Value Distribution in Ultrametric Analysis and Applications

Value Distribution in Ultrametric Analysis and Applications

Alain Escassut
Université Clermont Auvergne,
UMR CNRS 6620, France

World Scientific

NEW JERSEY • LONDON • SINGAPORE • BEIJING • SHANGHAI • HONG KONG • TAIPEI • CHENNAI

Published by

World Scientific Publishing Co. Pte. Ltd.
5 Toh Tuck Link, Singapore 596224
USA office: 27 Warren Street, Suite 401-402, Hackensack, NJ 07601
UK office: 57 Shelton Street, Covent Garden, London WC2H 9HE

Library of Congress Cataloging-in-Publication Data
Names: Escassut, Alain, author.
Title: Value distribution in ultrametric analysis and applications / Alain Escassut.
Description: New Jersey : World Scientific, [2025] | Includes bibliographical references.
Identifiers: LCCN 2024038421 | ISBN 9789811298325 (hardcover) |
ISBN 9789811298332 (ebook for institutions) | ISBN 9789811298349 (ebook for individuals)
Subjects: LCSH: Value distribution theory. | Theory of distributions (Functional analysis)
Classification: LCC QA331 .E7155 2025 | DDC 515/.782--dc23/eng/20241029
LC record available at https://lccn.loc.gov/2024038421

British Library Cataloguing-in-Publication Data
A catalogue record for this book is available from the British Library.

For any available supplementary material, please visit
https://www.worldscientific.com/worldscibooks/10.1142/13988#t=suppl

Desk Editors: ACES-Selas Hamilton/Rok Ting Tan

Typeset by Stallion Press
Email: enquiries@stallionpress.com

About the Author

Alain Escassut after studies in Licence and Master at the University of Bordeaux obtained a PhD in 1970 and a Doctorat d'Etat in Mathematics in 1972, working in this university as Maître-Assistant and Maître de conférence until 1987 (he was also Visiting-Assistant Professor at Princeton University in Spring 1981). Next, he was Professor at Université Blaise Pascal (from Clermont-Ferrand) until 2012 and has been Emeritus Professor since 2012 at this university which become now Université Clermont Auvergne.

Alain Escassut is a specialist in ultrametric analysis, particularly p-adic analysis and first studied algebras of analytic elements (long ago defined by Marc Krasner). He first examined the problem of analytical extension where the rôle of T-filters first appeared, explaining also algebraic properties of algebras, and he characterized Krasner–Tate algebras. He constructed a holomorphic functional calculus where T-filters are essential, again. He then examined the rôle of the multiplicative spectrum of an ultrametric Banach algebra.

With Abdelbaki Boutabaa and Labib Haddad, he obtained applications of the p-adic Nevanlinna theory to problems of unicity. Joint works with Yang and Ta Thi Hoai An let them obtain new results. He obtained results on small functions and solved the problem of the Hayman conjecture in a p-adic field. He also examined Valiron-like results on the growth of entire functions in a p-adic field. He examined the transcendence type in $\mathbb{C}_p$ and gave a new proof of the Hermite–Lindemann theorem in an ultrametric field, including the case of a field of residue characteristic 0. All these studies are detailed in this book.

Contents

Introduction

Value distribution for complex meromorphic functions is a topic that knew an impressive development during the 20th century and the early 2000, particularly from the eighties until now. The Nevanlinna theory has been very fruitful and has many applications, for instance, in algebraic geometry, problems of uniqueness, and value sharing values and small functions. On the other hand, during the second part of the 20th century and later on, ultrametric analysis became a new domain of research in value distribution and many problems once considered on the complex field were similarly considered on an ultrametric field. That needed the use of a complete algebraically closed field. That was done with the field $\mathbb{C}_p$ which is the completion of the algebraic closure of $\mathbb{Q}_p$, the completion of $\mathbb{Q}$ with respect to the p-adic absolute value. Thus, $\mathbb{C}_p$ is a field having certain properties similar to those of $\mathbb{C}$ but many other properties are much different, particularly, it has no connected subset except singletons. However, there exist other complete ultrametric fields, particularly the Levi-Civita field $\mathcal{R}$ [88] and its algebraic closure that is complete.

A field is said to be spherically complete when every decreasing sequence of disks has a not empty intersection. A field may be complete for an ultrametric absolute value but not spherically complete. For instance, $\mathbb{C}_p$ is not spherically complete and several serious problems happen because of that, particularly the well-known Lazard's problem. However, following, for example, a method proposed by Bertin Diarra, we show how to construct a spherically complete extension of a field. This technique enables us to solve certain problems when a spherically complete field is necessary.

This book is aimed at gathering classical results on analytic elements and meromorphic functions known for almost 70 or 80 years but also many recent works in p-adic value distribution and also some results in number theory, concerning problems of transcendence and exponential polynomials. Overall, it is aimed at giving correct proofs for all properties of the functions we consider, while several times we can find in the literature approximative citations or proofs putting in doubt certain statements. Particularly, here as in [44, 53] and in [48, 49], the role of circular filters is emphasized because it is the best way to obtain rigorous explanations. Concerning the famous Corona problem, here we give a description of all maximal ideals and of most of continuous multiplicative semi-norms, even though it is not known whether there exist continuous multiplicative semi-norms other than those we describe.

Most of the results presented in [49] are also presented here with sometimes some small corrections but several new results are given, particularly on p-adic transcendence, exponential polynomials, growth order of analytic functions, Picard's values of meromorphic functions, meromorphic functions sharing small functions, and every application of the Nevanlinna theory.

In the first chapters, we define affinely rigid sets, a basic notion indispensable in problems of uniqueness [22], and we present the main properties of p-adic fields, defining fields such as $\mathbb{Q}_p$, $\mathbb{C}_p$ and examining some classical properties. Thanks to the help of Bertin Diarra, finite extensions of $\mathbb{Q}_p$ are examined, particularly normal extensions and ramification index, in order to study the notion of transcendence order over $\mathbb{Q}_p$: this transcendence order is stable through an algebraic extension and we construct p-adic numbers of order $\leq 1 + \epsilon \ \forall \epsilon > 0$. The help of Bertin Diarra was very useful again. Similar to the transcendence type of complex numbers, the transcendence type of a p-adic transcendental number over $\mathbb{Q}$ is also defined and examined: we show that it is also stable through an algebraic extension and I am very grateful to Michel Waldschmidt for his help in that study [55]. The numbers of zeros of exponential polynomials have a good estimation through a computation that is not too complicated [57]. The ultrametric versions of the Hermite–Lindemann theorem, the Gelfond–Schneider theorem, and the exponentials theorem are given in residue characteristic p and 0. The case of residue characteristic 0 may apply to Levi-Civita fields [88].

Explaining correctly proofs requires to know perfectly all properties of analytic elements because all analytic or meromorphic functions are at least locally, analytic elements, as defined by Marc Krasner [69]. This is why we first recall them in order to make an autocontained book. Particularly, any student in master of mathematics, having a knowledge in general topology, can understand definitions and proofs.

Defining holomorphic functions in a p-adic field was not easy, except if we restrict ourselves to disks or sets with finitely many holes. The foundation of a veritable theory of holomorphic functions is due to Marc Krasner who defined analytic elements as limits of sequences of rational functions with respect to the topology of uniform convergence. Since the only connected sets are singletons, this requires the use of sets enjoying a property that can replace connexity. Marc Krasner defined quasi-connected sets [69]. But most of the useful properties of quasi-connected sets actually hold in a larger class of sets: the infraconnected sets [36, 44, 45].

Thus, we must place our study in an algebraically closed field $\mathbb{K}$, complete with respect to an ultrametric absolute value. For a study of Krasner's analytic functions or the generalization of that theory to infraconnected analytic sets made by Philippe Robba, readers can refer [83]. On the other hand, another theory of analytic functions was made by John Tate with the use of a sheaf, considering Banach algebras, usually called affinoid algebras [89]. The comparison between the two theories through their algebras of functions is made in [37, 45]. Here, these theories won't be developed again because they are not necessary to examine properties of meromorphic functions and problems of value distribution in p-adic analysis.

Mittag-Leffler's theorem due to Marc Krasner is another important tool to describe the behavior of analytic elements. The derivative of an analytic element f is examined with its Mittag-Leffler series relatively to those of f. Overall, the maximum principle satisfied by an analytic element comes from its behavior on circular filters.

A key problem for analytic elements is whether it can vanish along a circular filter. This was thoroughly explained in [44, 53]. Since it is a quite substantial problem involving the characterization of T-filters [38, 53], we do not describe that in this book, but one can find it in [44, 53]: here, we only recall sufficient conditions, on infraconnected sets, to avoid that any analytic element vanishes along a large circular filter, which is sufficient for our further studies. However, the description of that problem of filters is essential to understand the behavior of analytic and meromorphic functions in any context. Moreover, we give a very simple example of T-filter which is indispensable for the solution of the Corona problem.

An important particularity of the theory of analytic functions on $\mathbb{K}$ holds in the role played by multiplicative semi-norms, characterized by circular filters on $\mathbb{K}$. Their role in ultrametric Banach algebras was discovered by Bernard Guennebaud in the seventies [62] and they particularly apply to all kinds of analytic functions. Several properties of the spectral semi-norms are examined in [40, 45, 53] with the use of multiplicative semi-norms and circular filters. On the other hand, Berkovich's theory [9], made in the late eighties, is entirely based on continuous multiplicative semi-norms. Here we recall their description and their intervention to determine, for instance, what we denote by $|f|(r)$, a kind of "maximum modulus principle".

We characterize the continuous multiplicative norms on the algebra A of bounded analytic functions inside an "open" disk. A major problem posed in ultrametric analysis is the Corona problem, through a kind of analogy with the well-known Kakutani problem in $\mathbb{C}$, solved by Carlesson [28]. Concerning the algebra A, we explain why it cannot be considered in a similar way. However, by considering continuous multiplicative semi-norms, we can define an almost similar problem and show that the set of continuous multiplicative semi-norms whose kernel is a maximal ideal of codimension 1 is dense inside the whole set of all continuous multiplicative semi-norms whose kernel is a maximal ideal. That result, first proved when $\mathbb{K}$ is spherically complete [48], is generalized here. The characterization of all continuous multiplicative semi-norms of that algebra is far from over, but we present examples of continuous multiplicative semi-norms called *Araujo's semi-norms*, whose kernel is neither a maximal ideal nor the null ideal, and I have been grateful to Jesus Araujo for communicating me his results.

The Shilov boundary of an algebra of analytic elements is described with the help of circular filters, in connection with the maximum principle [15]. Mappings between the multiplicative spectra of such algebras are studied by using the hyperbolic distance in order to characterize injective analytic elements, a property also stated by Rivera-Letelier in [82] in connection with previous works

[42, 44]. The idea consists of showing that an injective meromorphic function has an absolute value that looks like this of a Moebius function (also called a linear fractional function). A correct proof requires first to show that an analytic element f maps the Shilov boundary of a set D into that of $f(D)$.

The divisor of an analytic function is introduced and Lazard's problem for analytic functions inside an "open" disk is described. The very complicated construction made by Michel Lazard in a spherically complete field is explained in a way that is more comprehensible [72]. In a field that is not spherically complete, a counter-example due to Michel Lazard shows that the theorem may not be generalized and this is a gap for certain problems. However, we can provide a result that is less accurate but sufficient for certain problems. Yet, sometimes we can place ourselves in a spherically complete extension. All closed ideals of algebras of entire functions or algebras of analytic functions inside an open disk are principal, we show that the algebra of entire functions is a Bezout ring, and when $\mathbb{K}$ is spherically complete, so is the algebra of analytic functions inside an open disk.

Similar to definitions well known in complex analysis, we have notions of order of growth and type of growth for an entire function in $\mathbb{K}$. But here, another notion appears essential: *the cotype of growth* [16]. Thus, we can obtain similar formulas for the growth order and the growth type. Let f be an entire function in $\mathbb{K}$ whose order of growth is finite. We give bounds for the cotype of growth and also for the lower cotype of growth. We show that the type of growth of f is equal to its lower type if and only if its cotype is equal to its lower cotype, and when these are realized, then the cotype is the product of the type by the order of growth. Such entire functions are called *clean functions*. But we construct an example of non-clean entire function f where the cotype is strictly superior to the product of the type by the order, which answers negatively a question asked long ago.

Suppose that $\mathbb{K}$ is of characteristic 0. Given a meromorphic function $f = \frac{g}{h}$, if the type or the cotype of h is finite, then f assumes all values infinitely many times. A counter-example is constructed showing that if the lower order of growth is equal to the order, that does not imply that the lower type of growth is equal to the type of growth, although such a claim was made for complex meromorphic functions [32] and we contest the proof.

Next, we examine notions of order of growth and type of growth for analytic functions inside an open disk [19].

Meromorphic functions in $\mathbb{K}$ and inside a disk are introduced and their basic properties are examined. Next, we examine meromorphic functions out of a hole, using Motzkin products. A meromorphic function admits primitives if and only if all residues are null. This property seems obvious but actually requires the use a form of Mittag-Leffler theorem to be correctly proved and does not seem to appear in other work. We study exceptional values of a function or its derivative, particularly we recall why a meromorphic function in $\mathbb{K}$ or an "unbounded" meromorphic function in an open disk admits at most one exceptional value, which suggests a conjecture.

In a field such as $\mathbb{K}$, the first fundamental theorem is almost trivial, but the p-adic Nevanlinna theory entirely holds in the so-called second fundamental theorem that we call here the main fundamental theorem. It was completely established in [21] by Abdelbaki Boutabaa for meromorphic functions in the whole field $\mathbb{K}$, after the first study by Ha Huy Khoai [63]. The generalization for meromorphic functions in an open disk was made further [23], taking into account Lazard's problem: this theorem is useful when it applies to functions which are not a quotient of two bounded analytic functions in that disk. Next, a similar statement was established for meromorphic functions out of a hole [54] which lets us obtain applications for these functions.

This is the beginning of many applications to the behaviour of meromorphic functions. Using the main fundamental theorem, we can show that many functional equations have no solution, particularly, elliptic and hyperelliptic curves admit no parametrization by meromorphic functions (more generally, Berkovich theory shows that no curve of genus ≥ 1 admits any parametrization by non-constant meromorphic functions). Similarly, equations of the form $f^m + g^n = 1$ admit no non-constant meromorphic solutions in the whole field, provided $\min(m, n) \geq 2$, $\max(m, n) \geq 3$ (results inside a disk are also given).

As it was done for complex functions, we define small functions. Generalizing Picard's values for meromorphic functions, given a meromorphic function f in $\mathbb{K}$ or an unbounded meromorphic function inside an open disk, there exists at most one small function w such that $f - w$ has finitely many zeros. One can easily obtain the main fundamental theorem on three small functions, as in $\mathbb{C}$, which was first noticed by Peichu Hu and Yang [67]. This is a useful tool in certain problems of uniqueness. Unfortunately, contrary to complex analysis, we know no main fundamental theorem on n small functions ($n > 3$) as it was made in $\mathbb{C}$ by Yamanoi [94]. However, a result has just been obtained by Ta Thi Hoai An and An-Nguyen Phuong, which we explain here.

When two meromorphic functions share (ignoring multiplicity) seven small functions, in a work with Yang, we showed that they are equal [56]. As corollary, if two entire functions share three small functions, then they are equal. But a much better result has been obtained by Ta Thi Hoai An and An-Nguyen Phuong, showing, through big computations, that it is sufficient to assume the two meromorphic functions to share five small ones [5]. All results in that domain hold in $\mathbb{K}$ like in an open disk and out of a hole.

As in complex analysis [29], we can also define and study perfectly and totally branched values, showing that in a p-adic field, a meromorphic function in $\mathbb{K}$ has at most three totally branched values and four perfectly branched values. But we can also prove that a meromorphic function having finitely many poles has at most one perfectly branched value, which is obtained without the use of the Nevanlinna main theorem but with just the use of square roots of certain analytic functions [17]. This last result has a generalization to branched rational functions.

Concerning functions inside a disk, we obtain results that look like those obtained for complex functions in the whole field [33]. We can also show that a meromorphic function whose numerator and denominator have different growth order admits at most two perfectly branched values. Moreover, a meromorphic function whose numerator and denominator have same finite growth order but different growth type admits at most three perfectly branched values [17].

We examine the Hayman conjecture in the p-adic context: given $f \in \mathcal{M}(\mathbb{K})$, can we state that $f'f^n$ takes every value $b \neq 0$ infinitely many times? We first give the proof by Jacqueline Ojeda obtained for every $n \geq 3$ [80]. Next we give the proof for $n = 2$ [52]. Moreover, all results given for $n \geq 3$ also apply to meromorphic functions inside an open disk and to meromorphic functions out of a hole. Many particular cases are also examined around such problems.

Concerning entire functions, a link then appears between growth order and small functions. Let $f,\ h \in \mathcal{A}(\mathbb{K})$ be such that $0 < \rho(f) < +\infty$ and $\rho(h) < \widetilde{\rho}(f)$. Then, h is a small function with respect to f. Let $f,\ h \in \mathcal{A}(\mathbb{K})$ be such that $\rho(h) = \rho(f)$ and such that $0 < \rho(f) < +\infty$. If f is clean and if $\sigma(h) = 0 < \sigma((f)$, then h is a small function with respect to f.

Let $f,\ g \in \mathcal{A}(\mathbb{K})$ be transcendental, regular and share three distinct functions, ignoring multiplicity, $\omega_j \in \mathcal{A}(\mathbb{K}),\ j = 1, 2, 3$, such that $\max_{1\leq j\leq 3} \rho(\omega_j) < \min(\rho(f), \rho(g))$. Then, $f = g$.

Let $f,\ g \in \mathcal{A}(\mathbb{K})$ be clean and share $h_1, h_2, h_3 \in \mathcal{A}(\mathbb{K})$, ignoring multiplicity, such that $\rho(h_j) = \rho(f) = \rho(g),\ j = 1, 2, 3,\ \sigma(h_j) = 0$, $j = 1, 2, 3$, and $0 < \min(\sigma(f), \sigma(g))$. Then, $f = g$.

We show that a meromorphic function admitting finitely many poles of order ≥ 3 and admitting primitives must have infinitely many zeros. That suggests the conjecture: *a meromorphic function admitting primitives must have infinitely many zeros.* Thanks to an idea due to Bézivin [10], the conjecture is proved for all meromorphic functions f having a total number of poles $t(r, f)$ in the disk of center 0 and radius r satisfying: $\text{Log}(t(r, f)) \leq O(\text{Log}(r)$, which makes the conjecture much likely. That suggests an interpretation by using type and cotype of the denominator of a meromorphic function: *given a meromorphic function* $f = \frac{g}{h}$ *admitting primitives, if the type of growth of h is finite, then f should have infinitely many zeros.*

We examine functions of uniqueness and unique range sets, counting or ignoring multiplicity, for p-adic analytic and meromorphic functions in the whole field and inside a disk. The Nevanlinna theory is the main tool for most of the results. Unique range sets counting multiplicity for entire functions in $\mathbb{K}$ and for polynomials are characterized as the affinely rigid subsets of $\mathbb{K}$ [22, 45] (such a characterization has no equivalent for complex functions). The result, first proved for polynomials and for entire functions and sets of three points [22], was generalized to entire functions on $\mathbb{K}$ by Cherry and Yang [31].

It is shown that the situation is much more complicated concerning meromorphic functions. Examples of sets which are or are not unique range sets counting multiplicity for meromorphic function are presented. The situation is examined in the whole field and inside a disk.

The problem of unique range sets counting multiplicity for entire functions in $\mathbb{K}$ is closely linked to this of polynomials of uniqueness. That suggests a more general study of meromorphic functions of uniqueness for various kinds of functions (in the whole field and inside a disk) [47], in connection with zeros of the derivative, a notion first defined by Fujimoto [60].

A Nevanlinna theory in characteristic $p \neq 0$ is made and applications are given to problems of the form $P(f) - Q(g)$, with P, Q polynomials [25, 26]. Finally, the Malmquist–Yoshida equation is studied in characteristic $p \geq 0$ and all solutions in the whole field are characterized [24]. The problem is finally examined in characteristic 0 for functions in an open disk following a work by Abdelbaki Boutabaa [27].

I would also like to thank many colleagues for many exchanges, particularly Ta Thi Hoai An, Kamal Boussaf, Jesus Araujo, Bézivin, and Yang. Abdelbaki Boutabaa passed away in 2022. He made the p-adic Nevanlinna theory and took part in most of our work, then.

Chapter 1

Affinely Rigid Sets

Affinely rigid sets have proven to play a crucial role in studying problems of uniqueness. They were introduced in [22] and were first called affinely rigid sets. Next, fearing a confusion with rigid geometry and affinoid sets, they were called *stiff sets* in [22]. However, the term affinely rigid was already known and popular. This is why it has become the one currently used in all further papers [3, 4, 93]. It is a very basic notion that requires to be thoroughly examined.

Definition. In this chapter, we denote by $\mathbb{F}$ an algebraically closed field. We call *centered similarity in the field* $\mathbb{F}$ a similarity in $\mathbb{F}$ of the form $\phi(x) = a + \alpha(x - a)$, and if ϕ is not the identity, the point a is called *the center of* ϕ.

We call *translation* a similarity of the form $\phi(x) = x + b$ with $b \neq 0$.

A subset S of $\mathbb{F}$ is said to be *affinely rigid* if there exists no affine mapping ϕ from $\mathbb{F}$ to $\mathbb{F}$, other than the identity, such that $\phi(S) = S$.

Let S be a non-affinely rigid set. S is said to be *a centered non-affinely rigid set* (respectively *non-centered non-affinely rigid set*) *with respect to a centered similarity* ϕ if it is preserved by a centered similarity ϕ and if the center of ϕ belongs to S (respectively does not belong to S).

Propositions 1.1 and 1.2 are immediate:

Proposition 1.1. *A similarity ϕ is not centered if and only if it is a translation.*

Proposition 1.2. *If a finite subset S of $\mathbb{F}$ is preserved by a similarity ϕ, that ϕ is a centered similarity.*

Proof. Let ϕ be a non-centered similarity preserving a finite set S. Then, ϕ^n also preserves S. By Proposition 1.1, $\phi(x)$ is of the form $x+b$ with $b \neq 0$. Then, $\phi^n(x) = x + nb$. Since $\mathbb{F}$ is of characteristic 0, the sequence $(x+nb)_{n\in\mathbb{N}}$ is injective. Particularly, taking $u \in S$, we have $u + nb \in S\ \forall n \in \mathbb{N}$, hence S has infinitely many elements, a contradiction. $\square$

Lemma 1.3 is immediate:

Lemma 1.3. *Let ϕ be a similarity and let S be a non-affinely rigid set. Then $\phi(S)$ is a non-affinely rigid set. Moreover, if S is centered (respectively non-centered), so is $\phi(S)$.*

Theorem 1.4. *Let $P \in \mathbb{F}[x]$ be of degree n and have all its zeros of order 1 and let S be the set of zeros of P. Then, S is not affinely rigid if and only if there exists a centered similarity of center a: $\phi(x) = a + \alpha(x-a)$ such that $P(\phi(x)) = \alpha^n P(x)$, with $\alpha \neq 1$. Let S be non-affinely rigid and let ϕ be such a centered similarity $\phi(x) = a + \alpha(x-a)$ preserving S. Then putting $u = x - a$, $Q(u) = P(x)$, Q is of one of the following two forms:*

(i) *$Q(u) = \sum_{k=0}^{q} a_{kd}u^{kd}$, with $\alpha^d = 1$ and $d \geq 2$ and then S is non-centered.*
(ii) *$Q(u) = \sum_{k=0}^{q} a_{kd+1}u^{kd+1}$, with $\alpha^d = 1$ and $d \geq 2$ and then S is centered.*

Moreover, if two centered similarities preserve S, they have the same center. Further, S is never both centered and non-centered.

Proof. Considering the coefficients of degree n, it is obvious that S is not affinely rigid if and only if there exists a centered similarity of center a: $\phi(x) = a + \alpha(x-a)$ such that $P(\phi(x)) = \alpha^n P(x)$, with $\alpha \neq 1$.

If (i) or (ii) is satisfied, it is obviously seen that S is preserved by the centered similarity ϕ of center a. Moreover, in Case (i), a does not lie in S, hence S is non-centered. In Case (ii), a lies in S, hence S is centered.

Now, suppose S is not affinely rigid and let $\phi(x) = a + \alpha(x-a)$ be a centered similarity preserving S. Then, $P(\phi(x)) = P(a + \alpha u) = Q(\alpha u)$ and $\alpha^n P(x) = \alpha^n Q(u)$, hence $Q(\alpha u) = \alpha^n Q(u)$.

Let $Q(u) = \sum_{j=0}^{n} a_j u^j$. We have $\alpha^j a_j = \alpha^n a_j \ \forall j = 0, \ldots, n$. Consequently, (1) $a_j(\alpha^{n-j} - 1) = 0 \ \forall j = 0, \ldots, n$.

Suppose first $a_0 \neq 0$. Clearly, we have $\alpha^n = 1$. Let d be the order of α as a nth root of 1. Since ϕ is not the identity, α is $\neq 1$. Consequently, by (1), we note that $d \geq 2$ because $\alpha^{n-1} \neq 1$. Then, n is of the form qd ($q \in \mathbb{N}^*$), and then, by (1), we have $a_j = 0$ for every j which is not multiple of d. Consequently, we have obtained $Q(u) = \sum_{k=0}^{q} a_{kd} u^{kd}$, with $\alpha^d = 1$ and $d \geq 2$.

Suppose now $a_0 = 0$. Since the zeros of P are not multiple, neither are those of Q. Consequently, $a_1 \neq 0$, hence, by (1), we have $\alpha^{n-1} = 1$. Let d be the order of α as an $(n-1)$th root of 1. Since $\alpha \neq 1$, by (1), we have $d \geq 2$ because $\alpha^{n-2} \neq 1$. Then, n is of the form $qd+1$ ($q \in \mathbb{N}^*$), and by (1), we have $a_j = 0$ for every j which is not of the form $kd+1$. Consequently, we obtain $Q(u) = \sum_{k=0}^{q} a_{kd+1} u^{kd+1}$, with $\alpha^d = 1$ and $d \geq 2$.

Suppose S is non-centered with respect to ϕ and consider another centered similarity ψ preserving S. Let Q be a polynomial defined as in (i) with respects to ϕ, admitting S for set of zeros, all of order 1. Then, we have $\psi(u) = \lambda u + \mu$ and $Q(\psi(u)) = \lambda^n Q(u)$. Consequently,

$$\sum_{k=0}^{q} a_{kd+1}(\lambda u + \mu)^{kd+1} = (\lambda)^{qd+1} \sum_{k=0}^{q} a_{kd+1} u^{kd+1}.$$

Examining terms of degree qd, we can check that $a_{qd+1}(qd+1)\lambda^{qd} \mu = 0$ because $d \geq 2$. Consequently, $\mu = 0$. Therefore, $Q(\psi(u)) = \sum_{k=0}^{q} a_{kd}(\lambda u)^{kd}$ and then, the polynomial associated with Q with respect to ψ, as defined in (i), is Q itself. Consequently, there exists no centered similarity ψ preserving S such that the polynomial associated with Q with respect to ψ is of the form (ii). This proves that S is non-centered with respect to ψ. That ends the proof of Theorem 1.4. □

Theorem 1.5. *Let S be a finite centered (respectively non-centered) non-affinely rigid set of n elements and let $\phi(x) = a + \alpha(x - a)$ be a similarity that preserves S. The order of ϕ divides $n-1$ (respectively divides n) and α is some $(n-1)$th root of $n-1$ (respectively α is some $(n-1)$th root of n).*

Proof. Let $\phi(x) = a + \alpha(x - a)$. Without loss of generality, we can assume that $a = 0$ and hence $\phi(x) = \alpha x$.

We first note that the restriction $\overline{\phi}$ of ϕ to S is a bijection from S to S and hence is a permutation of S.

Suppose first that S is centered. So, 0 is the center of ϕ. Then, $\overline{\phi}$ preserves $S\setminus\{0\}$ and hence its order divides $n-1$. Consequently, α is an $(n-1)$th root of 1.

Suppose now that S is non-centered. Then, $\overline{\phi}$ preserves S but does not preserve any point of S. Consequently, its order divides n and hence α is an nth root of 1. □

Corollary 1.6. *Let S be a finite centered non-affinely rigid set of n elements. If $n-1$ is prime, then S is of the form $\{a,\ a+u^k h \mid k = 0,\dots,n-2\}$ with $a,\ h \in \mathbb{F}$ and u some $(n-1)$th root of 1.*

Let S be a finite non-centered non-affinely rigid set of n elements. If n is prime, then S is of the form $\{a+u^k h \mid k = 0,\dots,n-1\}$ with $a,\ h \in \mathbb{F}$ and u is some nth root of 1.

Corollary 1.7. *Let S be a finite centered non-affinely rigid set of three elements. Then, S is of the form $\{a,\ a-h\ ,a+h\}$ with $a,\ h \in \mathbb{F}$.*

Let S be a finite non-centered non-affinely rigid set of three elements. Then, S is of the form $\{a+u^k h \mid k = 0,\ 1,\ 2\}$ with $a,\ h \in \mathbb{F}$ and u is a primitive third root of 1.

Theorem 1.8. *Let S be a subset of $\mathbb{F}$ of four elements. The following two conditions are equivalent. Then:*

(1) *S is a centered non-affinely rigid set if and only if it is of the form $\{a, a+h, a+uh, a+u^2h\}$, with $h \in \mathbb{F}$ and $u^3 = 1,\ u \neq 1$.*
(2) *S is a non-centered non-affinely rigid set if and only if it is of the form $\{a-h, a+h, a-k, a+k\}$, with $h,\ k \in \mathbb{F}^*$.*

Proof. If S is of the form $\{a, a+h, a+uh, a+u^2h\}$, with $h \in \mathbb{F}$ and $u^3 = 1,\ u \neq 1$, it is obvious that the similarity $\phi(x) = a + u(x-a)$ preserves S. Conversely, suppose that S is a centered non-affinely rigid set and let ϕ be a similarity of center a preserving $S\setminus\{a\}$ and different from the identity. Then, u is a third- root of 1 different from 1.

If S is of the form $\{a-h, a+h, a-k, a+k\}$, with $h,\ k \in \mathbb{F}^*$, then the similarity $\phi(x) = a - (x-a)$ preserves S and does not preserve any point of S, hence S is a non-centered non-affinely rigid set. Conversely, suppose that S is a non-centered non-affinely rigid set and let $\phi(x) = a + u(x - a)$. For convenience, suppose $a = 0$. Then, u is a root of 1 of order 2 or 4. Suppose first u is a root of 1 of order 2 i.e. $u = -1$. Then, $\phi(x) = -x$. Let $S = \{b_1, b_2, b_3, b_4\}$. Then, up to a reordering, we have $\phi(b_1) = b_3$, $\phi(b_2) = b_4$, hence $b_3 = -b_1$, $b_4 = -b_2$ which proves the claim.

Suppose now that u is root of 1 of order 4. Then, up to a reordering, we must have $\phi(b_j) = b_{j+1}$ $\forall j = 1, 2, 3$ and of course, $\phi(b_4) = b_1$. But then we note that $b_3 = -b_1$, $b_4 = -b_2$ again, which finishes proving the claim. □

The following condition (F) was introduced by Yang and Hirotaka Fujimoto.

Definition. A polynomial $P \in \mathbb{F}[x]$ is said *to satisfy Condition (F)* [60] if for any two distinct zeros a, b of P', we have $P(a) \neq P(b)$ (i.e. the restriction of P to the set of zeros of P' is injective).

Lemma 1.9. *Let $P \in \mathbb{F}[x]$ be of degree 4 and assume it does not satisfy Condition (F). Then, P' admits three distinct zeros.*

Proof. Of course, P' can't have a unique triple zero. Suppose P' only has two distinct zeros. Up to an affine change of variable, P' is of the form $\lambda(x^3 - bx^2)$. Since P does not satisfy Condition (F), we can conclude $b = 0$, hence P' has a triple zero, a contradiction. □

Theorem 1.10. *Let $P \in \mathbb{F}[x]$ be monic of degree 4. The following three statements are equivalent:*

(i) *P does not satisfy Condition (F).*
(ii) *P is of the form $[(x-a+l)(x-a-l)]^2 + B$ with $B \in \mathbb{F}$, $l \in \mathbb{F}^*$.*
(iii) *There exists an affine change of variable transforming P into an even function.*

Moreover, if those conditions are satisfied, assuming $P(c_1) = P(c_2)$, then we have $P(c_3) = -P(c_1)$ if and only if $B = -\frac{l^4}{2}$.

Proof. By putting $t = x - a$, it is obvious that (ii) implies (iii) and that (iii) implies (i). Suppose that (i) is satisfied. By Lemma 1.9, P' has three distinct zeros. Suppose first $P(c_3) = P(c_2) = P(c_1)$. Then, $P - P(c_1)$ has three distinct zeros of order ≥ 2, which contradicts the hypothesis $\deg(P) = 4$. Hence, we can assume $P(c_3) \neq P(c_1)$.

Let $a = \frac{c_1+c_2}{2}$, $c_1 = a + l$, $c_2 = a - l$. Thus, P is of the form $[(x-a+l)(x-a-l)]^2 + P(c_1)$, with $P(c_1) \neq 0$. Moreover, since P' has three distinct zeros, we see that $l \neq 0$, which also proves that (ii) is satisfied.

Now, suppose that the above three conditions are satisfied. By (ii), we have $P(c_1) = P(c_2) = B$ and $P'(x) = 4(x-a)[(x-a)^2 - l^2]$. So the zeros of P' are $a,\ a+l,\ a-l$. Since $P(c_1) = P(c_2)$, the pair $\{c_1, c_2\}$ is $\{a-l, a+l\}$. Therefore, $c_3 = a$ and hence $P(c_3) = l^4 + B = l^4 + P(c_1)$. Consequently, $P(c_3) = -P(c_1)$ holds if and only if $l^4 = -2P(c_1) = -2B$, which ends the proof. □

Corollary 1.11. *Let $P \in \mathbb{F}[x]$ be of degree 4 such that its set of zeros is affinely rigid. Then, P satisfies Condition (F).*

Proof. Indeed, suppose that P does not satisfy Condition (F). By Theorem 1.10, through a change of variable, P is an even function, hence for each zero, h and $-h$ are other zeros. Consequently, the set of zeros is not affinely rigid. □

Theorem 1.12. *Let $P(x) = \sum_{k=0}^{q} a_{kd+1}x^{kd+1}$, with $a_1 \neq 0$, $d \geq 2$, and let Z be the set of the zeros of P'. For all $c \in \mathbb{F}^*\setminus P(Z)$, $P - c$ admits $qd+1$ distinct zeros and its set of zeros is affinely rigid.*

Proof. Let $c \in \mathbb{F}^*\setminus P(Z)$, let $G(x) = P(x) - c$, and let T be the set of zeros of G. Suppose that T is not affinely rigid. By Theorem 1.3, there exists a centered similarity ϕ defined as $\phi(x) = \lambda x + \mu$ such that $G(\phi(x)) = \lambda^n G(x)$. So, we have $\sum_{k=0}^{q} a_{kd+1}(\lambda x + \mu)^{kd+1} - c = (\lambda)^{qd+1}(\sum_{k=0}^{q} a_{kd+1}x^{kd+1} - c)$ with $d \geq 2$. Examining terms of degree qd, we can check that $a_{qd+1}(qd+1)\lambda^{qd}\mu = 0$ because $d \geq 2$. Consequently, $\mu = 0$. Therefore, ϕ has center 0. Next, examining terms of degree 0 shows that $\lambda^n = 1$ because $c \neq 0$. Hence, $G(x) = \sum_{k=0}^{q} a_{kd+1}x^{kd+1} - c$. But since the center of ϕ is 0, by Theorem 1.3, $G(x)$ is either of the form $\sum_{j=0}^{s} b_{jt}x^{jt}$ or of the form $\sum_{j=0}^{s} b_{jt+1}x^{jt+1}$ with $t \geq 2$. Now, since $c \neq 0$, the second case is excluded. Consequently, we have $G(x) = \sum_{j=0}^{s} b_{jt}x^{jt}$ and therefore $\sum_{j=0}^{s} b_{jt}x^{jt} = \sum_{k=0}^{q} a_{kd+1}x^{kd+1} - c$.

In particular, since $t \geq 2$, we note that the coefficient of x on the right-hand side is null, a contradiction to $a_1 \neq 0$. Therefore, the assumption "T is not affinely rigid" is not true, which ends the proof. □

Theorem 1.13. *$P \in \mathbb{F}[x]$ be such that P' admits exactly two distinct zeros c_1, c_2. Then, $P(c_1) \neq P(c_2)$. Assume that all zeros of P are distinct, of order 1. If the set of zeros S of P is not affinely rigid, then P' is of the form $B((x - c_1)(x - c_2))^m$, $B \in \mathbb{F}$, and the unique similarity preserving S, other than the identity, is the mapping $\phi(x) = -x + c_1 + c_2$.*

Proof. Let $P'(x) = B(x - c_1)^{m_1}(x - c_2)^{m_2}$. Of course, $\deg(P) = m_1 + m_2 + 1$. Without loss of generality, we may obviously assume $B = 1$. Suppose first that $P(c_1) = P(c_2)$. Then, $P - P(c_1)$ admits each c_j as a zero of order $m_j + 1$ $(j = 1, 2)$ and therefore $\deg(P) = m_1 + m_2 + 2$, a contradiction. Hence, $P(c_1) \neq P(c_2)$.

Suppose that S is not affinely rigid and let $\phi(x) = ax + b$ preserve S, ϕ being different from the identity. Since ϕ preserves S and since all zeros of P are of order 1, $P \circ \phi$ is a polynomial of same degree as P, admitting the same zeros, all of order 1 and therefore $P \circ \phi$ is of the form λP with $\lambda \in \mathbb{F}^*$. Consequently, $P'(x) = \lambda a(ax + b - c_1)^{m_1}(ax + b - c_2)^{m_2}$.

Suppose first that $m_1 \neq m_2$. Then, we can identify c_1 with $\frac{c_1 - b}{a}$ and c_2 with $\frac{c_2 - b}{a}$. Consequently, $b = 0$, $a = 1$, a contradiction since ϕ is not the identity. Thus we are led to assume that $m_1 = m_2$. Put $m = m_1 = m_2$. Thus, $P'(x) = [(x - c_1)(x - c_2)]^m$. We may now write $P'(x) = a[(ax + b - c_1)(ax + b - c_2)]^m$ and we see that either $\frac{c_1 - b}{a} = c_1$, $\frac{c_2 - b}{a} = c_2$ which yields $a = 1, b = 0$ again or $\frac{c_1 - b}{a} = c_2$, $\frac{c_2 - b}{a} = c_1$. And since ϕ is not the identity, the second conclusion is the only possible. Then $a = -1$ and $b = c_1 + c_2$. □

Chapter 2

Properties of Ultrametric Fields

In this chapter, we recall basic definitions and properties on ultrametric fields: ultrametric absolute values, valuation rings, and residue fields. We must define holes of a subset and infraconnected subsets that are essential for the behavior of analytic functions (certain authors improperly call such sets "connected sets" which makes no sense in topology since there are no connected sets except singletons in an ultrametric field). A major interest of the class of infraconnected sets is that it is the biggest class of sets in an ultrametric complete algebraically closed field where the famous Krasner Mittag-Leffler theorem applies.

Definition and notation. Throughout this book, we denote by $\mathbb{N}$ the set of integers ≥ 0, by $\mathbb{Z}$ the ring of relative integers, by $\mathbb{Q}$ the field of rational numbers, by $\mathbb{R}$ the field of real numbers, and by $\mathbb{C}$ the field of complex numbers.

Given a topological space T and a subset S of T, we denote by $\overline{S}$ its closure (also called adherence) and by $\overset{\circ}{S}$ its interior (also called opening).

Let E be a field provided with an absolute value $|\ .\ |$ and let log be a real logarithm function of basis $\theta > 1$. We call *valuation associated with that absolute value* the mapping v from E to $\mathbb{R}$ defined as $v(x) = -\log|x|$ and here, we set $\Psi(x) = \log|x|$ and $|E| = \{|x| \mid x \in E\}$. If a set F contains the zero of a ring, we denote by F^* the set $F\setminus\{0\}$. An absolute value is said to be *trivial* if $|x| = 1\ \forall x \in E\setminus\{0\}$.

Throughout this chapter, we denote by $\mathbb{L}$ a field complete with respect to a non-trivial ultrametric absolute value, and throughout this book, we denote by $\mathbb{K}$ an algebraically closed field complete with respect to a non-trivial ultrametric absolute value. We denote by $| \,.\, |_\infty$ the archimedean absolute value defined on $\mathbb{R}$.

Lemma 2.1. *Let $\mathbb{E}$ be a field provided with an ultrametric absolute value $| \,.\, |$. The completion of $\mathbb{E}$ with respect to that absolute value is provided with an ultrametric absolute value which continues that of $\mathbb{E}$. The set $\{|x| \mid x \in \mathbb{E}^*\}$ is a subgroup of the multiplicative group $\mathbb{R}_+^*$.*

Definition. Given a field $\mathbb{E}$ provided with an ultrametric absolute value $| \,.\, |$, the multiplicative group $\{|x| \mid x \in \mathbb{E}^*\}$ is called *the value group of* $\mathbb{E}$ and the additive group $\{v(x) \mid x \in \mathbb{E}\}$ is called *valuation group of* $\mathbb{E}$.

Similarly, the set $\{\Psi(x) \mid x \in \mathbb{E}^*\}$ is a subgroup of $\mathbb{R}$ called *valuation group of* $\mathbb{E}$.

The field $\mathbb{E}$ is said to have *discrete valuation* or to have *discrete absolute value* if its valuation group is a discrete subgroup of $\mathbb{R}$ and hence is isomorphic to $\mathbb{Z}$. Else, the valuation group is dense in $\mathbb{R}$ and $\mathbb{E}$ is said to have *dense valuation* or to have a *dense absolute value.*

Lemma 2.2 is classical and proven in the same way no matter what the absolute value of $\mathbb{E}$.

Lemma 2.2. *Let $\mathbb{E}$ be a field provided with two absolute values whose associated valuations are v and w, respectively. They are equivalent if and only if there exists $r > 0$ such that $w(x) = rv(x)$ whenever $x \in \mathbb{E}$.*

Proof. If such an r exists, the two absolute values are seen to be equivalent. Reciprocally, we assume them to be equivalent and take $a \in \mathbb{E}$ such that $v(a) \geq 0$. It is seen that $w(a) \geq 0$. On the other hand, for all $x \in \mathbb{E}$ and for all $m, n \in \mathbb{N}$, we have $v(\frac{x^m}{a^n}) > 0$ if and only if $w(\frac{x^m}{a^n}) > 0$. Therefore, we see that $\frac{v(x)}{v(a)} > \frac{n}{m}$ is equivalent to $\frac{w(x)}{w(a)} > \frac{n}{m}$. Then, since $\mathbb{Q}$ is dense in $\mathbb{R}$, we have $\frac{v(x)}{v(a)} = \frac{w(x)}{w(a)}$ whenever $x \in \mathbb{E}$, and therefore, $\frac{w(x)}{v(x)} = \frac{w(a)}{v(a)}$. □

Notation. The set of the $x \in E$ such that $|x| \leq 1$ will be denoted by U_E and the set of the $x \in E$ such that $|x| < 1$ will be denoted by M_E.

Then Lemma 2.3 is immediate:

Lemma 2.3. *$U_{\mathbb{E}}$ is a local subring of E whose maximal ideal is $M_{\mathbb{E}}$.*

Definition and notation. Henceforth, U_E is called *the valuation ring of E*. The maximal ideal M_E of U_E is called the *valuation ideal* and the field $\mathcal{E} = \frac{U_E}{M_E}$ is called *the residue class field of* $\mathbb{E}$. For any $a \in \mathbb{E}$, the residue class of a is denoted by $\overline{a}$.

The characteristic of $\mathcal{E}$ is named *the residue characteristic of* $\mathbb{E}$ and is denoted by p.

Lemma 2.4. *Let F be a subfield of $\mathbb{E}$, let $\mathcal{E}$ (respectively $\mathcal{F}$) be the residue class field of $\mathbb{E}$ (respectively of F). Then, $\mathcal{F}$ is a subfield of $\mathcal{E}$. If $\mathbb{E}$ is algebraically closed and if its valuation is not trivial, it is dense.*

Proof. The first statement is immediate. Next, given $\alpha \in \mathbb{E}$ such that $0 < |\alpha| < 1$ and $\beta \in \mathbb{E}$ such that $\beta^q = \alpha^s$, we have $v(\beta) = \frac{q}{s}v(\alpha)$ whenever $s \in \mathbb{N}^*$ and $q \in \mathbb{Z}$. □

Lemma 2.5. *Let V be an $\mathbb{L}$-vector space of finite dimension provided with two norms. Then the two norms are equivalent.*

Proof. Let $\| \,.\, \|$ and $\| \,.\, \|'$ be the two norms on V. We proceed by induction on the dimension of V and assume the equivalence true for subspaces of dimension $n < q$. Let V have dimension q. Let $e_1, \dots, e_q$ be a base of V. Let us suppose that the two norms are not equivalent on V. Then there exists a sequence $(u_n)_{n\in\mathbb{N}}$ of the form $u_n = \sum_{j=1}^{q} a_{j,n}e_j$, with $\|u_n\| \geq 1$, such that $\lim_{n\to\infty} \|u_n\|' = 0$. Let S be the subspace of V generated by $\{e_1, \dots, e_{q-1}\}$. For every $n \in \mathbb{N}$, we put $v_n = \sum_{j=1}^{q-1} a_{j,n}e_j$.

First, we suppose that

(1) $\lim_{n\to\infty} |a_{q,n}| = 0$.

Since $\lim_{n\to\infty} \|u_n\|' = 0$, we have $\lim_{n\to\infty} \|v_n\|' = 0$. By hypothesis, the restrictions of the two norms to S are equivalent, hence we have $\lim_{n\to\infty} \|v_n\| = 0$. But since $\|u_n\| \geq 1$ for all $n \in \mathbb{N}$, this contradicts (1).

Now, since (1) is not true, there exists a subsequence of the sequence $(|a_{q,n}|)_{n\in\mathbb{N}}$ that admits a strictly positive lower bound and therefore, without loss of generality, we can clearly assume that there exists $r > 0$ such that $|a_{q,n}| \geq r$ for all $n \in \mathbb{N}$. Let $(x_n)_{n\in\mathbb{N}}$ be the sequence defined as $x_n = \frac{u_n}{a_{q,n}}$. It is seen that

(2) $\lim_{n\to\infty} \|x_n\|' = 0$.

The two norms $\| \, . \, \|$ and $\| \, . \, \|'$ are equivalent on S and they both are equivalent to the product norm $\| \, . \, \|''$ defined as $\|\sum_{j=1}^{q-1} b_j e_j\|'' = \max_{1\leq j\leq q-1} |b_j|$. Since $\mathbb{L}$ is complete, S is complete with respect to $\| \, . \, \|''$, hence S is closed in V with respect to the two norms $\| \, . \, \|$ and $\| \, . \, \|'$. Hence, by (2), e_q belongs to S, which is absurd and finishes the proof. □

Theorem 2.6. *Let F be an algebraic extension of $\mathbb{L}$, provided with two absolute values extending the one $\mathbb{L}$. These absolute values are equal.*

Proof. Let v, w be the valuations associated with these absolute values. Let $a \in F$. By Lemma 2.5, the two absolute values are equivalent on $\mathbb{L}[a]$. Hence, by Lemma 2.2, there exists $r > 0$ such that $w(x) = rv(x)$ whenever $x \in \mathbb{L}[a]$. But since $v(x) = w(x)$ whenever $x \in \mathbb{L}$ and since there exists $u \in \mathbb{L}$ such that $v(u) \neq 0$, we have $r = 1$. □

Lemma 2.7. *Let A be an $\mathbb{L}$-algebra. Let ϕ be a semi-norm of $\mathbb{L}$-algebra satisfying $\phi(x^n) = (\phi(x))^n \; \forall x \in A$. Then, ϕ is ultrametric.*

Proof. Let $a, b \in A$ satisfy $\phi(a) \geq \phi(b)$. We just have to show that $\phi(a + b) \leq \phi(a)$. Obviously, we have $\phi((a + b)^n) = \phi(\sum_{k=0}^{n} C_n^k a^k b^{n-k})$. For each $k = 0, \ldots, n$, we have $\phi(C_n^k a^k b^{n-k}) = |C_n^k|\phi(a^k b^{n-k}) \leq \phi(a)^k \phi(b)^{n-k} \leq \phi(a)^n$, hence $\phi((a + b)^n) \leq (n + 1)\phi(a)^n$, and therefore, $\phi(a + b) \leq \sqrt[n]{n+1} \; \phi(a)$ for all $n \in \mathbb{N}^*$. Finally, we obtain $\phi(a + b) \leq \phi(a)$. □

The most classical example of an ultrametric complete algebraically closed field is the field $\mathbb{C}_p$ that is described later.

Definition and notation. Consider the field $\mathbb{E}$ provided with an ultrametric absolute value. Let $a \in \mathbb{E}$ and let $r \in \mathbb{R}_+$. We denote by $d(a, r)$ the disk $\{x \in \mathbb{E} | \; |x - a| \; \leq r\}$, by $d(a, r^-)$ the disk $\{x \in \mathbb{E} | \; |x - a| \; < r\}$, and we call *circle of center a, of radius r* the set $C(a, r) = d(a, r) \setminus d(a, r^-)$.

Given r_1 and r_2 such that $0 < r_1 < r_2$, we denote by $\Gamma(a, r_1, r_2)$ the annulus $\{x \in E | \; r_1 < |x - a| \; < r_2\}$ and by $\Delta(a, r_1, r_2)$ the annulus $\{x \in E | \; r_1 \; \leq |x - a| \; \leq \; r_2\}$.

We know that if $b \in d(a,r)$, then $d(b,r) = d(a,r)$. In the same way, if $b \in d(a,r^-)$, then $d(b,r^-) = d(a,r^-)$. Moreover, given two disks T and T' such that $T \cap T' \neq \emptyset$, then either $T \subset T'$ or $T' \subset T$.

We denote by δ the distance defined on $\mathbb{E}$ by $\delta(a,b) = |a-b|$. Given $a \in \mathbb{E}$ and a subset D of $\mathbb{E}$, we set $\delta(a,D) = \inf\{|x-a| \mid x \in D\}$, and given two subsets D, F of $\mathbb{E}$, we set $\delta(D,F) = \inf\{|x-y| \mid x \in D, y \in F\}$.

We set $\text{diam}(D) = \sup\{|x-y| \mid |x \in D, y \in D\}$ and $\text{diam}(D)$ is named the *diameter* of D.

Similarly, we set $\text{codiam}(D) - \sup\{|x-y| \; |x \in D, y \notin D\}$ and $\text{codiam}(D)$ is named the *codiameter of D*.

Of course, the following three statements are seen to be equivalent:

(i) $d(a,r) = d(a,r^-)$.
(ii) $C(a,r) = \emptyset$.
(iii) $r \notin | E |$.

Further, the disks $d(b,r^-)$ included in $C(a,r)$ (respectively in $d(a,r)$) are the disks $d(b,r^-)$ such that $b \in C(a,r)$ (respectively in $d(a,r)$). They are called *the classes* of $C(a,r)$ (respectively of $d(a,r)$).

Henceforth, D will denote a subset of the field $\mathbb{L}$.

The closure of D (also called *adherence* of D) is denoted by $\overline{D}$ and the interior of D (also called *opening* of D) is denoted by $\overset{\circ}{D}$.

Given a point $a \in L$, we put $\delta(a,D) = \inf\{|x-a| \; |x \in D\}$. Then, $\delta(a,D)$ is named *the distance of a to D*.

Given two subsets D, D' of $\mathbb{L}$, we put $\delta(D,D') = \inf\{|x-y| \mid x \in D, \; u \in D'\}$. Then, $\delta(D,D')$ is called *the distance between D and D'*.

Given a bounded subset D of $\mathbb{L}$ of diameter R, we denote by $\widetilde{D}$ the disk $d(a,R)$, whenever $a \in D$. Given an unbounded subset D of $\mathbb{L}$, we put $\widetilde{D} = \mathbb{L}$.

We denote by $\widehat{\mathbb{L}}$ an extension of $\mathbb{L}$ provided with an absolute value that extends that of $\mathbb{L}$. Given $a \in \widehat{\mathbb{L}}$, $r > 0$, $\widehat{d}(a,r)$ (respectively $\widehat{d}(a,r^-)$) denotes the disk $\{x \in \widehat{\mathbb{L}} \mid \; |x-a| \leq r\}$ (respectively $\{x \in \widehat{\mathbb{L}} \mid |x-a| < r\}$).

Lemma 2.8. *Let $d(a,r)$, $d(b,s)$ be disks such that $d(a,r) \cap d(b,s) \neq \emptyset$ with $r \leq s$. Then, $d(a,r) \subset d(b,s)$.*

Let us also note the following basic lemma:

Lemma 2.9. *Suppose that the residue class field $\mathcal{E}$ of the field $\mathbb{E}$ is finite, of cardinal q. Then, for every disk $d(a,r)$ with $a \in \mathbb{E}$ and $r \in |\mathbb{E}|$ admits only q classes.*

Lemma 2.10. *$\widetilde{D}\backslash\overline{D}$ admits a unique partition of the form $(T_i)_{i\in I}$, whereas each T_i is a disk of the form $d(a_i, r_i^-)$ with $r_i = \delta(a_i, D)$.*

Proof. For every $a \in \widetilde{D}\backslash\overline{D}$, let $r(a) = \delta(a, D)$. Let α and β be two points in $\widetilde{D}\backslash\overline{D}$ such that $\mid \beta - \alpha \mid < r(\alpha)$. It is easily seen that for every $x \in D$, we have $|x - \beta| = |x - \alpha \mid$, and then the family of the disks $T(\alpha) = d(\alpha, r(\alpha)^-)$ $(\alpha \in \widetilde{D}\backslash\overline{D})$ makes a partition of $\widetilde{D}\backslash D$ because given α and $\beta \in \widetilde{D}\backslash\overline{D}$, either $|\alpha - \beta| < r(\alpha)$ and then $T(\alpha) = T(\beta)$, or $|\alpha - \beta| \geq r(\alpha)$ and then $|\alpha - \beta| \geq r(\beta)$, hence $T(\alpha) \cap T(\beta) = \emptyset$. □

Definition and notation. Such disks $d(a_i, r_i^-)$ are called *the holes* of D.

Example 1. The holes of a disk $d(a, r^-)$, with $r \in \mid \mathbb{L} \mid$, are the classes of $C(a,r)$.

Example 2. The only one hole of $\mathbb{L}\backslash d(0,1^-)$ is $d(0,1^-)$.

Example 3. The holes of $\mathbb{L}\backslash d(0,1)$ are the disks $d(a,1^-)$ with $a \in d(0,1)$.

Definition. D is said to be *infraconnected* [35, 36, 44, 49] if for every $a \in D$, the mapping I_a from D to $\mathbb{R}_+$ defined by $I_a(x) = |x - a|$ has an image whose closure in $\mathbb{R}_+$ is an interval. In other words, D is not infraconnected if and only if there exist a and $b \in D$ and an annulus $\Gamma(a, r_1, r_2)$ with $0 < r_1 < r_2 < |a - b|$ such that $\Gamma(a, r_1, r_2) \cap D = \emptyset$.

Lemma 2.11 is immediate:

Lemma 2.11. *If D is infraconnected of diameter $R \in \mathbb{R}$ (respectively $+\infty$), then $\overline{I_a(D)} = [0, R]$ (respectively $\overline{I_a(D)} = [0, +\infty[$).*

Lemma 2.12 gives a point of view from a hole of D.

Lemma 2.12. *Let D be infraconnected and let α belong to a hole T of diameter ρ. The closure of the set $\{|x - \alpha| \; |x \in D\}$ is an interval whose lower bound is ρ.*

Proof. We just have to show that for every r and r' such that $\rho < r < r' < \text{diam}(D)$, there exists $\beta \in D$ such that $r < |\beta - \alpha| < r'$. By definition of the holes, there exists $b \in D$ such that $|\alpha - b| < r$ and then, since D is infraconnected, there exists $\beta \in D$ such that $r < |b - \beta| < r'$. But it is seen that $|\beta - \alpha| = |b - \beta|$. □

Given two infraconnected sets A and B, we may prove $A \cup B$ to be infraconnected in the following two hypothesis (Theorems 2.13 and 2.15).

Theorem 2.13. *Let A and B be two infraconnected sets such that $A \cap B \neq \emptyset$. Then, $A \cup B$ is infraconnected.*

Proof. If A and B are not bounded, the statement is obvious because for every $a \in A$, $\overline{I_a(A)} = \mathbb{R}_+$ and for every $a \in B$, we have $\overline{I_a(B)} = \mathbb{R}_+$. Now, we may assume A to be bounded, of diameter R, while B has diameter $R' \geq R$ (respectively is not bounded). Then, $A \cup B$ has diameter R' (respectively is not bounded). Let $c \in A \cap B$, let $a \in A \cup B$, and let us show that $\overline{I_a(A \cup B)} = [0, R']$ (resp. $[0, +\infty[$).

For convenience, we first assume B to be bounded. Since $c \in A \cap B$, we see that $|x - a| \leq \max(|x - c|, |c - a|) \leq R'$ whenever $x \in A \cup B$, hence $\overline{I_a(A \cup B)} \subset [0, R']$. Hence, we just have to show that $\overline{I_a(A \cup B)} \supset [0, R']$. Obviously, $\overline{I_a(A \cup B)} = \overline{I_a(A)} \cup \overline{I_a(B)} = [0, R] \cup \overline{I_a(B)}$. Hence, we have to show that $\overline{I_a(B)} \supset [R, R']$. But when $x \in B$ with $|x - a| > R$, we see that $|x - a| = |x - c|$ (because $|c - a| \leq R$), hence $I_a(B) \cap]R, R'] = I_c(B) \cap]R, R']$, and finally, $\overline{I_a(B)} \supset [R, R']$ because $\overline{I_c(B)} \supset [R, R']$.

When B is not bounded, in the same way it is seen that $\overline{I_a(A \cup B)} = [0, +\infty[$. This finishes showing that $A \cup B$ is infraconnected. □

Corollary 2.14. *The relation $\mathcal{R}$ defined by $x\mathcal{R}y$ if there exists an infraconnected subset of D that contains x and y is an equivalence relation.*

Proof. $\mathcal{R}$ is obviously reflexive and symmetric. It is transitive by Theorem 2.13. □

Definition. The equivalence classes with respect to this relation are called *the infraconnected components*.

Examples.

(1) $d(0, 1^-) \cup d(1, 1^-)$ is infraconnected. Its holes are the disks $d(\alpha, 1^-)$ with $|\alpha| = |\alpha - 1| = 1$.
(2) Let $r \in]0, 1[$ and let $D = d(0, 1^-) \cup d(1, r)$. Then, D is not infraconnected; its infraconnected components are $d(0, 1^-)$ and $d(1, r)$. The holes of D are the disks $d(\alpha, 1^-)$ with $|\alpha| = |\alpha - 1| = 1$ and the disks $d(\alpha, |\alpha - 1|^-)$ with $r < |\alpha - 1| < 1$.

Theorem 2.15. *Let A and B be infraconnected sets such that $\widetilde{A} = \widetilde{B}$. Then, $A \cup B$ is infraconnected.*

Proof. Obviously, $\widetilde{A \cup B} = \widetilde{A}$. If A is bounded, let $\widetilde{A} = d(\alpha, R)$, and otherwise, let $\widetilde{A} = L$. First, let us assume A to be bounded. For $a \in A$, the set $\{|x - a| \mid x \in A\}$ is dense in $[0, R]$, hence so is the set $\{|x - a\ | \mid x \in A \cup B\}$. In the same way, B plays the same role, hence this still holds for $a \in B$. Finally, if A is not bounded, we just replace $[0, R]$ with $[0, +\infty[$. That finishes proving Theorem 2.15. □

Definition. An infraconnected subset D of $\mathbb{L}$ is said to be *affinoid* if it is of the form $d(a, R) \setminus \bigcup_{k=1}^{q} d(b_k, r_k^-)$ with R and $r_k \in |\mathbb{L}|\ \forall k$. A subset D of $\mathbb{L}$ is said to be *affinoid* if it is a finite union of infraconnected affinoid subsets.

Proposition 2.16. *Let D_1, D_2 be two infraconnected affinoid subsets of $\mathbb{L}$ such that $D_1 \cap D_2 \neq \emptyset$ and set $D = D_1 \cup D_2$ and $E = D_1 \cap D_2$. Then both D and E are infraconnected affinoid. Moreover, $\widetilde{D}$ is either $\widetilde{D_1}$ or $\widetilde{D_2}$ and each hole of D is either a hole of D_1 or a hole of D_2.*

Proof. By Theorem 2.13, D is infraconnected. Consider now D_1 of the form $d(a, r) \setminus (\cup_{i=1}^{m} d(a_i, r_i^-))$ and D_2 of the form $d(b, s) \setminus (\cup_{i=1}^{n} d(b_i, s_i^-))$. Suppose, for instance, $r \leq s$ and let $c \in D_1 \cap D_2$. Then, we can check that

$$E = d(c, r) \setminus \Big(\big(\cup_{i=1}^{d} (a_i, r_i^-) \big) \cup \big(\cup_{i=1}^{n} d(b_i, s_i^-) \big) \Big),$$

which is an infraconnected affinoid again. Since $D_1 \cap D_2 \neq \emptyset$, we have $\widetilde{D_1} \cap \widetilde{D_2} \neq \emptyset$, hence $\widetilde{D}$ is either $\widetilde{D_1}$ or $\widetilde{D_2}$, hence $\text{diam}(D) \in |\mathbb{L}|$. Next, since the holes of both sets are in finite number, each hole of

D is either a hole of D_1 or a hole of D_2, so each hole of D has a diameter in $|\mathbb{L}|$ and of course they are in finite number. □

Definition. We call *empty annulus of* D an annulus $\Gamma(a, r_1, r_2)$ such that

(i) $r_1 = \sup\{|x-a| \quad |x \in D, |x-a| \leq r_2\}$,
(ii) $r_2 = \inf\{\ |x-a| \quad |x \in D, |x-a| \geq r_1\}$.

The set $d(a, r_1) \cap D$ is denoted by $\mathcal{I}_D(\Gamma(a, r_1, r_2))$, while the set $(\mathbb{L} \backslash d(a, r_2^-)) \cap D$ is denoted by $\mathcal{E}_D(\Gamma(a, r_1, r_2))$. When there is no risk of confusion about the set D, we just write $\mathcal{I}(\Gamma(a, r_1, r_2))$ (respectively $\mathcal{E}(\Gamma(a, r_1, r_2))$) instead of $\mathcal{I}_D(\Gamma(a, r_1, r_2))$ (respectively $\mathcal{E}_D(\Gamma(a, r_1, r_2))$).

Remark 1. By definition, D is not infraconnected if and only if it admits an empty annulus.

Remark 2. By definition, $\{\mathcal{I}(\Gamma(a, r_1, r_2)), \mathcal{E}(\Gamma(a, r_1, r_2))\}$ is a partition of D.

Example. Let $r \in]0, 1[$, let $D = d(0, r) \cup d(1, 1^-)$, and let $D' = d(0, r^-) \cup d(1, r)$. Then, $\Gamma(0, r, 1)$ is an empty annulus of D and also of D'. In the same way, $\Gamma(1, r, 1)$ is also an empty annulus of D'.

Notation. Let $\mathcal{X}(D)$ be the set of the empty annuli of D. Given Λ_1 and $\Lambda_2 \in \mathcal{X}(D)$, it is easily seen that $\mathcal{I}(\Lambda_1) \subset \mathcal{I}(\Lambda_2)$ is equivalent to $\mathcal{E}(\Lambda_1) \supset \mathcal{E}(\Lambda_2)$. We denote by $\leq$ the relation defined on $\mathcal{X}(D)$ by $\Lambda_1 \leq \Lambda_2$ if $\mathcal{I}(\Lambda_1) \subset \mathcal{I}(\Lambda_2)$ and we set $\Lambda_1 < \Lambda_2$ if $\Lambda_1 \leq \Lambda_2$ and $\Lambda_1 \neq \Lambda_2$.

Lemmas 2.17 and 2.18 are easily seen.

Lemma 2.17. *The relation $\leq$ is a relation of order on $\mathcal{X}(D)$. Let Λ_1 and Λ_2 be two empty annuli of D. The following assertions are equivalent:*

(i) *Λ_1 and Λ_2 are not comparable with respect to the order $\leq$.*
(ii) *$\mathcal{I}(\Lambda_1) \subset \mathcal{E}(\Lambda_2)$.*
(iii) *$\mathcal{I}(\Lambda_2) \subset \mathcal{E}(\Lambda_1)$.*
(iv) *$\mathcal{I}(\Lambda_1) \cap \mathcal{I}(\Lambda_2) = \emptyset$.*

Lemma 2.18. *Let $\Lambda \in \mathcal{X}(D)$ and let $x \in \mathcal{I}(\Lambda)$ (respectively $x \in \mathcal{E}(\Lambda)$). The infraconnected component of x is included in $\mathcal{I}(\Lambda)$*

(respectively in $\mathcal{E}(\Lambda)$). If $\Lambda' \in \mathcal{X}(D)$ is such that $\Lambda < \Lambda'$, then $\mathcal{I}(\Lambda') \cap \mathcal{E}(\Lambda) \neq \emptyset$.

Lemma 2.19 is a direct consequence of Lemmas 2.17 and 2.18.

Lemma 2.19. *Let Θ be an empty annulus of D. The family of the empty annuli $\Lambda \geq \Theta$ is totally ordered.*

Proof. Let Λ_1 and $\Lambda_2 \in \mathcal{X}(D)$ satisfy $\Lambda_1 \geq \Theta, \Lambda_2 \geq \Theta$. Then, $\mathcal{I}(\Lambda_1) \cap \mathcal{I}(\Lambda_2) \supset \mathcal{I}(\Theta) \neq \emptyset$, hence $\mathcal{I}(\Lambda_1)$ is not included in $\mathcal{E}(\Lambda_2)$, hence Λ_1 and Λ_2 are comparable. □

Lemma 2.20. *Let Θ be a minimal element of $\mathcal{X}(D)$ for the order $\leq$. Then, $\mathcal{I}(\Theta)$ is an infraconnected component of D.*

Proof. Suppose that $\mathcal{I}(\Theta)$ is not infraconnected. By definition, $\mathcal{I}(\Theta)$ is of the form $d(a, R) \cap D$, hence there exists an empty annulus $\Lambda = \Gamma(\alpha, r_1, r_2)$ of $\mathcal{I}(\Theta)$ with $\alpha \in d(a, R), r_1 < r_2 \leq R$ and some $\beta \in \mathcal{I}(\Theta)$ such that $r_2 \leq |\alpha - \beta| \leq R$. Since $\Lambda \subset d(a, R)$, we see that $\Lambda \cap D = \emptyset$, hence Λ is an empty annulus of D, and therefore, $\Lambda < \Theta$. This ends the proof of Lemma 2.20. □

Theorem 2.21. *D has finitely many infraconnected components if and only if it has finitely many empty annuli. Moreover, if so does D, then one of the infraconnected components is $A_0 = \bigcap_{\Theta \in \mathcal{X}(D)} \mathcal{E}(\Theta)$, while the others are of the form $A_i = \mathcal{I}(\Lambda_i) \bigcap \left(\bigcap_{\Theta < \Lambda_i} \mathcal{E}(\Theta) \right)$, with $\Lambda_i \in \mathcal{X}(D)$.*

Proof. We first assume $\mathcal{X}(D)$ to be finite and we prove that the infraconnected components are in the form A_i, above, so that there will be finitely many ones.

Let $\Lambda_1, \ldots, \Lambda_n$ be these empty annuli of D, and for every $i = 0, \ldots, n$, let A_i be the subsets of D defined from $\Lambda_1, \ldots, \Lambda_n$ as above. For every $x \in D$, for every $i = 1, \ldots, n$, either $x \in \mathcal{I}(\Lambda_i)$ or $x \in \mathcal{E}(\Lambda_i)$, hence it is easily seen that x belongs to one of the A_i, hence $D = \bigcup_{i=0}^{n} A_i$. We check that $A_i \cap A_j = \emptyset$ whenever $i \neq j$. First, we assume $i = 0, j > 0$. Hence, $A_0 \subset \mathcal{E}(\Lambda_j)$, while $A_j \subset \mathcal{I}(\Lambda_j)$, hence $A_0 \cap A_j = \emptyset$. Now, we suppose $i > 0, j > 0$. If $\Lambda_i < \Lambda_j$, then $a_j \subset \mathcal{E}(\Lambda_i)$, while $A_i \subset \mathcal{I}(\Lambda_i)$ and then $A_i \cap A_j = \emptyset$. Hence, we may assume that Λ_1 and Λ_2 are not comparable, and then, by Lemma 2.17, we have $\mathcal{I}(\Lambda_i) \cap \mathcal{I}(\Lambda_j) = \emptyset$, hence $A_i \cap A_j = \emptyset$. Consequently, the family $(A_i)_{0 \leq i \leq n}$ makes a partition of D.

Now, we show that each A_i is infraconnected. Suppose that a certain A_h is not infraconnected for some $h > 0$ (respectively $h = 0$). Then, it admits an empty annulus $\Lambda = \Gamma(a, r_1, r_2)$. First, we note that if $h = 0$, then $\Lambda_h > \Lambda$ because both a, b are centers of Λ_h. Now, if $h = 0$ (respectively $h > 0$), let $\Theta \in \mathcal{X}(D)$ (respectively let $\Theta \in \mathcal{X}(D)$ be such that $\Theta < \Lambda_h$). Since both a, b belong to $\mathcal{E}(\Theta)$, it is seen that all Λ is included in $\mathcal{E}(\Theta)$ and therefore is included in A_h. This contradicts the hypothesis and finishes proving that A_h is infraconnected.

Next, we check that each A_j is maximal in the set of the infraconnected subsets of D. Indeed, let B be a subset of D that strictly contains a certain A_h and let $a \in B \setminus A_h$. If $h = 0$, there exists $\Theta \in \mathcal{X}(D)$ such that $a \in \mathcal{I}(\Theta)$, but $A_h \subset \mathcal{E}(\Theta)$, and therefore, Θ is included in an empty annulus of B. If $h > 0$, either a belongs to $\mathcal{E}(\Lambda_h)$ whereas $A_h \subset \mathcal{I}(\Lambda_h)$ and then Λ_h is included in an empty annulus of B, or there exists $\Theta \in \mathcal{X}(D)$ satisfying $\Theta < \Lambda_h$ and $a \in \mathcal{I}(\Theta)$, but then $A_h \subset \mathcal{I}(\Theta)$ and therefore Θ is included in an empty annulus of B. Thus, in each case, B is not infraconnected and this finishes showing that each A_i is maximal in the set of the infraconnected subsets of D. As a consequence, the infraconnected components of D are the A_i.

Now, conversely, we assume D to have infinitely many empty annuli. First, let us suppose that D has a sequence of empty annuli $(\Lambda_n)_{n \in \mathbb{N}}$ such that $\Lambda_n < \Lambda_{n+1}$ (respectively $\Lambda_n > \Lambda_{n+1}$) for all $n \in \mathbb{N}$. By Theorem 2.15, for every $n \in \mathbb{N}$, there exists $x_n \in \mathcal{E}(\Lambda_n) \cap \mathcal{I}(\Lambda_{n+1})$ (respectively $x_n \in \mathcal{I}(\Lambda_n) \cap \mathcal{E}(\Lambda_{n+1})$) and then the infraconnected component X_n of x_n satisfies $X_n \subset \mathcal{E}(\Lambda_n) \cap \mathcal{I}(\Lambda_{n+1})$ (respectively $X_n \subset \mathcal{I}(\Lambda_n) \cap \mathcal{E}(\Lambda_{n+1})$), hence $X_n \cap X_m = \emptyset$ for all $n \neq m$, hence D has infinitely many infraconnected components.

Finally, we may assume that every totally ordered set of empty annuli is finite. Hence, there exists a sequence of empty annuli Λ_n that are minimal elements for the order $\leq$ on $\mathcal{X}(D)$ and then $\mathcal{I}(\Lambda_n) \cap \mathcal{I}(\Lambda_m) = \emptyset$ whenever $n \neq m$. By Lemma 2.19, $\mathcal{I}(\Lambda_n)$ is an infraconnected component D_n of D such that $D_n \cap D_m = \emptyset$ whenever $n \neq m$. This finishes proving that D has infinitely many infraconnected components and this ends the proof. $\square$

Chapter 3

Monotonous and Circular Filters

Monotonous and circular filters are essential on an ultrametric field, mainly because for any rational function, its absolute value admits a limit along each circular filter [36, 44, 45, 49, 53] and circular filters are the least thin filters having this property. Most of properties of analytic functions of all kinds are derived from that property of circular filters. Certain authors call "generic disk" a notion which is not clearly defined but actually represents a circular filter. We see that, given a bounded sequence, there exists a subsequence thinner than a circular filter.

For certain problems, we can reduce ourselves to consider monotonous filters instead of circular filters. Monotonous filters are linked to sequences (a_n) such that $|a_{n+1} - a_n|$ is strictly monotonous. Moreover, decreasing filters let us define spherically complete fields.

Definition. Let J be set. A filter $\mathcal{F}$ on J is said to be *thinner* than a filter $\mathcal{G}$ if every element of $\mathcal{G}$ belongs to $\mathcal{F}$. In such a case, $\mathcal{G}$ is said to be *less thin than* $\mathcal{F}$. Two filters $\mathcal{F}$, $\mathcal{G}$ are said *to be secant* if for all $A \in \mathcal{F}$, $B \in \mathcal{G}$ we have $A \cap B \neq \emptyset$.

A filter $\mathcal{F}$ is said to be *secant* to a subset $B \subset J$ if $\{F \cap B \mid F \in \mathcal{F}\}$ is a filter.

A sequence $(u_n)_{n\in\mathbb{N}}$ in J is said to be *thinner* than a filter $\mathcal{G}$ if so is the filter defined by the sets $A_q = \{u_n | n \geq q\}$ $(q \in \mathbb{N})$. In such a case, $\mathcal{G}$ is said to be *less thin than* the sequence $(u_n)_{n\in\mathbb{N}}$.

A sequence $(u_n)_{n\in\mathbb{N}}$ in $\mathbb{L}$ will be said to be *an increasing distances sequence* (respectively *a decreasing distances sequence*) if the sequence $|u_{n+1} - u_n|$ is strictly increasing (respectively decreasing) and has a limit $\ell \in \mathbb{R}_+^*$.

The sequence $(u_n)_{n\in\mathbb{N}}$ will be said to be *a monotonous distances sequence* if it is either an increasing distances sequence or a decreasing distances sequence.

A sequence $(u_n)_{n\in\mathbb{N}}$ in $\mathbb{L}$ will be said to be *an equal distances sequence* if $|u_n - u_m| = |u_m - u_q|$ whenever $n, m, q \in \mathbb{N}$ such that $n \neq m \neq q$.

Theorem 3.1. *Let $\mathbb{E}$ be a field provided with an ultrametric absolute value. Let $(u_n)_{n\in\mathbb{N}}$ be a bounded sequence in $\mathbb{E}$. Either we may extract a Cauchy subsequence or we may extract a monotonous distances subsequence or we may extract an equal distances subsequence from the sequence $(u_n)_{n\in\mathbb{N}}$. Further, if the absolute value of $\mathbb{E}$ is discrete, there is no monotonous distances sequence in $\mathbb{E}$. And if the residue class field of $\mathbb{E}$ is finite, there is no equal distances sequence in $\mathbb{E}$.*

Proof. Suppose Theorem 3.1 to be false. For every $q \in \mathbb{N}$, the set of the circles $C(u_q, r)$ that contain some u_n is then finite.

Suppose that we have already defined integers n_q for $q \leq t$ satisfying (1) $|u_{n_q} - u_{n_{q-1}}| < |u_{n_{q-1}} - u_{n_{q-2}}|$ for $2 \leq q \leq t$ and such that $d(u_{n_q}, |u_{n_q} - u_{n_{q-1}}|^-)$ contains infinitely many terms of the sequence (u_n). For every $q = 2, \ldots, t$, let $r_q = |u_{n_q} - n_{n_{q-1}}|$. Obviously, at least one of the circles $C(u_{n_t}, r)$, with $r < r_t$, contains infinitely many terms of the sequence $(u_n)_{n\in\mathbb{N}}$. Let $C(u_{n_t}, r_{t+1})$ be such a circle. It is seen that at least one class Λ of this circle contains infinitely many terms of the sequence because otherwise we would have a sequence of classes (Λ_j) each one containing at least one term $u_{\tau(j)}$ and then they should satisfy $|u_{\tau(j)} - u_{\tau(i)}| = r_{t+1}$ whenever $i \neq j$. Hence, the sequence $(u_n)_{n\in\mathbb{N}}$ should admit an equal distances subsequence. Then we may pick up one term $u_{n_{t+1}}$ in Λ and we have constructed the finite subsequence up to the rank $t + 1$, satisfying the properties mentioned above. In the same way, we may initiate the induction by defining n_2 from arbitrary n_0, n_1. The sequence $(u_{n_t})_{t\in\mathbb{N}}$ is then defined for every $t \in \mathbb{N}$ and satisfies (1) for $t > 1$. Let $\ell = \lim_{t\to\infty}|u_{n_t} - u_{n_{t+1}}|$. If $\ell = 0$, the subsequence $(u_{n_t})_{t\in\mathbb{N}}$ is a Cauchy subsequence. If $\ell > 0$, this is a decreasing distances subsequence. Thus, we have proven that we can extract a sequence which is either a convergent sequence, or a monotonous distances sequence, or an equal distances sequence.

Now, suppose that the absolute value is discrete and suppose that we have extracted monotonous distances sequences $(b_m)_{m\in\mathbb{N}}$ from the sequence (u_n). Then the strictly monotonous sequence $|b_{m+1} - b_m|$ must tend to 0, a contradiction. Finally, suppose that the residue class field $\mathcal{E}$ of $\mathbb{E}$ is finite and suppose that we have extracted an equal distances sequences $(b_m)_{m\in\mathbb{N}}$. So, $|b_0 - b_m| = |b_0 - b_n| = |b_m - b_n|$ $\forall m \neq n,\ m \neq 0, n \neq 0$. Let q be the cardinal of $\mathcal{E}$. Then, by Lemma 2.9, the set of terms b_m is at most q, a contradiction. □

Henceforth, throughout this chapter, the field $\mathbb{L}$ *is supposed to have a dense valuation and* D *is an infraconnected subset of* $\mathbb{L}$.

Definition and notation. Let $a \in \widetilde{D}$ and $R \in \mathbb{R}_+^*$ be such that $\Gamma(a, r, R) \cap D \neq \emptyset$ whenever $r \in]0, R[$ (respectively $\Gamma(a, R, r) \cap D \neq \emptyset$ whenever $r > R$). We call *an increasing* (respectively *a decreasing*) *filter of center a and diameter R, on D* the filter $\mathcal{F}$ on D that admits for basis the family of sets $\Gamma(a, r, R) \cap D$ (respectively $\Gamma(a, R, r) \cap D$). For every sequence $(r_n)_{n\in\mathbb{N}}$ such that $r_n < r_{n+1}$ (respectively $r_n > r_{n+1}$) and $\lim_{n\to\infty} r_n = R$, it is seen that the sequence $\Gamma(a, r_n, R) \cap D$ (respectively $\Gamma(a, R, r_n) \cap D$) is a basis of $\mathcal{F}$, and such a basis is called *a canonical basis.* We call *a decreasing filter with no center of canonical basis* $(D_n)_{n\in\mathbb{N}}$ *and diameter* $R > 0$, *on* D a filter $\mathcal{F}$ on D that admits for basis a sequence $(D_n)_n \in \mathbb{N}$ in the form $D_n = d(a_n, r_n) \cap D$ with $D_{n+1} \subset D_n$, $r_{n+1} < r_n, \lim_{n\to\infty} r_n = R$, and $\bigcap_{n\in\mathbb{N}} d(a_n, r_n) = \emptyset$.

Given an increasing (respectively a decreasing) filter $\mathcal{F}$ on D of center a and diameter r, we denote by $\mathcal{B}_D(\mathcal{F})$ the set $\{x \in D|\ |x - a| \geq r\}$ (respectively the set $\{x \in D|\ |x - a| \leq r\}$ and by $\mathcal{C}_D(\mathcal{F})$ the set $\{x \in D|\ |x - a| < r\}$ (respectively the set $\{x \in D|\ |x - a| > r\}$. When there is no risk of confusion, we only write $\mathcal{B}(\mathcal{F})$ instead of $\mathcal{B}_D(\mathcal{F})$ and $\mathcal{C}(\mathcal{F})$ instead of $\mathcal{C}_D(\mathcal{F})$. Next, $\mathcal{C}_D(\mathcal{F})$ is named *the body of* $\mathcal{F}$ and $\mathcal{B}_D(\mathcal{F})$ is named *the beach of* $\mathcal{F}$.

We call *a monotonous filter on D* a filter which is either an increasing filter or a decreasing filter (with or without a center). Given a monotonous filter $\mathcal{F}$, we denote by $\operatorname{diam}(\mathcal{F})$ its diameter.

The field $\mathbb{L}$ is said to be *spherically complete* if every decreasing filter on $\mathbb{L}$ has a center in $\mathbb{L}$. The field $\mathbb{C}_p$, for example, is not spherically complete (see Chapter 6). However, every algebraically closed complete ultrametric field admits a spherically complete algebraically closed extension and this is recalled in Chapter 8.

Lemma 3.2. *Let $(a_n)_n \in \mathbb{N}$ be an increasing distances (respectively a decreasing distances) sequence in D. There exists a unique increasing (respectively decreasing) filter $\mathcal{F}$ on D such that the sequence $(a_n)_n \in \mathbb{N}$ is thinner than $\mathcal{F}$.*

Proof. Let $r_n = |a_{n+1} - a_n|$ and let $R = \lim_{n\to\infty} r_n$.

We first suppose $(a_n)_{n\in\mathbb{N}}$ to be an increasing distances sequence. The increasing filter $\mathcal{F}$ of center a_0, of diameter R is obviously less thin than the sequence $(a_n)_{n\in\mathbb{N}}$. We show that $\mathcal{F}$ is unique. Let $\mathcal{G}$ be an increasing filter of center a, of diameter R', less thin than the sequence $(a_n)_{n\in\mathbb{N}}$. For every $r < R'$, there exists $q \in \mathbb{N}$ such that $a_n \in \Gamma(a, r, R')$ whenever $n \geq q$. If $a \in d(a_0, R^-)$, this clearly requires that $R = R'$ and then $\mathcal{G} = \mathcal{F}$. Let us suppose that $a \notin d(a_0, R^-)$. Then we have $|a - a_n| = |a_n - a_m| = C$ whenever $n \neq m$ so $R' > R$ and then $\Gamma(a, r, R')$ does not contain the a_n whenever $r > R$. Finally, $\mathcal{G} = \mathcal{F}$.

We now suppose the sequence $(a_n)_{n\in\mathbb{N}}$ to be a decreasing distances sequence with a point a such that $|a - a_n| = |a_{n+1} - a_n|$ whenever $n \in \mathbb{N}$. Then the decreasing filter of center a, of diameter R is a decreasing filter less thin than the sequence $(a_n)_{n\in\mathbb{N}}$. We show it to be the only decreasing filter less thin than the sequence $(a_n)_{n\in\mathbb{N}}$. Indeed, given a decreasing filter $\mathcal{G}$ less thin than the sequence $(a_n)_{n\in\mathbb{N}}$, it must have a center because if it had no center, the sequence $d(a_{n+1}, |a_{n+1} - a_n|)$ would be one of its canonical basis, but by definition, it has an intersection that contains a. Then, symmetrical to the case when $\mathcal{F}$ is increasing, it is easily seen that $\mathcal{F}$ is unique.

Now, we suppose that the sequence $(a_n)_{n\in\mathbb{N}}$ is a decreasing distances sequence and that there does not exist $a \in L$ such that $|a - a_n| = |a_{n+1} - a_n|$ whenever $n \in \mathbb{N}$. We put $|a_{n+1} - a_n| = r_n$. Hence, the sequence of disks $d(a_{n+1}, r_n)$ has empty intersection and then the filter $\mathcal{F}$, a basis of which is the sequence $(D_n)_{n\in\mathbb{N}}$ with $D_n = d(a_{n+1}, r_n) \cap D$, is a decreasing filter with no center, of diameter R. There is no decreasing filter with center $a \in \mathbb{L}$, less thin than the sequence (a_n) because we should have $|a - a_n| = r_n$ whenever $n \in \mathbb{N}$. Hence, it just remains to show that $\mathcal{F}$ is the only decreasing filter with no center less thin than the sequence (a_n). Let us suppose that there exists another decreasing filter $\mathcal{G}$ of diameter R' with no center, of canonical basis $(D'_m)_{m\in\mathbb{N}}$ less thin than the sequence

(a_n). If $R' > R$, since every D'_m contains points a_n, it is seen that all the a_n lie in $D \cap d(a_0, R) \subset D'_m$ whenever $m \in \mathbb{N}$ and this contradicts that $\mathcal{G}$ has no center. Hence, we have $R' \le R$. But symmetrically, we have $R \le R'$. Hence, $R = R'$. We show that $\mathcal{G} = \mathcal{F}$. For every $m \in \mathbb{N}$, let ρ_m be the diameter of D'_m and let $a_q \in D'_m$ be such that $r_q \le \rho_m$. Clearly, $a_n \in D'_m$ whenever $n \ge q$, hence $D_n \subset D'_m$ whenever $n > q$. In the same way, let $n \in \mathbb{N}$ and $t \in \mathbb{N}$ be such that $\rho_m < r_n$ whenever $m \ge t$. Then, D'_m contains some a_s which belongs to $d(a_{n+1}, r_n) \cap D = D_n$, hence $D'_m \subset D_n$ whenever $m \ge t$. That finishes showing that $\mathcal{G} = \mathcal{F}$ and that ends the proof of Lemma 3.2. □

Lemma 3.3. *Let $\mathcal{F}$ be an increasing filter (respectively a decreasing filter) on $\mathbb{L}$, of center $a \in \widetilde{D}$ and diameter $R \le \mathrm{diam}(D)$ (respectively $R < \mathrm{diam}(D)$) such that a does not belong to a hole of diameter $\rho \ge R$ (respectively $\rho > R$). Then, $\mathcal{F}$ is secant with D and induces on D an increasing filter (respectively a decreasing filter) of center a and diameter R, on D.*

Proof. We just have to check that $\Gamma(a, r, R) \cap D \neq \emptyset$ whenever $r \in]0, R[$ (respectively $\Gamma(a, R, r) \cap D \neq \emptyset$ whenever $r > R$) and this is obvious when $a \in \overline{D}$ because D is infraconnected and $R \le \mathrm{diam}(D)$ (respectively $R < \mathrm{diam}(D)$). Now, let us assume a to belong to a hole T of diameter $\rho < R$ (respectively $\rho \le R$). Since $|\mathbb{L}|$ is dense in $[0, +\infty[$, for every $r < R$ (respectively $r > R$), D has points α such that $r < |a - \alpha| < R$ (respectively $R < |a - \alpha| < r$) and this ends the proof. □

Definition. Let $\mathcal{F}$ be an increasing (respectively a decreasing) filter of center a and diameter R on D. $\mathcal{F}$ is said *to be pierced* if for every $r \in]0, R[$ (respectively $r > R$), $\Gamma(a, r, R)$ (respectively $\Gamma(a, R, r)$) contains some hole T_m of D.

A decreasing filter with no center $\mathcal{F}$ of canonical basis $(D_m)_{m \in \mathbb{N}}$ on D is said *to be pierced* if for every $m \in \mathbb{N}$, $\widetilde{D}_m \backslash \widetilde{D}_{m+1}$ contains some hole T_m of D.

Remarks. The definition of a pierced filter with no center also applies to a decreasing filter with a center and then is equivalent to that given just above for such a filter.

If $\mathcal{F}$ is an increasing (respectively a decreasing) filter of center a, of diameter R, $\mathcal{F}$ is pierced if and only if there exists a sequence of holes $(T_n)_{n\in\mathbb{N}}$ of D such that $\delta(a, T_n) < \delta(a, T_{t+1})$ (respectively $\delta(a, T_n) > \delta(a, T_{n+1})$), $\lim_{n\to\infty}\delta(a, T_n) = R$.

Given a Cauchy filter $\mathcal{F}$ on D, of limit a in $|L$, we call *a canonical basis of* $\mathcal{F}$ a sequence D_m in the form $d(a, r_m) \cap D$ with $0 < r_m < r_{m+1}$ and $\lim_{m\to\infty} r_m = 0$. The filter $\mathcal{F}$ is said *to be pierced* if for every $m \in \mathbb{N}, \widetilde{D}_m$ contains some hole of D.

Let $a \in \widetilde{D}$. Let $(T_{m,i})_{\substack{1\leq i\leq s(m)\\ m\in\mathbb{N}}}$ be a sequence of holes of D which satisfies $\delta(a, T_{m,i}) = d_m \quad (1 \leq i \leq h_m)$, $d_m < d_{m+1}$ (respectively $d_m > d_{m+1}$), $\lim_{m\to\infty} d_m = S > 0$.

The sequence $(T_{m,i})_{\substack{1\leq i\leq s(m)\\ m\in\mathbb{N}}}$ is called *an increasing (respectively a decreasing) distances holes sequence that runs the increasing (respectively decreasing) filter of center a, of diameter R.*

Now, let $(T_{m,i})_{\substack{1\leq i\leq s(m)\\ m\in\mathbb{N}}}$ be a sequence of holes of D that satisfies $\delta(a_m, T_{m,i}) = d_m \ (1 \leq i \leq s(m))$, $d_m > d_{m+1}$, $\lim_{m\to\infty} d_m = R > 0$, where the filter $\mathcal{F}$ of basis $D_m = d(a_m, d_m) \cap D$ is a decreasing filter with no center. The sequence $(T_{m,i})_{\substack{1\leq i\leq s(m)\\ m\in\mathbb{N}}}$ is called *a decreasing distances holes sequence that runs* $\mathcal{F}$.

Summarizing these definitions, an increasing (respectively decreasing) distances holes sequence that runs an increasing (respectively decreasing) filter $\mathcal{F}$ is just named *an increasing (respectively decreasing) distances holes sequence* and the filter $\mathcal{F}$ is named *the increasing (respectively decreasing) filter associated with the sequence* $(T_{m,i})_{\substack{1\leq i\leq s(m)\\ m\in\mathbb{N}}}$. The diameter of $\mathcal{F}$ is called *the diameter of the sequence* $(T_{m,i})_{\substack{1\leq i\leq s(m)\\ m\in\mathbb{N}}}$. If $\mathcal{F}$ has a center a, a is named *the center of the sequence* $(T_{m,i})_{\substack{1\leq i\leq s(m)\\ m\in\mathbb{N}}}$. If $\mathcal{F}$ has no center, the sequence $(T_{m,i})$ is called *a decreasing distances holes sequence with no center.*

Finally, an increasing (respectively decreasing) distances holes sequence is called *a monotonous distances holes sequence* and the sequence $(d_m)_{m\in\mathbb{N}}$ is called *the monotony* of the monotonous distances holes sequence.

Let $(T_{m,i})_{\substack{1\le i\le s(m)\\ m\in\mathbb{N}}}$ be a monotonous distances holes sequence and for every $(m,i)_{\substack{1\le i\le s(m)\\ m\in\mathbb{N}}}$ let $\rho_{m,i} = \text{diam}(T_{m,i})$. The number $\inf_{\substack{1\le i\le s(m)\\ m\in\mathbb{N}}} \rho_{m,i}$ is called *piercing of the sequence* $(T_{m,i})_{\substack{1\le i\le s(m)\\ m\in\mathbb{N}}}$.

If a monotonous holes sequence has a piercing $\rho > 0$, it is said to be *well pierced.* If a monotonous filter $\mathcal{F}$ is run by a well-pierced monotonous holes sequence, $\mathcal{F}$ is said to be well pierced.

In each case, the sequence of circles $C(a, d_m)$ when $\mathcal{F}$ has center a (respectively $C(a_{m+1}, d_m)$ when $\mathcal{F}$ has no center) is said *to run the filter* $\mathcal{F}$ and *to carry* the monotonous distances holes sequence $(T_{m,i})_{\substack{1\le i\le s(m)\\ m\in\mathbb{N}}}$.

A monotonous distances holes sequence $(T_{m,i})_{\substack{1\le i\le s(m)\\ m\in\mathbb{N}}}$ is said to be *simple* if $s(m) = 1$ for all $m \in \mathbb{N}$.

Next, a sequence of holes $(T_m)_{m\in\mathbb{N}}$ of D is called *a Cauchy sequence of holes of limit* $a \in \mathbb{L}$ if $\lim_{m\to\infty} \delta(a, T_m) = 0$. Such a sequence is said *to run* the Cauchy filter of basis $\{d(a,r) \cap D | r > 0\}$.

Notation. In all the propositions, theorems, corollaries, and Lemmas 3.4–3.10, γ is the Moebius function $b + \frac{1}{x-a}$ with $a, b \in \mathbb{L}$.

Proposition 3.4. *Let* $\alpha \in D, r > 0$ *be such that* $|a - \alpha| < t$. *Then,* $\gamma(C(\alpha, r)) = C(b, \frac{1}{r})$.

Proof. We may assume $b = 0$ and then the proof is immediate. □

Corollary 3.5. *Let* $\alpha \in \mathbb{L}, r_1, r_2 \in]0, +\infty[$ *with* $|a - \alpha| < r_1 < r_2$. *Then,* $\gamma(\Gamma(\alpha, r_1, r_2)) = \Gamma\left(b, \frac{1}{r_2}, \frac{1}{r_1}\right)$.

Corollary 3.6. *Let* $\mathcal{F}$ *be the increasing (respectively decreasing) filter of center* α *and diameter* $R > |a-\alpha|$, *on* $\mathbb{L}\setminus\{a\}$. *Then,* $\gamma(\mathcal{F})$ *is the decreasing (respectively increasing) filter of center* b *and diameter* $\frac{1}{R}$.

Lemma 3.7. *Let* $\alpha \in \mathbb{L}$ *be such that* $|\alpha - a| \neq r$. *Then,*

$$\gamma(C(\alpha, r)) = C\Big(\gamma(\alpha), \frac{r}{|a-\alpha|^2}\Big).$$

Proof. When x belongs to $C(\alpha, r)$, we have

$$\gamma(x) - \gamma(\alpha) = \left|\frac{\alpha - x}{(x-\alpha)(a-\alpha)}\right| = \frac{r}{|a-\alpha|^2},$$

hence $\gamma(C(\alpha, r)) \subset C\Big(\gamma(\alpha), \frac{r}{|a-\alpha|^2}\Big)$. Now, let $\xi(u) = \gamma^{-1}(u) = a + \frac{1}{u-b}$. We see that $C\Big((\gamma(\alpha), \frac{r}{|a-\alpha|^2}\Big) \subset C(\alpha, r)$. Since γ and ξ are injective, we see that γ must be a surjection onto $C\Big(\gamma(\alpha), \frac{r}{|a-\alpha|^2}\Big)$. □

Corollary 3.8. *Let $\alpha \in \mathbb{L}$ and $r, r' \in]0, +\infty[$ be such that $0 < r < r' < |a - \alpha|$. Then, we have* $\gamma\Big(\Gamma(\alpha, r, r')\Big) = \Gamma\Big(\gamma(\alpha), \frac{r}{|a-\alpha|^2}, \frac{r'}{|a-\alpha|^2}\Big)$, $\gamma(d(\alpha, r)) = d\Big(\gamma(\alpha), \frac{r}{|a-\alpha|^2}\Big)$, $\gamma(d(\alpha, r^-)) = d\Big(\gamma(\alpha), \Big(\frac{r}{|a-\alpha|^2}\Big)^-\Big)$.

Corollary 3.9. *Let $\mathcal{F}$ be the increasing (respectively decreasing) filter of center α and diameter R on $\mathbb{L}\setminus\{a\}$ with $|a - \alpha| > R$. Then, $\gamma(\mathcal{F})$ is an increasing (respectively a decreasing) filter of center $\gamma(\alpha)$, of diameter $\frac{R}{|a-\alpha|^2}$ on $\mathbb{L}\setminus\{b\}$.*

Corollary 3.10. *Let $\mathcal{F}$ be a decreasing filter with no center, of basis $(D_n)_{n\in\mathbb{N}}$ on $\mathbb{L}\setminus\{a\}$ such that $a \notin D_0$. Then, $\gamma(\mathcal{F})$ is a decreasing filter with no center, of canonical basis $\Big(\gamma(D_n)\Big)_{n\in\mathbb{N}}$ on $\mathbb{L}\setminus\{b\}$.*

Theorem 3.11. *We suppose $a \in D$. Let $D' = \gamma(D)$. Let $\mathcal{F}$ be a filter on D which is either a monotonous filter or a Cauchy filter. Then, $\mathcal{F}$ is pierced if and only if $\gamma(\mathcal{F})$ is a pierced filter on D'.*

Proof. For example, we suppose first $\mathcal{F}$ to be a monotonous filter. By definition, $\mathcal{F}$ is the intersection with D of a monotonous filter $\mathcal{G}$ of $\mathbb{L}\setminus\{a\}$. Hence, $\mathcal{F}$ is pierced if and only if $\mathcal{G}$ is secant with $(\mathbb{L}\setminus\{a\})\setminus\overline{D}$. Since γ is bicontinuous in $\mathbb{L}\setminus\{a\}$, we see that $\gamma(\mathcal{G})$ is secant with $(\mathbb{L}\setminus\{b\})\setminus\overline{D'}$ if and only if $\mathcal{G}$ is secant with $(\mathbb{L}\setminus\{a\})\setminus\overline{D}$ because $\gamma(\overline{D})\setminus\{a\} = \overline{D'}\setminus\{b\}$. Hence, the conclusion is clear. In the same way if $\mathcal{F}$ is a Cauchy filter of limit $\alpha \in \mathbb{L}$, we consider the filter $\mathcal{G}'$ of the neighborhoods of $\gamma(\alpha)$ in $\mathbb{L}\setminus\{b\}$ and we see that $\mathcal{G}$ is secant with $(L\backslash\{a\})\backslash\overline{D}$ if and only if $\mathcal{G}'$ is secant with $(\mathbb{L}\backslash\{b\})\backslash\overline{D'}$. □

We are now going to define circular filters, which roughly characterize the absolute values on $\mathbb{L}(x)$ when $\mathbb{L}$ is algebraically closed.

Definition. Let $a \in \mathbb{L}$ and let $R \in]0, +\infty[$. We call *circular filter of center a and diameter R on L* the filter $\mathcal{F}$ which admits as a generating system the family of sets $\Gamma(\alpha, r', r'') \cap D$ with $\alpha \in d(a, R), r' < R < r''$, i.e. $\mathcal{F}$ is the filter which admits for basis the family of sets of the form $\left(\bigcap_{i=1}^{q} \Gamma(\alpha_i, r'_i, r''_i)\right)$ with $\alpha_i \in d(a, R), r'_i < R < r''_i$ $(1 \leq i \leq q\ ,\ q \in \mathbb{N})$.

For reasons that appear when characterizing the absolute values of $\mathbb{L}(x)$ when $\mathbb{L}$ is algebraically closed, a decreasing filter with no center on $\mathbb{L}$, of canonical basis $(D_n)_{n\in\mathbb{N}}$ is also called *circular filter on $\mathbb{L}$ with no center, of canonical basis $(D_n)_{n\in\mathbb{N}}$*.

Finally, the filter of the neighborhoods of a point $a \in \mathbb{L}$ is called *circular filter of the neighborhoods of a on $\mathbb{L}$*. It is also named *circular filter of center a and diameter* 0. A circular filter on $\mathbb{L}$ is said to be *large* if it has a diameter different from 0.

A circular filter on $\mathbb{L}$ secant with D is called *circular filter* on D. Given a circular filter $\mathcal{F}$ on $\mathbb{L}$, its diameter is denoted by $\mathrm{diam}(\mathcal{F})$ and we call *$\mathcal{F}$-affinoid* any infraconnected affinoid subset of $\mathbb{L}$ lying in $\mathcal{F}$.

Lemma 3.12 lets us describe circular filters on an infraconnected subset of $\mathbb{L}$.

Lemma 3.12. *Let $a \in \widetilde{D}$, let ρ be the distance from a to D, and let R be such that $\rho \leq R \leq \mathrm{diam}(D)$. For $j = 1, \ldots, q$, let $\alpha_j \in d(a, R)$ and let $r'_j, r''_j \in \mathbb{R}_+$ be such that $r'_j < R < r''_j$. Then, $\bigcap_{j=0}^{q}(\Gamma(\alpha_j, r'_j, r''_j) \cap D) \neq \emptyset$.*

Proof. If $\rho < R$, we put $r' = \max_{1\leq j\leq q} r'_j$ and we see that $\Gamma(a, r', R) \cap D$ is not empty (because D is infraconnected) and is included in every set $\Gamma(\alpha_j, r'_j, r''_j) \cap D$. If $R < \mathrm{diam}(D)$, we put $r'' = \min_{1\leq j\leq q} r''_j$, and in the same way, $\Gamma(a, R, r'') \cap D$ is not empty and is included in every set $\Gamma(\alpha_j, r'_j, r''_j) \cap D$.

Now, if $\rho = R = \mathrm{diam}(D)$, let $b \in D$ and let $r' = \max_{1\leq\ j\leq\ q} r'_j$. Then, $\Gamma(b, r', R) \cap D$ is not empty and is included in every set $\Gamma(\alpha_j, r'_j, r''_j) \cap D$. □

Corollary 3.13. *Let $a \in \widetilde{D}$, let ρ be the distance from a to D, and let R be such that $\rho \leq R \leq \mathrm{diam}(D)$. The circular filter on $\mathbb{L}$ of center a and diameter R is secant with D.*

Proposition 3.14 is immediate according to definitions:

Proposition 3.14. *Let $\mathcal{F}$ be an increasing filter (respectively a decreasing filter) of center a and diameter R, on D. Then, the circular filter of center a and diameter R on $\mathbb{L}$ is secant with D and is the only circular filter on D less thin than $\mathcal{F}$.*

Conversely, let $\mathcal{F}$ be a circular filter of center a and diameter R, on D secant with $d(a, R^-)$ (respectively $\mathbb{L}\backslash d(a, R)$). Then, the increasing filter (respectively decreasing filter) of center a and diameter R on $\mathbb{L}$ is secant with D and thinner than $\mathcal{F}$.

Lemma 3.15. *Let $\mathcal{F}$ be a circular filter. Then, $\mathcal{F}$ admits a basis consisting of the family of all $\mathcal{F}$-affinoids. If $\mathcal{F}$ does not admit a countable basis, it has a center and its diameter belongs to $|\mathbb{L}|$. If $\mathcal{F}$ has no center and is secant with an infraconnected affinoid subset B of $\mathbb{L}$, then B lies in $\mathcal{F}$. If $\mathcal{F}$ has center a and diameter r, then an infraconnected affinoid set B lies in $\mathcal{F}$ if and only if satisfy $E \cap (\mathbb{L}\backslash d(a, r)) \neq \emptyset$, $B \cap d(b, r^-) \neq \emptyset\ \forall b \in d(a, r)$.*

Proof. By definition, a circular filter with no center has a countable basis and of course so does a Cauchy circular filter. In both cases, it admits a basis consisting of a family of disks which are $\mathcal{F}$-affinoid sets.

Now, consider a circular filter of center a and diameter $r > 0$. Then, $\mathcal{F}$ admits for basis the family of sets of the form $d(a, r + \frac{1}{n})\backslash(\bigcup_{i=1}^{q} d(a_i, (r - \frac{1}{n})^-)$, where the a_i are centers of $\mathcal{F}$ satisfying $|a_i - a_j| = r$. In particular, if $r \notin |L|$, we have $q = 1$ and we obtain a basis of the form $\Gamma(a, r - \frac{1}{n}, r + \frac{1}{n})$ which is countable.

Now, suppose that $\mathcal{F}$ is secant with an infraconnected affinoid subset B of $\mathbb{L}$. Suppose first that $\mathcal{F}$ has no center. Let $(A_n)_{n\in\mathbb{N}}$ be a canonical basis of $\mathcal{F}$. Since each A_n admits common points with B, each is included in $\widetilde{B}$, and therefore, it is included in B if and only if it contains no hole of B. But since $\mathcal{F}$ has no center, $\bigcap_{n=0}^{\infty} A_n = \emptyset$, hence there exists $q \in \mathbb{N}$ such that $A_n \subset B\ \forall n \geq q$, and therefore, $B \in \mathcal{F}$.

Now, suppose that $\mathcal{F}$ has center a and diameter r. If $B \in \mathcal{F}$, it obviously satisfies $B \cap (\mathbb{L} \setminus d(a,r)) \neq \emptyset$, $B \cap d(b, r^-) \neq \emptyset\ \forall b \in d(a,r)$. Since B has finitely many holes, on one hand, there exists $s > r$ such that $\Gamma(a,r,s) \subset E$, and on the other hand, all classes of $d(a,r)$ are included in B, except finitely many: $d(b_j, r^-)$, $1 \leq j \leq n$. And for each $j = 1, \ldots, n$, there exists $r_j < r$ such that $\Gamma(b_j, r_j, r) \subset B$. Finally, B contains the set $d(a,s) \setminus \bigcup_{j=1}^{n} d(b_j, r_j)$ which obviously lies in $\mathcal{F}$. □

Corollary 3.16. *Let $\mathcal{F}$ be a circular filter of diameter r. For every $s \in]0, r[$, the family of $\mathcal{F}$-affinoid of codiameter $\rho > s$ is a basis of $\mathcal{F}$. If two disks $d(a,r)$ and $d(b,s)$ have no common points and if $\mathcal{F}$ is secant with $d(a,r)$, it is not secant with $d(b,s)$.*

Proposition 3.17. *Let $a \in \widetilde{D}$ and let S be the closure of $\{|x - a| \mid x \in D\}$ in $\mathbb{R}$. For every $r \in S$, the circular filter $\mathcal{F}$ of center a and diameter r is secant with D.*

Proof. Let $a \in \widetilde{D}$. We first suppose that $C(a,r) \cap D = \emptyset$. Then, either there exists a sequence $(x_n)_{n\in\mathbb{N}}$ in D such that $r < |x_{n+1} - a| < |x_n - a|$, $\lim_{n\to\infty} |x_n - a| = r$, or there exists a sequence $(x_n)_{n\in\mathbb{N}}$ in D such that $|x_n - a| < |x_{n+1} - a| < r$, $\lim_{n\to\infty} |x_n - a| = r$. In both cases, the circular filter $\mathcal{F}$ of center a and diameter r is clearly secant with D.

Now, we may suppose that $C(a,r) \cap D \neq \emptyset$. Let $b \in C(a,r) \cap D$. We see that b is also a center of $\mathcal{F}$. Since D is infraconnected and since $|a - b| = r \leq \operatorname{diam}(D)$, there does exist a sequence $(x_n)_{n\in\mathbb{N}}$ in D such that $\lim_{n\to\infty} |x_n - b| = r$. Hence, $\mathcal{F}$ (that has center b and diameter r) is secant with D. □

Proposition 3.18. *Let $(a_n)_{n\in\mathbb{N}}$ be a sequence in D that is either a monotonous distances sequence or a constant distances sequence. Then, there exists a unique circular filter on D less thin than the sequence (a_n).*

Proof. First, we suppose that the sequence $(a_n)_{n\in\mathbb{N}}$ is an increasing (respectively a decreasing) distances sequence. By Lemma 3.2, there exists a unique increasing (respectively decreasing) filter $\mathcal{F}$ on D less thin than the sequence $(a_n)_{n\in\mathbb{N}}$. If $\mathcal{F}$ has center a and diameter R, by Proposition 3.14, $\mathcal{F}$ is less thin than the circular filter of center a, of diameter R on D. If $\mathcal{F}$ is decreasing with no center, $\mathcal{F}$ is a circular filter.

Now, we suppose that $(a_n)_{n\in\mathbb{N}}$ is a constant distances sequence. We put $R = |a_n - a_m|$ for $n \neq m$ and $a = a_0$. The circular filter $\mathcal{F}$ of center a of diameter R on $\mathbb{L}$ is clearly secant with D because each set $\Lambda_n = \Gamma(a_n, r', r'')$ with $r' < R < r''$ belongs to a generating system of $\mathcal{F}$ and contains a_m for every $m > n$, hence its intersection with D is a circular filter $\mathcal{C}$ on D less thin than the sequence $(a_n)_{n\in\mathbb{N}}$. That ends the proof. □

Corollary 3.19. *Let $\mathcal{F}$ and $\mathcal{G}$ be two circular filters that are secant. Then, they are equal.*

Proof. We can find a monotonous sequence thinner than $\mathcal{F}$. Then, the sequence is thinner than $\mathcal{G}$ and hence $\mathcal{G} = \mathcal{F}$. □

Notation. Let $\mathbb{L}'$ be an extension of $\mathbb{L}$ provided with an absolute value that extends the one of $\mathbb{L}$. Let D be a set in $\mathbb{L}$, let $\mathcal{F}$ be a monotonous filter on D, and let D' be a set in $\mathbb{L}'$ that contains D. Let $(a_n)_{n\in\mathbb{N}}$ be a monotonous distances sequence that runs $\mathcal{F}$. In $\widehat{D}$, there is a unique monotonous filter less thin than the sequence $(a_n)_{n\in\mathbb{N}}$. This filter is denoted by $\widehat{\mathcal{F}}$.

In the same way, let $\mathcal{G}$ be a circular filter of center a and diameter r on D. We denote by $\widehat{\mathcal{G}}$ the filter of center a and diameter r on D'. Finally, let $\mathcal{G}$ be a circular filter with no center. Then, it is a decreasing filter, hence we have already previously defined $\widehat{\mathcal{G}}$.

Corollary 3.20. *Let $(a_n)_{n\in\mathbb{N}}$ be a bounded sequence in $\mathbb{L}$. Then, there exists a subsequence $(a_{n_t})_{t\in\mathbb{N}}$ and a unique circular filter $\mathcal{F}$ on $\mathbb{L}$ less thin than the subsequence $(a_{n_t})_{t\in\mathbb{N}}$.*

Proof. Since the sequence $(a_n)_{n\in\mathbb{N}}$ is bounded, by Theorem 3.1, we can extract either a monotonous distances subsequence, or a constant distances subsequence, or a converging subsequence. In all cases, once such a subsequence is chosen, there exists a unique circular filter $\mathcal{F}$ on $\mathbb{L}$ less thin than the subsequence. □

Theorem 3.21. *Let $(a_n)_{n\in\mathbb{N}}$, $(b_n)_{n\in\mathbb{N}}$ be two sequences such that $|a_n - b_n| \leq t < r$ $\forall n \in \mathbb{N}$. Suppose that the sequence $(a_n)_{n\in\mathbb{N}}$ is thinner than a circular filter $\mathcal{F}$ of diameter r. Then, the sequence $(b_n)_{n\in\mathbb{N}}$ also is thinner than $\mathcal{F}$.*

Proof. By Corollary 3.16, $\mathcal{F}$ admits a basis consisting of $\mathcal{F}$-affinoids S of codiameter $\rho > t$. Consider such a $\mathcal{F}$-affinoid S. Then, if a_n belongs to S, so does b_n. Now, when n is big enough, all a_n belong to S and hence so do all b_n. And since $\mathcal{F}$ admits a basis of $\mathcal{F}$-affinoids with a codiameter $s > t$, we see that the sequence $(b_n)_{n\in\mathbb{N}}$ is thinner than $\mathcal{F}$. □

Chapter 4

Ultrametric Absolute Values for Rational Functions

Notation. As mentioned in Chapter 1, log denotes a real logarithm function of basis $\theta > 1$ (eventually, we can take for θ an integer p that is the residue characteristic of $\mathbb{K}$). When a function f from an interval I to $\mathbb{R}$ admits a right side (respectively a left side) derivative at a point $a \in I$, we denote it by $f'^r(a)$ (respectively $f'^l(a)$). If the variable is μ, we also denote it by $\frac{d_r f}{d\mu}$ (respectively $\frac{d_l f}{d\mu}$).

Moreover, though out this chapter, we denote by $\mathbb{L}$ be a field provided with an ultrametric absolute value $| \,.\, |$.

The set of circular filters on $\mathbb{K}$ secant with a subset D of $\mathbb{K}$ is denoted by $\Phi(D)$ and the subset of large circular filters on $\mathbb{K}$ secant with D is denoted by $\Phi^\circ(D)$. We show the absolute values on the field $\mathbb{K}(x)$ of rational functions to be characterized by the circular filters on $\mathbb{K}$. Actually, the most important property of such absolute values comes from the fact that the logarithm of an absolute value is a piecewise affine function of the logarithm of the absolute value of the variable. And next, a valuation function is then defined for any $h \in \mathbb{K}(x)$ in the following way: let $r \in]0, +\infty[$ be such that $\mu = -\log r$ and let $\mathcal{F}$ be the circular filter of center 0 and diameter r. Following classical notations [1, 14], one sets

$$v(h, \mu) = -\log\left(\lim_{\mathcal{F}} |h(x)|\right).$$

This function $v(h, \mu)$, called *the valuation function of h*, is convenient mainly because it is piecewise affine. However, in order to avoid many changes of sign, here we consider $\Psi(h, \log r) = \log(\lim_{\mathcal{F}} |h(x)|)$ and we show that when $\frac{d_r\Psi(h,\mu)}{d\mu} \neq \frac{d_l\Psi(h,\mu)}{d\mu}$ then $\frac{d_r\Psi(h,\mu)}{d\mu} - \frac{d_l\Psi(h,\mu)}{d\mu}$ is equal to the difference between the number of zeros and the number of poles of h (taking multiplicity into account) on the circle $C(0, r)$ such that $\log r = \mu$. This translates properties of $|h(x)|$ into terms of piecewise affine functions.

However, this kind of definition presents the inconvenient of changing the sense of monotony for both $|x|$ and $|h(x)|$. Moreover, its sign is opposite to this of the counting function of zeros for entire functions in the Nevanlinna theory. This is why, here we adopt another set of notation and put $\Psi(x) = \log|x|$ $\forall x \in \mathbb{K}$. First, we have to state several basic properties that work not only in an algebraically closed field such as $\mathbb{K}$ but more generally in a field $\mathbb{E}$ that is just provided with an ultrametric absolute value.

Let $\mathbb{L}[x_1, \ldots, x_q]$ be an algebra of polynomials in q indeterminates, with coefficients in $\mathbb{L}$. For each $P(x_1, \ldots, x_q) = \sum_{j_1+\cdots+j_q\leq t} a_{i_1,\ldots,i_q} x_1^{j_1} \ldots\ldots x_q^{j_q}$ we set

$\overline{P(x_1, \ldots, x_q)} := \sum_{j_1+\cdots+j_q\leq t} \overline{a_{i_1,\ldots,i_q}} x_1^{j_1} \ldots\ldots x_q^{j_q}$.

On $\mathbb{L}[x_1, \ldots, x_q]$, we put $\|P\| := \sup_{j_1+\cdots+j_q\leq t} |a_{i_1,\ldots,i_q}|_0$.

However, when there is no risk of confusion, we just write $\| \, . \, \|$ instead of $\| \, . \, \|_0$.

Lemma 4.1. *$\| \, . \, \|$ is a multiplicative norm of $\mathbb{L}$-algebra.*

Proof. Let $B = \mathbb{L}[x_1, \ldots, x_q]$. Clearly, $\| \, . \, \|$ is an ultrametric norm of $\mathbb{L}$-vector space on B. We check that $\|PQ\| = \|P\|\|Q\|$ whenever $P, Q \in B$. Both $\|P\|$, $\|Q\|$ belong to $|\mathbb{L}|$. Hence, without loss of generality, we may clearly assume $\|P\| = \|Q\| = 1$. Thus, we have $\overline{P} = \overline{Q} = \overline{1}$. Let $\mathcal{L}$ be the residue field of $\mathbb{L}$. Since $\mathcal{L}[x_1, \ldots, x_q]$ is a ring without divisors of zeros, we have $\overline{PQ} \neq \overline{0}$, and therefore, $\|PQ\| = 1$. This ends the proof. □

Definition and notation. The norm $\| \, . \, \|$ on $\mathbb{L}[x_1, \ldots, x_q]_0$ is called *the Gauss norm*. Given a polynomial $P(x) = \sum_{j=1}^{q} a_j x^j \in \mathbb{L}[x]$, for any $r > 0$, we set $|P|(r) = \max_{0\leq j\leq q}(|a_j| r^j)$.

By ultrametricity, Lemma 4.2 is then immediate:

Lemma 4.2. *Let $P(x) \in \mathbb{L}[x]$. For all $x \in \mathbb{L}$, one has $|P(x)| \leq |P|(|x|)$.*

Lemma 4.3. *Suppose $\mathbb{L}$ is algebraically closed. Let $P(x) = \sum_{j=0}^{n} a_j x^j \in \mathbb{L}[x]\setminus\{0\}$ and let $r \in \mathbb{R}_+$. Then, $|P(x)|$ admits a limit equal to $|P|(r)$ when $|x|$ approaches r but remains different from r. Let $x \in d(0,r)$. Then, $|P(x)| \leq |P|(r)$. If P has no zero in the class of x in $d(0,r)$, then $|P(x)| = |P|(r)$. If P has at least one zero in that class, then $|P(x)| < |P|(r)$.*

Proof. Let $r' < r$ and $r'' > r$ be such that P has no zero in $\Gamma(0,r',r) \cup \Gamma(0,r,r'')$. We may obviously assume P to be monic. Let $P(x) = \prod_{i=1}^{n}(x-\alpha_i)$ be the factorization of P in an algebraic closure of $\mathbb{L}$, with

$|\alpha_i| < r'$ for $i \leq h$,
$|\alpha_i| > r''$ for $i \geq \ell$,
$|\alpha_i| = r$ for $i = h, \dots, \ell$.

Now, let $x \in \Gamma(0,r',r)$. Clearly, $|x-\alpha_i| = |x|$ whenever $i \leq h$, while $|x-\alpha_i| = |\alpha_i|$ whenever $i > h$, hence $|P(x)| = |x|^h \prod_{i=h+1}^{n}|\alpha_i|$, hence $\lim_{|x|\to r^-}|P(x)| = r^h \prod_{i=h+1}^{n}|\alpha_i|$.

Symmetrically, we show that $\lim_{|x|\to r^+}|P(x)| = r^{\ell-1}\prod_{i=\ell}^{n}|\alpha_i| = r^h \prod_{i=h+1}^{n}|\alpha_i|$. But the terms $|a_j x^j|$ are all different for every $|x|$ except for finitely many values so that there exist $\rho' \in [r',r[$ and $\rho'' \in]r,r'']$ such that the $|a_j x^j|$ are all different when $x \in \Gamma(0,\rho',r) \bigcup \Gamma(0,r,\rho'')$. Then, we have $|P(x)| = \max_{0\leq j\leq n}|a_j||x|^j$ and hence $\lim_{\substack{|x|\to r \\ |x|\neq r}} |P(x)| = \max_{0\leq j\leq n}|a_j|r^j$.

Now, let $x \in d(0,r)$ and let us assume P to have no zero in the class Λ of x in $d(0,r)$. This means that $|x-\alpha_i| = r$ whenever $i = 0,\dots,\ell-1$ and $|x-\alpha_i| = |\alpha_i|$ whenever $i \geq \ell$. Thus, $|P(x)| = r^{\ell-1}\prod_{i=\ell}^{n}|\alpha_i|$. If P has at least one zero α_{h+1} in the class of x, we see that $|x-\alpha_{h+1}| < r$, while $|x-\alpha_i| \leq r$ for every $\imath = h+2,\dots,\ell-1$, hence $|P(x)| < r^{\ell-1}\prod_{i=\ell}^{n}|\alpha_i|$, and finally, $|P(x)| < |P|(r)$. □

Theorem 4.4. *Let $P \in \mathbb{L}[X]$ and let U be the unit disk $\{x \in \mid |x| \leq 1\}$ of $\mathbb{K}$. Then, $\|P\| = \sup_{x\in U}(|P(x))$.*

Proof. On one hand, $|P(x)| \leq \|P\| \ \forall x \in U$. On the other hand, by Theorem 4.3, $\lim_{\substack{|x|\to 1,\ x\in U,\\ |x|\neq 1}} |P(x)| = \|P\|$. Consequently, the equality holds. □

Corollary 4.5. *Let $P \in \mathbb{L}[X]$ and let $t \in \mathbb{L}$ be such that $|t| \leq 1$. Let $Q(X) = P(X+t)$. Then, $\|P\| = \|Q\|$. If $\|P\| \leq 1$, then P is 1-Lipschitzian.*

Theorem 4.6. *For every $r > 0$, the mapping from $\mathbb{L}[x]$ to $\mathbb{R}_+$ defined by $P \to |P|(r)$ is an absolute value on $\mathbb{L}[x]$ such that $|P(a)| \leq |P|(|a|) \ \forall a \in \mathbb{L}$.*

Proof. Due to the definition of $|P|(r)$, it is easily checked that $|P|(r) = 0$ if and only if $P = 0$ and that

$$|P+Q|(r) \leq \max(|P|(r), |Q|(r)).$$

We can also check that $|P(a)| \leq |P|(|a|) \ \forall a \in \mathbb{L}$. Now, set $P(x) = \sum_{j=1}^{m} a_j x^j$, $Q(x) = \sum_{j=1}^{n} b_j x^j$, and let $P(x)Q(x) = \sum_{j=1}^{m+n} c_j x^j$. Let s (respectively t) be the biggest of the integers such that $|P|(r) = |a_s|r^s$ (respectively $|Q|(r) = |b_t|r^t$). Then, $|P|(r)|Q|(r) = |a_s b_t| r^{s+t}$. On the one hand, we can check that, obviously, $|c_j|r^j \leq |a_s b_t| r^{s+t} \ \forall j = 0, \ldots, m+n$, hence $|PQ|(r) \leq |P|(r)|Q|(r)$. On the other hand, since $|a_j|r^j < |a_s|r^s \ \forall j > s$ and $|b_j|r^j < |b_t|r^t \ \forall j > t$, we have $|c_{s+t}|r^{s+t} = |a_s b_t| r^{s+t}$, which proves that $|PQ|(r) \geq |P|(r)|Q|(r)$ and hence ends the proof. □

Now, Lemma 4.7 shows that we can change the origin, inside the disk $d(0, r)$.

Lemma 4.7. *Suppose $\mathbb{L}$ is algebraically closed. Let $r \in \mathbb{R}_+$ and let $a \in \mathbb{L}$ be such that $|a| \leq r$. Then, $|P(x)|$ has a limit when $|x - a|$ approaches r but remains different from r. Further, that limit does not depend on $a \in d(0, r)$ and it belongs to $|\mathbb{L}|$ if and only if so does r.*

Proof. We set $x = a + u$ and $P_a(u) = P(a+u)$. For every $P \in \mathbb{L}[x]$, we have $\lim_{\substack{|u|\to r\\ |u|\neq r}} |P_a(u)| = |P_a|(r)$. In particular, $|P_a|(r) = \lim_{\substack{|u|\to r^+\\ |u|<r}} |P_a(u)|$. But for every $\rho > r, C(0, \rho) = C(a, \rho)$, hence

$\lim_{|x|\to r^+}|P(x)| = \lim_{|u|\to r^+}|P(a+u)| = \lim_{|u|\to r^+}|P_a(u)|$. That ends the proof of Lemma 4.7. $\square$

Theorem 4.8. *Suppose $\mathbb{L}$ is algebraically closed. Let $P(x) = \sum_{j=0}^{q} a_j x^j \in \mathbb{L}[x]$ be a monic polynomial such that $a_j \in d(0,1)$ whenever $j = 0, \ldots, q$. Then, the q zeros of P belong to $d(0,1)$.*

Proof. Let ψ be the absolute value defined on $\mathbb{L}[x]$ by $\psi(P) = \lim_{|u|\to 1, |u|\neq 1}|P(u)|$ and let $P(x) = \prod_{j=1}^{q}(x - c_j)$. By Lemma 4.3, for each $j = 1, \ldots, q$, it is seen that $\psi(x - c_j) \geq 1$, while $\psi(P) = 1$. Hence, $\psi(x - c_j) = 1$ for every $j = 1, \ldots, q$, and therefore, by Lemma 4.3 again, we have $c_j \leq 1$ whenever $j = 1, \ldots, q$. $\square$

Notation. These ultrametric absolute values defined on $\mathbb{L}[x]$ are immediately extended to rational functions by $\left|\frac{P}{Q}\right|(r) := \frac{|P|(r)}{|Q|(r)}$.

Then, Lemma 4.9 is immediate:

Lemma 4.9. *Suppose $\mathbb{L}$ is algebraically closed. Let $h \in \mathbb{L}(x)$ and let $r \in \mathbb{R}_+$. For every $a \in d(0,r)$, we have $\lim_{\substack{|x-a|\to r \\ |x-a|\neq 0}} |h(x)| = |h|(r)$. Let $x \in C(0,r)$. If h has no zero and no pole in the class of x in $C(0,r)$, then $|h(x)| = |h|(r)$. Further, $|h|(r)$ belongs to $|\mathbb{K}|$ if and only if so does r.*

Circular filters characterize the multiplicative norms defined on $\mathbb{K}(x)$ [61, 36, 49].

Theorem 4.10 (Guennebaud). *For every circular filter $\mathcal{F}$ on $\mathbb{K}$, for every polynomial $P(x) \in \mathbb{K}[x], |P(x)|$ has a limit $\varphi_{\mathcal{F}}(P)$ along the filter $\mathcal{F}$. The mapping $\mathcal{F} \to \varphi_{\mathcal{F}}$ from $\Phi(\mathbb{K})$ into the set of the multiplicative semi-norms on $\mathbb{K}[x]$ is a bijection. Moreover, for every large circular filter on $\mathbb{K}$, $\varphi_{\mathcal{F}}$ has continuation to $\mathbb{K}(x)$ and the mapping $\mathcal{F} \to \varphi_{\mathcal{F}}$ from $\Phi^{\circ}(\mathbb{K})$ into the set of the multiplicative norms on $\mathbb{K}(x)$ is a bijection.*

If $\mathcal{F}$ has center 0 and diameter r, then $\varphi_{\mathcal{F}}(h) = |h|(r)$.

Proof. We first suppose that $\mathcal{F}$ has center $a \in \mathbb{K}$ and diameter $r > 0$. With no loss of generality, we may obviously assume $a = 0$ by means of the change of variable $x = a + u$. Then, by Lemma 4.3,

$|h(x)| = |h|(r)$ holds in every class of $C(0,r)$ but finitely many ones $\Lambda_1, \dots, \Lambda_q$. For every $j = 1, \dots, q$, we take $\alpha_j \in \Lambda_j$ and set $\alpha_0 = 0$. Let ϵ be > 0. By Lemma 4.3, there exist $\rho' \in]0, r[$ and $\rho'' > r$ such that $|\ |h(x)| - |h|(r)\ |_\infty \le \epsilon$ for every $x \in \bigcap_{j=0}^{q} \Gamma(\alpha_j, \rho', \rho'')$, so $\lim_{\mathcal{F}} |h(x)| = |h|(r)$.

Now, we suppose that $\mathcal{F}$ has no center in $\mathbb{K}$. It admits a canonical basis $(D_n)_{n\in\mathbb{N}}$, and then given $h \in \mathbb{K}(x)$, there exists $q \in \mathbb{N}$ such that h has neither any zero nor any pole inside D_q. Hence, $|h(x)|$ is equal to a constant l in D_q, and therefore, we have $\lim_{\mathcal{F}} |h(x)| = l$. By the same kind of reasoning as in Theorem 4.6, it is easily seen that $\varphi_{\mathcal{F}}$ is an absolute value on $\mathbb{K}(x)$.

Now, we check that the mapping $\mathcal{F} \to \varphi_{\mathcal{F}}$ is injective. Indeed, let $\mathcal{F}_1, \mathcal{F}_2$ be two different circular filters and let r_1 (respectively $r_2 \ge r_1$) be the diameter of $\mathcal{F}_1$ (respectively $\mathcal{F}_2$). We first suppose that we may find disks $\Lambda_1 = d(a_1, \rho_1)$ and $\Lambda_2 = d(a_2, \rho_2)$ such that $\Lambda_1 \cap \Lambda_2 = \emptyset$ and such that $\mathcal{F}_1$ (respectively $\mathcal{F}_2$) is secant with Λ_1 (respectively Λ_2). Then, it is seen that $|a_1 - a_2| > \rho_1 \ge r_1$, hence $\varphi_{\mathcal{F}_1}(x - a_1) \le r_1$, while $\varphi_{\mathcal{F}_2}(x - a_1) = |a_1 - a_2| > r_1$, and therefore, $\varphi_{\mathcal{F}_1} \ne \varphi_{\mathcal{F}_2}$.

We now suppose that we cannot find disks Λ_1, Λ_2 defined as above. Since $r_1 \le r_2$, any disk Λ which belongs to $\mathcal{F}_1$ is included in any disk that belongs to $\mathcal{F}_2$, and therefore, any point of Λ_1 is a center of $\mathcal{F}_2$. Thus, $\mathcal{F}_2$ admits a center $a \in \Lambda$, and then, $\mathcal{F}_1$ is secant with $d(a, r_2)$. Hence, we have $r_1 < r_2$ because otherwise $\mathcal{F}_1$ would be equal to $\mathcal{F}_2$. In particular, $\mathcal{F}_2$ is secant with one class $d(\alpha, r_2^-)$ of $d(a, r_2)$. Then, we have $\varphi_{\mathcal{F}_1}(x - \alpha) \le r_1$, while $\varphi_{\mathcal{F}_2}(x - \alpha) = r_2$. This finishes showing that the mapping $\mathcal{F} \to \varphi_{\mathcal{F}}$ is injective.

Now, we show that this mapping defined on $\Phi^\circ(\mathbb{K})$ is also surjective onto the set of multiplicative norms i.e. the absolute values on $\mathbb{K}[x]$ continuing these of $\mathbb{K}$. Indeed, let ψ be such an absolute value on $\mathbb{K}[x]$ and let $r = \inf_{\lambda \in \mathbb{K}} \psi(x - \lambda)$.

We first suppose that there exists $a \in \mathbb{K}$ such that $\psi(x - a) = r$. Since ψ is an absolute value, we check that $r > 0$ because if $r = 0$, we have $\psi(h) = h(a)$ for every $h \in \mathbb{K}[x]$ and then ψ is not an absolute value. Hence, we can assume $r > 0$. Let $\mathcal{F}$ be the circular filter of center a, of diameter r. By Lemma 2.7, we know that ψ is ultrametric, and then, for every $b \in \mathbb{K}$, we have $\psi(x-b) \le \max\ (\psi(x-a), |a-b|) = \max(|a - b|, r)$. But, by definition, we have $\psi(x - b) \ge r$, hence (1) $r \le \psi(x - b) \le \max(r, |a - b|)$.

If $|a-b| > r$, then both $\psi(x-b), \varphi_{\mathcal{F}}(x-b)$ are equal to $|a-b|$. If $|a-b| \leq r$, then b is another center of $\mathcal{F}$ and we have $\varphi_{\mathcal{F}}(x-a) = \varphi_{\mathcal{F}}(x-b) = r$. But, by (1), we see that $\psi(x-b) = r$. So, we have shown that $\varphi_{\mathcal{F}}(x-b) = \psi(x-b)$ for all $b \in \mathbb{K}$, and since $\mathbb{K}$ is algebraically closed, this finishes proving that $\psi = \varphi_{\mathcal{F}}$.

We now suppose that there does not exist $a \in \mathbb{K}$ such that $r = \psi(x-a)$. There exists $\alpha_n \in \mathbb{K}$ such that $r < \psi(x-\alpha_n) < r + \frac{1}{n}$. Let $\rho_n = \psi(x-\alpha_n)$. For $b \in \mathbb{K}\backslash d(\alpha_n, \rho_n)$, clearly we have

(2) $\psi(x-b) = |b-\alpha_n| > \rho_n$

because $\psi(x-\alpha_n) < |\alpha_n - b|$. Further, if $\mathcal{F}_n$ is the circular filter of center α_n and diameter ρ_n, we have

(3) $\psi(x-b) = |b-\alpha_n| = \varphi_{\mathcal{F}_n}(x-b)$.

However, there exists $\alpha_{n+1} \in \mathbb{K}$ such that $r < \psi(x-\alpha_{n+1}) < \min(\rho_n, r + \frac{1}{n+1})$. Hence, by (2), we see that $\alpha_{n+1} \in d(\alpha_n, \rho_n)$. That way, we may define a decreasing sequence of disks $D_n = d(\alpha_n, \rho_n)$ such that $r < \rho_n < r + \frac{1}{n}$ and $\psi(x-\alpha_n) = \rho_n$. Let $D'_n = D_n \cap D$. Then the decreasing filter $\mathcal{F}$ of basis $(D'_n)_{n\in\mathbb{N}}$ satisfies $\lim_{\mathcal{F}} \psi(x-\alpha_n) = r$. It is easily seen that $\mathcal{F}$ has no center because if α is a center of $\mathcal{F}$ then $\psi(x-\alpha) \leq \max\left(|\psi - \alpha_n|, |\alpha_n \ - \alpha|\right)$ hence $\psi(x-\alpha) = r$. We show that $\psi = \varphi_{\mathcal{F}}$. Let $b \in \mathbb{K}$ and let $q \in \mathbb{N}$ be such that $b \notin D_q$. Then, by (3), for $n \geq q$, we have $\psi(x-b) = \varphi_{\mathcal{F}_n}(x-b)$. On the other hand, it is easily seen that $\varphi_{\mathcal{F}_n}(x-b) = \varphi_{\mathcal{F}}(x-b)$. Thus, $\psi(x-b) = \varphi_{\mathcal{F}}(x-b)$ whenever $b \in \mathbb{K}$ and then $\psi = \varphi_{\mathcal{F}}$.

Finally, let ψ be a multiplicative semi-norm that is not a norm: there exists a polynomial P such that $\psi(P) = 0$, and hence, there exists $a \in \mathbb{K}$ such that $\psi(x-a) = 0$. Let $b \in \mathbb{K}$. Then, $\psi(x-b) = \psi((x-a)+(a-b)) \leq \max \psi(x-a), \psi(a-b))$. But since $\psi(t) = |t|\ \forall t \in \mathbb{K}$, we have $\psi(x-b) = |a-b|$, hence, putting $h(x) = x-b$, we have $\psi(h) = |h(a)|$, which shows that the equality $\psi(h) = |h(a)|$ holds for every polynomial of degree 1, and therefore, it holds in all $\mathbb{K}[x]$. This ends the proof of Theorem 4.10. □

Theorem 4.11. *Let $\mathcal{H}$ be a filter on $\mathbb{K}$ and let $\mathcal{F}, \mathcal{G} \in \Phi(\mathbb{K})$ be less thin than $\mathcal{H}$. Then, $\mathcal{F} = \mathcal{G}$.*

Proof. Since $\mathcal{H}$ is thinner than $\mathcal{F}$ and $\mathcal{G}$, we have $\lim_{\mathcal{H}} |P(x)| = \varphi_{\mathcal{F}}(P) = \varphi_{\mathcal{G}}(P)\ \ \forall P \in \mathbb{K}[x]$, hence $\varphi_{\mathcal{F}} = \varphi_{\mathcal{G}}$. But, by Theorem 4.10, the mapping that associates with each circular filter $\mathcal{F}$ the multiplicative semi-norm $\varphi_{\mathcal{F}}$ is injective and hence $\mathcal{F} = \mathcal{G}$. □

Notation. When $\mathcal{F}$ is the circular filter of center a, of diameter r, we also denote by $\varphi_{a,r}$ the absolute value $\varphi_{\mathcal{F}}$. Hence, by definition, we have $\varphi_{a,r}(h) = \lim_{\substack{|x-a|\to r \\ |x-a|\neq a}} |h(x)|$. In particular, we note that $\varphi_{0,r}(h) = |h|(r)$.

Finally, we denote by φ_a the multiplicative semi-norm defined on rational functions with no pole at a as $\varphi_a(h) = |h(a)|$.

Now, let us go back to the field $\mathbb{L}$. For $\mu \in \mathbb{R}$, we set $\Psi(h,\mu) = \log(|h|(\theta^{\mu}))$, for simplicity, we set $\Psi(h) = \Psi(h,0)$. Thus, comparative to the valuation function $v(h,\mu)$ defined and used in previous works [1, 14, 49], we have $\Psi(h,\mu) = -v(h,-\mu)$. The advantage of the function Ψ is to respect the sense of variation of $|h|(r)$.

The translation of statements 4.1, 4.2, 4.3, 4.6, 4.7, 4.8, and 4.9 into terms of valuation allows us to obtain Lemmas 4.12 and 4.13.

Lemma 4.12. *Let* $P(x) = \sum_{j=0}^{n} a_j x^j \in \mathbb{K}[x]\setminus\{0\}$. *For every* $\mu \in \mathbb{R}$, *we have* $\Psi(P,\mu) = \max_{0\leq j\leq n} \Psi(a_j) + j\mu$. *Moreover,* $\Psi(P(a)) \leq \Psi(P,\Psi(a))$ $\forall a \in L$.

Suppose $P \in \mathbb{K}[x]$. *The equality* $\Psi(P(x)) = \Psi(P,\Psi(x))$ *holds if and only if* P *has no zero* α *such that* $\Psi(x-\alpha) < \Psi(x)$.

Lemma 4.13. *Let* $h \in \mathbb{K}(x)\setminus\{0\}$. *We have* $\Psi(h(x)) = \Psi(h,\Psi(x))$ *for every* $x \in \mathbb{K}$ *such that* h *has no zero* α *satisfying* $\Psi(x-\alpha) < \Psi(x)$ *and no pole* β *satisfying* $\Psi(x-\beta) < \Psi(x)$.

Lemma 4.14. *Let* $h_1, h_2 \in \mathbb{K}(x)\setminus\{0\}$. *Then, we have*

$$\Psi(h_1 + h_2, \mu) \leq \max\big(\Psi(h_1,\mu), \Psi(h_2,\mu)\big).$$

When $\Psi(h_1,\mu) > \Psi(h_2,\mu)$, *then we have* $\Psi(h_1+h_2,\mu) = \Psi(h_1,\mu)$. *Moreover,* $\Psi(h_1.h_2,\mu) = \Psi(h_1,\mu) + \Psi(h_2,\mu)$.

Notation. In order to perform easily any change of origin, for every $a \in \mathbb{K}$ and $h \in \mathbb{K}(x)\setminus\{0\}$, we put $\Psi_a(h,\mu) = \Psi(h_a,\mu)$ with $h_a(u) = h(a+u)$. Thus, if $\mathcal{F}$ denotes the circular filter of center a and diameter θ^{μ}, then $\Psi_a(h,\mu) = \log(\varphi_{\mathcal{F}}(h))$.

We now consider again a polynomial $P(x) = \sum_{j=0}^{n} a_j x^j \neq 0$. We denote by $\nu^{+}(P,\mu)$ (respectively $\nu^{-}(P,\mu)$) the biggest (respectively the smallest) index j such that $\Psi(a_j) + j\mu = \Psi(P,\mu)$.

Lemma 4.15 is a consequence of Lemma 4.9:

Lemma 4.15. *Let $h \in \mathbb{K}(x)\setminus\{0\}$ and let $a, b \in \mathbb{K}$. For every $\mu \geq \Psi(a-b)$, we have $\Psi_a(h,\mu) = \Psi_b(h,\mu)$.*

Theorem 4.16. *Let $P(x) = \sum_{j=0}^{n} a_j x^j \in \mathbb{K}[x]$. For every $\mu \in \mathbb{R}$, $\nu^+(P,\mu) - \nu^-(P,\mu)$ is equal to the number of zeros admitted by P in the circle $C(0,\theta^\mu)$ in $\mathbb{K}$. The function $\nu^+(P,.)$ (respectively $\nu^-(P,.)$) is increasing and continuous on the right (respectively on the left). Moreover, given $Q \in \mathbb{K}[x]$, then ν^+, ν^- satisfy $\nu^+(PQ,\mu) = \nu^+(P,\mu) + \nu^+(Q,\mu)$, $\nu^-(PQ,\mu) = \nu^-(P,\mu) + \nu^-(Q,\mu)$. Further, if $\nu^+(P,\mu) = \nu^-(P,\mu)$, then both are constant in a neighborhood of μ.*

The function $\Psi(P,.)$ is continuous, piecewise affine, increasing, and convex and has a right-side derivative (respectively a left-side derivative) equal to $\nu^+(P,\mu)$ (respectively $\nu^-(P,\mu)$).

Proof. It is easily seen that the equality (1) $\nu^+(P,\mu) = \nu^-(P,\mu)$ holds for every μ but finitely many values, at most n. It is also clear that the functions $\nu^+(P,.)$ and $\nu^-(P,.)$ are increasing. By continuity, we see that the function $\nu^+(P,\mu)$ is continuous on the right at each point, while $\nu^-(P,\mu)$ is continuous on the left at each point. Finally, if (1) holds in an interval $]\mu',\mu''[$, then the functions $\nu^+(P,.)$ and $\nu^-(P,.)$ are constant and equal. Consider an interval $]\mu',\mu''[$ such that $\nu^+(P,\mu) = \nu^-(P,\mu)$ for all $\mu \in]\mu',\mu''[$ and let $j = \nu^+(P,\mu)$ whenever $\mu \in]\mu',\mu''[$. Then, $\Psi(P,\mu) = \Psi(a_j) + j\mu$ so that the function $\Psi(P,.)$ is in the form $A + i\mu$ in this interval.

Now, let μ be such that $\nu^+(P,\mu) < \nu^-(P,\mu)$. We see that $\Psi(P,.)$ is still continuous at μ and has a left-side derivative equal to $\nu^-(P,\mu)$ and a right-side derivative equal to $\nu^+(P,\mu)$. Finally, the function $\Psi(P,.)$ is continuous, piecewise affine, convex, and largely increasing.

If P and $Q \in \mathbb{K}(x)\setminus\{0\}$, then $\nu^+(PQ,\mu)$ is the right-side derivative of the function $\Psi(PQ,.)$. But $\Psi(PQ,.) = \Psi(P,.) + \Psi(Q,.)$, hence its right-side derivative at μ is just $\nu^+(P,\mu) + \nu^+(Q,\mu)$. In the same way, we have $\nu^-(PQ,\mu) = \nu^-(P,\mu) + \nu^-(Q,\mu)$ by considering left-side derivatives.

Then, to prove that $\nu^+(P,\mu) - \nu^-(P,\mu)$ is the number of zeros of P in $C(0,\theta^\mu)$, it is sufficient to show this when P is a binomial $x - a$. But then, this is obvious because $\nu^+(P,\mu) = \nu^-(P,\mu) = 0$ whenever $\mu < \Psi(\alpha)$, $\nu^+(P,\mu) = \nu^-(P,\mu) = 1$ whenever $\mu > \Psi(\alpha)$, while $\nu^+(P, \Psi(\alpha)) = 1, \nu^-(P, \Psi(\alpha)) = 0$. So all the statements of Theorem 4.16 have been proven. □

Applying Lemma 4.12 and Theorem 4.16 to the numerator and the denominator of a rational function, we obtain Corollary 4.17.

Corollary 4.17. *Let $P(x) = \sum_{j=0}^{n} a_j x^j \in \mathbb{K}[x]$. For every $\mu \in \mathbb{R}$, $\nu^+(P, \log(r))$ is equal to the number of zeros admitted by P in $d(0, r)$.*

Corollary 4.18. *Let $h \in \mathbb{K}(x)\setminus\{0\}$. The function in μ $\Psi(h, \mu)$ is continuous and piecewise affine.*

If μ is such that $d(0, \theta^\mu)$ contains s zeros and t poles of h (taking multiplicity into account), but neither any zero nor any pole in $C(0, \theta^\mu)$, then $\Psi(h, .)$ has a derivative equal to $s - t$ at μ.

If μ is such that $C(0, \theta^\mu)$ contains s zeros and t poles of h (taking multiplicity into account), then we have $\frac{d_l \Psi}{d\mu}(h, \mu) - \frac{d_r \Psi}{d\mu}(h, \mu) = s - t$. Further, if the function $\Psi(f, \mu)$ is not derivable at μ, then μ lies in $\Psi(\mathbb{K})$.

Corollary 4.19. *Let $h \in \mathbb{K}(x)\setminus\{0\}$ have no pole (respectively no zero) in an annulus $\Gamma(0, r', r'')$. Then, $\Psi(h, \mu)$ is convex (respectively concave) in $[\log r', \log r'']$.*

Corollary 4.20. *Let $h \in \mathbb{K}(x)\setminus\{0\}$ have s zeros and t poles in $d(0, r')$ and have neither any zero nor any pole in an annulus $\Gamma(0, r', r'')$. Then in $\Gamma(0, r', r'')$, $\Psi(h, \log|x|)$ is of the form $A + (s - t)\log|x|$. Moreover, $\nu^+(h, \mu) - \nu^-(h, \mu) = s - t \ \forall \mu \in]\log r', \log r''[$ and $\nu^+(h, \mu)$ (respectively $\nu^-(h, \mu)$) is continuous on the right (respectively on the left). Finally, given $g \in \mathbb{K}(x)$, we have $\nu^+(gh, \mu) = \nu^+(g, \mu) + \nu^+(h, \mu)$, $\nu^-(gh, \mu) = \nu^-(g, \mu) + \nu^-(h, \mu)$.*

Theorem 4.21. *Let $\mathcal{G} \in \Phi(\mathbb{K})$. Let $f \in \mathbb{K}(x)$ and take $\epsilon > 0$. There exists a $\mathcal{G}$-affinoid E such that $\big| |f(x)| - \varphi_{\mathcal{G}}(f)\big|_\infty \leq \epsilon$, $\forall x \in E$.*

Proof. Let $r = \text{diam}(\mathcal{G})$ and let $l > r$. If $\mathcal{G}$ has no center, there exists a disk $d(a, l) \in \mathcal{G}$, with $r \in |\mathbb{K}|$, containing neither zeros nor poles of f, therefore, by Lemma 4.9, $|f(x)|$ is a constant equal to $\varphi_{\mathcal{F}}(f)$ in $d(a, r)$, so our claim is obvious. Now, suppose that $\mathcal{G} = \mathcal{F}_{a,r}$. Let $\Lambda_1, \ldots, \Lambda_q$ be the classes of $d(a, r)$ containing at least one zero or one pole of f. By Lemma 4.9, $|f(x)|$ is a constant equal to $\varphi_{\mathcal{G}}(f)$ in $d(a, r)\setminus\left(\bigcup_{j=1}^{q} \Lambda_j\right)$. Consider a class $\Lambda_j = d(a_j, r^-)$

Lemma 3.2. *Let $(a_n)_n \in \mathbb{N}$ be an increasing distances (respectively a decreasing distances) sequence in D. There exists a unique increasing (respectively decreasing) filter $\mathcal{F}$ on D such that the sequence $(a_n)_n \in \mathbb{N}$ is thinner than $\mathcal{F}$.*

Proof. Let $r_n = |a_{n+1} - a_n|$ and let $R = \lim_{n\to\infty} r_n$.

We first suppose $(a_n)_{n\in\mathbb{N}}$ to be an increasing distances sequence. The increasing filter $\mathcal{F}$ of center a_0, of diameter R is obviously less thin than the sequence $(a_n)_{n\in\mathbb{N}}$. We show that $\mathcal{F}$ is unique. Let $\mathcal{G}$ be an increasing filter of center a, of diameter R', less thin than the sequence $(a_n)_{n\in\mathbb{N}}$. For every $r < R'$, there exists $q \in \mathbb{N}$ such that $a_n \in \Gamma(a, r, R')$ whenever $n \geq q$. If $a \in d(a_0, R^-)$, this clearly requires that $R = R'$ and then $\mathcal{G} = \mathcal{F}$. Let us suppose that $a \notin d(a_0, R^-)$. Then we have $|a - a_n| = |a_n - a_m| = C$ whenever $n \neq m$ so $R' > R$ and then $\Gamma(a, r, R')$ does not contain the a_n whenever $r > R$. Finally, $\mathcal{G} = \mathcal{F}$.

We now suppose the sequence $(a_n)_{n\in\mathbb{N}}$ to be a decreasing distances sequence with a point a such that $|a - a_n| = |a_{n+1} - a_n|$ whenever $n \in \mathbb{N}$. Then the decreasing filter of center a, of diameter R is a decreasing filter less thin than the sequence $(a_n)_{n\in\mathbb{N}}$. We show it to be the only decreasing filter less thin than the sequence $(a_n)_{n\in\mathbb{N}}$. Indeed, given a decreasing filter $\mathcal{G}$ less thin than the sequence $(a_n)_{n\in\mathbb{N}}$, it must have a center because if it had no center, the sequence $d(a_{n+1}, |a_{n+1} - a_n|)$ would be one of its canonical basis, but by definition, it has an intersection that contains a. Then, symmetrical to the case when $\mathcal{F}$ is increasing, it is easily seen that $\mathcal{F}$ is unique.

Now, we suppose that the sequence $(a_n)_{n\in\mathbb{N}}$ is a decreasing distances sequence and that there does not exist $a \in L$ such that $|a - a_n| = |a_{n+1} - a_n|$ whenever $n \in \mathbb{N}$. We put $|a_{n+1} - a_n| = r_n$. Hence, the sequence of disks $d(a_{n+1}, r_n)$ has empty intersection and then the filter $\mathcal{F}$, a basis of which is the sequence $(D_n)_{n\in\mathbb{N}}$ with $D_n = d(a_{n+1}, r_n) \cap D$, is a decreasing filter with no center, of diameter R. There is no decreasing filter with center $a \in \mathbb{L}$, less thin than the sequence (a_n) because we should have $|a - a_n| = r_n$ whenever $n \in \mathbb{N}$. Hence, it just remains to show that $\mathcal{F}$ is the only decreasing filter with no center less thin than the sequence (a_n). Let us suppose that there exists another decreasing filter $\mathcal{G}$ of diameter R' with no center, of canonical basis $(D'_m)_{m\in\mathbb{N}}$ less thin than the sequence

By ultrametricity, Lemma 4.2 is then immediate:

Lemma 4.2. *Let $P(x) \in \mathbb{L}[x]$. For all $x \in \mathbb{L}$, one has $|P(x)| \leq |P|(|x|)$.*

Lemma 4.3. *Suppose $\mathbb{L}$ is algebraically closed. Let $P(x) = \sum_{j=0}^{n} a_j x^j \in \mathbb{L}[x] \setminus \{0\}$ and let $r \in \mathbb{R}_+$. Then, $|P(x)|$ admits a limit equal to $|P|(r)$ when $|x|$ approaches r but remains different from r. Let $x \in d(0,r)$. Then, $|P(x)| \leq |P|(r)$. If P has no zero in the class of x in $d(0,r)$, then $|P(x)| = |P|(r)$. If P has at least one zero in that class, then $|P(x)| < |P|(r)$.*

Proof. Let $r' < r$ and $r'' > r$ be such that P has no zero in $\Gamma(0,r',r) \cup \Gamma(0,r,r'')$. We may obviously assume P to be monic. Let $P(x) = \prod_{i=1}^{n}(x - \alpha_i)$ be the factorization of P in an algebraic closure of $\mathbb{L}$, with

$|\alpha_i| < r'$ for $i \leq h$,
$|\alpha_i| > r''$ for $i \geq \ell$,
$|\alpha_i| = r$ for $i = h, \dots, \ell$.

Now, let $x \in \Gamma(0,r',r)$. Clearly, $|x - \alpha_i| = |x|$ whenever $i \leq h$, while $|x - \alpha_i| = |\alpha_i|$ whenever $i > h$, hence $|P(x)| = |x|^h \prod_{i=h+1}^{n}|\alpha_i|$, hence $\lim_{|x|\to r^-}|P(x)| = r^h \prod_{i=h+1}^{n}|\alpha_i|$.

Symmetrically, we show that $\lim_{|x|\to r^+}|P(x)| = r^{\ell-1}\prod_{i=\ell}^{n}|\alpha_i| = r^h \prod_{i=h+1}^{n}|\alpha_i|$. But the terms $|a_j x^j|$ are all different for every $|x|$ except for finitely many values so that there exist $\rho' \in [r', r[$ and $\rho'' \in]r, r'']$ such that the $|a_j x^j|$ are all different when $x \in \Gamma(0,\rho',r) \bigcup \Gamma(0,r,\rho'')$. Then, we have $|P(x)| = \max_{0\leq j\leq n}|a_j||x|^j$ and hence $\lim_{\substack{|x|\to r \\ |x|\neq r}} |P(x)| = \max_{0\leq j\leq n}|a_j|r^j$.

Now, let $x \in d(0,r)$ and let us assume P to have no zero in the class Λ of x in $d(0,r)$. This means that $|x - \alpha_i| = r$ whenever $i = 0, \dots, \ell - 1$ and $|x - \alpha_i| = |\alpha_i|$ whenever $i \geq \ell$. Thus, $|P(x)| = r^{\ell-1}\prod_{i=\ell}^{n}|\alpha_i|$. If P has at least one zero α_{h+1} in the class of x, we see that $|x - \alpha_{h+1}| < r$, while $|x - \alpha_i| \leq r$ for every $i = h+2, \dots, \ell - 1$, hence $|P(x)| < r^{\ell-1}\prod_{i=\ell}^{n}|\alpha_i|$, and finally, $|P(x)| < |P|(r)$. □

Theorem 4.4. *Let $P \in \mathbb{L}[X]$ and let U be the unit disk $\{x \in | \ |x| \leq 1\}$ of $\mathbb{K}$. Then, $\|P\| = \sup_{x\in U}(|P(x))$.*

and let s_j (respectively t_j) be the number of zeros (respectively poles) of f inside Λ_j and let s_0 (respectively t_0) be the number of zeros (respectively poles) of f in all $d(a,r)$. Let $\rho \in]0,r[\cap|\mathbb{K}|$ be such that $\left|\left(\frac{r}{\rho}\right)^{s_j-t_j} - 1\right|\varphi_{\mathcal{G}}(f) \leq \epsilon \ \forall j = 0,\dots,q$. Let $l = \frac{r^2}{\rho}$ and let $E = d(a,l)\setminus\bigcup_{j=1}^{q} d(a_j,\rho^-)$. By Corollary 4.19, we can check that the inequality $\|\,|f(x)| - \varphi_{\mathcal{G}}(f)|_\infty \leq \epsilon$ holds in all E. Since $\rho < r$, E is an infraconnected affinoid set which belongs to $\mathcal{G}$. Moreover, by definition, $l > r$. □

Theorem 4.22. *Let $\mathcal{F}$ be a filter on $\mathbb{K}$ such that for every $h \in \mathbb{K}(x)$, $|h(x)|$ admits a limit along $\mathcal{F}$. Then there exists a circular filter $\mathcal{H}$ less thin than $\mathcal{F}$.*

Proof. For every $h \in \mathbb{K}(x)$, set $\phi(h) = \lim_{\mathcal{F}} |h(x)|$. Then, ϕ belongs to Mult($|K(x)$), and hence, by Theorem 4.10, there exists a unique circular filter $\mathcal{H}$ such that $\phi = \varphi_{\mathcal{H}}$. Suppose that $\mathcal{F}$ is not thinner than $\mathcal{H}$. There exists a subset B of $\mathbb{K}$ such that $\mathcal{F}$ is secant with B but $\mathcal{H}$ is not. Since $\mathcal{H}$ admits a basis consisting of affinoid subsets, there exists an $\mathcal{H}$-affinoid D such that $D \cap B = \emptyset$. Since D is affinoid, it is of the form $d(a,R)\setminus\left(\bigcup_{k=1}^{q} d(a_i,r_i^-)\right)$ and $\mathcal{H}$ also admits an $\mathcal{H}$-affinoid E of the form $d(a,S)\setminus\left(\bigcup_{k=1}^{q} d(a_i,s_i^-)\right)$ with $S < R$ and $r_i < s_i \ \forall i = 1,\dots,q$. Let $b \in \Gamma(a,S,R)$ and let $h(x) = \frac{\prod_{i=1}^{q}(x-a_i)^{m_i}}{(x-b)^n}$. Then with integers m_i and n big enough, we can get

$$\inf\{|h(x)| \mid x \in E\} > \sup\{|h(x)| \mid x \in B\},$$

a contradiction to the hypothesis: $|h(x)|$ admits a limit along $\mathcal{F}$. □

Chapter 5

Hensel Lemma

The Hensel lemma is a classical tool for studying the factorization of analytic functions on a circle [1, 44] and is indispensable in Chapter 6. It is a strong result that roughly says, "In a complete field $\mathbb{L}$, if $\overline{P}$ splits in the form $\gamma\eta$ with $(\gamma, \eta) = \overline{1}$, then P also splits in $\mathbb{L}[x]$ in the form gh with $\overline{g} = \gamma,\ \overline{h} = h, \deg(\overline{g}) = \deg(\gamma)$". The proof is not very easy and requires a serious preparation. Here, we roughly follow the same process as in [1], [44].

Notation. According to Chapter 2, we denote by $U_{\mathbb{L}}$ the unit ball of $\mathbb{L}$ and by $M_{\mathbb{L}}$ its maximal ideal. We denote by $\mathcal{L}$ the residue class field of $\mathbb{L}$: $\frac{U_{\mathbb{L}}[x]}{M_{\mathbb{L}}[x]}$.

Given $g, h \in \mathbb{L}[x]$, here (g, h) denotes the monic greatest common divisor of g and h. Given $Q(x) = \sum_{j=0}^{n} b_j x^j \in U_{\mathbb{L}}[x]$, as in Chapter 2, we denote by $\overline{Q}$ the polynomial $\sum_{j=0}^{q} \overline{a_j}\ x^j \in \mathcal{L}[x]$. In this chapter, $P(x) = \sum_{j=0}^{q} a_j x^j \in \mathbb{L}[x]$ denotes a polynomial of degree q.

Lemma 5.1. *For all $\alpha \in \mathbb{L}$, we have $\Psi(P(\alpha)) \leq \Psi(P, 0) + \max(0, q\Psi(\alpha))$.*

Proof. By Lemma 4.12, we have $\Psi(P(\alpha)) \leq \max_{0 \leq j \leq q}(\Psi(a_j) + j\Psi(\alpha)) \leq \max_{0 \leq j \leq q} \Psi(a_j) + \max_{0 \leq j \leq q} j\Psi(\alpha)$. But $\max_{0 \leq j \leq q} \Psi(a_j) = \Psi(P, 0)$ and $\max_{0 \leq j \leq q} j\Psi(\alpha) = \max(0, q\Psi(\alpha))$. □

Definition. A polynomial $\sum_{j=0}^{q} a_j x^j \in L[x]$ is said to be *quasi-monic* if $|a_q| = 1$.

Lemma 5.2. *Let $F, D \in U_{\mathbb{L}}[x]$ with D quasi-monic. Let $Q, R \in U_{\mathbb{L}}[x]$ satisfy $F = DQ + R$ and $\deg(R) < \deg(D)$. Then, we have $\Psi(Q, 0) \leq \Psi(F, 0)$ and $\Psi(R, 0) \leq \Psi(F, 0)$.*

Proof. We can clearly assume $F \neq 0$. Then, by multiplying F by a suitable constant λ, we can also assume $\Psi(F, 0) = 0$. Since D is quasi-monic, the Euclidean division of F by D is clearly possible in $U_{\mathbb{L}}[x]$, and therefore, Q is the quotient and R is the rest of this division, due to the fact that $\deg(R) < \deg(D)$. So we have $\Psi(Q, 0) \leq 0$, $\Psi(R, 0) \leq 0$ because both Q, R belong to $U_{\mathbb{L}}[x]$. □

Corollary 5.3. *Let $F, D \in \mathbb{K}[x]$ with D having all its zeros in $d(0, 1)$. Let $Q, R \in U_{\mathbb{K}}[x]$ satisfy $F = DQ+R$ and $\deg(R) < \deg(D)$. Then we have $\Psi(Q, 0) \leq \Psi(F, 0) - \Psi(D, 0)$ and $\Psi(R, 0) \leq \Psi(F, 0)$. Moreover, if F has all its zeros in $d(0, 1)$, then $\Psi(Q, 0) = \Psi(F, 0) - \Psi(D, 0)$.*

Proof. The first statement is just an application of Lemma 5.2. Next, if F has all its zeros in $d(0, 1)$, we can assume that both F, D are monic and satisfy $\|F\| = \|D\| = 1$. Consequently, Q also must be monic and hence $\|Q\| = 1$, which ends the proof. □

Lemma 5.4. *Let $g, h \in U_{\mathbb{L}}[x]$ be quasi-monic such that $(\overline{g}, \overline{h}) = \overline{1}$ and $\deg(P) < \deg(g) + \deg(h)$. There exist $V, W \in \mathbb{L}[x]$ satisfying $\Psi(Vg + Wh - P, 0) < \Psi(P, 0), \Psi(V, 0) \leq \Psi(P, 0)$, $\Psi(W, 0) \leq \Psi(P, 0)$, $\deg(V) < \deg(h)$, $\deg(W) < \deg(g)$.*

Proof. Since $(\overline{g}, \overline{h}) = \overline{1}$, by Bezout's theorem, there exists υ and $\tau \in \mathcal{L}[x]$ such that $\upsilon\overline{g} + \tau\overline{h} = \overline{1}$, $\deg(\upsilon) < \deg(\overline{h})$, $\deg(\tau) < \deg(\overline{g})$. Let $S, T \in U\mathbb{L}[x]$ satisfy $\overline{S} = \upsilon$, $\overline{T} = \tau$, $\deg(S) = \deg(\upsilon)$, $\deg(T) = \deg(\tau)$. Thus, we have $\overline{Sg + Th - 1} = \overline{0}$ i.e.

(1) $\Psi(Sg + Th - 1) < 0$.

We now consider the Euclidean division of SP by h and TP by g, respectively. We obtain $SP = S_0h + V$and $TP = T_0g + W$. By Lemma 5.1, it is seen that $\max(\Psi(V, 0), \Psi(S_0, 0)) \leq \Psi(SP, 0) \leq \Psi(P, 0)$. Moreover, by hypothesis, we have

(1) $\deg(V) < \deg(h)$ and
(2) $\deg(W) < \deg(g)$.

Let $M = Sg + Th - 1$. Then, we have $MP = (S_0 + T_0)gh + Vg + Wh - P$. Since $\deg(P) < \deg(g) + \deg(h)$, by (1) and (2), we see that

$\deg(Vg+Wh-P) < \deg(g)+\deg(h)$, and therefore, $Vg+Wh-P$ is just the remainder of the Euclidean division of BP by gh. But then, by Lemma 5.2, we have $\Psi(Vg+Wh-P,0) \leq \Psi(MP,0) = \Psi(M,0)+\Psi(P,0)$, and therefore, by (1) and by definition of M, it is seen that $\Psi(M,0) < 0$. This finishes proving that $\Psi(Vg+Wh-P,0) < \Psi(P,0)$ and this ends the proof of Lemma 5.4. □

Notation. Let $g, h \in U_{\mathbb{L}}[x]$ be monic and satisfy $(\overline{g}, \overline{h}) = \overline{1}$. We denote by $B(f,g)$ the set of constants $c \in \mathbb{R}_+$ such that, for every polynomial $Q \in \mathbb{L}[x]$ satisfying $\deg(Q) < \deg(g)+\deg(h)$, there exist $V, W \in \mathbb{L}[x]$ satisfying $\Psi(Vg+Wh-Q,0) \leq \Psi(Q,0)+c$, $\Psi(V,0) \leq \Psi(Q,0)$, $\Psi(W,0) \leq \Psi(Q,0)$, $\deg(V) < \deg(h)$, $\deg(W) < \deg(g)$.

Lemma 5.5. *Let $g, h \in U_{\mathbb{L}}[x]$ be quasi-monic and satisfy $(\overline{g}, \overline{h}) = \overline{1}$ and let $d = \deg(g) + \deg(h)$. Then, $B(f,g)$ is a not empty interval whose lower bound is 0. Moreover, given $\lambda \in B(f,g)$ and monic polynomials s, $t \in U_{\mathbb{L}}[x]$ such that $\Psi(g-s,0) \leq \lambda$, $\Psi(h-t,0) \leq \lambda$, then $B(s,t) = B(g,h)$.*

Proof. Let $d = \deg(g)+\deg(h)$. We can apply Lemma 5.4 to each polynomial $Q_n = x^n$ for every $n = 0, \ldots, d-1$. Thus, we have polynomials V_n, W_n satisfying $\Psi(V_ng + W_nh - x^n, 0) < 0$, $\Psi(V_n, 0) \leq 0$, $\Psi(W_n, 0) \leq 0$, $\deg(V_n) < \deg(h)$, $\deg(W_n) < \deg(g)$. We put $\lambda_n = \Psi(V_ng + W_nh - x^n, 0)$, $(0 \leq n \leq d-1)$. Now, let $Q = \sum_{n=0}^{d-1} a_nx^n$, let $V = \sum_{n=0}^{d-1} a_nV_n$, $W_n = \sum_{n=0}^{d-1} a_nW_n$, and let $\lambda(g,h) = \max_{0\leq n\leq d-1} \lambda_n$. Clearly, we have $\Psi(Vg+Wh-Q,0) \leq \max_{0\leq n\leq d-1}(\Psi(a_n)+\lambda_n) \leq \max_{0\leq n\leq d-1} \Psi(a_n) + \max_{0\leq n\leq d-1} \lambda_n = \Psi(Q,0)+\lambda(g,h)$.
But trivially,

$$\Psi(V,0) \leq \max_{0\leq n\leq d-1} \Psi(a_n), \ \Psi(W,0) \leq \Psi(Q,0),$$

$$\deg(V) \leq \max_{0\leq n\leq d-1}(\deg(V_n))$$

$$< \deg(h), \ \deg(W) \leq \max_{0\leq n\leq d-1}(\deg(W_n)) < \deg(h).$$

So, $\lambda(g,h)$ lies in $B(f,g)$, and hence, it is obviously seen that $B(f,g)$ is a not empty interval and that its lower bound is 0.

Now, let $c \in B(f,g)$ and let s, $t \in U_{\mathbb{L}}[x]$ be monic and satisfy $\Psi(g-s,0) < c$, $\Psi(h-t,0) < c$. Since $\Psi(V,0) \leq \Psi(Q,0)$, $\Psi(W,0) \leq$

$\Psi(Q,0)$, it is easily seen that $\Psi(V(g-s)+W(h-t),0) < c+\Psi(Q,0)$, and therefore, $\Psi(Vs+Wt-Q,0) < c+\Psi(Q,0)$. This shows that $\lambda(s,t) \leq c$, and therefore, $B(f,g) \subset B(s,t)$. But similarly we have $B(s,t) \subset B(g,h)$ and this ends the proof of Lemma 5.5. $\square$

Lemma 5.6. *Let $Q \in L[x]$ and let $g, h \in U_{\mathbb{L}}[x]$ be quasi-monic and satisfy $(\overline{g}, \overline{h}) = \overline{1}$. Let $c \in B(g,h)$. There exist monic polynomials V, $W \in L[x]$ satisfying*
$\Psi(Vg+Wh-Q,0) \leq c+\Psi(Q,0)$,
$\deg(W) < \deg(g)$, $\deg(V) \leq \max(\deg(h), \deg(Q)-\deg(g))$,
$\Psi(V,0) \leq \Psi(Q,0)$, $\Psi(W,0) \leq \Psi(Q,0)$.

Proof. We consider the Euclidean division of Q by $gh : Q = \ell gh + Q_1$. Hence, $\deg(Q_1) < \deg(g) + \deg(h)$. By Lemma 5.2, we have

(1) $\Psi(Q_1,0) \leq \Psi(Q,0)$,
(2) $\Psi(\ell,0) \leq \Psi(Q,0)$.

By Lemma 5.5, there exist $V_1, W_1 \in \mathbb{L}[x]$ satisfying

(3) $\Psi(V_1 g + W_1 h - Q_1, 0) \leq \Psi(Q_1,0) + c$,
(4) $\Psi(V_1,0) \leq \Psi(Q_1,0)$,
(5) $\Psi(W_1,0) \leq \Psi(Q_1,0)$,
(6) $\deg(V_1) < \deg(h)$,
(7) $\deg(W_1) < \deg(g)$,

Now, we put $V = V_1 + \ell h$, $W = W_1$. So, we have $Vg + Wh - Q = V_1 + \ell gh + W_1 h - \ell gh - Q_1$, and therefore, by (3), we obtain $\Psi(Vg+Wh-Q,0) \leq \Psi(Q_1,0)+c$. Hence, by (1), we obtain

(8) $v(Vg+Wh-Q) \geq v(Q)+c$.

Now, by (1), (5), it is seen that

(9) $\Psi(W,0) \leq \Psi(Q,0)$.

We check

(10) $\Psi(V,0) \leq \Psi(Q,0)$.

Indeed, we have $\Psi(h,0) = 0$, hence, by (2), we see

(11) $\Psi(\ell h,0) \leq \Psi(Q,0)$.

But, by (4), we have $\Psi(V_1, 0) \le \Psi(Q, 0)$, and therefore, by (11), we obtain (10). Finally, by definition, we have $\deg(\ell) = \deg(Q) - \deg(gh)$, and therefore,

(12) $\deg(V) \le \max(\deg(V_1), \deg(\ell h)) \le \max(\deg(h), \deg(Q) - \deg(g))$. Thanks to (7), (8), (9), (10), and (12), Lemma 5.6 is now proven. □

Theorem 5.7 (Hensel Lemma). $\mathbb{L}$ *is supposed to be complete. Let* $P \subset U_{\mathbb{L}}[x]$ *be such that* $\overline{P}$ *splits in* $\mathcal{L}[x]$ *in the form* $\gamma\eta$ *with* γ, η *relatively prime. There exists* $g, h \in U_{\mathbb{L}}[x]$ *such that* $P = gh, \overline{g} = \gamma, \overline{h} = \eta, \deg(g) = \deg(\gamma)$.

Proof. We can obviously take quasi-monic polynomials $g_0, h_0 \in U_{\mathbb{L}}[x]$ such that $\overline{g_0} = \gamma$, $\overline{h_0} = \eta$. We put $\xi = \Psi(P - g_0h_0, 0)$ and take $\zeta \in B(g_0, h_0)$ satisfying $\zeta \le \xi$. We construct sequences $(g_n)_{n\in\mathbb{N}}, (h_n)_{n\in\mathbb{N}}$ in $\mathbb{L}[x]$ satisfying for all $n \ge 0$:

(i_n) $\Psi(P - g_nh_n, 0) \le (n+1)\zeta$,
(ii_n) $\Psi(g_n - g_{n-1}, 0) \le n\zeta,\ \Psi(h_n - h_{n-1}, 0) \le n\zeta$,
(iii_n) $\deg(h_n) \le \deg(P) - \deg(g_0),\ \deg(g_n) = \deg(g_0)$,
(iv_n) $\overline{g_n} = \gamma,\ \overline{h_n} = \eta$,
(v_n) $\zeta \in B(g_n, h_n)$.

First, we put $P_1 = P - g_0h_0$. We note that $\deg(P_1) = \deg(P)$. We now apply Lemma 5.6 to the case when $(Q, g, h) = (P_1, g_0, h_0)$: there exist $V_1, W_1 \in \mathbb{L}[x]$ satisfying (1) $\deg(W_1) < \deg(g_0)$, (2) $\deg(V_1) < \deg(P) - \deg(g_0)$, (3) $\Psi(V_1, 0) \le \zeta$, (4) $\Psi(W_1, 0) \le \zeta$, (5) $\Psi(V_1g_0 + W_1h_0 - P_1, 0) \le \zeta + \Psi(P_1, 0)$. Next, we put $g_1 = g_0 + W_1$, $h_1 = h_0 + V_1$. We check that $(i_1), (ii_1), (iii_1)$, and (iv_1) are satisfied. Moreover, by (3) and (4) and by Lemma 5.5, ζ lies in $B(g_1, h_1)$, hence v_1 is satisfied.

Now, we suppose we have already constructed the pairs (g_m, h_m) satisfying $(i_m), (ii_m), (iii_m), (iv_m)$, and (v_m) for every $m = 0, \dots, n$. Then, we put $P_{n+1} = P - g_nh_n$.

We can apply Lemma 5.6 to the case when (Q, g, h) is equal to (P_{n+1}, g_n, h_n). So, we can obtain $V_{n+1}, W_{n+1} \in L[x]$ satisfying (6) $\Psi(W_{n+1}g_n + V_{n+1}h_n - P_{n+1}, 0) \le \zeta + \Psi(P_{n+1}, 0)$, (7) $\deg(W_{n+1}) < \deg(g_n)$, (8) $\deg(V_{n+1}) \le \max(\deg(h_n), \deg(P_{n+1}) - \deg(g_n))$, (9) $\Psi(V_{n+1}, 0) \le \Psi(P_{n+1}, 0)$, $\Psi(W_{n+1}, 0) \le \Psi(P_{n+1}, 0)$. By (6) and by (v_n), we obtain (10) $\Psi(W_{n+1}g_n + V_{n+1}h_n - P_{n+1}, 0) \le (n+2)\zeta$.

Now we put $g_{n+1} = g_n + W_{n+1}$, $h_{n+1} = h_n + W_{n+1}$. We check that $P - g_{n+1}h_{n+1} = (P_{n+1} - h_nW_{n+1} - g_nV_{n+1}) - V_{n+1}W_{n+1} = P_{n+1} - hW_{n+1} - g_{n+1}V_{n+1} + (h_{n+1} - h_n)W_{n+1} + (g_{n+1} - g_n)V_{n+1} + V_{n+1}W_{n+1}$. By (ii_m) true for $m \leq n$, we note that $(11)\Psi(g_n - g_{n+1}, 0) \leq (n+1)\zeta$, $\Psi(h_n - h_{n+1}, 0) \leq (n+1)\zeta$, and then, by (9) and (10), we obtain $(i)_{n+1}$ $\Psi(P - g_{n+1}h_{n+1}, 0) \leq (n+2)\zeta$.

Relation (i_{n+1}) is true by definition and (iii_{n+1}) and (iv_{n+1}) are easily checked. By (11) and by Lemma 5.5, Relation (v_{n+1}) is also clear.

Therefore, the sequences $(g_n)_{n\in\mathbb{N}}$, $(h_n)_{n\in\mathbb{N}}$ satisfying (i_n), (ii_n), (iii_n), (iv_n), and (v_n) are now constructed. Since $\mathbb{L}$ is complete, the vector space $\mathbb{L}_q[x]$ of polynomial of degree $m \leq q$ is obviously complete with respect to the Gauss norm $\|\cdot\|$ which is characterized by $\log\|Q\| = \Psi(Q, 0)$.

Then, by Relation (ii_n), the sequences $(g_n)_{n\in\mathbb{N}}, (h_n)_{n\in\mathbb{N}}$ converge in $\mathbb{L}_q[x]$. We put $g = \lim_{n\to\infty} g_n$, $h = \lim_{n\to\infty} h_n$. By (iii_n), we have $\deg(g) = \deg(g_0) = \deg(\gamma)$. By (iv_n), we have $\overline{g} = \gamma$, $\overline{h} = \eta$, and finally, by (i_n), we have $\Psi(P - gh, 0) = +\infty$, hence $P = gh$. That ends the proof of Theorem 5.7. □

Chapter 6

Extensions of Ultrametric Fields: The Field $\mathbb{C}_p$

All considerations on analytic and meromorphic functions require considering a complete ultrametric algebraically closed field $\mathbb{K}$. Here we construct the field $\mathbb{C}_p$ and study finite extensions of $\mathbb{Q}_p$ [70]. And we show that $\mathbb{C}_p$ is not spherically complete.

Notation. As in the previous chapters, $\mathbb{L}$ denotes a complete ultrametric field whose absolute value is not trivial and whose residue class field is $\mathcal{L}$. We denote by F an algebraically closed ultrametric field whose absolute value is not trivial.

Let $\mathbb{E}$ be a field, let $\mathbb{B}$ be a finite algebraic extension of $\mathbb{E}$, and let $q = [\mathbb{B} : \mathbb{E}]$. We denote by $\mathcal{N}$ the algebraic norm of $\mathbb{B}$ over $\mathbb{E}$. Given $a \in \mathbb{B}$, we denote by $\mathrm{irr}(a, \mathbb{E})$ the minimal polynomial of a over $\mathbb{E}$.

Lemma 6.1 is classical in algebra:

Lemma 6.1. *Let $q = [\mathbb{B} : \mathbb{E}]$, and let $\mathcal{N}$ be the norm of $\mathbb{B}$ over $\mathbb{E}$. Let $a \in \mathbb{B}$, let $P_a = \mathrm{irr}(a, \mathbb{E})$, and let $d = \deg(P_a)$. Then $\mathcal{N}$ satisfies $\mathcal{N}(a) = \big((-1)^d P_a(0)\big)^{\frac{q}{d}}$ and $\mathcal{N}(ab) = \mathcal{N}(a)\mathcal{N}(b)$, $\forall b \in \mathbb{B}$.*

Theorem 6.2. *Let $\mathbb{E}$ be an algebraic extension of $\mathbb{L}$, let $a \in \mathbb{E}$, and let $P = \mathrm{irr}(a, \mathbb{L})$. Then a is integral over $U_{\mathbb{L}}$ if and only if $|P(0)| \leq 1$. Moreover, if $|P(0)| \leq 1$, then $\|P\| = 1$. Finally, if $|P(0)| = 1$, then $\mathrm{irr}(\overline{a}, \mathcal{L}) = \overline{P}$.*

Proof. First, we assume a to be integral over $U_{\mathbb{L}}$. Then there exists a monic polynomial $Q \in U_{\mathbb{L}}[x]$ such that $Q(a) = 0$. Therefore,

P divides Q in $\mathbb{L}[x]$. Let $Q(x) = P(x)T(x)$. Since both P, Q are monic, so is T. Therefore, $\Psi(P(,0) \geq 0$, $\Psi(T(,0) \geq 0$. But since $\Psi(Q,0) = 0$ and since $\Psi(Q,0) = \Psi(P,0) + \Psi(T,0)$, P, T must satisfy $\Psi(P,0) = \Psi(T,0) = 0$, and therefore, $|P(0)| \leq 1$.

Now, we assume $|P(0)| \leq 1$. Suppose $\Psi(P,0) > 0$. There exists $b \in M_{\mathbb{L}}$ such that $\Psi(bP,0) = 0$ and then $|bP(0)| < 1$, hence $\overline{0}$ is a zero of $\overline{bP}$. Further, we note

(1) $\deg(\overline{bP}) < \deg(P)$.

Let $\overline{bP} = x^d\phi$ with $\phi(\overline{0}) \neq \overline{0}$. Then, x^d and ϕ are relatively prime in $\mathcal{L}[x]$. Therefore, by Theorem 5.7, there exist g, $h \in U_{\mathbb{L}}[x]$ such that $\overline{g} = x^d$, $\overline{h} = \phi$, $\deg(g) = d$, and $P = gh$. But since P is irreducible in $\mathbb{L}[x]$ and since $d > 0$, h must be a constant, and therefore, $\deg(P) = d$, a contradiction to (1). Consequently, $\Psi(P,0) = 0$. Now, suppose $|P(0)| = 1$. Since P is irreducible in $\mathbb{L}[X]$, by Theorem 5.7, so is $\overline{P}$ in $\mathcal{L}[X]$, hence $\mathrm{irr}(\overline{\mathrm{a}}, \mathcal{L}) = \overline{\mathrm{P}}$. □

Corollary 6.3. *Let $\mathbb{E}$ be an algebraic extension of $\mathbb{L}$ equipped with the unique extension of the absolute value of $\mathbb{L}$. Let $a \in \mathbb{E}$ be such that $|a| = 1$, of degree l over $\mathbb{L}$. Then the residue class $\overline{a}$ of a in the residue class field of $\mathbb{E}$ is algebraic, of degree l over $\mathcal{L}$.*

Proof. Let $P(x) = \sum_{j=0}^{m} a_j x^j = \mathrm{irr}(\mathrm{a}, \mathbb{L})$. Since $|a| = 1$, we have $|a_0| = 1$, hence, by Theorem 6.2, we have $\|P\| = 1$ and $\overline{a}$ satisfies $\mathrm{irr}(\overline{\mathrm{a}}, \mathcal{L}) = \overline{\mathrm{P}}$, which ends the proof. □

Theorem 6.4. *Let $\mathbb{E}$ be an algebraic extension of $\mathbb{L}$. There exists a unique absolute value φ on $\mathbb{E}$ that extends the one of $\mathbb{L}$. Further, this absolute value is ultrametric and defined as follows: given $a \in \mathbb{E}$, $Q = \mathrm{irr}(\mathrm{a}, \mathbb{L})$, and $t = \deg(Q)$, then $\varphi(a) = \sqrt[t]{|Q(0)|}$.*

Proof. We first note that $\varphi(a) = |a|$ whenever $a \in \mathbb{L}$. We show that φ is an ultrametric absolute value on $\mathbb{E}$. Clearly, we have $\varphi(a) \neq 0$ whenever $a \in \mathbb{E}$. By Lemma 6.1, it is easily seen that we have $\varphi(ab) = \varphi(a)\varphi(b)$ whenever a, $b \in \mathbb{E}$, and therefore, $\varphi(a)^{-1} = \varphi(a^{-1})$. So it remains to show the ultrametric inequality. For this, we show $\varphi(1+z) \leq 1$ for every $z \in U_{\mathbb{E}}$. For convenience, we put again $P_z = \mathrm{irr}(\mathrm{z}, \mathrm{E})$ whenever $z \in \mathbb{E}$. Let $z \in U_{\mathbb{E}}$. So, we have $|P_z(0)| \leq 1$, and then by Theorem 6.2, z is integral over $U_{\mathbb{L}}$, hence so is $1+z$. Hence, by Theorem 6.2, we have

$|P_{1+z}(0)| \leq 1$, and therefore, $\varphi(1 + z) \leq 1$. Now, the ultrametric inequality is easily derived. Let $a, b \in \mathbb{E}$ satisfy $0 < |a| \leq |b|$. We have $\varphi(a+b) = \varphi\big(b(1+\frac{a}{b})\big) = \varphi(b)\varphi(1+\frac{a}{b})$. But $\varphi(1+\frac{a}{b}) \leq 1$, hence finally, $\varphi(a+b) \leq \varphi(b)$. Thus, we have now proven φ to be an ultrametric absolute value that extends that of $\mathbb{L}$. Then, by Theorem 2.6, this absolute value on $\mathbb{E}$ is unique, which ends the proof. □

Corollary 6.5. *Let Ω be an algebraic closure of $\mathbb{L}$. There exists a unique absolute value φ on Ω that extends the one of $\mathbb{L}$. Further, this absolute value is ultrametric and defined as follows: given $a \in \Omega$, $Q = \mathrm{irr}(\mathrm{a}, \mathbb{L})$ and $t = \deg(Q)$, then $\varphi(a) = \sqrt[t]{|Q(0)|}$.*

Corollary 6.6. *Let $P(x) \in \mathbb{L}[x]$ be irreducible over $\mathbb{L}$, let Ω be an algebraic closure of $\mathbb{L}$ provided with the absolute value extending that of $\mathbb{L}$, and let $b_1, \dots, b_q$ be the zeros of P in Ω. Then, $|b_i| = |b_j|$ $\forall i$, $j \leq q$.*

Corollary 6.7. *Let Ω be an algebraic closure of $\mathbb{L}$ provided with the unique absolute value $|\ .\ |$ that extends the one of $\mathbb{L}$. Then, U_Ω is equal to the integral closure of $U_{\mathbb{L}}$. Moreover $|\Omega| = \{\sqrt[n]{r} \mid r \in |\mathbb{L}|,\ n \in \mathbb{N}^*\}$.*

Corollary 6.8. *Suppose that the value group of $\mathbb{L}$ is $\mathbb{Z}$. Let $\mathbb{E}$ be a finite algebraic extension of $\mathbb{L}$ of degree t provided with the unique absolute value $|\ .\ |$ that extends the one of $\mathbb{L}$. There exists a rational r of the form $\frac{s}{t}$ such that the value group of $\mathbb{E}$ is $r\mathbb{Z}$.*

Lemma 6.9. *Let $\mathbb{B}$ be an algebraic extension of $\mathbb{E}$ provided with an absolute value extending that of $\mathbb{E}$. Then the residue class field of $\mathbb{B}$ is algebraic over the residue class field of $\mathbb{E}$.*

Proof. Let $t \in U_{\mathbb{B}}$ and let $\widehat{\mathbb{E}}$ be the completion of $\mathbb{E}$ with respect to the absolute value of $\mathbb{E}$. Since t is algebraic over $\mathbb{E}$, so much the more it is algebraic over $\widehat{\mathbb{E}}$. Then, by Corollary 6.5, the residue class $\overline{t}$ of t is algebraic over the residue class field of $\widehat{\mathbb{E}}$. But $\widehat{\mathbb{E}}$ obviously has the same residue class field as $\mathbb{E}$. □

Corollary 6.10. *Let $\mathbb{B}$ be an algebraic extension of $\mathbb{L}$ provided with the unique absolute value that extends the one of $\mathbb{L}$. Then the residue class field of $\mathbb{B}$ is an algebraic extension of the residue class field of $\mathbb{L}$. Moreover, if $\mathbb{B}$ is finite over $\mathbb{L}$, then residue class field of $\mathbb{B}$ is finite over the residue class field of $\mathbb{L}$.*

Proof. Suppose first that $\mathbb{B}$ is of the form $\mathbb{L}[u]$. Without loss of generality, we may assume that $|u| = 1$. Then the residue class field of $\mathbb{B}$ is $\mathcal{L}[\overline{u}]$. Next, we can generalize by induction. $\square$

Theorem 6.11. *Let Ω be an algebraic closure of $\mathbb{L}$ provided with the unique absolute value extending the one of $\mathbb{L}$. Then the residue class field of Ω is an algebraic closure of $\mathcal{L}$.*

Proof. Let $\mathcal{T}$ be the residue class field of Ω. Let $u \in U_\Omega$ and let $P = \text{irr}(u, \mathbb{L})$. By Corollary 6.7, P belongs to $U_{\mathbb{L}}[x]$ and obviously satisfies $\overline{P}(\overline{u}) = \overline{0}$, hence $\overline{u}$ is algebraic over $\mathcal{L}$. So, $\mathcal{T}$ is an algebraic extension of $\mathcal{L}$. Now, let $q \in \mathcal{L}[x]$, and let $Q \in U_{\mathbb{L}}[x]$ be a polynomial such that $\overline{Q} = q$. Then, Q factorizes in $\Omega[x]$ in the form $\prod_{j=1}^{s}(x - a_j)$ with $|a_j| \leq 1\ \forall j = 1, \ldots, s$, hence $\overline{a}_j$ belongs to $\mathcal{T}$ and $q(x) = \overline{a}_j = \prod_{j=1}^{s}(x - \overline{a}_j)$. So, $\mathcal{T}$ contains the algebraic closure of $\mathcal{L}$. And since it is an algebraic extension of $\mathcal{L}$, then it is the algebraic closure of $\mathcal{L}$. $\square$

Corollary 6.12. *The residue class field $\mathcal{K}$ of $\mathbb{K}$ is algebraically closed.*

Lemma 6.13. *Let $P(x) = \sum_{j=0}^{t} a_j x^j$, $Q(x) = \sum_{j=0}^{t} b_j x^j$ be monic, belong to $\mathbb{B}[x]$ and satisfy $\|P\|\|Q\| = 1$. For each zero α of P, Q admits at least one zero β such that $|\alpha - \beta|^t \leq \max_{0 \leq j \leq t} |a_j - b_j|$.*

Proof. Let $s = \max_{0 \leq j \leq t} |a_j - b_j|$ and let α be a zero of P. By Theorem 4.6, we have $|(P - Q)(x)| \leq s$ whenever $x \in d(0, 1)$, hence, in particular, $|Q(\alpha)| \leq s$. Let $\beta_1, \ldots, \beta_t$ be the zeros of Q (taking multiplicities into account). So, we have $\prod_{j=1}^{t} |\beta_j - \alpha| \leq s$ and then that at least one of the β_j satisfies $|\beta_j - \alpha| \leq s^{\frac{1}{t}}$. $\square$

Theorem 6.14. *Let $\mathbb{B}$ be an algebraically closed extension of $\mathbb{L}$ provided with the unique absolute value that extends the one of $\mathbb{L}$. The completion of $\mathbb{B}$ also is algebraically closed.*

Proof. Let $\widetilde{\mathbb{B}}$ be the completion of $\mathbb{B}$ and let $P(x) = \sum_{j=0}^{t} a_j x^j \in \widetilde{\mathbb{B}}[x]$ be monic. Let Ω be an algebraic closure of $\widetilde{\mathbb{B}}$, provided with the unique absolute value extending that of $\widetilde{\mathbb{B}}$ and let $\alpha_1, \ldots, \alpha_t$ be the zeros of P in Ω. Up to a change of variable, we may assume that $|\alpha_j| \leq 1\ \forall j = 1, \ldots, t$. Let $\varepsilon \in]0, 1[$ and let $Q(x) = \sum_{j=0}^{t} b_j x^j \in \mathbb{B}[x]$ be such that $\max_{0 \leq j \leq t} |a_j - b_j| \leq \varepsilon^t$. For each $j = 1, \ldots, t$, by Lemma 6.13, Q admits a zero β such that $|\alpha_j - \beta|^t \leq \varepsilon$. Since $Q \in \mathbb{B}[x]$, obviously β belongs to $\mathbb{B}$, and therefore, we see that α_j belongs to $\widetilde{\mathbb{B}}$. That ends the proof. □

The following theorem is due to Marc Krasner [70].

Theorem 6.15 (Krasner). *Let* $\mathbb{L}$ *have characteristic zero. Let* Ω *be an algebraic closure of* $\mathbb{L}$ *provided with the unique absolute value extending the one of* $\mathbb{L}$. *Let* $a \in \Omega$, *let* $a_2, \ldots, a_n$ *be the conjugates of* a *in* Ω, *and let* $b \in \Omega$ *satisfy* $|b-a| < |b-a_j|$ *for every* $j = 2, \ldots, n$. *Then, we have* $\mathbb{L}[a] \subset \mathbb{L}[b]$.

Proof. Let $a_1 = a$ and let $P(x) = \mathrm{irr}(\mathrm{a}, \mathbb{L})$. In $\Omega[x]$, the polynomial P splits in the form $\prod_{j=1}^{n}(x - a_j)$. Let $Q(x) = \mathrm{irr}(\mathrm{a}, \mathbb{L}[\mathrm{b}])$. Then, Q divides P. Let $t = \deg(Q)$ and suppose the a_j ranged in such a way that $Q(x) = \prod_{j=1}^{t}(x - a_j)$. Let $R(y) = Q(b + y)$. Then $R(y)$ is seen to be irreducible in $\mathbb{L}[b][y]$ like $Q(x)$ in $\mathbb{L}[b][x]$. Moreover, the zeros of R are just the $a_j - b$, with $1 \leq j \leq t$. Thus, we have $R = \mathrm{irr}(\mathrm{a} - \mathrm{b}, \mathbb{L}[\mathrm{b}])$. But since $\mathbb{L}$ is complete, by Corollary 6.5, we have $|a_j - b| = \sqrt[q]{|R(0)|}$ for every $j = 1, \ldots, t$. In particular, for $j = 2, \ldots, t$, we have $|a_j - b| = |a - b|$ and this contradicts the hypothesis. Finally, we have $t = 1$, and therefore, a lies in $\mathbb{L}[b]$. □

Corollary 6.16. *Let* $\mathbb{L}$ *have characteristic zero. Let* Ω *be an algebraic closure of* $\mathbb{L}$ *provided with the unique absolute value extending the one of* $\mathbb{L}$. *Let* $a \in \Omega$, *let* $a_2, \ldots, a_n$ *be the conjugates of* a *in* Ω, *and let* $b \in \Omega$ *satisfy* $|b - a| < |b - a_j|$ *for every* $j = 2, \ldots, n$ *and* $[\mathbb{L}[b] : \mathbb{L}] \leq n$. *Then, we have* $\mathbb{L}[a] = \mathbb{L}[b]$.

We can now recall the construction of p-adic fields.

Definition and notation. Let p be a prime number. On $\mathbb{Z}$, the *p-adic absolute value* is defined as follows: given $n \in \mathbb{Z}^*$, it factorizes in a unique way in the form $p^s q$, with $q \in \mathbb{Z}^*$, prime to p. So, here we take $\theta = p$ and set $|n|_p = p^{-s}$.

Lemma 6.17 is immediate.

Lemma 6.17. $|\ .\ |_p$ *is an ultrametric absolute value on* $\mathbb{Z}$ *that has continuation to* $\mathbb{Q}$ *and defines an ultrametric absolute value on* $\mathbb{Q}$

and $\mathbb{N}$ is dense in $\mathbb{Z}$. Then, $|n| \geq \frac{1}{n}$, $\forall n \in \mathbb{N}^*$,

$$U_{\mathbb{Q}} = \left\{p^n\left(\frac{a}{b}\right) \mid n \in \mathbb{N},\ a \in \mathbb{Z}, b \in \mathbb{Z}^*,\ \text{g.c.d.}(a,p) = \text{g.c.d.}(b,p) = 1\right\},$$

$$M_{\mathbb{Q}} = \left\{p^n\left(\frac{a}{b}\right) \mid n \in \mathbb{N}^*,\ a \in \mathbb{Z}, b \in \mathbb{Z}^*,\ \text{g.c.d.}(a,p) = \text{g.c.d.}(b,p) = 1\right\},$$

and the residue characteristic of $\mathbb{Q}$ is p. The residue class field of $\mathbb{Q}$ is the field of p elements $\mathbb{F}_p$. The valuation group of $\mathbb{Q}$ is isomorphic to the additive group $\mathbb{Z}$.

Remark and notation. Now, $\mathbb{Q}$ admits a completion with respect to the p-adic absolute value and its completion is denoted by $\mathbb{Q}_p$. The closure of $\mathbb{Z}$ in $\mathbb{Q}_p$ is denoted by $\mathbb{Z}_p$.

On $\mathbb{Q}_p$, we extend the valuation and the absolute value $|\ .\ |_p$ defined on $\mathbb{Q}$ and we set again $\Psi_p(x) = -v_p(x)$.

An algebraic closure Ω_p of $\mathbb{Q}_p$ is equipped with the unique extension of the p-adic absolute value defined on $\mathbb{Q}_p$ and we again denote it by $|\ .\ |_p$. The valuation group of $\mathbb{Q}_p$ is obviously equal to the one of $\mathbb{Q}$. Next, the valuation group of Ω_p is easily seen to be isomorphic to $(\mathbb{Q}, +)$. In Chapter 8, we see that Ω_p is not complete.

By Theorem 6.14, Ω_p has a completion denoted by $\mathbb{C}_p$ that is algebraically closed. The valuation group of $\mathbb{C}_p$ is then isomorphic to $(\mathbb{Q}, +)$ like the one of Ω_p. Moreover, by Theorem 6.11, the residue class field of Ω_p is an algebraic closure of $\mathbb{F}_p$ and the one of $\mathbb{C}_p$ is seen to be the same. The absolute value $|\ .\ |_p$ defined on Ω_p has a natural extension to $\mathbb{C}_p$ and the associated valuation is denoted by v_p again, and we set again $\Psi_p(x) = -v_p(x)$, $\forall x \in \mathbb{C}_p$. However, when there is no risk of confusion, we just write Ψ instead of Ψ_p.

Theorem 6.18. *Let a be integral over $\mathbb{Z}$ and let $a_2, \dots, a_q$ be the conjugates of a over $\mathbb{Q}$. Then, $|a| \leq 1$ and $a_j| \leq 1$, $\forall j = 2, \dots, q$.*

Proof. Since a is integral over $\mathbb{Z}$, it is integral over $\mathbb{Z}_p$. Let $P(X) = irr(a, \mathbb{Q})$ and let $B(X) = irr(a, \mathbb{Q}_p)$. Let $a_1, \dots, a_h$ be the conjugates of a over $\mathbb{Q}_p$ (with $a_1 = a$). Then, $P(0) = \prod_{j=1}^q a_j$. Then, B divides P in $\mathbb{Q}_p[X]$. Moreover, $\|P\| = 1$ and $\|B\| = 1$. Next, $B(0) = \prod_{j=1}^h a_j$. By Corollary 6.6, we have $|a_1| =, \dots, = |a_h|$, while $|B(0)| \leq 1$, hence $|a_j| \leq 1$, i.e. $|a| \leq 1$. Next, what is true for a_1 also holds for every $a_j, \dots, j = 2, \dots, q$, hence $|a_j| \leq 1$. □

In the future, we use Lemma 6.19:

Lemma 6.19. *Let* $a \in \mathbb{C}_p$ *be algebraic over* $\mathbb{Q}_p$ *such that* $\log_p(|a|)$ *is of the form* $\frac{\lambda}{t}$, *with* $\lambda \in \mathbb{Z}$ *and* t *in* $\mathbb{N}^*$. *Take* m, $n \in \mathbb{N}$ *and* $b \in \mathbb{C}_p$ *such that* $\log_p(|b|)$ *is of the form* $\frac{u}{w}$ *with* $u \in \mathbb{N}$ *and* $w \in \mathbb{N}$ *prime, prime with* u *and such that* $w > \max(m, n, t)$.

Let f, $g \in \mathbb{Q}_p[a]$ *be such that* $|fb^m|_p = |gb^n|_p$. *Then,* $m = n$.

Proof. We note that for every $x \in \mathbb{Q}_p[a]$, $\Psi_p(|x|)$ is of the form $\frac{\ell}{t}$ with $\ell \in \mathbb{Z}$. Consequently, $\Psi_p(|f|)$ is of the form $\frac{h}{t}$ and $\log_p(|g|)$ is of the form $\frac{k}{t}$ with h and $k \in \mathbb{Z}$. Consequently, $\Psi_p(|fb^m|) = \frac{h}{t} + \frac{mu}{w}$ and $\Psi_p(|gb^n|) = \frac{k}{t} + \frac{nu}{w}$, and therefore, due to the equality $|fb^m|_p = |gb^n|_p$, we have $(h - k)w = ut(n - m)$. But since $w > t$, it is prime with ut, hence it must divide $n - m$, which is impossible because $\max(m, n) < w$, except if $m = n$. □

Lemma 6.20. *Let* $|\ .\ |$ *be an ultrametric absolute value on* $\mathbb{Q}$. *If this absolute value is trivial, the residue characteristic is zero. If the absolute value is not trivial, there exists a prime number* q *such that* $|\ .\ |$ *is equivalent to* $|\ .\ |_q$.

Proof. If this absolute value is trivial, it is clear that the residue characteristic is zero. So, we suppose that the absolute value is not trivial. For every $n \in \mathbb{N}^*$, we have $|n| \leq 1$. Since $|\ .\ |$ is not trivial, there certainly exists $s \in \mathbb{N}^*$ such that $|s| < 1$. Let q be the smallest $s \in \mathbb{N}^*$ such that $|s| < 1$. It is easily checked that q is prime. Since $M_{\mathbb{Q}}$ is a principal ideal of $\mathbb{Z}$, we have $M_{\mathbb{Q}} = q\mathbb{Z}$.

Let $t = \left|\frac{1}{q}\right|$. It is easily checked that given $m \in \mathbb{Z}^*$, of the form $q^s n$, with $n \in \mathbb{Z}^*$, prime to q, we have $|m| = t^{-s}$. Let w be the valuation associated with this absolute value. Then, w is clearly proportional to v_q and by Lemma 2.2 is equivalent to v_q. This ends the proof. □

Lemma 6.21 is easily seen.

Lemma 6.21. $\mathbb{N}$ *is dense in* $\mathbb{Z}_p$, *the invertible elements in* $\mathbb{Z}_p$ *are the ones whose absolute value is* 1, $\mathbb{Z}_p$ *is compact, equal to* $U_{\mathbb{Q}_p}$, *and* $p\mathbb{Z}_p$ *is equal to* $M_{\mathbb{Q}_p}$. $\mathbb{Q}_p$ *is locally compact. The residue class field of* $\mathbb{Q}_p$ *is equal to the field of* p *elements* $\mathbb{F}_p$. *Finally,* U_p *is the union of* p *disks* $d(u, \frac{1}{p})$.

Proof. All statements are immediate except the compacity of $\mathbb{Z}_p$. Consider a sequence (a_n) in $\mathbb{Z}_p$. Since it is bounded, by Theorem 3.1, we can extract either a monotonous distances sequence or an equal distances sequence, or a converging sequence. But since each circle $d(a,r)$ with $r \in |\mathbb{Q}_p|$ only has p classes, there are no equal distances sequences in $\mathbb{Q}_p$. And since the absolute value is discrete, there is no monotonous distances sequence in $\mathbb{Q}_p$. Hence, we can extract a converging sequence from the sequence (a_n). □

Corollary 6.22. *For each $m \in \mathbb{N}$, U_p is the union of p^m distinct disks $d(u, \frac{1}{p^m})$. Let E be a finite algebraic extension of $\mathbb{Q}_p$. There exists a constant $B > 0$ such that for all $m \in \mathbb{N}$ the number of distinct $d(u, \frac{1}{p^m})$ is inferior or equal to Bp^m.*

By definition and construction of $\mathbb{C}_p$, we have this corollary:

Corollary 6.23. *The field of algebraic numbers is dense in $\mathbb{C}_p$. Hence, $\mathbb{C}_p$ contains a dense countable subset.*

In Theorem 6.24, we follow the method of [84].

Theorem 6.24. *$\mathbb{C}_p$ is not spherically complete.*

Proof. Let $(r_n)_{n\in\mathbb{N}}$ be a real sequence such that $0 < r < r_{n+1} < r_n < 1$, $\forall n \in \mathbb{N}$. Let $\mathcal{S}$ be the set of sequences of the set $\{0,1\}$. Suppose that for each $n = 0,\dots,q$ we have defined 2^n disks $d(a_{n,k}, r_n), k = 0,1$, such that for each $n = 1,\dots,q$ and for each $k = 0,1$ the disks $d(a_{n,0}, r_n)$ and $d(a_{n,1}, r_n)$ are included in some $d(a_{n-1,k}, r_{n-1})$ (with $k = 0$ or $k = 1$) and have an empty intersection. It is then immediate to define in each disk $d(a_{q,k}r_q)$ two disks $d(a_{q+1,0}, r_{q+1})$ and $d(a_{q+1,1}, r_{q+1})$ having an empty intersection. So, the family is defined for every $n \in \mathbb{N}$.

Now, let $(u_n) \in \mathcal{S}$ and let (V_n) be a decreasing sequence of disks defined as follows: suppose defined (V_n) for $n \leq q$. If $u_{q+1} = 0$, we set $V_{q+1} = d(a_{q+1,0}, r_{q+1})$, and if $u_{q+1} = 1$, we set $V_{q+1} = d(a_{q+1,1}, r_{q+1})$. That mapping which associates with each sequence (u_n) the decreasing sequence of disks (V_n) is clearly injective. Now, consider two distinct such sequences (u_n) and (u'_n). Let q be the smallest integer such that $u_q \neq u'_q$. The distance from V_q to V'_q is at least r_q. Consequently, for every $n \geq q$, the distance between V_n and V'_n is at least $r_q \geq r$.

Suppose now that $\mathbb{C}_p$ is spherically complete. For each sequence (u_n) the intersection of the decreasing sequence of disks (V_n) contains a point $\alpha(u_n)$ and hence, by the last conclusion, if (u_n) and (u'_n) are two different sequences, we have $|\alpha(u_n) - \alpha(u'_n)| \geq r$. But we know that the set of sequences (u_n) is not countable and hence the set of the $\alpha(u_n)$, $((u_n) \in \mathcal{S})$ is not countable. Consequently, $\mathbb{C}_p$ contains an uncountable subset Σ such that $|x - y| \geq r$, $\forall x \neq y$, x, $y \in \Sigma$. This contradicts the fact that $\mathbb{C}_p$ contains a dense countable subset. $\square$

Chapter 7

Normal Extensions of $\mathbb{Q}_p$ Inside $\mathbb{C}_p$

Notation. Recall that $\mathbb{L}$ is a complete field with respect to an ultrametric absolute value. For every $s \in \mathbb{N}^*$, we put $u_s = \frac{1}{p^{s-1}(p-1)}$ and $r_s = p^{-u_s}$. We study the p^sth roots of 1 and we show that they lie in circles of center 1 and radius r_s. We examine normal extensions of $\mathbb{Q}_p$ and totally ramified extensions and show the role of Eisenstein polynomials.

We need certain technical lemmas.

Lemma 7.1. *Let $s \in \mathbb{N}^*$. For every $n \in \mathbb{N}^*$ such that $n < p^s$, we have* $\left|\binom{p^s}{n}\right|_p = \frac{1}{p^s|n|_p}$.

Proof. We note that for any $n \in \mathbb{N}^*$ such that $n < p^s$, n is a multiple of p^h for some $h < s$, if and only if so is $p^s - n$. Now, let B be the bijection from $\{1, \ldots, n-1\}$ onto $\{(p^s - n + 1), \ldots, p^s - 1\}$, defined as $B(j) = p^s - j$. Thus, for every $j = 1, \ldots, n-1$, we have $|j|_p = |B(j)|_p$. Now, obviously, $\left|\binom{p^s}{n}\right|_p = \left|\frac{p^s}{n}\right|_p \cdot \left|\prod_{j=1}^{n-1} \frac{B(j)}{j}\right|_p$. But as we just saw, each factor $\left|\frac{B(j)}{j}\right|_p$ is equal to 1, and therefore, the conclusion is clear. □

Notation. Let $s \in \mathbb{N}$. We denote by W_s the group of the p^sth roots of 1 in $\mathbb{C}_p$ i.e. the $\zeta \in \mathbb{C}_p$ such that $\zeta^{p^s} = 1$ and we denote by B_s the set $W_s \setminus W_{s-1}$ and we set $W = \bigcup_{s \in \mathbb{N}} W_s$.

F_s denotes the polynomial $\sum_{j=0}^{p-1} x^{jp^{s-1}}$ and we put $G_s(x) = F_s(1+x)$.

Definition. A monic polynomial $P(x) = \sum_{j=0}^{q} a_j x^j \in \mathbb{L}[x]$ is called *an Eisenstein polynomial* if it satisfies $a_j \in M_{\mathbb{L}}$ whenever $j = 0, \dots, q-1$ and $a_0 \notin (M_{\mathbb{L}})^2$.

Theorem 7.2 (Eisenstein). *Let $\mathbb{L}$ have a discrete valuation. Let $P(x) = \sum_{j=0}^{q} a_j x^j \in \mathbb{L}[x]$ be an Eisenstein polynomial. Then, P is irreducible in $\mathbb{L}[x]$.*

Proof. We suppose P not irreducible. Then, P splits in $\mathbb{L}[x]$ in the form $S(x)T(x)$ with $S(x) = \sum_{j=0}^{m} \alpha_j x^j$, $T(x) = \sum_{j=0}^{n} \beta_j x^j$, and $\alpha_m = \beta_n = 1$. Since S, T are monic, we have $\|S\| \geq 1$, $\|T\| \geq 1$, and since $\|S\|\|T\| = \|ST\| = \|P\| = 1$, we have $\|S\| = \|T\| = 1$. Hence, both S, T belong to $U_{\mathbb{L}}[x]$. First, we note that if α_0 belongs to $M_{\mathbb{L}}$, then β_0 does not because $a_0 \notin (M_{\mathbb{L}})^2$. Hence, we may assume $\alpha_0 \in M_{\mathbb{L}}$ and $\beta_0 \notin M_{\mathbb{L}}$. Further, we have $\alpha_j \in M_{\mathbb{L}}$ for every $j = 0, \dots, m-1$. Indeed, let ℓ be the smallest of the integers h such that $|\alpha_h| = 1$. Then we have $|a_l| = |\beta_0 \alpha_\ell + \sum_{j=1}^{\ell} \beta_j \alpha_{\ell-j}| = 1$ because $|\beta_0 \alpha_\ell| = 1$ and $\sum_{j=1}^{\ell} \beta_j \alpha_{\ell-j} \in M_{\mathbb{L}}$. Consequently, $\ell = q$, therefore P is irreducible. $\square$

The following lemmas are useful in the sequel.

Lemma 7.3. *Let $\mathcal{G}$ be a subgroup of the multiplicative group $(\mathbb{K}^*, \cdot)$ included in $C(0,1)$ and let $u \in \mathcal{G}$. The bijection γ from $\mathcal{G}$ onto $\mathcal{G}$ defined as $\gamma(x) = ux$ is isometric.*

Lemma 7.4. *Let $(j, n) \in \mathbb{N} \times \mathbb{N}^*$ be such that $j < n$. Then, we have*

$$\binom{n}{j+1} = \sum_{h=j}^{n-1} \binom{h}{j}.$$

Lemma 7.5. *For every $s \in \mathbb{N}$, G_s is an Eisenstein polynomial.*

Proof. First, we suppose $s = 1$. We have

$$G_1(x) = \sum_{h=0}^{p-1} \left(\sum_{j=0}^{h} \binom{h}{j} x^j \right) = \sum_{j=0}^{p-1} \left(\sum_{h=j}^{p-1} \binom{h}{j} \right) x^j.$$

Hence, by Lemma 7.4, we have $G_1(x) = \sum_{j=0}^{p-1} \binom{p}{j+1} x^j$. Moreover, by Lemma 7.1, we have $\left|\binom{p}{j+1}\right|_p = \frac{1}{p}$ for every $j = 0, \ldots, p-2$, and therefore, G_1 is an Eisenstein polynomial.

Now we consider the general case $s \in \mathbb{N}^*$. First, we put $T_s(x) = (1+x)^{p^s}$. By Lemma 7.1, it is seen that $T_1(x)$ is of the form $1 + x^p + \gamma_1(x)$ with $\gamma_1(x) \in px\mathbb{Z}_p[x]$ and $\deg(\gamma_1) = p-1$. Then by an immediate induction, we see that $T_s(x)$ is of the form $1 + x^{p^s} + \gamma_s(x)$ with $\gamma_s(x) \in px\mathbb{Z}_p[x]$ and $\deg(\gamma_s) = p^s - 1$.

As a consequence, it is easily seen that G_s is an Eisenstein polynomial if and only if so is the polynomial $g_s(x) = \sum_{j=0}^{p-1}(1 + x^{p^s})^j$. But we have $g_s(x) = G_1(x^{p^s})$. Since G_1 is an Eisenstein polynomial, so is g_s. This ends the proof. □

Theorem 7.6. *For each* $s \in \mathbb{N}^*$, B_s *consists of* $p^s - p^{s-1}$ *roots of* 1 *of order* p^s *that lie in* $C(1, r_s)$. *For every* $t \in B_s, \operatorname{irr}(\mathrm{t}, \mathbb{Q}_\mathrm{p})$ *is equal to* F_s *and* $B_s \cap d(t, r_s^-)$ *is equal to* tW_{s-1}.

Proof. Let $t \in B_s$ and let $F(x) = \operatorname{irr}(\mathrm{t}, \mathbb{Q}_\mathrm{p})$. Then, F divides $x^{p^s} - 1$ and has a degree $d > p^{s-1}$. Since $x^{p^s} - 1 = (x^{p^{s-1}} - 1)F_s$, F divides F_s. But by Lemma 7.5 and Theorem 6.1, G_s is irreducible over $\mathbb{Q}_p$, hence so is F_s, and therefore, $F = F_s$. Then, by Corollary 6.5, we have $\Psi_p(t-1) = \frac{\Psi_p(G_s(0))}{p^s - p^{s-1}} = -u_s$. Hence, B_s is included in $C(1, r_s)$ and obviously consists of $p^s - p^{s-1}$ different points in this circle. Let ϕ be the mapping defined in W_{s-1} as $\phi(\xi) = t\xi$. Since t is of order p^s and any element of W_{s-1} is of order $< p^s$, one sees that ϕ is an injection from W_{s-1} into B_s, and then, by Lemma 7.3, we have $\phi(W_{s-1}) \subset B_s \cap d(t, {r_s}^-)$. Conversely, given $u \in B_s \cap d(t, {r_s}^-)$, then $t^{-1}u$ lies in W_{s-1} which shows that ϕ is a bijection from W_{s-1} onto $B_s \cap d(t, {r_s}^-)$. □

Corollary 7.7. *For every* $\xi, \zeta \in W_1$ *of* $\xi \neq \zeta$, *we have* $\Psi_p(\xi - \zeta) = -\frac{1}{p-1}$.

Theorem 7.8. *Let* $n \in \mathbb{N}^*$ *and let* ζ *be an* n*th root of 1 of order* n. *Then* ζ *belongs to* $d(1, 1^-)$ *if and only if* n *is of the form* p^s $(s \in \mathbb{N})$.

Proof. By Theorem 7.6, we know that if n is of the form p^s, then ζ belongs to $d(1, 1^-)$. Now we suppose $\zeta \in d(1, 1^-)$ and put $n = q\, p^s$

with q prime to p. Let $\xi = \zeta^{(p^s)}$. It is seen that ξ also belongs to $d(1,1^-)$ because $|\xi - 1|_p = |\zeta - 1|_p \left| \sum_{j=0}^{p^s-1} \zeta^j \right|_p$. Let $P(x) = x^q - 1$. If $\xi \neq 1$, then $\overline{P}(x)$ admits 1 as a zero of order $t \geq 2$. But $\overline{P}'(\overline{1}) = \overline{q} \neq \overline{0}$, hence $\xi = 1$. Therefore, we have $q = 1$ and this ends the proof. □

Definition. Let $\mathbb{E}$ be a finite extension of $\mathbb{Q}_p$. Recall that the residue class field of $\mathbb{Q}_p$ is $\mathbb{F}_p$. Let $\mathcal{E}$ be the residue class field of $\mathbb{E}$. Since $\mathbb{E}$ is finite over $\mathbb{Q}_p$, it is locally compact, hence $\sup\{|x| \ |x \in \mathbb{E} \ |x| < 1\} < 1$. So, we can choose an element $s \in \mathbb{E}$ such that $|s| < 1$ and such that $|s| = \sup\{|x| \ |x \in \mathbb{E} \ |x| < 1\}$. Such an element s is called *a uniformizer* of $\mathbb{E}$. Since s is algebraic over $\mathbb{Q}_p$, $|p|$ is of the form $|s|^c$ with $c \in \mathbb{N}^*$. The number c is called *ramification index of* $\mathbb{E}$.

Next, by Corollary 6.3, we know that if $a \in \mathbb{E}$ is algebraic over $\mathbb{Q}_p$ of degree q and such that $|a| = 1$, then its residue class $\overline{a}$ is algebraic over $\mathbb{F}_p$, of degree $\leq q$. Consequently, if $\mathbb{E}$ is finite over $\mathbb{Q}_p$, then $\mathcal{E}$ is finite over $\mathbb{F}_p$. The number $[\mathcal{E} : \mathbb{F}_p]$ is called residual degree of $\mathbb{E}$ and is denoted by f.

The extension $\mathbb{E}$ is said to be *ramified* if $c > 1$ and *unramified* if $c = 1$.

Lemma 7.9 is just a remark:

Lemma 7.9. *Let T be an unramified extension of $\mathbb{Q}_p$ and let $\mathbb{E}$ be an extension of $\mathbb{Q}_p$ such that $\mathbb{E} \subset T$. Then, $\mathbb{E}$ is unramified.*

Theorem 7.10. *Let $n \in \mathbb{N}^*$ be prime to p and let u, w be two distinct roots of 1 in Ω_p of order n. Then, in Ω_p, we have $|u - w| = 1$.*

Proof. Let $h = \frac{u}{w}$. Then, $h^n = 1$. Suppose $|h - 1| < 1$. By Theorem 7.8, n is of the form p^s, a contradiction. □

Chapter 8

Spherically Complete Extensions

Several problems on p-adic analytic functions require one to consider an ultrametric algebraically closed extension of $\mathbb{K}$ which is spherically complete, in order to give every circular filter a center. Others require to have a complete algebraically closed extension which admits a non-countable residue class field. Proving the existence of a spherically complete algebraically closed extension of the ground field $\mathbb{K}$ isn't easy, most of the ways involving basic considerations in logic. Here we follow the method proposed by Bertin Diarra that is only based on the notion of ultraproducts [33].

Definition and notation. Here we denote by $(\mathbb{E}_i)_{i\in I}$ an infinite family of field extensions of $\mathbb{L}$, provided each with an ultrametric absolute value $|\,.\,|_i$ extending that of $\mathbb{L}$. Next, $\mathcal{U}$ denotes an ultrafilter on I. We remember that $\mathcal{U}$ is said to be *principal* if there exists $\alpha \in I$ such that $\mathcal{U}$ is the set of the subsets of I that contain α. Then, $\mathcal{U}$ is said to be *incomplete* if there exists a decreasing sequence $(X_n)_{n\in\mathbb{N}}$ of elements of $\mathcal{U}$ such that $\bigcap_{n\in\mathbb{N}} X_n = \emptyset$. Since I is infinite, there obviously exist incomplete ultrafilters on I. In particular, any incomplete ultrafilter is not principal. $\mathcal{R}$ denotes the subring of $\prod_{i\in I}\mathbb{E}_i$ that consists of the set $(a_i)_{i\in I} \in \prod_{i\in I}\mathbb{E}_i$ such that $\sup_{i\in I}|a_i|_i < +\infty$. Of course, $\mathcal{R}$ is an $\mathbb{L}$-algebra.

We denote by φ the mapping from $\mathcal{R}$ into $\mathbb{R}_+$ defined as $\varphi((a_i)_{i\in I}) = \lim_{\mathcal{U}} |a_i|_i$. Then, φ is seen to be a multiplicative semi-norm of the $\mathbb{L}$-algebra $\mathcal{R}$.

We put $\mathcal{J} = Ker(\varphi)$ and $\mathcal{S} = \frac{\mathcal{R}}{\mathcal{J}}$ and we denote by ψ the canonical surjection from $\mathcal{R}$ onto $\mathcal{S}$. Then, $\mathcal{S}$ is obviously provided with an absolute value $|\ .\ |$ defined as $|\psi(a)| = \varphi(a),\ (a \in \mathcal{R})$.

On the other hand, $\mathcal{R}$ is seen to be provided with a norm of $\mathbb{L}$-algebra $\|\ .\ \|$ defined as $\|(a_i)_{i\in I}\| = \sup_{i\in I} |a_i|$. Next, we denote by $|||\ .\ |||$ the semi-norm quotient of the norm of $\mathbb{L}$-algebra by the ideal $\mathcal{J}$, defined on $\mathcal{R}$ as $|||a||| = \inf_{t\in\mathcal{J}} \|a - t\|$.

Theorem 8.1 (Diarra). *$\mathcal{S}$ is a field extension of $\mathbb{L}$ and its absolute value $|\ .\ |$ extends the one of $\mathbb{L}$. Moreover, if $\mathcal{U}$ is non-principal and if each $\mathbb{E}_i$ has a dense valuation group, then $\mathcal{S}$ has a valuation group equal to $(\mathbb{R}, +)$. Further, if each $\mathbb{E}_i$ is algebraically closed, then so is $\mathcal{S}$.*

Proof. Let $\alpha \in \mathcal{S}\setminus\{0\}$ and let $a = (a_i)_{i\in I} \in \mathcal{R}$ be such that $\psi(a) = \alpha$. By definition, we have $\lim_{\mathcal{U}} |a_i|_i \neq 0$. Hence, there exists $J \in \mathcal{U}$ such that for every $i \in J$ we have $\frac{\varphi(a)}{2} < |a_i|_i < \frac{3\varphi(a)}{2}$, hence $a_i \neq 0$, and therefore,

(1) $\frac{2}{3\varphi(a)} < |a_i^{-1}|_i < \frac{2}{\varphi(a)}$ whenever $i \in J$.

Now, let $b = (b_i)_{i\in I} \in \prod_{i\in I} \mathbb{E}_i$ be defined as $b_i = a_i^{-1}$ whenever $i \in J$ and $b_i = 1$ whenever $i \in I\setminus J$. By (1), it is seen that b does belong to $\mathcal{R}$. But now, $ab - 1$ is an element $(c_i)_{i\in I}$ of $\mathcal{R}$ that satisfies $c_i = 0$ whenever $i \in J$, hence $\lim_{\mathcal{U}} |c_i| = 0$. Therefore, $ab - 1$ belongs to $\mathcal{J}$, and finally, in $\mathcal{S}$ we have $\psi(a)\psi(b) = 1$. This shows $\mathcal{S}$ to be a field.

Next, we suppose that for each $i \in I$ the valuation group of $\mathbb{E}_i$ is dense. Let $r \in]0, +\infty[$. We can obviously find a family $(\varepsilon_i)_{i\in I}$ in $]0, +\infty[$ such that $\lim_{\mathcal{U}} \varepsilon_i = 0$. For every $i \in I$, let $a_i \in \mathbb{E}_i$ satisfy $r - \varepsilon_i < |a_i| < r$ and let $a = (a_i)_{i\in I}$. Of course, a belongs to $\mathcal{R}$ and satisfies $\lim_{\mathcal{U}} |a_i|_i = r$, hence r belongs to $|\mathcal{S}|$. This shows that the valuation group of $\mathcal{S}$ is equal to $\mathbb{R}$.

Finally, we suppose that each field $\mathbb{E}_i$ is algebraically closed. Let

$$P(x) = \sum_{n=0}^{q} \lambda_n x^n \in \mathcal{S}[x]$$

with $\lambda_q = 1$, and $q > 0$. We show that P admits at least one zero in $\mathcal{S}$. For every $n = 0, 1, \ldots, q-1$, let $(a_{i,n})_{i\in I} \in \mathcal{R}$ satisfy $\psi((a_{i,n})_{i\in I}) = \lambda_n$ and let $a_{i,q} = 1$ whenever $i \in I$. So, we have $\psi((a_{i,n})_{i\in I}) = \lambda_n$ whenever $n = 0, \ldots, q$. For every $i \in I$, we put $T_i(x) = \sum_{n=0}^{q} a_{i,n} x^n$.

Since $\mathbb{E}_i$ is algebraically closed, and since $a_{i,q} = 1$ for every $i \in I$, at least one of the zeros α_i of T_i in $\mathbb{E}_i$ satisfies $|\alpha_i|^q \leq |a_{i,0}|$. But by hypothesis, $(a_{i,0})_{i\in I}$ belongs to $\mathcal{R}$, hence so does $(\alpha_i)_{i\in I}$. Hence, we can put $\alpha = \psi((\alpha_i)_{i\in I})$ and then we have $P(\alpha) = 0$. This finishes showing that $\mathcal{S}$ is algebraically closed and this ends the proof of Theorem 8.1. □

Lemma 8.2 (Diarra). *Let $a = (a_i)_{i\in I} \in \mathcal{R}$. Then we have $|||a||| = \varphi(a)$.*

Proof. Let $J \in \mathcal{U}$ and let $e = (e_i)_{i\in I} \in \mathcal{R}$ be defined as $e_i = 0$ whenever $i \in J$ and $e_i = 1$ whenever $i \notin J$. For convenience, we put $b = ae$. Clearly, b belongs to $\mathcal{J}$, hence, we have $|||a||| = \inf_{t\in\mathcal{J}} \|a - t\| \leq \|a - b\|$. But now, we check that $\|a - b\| = \sup_{i\in J} |a_i|_i$. Further, this is true for every $J \in \mathcal{U}$. Hence, we obtain $|||a||| \leq \inf_{J\in\mathcal{U}}(\sup_{i\in J} |a_i|_i) = \lim_{\mathcal{U}} |a_i|_i = \varphi(a)$. On the other hand, for all $t \in \mathcal{J}$, we have $\varphi(a-t) = \varphi(a) \leq \|a-t\|$, hence $\varphi(a) \leq |||a|||$. This ends the proof of Lemma 8.2. □

Theorem 8.3 (Diarra). *If $\mathcal{U}$ is incomplete, then $\mathcal{S}$ is spherically complete.*

Proof. Let $(\alpha_n)_{n\in\mathbb{N}}$ be a decreasing distances sequence in $\mathcal{S}$ and for every $n \in \mathbb{N}$, let $a_n \in \mathcal{R}$ be such that $\psi(a_n) = \alpha_n$. By induction, we can easily construct another sequence $(b_n)_{n\in\mathbb{N}}$ in $\mathcal{R}$ satisfying

$(\mathcal{V}_n)$ $\psi(b_n) = \alpha_n$ for every $n \in \mathbb{N}$,

$(\mathcal{W}_n)$ $\|b_n - b_{n-1}\| < r_{n-2}$ whenever $n > 1$.

Indeed, let $q \in \mathbb{N}^*$ suppose that we have defined $b_0, \dots, b_q$ satisfying $(\mathcal{V}_n)$ for every $n = 0, \dots, q$. Of course, we have

(1) $\varphi(b_q - a_{q+1}) = \varphi(a_q - a_{q+1}) = r_q < r_{q-1}$.

By Lemma 8.2, we have $\varphi(b_q - a_{q+1}) = |||b_q - a_{q+1}|||$, hence by (1), there exists $c \in \mathcal{J}$ such that $\|b_q - a_{q+1} - c\| < r_{q-1}$. So, we put $b_{q+1} = a_{q+1} + c$ and then $(\mathcal{V}_{q+1})$, $(\mathcal{W}_{q+1})$ are satisfied. In order to begin the induction, we put $b_0 = a_0$, $b_1 = a_1$ and then, we can define the sequence $(b_n)_{n\in\mathbb{N}}$ for every $n \in \mathbb{N}$, satisfying $(\mathcal{V}_n)$ and $(\mathcal{W}_n)$.

Now, for each $n \in \mathbb{N}$, we put $b_n = (b_{i,n})_{i\in I}$. Since $\mathcal{U}$ is incomplete, we can take a decreasing sequence $(X_n)_{n\in\mathbb{N}}$ of elements of $\mathcal{U}$ such that $\bigcap_{n\in\mathbb{N}} X_n = \emptyset$. We put $I_0 = I \setminus X_0$ and for each $n \in \mathbb{N}^*$, $I_{n+1} = X_n \setminus X_{n+1}$. Thus, the family $(I_n)_{n\in\mathbb{N}}$ makes a partition of I. Further, for each $q \in \mathbb{N}$, $(I_n)_{n\geq q}$ makes a partition of X_q. Hence, we can define a surjective mapping g from I onto $\mathbb{N}$ as $g(i) = n$ whenever $i \in I_n$. Now, for every $i \in I$, we put $h_i = b_{i,g(i)}$. By $(\mathcal{W}_n)$, we have $\|b_n\| \leq r_0$ whenever $n \in \mathbb{N}$, hence $|h_i|_i \leq r_0$ for each $i \in I$, and therefore, $(h_i)_{i\in I}$ belongs to $\mathcal{R}$. We put $h = (h_i)_{i\in I}$ and $w = \psi(h)$. We show

(2) $|w - \alpha_n| \leq r_{n-1}$ whenever $n \in \mathbb{N}^*$.

Let $n \in \mathbb{N}^*$ be fixed. It is seen that for every $m > n$, we have $\|b_n - b_m\| < r_{n-1}$, hence for every $i \in I$, we have $|b_{i,n} - b_{i,m}|_i < r_{n-1}$. Moreover, since $(I_m)_{m>n}$ makes a partition of X_{n+1}, for every $i \in X_{n+1}$, there exists $m > n$ such that $i \in I_m$ and then we have $|b_{i,n} - h_i|_i = |b_{i,n} - b_{i,g(i)}|_i = |b_{i,n} - b_{i,m}|_i < r_{n-1}$ whenever $i \in X_{n+1}$. But X_{n+1} belongs to $\mathcal{U}$, and therefore, in $\mathcal{S}$ we have $|b_n - h| = \lim_{\mathcal{U}} |b_{i,n} - h_i|_i \leq r_{n-1}$. This is true for every $n \in \mathbb{N}^*$ and finally this shows (2). Hence, w belongs to $\bigcap_{n\in\mathbb{N}} d_{\mathcal{S}}(\alpha_{n+1}, r_n)$ and this finishes proving that $\mathcal{S}$ is spherically complete. □

Theorem 8.4. *$\mathbb{K}$ admits a spherically complete algebraically closed extension whose residue class field is not countable and whose valuation group is equal to $\mathbb{R}$.*

Proof. First, we construct a complete algebraically closed extension of $\mathbb{K}$ whose residue class field is not countable. Let T be a transcendental extension of the form $\mathbb{K}((x_j)_{j\in\mathbb{R}})$ provided with the absolute value $|\ .\ |$ defined on $\mathbb{K}[(x_j)_{j\in\mathbb{R}}]$ by

$$\left|\sum_{j_1,\dots,j_q\leq N} a_{j_1,\dots,j_q} x_{j_1}^{t_1}\cdots x_{j_q}^{t_q}\right| = \max_{j_1,\dots,j_q\leq N} |a_{j_1,\dots,j_q}|.$$

It is seen that $|x_j - x_h| = 1$ whenever $j, h \in \mathbb{R}$ such that $j \neq h$, and therefore, the residue class field of T is not countable. Let T' be the completion of T and let $\mathbb{E}$ be an algebraic closure of T', provided with the unique absolute value that extends the one of T'. Let $\mathbb{E}'$ be the completion of $\mathbb{E}$. By Theorem 6.14, $\mathbb{E}'$ is algebraically closed. Obviously, its residue class field contains one of T and therefore is not countable.

Now, we can construct $\mathcal{S}$ by taking $I = \mathbb{N}$ and $\mathbb{E}_i = \mathbb{E}'$ for every $i \in \mathbb{N}$. Since $\mathbb{E}'$ is algebraically closed, by Theorem 8.1, so is $\mathcal{S}$. Moreover, the valuation group of $\mathbb{E}'$ obviously is dense and therefore, by Theorem 7.1, $\mathcal{S}$ has a valuation group equal to $\mathbb{R}$. Finally, by Theorem 8.3, $\mathcal{S}$ is spherically complete. That ends the proof of Theorem 8.4. $\square$

Thanks to Theorem 8.3, we can generalize Corollary 5.3:

Theorem 8.5. *Let $P, F \in \mathbb{K}[x]$ with F having all its zeros in $d(0, r)$. Let $Q, R \in \mathbb{K}[x]$ satisfy $P = FQ + R$ and $\deg(R) < \deg(F)$. Then we have $\Psi(Q, \log r) \leq \Psi(P, \log r) - \Psi(F, \log r)$ and $\Psi(R, \log r) \leq \Psi(P, \log r)$. Moreover, if P has all its zeros in $d(0, r)$, then $\Psi(Q, \log r) = \Psi(P, \log r) - \Psi(F, \log r)$.*

Proof. Without loss of generality, we can assume that the valuation group of $\mathbb{K}$ is $\mathbb{R}$. Consequently, up to a change of variable, we can suppose that $r = 1$. We can also assume that F is monic. Now, since F has all zeros in $d(0, r)$, this means that $\Psi(F, 0) = 0$ and hence Theorem 8.5 is reduced to Corollary 5.3. $\square$

Notation. Henceforth, $\widehat{\mathbb{K}}$ denotes an algebraically closed spherically complete extension of $\mathbb{K}$.

For every disk $d(a, r^-)$ (respectively $d(a, r)$) in $\mathbb{K}$, we denote by $\widehat{d}(a, r^-)$ (respectively $\widehat{d}(a, r)$) the disk of same center and diameter in $\widehat{\mathbb{K}}$. Similarly, we denote by $\widehat{C}(a, r)$ the circle $\{x \in \widehat{\mathbb{K}} \mid |x - a| = r\}$.

Remark. There exists another way to construct a spherically complete extension, due to Irving Kaplansky [68].

Definition. Let $\mathbb{E}$ be an extension of $\mathbb{L}$ provided with an ultrametric absolute value that extends that of $\mathbb{L}$. The extension $\mathbb{E}$ is said to be *immediate* if its residue class field is identical to that of $\mathbb{L}$ and its value group also is identical to that of $\mathbb{L}$.

The following theorem is due to Kaplansky [68]:

Theorem 8.6. $\mathbb{L}$ *admits an immediate extension that is maximal with respect to the inclusion.*

An immediate extension of $\mathbb{L}$ *is spherically complete if and only if it is maximal with respect to the inclusion.*

A maximal immediate extension of $\mathbb{L}$ *is unique up to an* $\mathbb{L}$*-isomorphism.*

The proof of this theorem represents a very big work. In what follows, we do not need that theorem.

Chapter 9

Transcendence Order and Transcendence Type

In $\mathbb{C}_p$, we can define a notion of transcendence order stating that if a is transcendental over $\mathbb{Q}_p$ and has a transcendence order $\leq t$ and if b is transcendental over $\mathbb{Q}_p$ but algebraic over $\mathbb{Q}_p[a]$, then b also has a transcendence order $\leq t$. We show the existence of numbers of order less than $1+\epsilon$ for every $\epsilon > 0$ [41, 55].

Definition. Let $\tau \in]0, +\infty[$. Let F be a transcendental extension of $\mathbb{Q}_p$ provided with an absolute value $|\ .\ |$ extending that of $\mathbb{Q}_p$. An element $a \in F$ is said *to have transcendence order* $\leq \tau$ or *order* $\leq \tau$ in brief if there exists a constant $C_a \in]0, +\infty[$ such that every polynomial $P \in \mathbb{Q}_p[x]$ satisfies $\log_p(|P(a)|) \geq \log(\|P\|) - C_a(\deg(P))^\tau$. Moreover, a is said *to have weak transcendence order* $\leq \tau^+$ or *weak order* $\leq \tau^+$ in brief if a has transcendence order $\leq \tau + \epsilon$ for every $\epsilon > 0$.

Notation. We denote by $\mathcal{S}(\tau)$ the set of numbers $x \in \mathbb{C}_p$ having transcendence order $\leq \tau$ and by $\mathcal{S}(\tau^+)$ the set of numbers $x \in \mathbb{C}_p$ having weak transcendence order $\leq \tau^+$.

Finally, we say that a number $x \in \mathbb{C}_p$ is *of infinite order* if it does not belong to $\mathcal{S}(\tau)$ for all $\tau \in \mathbb{R}_+^*$.

Remark. By definition, an element $a \in \mathbb{C}_p$ having transcendence order $\leq \tau$ or weak transcendence order $\leq \tau^+$ is transcendental over $\mathbb{Q}_p$.

Theorem 9.1. *Let $\tau \in]0, +\infty[$. If $\mathcal{S}(\tau) \neq \emptyset$ then $\tau \geq 1$.*

Proof. Let $a \in \mathbb{C}_p$, $a \neq 0$ be transcendental over $\mathbb{Q}_p$ and have transcendence order $\leq \tau$. We can find $b \in \Omega_p$ ($b \neq 0$) such that $|a - b|_p < 1$. Consider the minimal polynomial Q of b over $\mathbb{Q}_p$. Let $b_2, \dots, b_q$ be the conjugates of b over $\mathbb{Q}_p$ and set $b_1 = b$. We note that by Corollary 6.6 all conjugates b_j of b over $\mathbb{Q}_p$ satisfy $|b_j|_p = |b|_p$.

Suppose first that $|a|_p \leq 1$. Since $|b_j|_p = |b|_p = |a|_p \leq 1$, all coefficients of Q belong to $\mathbb{Z}_p$. Obviously, Q is monic, hence $\|Q\| = 1$. By hypothesis, there exists $C_a \in]0, +\infty[$ such that $\Psi_p(P(a)) \geq \log_p(\|P\|) - C_a(\deg(P))^\tau$, $\forall P \in \mathbb{Q}_p[x]$. Consequently, $-n\Psi_p(Q(a)) = -\Psi_p((Q(a))^n) \leq C_a(n \deg(Q))^\tau$, $\forall n \in \mathbb{N}^*$. Since $Q(b) = 0$ and since, by Corollary 4.5, Q is 1-Lipschitzian in U, we have $-\Psi_p(Q(a)) > 0$, and therefore, if $\tau < 1$, the inequality $-n\Psi_p(Q(a)) \leq C_a(n \deg(Q))^\tau$ $\forall n \in \mathbb{N}^*$ is impossible when n tends to $+\infty$.

Suppose now $|a|_p > 1$. Set $Q(X) = \sum_{k=0}^{q} c_k X^k$. Since the b_j satisfy $|b_j|_p = |a|_p$ ($1 \leq j \leq q$), we have $|c_k|_p \leq (|a|_p)^{q-k}$ and particularly $|c_0|_p = \prod_{j=1}^{q} |b_j|_p = (|a|_p)^q$. Consequently, $\|Q\| = (|a|_p)^q$, and therefore, considering the sequence $(Q^n)_{n \in \mathbb{N}}$, for every $n \in \mathbb{N}^*$, we have

(1) $-n\Psi_p(Q(a)) \leq -nq\Psi_p(a) + C_a(nq)^\tau$.

On the other hand, $Q(a) = Q(a) - Q(b) = (a - b)\sum_{k=1}^{q} c_k \sum_{j=0}^{k} a_j b^{k-j-1}$ and hence $|Q(a)|_p \leq |a - b|_p(|a|_p)^{q-1}$. Consequently, we obtain $-n\Psi_p(a - b) - n(q - 1)\Psi_p(a) \leq \Psi_p(Q(a)$, and hence, by (1), $-n\Psi_p(a - b) - n(q - 1)\Psi_p(a) \leq -nq\Psi_p(a) + C_a(nq)^\tau$. Finally, $n(\Psi_p(a) - \Psi_p(a - b)) \leq C_a(nq)^\tau$. Since $|a|_p > 1$ and $|a - b|_p < 1$, this inequality is impossible again when n tends to $+\infty$, which ends the proof. □

Theorem 9.2. *There exists $b \in \mathbb{C}_p$ transcendental over $\mathbb{Q}_p$, of order $\leq 1 + \epsilon$, for every $\epsilon > 0$.*

Proof. Consider first a strictly decreasing sequence $(\epsilon_n)_{n \in}$ such that $\lim_{n \to +\infty} \epsilon_n = 0$ and $\lim_{n \to +\infty} \epsilon_n \log(n) = +\infty$.

We can always divide any polynomial $P \in \mathbb{Q}_p[x]$ by some $\lambda \in \mathbb{Q}_p$ such that $|\lambda|_p = \|P\|$ and hence we go back to the hypothesis $\|P\| = 1$. So, if we can find some $b \in \mathbb{C}_p$ and, for every $\omega > 0$, a constant $C(\omega) > 0$ and show that for every $P \in \mathbb{Q}[X]$ such that

$\|P\| = 1$, we have $-\log(|P(b)|_p) \leq C(\omega)(\deg(P))^{1+\omega}$, Theorem 9.2 will be proven.

By induction, we can define a strictly increasing sequence $(r_n)_{n\in\mathbb{N}}$ of $\mathbb{Q}$ and a sequence $(a_n)_{n\in\mathbb{N}}$ of $\mathbb{C}_p$ with $r_n = \frac{u_n}{v_n}$, irreducible and $(v_n)_{n\in\mathbb{N}}$ a strictly increasing sequence of prime numbers satisfying further the following properties:

(i) $\lim_{n\to+\infty} r_n = +\infty$,
(ii) for every $n \in \mathbb{N}$, $n^{\epsilon_n} < r_n < (n+1)^{\epsilon_n}$,
(iii) $v_n > \prod_{j=1}^{n-1} v_j$,
(iv) $(a_n)^{v_n} = p^{u_n}$.

By construction, the sequence $(|a_n|_p)_{n\in\mathbb{N}}$ is strictly decreasing and tends to 0 and all terms belong to U. Set $b = \sum_{n=1}^{\infty} a_n$. Now, let us fix $\varepsilon > 0$. We show that b is transcendental over $\mathbb{Q}_p$ and has a transcendence order $\leq 1+\varepsilon$.

Since the sequence (ϵ_n) tends to 0, we can find an integer $t(\varepsilon)$ such that $\epsilon_m < \varepsilon$ $\forall m \geq t(\varepsilon)$. Thus, as the first step, let us take $q \geq t(\varepsilon)$ and let us find a constant $C(\varepsilon) > 0$, not depending on b, such that for every $P \in \mathbb{Q}[X]$ satisfying $\|P\| = 1$ and $\deg(P) = q$, we have $-\log_p(|P(b)|_p) \leq C(\varepsilon)q^{1+\varepsilon}$.

For each $n \in \mathbb{N}^*$, set $b_n = \sum_{m=1}^{n} a_m$. Since the sequence $(|a_m|_p)_{m\in\mathbb{N}}$ is strictly decreasing, we have $|b - b_n|_p = |a_{n+1}|_p$, and since P is obviously 1-Lipschitzian in the disk U, we have $|P(b) - P(b_n)|_p \leq |a_{n+1}|_p$, hence

(1) $\log_p(|P(b) - P(b_n)|_p) \leq \log_p(|a_{n+1}|_p) = -r_{n+1}$.

Now, since the sequence $\epsilon_n \log_p(n)$ tends to $+\infty$, we can choose $n(q)$ such that $(n(q)+1)^{\epsilon_{n(q)+1}} > (q+1)^{1+\varepsilon}$. Then, by (1), we have

(2) $\log_p(|P(b) - P(b_{n(q)})|_p) < \log_p(|a_{n(q)+1}|_p) = -(r_{n(q)+1}) < -(n(q)+1)^{\epsilon_{n(q)+1}} < -(q+1)^{1+\varepsilon}$.

We show the following inequality:

(3) $-\log_p(|P(b_{n(q)})|_p) \leq (q+1)^{1+\varepsilon}$.

Thus, suppose (3) is wrong. Set $h_q = \sum_{m=q}^{n(q)} a_m$. Then, $b_{n(q)} = b_{q-1} + h_q$. Now, developing P at the point b_{q-1}, we have

(4)
$$\log_p(|P(b_{n(q)})|_p) = \log_p\left(\left|\sum_{m=0}^{q} \frac{P^{(m)}(b_{q-1})}{m!}(h_q)^m\right|_p\right) < -(q+1)^{1+\varepsilon}.$$

Consider now the sum $\sum_{m=0}^{q} \frac{P^{(m)}(b_{q-1})}{m!}(h_q)^m$. Since the sequence $|a_m|_p$ is strictly decreasing, we have $|h_q|_p = |a_q|_p$, hence $\log_p(|h_q|_p) = -r_q$. We note that $\mathbb{Q}_p[a_1, \dots, a_{q_1}]$ is an algebraic extension of $\mathbb{Q}_p$ of degree at most $\prod_{j=1}^{q-1} v_j$. Consequently, by Corollary 6.8, the extension $\mathbb{Q}_p[a_1, \dots, a_{q_1}]$ has a value group of the form $\frac{s}{t}\mathbb{Z}$ with $t \leq \prod_{j=1}^{q-1} v_j$. On the other hand, due to the hypothesis $r_q = \frac{u_q}{v_q}$, it appears that v_q is a prime integer, prime to u_q and bigger than q and $\prod_{j=1}^{q-1} v_j$. Consequently, we can apply Lemma 6.19 with h_q in the role of c and b_{q-1} in the role of a. Therefore, for each $m = 0, \dots, q-1$, all the $\left|\frac{P^{(m)}(b_{q-1})}{m!}(h_q)^m\right|_p$ are pairwise distinct. Consequently, we have

(5) $\left|\sum_{m=0}^{q} \frac{P^{(m)}(b_{q-1})}{m!}(h_q)^m\right|_p = \max_{1\leq m\leq q} \left|\frac{P^{(m)}(b_{q-1})}{m!}(h_q)^m\right|_p.$

Next, since $-\Psi_p(h_q) = r_q < (q+1)^\omega$, for each integer $m = 1, \dots, q$, we have $\Psi_p(h_q)^m) = -mr_q > -m(q+1)^\omega \geq -q(q+1)^\omega$, hence

(6) $\Psi_p((h_q)^m) \geq -q(q+1)^\omega > -(q+1)^{1+\omega}\ \forall m \leq q.$

Consequently, by (4), (5), and (6), the polynomial $Q(X) = \sum_{m=0}^{q} \frac{P^{(m)}(b_{q-1})}{m!}(X)^m$ has all coefficients in $d(0, 1^-)$ and hence we have $\|Q\| < 1$. But since $|b_{q-1}|_p < 1$, by Theorem 4.5, we have $\|P\| = \|Q\| < 1$, a contradiction to the hypothesis $\|P\| = 1$. Therefore, Relation (3) is proven for every polynomial $P \in \mathbb{Q}_p[X]$ of degree $q \geq t(\omega)$ such that $\|P\| = 1$. Consequently, by (3), we obviously have a constant $C > 0$, not depending on b, such that $-\Psi_p(P(b)) \leq C(\deg(P))^{1+\omega}$ for every $P \in \mathbb{Q}_p[X]$ such that $\deg(P) \geq t(\omega)$ and $\|P\| = 1$.

Particularly, b is transcendental over $\mathbb{Q}_p$ because if it were algebraic, the degrees of polynomial $P \in \mathbb{Q}_p[X]$ such that $P(b) = 0$ wouldn't be bounded. Finally, it is easily seen that there exists a constant $m > 0$ such that $|Q(b)|_p \geq m$ for every polynomial $Q \in \mathbb{Q}_p[X]$ of degree $q \leq t(\omega)$ and $\|Q\| = 1$. Therefore, b is clearly of order $\leq 1 + \omega$. □

Corollary 9.3. *Ω_p is not complete.*

The transcendence type is defined in $\mathbb{C}_p$ in the same way as in $\mathbb{C}$ [41].

Definition and notation. Given a complex number z, we denote by $|z|_\infty$ its modulus. Throughout this chapter, a number $a \in \mathbb{C}_p$ is just said to be *algebraic* (respectively *transcendental*) if it is algebraic (respectively transcendental) over $\mathbb{Q}$. When a is algebraic or transcendental over $\mathbb{Q}_p$, we precise this. Throughout this chapter, we denote by Ω the field of algebraic numbers and by $\mathbf{A}$ the ring of algebraic integers.

Let $a \in \Omega$. We call *denominator* of a any strictly positive integer n such that na and we denote by $\text{den}(a)$ the smallest denominator of a. Let $a_2, \ldots, a_n$ be the conjugates of a over $\mathbb{Q}$ in $\mathbb{C}$ and put $a_1 = a$. For convenience, we use the logarithm of base p denoted by $\log_p$. We set $\overline{|a|} = \max_{j=1,\ldots,n}(|a_j|_\infty)$ and $s(a) = \max(\log_p(\overline{|a|}, \log(\text{den}(a)))$.

The following relations are classical and immediate:

Lemma 9.4. *Let $a, b \in \Omega$ and let $m \in \mathbb{N}$. Then,* $\text{den}(ab) \leq \text{den}(a)$ $\text{den}(b)$, $\text{den}(a+b) \leq \text{den}(a)\text{den}(b)$, $\text{den}(ma) \leq m\text{den}(a), \text{den}(a^m) \leq (\text{den}(a))^m$ *and* $\overline{|ab|} \leq \overline{|a|}.\overline{|b|}$, $\overline{|a+b|} \leq \overline{|a|} + \overline{|b|}$, $\overline{|ma|} \leq m\overline{|a|}$, $\overline{|a^m|} = (\overline{|a|})^m$.

Definition and notation. Let

$$P(X_1, \ldots, X_n) = \sum_{i_1,\ldots,i_n} a_{i_1,\ldots,i_n}(X_1)^{i_1} \cdots (X_n)^{i_n} \in \mathbb{C}[X_1, \ldots, X_n].$$

We put $H(P) = \max_{i_1,\ldots,i_n} |a_{i_1,\ldots,i_n}|_\infty$ and $t(P) = \max(\log_p(H(P), \deg(P)+1)$.

A number $a \in \mathbb{C}_p$ is said *to have transcendence type less than* α if there exists a constant $C_a > 0$ such that, for every $Q \in \mathbb{Z}[X]$, we have either $Q(a) = 0$ or $-\Psi_p(Q(a)) \leq C_a(t(Q))^\alpha$. We denote by $\mathcal{T}(\alpha)$ the set of numbers $a \in \mathbb{C}_p$ having a transcendence type less than or equal to α.

If a number $a \in \mathbb{C}_p$ does not belong to $\mathcal{T}(\alpha)$ for all $\alpha > 0$, we say that a is of *infinite type*.

By Lemma 6.17, Lemmas 9.5, 9.6, and 9.7 are immediate:

Lemma 9.5. *Let $P \in \mathbb{Z}[X]$. Then,* $-\Psi_p(P, 0) \leq \log_p(H(P))$.

Lemma 9.6. *Let $P \in \mathbb{Z}[X]$ be of degree k and let $a \in \Omega$. Then,*

$$\overline{|P(a)|} \leq H(P)(k+1)\big(\max(\overline{|a|}, 1)\big)^k.$$

Lemma 9.7. *Let* $\alpha_1, \alpha_2 \in]0, +\infty[$ *satisfy* $\alpha_1 \leq \alpha_2$. *Then,* $\mathcal{T}(\alpha_1) \subset \mathcal{T}(\alpha_2)$.

There exists a link between transcendence order over $\mathbb{Q}_p$ and transcendence type over $\mathbb{Q}$ [41, 50].

Theorem 9.8. *Let* $\alpha \in [1, +\infty[$. *Then,* $\mathcal{S}(\alpha) \subset \mathcal{T}(\alpha)$.

Proof. Let $a \in \mathcal{S}(\alpha)$. By hypothesis, there exists $C > 0$ such that

$$-\Psi_p(Q(a)) \leq -\Psi_p(Q, 0) + C\Big(\deg(Q)\Big)^{\alpha} \ \forall Q \in \mathbb{Q}_p[X].$$

Hence, by Lemma 9.5, we have $-\Psi_p(Q(a)) \leq \log_p(H(Q)) + C\Big(\deg(Q)\Big)^{\alpha}$ $\forall Q \in \mathbb{Q}[X]$. Then, taking $C \geq 1$, we can derive

$$-\Psi_p(Q(a)) \leq C\Big(\log_p(H(Q)) + \big(\deg(Q)\big)^{\alpha}\Big) \leq 2C(t(Q))^{\alpha} \ \forall Q \in \mathbb{Q}[X],$$

which proves that $a \in \mathcal{T}(\alpha)$. □

By Theorem 9.2, we can now state the following corollary [41, 49]:

Corollary 9.9. *There exists* $b \in \mathbb{C}_p$, *transcendental over* $\mathbb{Q}_p$, *such that* $b \in \mathcal{T}(1+\epsilon)$ *for every* $\epsilon > 0$.

Proof. Indeed, in Theorem 9.2, we saw that there exists $b \in \mathbb{C}_p$ that belongs to $\mathcal{S}(1+\epsilon)$ for all $\epsilon > 0$. □

By Lemma 6.17, we can immediately derive the following inequality:

Theorem 9.10. *Let* $a \in \Omega^*$ *be integral of degree* q, *over* $\mathbb{Z}$. *Then,* $|a|_p \geq \frac{1}{\left(\overline{|a|}\right)^q}$.

Proof. Let $Q(X) = irr(a, \mathbb{Q})$ and let $a_1, \ldots, a_q$ be the conjugates of a over $\mathbb{Z}$, with $a_1 = a$. Then, $\prod_{j=1}^{q} a_j$ belongs to $\mathbb{Z}^*$, hence, by

Lemma 6.18, we have $|a_j| \leq 1\ \forall j = 2, \dots, q$:

$$\left|\prod_{j=1}^{q} a_j\right|_p \geq \frac{1}{|\prod_{j=1}^{q} a_j|_\infty}. \tag{1}$$

Consequently,

$$|a|_p \geq \left|\prod_{j=1}^{q} a_j\right|_p \geq \frac{1}{\prod_{j=1}^{q} |a_j|_\infty}. \tag{2}$$

Now, $|\prod_{j=1}^{q} a_j|_\infty = \prod_{j=1}^{q} |a_j|_\infty \leq \left(\overline{|a|}\right)^q$. Thus, by (2), we obtain

$$|a|_p \geq \frac{1}{\left(\overline{|a|}\right)^q}.$$

□

Corollary 9.11. *Let $a \in \Omega^*$ be of degree q over $\mathbb{Q}$ and let $t = \mathrm{den}(a)$. Then, $|a|_p \geq \frac{1}{t^q\left(\overline{|a|}\right)^q}$.*

Corollary 9.12. *Let $a \in \Omega^*$ be of degree q over $\mathbb{Q}$ and let $t = \mathrm{den}(a)$. Then, $\log(|a|_p) \geq -2q(s(a))$.*

Chapter 10

Algebras $R(D)$

The idea of considering rational functions with no pole inside a domain D, in order to define analytic functions in D, is due to Marc Krasner [69]. The behavior of rational functions in $\mathbb{K}$ is determined by circular filters which characterize all multiplicative norms on rational functions. We make a general study of the set of multiplicative semi-norms of a normed algebra which is locally compact with respect to the topology of pointwise convergence. Results are first due to Guennebaud and Garandel [61, 62]. Henceforth, the idea of considering the topologic space of multiplicative semi-norms continuous with respect to the topology of a normed algebra was used in many works on Banach algebra.

Notation. In this chapter, we denote by D an infinite subset of $\mathbb{K}$ and then $\mathbb{K}^D$ is provided with the topology $\mathcal{U}_D$ of uniform convergence on D.

We denote by $R(D)$ the $\mathbb{K}$-algebra of rational functions $h(x) \in \mathbb{K}(x)$ with no pole in D. Since D is infinite, $R(D)$ is clearly a $\mathbb{K}$-subalgebra of $\mathbb{K}^D$ and is provided with the topology induced by $\mathcal{U}_D$ that makes it a topological subgroup of $\mathbb{K}^D$. Algebraically, $R(D)$ is a $\mathbb{K}$-subalgebra of $\mathbb{K}(x)$ and more precisely is of the form $S(D)^{-1}\mathbb{K}[x]$ with $S(D)$ the multiplicative set of polynomials whose zeros do not belong to D.

We denote by $R_b(D)$ the $\mathbb{K}$-subalgebra of $R(D)$ consisting of the $f \in R(D)$ which are bounded in D. Finally, if D is not bounded, we denote by $R_0(D)$ the $\mathbb{K}$-subalgebra of $R(D)$ that consists of the $f \in R(D)$ such that $\lim_{|x|\to+\infty,\ x\in D} f(x) = 0$.

For every $f \in \mathbb{K}^D$, we set $\|f\|_D = \sup_{x\in D}|f(x)| \in [0, +\infty]$.

Recall that an algebra-semi-norm ψ of a $\mathbb{K}$-algebra A is said to be *semi-multiplicative* or *power multiplicative* if it satisfies $\psi(x^n) = (\psi(x))^n$ $\forall x \in A$ and is said to be *multiplicative* if $\psi(xy) = \psi(x)\psi(y)$ $\forall x,\ y \in A$.

Lemma 10.1 is then immediate.

Lemma 10.1. *$R(D)$ is a principal ideal ring. Every ideal is of the form $P(x)R(D)$ with P a polynomial whose zeros belong to D.*

Lemma 10.2 is an immediate application of general properties of the supremum, once the set $[0, +\infty]$ is provided with the classical extensions of the addition and the multiplication:

$a + (+\infty) = +\infty$ for every $a \in [0, +\infty]$,
$a\,.\,(+\infty) = +\infty$ for every $a \in]0, +\infty]$.

Lemma 10.2. *For every $g, h \in R(D)$, we have*

(i) $\|h\|_D = 0$ *if and only if* $h = 0$,
(ii) $\|\lambda h\|_D = |\lambda|\ \|h\|_D$ *for every* $\lambda \in \mathbb{K}^*$,
(iii) $\|h + g\|_D \le \max(\|h\|_D, \|g\|_D)$,
(iv) *If* $(\|f\|_D, \|g\|_D)$ *is different from* $(0, +\infty)$ *and from* $(+\infty, 0)$, *then* $\|hg\|_D \le \|h\|_D.\|g\|_D$,
(v) $\|h^n\|_D = (\|h\|_D)^n$ *whenever* $n \in \mathbb{N}$.

Theorem 10.3. *$R_b(D) = R(D)$ if and only if D is closed and bounded. Moreover, if D is closed and bounded, then $\|\ .\ \|_D$ is a semi-multiplicative ultrametric norm of $\mathbb{K}$-algebra.*

Proof. We first suppose D to be bounded. By Lemma 10.2, we just have to show that $\|h\|_D < +\infty$ for every $h \in R(D)$ in order to show that $\|\ .\ \|_D$ is a norm of $\mathbb{K}$-algebra such that $\|h^n\|_D = \|h\|_D^n$. Since D is bounded, obviously every polynomial P satisfies $\|P\|_D < +\infty$, hence, by Lemma 10.2 (iv), we just have to check that $\left\|\frac{1}{Q}\right\|_D < +\infty$. To show this, it is sufficient to prove that $\left\|\frac{1}{x-a}\right\|_D < +\infty$ for every $a \in \mathbb{K}\setminus D$. Since D is closed, the distance r from a to D is not zero, hence $\left\|\frac{1}{x-a}\right\|_D \le \frac{1}{r}$.

Now, if D is not bounded, obviously $\|x\|_D = +\infty$. If D is not closed, there exists at least one point $a \in \mathbb{K}\setminus D$ with a sequence

$(a_n)_{n\in\mathbb{N}}$ in D which converges to a, hence $\left\|\frac{1}{x-a}\right\|_D = +\infty$. That ends the proof of Theorem 10.3. □

Theorem 10.4. (Garandel, Guennebaud) [61, 62]: *Let $\mathcal{F}$ be a large circular filter on $\mathbb{K}$ of diameter $s > 0$. The following three assertions are equivalent:*

(i) $\varphi_{\mathcal{F}}(h) \leq \|h\|_D$ *whenever* $h \in R(D)$.
(ii) $\varphi_{\mathcal{F}}$ *is a continuous ultrametric multiplicative norm on* $R(D)$ *with respect to the topology of uniform convergence.*
(iii) $\mathcal{F}$ *is secant with* D.

Proof. First, (i) and (ii) are obviously equivalent. Next, (iii) clearly implies (i) because if $\mathcal{F}$ is secant with D, then $\lim_{\mathcal{F}}|h(x)| = \lim_{\mathcal{F}\cap D}|h(x)| \leq \|h\|_D$.

Hence, we just have to show that (i) implies (iii). For this, we assume (iii) to be false and prove that (i) is false. We first assume $\mathcal{F}$ to have center a. There exist annuli $\Gamma(a_i, r'_i, r''_i)$ $(1 \leq i \leq q)$ with $|a_i - a_j| = s$ whenever $i \neq j$ and $r'_i < s < r''_i$ such that the set $B = \bigcap_{i=1}^{q}\Gamma(a_i, r'_i, r''_i)$ belongs to $\mathcal{F}$ and satisfies $B \cap D = \emptyset$. We put $r' = \max_{1\leq i\leq q} r'_i$ and $r'' = \min_{1\leq i\leq q} r''_i$. Let $\rho' \in]r', s[$, let $\rho'' \in]s, r''[$, and for every $i = 1, \dots, q$ set $b_i \in \Gamma(a_i, \rho', s)$ and set $b \in \Gamma(a, s, \rho'')$.

We put $h(x) = \left(\prod_{i=1}^{q}\left(\frac{x-a_i}{x-b_i}\right)\right)\left(\frac{\lambda}{x-b}\right)$ with $|\lambda| = b$. We first note that $\varphi_{\mathcal{F}}(h) = 1$ because $\varphi_{\mathcal{F}}\left(\frac{x-a_i}{x-b_i}\right) = 1$ whenever $i = 1, \dots, q$ and $\varphi_{\mathcal{F}}\left(\frac{\lambda}{x-b}\right) = 1$. Next, it is easily seen that $\|h\|_D \leq \max\left(\frac{r'}{\rho'}, \frac{\rho''}{r''}\right)$. Indeed, $|h(x)| = |\frac{x}{b_i}| \leq \frac{r'}{\rho'}$ when $|x - a_i| \leq r'$ and $|h(x)| = |\frac{\lambda}{x-b}| \leq \frac{\rho''}{r''}$ when $|x - a| \geq r''$. Hence, we have $\|h\|_D < 1$ while $\varphi_{\mathcal{F}}(h) = 1$ and that contradicts the assertion (i).

We now suppose that $\mathcal{F}$ has no center. Let $d(a, r)$ belong to $\mathcal{F}$ such that $d(a, r) \cap D = \emptyset$ and let $\rho \in]s, r[$. There still exists a disk $d(\alpha, \rho) \in \mathcal{F}$ such that $d(\alpha, \rho) \subset d(a, r)$. Let us take $b \in \Gamma(\alpha, \rho, r)$ and $\lambda \in \mathbb{K}$ such that $|\lambda| = |b|$. We just put $h(x) = \frac{\lambda}{x-b}$ and we have $\varphi_{\mathcal{F}}(h) = 1$ because $|h(x)| = 1$ whenever $x \in d(\alpha, \rho)$, while $|h(x)| \leq \frac{|b-\alpha|}{r} < 1$ whenever $|x - \alpha| \geq r$, hence finally $\|h\|_D < 1$. That ends the proof of Theorem 10.4. □

In order to describe properties of the multiplicative semi-norms on $R(D)$ and next on analytic elements, we must recall a classical result on continuous multiplicative semi-norms on a normed $\mathbb{K}$-algebra A.

Notation. We denote by $\text{Mult}(A)$ the set of $\mathbb{K}$-algebra multiplicative semi-norms of a $\mathbb{K}$-algebra A. Given $\psi \in \text{Mult}(A)$, we denote by $Ker(\psi)$ the set of $x \in A$ such that $\psi(x) = 0$ and $Ker(\psi)$ is called *the kernel of* ψ.

Suppose now A is a normed $\mathbb{K}$-algebra whose norm is denoted by $\| \, . \, \|$. We denote by $\text{Mult}(A, \| \, . \, \|)$ the set of $\mathbb{K}$-algebra multiplicative semi-norms of A that are continuous functions on A with respect to the norm $\| \, . \, \|$ of A. Similarly, we denote by $\text{Mult}_m(A, \| \, . \, \|)$ the set of $\mathbb{K}$-algebra multiplicative semi-norms of A whose kernel is a maximal ideal that are continuous functions on A and by $\text{Mult}_1(A, \| \, . \, \|)$ the set of $\mathbb{K}$-algebra multiplicative semi-norms of A whose kernel is a maximal ideal of codimension 1 that are continuous functions on A.

Lemma 10.5. *Let A be a $\mathbb{K}$-algebra provided with a $\mathbb{K}$-algebra norm $\| \, . \, \|$ and let $\varphi \in \text{Mult}(A)$. Then, φ belongs to $\text{Mult}(A, \| \, . \, \|)$ if and only if $\varphi(x) \leq \| \, x \, \|$ whenever $x \in A$. Moreover, if A has a unity u and if φ is not identically 0, then $\varphi(\lambda u) = |\lambda|$ whenever $\lambda \in \mathbb{K}$. Further, if φ belongs to $\text{Mult}(A)$, $Ker(\varphi)$ is a prime ideal, and if $\text{Mult}(A, \| \, . \, \|)$, then $Ker(\varphi)$ is a closed prime ideal.*

Proof. Suppose that for some $x \in A$ we have $\varphi(x) > \| \, x \, \|$. Since the valuation group of $\mathbb{K}$ is dense, it contains a subgroup of the form $a\mathbb{Z}$, with $a > 0$. Let $q \in \mathbb{N}$ be such that $q(\log(\varphi(x)) - \log(\|x\|)) > a$. Then there clearly exists $\lambda \in \mathbb{K}$ satisfying $\| \, x \, \|^q < |\lambda| < \varphi(x)^q$. So, much the more, we have $\| \, x^q \, \| < |\lambda| < \varphi(x^q)$. Let $t = x^q$. Then, $\lim_{n\to\infty} \left(\frac{t}{\lambda}\right)^n = 0$ but $\lim_{n\to\infty} \varphi\left(\left(\frac{t}{\lambda}\right)^n\right) = +\infty$, and then φ is not continuous.

Now, let u be a unity in A. Either $\varphi(u) = 0$ and then $\varphi(x) = 0$ whenever $x \in A$, or $\varphi(u) = 1$ and then we have $\varphi(\lambda u) = |\lambda|\varphi(u) = |\lambda|$ whenever $\lambda \in L$. The last statement is immediate. This ends the proof of Lemma 10.5. □

Theorem 10.6 is well known and may be found in [62] and in Theorems 6.9 and 6.19 of [45] (see also [53]).

Theorem 10.6. *Let A be a $\mathbb{K}$-algebra provided with a $\mathbb{K}$-algebra norm $\| \, . \, \|$. For every $x \in A$, the sequence $\left(\| \, x^n \, \|^{\frac{1}{n}}\right)_{n\in\mathbb{N}}$ has a limit denoted by $\| \, x \, \|_{si}$, satisfying $\|x\|_{si} \leq \|x\| \ \forall x \in A$ and $\|x\|_{si} = \sup\{\phi(x) \mid \phi \in \text{Mult}(A, \|. \, .\|)$. Moreover, $\|x^n\|_{si} = (\|x\|_{si})^n \ \forall f \in A$, $\forall n \in \mathbb{N}$.*

Theorem 10.7. *Let A be a $\mathbb{K}$-algebra provided with a $\mathbb{K}$-algebra norm $\| \,.\, \|$. Then, Mult($A, \| \,.\, \|$) is compact with respect to the topology of pointwise convergence.*

Proof. Let B be the unit ball of A. By Lemma 10.5, each $\varphi \in \text{Mult}(A, \| \,.\, \|)$ has a restriction $\widehat{\varphi}$ to B which satisfies $\widehat{\varphi}(B) \subset [0, 1]$. Hence, $\text{Mult}(B, \| \,.\, \|)$ is a closed subset of $[0, 1]^B$ provided with the topology of pointwise convergence on B. But by Tykhonov's theorem, $[0,\ 1]$ is compact for this topology and then so is $\text{Mult}(B, \| \,.\, \|)$. Moreover, the mapping $\varphi \to \widehat{\varphi}$ from $\text{Mult}(A, \| \,.\, \|)$ into $\text{Mult}(B, \| \,.\, \|)$ is a bijection. Indeed, it is clearly injective and it is surjective because given $\psi \in \text{Mult}(B, \| \,.\, \|)$, we may extend ψ to A by putting $\overline{\psi}(x) = |\lambda|\psi(\frac{x}{\lambda})$ with $\lambda \in \mathbb{K},\ |\lambda| \geq \|x\|$. Finally, this bijection is bicontinuous with respect to the pointwise convergence on both $\text{Mult}(A, \| \,.\, \|)$ and $\text{Mult}(B, \| \,.\, \|)$ and this ends the proof of Theorem 10.7. □

Theorem 10.8 is given in several works. This proof mainly is given in [53].

Theorem 10.8. *Let $\mathbb{B}$ be a field extension of $\mathbb{K}$ provided with a non-zero semi-norm of $\mathbb{K}$-algebra $\| \,.\, \|$. Then, $\| \,.\, \|$ is a norm of $\mathbb{K}$-algebra, and there exists an absolute value φ on $\mathbb{B}$ extending that of $\mathbb{K}$ such that $\varphi(x) \leq \|x\|$ whenever $x \in A$.*

Proof. Let $SM(A, \| \,.\, \|)$ denote the set of continuous semi-norms ϕ of A satisfying $\phi(f^n) = (\phi(f))^n\ \forall f \in A,\ \forall n \in \mathbb{N}$. It is seen that $\| \,.\, \|$ is a norm because $Ker\| \,.\, \| = \{0\}$. In the same way, so is the spectral semi-norm $\| \,.\, \|_{si}$ associated with $\| \,.\, \|$. Now, $SM(A, \| \,.\, \|_{si})$ is easily checked to be inductive with respect to the order $\geq$, i.e. given a totally ordered subset W of $SM(A, \| \,.\, \|_{si})$, the mapping ψ defined in A by $\psi(x) = \inf\{\theta(x) | \theta \in W\}$ belongs to $SM(A, \| \,.\, \|_{si})$. Then by Zorn's lemma, $SM(A, \| \,.\, \|_{si})$ admits a minimal element φ. As we just saw, φ is a norm of $\mathbb{K}$-algebra and we have $\varphi(x) \leq \|x\|_{si}$ whenever $x \in A$. We will prove that $\varphi(ab) = \varphi(a)\varphi(b)$ whenever $a,\ b \in A$. Let $a \in A\backslash\{0\}$. For every $x \in A$, we put $u_n(x) = \frac{\varphi(a^n x)}{\varphi(a)^n}$. The sequence $(u_n(x))_{n\in\mathbb{N}}$ is seen to be decreasing. We put $\sigma(x) = \lim_{n\to+\infty} \frac{\varphi(a^n x)}{\varphi(a)^n}$ whenever $x \in A$.

First, we check that σ is a norm of $\mathbb{K}$-algebra. Obviously, it is seen that for every $n \in \mathbb{N},\ u_n$ is a norm of $\mathbb{K}$-vector space, hence so is σ.

Next, we have

$$u_n(x)u_n(y) = \frac{\phi(a^n x)\phi(a^n y)}{\phi(a^n)\phi(a^n)}, = \frac{\phi(a^n x)\phi(a^n y)}{\phi(a^{2n})} \geq \frac{\phi(a^{2n}xy)}{\phi(a^{2n})} \geq \sigma(xy),$$

whenever $x,\ y \in A$, hence $\sigma(x)\sigma(y) \geq \sigma(xy)$. So, σ is a norm of $\mathbb{K}$-algebra. Now, we check that σ is semi-multiplicative, because

$$\lim_{n\to+\infty}\left(\frac{\varphi(a^n x^q)}{\varphi(a^n)}\right) = \lim_{n\to+\infty}\left(\frac{\varphi(a^{qn} x^q)}{\varphi(a^{qn})}\right) = \lim_{n\to+\infty}\left(\frac{\varphi(a^n x)}{\varphi(a^n)}\right)^q = \sigma(x)^q.$$

Then, since σ satisfies $\sigma(x) \leq \varphi(x) \leq \|x\|_{si}$ whenever $x \in A$, it clearly belongs to $SM(A, \| \, . \, \|_{si})$. But since φ is minimal in $SM(A, \| \, . \, \|_{si})$, actually φ is equal to σ.

Now, as the sequence $(u_n)_{n\in\mathbb{N}}$ is decreasing, we have $\sigma(x) \leq \frac{\varphi(ax)}{\varphi(a)}$, hence $\varphi(x)\varphi(a) \leq \frac{\varphi(ax)}{\varphi(a)}$ and hence $\varphi(a)\varphi(x) \leq \varphi(ax)$. But since φ is a norm of $\mathbb{K}$-algebra, $\varphi(ax) \leq \varphi(a)\varphi(x)$, which proves that φ is a multiplicative norm. □

Theorem 10.9 is immediate.

Theorem 10.9. *Let A be a Banach $\mathbb{K}$-algebra. For every maximal ideal $\mathcal{M}$ of A, there exists $\varphi \in Mult(A, \| \, . \, \|)$ such that $Ker(\varphi) = \mathcal{M}$. If $\mathcal{M}$ is of codimension 1, the mapping τ from A onto $\mathbb{K}$ admitting $\mathcal{M}$ for kernel satisfies $|\tau(f)| \leq \|f\| \ \forall f \in A$.*

Proof. Let $\mathcal{M}$ be a maximal ideal of A and let $\mathbb{B}$ be the field $\frac{A}{\mathcal{M}}$. Since A is complete, $\mathcal{M}$ is closed, therefore $\mathbb{B}$ is provided with the quotient norm. By Theorem 10.9, $\mathbb{B}$ admits an absolute value $| \, . \, |$ which extends that of $\mathbb{K}$. Let ψ be the canonical surjection from A to $\mathbb{B}$. On A we put $\varphi(x) = |\psi(x)|$. Then, φ is an element of $\mathrm{Mult}(A, \| \, . \, \|)$ such that $Ker(\varphi) = \mathcal{M}$.

Suppose now $\mathcal{M}$ is of codimension 1 and suppose $|\tau(f)| > \|f\|$. Then the series $\sum_{n=0}^{+\infty}\left(\frac{f}{\tau(f)}\right)^n$ converges and shows that $f - \tau(f)$ is invertible in A, a contradiction since $\tau(f - \tau(f)) = 0$. □

Corollary 10.10. *Let A be a Banach $\mathbb{K}$-algebra. Every $\mathbb{K}$-algebra homomorphism from A to $\mathbb{K}$ is continuous.*

The characterization of the continuous multiplicative norms of $(R(D), \| \, . \, \|_D)$ by means of the large circular filters secant with D

suggests us extending this characterization to the multiplicative semi-norms of $R(D)$.

Theorem 10.11. *Let D be a closed bounded subset of $\mathbb{K}$. The mapping Ξ from $\Phi(D)$ into* $\text{Mult}(R(D))$ *defined as $\Xi(\mathcal{F}) = \varphi_{\mathcal{F}}$ is a bijection from $\Phi(D)$ onto* $\text{Mult}(R(D), \mathcal{U}_D)$. *Moreover, $\varphi_{\mathcal{F}}$ is an absolute value if and only if $\mathcal{F}$ does not converge in D. Further,* $\text{Mult}(R(D, \| \,.\, \|_D))$ *is provided with the topology of pointwise convergence for which it is compact.*

Proof. On the one hand, by Theorems 4.10 and 10.4, it is clearly seen that the mapping defined on $\Phi^\circ(D)$ by $\mathcal{F} \longrightarrow \varphi_{\mathcal{F}}$ is a bijection from this set onto the set of continuous multiplicative norms on $R(D)$.

On the other hand, every $a \in D$ defines a multiplicative semi-norm ψ by $\psi(h) = |h(a)|$, the kernel of which is the maximal ideal $(x - a)R(D)$. Thus, we have a mapping from the set of convergent circular filters on D into the set of multiplicative semi-norms which are not norms: this mapping is obviously injective.

Finally, let ψ be a multiplicative semi-norm whose kernel is not zero. Then, $Ker(\psi)$ is a prime ideal hence a maximal ideal of $R(D)$, and therefore, it is of the form $(x - a)R(D)$ with $a \in D$. Then, $\psi(x-a) = 0$, hence $\psi(x-b) = |a-b|$ whenever $b \in \mathbb{K}$, and therefore, ψ is of the form $\varphi_{\mathcal{F}_a}$ with $\mathcal{F}_a$ the filter of the neighborhoods of a in D. Thus, Ξ is a bijection from $\Phi(D)$ onto $\text{Mult}(R(D), \mathcal{U}_D)$ and $\varphi_{\mathcal{F}}$ is a norm if and only if $\mathcal{F}$ is not convergent. Finally, by Theorem 10.7, $\text{Mult}(R(D, \| \,.\, \|_D))$ is compact with respect to the topology of pointwise convergence. That ends the proof of Theorem 10.11. □

Corollary 10.12. $\text{Mult}(\mathbb{K}[X])$ *is provided with the topology of pointwise convergence for which it is locally compact.*

Chapter 11

Analytic Elements

Due to the fact that any disk $d(a, r)$ is exactly the same as $d(b, r)$ for every $b \in d(a, r)$, it is easily seen that a power series $\sum_{n=0}^{\infty} a_n (x - a)^n$ which admits the disk $d(a, r)$ for disk of convergence may not be extended outside its convergence disk as it is done in complex analysis, by means of a change of origin.

However, by Runge's theorem, we remember that a holomorphic function in a compact subset D of $\mathbb{C}$ is equal to the limit of a sequence of rational functions with respect to the uniform convergence on D. This is why Marc Krasner introduced analytic elements on a subset D of $\mathbb{K}$ directly by considering limits of sequences of rational functions with respect to the uniform convergence on D [69].

Actually, Marc Krasner constructed a theory of analytic functions f defined on a quasi-connected set D equal to the union of a chained family of quasi-connected sets $(D_i)_{i \in I}$ such that the restriction of f to each D_i is an analytic element on D_i [69] (this construction was widen to the analytic infraconnected sets by Philippe Robba) [83].

Another theory was defined by John Tate, consisting (in one variable) of using infraconnected affinoid sets. Here, we only describe some basic properties of analytic elements on infraconnected sets in order to apply them to power series and various Laurent series that are used for studying meromorphic functions. A comparison between Krasner's theory and Tate's theory was made in [37]. Here, we aim at studying meromorphic functions in the field $\mathbb{K}$ and applications to problems of value distribution. This is why we do not repeat the study of Krasner–Tate algebras [37]. However, interesting properties of injective analytic elements can be proved with the use of Krasner–Tate algebra. In Chapter 42, we recall these properties letting us prove the famous relation satisfied by many kind of injective analytic functions in $\mathbb{K}$.

We examine algebras of analytic elements, particularly Banach algebras of bounded analytic elements. We see the characterization of sets D such that the space of analytic elements on D is a $\mathbb{K}$-algebra. We examine some basic properties of analytic elements such as poles when the set D is not closed and we see that analytic elements on a closed bounded set D are uniformly continuous. If D has finitely many infraconnected components, for each one, its characteristic function is an analytic element on D.

Definition and notation. Let D be an infinite subset of $\mathbb{K}$. We denote by $H(D)$ the completion of $R(D)$ for the topology $\mathcal{U}_D$ of uniform convergence on D. The elements of $H(D)$ are called the *analytic elements on* D [35, 44, 69].

The set $H(D)$ is then provided with the topology of uniform convergence on D for which it is complete and every $f \in H(D)$ defines a function on D which is the uniform limit (on D) of a sequence $(h_n)_{n\in\mathbb{N}}$ in $R(D)$. Thus, given two infinite sets D, D' such that $D \subset D'$, the restriction to D of elements of $H(D')$ enables us to consider that $H(D')$ is included in $H(D)$.

Next, $H(D)$ is a $\mathbb{K}$-vectorial space and a complete topological group with respect to the topology $\mathcal{U}_D$. The question of whether the product of two analytic elements on D is an analytic element on D is studied later. However, it is easily seen that given $f \in H(D)$, the function f^n also belongs to $H(D)$.

Lemma 11.1. *For every* $f, g \in H(D)$, *we have*

(i) $\|f\|_D = 0$ *if and only if* $f = 0$,
(ii) $\|\lambda f\|_D = |\lambda|\ \|f\|_D$ *whenever* $\lambda \in \mathbb{K}^*$,
(iii) $\|f + g\|_D \le \max(\|f\|_D, \|g\|_D)$,
(iv) *if* $(\|f\|_D, \|g\|_D)$ *is different from* $(0, +\infty)$ *and from* $(+\infty, 0)$, *then the function* fg *satisfies* $\|fg\|_D \le \|f\|_D\ \|g\|_D$,
(v) $\|f^n\|_D = \|f\|_D^n$ *whenever* $n \in \mathbb{N}^*$.

Notation. We denote by $H_b(D)$ the set of elements $f \in H(D)$ bounded on D. Then, $H_b(D)$ is clearly a $\mathbb{K}$-vectorial subspace of

$H(D)$ and is closed in $H(D)$. Moreover, $\| \,.\, \|_D$ is a norm on $H_b(D)$ that makes it a Banach $\mathbb{K}$-algebra. If D is unbounded, we denote by $H_0(D)$ the set of the $f \in H(D)$ such that $\lim_{\substack{|x|\to+\infty \\ x\in D}} f(x) = 0$.

Theorem 11.2 is an immediate consequence of Theorem 10.3.

Theorem 11.2. *$H_b(D)$ is a Banach $\mathbb{K}$-subalgebra of $\mathbb{K}^D$. The following three conditions are equivalent:*

(i) *$H_b(D) = H(D)$,*
(ii) *$H(D)$ is topological $\mathbb{K}$-vector space,*
(iii) *$(H(D), \| \,.\, \|_D)$ is a Banach $\mathbb{K}$-algebra,*
(iv) *D is closed and bounded.*

If these conditions are satisfied, then $\| \,.\, \|_D$ is a semi-multiplicative norm.

Definition. Let $f \in H(D)$ have no zero in D. The element f is said to be *invertible in* $H(D)$ if the function $\frac{1}{f}$ (also denoted by f^{-1}) belongs to $H(D)$. This definition holds even if $H(D)$ is not a ring.

The following lemma is classical.

Lemma 11.3. *Let $f \in H(D)$ be such that $\inf_{x\in D} |f(x)| > 0$. Then, f^{-1} belongs to $H(D)$. Moreover, if D is closed and bounded, f^{-1} belongs to $H(D)$ if and only if $\inf_{x\in D} |f(x)| > 0$.*

Let $g \in H(D)$ satisfy $|g(x)| = 1$ for all $x \in D$ and $\|f - 1\|_D > \|g - 1\|_D$. Then, we have $\|fg - 1\|_D = \|f - 1\|_D$.

Proof. We suppose $\inf_{x\in D} |f(x)| = \lambda > 0$. Let $(h_n)_{n\in\mathbb{N}}$ be a sequence in $R(D)$ such that $\lim_{n\to\infty} \|h_n - f\|_D = 0$. For n big enough, we have $|h_n(x)| \geq \lambda$ whenever $x \in D$, hence $\left|\frac{1}{h_n(x)} - \frac{1}{f(x)}\right| = \left|\frac{f(x)-h_n(x)}{h_n(x)f(x)}\right| \leq \frac{\|f-h_n\|_D}{\lambda^2}$, hence the sequence $\frac{1}{h_n}$ converges to $\frac{1}{f}$.

Conversely, if D is closed and bounded and if $\frac{1}{f} \in H(D)$, $\frac{1}{f}$ has to be bounded by some $M \in \mathbb{R}_+$, hence $|f(x)| \geq \frac{1}{M}$ whenever $x \in D$.

Now, let $g \in H(D)$ satisfy $|g(x)| = 1$ for all $x \in D$ and $\|f-1\|_D > \|g-1\|_D$. For every $x \in D$, we have $|f(x)-1||g(x)| > |g(x)-1|$, and therefore, $|f(x)g(x)-1| = |f(x)-1||g(x)| = |f(x)-1|$. This finishes proving Lemma 11.3. □

Theorem 11.4. *Let $f \in H(D)$ and let $h \in R_b(D)$. Then, fh belongs to $H(D)$.*

Proof. Let ε be > 0 and let $g \in R(D)$ satisfy $\|f - g\|_D < \varepsilon$. Then, we have $\|hf - hg\|_D < \varepsilon\|h\|_D$ and this clearly shows that $fh \in H(D)$. □

When D is not closed or is not bounded, we show how to split an element $f \in H(D)$.

Theorem 11.5. *The vector space $H(D)$ is equal to the direct sum $R_0(\mathbb{K}\setminus(\overline{D}\setminus D)) \oplus H(\overline{D})$. Moreover, if D is not bounded, then $H(\overline{D})$ is equal to the direct sum $\mathbb{K}[x] \oplus H_0(\overline{D})$.*

Proof. Let $(f_n)_{n\in\mathbb{N}}$ be a sequence in $R(D)$ such that $\lim_{n\to\infty} \|f_n - f\|_D = 0$. In particular, there exists $N \in \mathbb{N}$ such that $f_n - f_N$ is bounded when $n \geq N$. We put $g_n := f_n - f_N$. The sequence g_n converges in $H(D)$ to $f - f_N$. Let $g := f - f_N$. On the other hand, since each $f_n - f_N$ belongs to $R(D)$ and is bounded in D, $f_n - f_N$ belongs to $R(\overline{D})$. Obviously, $\|f_n - f_N\|_D = \|f_n - f_N\|_{\overline{D}}$, hence finally, g belongs to $H_b(\overline{D})$. Now, we may obviously split f_N in the form $E(x) + h_1(x) + h_2(x)$ with $E(x) \in \mathbb{K}[x], h_1 \in R_0(\overline{D}), h_2 \in R_0(\mathbb{K}\setminus(\overline{D}\setminus D))$. We put $f^* = h_2$ and $\overline{f} = E + h_1 + g$. We have clearly split f in the form $f^* + \overline{f}$ with $f^* \in R_0(\mathbb{K}\setminus(\overline{D}\setminus D))$, $\overline{f} \in H(\overline{D})$. Hence, we have proven that $H(D) = R_0(\mathbb{K}\setminus(\overline{D}\setminus D)) + H(\overline{D})$.

This sum is easily seen to be direct. Indeed, suppose that we have $h \in R_0(\mathbb{K}\setminus(\overline{D}\setminus D))$ and $g \in H(\overline{D})$ such that $h + g = 0$, with $h \neq 0$. Then, h has a pole $\alpha \in \overline{D}\setminus D$ and it may be written in the form $\sum_{i=1}^{q} \frac{\lambda_i}{(x-\alpha)^i} + h_\alpha$ with $\lambda_i \neq 0$ $(1 \leq i \leq q)$ and $h_\alpha \in R_0(\mathbb{K}\setminus(\overline{D}\setminus(D \cup \{\alpha\})))$. But g is obviously bounded around α, hence h has to be bounded when x approaches α, hence finally α does not exist. This shows that the sum is direct.

Now, we suppose D is unbounded. First, we prove that every element $f \in H_b(D)$ admits a limit when $|x|$ tends to $+\infty$. Let $\epsilon \in \mathbb{R}_+^*$ and let $h \in R(D)$ satisfy $\|f - h\|_D < \epsilon$. Since f is bounded in D, so is h. But then h is of the form $\frac{P}{Q}$, with $\deg(P) = \deg(Q)$, and therefore, h has a limit λ when $|x|$ tends to $+\infty$. Let $\epsilon > 0$ be such that $|h(x) - \lambda| < \epsilon$ whenever $x \in D\setminus d(0, r)$. Clearly, we have $|f(x) - \lambda| < \epsilon$ whenever $x \in D\setminus d(0, r)$. This proves that f does converge along the filter $\mathcal{F}$ which admits as a basis the family of sets $D\setminus d(0, r)$ $(r > 0)$.

Let $f \in H(\overline{D})$ be unbounded. Let $(x_n)_{n\in\mathbb{N}}$ be a sequence such that $\lim_{n\to\infty} |f(x_n)| = +\infty$. Suppose the sequence $(x_n)_{n\in\mathbb{N}}$ does not tend to $+\infty$. Then there exists a bounded subsequence $(x_{n_q})_{q\in\mathbb{N}}$ such that $\lim_{q\to\infty} |f(x_{n_q})| = +\infty$, but this is impossible due to Theorem 11.2 because such a sequence lies in a closed bounded set D' included in D. Now, there exists $h \in R(\overline{D})$ such that $f - h$ is bounded, and therefore, we have $\lim_{n\to\infty} |h(x_n)| = +\infty$. Let $h(x) = P(x) + u(x)$ with $P \in \mathbb{K}[x]$ and $u \in R_0(\overline{D})$. Since $f - h$ is bounded, clearly, $f - P$ belongs to $H_b(\overline{D})$, hence we have proven that $H(\overline{D}) = \mathbb{K}[x] + H_b(\overline{D})$. Moreover, since u has a limit when $|x|$ tends to $+\infty$ in D, we have $H_b(\overline{D}) = H_0(\overline{D}) + \mathbb{K}$, and therefore, $H(\overline{D}) = \mathbb{K}[x] + H_0(\overline{D})$. Finally, by considering elements when $|x|$ tends to $+\infty$, this sum is easily seen to be direct and this ends the proof. □

Definition. Let $\alpha \in \overline{D}\backslash D$ and $f \in H(D)$, and let $f = f^* + \overline{f}$, with $f^* \in R_0(\mathbb{K}\backslash(\overline{D}\backslash D))$ and $\overline{f} \in H(\overline{D})$. Let $\alpha \in \overline{D}\backslash D$ be a pole of f^* and let $f^*(x) = \sum_{j=1}^{q} \frac{\lambda_j}{(x-\alpha)^j} + u(x)$, with $u \in R_0(\mathbb{K}\backslash(\overline{D}\backslash(D\cup\{\alpha\})))$. The pole α of order q of f^* is called a *pole of order q of f* and λ_1 is called *the residue of f at α* and is denoted by $\operatorname{res}(f, \alpha)$.

Let $a_i, 1 \le i \le n$, be the poles of f and for each i let q_i be the order of a_i. The polynomial $\prod_{i=1}^{n}(x - a_i)^{q_i}$ is named *the polynomial of poles of f in $\overline{D}\backslash D$.*

Corollary 11.6. *Let $\alpha \in \overline{D}\backslash D$ and $f \in H(D)$ be such that $|f(x)|$ is bounded in $d(\alpha, r) \cap D$ with $r > 0$. Then $f \in H(D \cup \{\alpha\})$.*

Proof. Indeed, as $\overline{f}$ is obviously bounded in $D \cap d(\alpha, r)$, so is f^*, and therefore, f^* has no pole at α. □

Corollary 11.7. *$H(D) = H_b(\overline{D}) + R(\mathbb{K}\backslash(\overline{D}\backslash D))$, $H_b(D) = H_b(\overline{D}) \subset H(\overline{D})$ and $H_b(D) = H_b(\overline{D})$. If D is bounded, then $H_b(D) = H(\overline{D})$. If D is not bounded, for every unbounded $f \in H(D)$, there exists a unique $q \in \mathbb{N}^*$ such that $x^{-q}f(x)$ has a finite non-zero limit when $|x|$ tends to $+\infty$, $x \in D$. Let $d(a, r^-)$ be a hole of D. If f belongs to $H(\overline{D})$ and if $x^{-q}f(x)$ has a finite non-zero limit, then $\frac{f(x)}{(x-a)^q}$ belongs to $H_b(D)$.*

Corollary 11.8. *If $\overline{D} = \mathbb{K}$, then $H(D) = R(D)$.*

Corollary 11.9 comes from the definition of the poles and from Theorem 11.5.

Corollary 11.9. *Let $f \in H(D)$ and let $\alpha \in \overline{D}\backslash D$. Then α is a pole of order $n > 0$ for f if and only if $(x - \alpha)^n f(x)$ has a finite non-zero* *limit at α. If there exists no $r \in \mathbb{R}_+^*$ such that $|f(x)|$ is bounded in $d(\alpha, r)\setminus\{\alpha\}$, then α is a pole of order $n \geq 1$ for f and $(x-\alpha)^n f(x)$ has a finite non-zero limit at α.*

Theorem 11.10. *Let $f \in H(D)$ and let $\alpha \in \overline{D}\setminus D$. Either f belongs to $H(D \cup \{\alpha\})$ or α is a pole for f.*

Proof. If f does not belong to $H(D\cup\{\alpha\})$, by Corollary 11.6, f is unbounded in any disk $d(\alpha, r)$ whenever $r > 0$. Hence, by means of the notation of Theorem 11.5, α is clearly a pole of f^* and therefore is a pole of f. □

We must note Theorem 11.11:

Theorem 11.11. *Let D be closed and bounded and let $f \in H(D)$. Then, f is uniformly continuous in D.*

Proof. The claim is immediate when f is a polynomial. Suppose now that $f(x) = \frac{P(x)}{Q(x)} \in R(D)$. Since D is closed and bounded, there exists $m > 0$ such that $|Q(x)| \geq m$, $\forall x \in D$. Consequently, $\frac{P(x)}{Q(x)}$ also is uniformly continuous. Then, when $f \in H(D)$, since f is a uniform limit of rational functions, f is also uniformly continuous. □

Notation. We denote by Alg the family of sets E such that $H(E)$ is a $\mathbb{K}$-subalgebra of $\mathbb{K}^E$.

Theorem 11.12. *Let $f \in H(D)$. There exists $W \in R_b(D)$, whose zeros lie in $\overline{D}\setminus D$ and $h \in H(\overline{D})$ such that $f = \frac{h}{W}$. Further, if D is bounded or if $D \in$ Alg, then there exists $g \in H(\overline{D})$ such that $f = \frac{g}{Q}$ with Q the polynomial of poles of f in $\overline{D}\setminus D$.*

Proof. We may summarize Theorem 11.5 in this way: f is of the form $\widetilde{f}(x) + \widehat{f}(x)$ with $\widetilde{f} \in R(\mathbb{K}\setminus(\overline{D}\setminus D))$ and $\widehat{f} \in H_b(\overline{D})$. Indeed, if D is bounded, we just take $\widehat{f} = \overline{f}$, and if D is not bounded, $\widehat{f}$ is the one defined in Theorem 11.5. Thus, $\widetilde{f}(x)$ can be written in the form $\frac{P(x)}{Q(x)}$ with $Q(x) = \prod_{i=1}^n (x-a_i)^{q_i}$ i.e. the polynomial of the poles of f in $\overline{D}\setminus D$, and $P(x) \in \mathbb{K}[x]$. Let $q = \sum_{i=1}^n q_i$. Theorem 11.12 is obviously trivial if D has no hole, hence we may assume D to have at least one hole $T = d(a, r^-)$. Let $W(x) = \frac{Q(x)}{(x-a)^q}$. We know that $W \in R_b(\overline{D})$ hence $W\widehat{f} \in H_b(\overline{D})$. On the other hand, we see that

$W\widetilde{f} \in R(\overline{D})$, hence $Wf \in H(\overline{D})$. We just put $h = Wf$ and have the factorization $f = \frac{h}{W}$.

If D is bounded, we see that both Q, h are bounded in D, hence $Q\widehat{f}$ belongs to $H_b(\overline{D})$ and then Qf belongs to $H(\overline{D})$. In the same way, if $H(D)$ is supposed to be a ring, then $Q\widehat{f}$ belongs to $H(\overline{D})$ and then $Qf = P + Q\widehat{f}$ belongs to $H(\overline{D})$. This ends the proof. □

Corollary 11.13. *Let $S(D)$ be the set of polynomials whose zeros belong to $\overline{D}\backslash D$. If $D \in$ Alg, then $H(D) = S(D)^{-1}H(\overline{D})$.*

Theorem 11.14. *Let D be closed. Let $g \in H(D)$ and let $P \in \mathbb{K}[x]$ be such that Pg belongs to $H(D)$. For every $Q \in \mathbb{K}[x]$ such that $\deg(Q) \leq \deg(P)$, Qg also belongs to $H(D)$.*

Proof. Theorem 11.14 is clearly trivial when D belongs to Alg. Now, suppose that D does not belong to Alg, hence D is unbounded. If D has no hole, then $D = \mathbb{K}$, hence by Corollary 11.8, $H(D)$ is equal to $\mathbb{K}[x]$. Thus, we may assume that D admits at least one hole and then, without loss of generality, we can assume that this hole is $d(0, r^-)$. Let $q = \deg(P)$. Let $P_q = P$ and let $P_{q-1} = \frac{P(x)-P(0)}{x}$. Then, P_{q-1} is a polynomial of degree $q - 1$. We see that $(P(x) - P(a))g(x)$ belongs to $H(D)$. But $\left\|\frac{1}{x}\right\|_D$ is bounded and then by Theorem 11.4, $P_{q-1}(x)g(x)$ belongs to $H(D)$. Hence, by induction, it is seen that for each $j = 1, \ldots, q$ there exists a polynomial P_j of degree j such that $P_j g$ belongs to $H(D)$ and this clearly completes the proof. □

Now, when D is not infraconnected, we have to note an easy result on characteristic functions that shows how rich the algebra $H(D)$ is.

Proposition 11.15. *Let D have an empty annulus Λ. Let w_1, w_2 be the functions defined on D by $w_1(x) = 1, w_2(x) = 0$ if $x \in \mathcal{I}(\Lambda)$ and $w_1(x) = 0, w_2(x) = 1$ if $x \in \mathcal{E}(\Lambda)$. Then, w_1 and w_2 belong to $H(D)$.*

Proof. Let $\Lambda = \Gamma(a, r_1, r_2)$, with $a \in D$. With no loss of generality, we may obviously assume $a = 0$. Let $\alpha \in \Lambda$ be such that $r_1 < |\alpha| < r_2$ and for each $n \in \mathbb{N}^*$, let $u_n = \frac{1}{1-\left(\frac{x}{\alpha}\right)^n}$. Then, $\left\|1 - \left(\frac{x}{\alpha}\right)^n - 1\right\|_{\mathcal{I}(\Lambda)} \leq \left(\frac{r_1}{|\alpha|}\right)^n$, while $\left|1 - \left(\frac{x}{\alpha}\right)^n\right| \geq \left(\frac{r_2}{|\alpha|}\right)^n$ for every $x \in \mathcal{E}(\Lambda)$, hence finally, $\|u_n - w_1\|_D \leq \max\left(\left(\frac{r_1}{|\alpha|}\right)^n, \left(\frac{|\alpha|}{r_2}\right)^n\right)$. Thus, we see that $w_1 = \lim_{n\to\infty} u_n \in H(D)$ and $w_2 = 1 - w_1 \in H(D)$. □

Theorem 11.16. *Let E have finitely many infraconnected components $E_1, \dots, E_q$. For each $i = 1, \dots, q$, the characteristic function of E_i belongs to $H(E)$.*

Proof. Let A be one of the infraconnected components of E. By Theorem 2.21, there exist empty annuli $(\Lambda_j)_{0 \le j \le n}$ such that A is either of the form

$$(\alpha)\ \mathcal{I}_E(\Lambda_0) \bigcap \left(\bigcap_{j=1}^{n} \mathcal{E}_E(\Lambda_j) \right) \quad \text{or of the form} \quad (\beta)\ \bigcap_{j=1}^{n} \mathcal{E}_E(\Lambda_j).$$

But by Proposition 11.15, the characteristic function u_j of $\mathcal{E}_E(\Lambda_j)$ belongs to $H_b(E)$ $(1 \le j \le n)$ and so does the characteristic function u_0 of $\mathcal{I}_E(\Lambda_0)$. Since all the u_j belong to $H_b(E)$, we see that the products $u = \prod_{j=0}^{n} u_j$ and $w = \prod_{j=1}^{n} u_j$ belong to $H(E)$. Then, when A is of the form (α) (respectively (β)), its characteristic function is equal to u (respectively w) and therefore belongs to $H(E)$. □

Chapter 12

Composition of Analytic Elements

Given A and $B \subset \mathbb{K}$, $f \in H(A)$ such that $f(A) \subset B$ and $g \in H(B)$, a basic question is whether $g \circ f \in H(A)$. There is an immediate application to the study of homomorphisms from an algebra $H(D)$ to another $H(D')$.

Lemma 12.1. *Let A and B be subsets of $\mathbb{K}$ and let $f \in H(A)$ be such that $f(A) \subset B$. For every $\lambda \in \mathbb{K}\backslash\overline{B}$, $f - \lambda$ is invertible in $H(A)$. Moreover, if for every $h \in R(B)$, $h \circ f$ belongs to $H(A)$, then for every $g \in H(B)$, $g \circ f$ belongs to $H(A)$ and for every $\lambda \in \overline{B}\backslash B$, $f - \lambda$ is invertible in $H(A)$.*

Proof. Let $r = \delta(\lambda, B)$. If $\lambda \notin \overline{B}$, we have $r > 0$ and then $|f(x) - \lambda| \geq r$ whenever $x \in A$. Hence, by Lemma 11.3, $f - \lambda$ is invertible in $H(A)$.

Now, we assume that for every $h \in R(B)$, $h \circ f$ belongs to $H(A)$. Let $g \in H(B)$, let ϵ be > 0, and let $h \in R(B)$ satisfy $\|g - h\|_B \leq \epsilon$. It is seen that $\|g \circ f - h \circ f\|_A \leq \epsilon$. Since $h \circ f$ belongs to $H(A)$, then so does $g \circ f$.

Finally, let $\lambda \in \overline{B}\backslash B$ and let $h(u) = \frac{1}{u-\lambda}$. Since $h \circ f$ belongs to $H(A)$, $f - \lambda$ is invertible in $H(A)$. □

Theorem 12.2. *Let A, B be subsets of $\mathbb{K}$ and let $f \in H(A)$ satisfy $f(A) \subset B$:*

(i) *If $f \in R(A)$, then $g \circ f \in H(A)$ whenever $g \in H(B)$.*
(ii) *If $A \in \text{Alg}$, then $g \circ f \in H(A)$ for all $g \in H(B)$ if and only if $f - \lambda$ is invertible in $H(A)$ for all $\lambda \in \overline{B}\backslash B$.*

Proof. By Lemma 12.1, we just have to show that for every $g \in R(B)$, $g \circ f$ belongs to $H(A)$ in each one of these two hypotheses:

(H_1) $f \in R(A)$.
(H_2) $f - \lambda$ is invertible in $H(A)$ for all $\lambda \in \overline{B} \setminus B$.

So, we take $g(u) = \frac{P(u)}{Q(u)} \in R(B)$ and show that $g \circ f \in H(A)$ in each hypothesis. Let $\lambda_1, \ldots, \lambda_q$ be the poles of g in $\mathbb{K} \setminus B$.

(H_1) For every $j = 1, \ldots, q$, $f - \lambda_j$ is invertible in $R(A)$ because $f - \lambda_j$ has no zero in A. Hence, $Q \circ f$ is invertible in $R(A)$ and then $g \circ f$ belongs to $R(A)$.
(H_2) For each $j = 1, \ldots, q$, either λ_j belongs to $\overline{B} \setminus B$ or it belongs to $\mathbb{K} \setminus \overline{B}$. In both cases, by Lemma 12.1, each $f - \lambda_j$ is invertible in $H(A)$. Since A belongs to Alg, $Q \circ f$ is clearly invertible in $H(A)$ and $P \circ f$ belongs to $H(A)$. Hence, so does $g \circ f$. □

Corollary 12.3. *Let $A \in$ Alg and let B be a closed subset of $\mathbb{K}$. Let $f \in H(A)$ satisfy $f(A) \subset B$ and let $g \in H(B)$. Then, $g \circ f$ belongs to $H(A)$.*

Example. Let r, $s \in \mathbb{R}_+^*$, let $f \in H(d(0, r))$ be such that $f(d(0, r)) \subset d(0, s)$, and let $g \in H(d(0, s))$. Then, $g \circ f$ belongs to $H(d(0, r))$.

Lemma 12.4. *Let $h \in R(D)$ and let $D' = h(D)$. Let $f \in H(D')$. If f is invertible in $H(D')$, then $f \circ h$ is invertible in $H(D)$. If h is a Moebius function, f is invertible in $H(D')$ if and only if $f \circ h$ is invertible in $H(D)$.*

Proof. First, we suppose f invertible in $H(D')$. Let $g = \frac{1}{f}$. Then, by Theorem 12.2, $g \circ h$ belongs to $H(D)$ and is clearly equal to $\frac{1}{f \circ h}$. Now, we assume that h is a Moebius function and we put $\ell = h^{-1}$. If $f \circ h$ is invertible in $H(D)$, $(f \circ h) \circ \ell$ is invertible in $H(D')$ and this ends the proof of Lemma 12.4. □

We are now going to study the $\mathbb{K}$-algebra homomorphisms from $H(D)$ into $H(D')$. First, we consider homomorphisms from $R(D)$ into $R(D')$.

Proposition 12.5. *Let D, D' be subsets of $\mathbb{K}$ and let $\gamma \in R(D')$ satisfy $\gamma(D') \subset D$. Let ϕ_γ be the mapping from $R(D)$ into $R(D')$ defined as $\phi_\gamma(f) = f \circ \gamma$ $(f \in R(D))$. Then ϕ_γ is a homomorphism* *from $R(D)$ into $R(D')$ and this homomorphism is injective if and only if γ is not a constant. Every $\mathbb{K}$-algebra homomorphism is of this form and the mapping $\gamma \to \phi_\gamma$ is a bijection from the set of $\gamma \in R(D')$ such that $\gamma(D') \subset D$ onto the set of $\mathbb{K}$-algebra homomorphisms from $R(D)$ into $R(D')$.*

Proof. Let $\gamma \in R(D')$ satisfy $\gamma(D') \subset D$. Then it is seen that ϕ_γ takes values in $R(D')$, is a $\mathbb{K}$-algebra homomorphism, and is obviously injective if and only if γ is not a constant.

Conversely, let ψ be a $\mathbb{K}$-algebra homomorphism from $R(D)$ into $R(D')$ and let $\gamma = \psi(I_{D'})$ with $I_{D'}$ the identical mapping in D'. Then, we have $\psi(P) = P \circ \gamma$ for every polynomial P. On the other hand, if $\alpha \notin D$, then $(x - \alpha)$ is invertible in $R(D)$ and $\psi(\frac{1}{x-\alpha}) = (\psi(x-\alpha))^{-1} = (\gamma - \alpha)^{-1}$. Therefore, $\psi(h) = h \circ \gamma$ whenever $h \in R(D)$. The mapping $\gamma \to \phi_\gamma$ is obviously injective and hence is a bijection. □

Proposition 12.6. *Let D, D' be sets in $\mathbb{K}$ and let $\gamma \in H(D')$ satisfy $\gamma(D') \subset D$ and $f \circ \gamma \in H(D')$ for all $f \in H(D)$. Let ϕ_γ be the mapping from $H(D)$ into $H(D')$ defined as $\phi_\gamma(f) = f \circ \gamma$. Then ϕ_γ is a linear mapping from $H(D)$ into $H(D')$ continuous with respect to the topology of uniform convergence on D for $H(D)$ and on D' for $H(D')$. Moreover, given f, $g \in H(D)$ such that $fg \in H(D)$ we have $\phi_\gamma(fg) = \phi_\gamma(f)\phi_\gamma(g)$. The restriction of ϕ_γ to $H_b(D)$ is a Banach $\mathbb{K}$-algebra homomorphism from $H_b(D)$ into $H_b(D')$.*

If γ is a bijection from D' onto D and if $\gamma^{-1} \in H(D)$, then ϕ_γ is a $\mathbb{K}$-vector space isomorphism from $H(D)$ onto $H(D')$ bicontinuous with respect to the topology of uniform convergence on D for $H(D)$ and on $H(D')$ for $H(D')$, satisfying $(\phi_\gamma)^{-1} = \phi_{\gamma^{-1}}$ and the restriction of ϕ_γ to $H_b(D)$ is a Banach $\mathbb{K}$-algebra isomorphism from $H_b(D)$ onto $H_b(D')$. Further, if $\gamma(D') = D$, then the equality $\|\phi_\gamma(f)\|_{D'} = \|f\|_D$ is true for every $f \in H(D)$ and the restriction of ϕ_γ to $H_b(D)$ is an isometric Banach $\mathbb{K}$-algebra isomorphism from $H_b(D)$ onto $H_b(D')$.

Proof. It is easily seen that ϕ_γ is linear and satisfies $\phi_\gamma(fg) = \phi_\gamma(f)\phi_\gamma(g)$ when $fg \in H(D)$. Next, ϕ_γ is clearly continuous because $\|\phi_\gamma(f)\|_{D'} = \|f \circ \gamma\|_D = \sup_{x\in D'} |f(\gamma(x))| \le \sup_{u\in D} |f(u)| = \|f\|_D$. In particular, we note that if $\gamma(D') = D$, we have $\|\phi_\gamma(f)\|_{D'} = \|f \circ \gamma\|_D = \sup_{x\in D'} |f(\gamma(x))| = \sup_{u\in D} |f(u)| = \|f\|_D$.

If $f \in H_b(D)$, obviously, $f \circ \gamma \in H_b(D')$. Now, let γ be a bijection from D' onto D such that $\gamma^{-1} \in H(D)$. It is seen that $(\phi_{\gamma^{-1}}) \circ \phi_\gamma = I_{H(D)}$ while $\phi_\gamma \circ (\phi_{\gamma^{-1}}) = I_{H(D')}$, hence ϕ_γ is an isomorphism such that $\phi_{\gamma^{-1}} = (\phi_\gamma)^{-1}$. □

We study the $\mathbb{K}$-algebra homomorphisms from $H(D)$ into $H(D')$.

Notation. Given subsets D and D' of $\mathbb{K}$, we denote by $\Xi(D', D)$ *the set of the* $\gamma \in H(D')$ *such that* $\gamma(D') \subset D$ *and such that for every* $\lambda \in \overline{D} \backslash D$, $\gamma - \lambda$ *is invertible in* $H(D')$.

Given two $\mathbb{K}$-algebras A and B, we denote by $\mathcal{H}om(A, B)$ the set of $\mathbb{K}$-algebra homomorphisms from A into B.

Remark. In particular, $\Xi(D', D)$ contains the set of the $h \in R(D')$ such that $h(D') \subset D$.

Theorem 12.7. *Let* $D, D' \in \mathrm{Alg}$ *and let* $\gamma \in \Xi(D', D)$. *The mapping* ϕ_γ *defined in* $H(D)$ *by* $\phi_\gamma(f) = f \circ \gamma$ *has values in* $H(D')$ *and is a* $\mathbb{K}$*-algebra homomorphism from* $H(D)$ *into* $H(D')$. *Conversely, every* $\mathbb{K}$*-algebra homomorphism from* $H(D)$ *into* $H(D')$ *is continuous and of this form. Further, the mapping* $\gamma \to \phi_\gamma$ *from* $\Xi(D', D)$ *onto* $\mathcal{H}om(H(D), H(D'))$ *is a bijection.*

Let $D'' \in \mathrm{Alg}$ *and let* $\tau \in \Xi(D'', D')$. *Then,* $\gamma \circ \tau \in \Xi(D'', D)$ *and* $\phi_{\gamma \circ \tau} = \phi_\gamma \circ \phi_\tau$.

Further, a homomorphism ϕ_γ *from* $H(D)$ *into* $H(D')$ *is an isomorphism if and only if* γ *is a bijection from* D' *onto* D *such that* $\gamma^{-1} \in H(D)$ *and then, when it is satisfied, we have* $(\phi_\gamma)^{-1} = \phi_{\gamma^{-1}}$.

Proof. By Corollary 12.3, $f \circ \gamma$ belongs to $H(D')$ whenever $f \in H(D)$, and then by Proposition 12.6, ϕ_γ is a $\mathbb{K}$-algebra homomorphism from $H(D)$ into $H(D')$. Let $\psi \in \mathcal{H}om(H(D), H(D'))$ and first let us show that ψ satisfies $\|\psi(f)\|_{D'} \leq \|f\|_D$ whenever $f \in H(D)$. Indeed, suppose that for some $f \in H(D)$ we have $\|\psi(f)\|_{D'} > \|f\|_{D'}$. Let $g = \psi(f)$. There exists $\alpha \in D'$ such that $|g(\alpha)| > \|f\|_{D'}$. Let $\lambda = g(\alpha)$. The series $\frac{f}{\lambda} \sum_{n=0}^{\infty} \left(\frac{f}{\lambda}\right)^n$ does converge in $H(D)$ to $(\lambda - f)^{-1}$. Thus, $\lambda - f$ is invertible in $H(D)$ and then $\lambda - g = \psi(\lambda - f)$ is invertible in $H(D')$. But by hypothesis, α is a zero for $\lambda - g$, hence $\lambda - g$ is not invertible in $H(D')$ and this shows that $\|\psi(f)\|_{D'} \leq \|f\|_D$. Now, let $\gamma = \psi(I_D) \in H(D')$ and let us show that $\gamma \in \Xi(D', D)$. Let $\alpha \in \mathbb{K} \backslash D$. Since $\psi \in \mathcal{H}om(H(D), H(D'))$, ψ must satisfy $\psi\left(\frac{1}{x-\alpha}\right) = \frac{1}{\psi(x)-\alpha} = \frac{1}{\gamma-\alpha}$, hence $\gamma - \alpha$ has to be invertible in $H(D')$ for every $\alpha \in \mathbb{K}\setminus D$. But this just means that $\gamma \in \Xi(D', D)$.

In the same way, we see that for every $h \in R(D)$, we have $\psi(h) = h(\psi(x)) = h \circ \gamma$. Finally, since ψ is continuous, the equality $\psi(f) = f \circ \gamma$ holds in all $H(D)$. Obviously, given $\gamma, \tau \in \Xi(D', D)$, if $\phi_\gamma = \phi_\tau$, then $\phi_\gamma(I_D) = \phi_\tau(I_D)$, hence $\gamma = \tau$. The mapping $\gamma \to \phi_\gamma$ is then a bijection from $\Xi(D', D)$ onto $\mathcal{H}om(H(D), H(D'))$.

Now, let $D'' \in$ Alg and $\tau \in \Xi(D'', D')$. It is seen that $\gamma \circ \tau \in \Xi(D'', D)$ and $\phi_{\gamma\circ\tau}(I_D) = \gamma \circ \tau = \phi_\tau(\gamma) = \phi_\tau(\phi_\gamma(I_D)) = \phi_\tau \circ \phi_\gamma(I_D)$, hence $\phi_{\gamma\circ\tau} = \phi_\tau \circ \phi_\gamma$.

By Proposition 12.6, if γ is a bijection from D' onto D and such that $\gamma^{-1} \in H(D)$, then ϕ_γ is an isomorphism of $\mathbb{K}$-vector space, hence it is an isomorphism of $\mathbb{K}$-algebra, and then, by Proposition 12.6, we have $\left(\phi_\gamma\right)^{-1} = \phi_{\gamma^{-1}}$. Conversely, if ϕ_γ is an isomorphism, then $\left(\phi_\gamma\right)^{-1}$ is in the form ϕ_τ, with $\tau \in \Xi(D, D')$ and $\phi_\tau \circ \phi_\gamma(I_D) = \gamma \circ \tau(I_D) = I_D$ and $\phi_\gamma \circ \phi_\tau(I_{D'}) = \tau \circ \gamma(I_{D'}) = I_{D'}$. Hence, γ is a bijection from D' onto D such that $\gamma^{-1} = \tau \in H(D)$. That finishes showing Theorem 12.7. □

Definition. Let A, B be subsets of $\mathbb{K}$. If there exists a bijection $f \in H(A)$ from A onto B such that f^{-1} belongs to $H(B)$, then f is named *a bianalytic element from A onto B*.

Propositions 12.8 and 12.9 are often useful to transform unbounded domains into bounded domains.

Proposition 12.8. *Let $D \in$ Alg and let $h \in R(D)$ be a Moebius function. Let $D' = h(D)$. Then, D' belongs to Alg and $H(D')$ is isomorphic to $H(D)$ with respect to the mapping ψ defined in $H(D')$ as $\psi(f) = f \circ h$.*

Proof. By Theorem 12.2, for every $f \in H(D')$, $f \circ h$ belongs to $H(D)$, and by Proposition 12.6, this mapping is a $\mathbb{K}$-vector space isomorphism which satisfies $\psi(fg) = \psi(f)\psi(g)$ whenever $f, g \in H(D')$. Hence, the space $H(D') = \psi^{-1}(H(D))$ is a $\mathbb{K}$-algebra isomorphic to $H(D)$. In particular, D' belongs to Alg and ψ is a $\mathbb{K}$-algebra isomorphism. □

Proposition 12.9. *Let D be a set with a hole $T = d(a, r^-)$, let $\gamma(x) = \frac{1}{x-a}$, and let $D' = \gamma(D)$. Then, $\overline{D'} \in$ Alg and $H(\overline{D'})$ is isomorphic to $H_b(D)$.*

Proof. Without loss of generality, we may clearly assume D to be closed because by Corollary 11.7, $H_b(D)$ is equal to $H_b(\overline{D})$. For every $f \in H(D)$, let $\psi(f) = f \circ \gamma \in H(D')$. Then, $\psi(H_b(D))$ is a $\mathbb{K}$-algebra included in $H(D')$. If D is bounded, D' is bounded and closed like D, hence by Proposition 12.8, ψ is an isomorphism from $H(D)$ onto $H(D')$. Now, we suppose D unbounded. Then, D' is bounded and $\psi(H_b(D))$ is obviously included in $H_b(D')$ which, by Corollary 11.7, is just equal to $H(\overline{D'})$. On the other hand, γ clearly maps $R_b(D)$ onto $R(\overline{D'})$, hence $\psi(H_b(D)) = H(\overline{D'})$. □

Theorem 12.10. *Let $T = d(a, r^-)$ be a hole of D and let $\gamma(x) = \frac{1}{x-a}$. Let $D' = \gamma(D)$. The mapping ψ from $H_b(D')$ into $H_b(D)$ defined as $\psi(f) = f \circ \gamma$ is a $\mathbb{K}$-algebra isomorphism.*

Proof. D' is bounded, hence by Corollary 11.7, $H_b(D')$ is equal to the Banach $\mathbb{K}$-algebra $H(\overline{D'})$. Now, by Theorem 12.2, we see that $\gamma \in \Xi(D, D')$ and $\gamma^{-1} \in \Xi(D', D)$. Hence, ψ is clearly a $\mathbb{K}$-Banach space isomorphism from $H_b(D')$ onto $H_b(D)$. Now, ψ satisfies $\psi(fg) = \psi(f)\,\psi(g)$ whenever $f, g \in H(D)$ such that $fg \in H(D')$. But both $H_b(D)$, $H_b(D')$ are Banach $\mathbb{K}$-algebras, hence ψ is a Banach $\mathbb{K}$-algebra isomorphism. □

Theorem 12.11. *Let a be a point in D which is not isolated. Let $\gamma(x) = \frac{1}{x-a}$ and let $D' = \gamma(D \setminus \{a\})$. Then given $f \in H(D')$, $f \circ \gamma$ belongs to $H(D)$ if and only if $f(x)$ has a limit when $|x|$ tends to $+\infty$.*

Proof. If $f \circ \gamma$ belongs to $H(D)$, then we just have

$$\lim_{|x| \to +\infty} f(x) = \lim_{x \to a} f \circ \gamma(x) = f \circ \gamma(a).$$

Conversely, if f has a limit l when $|x|$ tends to $+\infty$, then $f \circ \gamma$ is bounded in certain disks $d(a, r) \setminus \{a\}$. Therefore, by Corollary 11.6, $f \circ \gamma$ belongs to $H(D)$. □

Corollary 12.12. *Let $D \in$ Alg, and let a be a point in D which is not isolated such that $(D \setminus \{a\})$ belongs to Alg. Let $\gamma(x) = \frac{1}{x-a}$ and let $D' = \gamma(D \setminus \{a\})$. Then, $H(D)$ is isomorphic to the subalgebra of $H(D')$ which consists of f such that $|f(x)|$ is bounded when $|x|$ approaches $+\infty$.*

Chapter 13

Multiplicative Spectrum of $H(D)$

In Chapter 10, we studied and characterized the multiplicative semi-norms on a $\mathbb{K}$-algebra $R(D)$ of rational functions. We apply these properties to the completion $H(D)$ of $R(D)$ by considering multiplicative semi-norms that are continuous with respect to the topology of $H(D)$. On $H(D)$ as on $R(D)$, the role of circular filters is obviously crucial: each continuous multiplicative semi-norm of $H(D)$ is defined by a circular filter secant with D exactly as it was explained for rational functions. However, circular filters that are not secant with D play no role with regards to $H(D)$.

Notation. Throughout this chapter, D is an infraconnected subset of $\mathbb{K}$. We denote by $\text{Mult}(H(D),\mathcal{U}_D)$ the set of continuous multiplicative semi-norms ψ of the $\mathbb{K}$-vector space $H(D)$ that satisfy $\psi(fg) = \psi(f)\psi(g)$ whenever $f, g \in H(D)$ such that $fg \in H(D)$.

Remark. This notation does not require $H(D)$ to be a $\mathbb{K}$-algebra, though it coincides with the notation already introduced for any topological algebra when $H(D)$ is a normed $\mathbb{K}$-algebra. Multiplicative semi-norms appeared to be the main tool for studying analytic elements [36, 44, 61]. They also are at the basis of Berkovich theory [9].

Theorem 13.1 ((Garandel) [61]). *For every $\mathcal{F} \in \Phi(D)$, the multiplicative semi-norm $\varphi_{\mathcal{F}}$ defined on $R(D)$ extends by continuity to $H(D)$ to a continuous semi-norm of $\mathbb{K}$-vector space ${}_D\varphi_{\mathcal{F}}$ of $H(D)$ that satisfies ${}_D\varphi_{\mathcal{F}}(f.g) = {}_D\varphi_{\mathcal{F}}(f)\ {}_D\varphi_{\mathcal{F}}(g)$ whenever $f, g \in H(D)$*

such that $fg \in H(D)$. Moreover, the mapping: $\mathcal{F} \to {}_D\varphi_{\mathcal{F}}$, from $\Phi(D)$ into $\text{Mult}(H(D),\mathcal{U}_D)$, *is a bijection.*

Proof. We may obviously extend $\varphi_{\mathcal{F}}$ by continuity to ${}_D\varphi_{\mathcal{F}}$ satisfying ${}_D\varphi_{\mathcal{F}}(fg) = {}_D\varphi_{\mathcal{F}}(f)\ {}_D\varphi_{\mathcal{F}}(g)$ whenever $f, g \in H(D)$ such that $fg \in H(D)$. We now check that the mapping $\mathcal{F} \to {}_D\varphi_{\mathcal{F}}$ from $\Phi(D)$ into $\text{Mult}(H(D),\mathcal{U}_D)$ is a bijection. It is obviously injective because if ${}_D\varphi_{\mathcal{F}_1} = {}_D\varphi_{\mathcal{F}_2}$, then $\varphi_{\mathcal{F}_1} = \varphi_{\mathcal{F}_2}$, hence by Corollary 10.10, $\mathcal{F}_1 = \mathcal{F}_2$. Now, let $\psi \in \text{Mult}(H(D),\mathcal{U}_D)$. The restriction of ψ to $R(D)$ is an element ψ_0 of $\text{Mult}(R(D),\mathcal{U}_D)$, hence by Corollary 10.10, ψ_0 is of the form $\varphi_{\mathcal{F}}$ and then, by continuity, we have $\psi = {}_D\varphi_{\mathcal{F}}$. □

From Theorem 10.4, the following theorem is immediate concerning a space $H(D)$.

Corollary 13.2. *Let $\mathcal{F}$ be a large circular filter on $\mathbb{K}$ of diameter $s > 0$. The following three assertions are equivalent:*

(i) *$\varphi_{\mathcal{F}}(h) \leq \|h\|_D$ whenever $h \in H(D)$.*
(ii) *$\varphi_{\mathcal{F}}$ is a continuous ultrametric multiplicative norm on $H(D)$ with respect to the topology of uniform convergence.*
(iii) *$\mathcal{F}$ is secant with D.*

By Corollary 10.10, we have Corollary 13.3:

Corollary 13.3. *Let D be a closed bounded subset of $\mathbb{K}$. Then,* $\text{Mult}(H(D), \|\cdot\|_D)$ *is compact with respect to the topology of pointwise convergence.*

Remark. If $\mathcal{F}$ is a large circular filter, we know that $\varphi_{\mathcal{F}}$ is a norm on $R(D)$. But we don't know whether ${}_D\varphi_{\mathcal{F}}$ is a norm on $H(D)$. This not trivial question is linked to the problem of T-filters which is a so big question that it would require another book [38, 44, 53].

Definition and notation. For convenience, for every $a \in D$, we put $\varphi_a(f) = |f(a)|$ whenever $f \in H(D)$ and so we define the semi-norms $\varphi_a \in \text{Mult}(H(D),\mathcal{U}_D)$.

An element $\psi \in \text{Mult}(H(D),\mathcal{U}_D)$ is said to be *punctual* if it is of the form φ_a with $a \in D$, i.e. if its circular filter is punctual.

Let $\mathcal{F}$ be a monotonous filter on D. By Proposition 3.14, there exists a unique circular filter $\mathcal{G}$ on D less thin than $\mathcal{F}$. Then, we put ${}_D\varphi_{\mathcal{F}}(f) = {}_D\varphi_{\mathcal{G}}(f)$ for all $f \in H(D)$.

For simplicity, when $\mathcal{G}$ has center 0 and diameter r, we set $|f|(r) = {}_D\varphi_{\mathcal{G}}(f)$. So, when f belongs to $R(D)$, this is the definition already given in Chapter 4.

Now, let D be infraconnected. Let $a \in \widetilde{D}$ and let r satisfy $\delta(a, D) \leq r \leq \text{diam}(D)$. The circular filter $\mathcal{F}$ of center a and diameter r is then secant with D. We put ${}_D\varphi_{a,r} = {}_D\varphi_{\mathcal{F}}$. Let A be a bounded subset of $\widetilde{D}$ and let $\widetilde{A} = d(a, r)$. If $\delta(a, D) \leq r \leq \text{diam}(D)$, we put ${}_D\varphi_A = {}_D\varphi_{a,r}$. In particular, this notation applies to holes of an infraconnected set D.

Let $\mathcal{F}$ be a circular filter or a monotonous filter on D. We denote by $\mathcal{J}(\mathcal{F})$ the set of the $f \in H(D)$ such that $\lim_{\mathcal{F}} f(x) = 0$. Hence, $\mathcal{J}(\mathcal{F})$ is equal to $Ker({}_D\varphi_{\mathcal{F}})$, and therefore, if $D \in \text{Alg}$, $Ker({}_D\varphi_{\mathcal{F}})$ is a closed prime ideal of $H(D)$.

If $\mathcal{F}$ is a monotonous filter on D, we denote by $\mathcal{J}_0(\mathcal{F})$ the set of the $f \in H(D)$ such that $f(x) = 0$ whenever $x \in \mathcal{B}(\mathcal{F})$.

Finally, given $a \in D$, we denote by $\mathcal{J}(a)$ the set of the $f \in H(D)$ such that $f(a) = 0$. Then, if $D \in \text{Alg}$, $\mathcal{J}_0(\mathcal{F})$ and $\mathcal{J}(a)$ are closed prime ideals of $H(D)$.

Among many ultrametric properties, we note the following.

Lemma 13.4. *Let $\mathcal{F}$ be a circular filter or a monotonous filter on D and let $f \in H(D)$. There exists $A \in \mathcal{F}$ such that $|f(x)|$ is bounded in A. Moreover, for every sequence $(a_n)_{n\in\mathbb{N}}$ thinner than $\mathcal{F}$, we have $\lim_{n\in\mathbb{N}} \varphi_{a_n} = \varphi_{\mathcal{F}}$. Given a, $b \in D$ and $r \in]0, \text{diam}(D)[$ such that $|a - b| \leq r$, we have ${}_D\varphi_{a,r} = {}_D\varphi_{b,r}$.*

Proof. Indeed, there does exist $A \in \mathcal{F}$ such that $|f(x)| \leq {}_D\varphi_{\mathcal{F}}(f) + 1$ for all $x \in A$. The last statements come from properties seen on $R(D)$. □

Lemma 13.5. *Let $f \in H(D)$ be invertible in $H(D)$. Then for every $\psi \in \text{Mult}(H(D), \mathcal{U}_D)$, we have $\psi(f) \neq 0$.*

Proof. Indeed, we have $\psi(f)\psi(\frac{1}{f}) = 1$. □

Lemma 13.6. *Let ${}_D\varphi_{\mathcal{F}} \in \text{Mult}(H(D), \mathcal{U}_D)$, let $f \in H(D)$, and let $g \in H(D)$ be such that $\|f - g\|_D < {}_D\varphi_{\mathcal{F}}(f)$. Then, ${}_D\varphi_{\mathcal{F}}(f) = {}_D\varphi_{\mathcal{F}}(g)$.*

Proof. Indeed, we know that ${}_D\varphi_{\mathcal{F}}(f - g) \leq \|f - g\|_D$, hence ${}_D\varphi_{\mathcal{F}}(f - g) < {}_D\varphi_{\mathcal{F}}(f)$, and therefore, ${}_D\varphi_{\mathcal{F}}(g) = {}_D\varphi_{\mathcal{F}}(f)$. □

Lemma 13.7. *Let D be unbounded and let $f \in H_b(D)$. Then, $|f(x)|$ has a limit ${}_D\varphi_D(f)$ when $|x|$ tends to $+\infty$, while x lies in D and ${}_D\varphi_D$ belongs to* $\text{Mult}(H_b(D), \|\ .\ \|_D)$.

Proof. By Corollary 11.7, $f(x)$ admits a limit λ when $|x|$ tends to $+\infty$ ($x \in D$). Hence, $\lim_{|x|\to+\infty,\ x\in D} |f(x)| = |\lambda|$. Thus, the mapping φ_∞, defined as $\varphi_\infty(f) = \lim_{|x|\to+\infty} |f(x)|$, belongs to $\text{Mult}(R_b(D), \|\ .\ \|_D)$ and obviously has continuation by continuity to an element ${}_D\varphi_D \in \text{Mult}(H_b(D), \|\ .\ \|_D)$ which satisfies ${}_D\varphi_D(f) = \lim_{|x|\to+\infty,\ x\in D} |f(x)|$. □

Notation. When there is no risk of confusion about the set D, we just write $\varphi_{\mathcal{F}}$ (respectively $\varphi_{a,r}$, respectively φ_D, and respectively φ_A), instead of ${}_D\varphi_{\mathcal{F}}$ (respectively ${}_D\varphi_{a,r}$, respectively ${}_D\varphi_D$, and respectively ${}_D\varphi_A$). Next, when D is unbounded, ${}_D\varphi_D$ is also denoted by ${}_D\varphi_\infty$.

Theorem 13.8. *Let $\psi \in \text{Mult}(H_b(D), \|\ .\ \|_D) \setminus \text{Mult}(H(D), \mathcal{U}_D)$. If D is bounded, ψ is of the form φ_a with $a \in \overline{D} \setminus D$. If D is not bounded, ψ is either of the form φ_a, with $a \in \overline{D} \setminus D$ or of the form ${}_D\varphi_\infty$.*

Proof. First, we suppose D bounded. By Corollary 11.7, we have $H_b(D) = H(\overline{D})$. Hence, ψ is equal to some ${}_{\overline{D}}\varphi_{\mathcal{F}}$, with $\mathcal{F}$ a circular filter on $\overline{D}$. If $\mathcal{F}$ is large, it is a large circular filter secant with D and then ψ belongs to $\text{Mult}(H(D), \mathcal{U}_D)$. If $\mathcal{F}$ is not large, it is the filter of neighborhoods of a point $a \in \overline{D}$. But if $a \in D$, obviously φ_a belongs to $\text{Mult}(H(D), \mathcal{U}_D)$. Hence, $a \in \overline{D} \setminus D$.

Now, we suppose D unbounded. If D has no hole, we just have $H_b(D) = R_b(D) = \mathbb{K}$, hence $\text{Mult}(H_b(D), \|\ .\ \|_D) = \text{Mult}(H(D), \mathcal{U}_D)$. Thus, we may assume D to have a hole $T = d(a, r^-)$. Without loss of generality, we may assume $a = 0$. Let $\gamma(x) = \frac{1}{x}$ and let $D' = \gamma(D)$. Then, D' is bounded and, by Proposition 12.9, we know that the algebra $H_b(D)$ is isomorphic to $H(D')$. By Proposition 12.8, the mapping $f \to f \circ \gamma$ defines a $\mathbb{K}$-vector space isomorphism from $H(D)$ onto $H(\overline{D'})$ and a $\mathbb{K}$-algebra isomorphism from $H_b(D)$ onto $H_b(D') = H(\overline{D'})$. Hence, we may define $\psi' \in \text{Mult}(H(\overline{D'}))$ by $\psi'(f \circ \gamma) = \psi(f)$ whenever $f \in H_b(D)$. If ψ' belonged to $\text{Mult}(H(D'), \mathcal{U}_{D'})$, then we would have $\psi(f) = \psi'(f \circ \gamma)$ whenever $f \in H(D)$, and therefore, $\psi \in \text{Mult}(H(D), \mathcal{U}_D)$. Hence, ψ' does not belong to $\text{Mult}(H(D'), \mathcal{U}_{D'})$ and then ψ' is of the form

φ_b with $b \in \overline{D'}\setminus D$. If $b \neq 0$, then $\psi = \varphi_{\frac{1}{b}}$. If $b = 0$, then $\psi = {}_D\varphi_D$ and this ends the proof. □

Theorem 13.9. *Let $(\alpha_n)_{n\in\mathbb{N}}$ be a bounded sequence in D such that no subsequence converges to any point of $\overline{D}\setminus D$. There exists a subsequence $(\alpha_{n_s})_{s\in\mathbb{N}}$ such that the sequence $(\varphi_{\alpha_{n_s}})_{s\in\mathbb{N}}$ converges in* Mult$(H(D),\mathcal{U}_D)$.

Proof. By Theorem 3.1, we may extract either a convergent subsequence, or a monotonous distances subsequence, or an equal distances subsequence from the sequence $(\alpha_n)_{n\in\mathbb{N}}$. Let $(\alpha_{n_s})_{s\in\mathbb{N}}$ be such a subsequence. If this subsequence converges to a point $\alpha \in \mathbb{K}$, then by hypothesis, α lies in D, hence $\lim_{s\to+\infty} \varphi_{\alpha_{n_s}} = \varphi_\alpha$. If this subsequence is a monotonous distances subsequence, or an equal distances subsequence, then by Proposition 3.18, on D there exists a large circular filter $\mathcal{F}$ less thin than the sequence (α_{n_s}) and then we see that for every $f \in H(D)$ we have $\lim_{s\to+\infty} |f(\alpha_{n_s})| = {}_D\varphi_{\mathcal{F}}(f)$, hence $\lim_{s\to+\infty} \varphi_{\alpha_{n_s}}(f) = \varphi_{\mathcal{F}}(f)$. Thus, in every case, we have proven that the subsequence $\varphi_{\alpha_{n_s}}$ converges in Mult$(H(D),\mathcal{U}_D)$. □

Chapter 14

Power and Laurent Series

A power series on a p-adic field admits a disk of convergence whose radius is defined in the same way as on $\mathbb{C}$. The difference in behavior between power series in $\mathbb{C}$ and in a field such as $\mathbb{K}$ concerns what happens when $|x|$ is equal to the radius of convergence. We show that the norm of uniform convergence in a disk $d(a,s) \subset d(0,R^-)$ is multiplicative and satisfies $\| \sum_{n=0}^{+\infty} a_n x^n \|_{d(0,s)} = \sup_{n\in\mathbb{N}} |a_n| s^n$. As a consequence, the product of two power series converging in $d(0,R^-)$ is bounded if and only if both are bounded. We show that the algebra of power series with a radius of convergence equal to R is equal to the intersection of algebras of analytic elements $H(d(0,s))$ when $s < R$. We show that all analytic elements in $d(0,R^-)$ are power series converging in $d(0,R^-)$. The converse is false. However, we see that the analytic elements in $d(0,R)$ are exactly the power series converging in this disk.

Definitions. Let $f(x) = \sum_{n=0}^{\infty} a_n x^n$ be a power series with coefficients in $\mathbb{K}$.

As usual, when $\limsup_{n\to\infty} \sqrt[n]{|a_n|} \neq 0$, we call *radius of convergence of* f the number $r = \frac{1}{\limsup_{n\to\infty} \sqrt[n]{|a_n|}}$ (with $r = 0$ when $\limsup_{n\to\infty} \sqrt[n]{|a_n|} = +\infty$).

When $\limsup_{n\to\infty} \sqrt[n]{|a_n|} = 0$, we define the radius of convergence of f as $+\infty$.

Example. Let $f(x) = \sum_{n=1}^{\infty} nx^n$. The radius of convergence of this series is 1. This function obviously defines the rational function $\frac{x}{(1-x)^2}$ in $d(1,1^-)$.

Remark. If a sequence of positive numbers $(u_n)_{n\in\mathbb{N}}$ is such that the sequence $\left(\frac{u_{n+1}}{u_n}\right)_{n\in\mathbb{N}}$ converges to a limit $l \geq 0$, then so does the sequence $(\sqrt[n]{u_n})_{n\in\mathbb{N}}$. On the field $\mathbb{K}$, as in Archimedean analysis, it is a way to compute easily many radii of convergence.

Lemma 14.1 is immediate.

Lemma 14.1. *Let $f = \sum_{n=0}^{\infty} a_n x^n$ be a power series with coefficients in $\mathbb{K}$. The series converges if and only if $\lim_{n\to\infty} |a_n x^n| = 0$. Let r be its radius of convergence. If $|x| < r$, then the series converges. If $|x| > r$, then the series diverges.*

Definition and notation. Power series whose radius of convergence is ∞ are called *entire functions on* $\mathbb{K}$ and the set of entire functions is denoted by $\mathcal{A}(\mathbb{K})$.

For every $a \in \mathbb{K}$, $r \in \mathbb{R}_+^*$, similarly we denote by $\mathcal{A}(d(a, r^-))$ the set of power series in $x - a$ whose radius of convergence is superior or equal to r and by $\mathcal{A}_b(d(a, r^-))$ the set of functions $f \in \mathcal{A}(d(a, r^-))$ that are bounded in $d(a, r^-)$. The set $\mathcal{A}(d(a, r^-))\backslash\mathcal{A}_b(d(a, r^-))$ is denoted by $\mathcal{A}_u(d(a, r^-))$.

Similarly, we denote by $\mathcal{A}(\mathbb{K}\backslash d(a, r))$ the set of Laurent series converging whenever $|x - a| > r$, by $\mathcal{A}_b(\mathbb{K}\backslash d(a, r))$ the set of bounded Laurent series converging whenever $|x - a| > r$, and by $\mathcal{A}_u(\mathbb{K}\backslash d(a, r))$ the set of unbounded Laurent series converging whenever $|x - a| > r$.

Finally, given r', r'' such that $0 < r' < r''$, we denote by $\mathcal{A}(\Gamma(a, r', r''))$ the set of Laurent series converging whenever $r' < |x - a| < r''$. And we denote by $\mathcal{A}_b(\Gamma(a, r', r''))$ the set of functions $f \in \mathcal{A}(\Gamma(a, r', r''))$ that are bounded in $\Gamma(a, r', r'')$.

From Lemma 14.1, we can derive Corollary 14.2:

Corollary 14.2. *Let $\sum_{-\infty}^{+\infty} a_n x^n$ be a Laurent series with coefficients in $\mathbb{K}$, let $r'' = \frac{1}{\limsup_{n\to\infty} \sqrt[n]{|a_n|}}$ with $r'' = 0$ whenever $\limsup_{n\to\infty} \sqrt[n]{|a_n|} = +\infty$, and let $r' = \frac{1}{\limsup_{n\to-\infty} \sqrt[-n]{|a_n|}}$ with $r' = 0$ whenever $\limsup_{n\to-\infty} \sqrt[-n]{|a_n|} = +\infty$. If $r' < |x| < r''$, the series converges. If $|x| > r''$ or if $|x| < r'$, the series diverges.*

Corollary 14.3. *Let $r', r'' \in \mathbb{R}_+$ satisfy $0 < r' < r''$. Then, $\mathcal{A}(\Gamma(0, r', r''))$ is the set of Laurent series $\sum_{-\infty}^{+\infty} a_n x^n$ such that $r' \leq \frac{1}{\limsup_{n\to-\infty} \sqrt[n]{|a_n|}} \leq r''$.*

Corollary 14.4. *Let r', $r'' \in \mathbb{R}_+$ be such that $r' < r''$ and let $f(x) = \sum_{-\infty}^{+\infty} a_n x^n \in \mathcal{A}(\Gamma(a, r', r''))$. For each $r \in]r', r''[$, one has*

$$\lim_{n\to+\infty} |a_n| r^n = \lim_{n\to-\infty} |a_n| r^n = 0.$$

Lemma 14.5 is useful in certain further problems:

Lemma 14.5. *Let $q \in \mathbb{N}^*$ and let $\mathbb{L}$ be a complete algebraically closed extension of $\mathbb{K}$. Let $f \in \mathcal{A}(d(a, R^-))$ and suppose that there exists a power series g with coefficients in $\mathbb{L}$, with radius of convergence $\geq R$ such that $(g(x))^q = f(x)$, $\forall x \in d(a, R^-)$. Then g has all coefficients in $\mathbb{K}$ and belongs to $\mathcal{A}(d(a, R^-))$.*

Proof. Without loss of generality, we can obviously suppose $a = 0$. Let $f(x) = \sum_{n=0}^{+\infty} b_n x^n$ (with $b_n \in \mathbb{K}$) and let $g(x) = \sum_{n=0}^{+\infty} a_n x^n$. Then, $(a_0)^q = b_0$, hence $a_0 \in \mathbb{K}$ because $\mathbb{K}$ is algebraically closed. Now, suppose we have proven that $a_n \in \mathbb{K}$, $\forall n \leq t-1$. We can see that b_t is of the form $a_t(a_0)^{q-t} + h$, where h is a polynomial in $a_0, a_1, \ldots, a_{t-1}$. Therefore, a_t also belongs to $\mathbb{K}$. Consequently, g has all coefficients in $\mathbb{K}$, which ends the proof. $\square$

Notation. Let $r \in \mathbb{R}_+^*$ and let $\sum_{-\infty}^{+\infty} a_n x^n \in H(C(0, r))$. By hypothesis, we have $\lim_{n\to+\infty} |a_n| r^n = \lim_{n\to-\infty} |a_n| r^n = 0$. Generalizing the notation already introduced for rational functions, we denote by $\nu^+(f, \log r)$ the highest of the integers $m \in \mathbb{Z}$ such that $|a_m| r^m = \sup_{n\in\mathbb{Z}} |a_n| r^n$ and by $\nu^-(f, \log r)$ the lowest of the integers $m \in \mathbb{Z}$ such that $|a_m| r^m = \sup_{n\in\mathbb{Z}} |a_n| r^n$. Next, when $\nu^+(f, \log r) = \nu^-(f, \log r)$, we just set $\nu(f, \log r)$.

Recall that we have, given a circular filter $\mathcal{F}$ of center 0 and diameter r, for every element of $H(d(0, r))$ and particularly for every analytic function $f \in \mathcal{A}(\mathbb{K})$, put $|f|(r) = \lim_{\mathcal{F}} |f(x)|$.

Theorem 14.6. *Let $r \in \mathbb{R}_+^*$, let $\mathcal{F}$ be the circular filter of center 0 and diameter r on $\mathbb{K}$, and let $E = d(0, r)$. Then, $H(E)$ is the set of power series $f(x) = \sum_{n=0}^{\infty} a_n x^n$ such that $\lim_{n\to\infty} |a_n| r^n = 0$ and we have $\|f\|_E = |f|(r) = \max_{n\in\mathbb{N}} |a_n| r^n = {}_E\varphi_{\mathcal{F}}(f) = \|f\|_{C(0,r)}$.*

For every $\alpha \in E$, $H(E)$ is also equal to the set of series $f(x) = \sum_{n=0}^{\infty} b_n (x - \alpha)^n$ such that $\lim_{n\to\infty} |b_n| r^n = 0$.

Let $B = \mathbb{K} \setminus d(0, r^-)$. Then, $H(B)$ is the set of Laurent series $f(x) = \sum_{n=0}^{\infty} \frac{a_n}{x^n}$ such that $\lim_{n\to\infty} |a_n| r^{-n} = 0$ and we have $\|f\|_B = \max_{n\in\mathbb{N}} |a_n| r^{-n} = {}_B\varphi_{\mathcal{F}}(f) = \|f\|_{C(0,r)}$.

For every $\alpha \in d(0, r^-)$, $H(B)$ *is also equal to the set of series* $f(x) = \sum_{n=0}^{\infty} \frac{b_n}{(x-\alpha)^n}$ *such that* $\lim_{n\to\infty} |b_n| r^{-n} = 0$.

Let $r' \geq r$ *and let* $D = \Delta(0, r, r'))$. *Then* $H(D)$ *is the set of Laurent series* $f(x) = \sum_{-\infty}^{\infty} a_n x^n$ *such that* $\lim_{n\to-\infty} |a_n| r^n = 0$ *and* $\lim_{n\to+\infty} |a_n|(r')^n = 0$ *and we have* $\|f\|_D = \max(\max_{n<0} |a_n| r^n, \max_{n\geq 0} |a_n|(r')^n)$. *Moreover, for every* $\alpha \in d(0, r^-)$, $H(D)$ *is also equal to the set of power series* $f(x) = \sum_{-\infty}^{\infty} b_n (x-\alpha)^n$ *such that* $\lim_{n\to-\infty} |b_n| r^n = 0$ *and* $\lim_{n\to+\infty} |b_n|(r')^n = 0$.

Proof. Let $S(r)$ be the set of power series $f(x) = \sum_{n=0}^{\infty} a_n x^n$ such that $\lim_{n\to\infty} |a_n| r^n = 0$. Such a power series obviously is a uniform limit of polynomials because $|f(x) - \sum_{n=0^q} a_n x^n| \leq \sup_{n\geq q} |a_n| r^n$ and hence it belongs to $H(E)$. Moreover, E is closed and bounded, hence by Theorem 11.2, $H(E)$ is a $\mathbb{K}$-Banach algebra with respect to the norm of uniform convergence on E. By Lemma 4.7, on $\mathbb{K}[x]$ the norm $\| \,.\, \|_E$ is $\varphi_{\mathcal{F}}$, and by Theorem 13.1, that equality has continuation to $H(E)$.

Now, for a polynomial $P(x) = \sum_{n=0}^{q} a_n x^n$, by Lemma 4.7, we have $\|P\|_E = \sup_{0\leq n\leq q} |a_n| r^n$, hence this equality also has continuation to f. Consequently, $\|f\|_E = \lim_{\mathcal{F}} |f(x)| = \varphi_{\mathcal{F}}(f)$. Particularly, $S(r)$ is a subset of $H(E)$.

In order to show that $S(r) = H(E)$, we first show that $S(r)$ is closed in $H(E)$. Since $\mathcal{F}$ is secant with $C(0, r)$, we have $\|f\|_{C(0,r)} = {}_E\varphi_{\mathcal{F}}(f)$. But we know that $\varphi_{\mathcal{F}}(P_n) \leq \max_{0\leq i\leq n} |a_i| r^i$. Since $\varphi_{\mathcal{F}}$ extends continuously to ${}_E\varphi_{\mathcal{F}} \in \text{Mult}(H(E), \| \,.\, \|_E)$, for n big enough, we have ${}_E\varphi_{\mathcal{F}}(f) =_E \varphi_{\mathcal{F}}(P_n) = |a_j| r^j$ with $j < n$ and $|a_n| r^m < |a_j| r^j$ whenever $m < j$, hence finally ${}_E\varphi_{\mathcal{F}}(f) = |a_j| r^j = \max_{n\in\mathbb{N}} |a_n| r^n$. Consequently, we have $\|f\|_{C(0,r)} = \|f\|_E \leq |a_j| r^j = \max_{0\leq n} |a_n| r^n$, and therefore, $\|f\|_{C(0,r)} = \|f\|_E = \max_{0\leq n} |a_n| r^n$. This finishes showing that $S(r)$ is a closed subset of $H(E)$.

Now, we show that $R(E)$ is included in $S(r)$. For this, we just have to show that, given any $\beta \in \mathbb{K}\backslash E$, $\left(\frac{1}{x-\beta}\right)^q = \left(-\frac{1}{\beta} \sum_{n=0}^{\infty} \left(\frac{x}{\beta}\right)^n\right)^q$ belongs to F. When developing $\left(\sum_{n=0}^{\infty} \left(\frac{x}{\beta}\right)^n\right)^q$, we see that for every fixed $q \in \mathbb{N}$, the coefficient A_q of x^q is a sum of terms of the form $\frac{s}{\beta^q}$, with $s \in \mathbb{N}$, hence finally, $|A_q| \leq \frac{1}{|\beta|^q}$, and therefore, $|A_q| r^q \leq \left(\frac{r}{|\beta|}\right)^q$. Since $|\beta| > r$, this shows that $\left(\frac{1}{x-\beta}\right)^q \in S(r)$. So, we have proven the inclusion $R(E) \subset S(r) \subset H(E)$. Since $S(r)$ is closed, we have $S(r) = H(E)$.

Now, let $\alpha \in E$. Since $d(\alpha, r) = d(0, r)$, after the change of variable $x = \alpha + u$, the same reasoning shows that a series $f(x) = \sum_{n=0}^{\infty} a_n\ x^n \in H(E)$ is also of the form $\sum_{n=0}^{\infty} b_n(\alpha)(x-\alpha)^n$ with $\lim_{n\to\infty} |b_n(\alpha)|r^n = 0$. Conversely, $H(E)$ is clearly equal to the set of series $\sum_{n=0}^{\infty} b_n(x+\alpha)^n$ such that $\lim_{n\to\infty} |b_n|r^n = 0$ because any series g of the form $\sum_{n=0}^{\infty} b_n\ u^n$, with $\lim_{n\to\infty} |b_n|r^n = 0$, can be written as $\sum_{n=0}^{\infty} a_n(u+\alpha)^n$.

The statements about $H(B)$ are an obvious consequence of those about $H(E)$ after the change of variable $y = \frac{1}{x}$ and more generally, $y = \frac{1}{x-\alpha}$. So, are the statements about $H(D)$. □

We can easily check the following corollaries:

Corollary 14.7. *Let $f \in \mathcal{A}(\mathbb{K})$. The following three statements are equivalent:*

(i) $\lim_{r\to+\infty} \frac{|f|(r)}{r^q} = +\infty,\ \forall q \in \mathbb{N}$.
(ii) *There exists no $q \in \mathbb{N}$ such that* $\lim_{r\to+\infty} \frac{|f|(r)}{r^q} = 0$.
(iii) *f is not a polynomial.*

Corollary 14.8. *Let $f,\ g \in \mathcal{A}\mathbb{K})$. Then $f.g$ is a polynomial if and only if both $f,\ g$ are polynomials.*

Corollary 14.9. *Let $r \in \mathbb{R}_+^*$ and let $D = d(0, r)$. Then, $H(D)$ is the set of power series $f(x) = \sum_{n=0}^{\infty} a_n x^n$ such that $\lim_{|n|_\infty\to\infty} |a_n|r^n = 0$ and we have*

$$\|f\|_D = \max_{n\in\mathbb{N}} |a_n|r^n = {}_D\varphi_{\mathcal{F}}(f).$$

Moreover, the norms $\|\ .\ \|_{C(0,r)}$, $|\ .\ |(r)$ and $\|\ .\ \|_{d(0,r)}$ are equal and are multiplicative.

Corollary 14.10. *Let $r \in \mathbb{R}_+^*$ and let $D = d(0, r)$ (respectively $D = d(0, r^-)$). Then the norm $\|\ .\ \|_D$ on $H(D)$ is multiplicative.*

Corollary 14.11. *Let $\alpha \in D$ and $r \in \mathbb{R}_+^*$ be such that $d(\alpha, r) \subset D$. Let $f \in H(D)$. In $d(\alpha, r)$, $f(x)$ is equal to a power series of the form $\sum_{n=0}^{\infty} a_n(x-\alpha)^n$ such that $\lim_{n\to\infty} |a_n|r^n = 0$. If $f(\alpha) = 0$ and if $f(x)$ is not identically zero in $d(\alpha, r)$, then there exists a unique* *integer $q \in \mathbb{N}^*$ such that $a_n = 0$ for every $n < q$ and $a_q \neq 0$ and α is an isolated zero of f in $d(\alpha, r)$.*

Proposition 14.12. *Let $r \in \mathbb{R}_+^*$ and let $f(x) = \sum_{n=0}^{\infty} a_n x^n$. The following statements are equivalent:*

(a) $f \in \mathcal{A}(d(0, r^-))$.
(b) $f \in \bigcap_{s<r} H(d(0, s))$.
(c) *The series f is convergent in all of $d(0, r^-)$.*

Proof. (b) and (c) are clearly equivalent to the condition $\lim_{n\to\infty} |a_n|s^n = 0$ whenever $s < r$ and, in the same way as in archimedean analysis, it is shortly checked that this is also equivalent to $\limsup_{n\to\infty} \sqrt[n]{|a_n|} \leq \frac{1}{r}$. □

Remark. If f is convergent for some $\alpha \in C(0, r)$, then $\lim_{n\to\infty} |a_n|r^n = 0$, hence f belongs to $H(d(0, r))$.

Corollary 14.13. *Let $r \in \mathbb{R}_+^*$ and let $f(x) = \sum_{-\infty}^{0} a_n x^n$. The following statements are equivalent:*

(a) $f \in \mathcal{A}(\mathbb{K}\setminus d(0, r))$.
(b) $f \in \bigcap_{s>r} H_b(\mathbb{K}\setminus d(0, s))$.
(c) *The series f is convergent in all of $\mathbb{K}\setminus d(0, r)$.*

Corollary 14.14. *Let r_1, $r_2 \in \mathbb{R}_+^*$ with $r_1 < r_2$ and let $f(x) = \sum_{-\infty}^{+\infty} a_n x^n$. The following statements are equivalent:*

(a) $f \in \mathcal{A}(\Gamma(0, r_1, r_2))$.
(b) $f \in \bigcap_{r_1<s_1<s_2<r_2} H(\Delta(0, s_1, s_2))$.
(c) *The series f is convergent in all of $\Gamma(0, r_1, r_2)$.*

Corollary 14.15. *Let $f \in \mathcal{A}(d(0, r^-))$ be not identically zero. For every $\alpha \in d(0, r^-)$, $f(x)$ is equal to a power series $\sum_{n=0}^{\infty} b_n(\alpha)(x-\alpha)^n$. If f is not identically zero and if α is a zero of f in $d(0, r^-)$, α is an isolated zero and f factorizes in $\mathcal{A}(d(0, r^-))$ in the form $(x-\alpha)^q g(x)$, with $g \in \mathcal{A}(d(0, r^-))$, $q \in \mathbb{N}^*$ $g(\alpha) \neq 0$.*

Definition. Let $f \in H(D)$ and let $\alpha \in \overset{\circ}{D}$, let $r > 0$ be such that $d(\alpha, r) \subset D$ and suppose $f(x) = \sum_{n=q}^{\infty} b_n(x-\alpha)^n$ whenever $x \in d(\alpha, r)$, with $b_q(\alpha) \neq 0$ and $q > 0$. Then, α is called *a zero of* *multiplicity order* q, or more simply, *a zero of order* q. In the same way, q is named *the multiplicity order of* α.

Remark. In particular, these definitions apply to functions $f \in \mathcal{A}(d(a, r^-))$, at any point $\alpha \in d(a, r^-)$.

Corollary 14.16. *Let* $a \in \mathbb{K}$, $R \in \mathbb{R}_+^*$ *and let* $f \in \mathcal{A}(d(a, R^-))$ *(respectively* $f \in \mathcal{A}(\mathbb{K})$*). Let* $a_1, \ldots, a_q$ *be zeros of* f *of respective order* s_j *and let* $P(x) = \prod_{j=1}^{q}(x - aj)^{s_j}$. *Then,* f *factorizes in the form* $P(x)u(x)$ *with* $u \in \mathcal{A}(d(a, R^-))$ *(respectively* $u \in \mathcal{A}(\mathbb{K})$.

Corollary 14.17. *Let* $a \in \mathbb{K}$, R, $R' \in \mathbb{R}_+^*$ *with* $R < R'$ *and let* $\Lambda = d(a, R^-)$ *(respectively* $\Lambda = \mathbb{K} \backslash d(a, R)$, *respectively* $\Lambda = \Gamma(a, R, R')$*)) and let* $C(b, r)$ *be a circle included in* Λ. *Then,* ${}_\Lambda\varphi_{b,r}$ *applies to* $\mathcal{A}(\Lambda)$.

Corollary 14.18. *Let* $f(x) = \sum_{n=0}^{\infty} a_n x^n \in \mathcal{A}(\mathbb{K})$. *If* f *is not a constant, then* $\lim_{r \to +\infty} |f|(r) = +\infty$.

Notation. Let $R \in \mathbb{R}_+^*$ and let $f \in \mathcal{A}(d(a, R^-))$. Given $r \in]0, R[$, by Proposition 14.12, f belongs to $H(d(a, r))$, hence for every circular filter $\mathcal{F}$ secant with $d(a, r)$, $\varphi_{\mathcal{F}}(f)$ is defined. Particularly, if $a = 0$, $|f|(r)$ is defined.

Theorem 14.19. *Let* $R \in \mathbb{R}_+^*$ *and let* $f \in \mathcal{A}(d(a, R^-))$. *Then,* f *is invertible in* $\mathcal{A}(d(a, R^-))$ *if and only if* f *has no zero in* $d(a, R^-)$.

Proof. Suppose that f has no zero in $d(a, R^-)$. For each $r \in]0, R[$, f belongs to $H(d(a, r))$, and hence by Lemma 11.3, it is invertible in $H(d(a, r))$. Consequently, the function defined in $d(a, r)$ as $g(x) = \frac{1}{f(x)}$ belongs to $H(d(a, r))$. This is true for all $r \in]0, R[$ and shows that f^{-1} belongs to $\mathcal{A}(d(a, R^-))$. The converse is obvious. □

Theorem 14.20. *Let* $R \in \mathbb{R}_+^*$. *The* $\mathbb{K}$*-subalgebra* $\mathcal{A}_b(d(0, R^-))$ *of* $\mathcal{A}(d(0, R^-))$ *is a Banach* $\mathbb{K}$*-algebra with respect to the norm* $\| \cdot \|_{d(0,R^-)}$. *Further, this norm is multiplicative and satisfies* $\|f\|_{d(0,R^-)} = \lim_{r \to R} |f|(r) = \sup_{n \in \mathbb{N}} |a_n| R^n$.

Let $f(x) = \sum_{n=0}^{\infty} a_n x^n \in \mathcal{A}(d(0, R^-))$. *Then,* f *is bounded in* $d(0, R^-)$ *if and only if so is the sequence* $(|a_n| R^n)_{n \in \mathbb{N}}$. *Moreover, if* f *is bounded, then* $\|f\|_{d(0,R^-)} = \sup_{n \in \mathbb{N}} |a_n| R^n$.

Proof. Let $f(x) = \sum_{n=0}^{\infty} a_n x^n \in \mathcal{A}_b(d(0, R^-))$. By Theorem 14.6, we have $\|f\|_{d(0,R^-)} = \sup_{n \in \mathbb{N}} |a_n| R^n$. The norm $\| \cdot \|_{d(0,R^-)}$ is a norm of $\mathbb{K}$-algebra, hence $\|f\ g\|_{d(0,R^-)} \leq \|f\|_{d(0,R^-)}\ \|g\|_{d(0,R^-)}$. On the other hand, by Theorem 14.6, the norm $\|\ .\ \|_{d(0,s)}$ is multiplicative on $H(d(0,s))$ for every $s < R$, hence $\|fg\|_{d(0,R^-)} \geq \|fg\|_{d(0,s)} = \|f\|_{d(0,s)}\ \|g\|_{d(0,s)}$ whenever $s < R$, and therefore, $\|\ .\ \|_{d(0,R^-)}$ is multiplicative on $\mathcal{A}_b(d(0,R^-))$. Now, let $(f_m)_{m\in\mathbb{N}}$ be a Cauchy sequence in $\mathcal{A}_b(d(0,R^-))$. We put $f_m(x) = \sum_{n=0}^{\infty} a_{n,m}x^n$. By hypothesis, for every $\epsilon > 0$, we have an integer $N(\epsilon)$ such that $|a_{n,m} - a_{n,q}|R^n \leq \epsilon$ for every $n \in \mathbb{N}$, whenever $m, q \geq N(\epsilon)$. Thus, it is easily seen that each sequence $(a_{n,m})_{m\in\mathbb{N}}$ converges in $\mathbb{K}$ to a limit a_n that satisfies $|a_n - a_{n,m}|R^n \leq \epsilon$ whenever $m \geq N(\epsilon)$ and then the series $f(x) = \sum_{n=0}^{\infty} a_n x^n$ satisfies $\|f - f_m\|_{d(0,R^-)} \leq \epsilon$. Obviously, f belongs to $H(d(0,s))$ for all $s < R$ and then the sequence (f_m) is proven to converge in $\mathcal{A}_b(d(0,R^-))$.

For every $s \in]0, R[$, we have

$$|f|(s) = \|f\|_{d(0,s)} = \sup_{n\in\mathbb{N}} |a_n|s^n \leq \sup_{n\in\mathbb{N}} |a_n|R^n = \|f\|_{d(0,R^-)},$$

and hence we can check that the real increasing bounded function h defined in $]0, R[$ as $h(s) = \sup_{n\in\mathbb{N}} |a_n|s^n$ is obviously continuous at R. Consequently, $\|f\|_{d(0,R^-)} = \lim_{r\to R} |f|(r) = \sup_n |a_n|R^n$. Therefore, obviously, f is bounded in $d(0,R^-)$ if and only if so is the sequence $(|a_n|R^n)_{n\in\mathbb{N}}$. □

Corollary 14.21. *Let $R \in \mathbb{R}_+^*$ and let $f,\ g \in \mathcal{A}(d(a,R^-))$. Then fg belongs to $\mathcal{A}_b(d(a,R^-))$ if and only if so do both f, g and $\mathcal{A}_b(d(a,R^-))$ is $\mathbb{K}$-subalgebra of $\mathcal{A}(d(a,R^-))$.*

Theorem 14.22. *Suppose that $\mathbb{K}$ has characteristic different from 2. Let $f,\ g \in \mathcal{A}(\mathbb{K})\setminus\mathbb{K}$ (respectively $f,\ g \in \mathcal{A}_u(d(0,r^-))$) be distinct. Then, $f^2 - g^2$ belongs to $\mathcal{A}(\mathbb{K})\setminus\mathbb{K}$ (respectively $f^2 - g^2 \in \mathcal{A}_u(d(0,r^-))$).*

Proof. Indeed, $f^2 - g^2 = (f-g)(f+g)$. Suppose that $|f+g|(r)$ is bounded when r tends to $+\infty$ (respectively to R). Since the characteristic of $\mathbb{K}$ is different from 2, $|f-g|(r)$ is obviously unbounded when r tends to $+\infty$ (respectively to R). Consequently, since the norm $|\ .\ |(r)$ is multiplicative, $|f^2 - g^2|(r)$ cannot be bounded when r tends to $+\infty$ (respectively to R). Therefore, $f^2 - g^2$ belongs to $\mathcal{A}(\mathbb{K})\setminus\mathbb{K}$ (respectively to $\mathcal{A}_u(d(0,r^-))$). Similarly, if $|f-g|(r)$ is bounded when r tends to $+\infty$ (respectively to R), we have the symmetric proof. □

Theorem 14.23. *For every $r \in \mathbb{R}_+^*$, $H(d(0,r^-))$ is included in $\mathcal{A}_b(d(0,r^-))$.*

Proof. Since $\mathcal{A}_b(d(0,r^-))$ is complete with respect to the norm $\| \cdot \|_{d(0,r^-)}$, we just have to show that $R(d(0,r^-)) \subset \mathcal{A}_b(d(0,r^-))$, hence finally, we just have to show that given $\alpha \in \mathbb{K}\setminus d(0,r^-)$, $\frac{1}{x-\alpha} \in \mathcal{A}_b(d(0,r^-))$. But we have $\frac{1}{x-\alpha} = -\frac{1}{\alpha(1-\frac{x}{\alpha})} = -\frac{1}{\alpha}\sum_{n=0}^{\infty}\left(\frac{x}{\alpha}\right)^n$ for all $x \in d(0,r^-)$ because $\left|\frac{x}{\alpha}\right| < 1$, hence $\frac{1}{x-\alpha} \in \mathcal{A}_b(d(0,r^-))$ and that finishes proving Theorem 14.23. □

Remarks. We see later that $H(d(0,r^-))$ is much smaller than $\mathcal{A}_b(d(0,r^-))$. In particular, we see that $\sqrt[q]{1+x}$ belongs to $\mathcal{A}_b(d(0,1^-)$ but does not belong to $H(d(0,1^-))$.

Let $\sum_0^{+\infty} a_n x^n$ be a power series whose radius of convergence is r. Suppose first that $r \in |\mathbb{K}|$. If there is at least one point $\alpha \in C(0,r)$ such that the series converges at α, then this implies that $\lim_{n\to+\infty} |a_n| r^n = 0$ and hence the series converges in all $C(0,r)$ and defines an element of $H(d(0,r))$. If r does not belong to $|\mathbb{K}|$, the power series converging in $d(0,r)$ are just the power series converging in $d(0,r^-)$. This is why we don't have to consider *analytic functions* inside a disk $d(a,r)$.

Theorem 14.24. *Let $f(x) = \sum_{n=0}^{\infty} a_n x^n \in \mathcal{A}(d(0,R^-))$ and suppose that $a_n \in \mathbb{Q}_p$, $\forall n \in \mathbb{N}$. Then, for every $x \in d(0,R^-)$, if x is algebraic over $\mathbb{Q}_p$, so is $f(x)$.*

Proof. Suppose x is algebraic, of degree q over $\mathbb{Q}_p$ and let $\mathbb{E} = \mathbb{Q}_p[x]$. For every $m \in \mathbb{N}$, $\sum_{n=0}^{m} a_n x^n$ belongs to $\mathbb{E}$. But since $\mathbb{E}$ is a finite extension of $\mathbb{Q}_p$, it is complete, hence $f(x)$ also belongs to $\mathbb{E}$. □

Chapter 15

Krasner Mittag-Leffler Theorem

The wonderful Mittag-Leffler theorem for analytic elements is due to Marc Krasner who showed it on quasi-connected sets [69]. The same proof holds on infraconnected sets as it was shown by Philippe Robba [83]. The theorem shows that a Banach space $H_b(D)$ is a direct topological sum of elementary subspaces and is indispensable to have a clear image of the space $H(D)$. Further, it appears necessary when studying meromorphic functions as we see later.

Throughout this chapter, D is supposed to be infraconnected.

We remember that if D is unbounded, $H_0(D)$ denotes the set of the $f \in H(D)$ such that $\lim_{\substack{|x|\to+\infty \\ x\in D}} f(x) = 0$.

Theorem 15.1 ((Krasner) [69, 83]). *Let D be closed and bounded (respectively unbounded) and let $f \in H_b(D)$. There exists a unique sequence of holes $(T_n)_{n\in\mathbb{N}^*}$ of D and a unique sequence $(f_n)_{n\in\mathbb{N}}$ in $H(D)$ such that $f_0 \in H(\widetilde{D})$ (respectively $f_0 \in \mathbb{K}$), $f_n \in H_0(\mathbb{K}\setminus T_n)$ $(n > 0)$, $\lim_{n\to\infty} f_n = 0$ satisfying further:*

(1) $f = \sum_{n=0}^{\infty} f_n$ *and* $\|f\|_D = \sup_{n\in\mathbb{N}} \|f_n\|_D$.

For every hole $T_n = d(a_n, r_n^-)$, we have

(2) $\|f_n\|_D = \|f_n\|_{\mathbb{K}\setminus T_n} = {}_D\varphi_{a_n,r_n}(f_n) \le {}_D\varphi_{a_n,r_n}(f) \le \|f\|_D$.

If D is bounded and if $\widetilde{D} = d(a, r)$, we have

(3) $\|f_0\|_D = \|f_0\|_{\widetilde{D}} = {}_D\varphi_{a,r}(f_0) \le {}_D\varphi_{a,r}(f) \le \|f\|_D$.

If D is not bounded, then $|f_0| = \lim_{\substack{|x|\to\infty \\ x\in D}} |f(x)| \le \|f\|_D$.

Let $D' = \widetilde{D}\setminus(\bigcup_{n=1}^{\infty} T_n)$. Then, f belongs to $H(D')$ (respectively $H_b(D')$) and its decomposition in $H(D')$ is given again by (1) *and f satisfies $\|f\|_{D'} = \|f\|_D$.*

Proof. Since $f \in H_b(D)$, by Corollary 11.7, we know that $f \in H(\overline{D})$. Hence, without loss of generality, we may assume that D is closed. Obviously, we may also assume $0 \in \widetilde{D}$.

First, we suppose $f \in R(D)$. Then, f has decomposition in the form $E(x) + \sum_{j=1}^{t} \frac{\lambda_j}{(x-\alpha_j)^{q_j}}$ with $E(x) \in \mathbb{K}[x]$ and $\alpha_j \in \mathbb{K}\setminus D$. Now, for each j, either α_j belongs to a hole T or α_j belongs to $\mathbb{K}\setminus\widetilde{D}$. Let $T_1, \ldots, T_s$ be the holes that contain some α_j. Then, $\sum_{j=1}^{t} \frac{\lambda_j}{(x-\alpha_j)^{q_j}}$ is of the form $\sum_{n=1}^{s} f_n + h_0$ with $f_i \in H_0(\mathbb{K}\setminus T_i)$ and $h_0 \in H_b(\widetilde{D})$. Finally, we put $f_0 = E(x) + h_0$ and we have the announced decomposition: $f = \sum_{n=0}^{s} f_i$ with $f_i \in H_0(\mathbb{K}\setminus T_i)$ and $f_0 \in H_b(\widetilde{D})$. In the case when D is unbounded, f_0 is just a constant.

For each $i = 1, \ldots, s$, $f - f_i$ clearly belongs to $H_b(D \cup T_i)$ and obviously f belongs to $H_b\big(\widetilde{D}\setminus\big(\bigcup_{i=1}^{s} T_i\big)\big)$.

First, we show that for any $n \in \mathbb{N}^*$, we have $\|f_n\|_D = \|f_n\|_{\mathbb{K}\setminus T_n}$. Let $\mathcal{F}_n$ be the circular filter on $\mathbb{K}$ of center α_n and diameter r_n. By Theorem 14.6, we have

(4) $\|f_n\|_{\mathbb{K}\setminus T_n} = \lim_{\mathcal{F}_n \cap (\mathbb{K}\setminus T_n)} |f_n(x)|$.

But by Proposition 3.17, $\mathcal{F}_n$ is secant with D, hence

(5) $\lim_{\mathcal{F}_n \cap (\mathbb{K}\setminus T_n)} |f_n(x)| = \lim_{\mathcal{F}_n \cap D} |f_n(x)|$

and obviously,

(6) $\lim_{\mathcal{F}_n \cap D} |f_n(x)| \le \|f_n\|_D \le \|f_n\|_{\mathbb{K}\setminus T_n}$.

Finally, by (4), (5), and (6), we obtain

(7) $\|f_n\|_{\mathbb{K}\setminus T_n} = \|f_n\|_D = {}_D\varphi_{\mathcal{F}_n}(f_n)$.

In the same way, when D is bounded, say $\widetilde{D} = d(0, r)$, we consider the circular filter $\mathcal{F}_0$ of center 0 and diameter r, in order to prove that

(8) $\|f_0\|_D = \|f_0\|_{\widetilde{D}} = {}_D\varphi_{\mathcal{F}_0}(f)$.

Now, let us show that $\|f\|_D \geq \|f_n\|_D$ for any $n \in \mathbb{N}^*$. Since $f \in R(D)$, there exists an annulus $\Gamma(a_n, r_n, r'_n)$ such that f has neither any zero nor any pole inside $\Gamma(a_n, r_n, r'_n)$. We put $I =]\log(r_n), \log(r'_n)[$. By hypothesis, f_n has no pole in $\mathbb{K}\setminus d(0, r_n)$. Hence, since $\lim_{|x|\to\infty} f_n(x) = 0$, by Corollary 3.17, we see that $\frac{d\Psi_{a_n}}{d\mu}(f_n, \mu) < 0$ whenever $\mu \in I$. Let $g_n = f - f_n \in R(D \cup T_n)$. Since g_n has no pole inside T_n, by Corollary 4.17, we see that $\frac{d_r\Psi_{a_n}}{d\mu}(g_n, \mu) \geq 0$ whenever $\mu < \log(r_n)$.

Therefore, the equation $\Psi_{a_n}(f_n, \mu) = \Psi_{a_n}(g_n, \mu)$ has at most one solution in I and then $\Psi_{a_n}(f, \mu)$ is equal to $\max(\Psi_{a_n}(f_n, \mu), \Psi_{a_n}(g_n, \mu))$ whenever $\mu \in I$, hence $\Psi_{a_n}(f, \mu) \geq \Psi_{a_n}(f_n, \mu)$ whenever $\mu \in I$. It follows that the multiplicative semi-norm $\varphi_{\mathcal{F}_n}$ defined on $R(D)$ satisfies $\log(\varphi_{\mathcal{F}_n}(f_n)) = \Psi_{a_n}(f_n, \log(r_n)) \leq \Psi_{a_n}(f, \log(r_n)) = \log(\varphi_{\mathcal{F}_n})(f)$, hence

(9) $\varphi_{\mathcal{F}_n}(f_n) \leq \varphi_{\mathcal{F}_n}(f)$.

But $\varphi_{\mathcal{F}_n}(f_n) = \|f_n\|_D$ and $\varphi_{\mathcal{F}_n}(f) \leq \|f\|_D$, hence by (9), we have $\|f\|_D \geq \|f_n\|_D$. Finally, by (7), we see that (2) is clearly proven.

When D is bounded, we put $\widetilde{D} = d(0, r)$ and we prove (3) in the same way as above when proving (2) by considering an annulus $\Gamma(0, r, r')$ such that f has neither any zero nor any pole inside $\Gamma(0, r, r')$. Then the element $g_0 = f - f_0$ is of the form $\frac{P}{Q}$ with $P, Q \in \mathbb{K}[x]$, all the zeros of Q in $\widetilde{D}$ and $\deg(P) < \deg(Q)$ because $g_0 \in R_0(\mathbb{K}\setminus\widetilde{D})$. Hence, we have $\frac{d\ \Psi}{d\mu}(g_0, \mu) < 0$, while $\frac{d\Psi}{d\mu}(f_0, \mu) \geq 0$ whenever $\mu \in]\log r, \log r'[$, so we have ${}_D\varphi_{\mathcal{F}_0}(f_0) \leq {}_D\varphi_{\mathcal{F}_0}(f)$, and hence, by (8), we obtain (3).

When D is not bounded, the inequality $|f_0| \leq \|f\|_D$ is obvious because $\lim_{\substack{|x|\to\infty \\ x\in D}} f(x) - f_0 = 0$. This finishes proving the Mittag-Leffler theorem when $f \in R(D)$.

Now, let $f \in H(D)$ and let $(h_m)_{m\in\mathbb{N}}$ be a sequence in $R(D)$ that converges to f in $H(D)$. The set of holes of D that contain at least one pole of some h_m is clearly countable. Hence, there exists a sequence of holes $(T_n)_{n\in\mathbb{N}^*}$ such that, denoting by D' the set $\widetilde{D}\setminus\big(\bigcup_{n\in\mathbb{N}^*}^{\infty} T_n\big)$, then h_m belongs to $H(D')$ whenever $m \in \mathbb{N}$. For each $m \in \mathbb{N}$, h_m splits in $H(D')$ in the form $h_m = \sum_{n=0}^{\infty} h_{m,n}$ with $h_{m,0} \in H(\widetilde{D}), h_{m,n} \in H_0(\mathbb{K}\setminus T_n)$. In particular, for each fixed $n \in \mathbb{N}$, we have $\|h_{m,n} - h_{q,n}\|_D \le \|h_m - h_q\|_D$. Thus, we see that the sequence $(h_{m,n})_{m\in\mathbb{N}}$ converges in $H(\mathbb{K}\setminus T_n)$ for $n > 0$ (respectively in $H(\widetilde{D})$ for $n = 0$) to a limit f_n and then we have $f = \sum_{n=0}^{\infty} f_n$ in $H(D')$. Obviously, $\|f\|_D = \sup_{n\in\mathbb{N}} \|f_n\|_D$, whereas $\|f_n\|_D = \|f_n\|_{D'}$ whenever $n \in \mathbb{N}$ and so, $\|f\|_D = \|f\|_{D'}$. This ends the proof of Theorem 15.1. □

Corollary 15.2. *Let $(T_i)_{i\in I}$ be the family of holes of D. Let J be a subset of I and let $S = I\setminus J$. Let $E = D\bigcup\big(\bigcup_{i\in J} T_i\big)$ and let $F = D\bigcup\big(\bigcup_{i\in S} T_i\big)$. Then we have $H(D) = H_0(E) \oplus H(F)$ and for each $g \in H_0(E)$, $h \in H(F)$, we have $\|g + h\|_D = \max(\|g\|_E, \|h\|_F)$.*

The Mittag-Leffler theorem suggests some new definitions.

Definition and notation. Let $f \in H_b(D)$. We consider the series $\sum_{n=0}^{\infty} f_n$ obtained in Theorem 15.1, whose sum is equal to f in $H(D)$, with $f_0 \in H(\widetilde{D}), f_n \in H(\mathbb{K}\setminus T_n)\setminus\{0\}$ and with the T_n holes of D. Each T_n is called an *f-hole* and f_n is called the *Mittag-Leffler term of f associated with T_n*, whereas f_0 is called the *principal term* of f. For each f-hole T of D, the Mittag-Leffler term of f associated with T is denoted by $\overline{\overline{f_T}}$, whereas the principal term of f is denoted by $\overline{\overline{f_0}}$. The series $\sum_{n=0}^{\infty} f_n$ is called *the Mittag-Leffler series of f on the infraconnected set D*. More generally, let E be an infraconnected set and let $f \in H(E)$. According to Theorem 11.5, f is of the form $g + h$ with $g \in R(\mathbb{K}\setminus(\overline{E}\setminus E))$ and $h \in H_b(\overline{E})$ whereas such a decomposition is unique, up to an additive constant. For every hole T of $\overline{E}$, we denote by $\overline{\overline{f_T}}$ the Mittag-Leffler term of h associated with T and $\overline{\overline{f_T}}$ is still named *the Mittag-Leffler term of f associated with T*.

Corollary 15.3. *Let $f \in H_b(D)$, let $(T_n)_{n\in\mathbb{N}^*}$ be the sequence of the f-holes, with $T_n = d(a_n, \rho_n^-)$, and let $f_0 = \overline{\overline{f_0}}$ and $f_n = \overline{\overline{f_{T_n}}}$ for every $n \in \mathbb{N}^*$. Let $\widetilde{D} = d(a, s)$ (respectively $\widetilde{D} = \mathbb{K}$). There exists $q \in \mathbb{N}$ such that $\|f\|_D = \|f_q\|_D$. If $q \ge 1$, then $\|f\|_D = {}_D\varphi_{a_q,r_q}(f) = {}_D\varphi_{a_q,r_q}(f_q)$. If $q = 0$ and if D is bounded (respectively is not bounded),* *then* $\|f\|_D = {}_D\varphi_{a,s}(f) = {}_D\varphi_{a,s}(f_0)$ *(respectively* $\|f\|_D = |f_0|$*). Further, given a hole* T *of* D*, if* f *belongs to* $H_b(D)$ *and if* g *belongs to* $H_0(\mathbb{K}\backslash T)$ *and satisfies* $f - g \in H(D \cup T)$*, then* $\overline{\overline{f_T}}$ *is equal to* g*.*

Definition. Let D be bounded, of center α and diameter r. A circular filter $\mathcal{F}$ on D is called *the specific filter of a hole* $T = d(a, r^-)$ if it is the circular filter on D of diameter r. If D is bounded and $\widetilde{D} = d(a, R)$, the circular filter of center a and diameter R is called *the specific filter of* $\widetilde{D}$. In general, a specific filter of a hole of D or of $\widetilde{D}$ is called *a specific filter of* D.

Corollary 15.4. *Let* $f \in H_b(D)$*. There exists a large circular filter* $\mathcal{F}$ *with center* $\alpha \in \widetilde{D}$ *secant with* D *such that* ${}_D\varphi_{\mathcal{F}}(f) = \|f\|_D$*. If* D *is bounded, there exists a specific filter* $\mathcal{F}$ *of* D *such that* ${}_D\varphi_{\mathcal{F}}(f) = \|f\|_D$*.*

Corollary 15.5. *Let* $f \in H(d(0, 1^-))$ *and let* $(d(\alpha_m, 1^-))_{m\in\mathbb{N}^*}$ *be the family of the f-holes. Then f is of the form*

$$(1) \quad \sum_{n=0}^{\infty} a_{n,0}x^n + \sum_{m,n\in\mathbb{N}^*} \frac{a_{n,m}}{(x-\alpha_m)^n}$$

with $\lim_{n\to\infty} a_n = 0, \lim_{n\to\infty} |a_{n,m}| = 0$ *whenever* $m \in \mathbb{N}^*$ *and* $\lim_{m\to\infty}(\sup_{n\in\mathbb{N}^*} |a_{n,m}|) = 0$*. On the other hand, f satisfies* $\|f\|_{d(0,1^-)} = \sup_{m,n\in\mathbb{N}^*} |a_{n,m}|$*.*

Conversely, every function of the form (1), with the α_m *satisfying* $|\alpha_m| = |\alpha_j - \alpha_m| = 1$ *whenever* $m \neq j$*, belongs to* $H(d(0, 1^-))$*. The norm* $\| \cdot \|_{d(0,1^-)}$ *is multiplicative and equal to* ${}_{d(0,1^-)}\varphi_{0,1}$*.*

Corollary 15.6. *Let* $r_1, r_2 \in \mathbb{R}_+$ *satisfy* $0 < r_1 < r_2$*. Then,* $H(\Delta(0, r_1, r_2))$ *is equal to the set of the Laurent series* $\sum_{-\infty}^{+\infty} a_n x^n$ *with* $\lim_{n\to-\infty} |a_n| r_1^n = \lim_{n\to\infty} |a_n| r_2^n = 0$ *and we have* $\|\sum_{-\infty}^{+\infty} a_n x^n\|_{\Delta(0,r_1,r_2)} = \max\left(\sup_{n\geq 0} |a_n| r_1^n,\ \sup_{n<0} |a_n| r_2^n\right)$*.*

Proof. On Theorem 14.6, we saw that $H(\Delta(0, r_1, r_2))$ is equal to the set of the Laurent series $\sum_{-\infty}^{+\infty} a_n x^n$ with $\lim_{n\to-\infty} |a_n| r_1^n = \lim_{n\to\infty} |a_n| r_2^n = 0$. Then, the conclusions on the norm come from Theorem 15.1. □

Theorem 15.7. *Let* $r \in \mathbb{R}_+$*. Then,* $H(C(0, r))$ *is equal to the set of the Laurent series* $\sum_{-\infty}^{+\infty} a_n x^n$ *with* $\lim_{|n|_\infty\to\infty} |a_n| r^n = 0$ *and we have*

$\|\sum a_n x^n\|_{C(0,r)} = \sup_{n\in\mathbb{Z}} |a_n| r^n$. *Next, the norm* $\|\ .\ \|_{C(0,r)}$ *is multiplicative and equal to* ${}_{C(0,r)}\varphi_{0,r}$.

Proof. We put $\Lambda = C(0,r)$. We may apply Theorem 14.6 by taking $r_1 = r_2 = r$ and we obtain all the conclusions but the fact that $\|\ .\ \|_\Lambda$ is multiplicative. Let us show this. Let $h \in R(\Lambda)$. Hence, h is of the form $\frac{P}{Q}$, with $P, Q \in \mathbb{K}[x]$ and $Q(x)$ has no zero in Λ. Let Θ be a class of Λ. By Lemma 4.9, we have $|Q(x)| = \varphi_{0,r}(Q)$ whenever $x \in \Theta$ and $|P(x)| \le \varphi_{0,r}(P)$ whenever $x \in \Theta$. Hence, we see that $\|\frac{P}{Q}\|_\Lambda \le \varphi_{0,r}(\frac{P}{Q})$, and therefore, $\|h\|_\Lambda = \varphi_{0,r}(h)$ whenever $h \in R(\Lambda)$. Consequently, we have $\|f\|_\Lambda = {}_\Lambda\varphi_{0,r}(f)$ whenever $f \in H(\Lambda)$. □

Proposition 15.8. *Let* $r_1, r_2 \in \mathbb{R}_+^*$, *with* $r_1 < r_2$*:*

(i) $\mathcal{A}(\Gamma(0,r_1,r_2)) = A(d(0,r_2^-)) \oplus \mathcal{A}_0(\mathbb{K}\setminus d(0,r_1))$ *and* $\mathcal{A}_b(\Gamma(0,r_1,r_2)) = \mathcal{A}_b(d(0,r_2^-)) \oplus \mathcal{A}_{0,b}(\mathbb{K}\setminus d(0,r_1))$.

(ii) *Let* $f(x) = \sum_{-\infty}^{+\infty} a_n x^n \in \mathcal{A}(\Gamma(0,r_1,r_2))$. *Then,* $f \in \mathcal{A}_b(\Gamma(0,r_1,r_2))$ *if and only if* $\max\left(\sup_{n\ge 0}|a_n|r_2^n, \sup_{n<0}|a_n|r_1^n\right) < +\infty$. *Moreover, if* $f \in \mathcal{A}_b(\Gamma(0,r_1,r_2))$, *then*

$$\|f\|_{\Gamma(0,r_1,r_2)} = \max\left(\sup_{n\ge 0}|a_n|r_2^n, \sup_{n<0}|a_n|r_1^n\right).$$

(iii) $\mathcal{A}_b(\Gamma(0,r_1,r_2))$ *is a Banach* $\mathbb{K}$*-algebra that contains* $H(\Gamma(0,r_1,r_2))$.

Proof. (i) is obvious. We show (ii). Let $f \in \mathcal{A}_b(\Gamma(0,r_1,r_2))$ and let $f = f_1 + f_2$ with $f_2 \in \mathcal{A}_b(d(0,r_2^-))$ and $f_1 \in \mathcal{A}_{0,b}(\mathbb{K}\setminus d(0,r_1))$. We put $\Lambda = \Gamma(0,r_1,r_2)$. It is obviously seen that $\|f\|_\Lambda \le \max\left(\|f_1\|_\Lambda, \|f_2\|_\Lambda\right) \le \max\left(\sup_{n\ge 0}|a_n|r_2^n, \sup_{n<0}|a_n|r_1^n\right)$. Now, for every s_1, s_2 such that $r_1 < s_1 < s_2 < r_2$, we know that f belongs to $H(\Delta(0,s_1,s_2))$ because so do both f_1, f_2. Then, by Theorem 15.1, we have

$$\|f\|_{\Delta(0,s_1,s_2)} = \max\left(\|f_2\|_{d(0,s_2)}, \|f_1\|_{\mathbb{K}\setminus d(0,s_1^-)}\right).$$

Finally, $\|f_2\|_{d(0,s_2)} = \sup_{n\in\mathbb{N}}|a_n|s_2^n$, while $\|f_1\|_{\mathbb{K}\setminus d(0,s_1^-)} = \sup_{n<0}|a_n|s_1^n$. Thus, we see that $\|f\|_{\Gamma(0,r_1,r_2)} \ge \|f\|_{\Delta(0,s_1,s_2)} = \max\left(\sup_{n\ge 0}|a_n|s_2^n, \sup_{n<0}|a_n|s_1^n\right)$. This is true for every $s_1, s_2 \in]r_1, r_2[$, hence finally, $\|f\|_\Lambda = \max\left(\sup_{n\ge 0}|a_n|r_2^n, \sup_{n<0}|a_n|r_1^n\right)$. All statements in (ii) are then proven.

We now prove (iii). By (ii), $\mathcal{A}_b(\Lambda)$ is just the Banach $\mathbb{K}$-algebra

$$\mathcal{A}_b(d(0, r_2^-)) \oplus \mathcal{A}_{0,b}(\mathbb{K}\backslash d(0, r_1))$$

provided with the norm $\|f_1+f_2\|_\Lambda = \max\left(\|f_2\|_{d(0,r_2^-)}, \|f_1\|_{\mathbb{K}\backslash d(0,r_1)}\right)$. We saw that $R(d(0, r_2^-)) \subset \mathcal{A}_b(d(0, r_2^-))$, hence $R(d(0, r_2^-)) \subset \mathcal{A}_b(\Lambda)$.

In the same way, we have $R(\mathbb{K}\backslash d(0, r_1)) \subset \mathcal{A}_b(\mathbb{K}\backslash d(0, r_1))$ and then $\mathcal{A}_b(\mathbb{K}\backslash d(0, r_1))$ is obviously included in $\mathcal{A}_b(\Lambda)$. Since $R(\Lambda) = R(d(0, r_2^-)) + R(\mathbb{K}\backslash d(0, r_1))$, $R(\Lambda)$ is included in $\mathcal{A}_b(\Lambda)$ which is complete for the norm $\| \,.\, \|_\Lambda$, hence $H(\Lambda) \subset \mathcal{A}_b(\Lambda)$. This finishes proving Proposition 15.8. □

Notation. Given a subset A of $\widehat{\mathbb{K}}$, we denote by $\widehat{H}(A)$ the set of the analytic elements on A, taking $\widehat{\mathbb{K}}$ as a ground field.
E^*

Now, we apply the Mittag-Leffler theorem to the analytic extension of analytic elements.

Theorem 15.9. *For all $f \in H(D)$, f has continuation to a unique element $\widehat{f} \in \widehat{H}(\widehat{D})$. Further, if $f \in H_b(D)$, the Mittag-Leffler series of $\widehat{f}$ in $\widehat{D}$ is the same as this of f in D.*

Proof. By Theorem 11.5, we may easily assume that f belongs to $H_b(\overline{D})$. The Mittag-Leffler series of f on D obviously converges on $\widehat{D}$, to an element of $\widehat{H}(\widehat{D})$. This is unique because so is the Mittag-Leffler series of f on D. □

Theorem 15.10. *Let E be an infraconnected set such that $D \cap E$ is infraconnected and such that every hole of $D \cap E$ is either a hole of D or a hole of E. Let $F \in H(D)$, $G \in H(E)$, satisfying $F(x) = G(x)$ whenever $x \in D \cap E$. Then there exists $h \in H(D \cup E)$ such that $h(x) = F(x)$ whenever $x \in D$, $h(x) = G(x)$ whenever $x \in E$, such that for every h-hole V of $D \cup E$, $\overline{\overline{h_V}}$ is either of the form $\overline{\overline{F_S}}$, when V is a F-hole S of D, or of the form $\overline{\overline{G_T}}$ when V is a G-hole T of E.*

Proof. By Theorem 11.5, it is easily seen that we may assume $F \in H_b(D)$, $G \in H_b(E)$ without loss of generality. Let $A = D \cup E$, $B = D \cap E$. Let h be the restriction of F and G to B. Let $(V_n)_{n\in\mathbb{N}^*}$ be the sequence of h-holes that are holes of D and let $(W_n)_{n\in\mathbb{N}^*}$ be the sequence of h-holes that are holes of E but not of D. For each $q \in \mathbb{N}^*$, as $h(x)$ is equal to $F(x)$ in B, $\overline{\overline{h_{W_q}}}$ is an element of $H_0(D)$ of the form $\sum_{m=1}^{\infty} \overline{\overline{F_{S_m^q}}}$ with S_m^q some F-holes of D included in W_q. We put $f_{q,m} = \overline{\overline{F_{S_m^q}}}$ for every $(q, m) \in (\mathbb{N}^{*2})$. In the same way, for each $q \in \mathbb{N}^*$, $\overline{\overline{h_{V_q}}}$ is an element of $H_0(E)$ of the form $\sum_{m=1}^{\infty} \overline{\overline{G_{T_m^q}}}$ with T_m^q the G-holes of E included in V_q. We put $g_{q,m} = \overline{\overline{G_{T_m^q}}}$ for every $(q, m) \in (\mathbb{N}^{*2})$. Without loss of generality, we may obviously assume $\widetilde{D} \subset \widetilde{E}$; we put $h_0(x) = \overline{\overline{G_0(x)}}$. We note that A is clearly included in the set $A' = \widetilde{E} \setminus \big((\bigcup_{(q,m)\in(\mathbb{N}^{*2})} S_m^q) \bigcup (\bigcup_{(q,m)\in(\mathbb{N}^{*2})} T_m^q) \big)$. Then, it is easily seen that the series $h_0(x) + \sum_{(m,q)\in\mathbb{N}^{*2}} f_{q,m} + \sum_{(m,q)\in\mathbb{N}^{*2}} g_{q,m}$ converges in $H(A')$ because by Corollary 15.2 we have $\|f_{q,m}\|_{A'} = \|\overline{\overline{F_{S_m^q}}}\|_D$ and $\|g_{q,m}\|_{A'} = \|\overline{\overline{G_{T_m^q}}}\|_E$, whereas $\lim_{q+m\to+\infty} \|\overline{\overline{F_{S_m^q}}}\|_E = \lim_{q+m\to+\infty} \|\overline{\overline{G_{T_m^q}}}\|_E = 0$. Further, by construction, $h(x)$ is equal to $F(x)$ and $G(x)$ in B and is such that for every h-hole V of $D \cup E$, $\overline{\overline{h_V}}$ is either of the form $\overline{\overline{F_S}}$, when V is a F-hole S of D, or of the form $\overline{\overline{G_T}}$ when V is a G-hole T of E. This clearly ends the proof of Theorem 15.10. □

Corollary 15.11. *Let E be an infraconnected set such that $D \subset E$ and such that each hole of D contains a unique hole of E. Let $f \in H(E)$ and let $f = \overline{\overline{f_0}} + \sum_{n=1}^{+\infty} \overline{\overline{f_{T_n}}}$ be the Mittag-Leffler series of f on the infraconnected E. For every $n \in \mathbb{N}^*$, let V_n be the hole of D containing T_n. Then the Mittag-Leffler series of f on the infraconnected D is of the form $\overline{\overline{f_0}} + \sum_{n=1}^{+\infty} \overline{\overline{f_{V_n}}}$ with $\overline{\overline{f_{V_n}}} = \overline{\overline{f_{T_n}}}$, $\forall n \in \mathbb{N}^*$.*

In the particular case of affinoid subsets, we can be more accurate for Theorem 15.10:

Theorem 15.12. *Let D_1, D_2 be infraconnected affinoid subsets of $\mathbb{K}$ such that $D_1 \cap D_2 \neq \emptyset$ and let $f_j \in H(D_j)$, $j = 1, 2$ be such that $f_1(x) = f_2(x)$, $\forall x \in D_1 \cap D_2$. Then the function f defined in $D_1 \cup D_2$ as $f(x) = f_j(x)$ $\forall x \in D_j$, $j = 1, 2$, belongs to $H(D_1 \cup D_2)$.*

Proof. Let $D = D_1 \cup D_2$. Without loss of generality, we can assume that $\widetilde{D_1}$ contains $\widetilde{D_2}$ and hence $\widetilde{D} = \widetilde{D_1}$. We can also assume that $0 \in D_1 \cap D_2$. Set $A = D_1 \setminus D_2$. Then A is included in a finite union of holes of D_1. Consider such a hole $T = d(a, r^-)$ of D_1 (with $r \in |\mathbb{K}|$) such that $T \cap D_2 \neq \emptyset$. Since $D_1 \cap D_2 \neq \emptyset$, both D_1, D_2 have points on $C(a, r)$. Moreover, since both are affinoid, D_1 contains all classes of $C(a, r)$ except maybe finitely many because T is a hole of D_1. On the other hand, D_2 also contains all classes of $C(a, r)$ except maybe finitely many because it has points on $C(a, r)$ and inside $d(a, r^-)$. Consequently, $f_1(x)$ and $f_2(x)$ coincide in all classes of $C(a, r)$ except maybe in finitely many $\Lambda_1, \dots, \Lambda_q$.

Let g be the Mittag-Leffler term of f_1 relative to T. Let S_k, $1 \leq k \leq t$, be the holes of D_2 included in T, and for each $k = 1, \dots, t$, let h_k be the Mittag-Leffler term of f_2 relative to the hole S_k. Consider now the restrictions $\underline{f_1}$ of f_1 and $\underline{f_2}$ of f_2 on the set $D' = D_1 \cap D_2$.

The two functions are equal in D' and of course have the same Mittag-Leffler term relative to the hole $d(a, r^-)$. Concerning h_2, this term is $\sum_{k=1}^{t} h_k$. Consequently, $g = \sum_{k=1}^{t} h_k$. Since g and the h_k are Laurent series converging in $\mathbb{K} \setminus d(a, r^-)$, g and $\sum_{k=1}^{t} h_k$ coincide in all this set. Consequently, in the Mittag-Leffler series of f_1, we can replace g by $\sum_{k=1}^{t} h_k$. Thus, f_1 becomes an element of $D_1 \cup \big((d(a, r^-) \cap D_2\big)$. We can do the same with each hole of D_1 containing points of D_2, and hence, after finitely many similar change, we obtain an element f of $H(D)$ such that $f(x) = f_j(x)\ \forall x \in D_j,\ j = 1, 2.$ $\square$

Notation. Let E be a $\mathbb{K}$-Banach space. We denote by E^{*} the $\mathbb{K}$-Banach space of continuous linear forms of E provided with its usual norm. The dual of a Banach space $H(D)$ was thoroughly studied by Yvette Amice [2].

Theorem 15.13 (Amice). *Let $r \in \mathbb{R}_+$. Given $h(t) = \sum_{n=0}^{\infty} \frac{b_n}{t^n} \in \mathcal{A}_b(\mathbb{K} \setminus d(0, r))$, there exists a unique $\phi_h \in H(d(0, r))^{*}$ satisfying $\phi_h(x^q) = b_q$, $(q \in \mathbb{N})$. Moreover, on the space $\mathcal{A}_b(\mathbb{K} \setminus d(0, r))$ provided with the norm $\| \cdot \|_{\mathbb{K} \setminus d(0,r)}$, the mapping $h \to \phi_h$ is an isometric isomorphism from $\mathcal{A}_b(\mathbb{K} \setminus d(0, r))$ onto $H(d(0, r))^{*}$.*

Proof. Let $F = \mathbb{K} \setminus d(0, r)$. First, let $h(t) = \sum_{n=0}^{\infty} \frac{b_n}{t^n} \in \mathcal{A}_b(F)$ and let $f(x) = \sum_{n=0}^{\infty} a_n x^n \in H(d(0, r))$. Since the sequence $\frac{|b_n|}{r^n}$ is bounded and $\lim_{n\to\infty} |a_n| r^n = 0$, it is seen that $\lim_{n\to\infty} a_n b_n = 0$ and then the series $\sum_{n=0}^{\infty} a_n b_n$ is convergent. Hence, we may put $\phi_h(f) = \sum_{n=0}^{\infty} a_n b_n$. Thus, we define a linear form ϕ_h of $H(d(0, r))$

that satisfies

$$|\phi_h(f)| \leq \sup_{n\in\mathbb{N}} |a_n b_n| \leq \left(\sup_{n\in\mathbb{N}} |a_n| r^n\right)\left(\sup_{n\in\mathbb{N}} \frac{|b_n|}{r^n}\right) = \|f\|_{d(0,r)}\|h\|_F.$$

Therefore, with respect to the norm $\| \ . \ \|$ of $H((d(0,r))^{\circledast}$, we have $\|\phi_h\| \leq \|h\|_F$. Now, we check that the equality is satisfied. Indeed, let $q \in \mathbb{N}$. We have

$$\frac{|\phi_h(x^q)|}{\|x^q\|_{d(0,r)}} = \frac{|\phi_h(x^q)|}{r^q} = \frac{|b_q|}{r^q} \leq \sup_{f\neq 0} \frac{|\phi_h(f)|}{\|f\|_{d(0,r)}}$$

for all $q \geq 0$. Hence, we have $\|h\|_F = \sup_{q\in\mathbb{N}} \frac{|b_q|}{r^q} \leq \|\phi_h\|$. So we obtain the announced equality. Thus, we have defined an isometric homomorphism from $\mathcal{A}_b(F)$ into $H(d(0,r))^{\circledast}$.

Now, we check that this mapping is surjective. Indeed, let $\psi \in H(d(0,r))^{\circledast}$, and for each $n \in \mathbb{N}$, let $b_n = \psi(x^n)$. Obviously, we have $\|\psi\| \geq \frac{|b_n|}{r^n}$ for every $n \in \mathbb{N}$, hence the sequence $(|b_n|r^{-n})_{n\in\mathbb{N}}$ is bounded and therefore defines a function $f(t) = \sum_{n=0}^{\infty} \frac{b_n}{t^n} \in \mathcal{A}_b(F)$. Thus, ψ is equal to ϕ_h, and therefore, the mapping $h \to \phi_h$ is surjective. This ends the proof of Theorem 15.13. □

Remark. There obviously exists an isometric homomorphism from $H(d(0,1))$ into $H(d(0,1))^{\circledast}$ defined as follows: let $f = \sum_{n=0}^{+\infty} a_n x^n \in H(d(0,1^-))$ and let $\underline{f}(x) = f(\frac{1}{x}) \in \mathcal{A}_b(\mathbb{K}\setminus d(0,1))$. Then we have an element $f^* \in H(d(0,1))^*$ equal to $\phi_{\underline{f}}$. The question of whether this homomorphism is surjective depends on the ground field $\mathbb{K}$. If $\mathbb{K}$ is spherically complete, this homomorphism is not surjective. If $\mathbb{K}$ is not spherically complete, this homomorphism is surjective [86].

Corollary 15.14. *Let $r \in \mathbb{R}_+$. For each $h(t) = \sum_{n=0}^{\infty} b_n t^n \in \mathcal{A}_b(d(0,r^-))$, there exists a unique $\phi_h \in H(\mathbb{K}\setminus d(0,r^-))^{\circledast}$ satisfying $\phi_h(x^{-q}) = b_q$ $(q \in \mathbb{N})$. Moreover, the space $\mathcal{A}_b(d(0,r^-))$ being provided with the norm $\| \ . \ \|_{d(0,r^-)}$, the mapping $h \to \phi_h$ is an isometric isomorphism from $\mathcal{A}_b(d(0,r^-))$ onto $H(\mathbb{K}\setminus d(0,r^-))^{\circledast}$.*

Corollary 15.15. *Let $r \in \mathbb{R}_+$. For each $h(t) = \sum_{n=0}^{\infty} b_n t^n \in \mathcal{A}_b(d(0,r^-))$ such that $h(0) = 0$, there exists a unique $\phi_h \in H_0(\mathbb{K}\setminus d(0,r^-))^{\circledast}$ satisfying $\phi_h(x^{-q}) = b_q$ $(q \in \mathbb{N}^*)$. Moreover, this mapping $h \to \phi_h$ from the subspace of the $h \in \mathcal{A}_b(d(0,r^-))$ such that $h(0) = 0$ into $H_0(\mathbb{K}\setminus d(0,r^-))^{\circledast}$ is an isometric isomorphism.*

Now, applying Theorem 15.13 to $H(\widetilde{D})$ and Corollary 15.15 to the spaces $H_0(\mathbb{K}\setminus T_i)$ for each hole T_i of an infraconnected set D, we obtain Corollary 15.16.

Corollary 15.16 (Amice). *Let D be closed bounded infraconnected. Let $(T_i)_{i\in J}$ be the family of its holes and for every $i \in J$, let $a_i \in T_i$. Let $M \in \mathbb{R}_+^*$. Let $h_0 \in \mathcal{A}_b(\mathbb{K}\setminus\widetilde{D})$ and let $(h_i)_{i\in J}$ be a family such that for each $i \in J$, h_i belongs to $\mathcal{A}_b(T_i)$ and satisfies*

(1) $h_i(a_i) = 0$

and

(2) $\|h_i\|_{T_i} \leq M$ *for all* $i \in J$.

There exists a unique $\psi \in H(D)^*$ satisfying

$\psi(f) = \phi_{h_0}(f)$ *for every* $f \in H(\widetilde{D})$,
$\psi(f) = \phi_{h_i}(f)$ *for every* $f \in H_0(\mathbb{K}\setminus T_i)$, *whenever* $i \in J$.

Further, for every element ψ of $H(D)^*$, there exists a unique $h_0 \in \mathcal{A}_b(\mathbb{K}\setminus d(0,r))$ and a unique family $(h_i)_{i\in J}$ satisfying (1) and (2) for some $M \in \mathbb{R}_+^*$ such that ψ is defined as above.

Now, we use the continuous linear forms to define the residue of an element on a hole.

Theorem 15.17. *Let $f \in H_b(\mathbb{K}\setminus d(a,r^-))$, and for each $\alpha \in d(a,r^-)$, let $f(x) = \sum_{n=0}^{\infty} \frac{b_n(\alpha)}{(x-\alpha)^n}$. Then, $b_1(\alpha)$ does not depend on α in $d(a,r^-)$.*

Proof. Let $E = \mathbb{K}\setminus d(a,r^-)$. We know that $\frac{|b_1(\alpha)|}{r} \leq \|f\|_E$, and therefore, fixing α in $d(a,r^-)$, the linear form ψ_α on $H_b(E)$ defined as $\psi_\alpha(f) = b_1(\alpha)$ is obviously continuous. We show that $\psi_\alpha(f) = \psi_a(f)$. First, for every $q \in \mathbb{N}$, we put $f_q(x) = \frac{1}{(x-\alpha)^q}$. We have $f_q(x) = \frac{1}{(x-a)^q}\left(\sum_{j=0}^{\infty}\left(\frac{\alpha-a}{x-a}\right)^j\right)^q$. Therefore, for every $q \geq 2$, we have $\psi_a(f_q) = 0$ and that $\psi_a(f_1) = 1$. Hence, $\psi_\alpha(f_q) = \psi_a(f_q)$ for every $q \in \mathbb{N}$. This shows that $\psi_\alpha(f) = \psi_a(f)$ for every $f \in H_b(E)$. □

Definition and notation. Let $f \in H_b(D)$, let T be a hole of D, and let $a \in T$. Let $f_T(x) = \sum_{n=1}^{\infty} \frac{b_n(a)}{(x-a)^n}$. By Theorem 15.17, $b_1(a)$

actually does not depend on a in T. We set $\text{res}(f, T) = b_1(a)$ and this number $\text{res}(f, T)$ is called *the residue of f on the hole T.*

By Theorem 15.1, Theorem 15.18 is obvious:

Theorem 15.18. *Let $f \in H(D)$ and let T be a hole of D of diameter r. Then,*

$$|\text{res}(f, T)| \le r\|\overline{\overline{f_T}}\|_{\mathbb{K}\setminus T} \le r\|f\|_D.$$

We can now characterize $\mathbb{K}$-algebra homomorphisms among the continuous linear forms.

Theorem 15.19. *Let D be a closed bounded infraconnected, let $a \in \widetilde{D}$, let $(T_i)_{i\in J}$ be the family of holes of D and for every $i \in J$, and let $a_i \in T_i$. Let $M \in \mathbb{R}_+^*$. Let $h_0 \in \mathcal{A}_b(\mathbb{K}\setminus\widetilde{D})$ and let $(h_i)_{i\in J}$ be a family such that for each $i \in J$, h_i belong to $\mathcal{A}_b(T_i)$ and satisfy Conditions (1) and (2):*

(1) $h_i(a_i) = 0$

and

(2) $\|h_i\|_{T_i} \le M$ *for all* $i \in J$.

Let $\psi \in H(D)^*$ satisfy

$\psi(f) = \phi_{h_0}(f)$ *for every* $f \in H(\widetilde{D})$,
$\psi(f) = \phi_{h_i}(f)$ *for every* $f \in H_0(\mathbb{K}\setminus T_i)$, *whenever* $i \in J$.

Then, ψ is a homomorphism of $\mathbb{K}$-algebra from $H(D)$ onto $\mathbb{K}$ if and only if there exists $\alpha \in D$ such that

(3) $h_0(t) = \frac{t-a}{t-\alpha}$,

and for every $i \in J$,

(4) $h_i(t) = \frac{t-a_i}{\alpha-t}$.

Moreover, every $\mathbb{K}$-algebra homomorphism from $H(D)$ to $\mathbb{K}$ is continuous and is of this form.

Proof. First, we suppose that ψ is a $\mathbb{K}$-algebra homomorphism from $H(D)$ onto $\mathbb{K}$ and we put $\psi(x) = a$. As h_0 is of the form

$h_0(t) = \sum_{n=0}^{\infty} \frac{b_n}{(t-a)^n}$, here for every $n \in \mathbb{N}$, we have $b_n = \psi((x-a)^n) = (\alpha - a)^n$, and therefore, $h_0(t) = \sum_{n=0}^{\infty} \left(\frac{\alpha-a}{t-a}\right)^n = \frac{t-a}{t-\alpha}$. Next, we fix $i \in J$. Then, h_i is of the form $h_i(t) = \sum_{n=1}^{\infty} b_{i,n}(t-a_i)^n$. Hence, for every $n \in \mathbb{N}^*$, we have $b_{i,n} = \psi\left(\frac{1}{(x-a_i)^n}\right) = \left(\frac{1}{\alpha-a_i}\right)^n$, and therefore, $h_i(t) = \sum_{n=1}^{\infty} \left(\frac{t-a_i}{\alpha-a_i}\right)^n = \frac{t-a_i}{\alpha-t}$.

Conversely, we suppose (3) and (4) are satisfied. Then, it is easily checked that $h_0(t) = \sum_{n=0}^{\infty} \left(\frac{\alpha-a}{t-a}\right)^n$, and therefore, for every $n \in \mathbb{N}$, we have $\psi((x-a)^n) = (\alpha - a)^n$. Hence, for every $f \in H(\widetilde{D})$, we have $\psi(f) = f(\alpha)$.

In the same way, we check that, fixing $i \in J$, we have $\psi\left(\frac{1}{(x-a_i)^n}\right) = \left(\frac{1}{\alpha-a_i}\right)^n$, hence $\psi(f) = f(\alpha)$ for every $f \in H_0(\mathbb{K}\setminus T_i)$. This clearly finishes proving that $\psi(f) = f(\alpha)$ for every $f \in H(D)$.

Now, let ψ be a $\mathbb{K}$-algebra homomorphism from $H(D)$ to $\mathbb{K}$. By Corollary 10.10, ψ is continuous and hence belongs to $H(D)^*$. Consequently, it is of the form defined by the theorem. □

Chapter 16

Factorization of Analytic Elements

In $\mathbb{C}$, it is well known that when a (not identically zero) holomorphic function admits a zero at a point α, this zero has a finite order of multiplicity. Actually, this is a generalization of a property of rational functions. In the non-Archimedean context, we find again that property among analytic elements and it is essential. In this chapter, D is just a subset of $\mathbb{K}$.

Lemma 16.1. *Let $\alpha \in \overset{\circ}{D}$. Let $q \in \mathbb{N}^*$ and let (g_n) be a sequence of $H(D)$ such that the sequence $(x-\alpha)^q g_n$ converges in $H(D)$. Then the sequence $(g_n)_{n\in\mathbb{N}}$ also converges in $H(D)$.*

Proof. Without loss of generality, we can assume $\alpha = 0$. Set $f_n = x^q g_n$, $n \in \mathbb{N}$. Since 0 lies in $\overset{\circ}{D}$, there exists a disk $d(0,r) \subset D$. Let $E = D\setminus d(0,r)$. Clearly, we have $\|f_s - f_n\|_{d(0,r)} = \|x^q\|_{d(0,r)}\|g_s - g_n\|_{d(0,r)} = r^q\|g_s - g_n\|_{d(0,r)}$ and hence $\|g_s - g_n\|_E \leq \frac{\|f_s-f_n\|}{r^q}$. Consequently, $\|g_s - g_n\|_D \leq \frac{\|f_s-f_n\|}{r^q}$, and therefore, the sequence (g_n) is a Cauchy sequence, which ends the proof. □

Theorem 16.2. *Let α belong to $D \cap \overline{\overset{\circ}{D}}$ and let $f \in H(D)$ be such that $f(\alpha) = 0$. Then, f has factorization in $H(D)$ in the form $(x-\alpha)g$ with $g \in H(D)$. If there is no neighborhood V of α such that $f(x) = 0$ whenever $x \in V$, then there exists a unique integer $q \in \mathbb{N}$ and $h \in H(D)$ such that $f(x) = (x-\alpha)^q\ h(x)$ and $h(\alpha) \neq 0$ and then α is a zero of order q of f.*

Proof. First, we prove the main factorization in the form $(x-\alpha)g$. We may obviously assume $\alpha = 0$. By hypothesis, there exists a disk $d(0,s)$ included in $\overline{D}$. And then, by Theorem 11.5, there exists a disk $d(0,r)$ included in $d(0,s)$ such that f has no pole in $d(0,r)$. Consequently, f belongs to $H(D \cup d(0,r))$. So, we can assume that 0 is interior to D and that $d(0,r) \subset D$ without loss of generality. By Corollary 14.11, the restriction of f to $d(0,r)$ is equal to a power series $\sum_{n=1}^{\infty} a_n x^n$ for all $x \in d(0,r)$.

Now, let t_n be a sequence in $R(D)$ such that $\lim_{n\to\infty} t_n = f$. Clearly, $|t_n(0)| \leq \|t_n - f\|_D$ because $f(0) = 0$, so we have $\|t_n - t_n(0) - f_n\|_D \leq \|t_n - f\|_D$. We put $h_n = t_n - t_n(0)$ $(n \in \mathbb{N})$. The sequence (h_n) of $R(D)$ approaches f in $H(D)$ and satisfies $h_n(0) = 0$ whenever $n \in \mathbb{N}$, hence h_n has factorization in $R(D)$ in the form xg_n. By Lemma 16.1, the sequence $(g_n)_{n\in\mathbb{N}}$ converges in $H(D)$. Let g be its limit. We show that $\lim_{q\to\infty}\|xg_q - xg\|_D = 0$. Let $\varepsilon > 0$ and let $N \in \mathbb{N}$ be such that $\|h_n - f\|_D \leq \varepsilon$ whenever $n \geq N$. We fix $q \geq N$. Then, $\|h_n - h_q\|_D \leq \varepsilon$, hence $|xg_n(x) - xg_q(x)| \leq \varepsilon$ whenever $x \in D$. So, when n tends to $+\infty$, we see that $|xg(x) - xg_q(x)| \leq \varepsilon$ whenever x in D. Thus, we have $\|xg - xg_q\|_D \leq \varepsilon$, and therefore, $\lim_{q\to\infty} \|xg_q - xg\|_D = 0$. But, by hypothesis, we have $\lim_{q\to\infty} \|f - xg_q\|_D = 0$ and then $f = xg$.

Now, we suppose that f is not identically zero in $d(0,r)$. Then at least one of the coefficients a_n of its power series is not zero. By Corollary 14.11, f admits 0 as a zero of order q and then q is the smallest integer such that $a_q \neq 0$. In $d(0,r)$, we have $f(x) = \sum_{n=q}^{\infty} a_n x^n = xg(x)$, hence $g(x) = \sum_{n=q}^{\infty} a_n x^{n-q}$ whenever $x \in d(0,r)$. Suppose that f has been proven to be factorized in the form $x^s g_s$ with $s < q$ and $g_s \in H(D)$. Clearly, $g_s(x) = \sum_{n=q}^{\infty} a_n x^{n-s}$ whenever $x \in d(0,r)$, hence $g_s(0) = 0$, and therefore, g_s has factorization in the form xg_{s+1} with $g_{s+1} \in H(D)$. Thus, by induction, we obtain $f = x^q g_q(x)$ with $g_q(x) = \sum_{n=q}^{\infty} a_n x^{n-q}$ and then $g_q(0) = a_q \neq 0$. That finishes proving Theorem 16.2. □

Notation. Let $a \in \overset{\circ}{D}$ and let $f \in H(D)$ be such that $f(a) = 0$, $f(x) \neq 0$ in a disk $d(a,r)$. The order of the zero a of f is denoted by $\omega_a(f)$.

Corollary 16.3. *Let $\overline{D}$ be open, let $f \in H(D)$, and let α be a zero of f in D. Either there exists a disk $d(\alpha, r)$ such that $f(x) \neq 0$*

whenever $x \in d(\alpha, r)\setminus\{\alpha\}$ *or there exists a disk* $d(\alpha, r)$ *such that* $f(x) = 0$ *whenever* $x \in d(\alpha, r)$.

Corollary 16.4. *Let* $f \in H(D)$ *have a zero of order* q *at a point* $\alpha \in \overset{\circ}{D}$. *Then for every* $s = 1, \ldots, q$, f *factorizes in the form* $(x-\alpha)^s g_s$, *with* $g_s \in H(D)$ *having a zero of order* $q - s$ *at* α.

Corollary 16.5. *Let* $\alpha \in \overset{\circ}{D}$ *and let* $f \in H(D)$. *Let* $\sum_{n=0}^{\infty} a_n(x-\alpha)^n$ *be its power series in a disk* $d(\alpha, r) \subset D$. *Let* $P(x) = \sum_{n=0}^{q} a_n(x-\alpha)^n$ *and let* $g(x) = f(x) - P(x)$. *Then* g *factorizes in the form* $(x-\alpha)^q h(x)$, *with* $h \in H(D)$.

Definition. Let $A \subset D$ be an open subset of $\mathbb{K}$ and let $f \in H(D)$ have finitely many zeros $a_1, \ldots, a_n$ in A of multiplicity order of $q_1, \ldots, q_n$, respectively. The polynomial $\prod_{i=1}^{n}(x - a_i)^{q_i}$ is named *the polynomial of zeros of* f *in* A.

We are now able to give the following corollary.

Corollary 16.6. *Let* A *be a subset of* D *open in* $\mathbb{K}$, *let* $f \in H(D)$ *have finitely many zeros in* A, *and let* P *be the polynomial of its zeros in* A. *Then* f *has a factorization in the form* $f = Pg$, *with* $g \in H(D)$ *and* $g(x) \neq 0$ *whenever* $x \in A$.

Definitions. An element $f \in H(D)$ is said to be *semi-invertible* (respectively *quasi-invertible*) if it factorizes in the form $P(x)\,g(x)$, with g invertible in $H(D)$ and with P a polynomial whose zeros belong to D (respectively to $D \cap \overline{\overset{\circ}{D}}$).

An element $f \in H(D)$ is said to be *quasi-minorated* if for every bounded sequence $(a_n)_{n\in\mathbb{N}}$ of D such that $\lim_{n\to\infty} f(a_n) = 0$ we can extract a subsequence that converges in $\mathbb{K}$.

Remarks. (1) If a semi-invertible element of $H(D)$ has no zero in D, it is invertible in $H(D)$. (2) Let D belong to Alg. If f_1, f_2 are semi-invertible (respectively quasi-invertible) elements of $H(D)$, then $f_1 f_2$ is also semi-invertible (respectively quasi-invertible). However, when D does not belong to Alg, counter-examples show that the product of two semi-invertible (respectively quasi-invertible) elements is not always semi-invertible (respectively quasi-invertible). Such counter-examples are given in a further remark.

Lemma 16.7. *Let $D \in$ Alg, let $f \in H(D)$ be quasi-invertible (respectively quasi-minorated), and let $h \in R(D)$ be a Moebius function. Let $D' = h(D)$ and let $g = f \circ h^{-1}$. Then, g is a quasi-invertible (respectively quasi-minorated) element of $H(D')$.*

Proof. Suppose first f is quasi-invertible in $H(D)$. Let $u = h(x)$. So, f is of the form $P(x)\phi(x)$ with P a polynomial whose zeros are interior to D and ϕ is an invertible element of $H(D)$. Then, $\phi \circ h^{-1}$ is invertible in $H(D')$ and $P \circ h^{-1}$ belongs to $R(D')$ and is of the form $\frac{Q(u)}{(u-b)^s}$, where b is the unique pole of h^{-1}. Consequently, g is of the form $Q(u)\frac{\phi(u)}{(u-b)^s}$. Thus, $\frac{\phi(u)}{(u-b)^s}$ is invertible in $H(D')$ and hence g is quasi-invertible.

Now, suppose f quasi-minorated. Let $(a_n)_{n\in\mathbb{N}}$ be a sequence in D' such that $\lim_{n\to\infty} g(a_n) = 0$ and let $b_n = h^{-1}(a_n)$, $(n \in \mathbb{N})$. Then, $\lim_{n\to\infty} f(b_n) = 0$. Since f is quasi-minorated, one can extract a subsequence $(b_{q(m)})_{m\in\mathbb{N}}$ that either converges or satisfies $\lim_{m\to\infty}|b_{q(m)}| = \infty$. But then, the sequence $(a_{q(m)})$ either converges or satisfies $\lim_{m\to\infty}|a_{q(m)}| = \infty$. Hence, f is quasi-minorated. □

Theorem 16.8. *Let D be bounded, open, closed and let $f \in H(D)$. If f is quasi-minorated, then it is quasi-invertible.*

Proof. We suppose f is not quasi-invertible and we prove that f is not quasi-minorated either.

First, we suppose f to have finitely many zeros. Since D is open, by Corollary 16.6, f has factorization in the form $P(x)\ g(x)$, with P a polynomial whose zeros are interior to D and g an element of $H(D)$ which has no zero in D but is not invertible in $H(D)$ since f is not quasi-invertible. Hence, there exists a bounded sequence $(\alpha_n)_{n\in\mathbb{N}}$ in D such that $\lim_{n\to\infty} g(\alpha_n) = 0$. If f were quasi-minorated, we could extract a convergent subsequence from the sequence $(\alpha_n)_{n\in\mathbb{N}}$ whose limit would belong to D and would be a zero of g. Hence, f is not quasi-minorated when it has finitely many zeros in D.

Now, we suppose that f has a sequence of (distinct) zeros (α_n) in D and that f is quasi-minorated. Hence, we may extract a convergent subsequence of limit α. Obviously, α is another zero of f, hence, by Corollary 16.3, $f(x)$ is equal to zero inside a disk $d(\alpha, r)$ and then

f is not quasi-minorated, a contradiction. That ends the proof of Theorem 16.8. □

Theorem 16.9. *Let D be closed and bounded. Let $f \in H(D)$ be quasi-minorated and have no zero in D. Then, f is invertible in $H(D)$.*

Proof. Assume that $\inf_{x \in D} |f(x)| = 0$ and let $(a_n)_{n \in \mathbb{N}}$ be a sequence in D such that $\lim_{n \to \infty} f(a_n) = 0$. Since D is bounded and since f is quasi-minorated, we can extract a subsequence (a_n) which converges in $\mathbb{K}$ to a point $a \in \overline{D}$. Since D is closed, a belongs to D and satisfies $f(a) = 0$, which contradicts the hypothesis. Thus, there exists $\lambda > 0$ such that $|f(x)| \geq \lambda$ whenever $x \in D$ and then by Lemma 11.3, f is invertible in $H(D)$. □

Chapter 17

Algebras $H(D)$

We have seen that $H(D)$ is a Banach $\mathbb{K}$-algebra if and only if D is closed and bounded. But studying analytic elements, analytic functions require to know algebras of analytic elements which are not necessarily bounded. Thus, we have to examine the class Alg of subsets D of $\mathbb{K}$ such that $H(D)$ is a $\mathbb{K}$-algebra with respect to usual laws [35, 37, 44, 50].

Notation. Throughout this chapter, D denotes a subset of $\mathbb{K}$. Let $f \in H(D)$. According to Theorem 11.5, f is of the form $f^* + \overline{f}$, with $f^* \in R_0(\mathbb{K}\setminus(\widetilde{D}\setminus D))$ and $\overline{f} \in H(\overline{D})$. We keep that notation throughout this chapter.

Proposition 17.1. *Let α belong to $\overset{\circ}{\overline{D}}$ and let $f \in H(D\setminus\{\alpha\})$. For every $q \in \mathbb{N}$, $\frac{f}{(x-\alpha)^q}$ belongs to $H(D\setminus\{\alpha\})$.*

Proof. Since α belongs to $\overset{\circ}{\overline{D}}$, there exists a disk $d(\alpha, s)$ included in $\overline{D}$. On the other hand, there exists $r \in]0, s[$ such that f^* has no pole in $d(\alpha, r)\setminus\{\alpha\}$. Hence, by Theorem 11.10, f is of the form $\frac{g}{(x-\alpha)^t}$ with $g \in H(D \cup d(\alpha, r))$ and $t \in \mathbb{N}$. Then, $\frac{f}{(x-\alpha)^q} = \frac{g}{(x-\alpha)^{q+t}}$. Thus, without loss of generality, we may assume that α belongs to D and that f belongs to $H(D \cup d(\alpha, r))$.

By Corollary 14.11, in $d(\alpha, r)$, $f(x)$ is equal to a power series $\sum_{n=0}^{\infty} a_n(x-\alpha)^n$. Let $P(x) = \sum_{n=0}^{q} a_n(x-\alpha)^n$. By Theorem 16.2, $f(x) - P(x)$ factorizes in the form $(x-\alpha)^q h$ with $h \in H(D)$.

Hence, we see that $\frac{f(x)}{(x-\alpha)^q} = \frac{P(x)}{(x-\alpha)^q} + h$. Since $\frac{P}{(x-\alpha)^q} \in R(D)$, it is clear that $\frac{f}{(x-\alpha)^q}$ belongs to $H(D)$. □

Corollary 17.2. *Let $f \in H(D)$ and let P be a polynomial whose zeros are interior to D. Let $a_1, \ldots, a_n$ be the zeros of P. Then, $\frac{f}{P}$ belongs to $H(D \setminus \{a_1, \ldots, a_n\})$.*

Proposition 17.3. *If D is bounded and satisfies $\overline{D} \setminus D \subset \left(\overset{\circ}{\overline{D}}\right)$, then $D \in \mathrm{Alg}$.*

Proof. Let $f, g \in H(D)$ and let us show that $fg \in H(D)$. By Theorem 11.5, we have $f = f^* + \overline{f}$, $g = g^* + \overline{g}$ with f^*, $g^* \in R_0(\mathbb{K} \setminus (\overline{D} \setminus D))$ and $\overline{f}$, $\overline{g} \in H(\overline{D})$. Since D is bounded, by Corollary 11.7, we have $H(\overline{D}) = H_b(D)$ and then $\overline{f}\overline{g}$ obviously belongs to $H(D)$ while $f^*g^* \in R_0(\mathbb{K} \setminus (\overline{D} \setminus D))$. Finally, by Corollary 17.2, both $f^*\overline{g}$ and $g^*\overline{f}$ belong to $H(D)$ and therefore so does fg. □

Definition. Let $\mathcal{F}$ be a filter in D. An element $f \in H(D)$ is said to be *vanishing along* $\mathcal{F}$ if $\lim_{\mathcal{F}} f(x) = 0$. Further, f is said to be *properly vanishing along* $\mathcal{F}$ if $\lim_{\mathcal{F}} f(x) = 0$ and if $\|f\|_A \neq 0$ whenever $A \in \mathcal{F}$.

Proposition 17.4 is a polyvalent result which helps us characterize the sets $D \in \mathrm{Alg}$ but also find conditions for $H(D)$ not to be a Noetherian algebra.

Proposition 17.4. *Let $\mathcal{F}$ be a pierced filter on D, let $(T_n)_{n\in\mathbb{N}}$ be a sequence of holes of D that runs $\mathcal{F}$, and let $E = \mathbb{K} \setminus (\bigcup_{n=0}^{\infty} T_n)$. Let $g_1, \ldots, g_q \in H_b(E)$ be vanishing along $\mathcal{F}$, with g_1 properly vanishing. For every $x \in E$ let $S(x) = \sup_{1 \leq i \leq q} |g_i(x)|$, let $\mathcal{J}$ be the ideal generated by $g_1, \ldots, g_q$ in $H_b(E)$ and let $\overline{\mathcal{J}}$ be its closure in $H_b(D)$.*

There exists a sequence $(z_n)_{n\in\mathbb{N}}$ in D, thinner than $\mathcal{F}$, such that $g_1(z_n) \neq 0$ and an element $G \in \overline{\mathcal{J}}$ such that $\lim_{n\to\infty} \frac{|G(z_n)|}{S(z_n)} = +\infty$.

Proof. Without loss of generality, we may assume $\mathcal{F}$ to be a decreasing filter or a Cauchy filter. Indeed, if $\mathcal{F}$ is an increasing filter of center α and diameter R, consider a hole of D $T(b, \rho)$ included in $d(\alpha, R^-)$, take $\gamma(x) = \frac{1}{x-b}$, and set $D' = \gamma(D)$. Then, by Theorem 3.11, D' admits a decreasing pierced filter $\mathcal{F}'$, the image of $\mathcal{F}$ by γ. Next, D' is clearly bounded. By Theorem 12.7, the mapping

ϕ from D onto D' defined by $\phi(f) = f \circ \gamma^{-1}$ is an isomorphism from $H(\overline{D'})$ onto $H_b(D)$. Hence, $\mathcal{J}$ is isomorphic to the ideal generated by $\{g_j \circ \gamma^{-1} | 1 \leq j \leq q\}$ in $H(\overline{D'})$. Hence, we assume $\mathcal{F}$ to be a decreasing pierced filter or a Cauchy pierced filter.

Without loss of generality, we may now clearly assume $D = E$. Since g_j are bounded, we may obviously assume $\|g_j\|_D \leq 1$ whenever $j = 1, \ldots, q$. Let $R = \operatorname{diam}(\mathcal{F})$ and let $(x_m)_{m\in\mathbb{N}}$ be a sequence in D thinner than $\mathcal{F}$, such that $g_1(x_m) \neq 0$ whenever $m \in \mathbb{N}$, with $|x_{m+2} - x_{m+1}| < |x_{m+1} - x_m|$. Since $\mathcal{J}$ is pierced, there exists a subsequence $(x_{m_q})_{q\in\mathbb{N}}$ of the sequence (x_m) together with a sequence of holes $(T_q)_{q\in\mathbb{N}}$ of D such that

$$T_q \subset d(x_{m_{q+1}}, d_{m_q}) \backslash d(x_{m_{q+2}}, d_{m_{q+1}}).$$

Hence, without loss of generality, we may assume that we have a sequence of holes $(T_m)_{m\in\mathbb{N}}$ of D such that $T_m \subset d(x_{m+1}, d_m) \backslash d(x_{m+2}, d_{m+1})$.

We put $D_m = d(x_{m+1}, d_m) \cap D$ and $A_n = D_{2n+1} \backslash D_{2n+3}$. For each n, let $u_n \in A_n$ be such that $|g_1(u_n)| \geq \|g_1\|_{A_n} (\frac{n}{n+1})$. For each $j = 1, \ldots, t$, let $M_n^j = \|g_j\|_{A_n}$ and let $M_n = \max_{1\leq j\leq t} M_n^j$. Since $g_1(x_m) \neq 0$, we have $M_n > 0$ whenever $n \in \mathbb{N}$, and since $\|g_j\|_D \leq 1$ for all j, we have $M_n \leq 1$ whenever $n \in \mathbb{N}$.

We construct a sequence (U_n) in $H_b(D)$ satisfying

$$(1) \qquad |U_n(x)| \leq \frac{1}{n+1} \text{ whenever } x \in D \backslash A_n,$$

$$(2) \qquad \sqrt{M_n}\Big(\frac{n+1}{n}\Big) > \|g_1 U_n\|_{A_n} > \sqrt{M_n}.$$

For every $n \in \mathbb{N}$, set $T_n = d(\beta_n, \rho_n^-)$, $u_n = x_{2n+2}$, $a_n = \beta_{n+1}$, $b_n = \beta_{n+2}$, $c_n = \beta_{2n+3}$, and $\epsilon_n \in d(0, \frac{1}{n})$. Let us fix $n \in \mathbb{N}$. It is seen that $|u_n - a_n| > |u_n - b_n|$, hence there exists $q_n \in \mathbb{N}$ such that

$$(3) \qquad |\epsilon_n| \left| \frac{u_n - a_n}{u_n - b_n} \right|^{q_n} g(u_n) > \sqrt{M_n}$$

and of course there exists q'_n such that

$$(4) \qquad \Big(\frac{d_{2n+1}}{d_{2n+2}}\Big)^{q_n} \Big(\frac{d_{2n+3}}{d_{2n+2}}\Big)^{q'_n} < 1.$$

We put $h_n(x) = \epsilon_n \big(\frac{x-a_n}{x-b_n}\big)^{q_n} \big(\frac{x-c_n}{x-b_n}\big)^{q'_n}$.

Then, by (4), we see that

when $|x - c_n| > d_{2n+1}$, we have $|h_n(x)| = |\epsilon_n| < \frac{1}{n}$, when $|x - c_n| \leq d_{2n+3}$, we have $|x - a_n| = |a_n - c_n| = d_{2n+1}$ and $|x - b_n| = |b_n - c_n| = d_{2n+2}$, hence $|h_n(x)| \leq |\epsilon_n|\big(\frac{d_{2n+1}}{d_{2n+2}}\big)^{q_n}$ $\big(\frac{d_{2n+3}}{d_{2n+2}}\big)^{q'_n} < \frac{1}{n}$.

But now we note that x belongs to $D\backslash A_n$ if and only if it satisfies: either $|x - c_n| > d_{2n+1}$ or $|x - c_n| \leq d_{2n+3}$, hence we have proven that $|h_n(x)| < \frac{1}{n}$ whenever $x \in D\backslash A_n$. This shows h_n satisfies (1).

When $x \in A_n$ i.e. when $d_{2n+3} < |x - c_n| < d_{2n+1}$, we see that $\|g_1 h_n\|_{A_n} \geq |g_1(u_n)h_n(u_n)|$, hence, by (3), we have $\|g_1 h_n\|_{A_n} \geq \sqrt{M_n}$. Hence, there trivially exists $\lambda_n \in d(0,1)$ such that $\big(\frac{n+1}{n}\big)\sqrt{M_n} > |\lambda_n g_1 h_n|_{A_n} > \sqrt{M_n}$.

Now, we put $U_n = \lambda_n h_n$ and we see that U_n satisfies (1) and (2). In particular, we have $\|g_1 U_n\|_D \leq \max\big(\sqrt{M_n}\big(\frac{n+1}{n}\big), \frac{\|g_1\|_D}{n+1}\big)$, hence $\lim_{n\to\infty} \|g_1 U_n\|_D = 0$. Let $T = \sum_{n=0}^{\infty} g_1 U_n$. By definition, T belongs to $\overline{F}$ because for every $t \in \mathbb{N}$, $g\sum_{n=0}^{t} U_n$ belongs to F.

By (2), there exists a sequence $(z_n)_{n\in\mathbb{N}}$ in D satisfying $z_n \in A_n$ and

$$(5) \qquad \sqrt{M_n} < |g_1(z_n)U(z_n)| < M_n\Big(\frac{n+1}{n}\Big),$$

hence we have

$$(6) \qquad |U_n(z_n)| > \frac{\sqrt{M_n}}{|g_1(z_n)|} \geq \frac{1}{\sqrt{M_n}} \text{ because } |g_1(z_n)| \leq M_n.$$

When $j \neq n$, z_n belongs to $D\backslash A_j$, hence by (1) and (6), we have $|U_j(z_n)| < \frac{1}{j+1} < \frac{1}{\sqrt{M_n}} < |U_n(z_n)|$ whenever $j \neq n$. Hence, we see that $|T(z_n)| = |g_1(z_n)U_n(z_n)|$ whenever $n \in \mathbb{N}$. But then, by (5), we see that $\frac{|T(z_n)|}{S(z_n)} = \frac{|T(z_n)|}{M_n} > \frac{1}{\sqrt{M_n}}$. Consequently, $\lim_{n\to\infty} \frac{|T(z_n)|}{S(z_n)} = +\infty$ and this finishes the proof of Lemma 17.4. □

Corollary 17.5. *Let $a \in D\backslash \overset{\circ}{D}$ and let $\mathcal{F}$ be the pierced filter of the neighborhoods of a. There exists a sequence $(z_n)_{n\in\mathbb{N}}$ in D of limit a and an element $G \in H_b(D)$ vanishing along $\mathcal{F}$ such that $\limsup_{n\to\infty} \frac{|G(z_n)|}{|z_n - a|} = +\infty$.*

Lemma 17.6. *Let D have a hole $T = d(a, r^-)$. Let $\gamma(x) = b + \frac{\lambda}{x-a}$ with $\lambda \in \mathbb{K}$ and let $D' = \gamma(D)$. For every $\alpha \in \overline{D}, \alpha$ belongs to $\left(\overset{\circ}{\overline{D}}\right)$ if and only if $\gamma(\alpha)$ belongs to $\left(\overset{\circ}{\overline{D'}}\right)$. Moreover, if D is not bounded, then $\mathbb{K}\setminus\overline{D}$ is bounded if and only if b belongs to $\left(\overset{\circ}{\overline{D'}}\right)$.*

Proof. γ is obviously a bicontinuous bijection from $\mathbb{K}\setminus\{a\}$ onto $\mathbb{K}\setminus\{b\}$. Let $\alpha \in \left(\overset{\circ}{\overline{D}}\right)$. There exists a disk $d(\alpha, r)$ included in $\overline{D}$. Since $a \notin d(\alpha, r), \gamma$ is bounded in $d(\alpha, r)$. Since γ is bicontinuous, $\gamma(d(\alpha, r))$ is open in $\mathbb{K}\setminus\{a\}$, hence it is clearly open in $\mathbb{K}$. So, $\gamma(\alpha)$ belongs to $\gamma\left(\overset{\circ}{\overline{D}}\right)$. But $\gamma(\overline{D}) \subset \overline{\gamma(D)}$, hence $\gamma(\alpha) \in \left(\overset{\circ}{\overline{D'}}\right)$. Let $\xi = \gamma^{-1}$. Then, $\xi(u) = a + \frac{\lambda}{u-b}$ and then what is true for γ is also true for ξ. Hence, conversely, if $\gamma(\alpha) \in \left(\overset{\circ}{\overline{D'}}\right)$, we see that $\alpha = \xi(\gamma(\alpha)) \in \overset{\circ}{\overline{D'}}$ because $D = \xi(D')$.

We now suppose D unbounded. If $\mathbb{K}\setminus\overline{D}$ is bounded, then $\overline{D}$ contains a set E of the form $\{|x| \ |x-a| \geq s\}$ with $s > |a-b|$ whose image E' is $d\big(a, \frac{|\lambda|}{s}\big)\setminus\{a\}$, hence $\overline{D'}$ contains $d\big(a, \frac{|\lambda|}{s}\big)$ and so does $\overset{\circ}{\overline{D'}}$.

Finally, we suppose that $\mathbb{K}\setminus\overline{D}$ is not bounded. Then, $\gamma((\mathbb{K}\setminus\{a\})\setminus\overline{D})$ is an open set E in $\mathbb{K}$ whose closure contains b. Since $b \in \overline{D'}\setminus D'$, it is easily seen that there is a sequence of holes of $D', (T_n)_{n\in\mathbb{N}}$, that approaches b (each one is obviously included in E), hence $b \notin \left(\overset{\circ}{\overline{D'}}\right)$. This ends the proof of Lemma 17.6. □

Corollary 17.7. *Let D have a hole $T = d(a, r^-)$, satisfy $\overline{D}\setminus D \subset \left(\overset{\circ}{\overline{D}}\right)$, and be such that $\mathbb{K}\setminus\overline{D}$ is bounded. Let $\gamma = \frac{1}{x-a}$ and let $D' = \gamma(D)$. Then, $\overline{D'}\setminus D' \subset \left(\overset{\circ}{\overline{D'}}\right)$.*

Lemma 17.8. *The following two conditions are equivalent:*

(A) $\widetilde{D}\setminus\overline{D}$ *is bounded.*
(A′) *Either D is bounded or $\mathbb{K}\setminus\overline{D}$ is bounded.*

Proof. If D is bounded, (A) and (A′) are clearly satisfied, hence we have nothing to show. Now, we suppose D to be unbounded.

Hence, $\widetilde{D} = \mathbb{K}$ and then $\widetilde{D}\setminus\overline{D} = \mathbb{K}\setminus\overline{D}$, so $\widetilde{D}\setminus\overline{D}$ is bounded if and only if $\mathbb{K}\setminus\overline{D}$ is bounded. Finally, (A) and (A′) are equivalent. □

Theorem 17.9. *D belongs to* Alg *if and only if it satisfies the following two conditions:*

(A) *$\widetilde{D}\setminus\overline{D}$ is bounded.*

(B) $\overline{D}\setminus D \subset \left(\overset{\circ}{\overline{D}}\right)$.

Proof. Suppose first that D satisfies A and B. By Lemma 17.8, it satisfies A′ and B. Suppose first that D is bounded. Since D satisfies B, by Proposition 17.3, $D \in$ Alg. Suppose now that $\mathbb{K}\setminus\overline{D}$ is bounded. We may obviously assume D to have a hole $T = d(a, r^-)$ because if D has no hole, then, by Corollary 11.8, we have $H(D) = R(D)$. Let $\gamma = \frac{1}{x-a}$ and let $D' = \gamma(D)$. The set D' is then bounded, and, by Corollary 17.7, D' satisfies $\overline{D'}\setminus D' \subset \left(\overset{\circ}{\overline{D'}}\right)$ and hence $D' \in$ Alg, therefore by Proposition 17.3, D belongs to Alg.

We now suppose that (B) is not satisfied and prove that $D \notin$ Alg. Indeed, let $\alpha \in (\overline{D}\setminus D)\setminus\left(\overset{\circ}{\overline{D}}\right)$. By definitions, D has a Cauchy pierced filter $\mathcal{F}$ that converges in $\mathbb{K}$ to α. Let $T = d(a, r^-)$ be a hole of D and let $f = \frac{x-\alpha}{x-a}$. Then, $f \in R_b(D)$. By Proposition 17.4, there exists $S \in H_b(D)$ such that $S(\alpha) = 0$ together with a sequence $(z_n)_{n\in\mathbb{N}}$ in D such that $\lim_{n\to\infty} z_n = \alpha$, while $\lim_{n\to\infty}\left|\frac{S(z_n)}{f(z_n)}\right| = +\infty$.

Let us assume $D \in$ Alg. Then, $\frac{S(x)}{f(x)} \in H(D)$ because $\frac{x-a}{x-\alpha} \in R(D)$. Since $\left|\frac{S(x)}{f(x)}\right|$ is not bounded in any neighbourhood of α, by Theorem 11.10 and Corollary 11.9, there exists an integer $n \geq 1$ such that $(x-\alpha)^n\frac{S(x)}{f(x)}$ has a non-zero limit ℓ at α. But when $x \in D \cap d(\alpha, |a|)$, we have $\left|(x-\alpha)^n\left(\frac{S(x)}{f(x)}\right)\right| = \left|(x-\alpha)^{n-1}(x-a)S(x)\right| = |a|\ |x-\alpha|^{n-1}\ |S(x)|$, hence $\ell = 0$, a contradiction. Consequently, $D \notin$ Alg.

Suppose now that D satisfies (B) but does not satisfy (A). Both D and $\mathbb{K}\setminus\overline{D}$ are unbounded. Since $\mathbb{K}\setminus\overline{D}$ is unbounded, D has a hole $T = d(\alpha, r^-)$, and then by Lemma 17.6, the inversion $\gamma(x) = \frac{1}{x-\alpha}$ maps D onto a bounded set D' such that $\alpha \in (\overline{D'}\setminus D') \subset \left(\overset{\circ}{\overline{D'}}\right)$. Hence, D' does not belong to Alg and neither does D. This ends the proof of Theorem 17.9. □

Corollary 17.10. *If D is bounded and if $\overline{D}$ is open, then D belongs to* Alg.

Notation. Throughout this book, Conditions (A) and (B) are those given in Theorem 17.9.

Theorem 17.11. *Let D belong to* Alg *and have a hole $T = d(a, r^-)$. Let $h(x) = \frac{1}{x-a}$ and let $D' = h(D)$. Then, D' is bounded and belongs to* Alg.

Proof. Since $D \in$ Alg, by Theorem 17.10, D satisfies Conditions (A) and (B) and hence we can we can check that D', being obviously bounded, satisfies (A) and (B) too. If D is bounded, this is immediate. If D is not bounded, then 0 is the unique point that might not belong to $\overline{D}'\setminus D' \subset \left(\overset{\circ}{\overline{D}'}\right)$. But if 0 is the limit of a sequence of holes of D', then $\widetilde{D}\setminus\overline{D}$ is not bounded, which contradicts Condition (A).

□

Lemma 17.12. *Let D be open. Then, D satisfies Condition (B) if and only if $\overline{D}$ is open.*

Proof. Since D is open, $\overline{D}$ is open if and only if for every $\alpha \in \overline{D}\setminus D$, α is interior to D. This is just equivalent to Condition (B). □

Notation. Let $D \in$ Alg and let $a \in D$. We denote by $\mathcal{J}(a)$ the ideal of the $f \in H(D)$ such that $f(a) = 0$.

Theorem 17.13. *Let $D \in$ Alg and let $a \in D$. If a belongs to $\overset{\circ}{\overline{D}}$, then $\mathcal{I}(a) = (x - a)H(D)$. Else, $\mathcal{J}(a)$ is not of finite type.*

Proof. Suppose first $a \in \overset{\circ}{\overline{D}}$. By Theorem 16.2, it is clearly seen that $\mathcal{J}(a) = (x - a)H(D)$. Now, let $a \notin \overset{\circ}{\overline{D}}$. Then the filter of the neighborhoods of a is a Cauchy pierced filter. We denote it by $\mathcal{F}$. Suppose that $\mathcal{J}(a)$ is of finite type and let $\{g_1, \dots, g_q\}$ be a system of generators. For each $j = 1, \dots s, q$, let Q_j be the polynomial of poles of g_j in $\overline{D}\setminus D$. Now, let $d(b, r^-)$ be a hole of D. By Corollary 11.7, for each $j = 1, \dots, q$, there exists a rational function of the form $\frac{1}{(x-b)^{t_j}}$ such that the function $h_j = \frac{g_j}{Q_j(x-b)^{t_j}}$ belongs to $H_b(\overline{D})$.

Of course, at least one of g_j is properly vanishing along $\mathcal{F}$, otherwise all the elements of $\mathcal{J}(a)$ would be equal to 0 inside a neighborhood of a and then $\mathcal{J}(a)$ couldn't contain $x-a$. Consequently, we can assume that g_1 is properly vanishing and so is h_1. For every $x \in D$, we put $S(x) = \max_{1 \leq j \leq q} |h_j(x)|$. We note that $\mathcal{J}(a)$ is obviously closed in $H(D)$, hence by Proposition 17.4, there exists $f \in \mathcal{J}(a)$ together with a sequence $(z_n)_{n \in \mathbb{N}}$ in D, of limit a, such that $S(z_n) \neq 0$ whenever $n \in \mathbb{N}$ and that $\lim_{n \to \infty} \frac{|f(z_n)|}{S(z_n)} = +\infty$. This obviously contradicts the fact that f should be of the form $\sum_{j=1}^{q} f_j g_j$ with the f_j in $H(D)$. Thus we have shown that $\mathcal{J}(a)$ is not of finite type and this ends the proof of Theorem 17.13. □

Corollary 17.14. *Let $D \in$ Alg. If $D \setminus \overset{\circ}{\overline{D}} \neq \emptyset$ then $H(D)$ is not noetherian.*

Chapter 18

Derivative of Analytic Elcments

Given an infraconnected set, the main question we consider here is whether an element f of $H(D)$ has a derivative that belongs to $H(D)$ and when it does, whether its Mittag-Leffler series is obtained by deriving that of f. Another question is whether an analytic element on D whose derivative is identically zero is a constant. Both questions are answered on an infraconnected closed set.

Throughout this chapter, D is a subset of $\mathbb{K}$ and is supposed to be open and infraconnected and we fix $R > 0$.

Theorem 18.1. *Let $f(x) = \sum_{n=0}^{+\infty} a_n x^n \in H(d(0,R))$. Then, f has a derivative $f'(\alpha)$ at each point $\alpha \in d(0,R)$ and $f'(0) = a_1$. Moreover, the function f' also belongs to $H(d(0,R))$, is equal to $\sum_{n=0}^{+\infty} n a_n x^{n-1}$, and satisfies*

$$|f'|(r) \leq \frac{|f|(r)}{r} \quad \forall r \in]0, R].$$

Further, f is indefinitely derivable in $d(0,R)$ and

$$f^{k)}(x) = \sum_{n=k}^{\infty} n(n-1)\cdots(n-k+1)a_n x^{n-k}.$$

Proof. Without loss of generality, we can suppose that $R \leq 1$. Obviously, $\lim_{x\to 0} \frac{f(x)-a_0}{x} = a_1$, hence $f'(0)$ exists and is equal to a_1.

More generally, take $\alpha \in d(0,R)\setminus\{0\}$ and consider

$$\frac{f(x)-f(\alpha)}{x-\alpha} = \frac{\sum_{n=1}^{\infty} a_n(x^n-\alpha^n)}{x-\alpha}$$
$$= \sum_{n=1}^{\infty} a_n(x^{n-1}+\alpha x^{n-2}+\cdots+\alpha^{n-1}).$$

Then, $|(x^{n-1}+\alpha x^{n-2}+\cdots+\alpha^{n-1})-n\alpha^{n-1}| \leq |x-\alpha| \big(\max(|\alpha|,|x|)\big)^{n-1}$. Particularly, when x is close enough to α, since $\alpha \neq 0$, we have $|x|=|\alpha|$, hence

$$|a_n||(x^{n-1}+\alpha x^{n-2}+\cdots+\alpha^{n-1})-n\alpha^{n-1}| \leq |a_n||\alpha|^{n-1}|x-\alpha|.$$

That proves that $(a_n(x^{n-1}+\alpha x^{n-2}+\cdots+\alpha^{n-1}))-na_n\alpha^{n-1}$ converges to 0 uniformly with respect to n and uniformly with respect to x inside a disk of center α. Consequently, $f'(\alpha)=\sum_{n=1}^{\infty} na_n(x-\alpha)^n \ \forall \alpha \in d(0,R)$. Then,

$$|f'|(r) = \sup_{n\in\mathbb{N}}(|na_n|r^{n-1}) \leq \sup_{n\in\mathbb{N}}(|a_n|r^{n-1}) = \frac{1}{r}|f|(r).$$

The last statement concerning $f^{(k)}$ is then immediate. □

More generally, we can derive the following:

Theorem 18.2. *Let $f \in H(\Delta(0,R,R'))$. Then, $f^{(k)}$ also belongs to $H(\Delta(0,R,R'))$ for every $k \in \mathbb{N}^*$ and satisfies $|f'|(r) \leq \frac{|f|(r)}{r}$ $\forall r \in]R,R'[$. Moreover, if the residue characteristic does not divide $\nu^+(f,\log r)$ or $\nu^-(f,\log r)$, then $|f'|(r) = \frac{|f|(r)}{r}$.*

Proof. $f(x)$ is equal to a Laurent series $\sum_{-\infty}^{+\infty} a_n x^n$ with $\lim_{n\to+\infty}|a_n|R'^n = 0$ and $\lim_{n\to-\infty}|a_n|R^n = 0$, hence obviously, $\lim_{n\to-\infty}|na_n|R^{n-1} = 0$ and $\lim_{n\to+\infty}|na_n|R'^{n-1} = 0$. Consequently, $f'(x)$ belongs to $H(\Delta(0,R,R'))$. Then,

$$|f'|(r) = \sup_{n\in\mathbb{Z}}(|na_n|r^{n-1}) \leq \sup_{n\in\mathbb{Z}}(|a_n|r^{n-1}) = \frac{1}{r}|f|(r).$$

Suppose now that the residue characteristic p does not divide $\nu^+(f,\log r)$ or $\nu^-(f,\log r)$. If $\nu^+(f,\log r)=\nu^-(f,\log r)$ is an integer q, it is obvious that $|f|(r)=|a_q|r^q$ and $|qa_q|=|a_q|$, hence $|f'|(r)=\frac{|f|(r)}{r}$, provided $q\neq 0$. Next, the property has continuation by continuity to the points μ such that $\nu^+(f,\mu)\neq\nu^-(f,\mu)$. □

Corollary 18.3. *Let* $R,\ R' \in]0,+\infty[\ (R < R')$ *and let* $f \in \mathcal{A}(d(0,R^-))$ *(respectively* $f \in \mathcal{A}(\Gamma((0,R,R'))$*. Then,* $|f'|(r) \leq \frac{|f|(r)}{r}\ \forall r \in]0,R[$ *(respectively* $\forall r \in]R,R'[$*).*

Corollary 18.4. *Suppose* $\mathbb{K}$ *has characteristic zero and* f *belongs to* $H(d(0,R))$*. Then,* $a_n = \frac{f^{(n)}(0)}{n!}$ *for every* $n \in \mathbb{N}$*, and if* α *is a zero of multiplicity order* q *of* f*, then we have* $f^{(j)}(\alpha) = 0$ *for every* $j < q$ *and* $f^{(q)}(\alpha) \neq 0$*.*

Corollary 18.5. *Let* $h \in \mathbb{K}(x)$*. Then, for all* $r > 0$*, we have* $|h'|(r) \leq \frac{|h|(r)}{r}$*.*

Theorem 18.6. *Let* $f(x) = \sum_{n=0}^{\infty} a_n x^n \in \mathcal{A}(d(0,r^-))$*. The power series* $\sum_{n=1}^{\infty} n a_n x^{n-1}$ *also belongs to* $\mathcal{A}(d(0,r^-))$ *and is equal to the derivative of* f *in* $d(0,r^-)$*. The radius of convergence of* f' *is superior or equal to the one of* f*. Further, if* $\mathbb{K}$ *has characteristic* 0*, the radius of convergence of* f' *is the same as that of* f*.*

Proof. By Theorem 18.1, the first statement is clear. Now, we suppose that $\mathbb{K}$ has characteristic zero. If $\mathbb{K}$ has residue characteristic zero, then $|n| = 1$ for all $n \in \mathbb{N}^*$, and therefore, the last statement is clear. Now, we assume that $\mathbb{K}$ has residue characteristic $p \neq 0$. By Lemma 6.17, we have $\frac{1}{n} \leq |n| \leq 1$ for every $n \in \mathbb{N}^*$, and therefore, $\lim_{n\to+\infty} \sqrt[n]{|n|} = 1$. Consequently, $\limsup_{n\to+\infty} \sqrt[n]{|a_n|} = \limsup_{n\to+\infty} \sqrt[n-1]{|na_n|}$, and finally, f' has the same radius of convergence as f. □

Corollary 18.7. *Suppose* $\mathbb{K}$ *has characteristic* 0*. Let* $f(x) = \sum_{n=0}^{\infty} a_n x^n \in \mathcal{A}(d(0,r^-))$*. The power series* $\sum_{n=0}^{\infty} \frac{a_n}{n+1} x^{n+1}$ *also belongs to* $\mathcal{A}(d(0,r^-))$ *and has the same radius of convergence as that of* f *and is a primitive of* f *in* $d(a,r^-)$*.*

Remark. Unlike in Archimedean analysis, when the characteristic p of $\mathbb{K}$ is not zero, there do exist power series f whose derivatives have a radius of convergence bigger than that of f. For example, let $f(x) = \sum_{n=0}^{\infty} x^{p^n}$: the radius of convergence of f is 1, while this of f' is $+\infty$.

Theorem 18.8. *Let* $a \in \mathbb{K}$ *and let* $R \in \mathbb{R}_+$*. Let* $(f_m)_{m\in\mathbb{N}^*}$ *be a sequence of* $H(d(a,R))$ *converging uniformly to a function* f*. Then* *the sequence* $(f'_m)_{m\in\mathbb{N}^*}$ *converges uniformly to* f' *in* $H(d(a,R))$ *and we have* $\|f'_m - f'\|_{d(a,R)} \leq \frac{\|f_m-f\|_{d(a,R)}}{R}$ $\forall m \in \mathbb{N}$.

Proof. For each $m \in \mathbb{N}$, set $f_m(x) = \sum_{n=0}^{\infty} a_{n,m}x^n$ and let $f(x) = \sum_{n=0}^{\infty} b_n x^n$. Then for each $m \in \mathbb{N}$, we have $\lim_{n\to+\infty} |a_{n,m}|R^n = 0$, $\lim_{n\to+\infty} |b_n|R^n = 0$. Now, the Banach norm of $f_m - f$ tends to zero when m goes to ∞, hence $\lim_{m\to+\infty} \big(\sup_{n\in\mathbb{N}}(|a_{n,m} - b_n|R^n)\big) = 0$. Consequently, considering the respective derivatives, we have

$$\lim_{m\to+\infty}\left(\sup_{n\in\mathbb{N}}(|na_{n,m} - nb_n|R^n)\right) = 0,$$

and therefore, by Theorem 18.1, we have $\|f'_m - f'\|_{d(a,R)} \leq \frac{\|f_m-f\|_{d(a,R)}}{R}$. We are done. □

Corollary 18.9. *Let* $a \in \mathbb{K}$ *and let* $R \in \mathbb{R}_+$. *Let* $(f_m)_{m\in\mathbb{N}^*}$ *be a sequence of* $H(\mathbb{K}\backslash d(a,R^-))$ *converging uniformly to a function* f. *Then the sequence* $(f'_m)_{m\in\mathbb{N}^*}$ *converges uniformly to* f' *in* $H(\mathbb{K}\backslash d(a,R^-))$ *and we have* $\|f'_m - f'\|_{\mathbb{K}\backslash d(a,R^-)} \leq R\|f_m - f\|_{\mathbb{K}\backslash d(a,R^-)}$ $\forall m \in \mathbb{N}$.

Theorem 18.10. *Suppose* $\mathbb{K}$ *has characteristic* 0. *Let* $a \in \mathbb{K}$ *and let* $R \in \mathbb{R}_+$. *Let* $(f_m)_{m\in\mathbb{N}^*}$ *be a sequence of* $\mathcal{A}(d(a,R^-))$ *such that the sequence* $(f'_m)_{m\in\mathbb{N}^*}$ *converges uniformly to a function* h *in* $H(d(a,r))$ $\forall r \in]0,R[$ *and such that the sequence* $f_m(a)$ *converges in* $\mathbb{K}$. *Then the sequence* $(f_m)_{m\in\mathbb{N}^*}$ *converges to a function* $f \in \mathcal{A}(d(a,R^-))$ *such that* $f' = h$ *and the convergence is uniform in* $d(a,r)$ *for every* $r \in]0,R[$.

Proof. Without loss of generality, we can assume $a = 0$. Let us fix $r \in]0,R[$ and let us show that the sequence $(f_m)_{m\in\mathbb{N}^*}$ converges uniformly to a function $f \in H(d(a,r))$ such that $f' = h$. For every $m \in \mathbb{N}$, let $f'_m = \sum_{n=0}^{+\infty} b_{n,m}x^n$ and let $h(x) = \sum_{n=0}^{+\infty} b_n x^n$. Take $s \in]r,R[$. By hypothesis, we have

(1) $\lim_{m\to+\infty} \sup_{n\in\mathbb{N}} |b_{n,m} - b_n|s^n = 0.$

But since $|n| \geq \frac{1}{n}$, we have $\lim_{n\to+\infty} \frac{\rho^{n+1}}{|n+1|} = 0$ $\forall \rho < 1$, therefore, by (1), we have $\lim_{m\to+\infty} \big(\sup_{n\in\mathbb{N}} \big|\frac{b_{n,m}-b_n}{n+1}\big| r^{n+1}\big) = 0$. Consequently, the sequence $(f_m - f_m(0))_{m\in\mathbb{N}}$ converges uniformly to the function $g(x) = \sum_{n=0}^{+\infty} \frac{b_n}{n+1} x^{n+1}$. Set $\lim_{m\to+\infty} f_m(0) = b_0$. Then the sequence $(f_m)_{m\in\mathbb{N}}$ converges to $g(x) + b_0$ uniformly in $d(0, r)$ for every $r < R$. □

Corollary 18.11. *Suppose $\mathbb{K}$ has characteristic 0. Let $\alpha \in \mathbb{K}$ and let $R \in \mathbb{R}_+$ and let $a \in \mathbb{K}\backslash d(\alpha, R)$. Let $(f_m)_{m\in\mathbb{N}^*}$ be a sequence of $\mathcal{A}(\mathbb{K}\backslash d(\alpha, R))$ such that the sequence $(f'_m)_{m\in\mathbb{N}^*}$ converges uniformly to a function h in $H(\mathbb{K}\backslash d(\alpha, r^-))$ $\forall r > R$ and that the sequence $(f_m(a))_{m\in\mathbb{N}}$ converges in $\mathbb{K}$. Then the sequence $(f_m)_{m\in\mathbb{N}^*}$ converges to a function $f \in \mathcal{A}(\mathbb{K}\backslash d(\alpha, R))$ such that $f' = h$ and the convergence is uniform in $\mathbb{K}\backslash d(\alpha, r^-)$ for every $r > R$.*

Theorem 18.12. *Let $r > 0$. If $\mathbb{K}$ has characteristic 0, then an element $f \in H(d(0, r))$ has a derivative identically equal to 0 if and only if it is equal to a constant.*

If $\mathbb{K}$ has a characteristic $p \neq 0$, then an element $f \in H(d(0, r))$ has a derivative identically equal to 0 if and only if there exists $g \in H(d(0, r))$ such that $f(x) = (g(x))^p$.

Proof. By Theorem 14.6, each element of $H(d(0, r)$ is a convergent power series, hence the statement about the case when $\mathbb{K}$ has characteristic zero is obvious. Now, suppose that $\mathbb{K}$ has a characteristic $p \neq 0$. If there exists $g \in H(d(0, r))$ such that $f(x) = (g(x))^p$, obviously we have $f'(x)$ identically equal to 0.

Now, we suppose $f'(x)$ identically equal to 0. Hence, $f(x)$ is of the form $\sum_{j=0}^{\infty} b_j x^{jp}$, with $\lim_{j\to\infty} |b_j| r^{jp} = 0$. For each $j \in \mathbb{N}$, we can take $c_j \in \mathbb{K}$ such that $(c_j)^p = b_j$. Then, it is seen that $\lim_{j\to\infty} |c_j| r^j = 0$. Now, we can put $g(x) = \sum_{n=0}^{\infty} c_n x^n$, and therefore, g belongs to $H(d(0, r))$. Since $\mathbb{K}$ has characteristic p, we have $(g(x))^p = f(x)$. This ends the proof. □

Corollary 18.13. *Let $r > 0$. If $\mathbb{K}$ has characteristic 0, then an element $f \in H(\mathbb{K}\backslash d(0, r^-))$ (respectively $f \in H_0(\mathbb{K}\backslash d(0, r^-))$) has a derivative identically equal to 0 if and only if it is equal to a constant (respectively to 0).*

If $\mathbb{K}$ has a characteristic $p \neq 0$, then an element $f \in H(\mathbb{K}\backslash d(0, r^-))$ (respectively $f \in H_0(\mathbb{K}\backslash d(0, r^-))$) has a derivative identically equal to 0 if and only if there exists $g \in H(\mathbb{K}\backslash d(0, r^-))$ (respectively $g \in H_0(\mathbb{K}\backslash d(0, r^-))$) such that $f(x) = (g(x))^p$.

Corollary 18.14. *Let $r > 0$. If $\mathbb{K}$ has characteristic 0, then a power series $f(x) \in \mathcal{A}(d(0, r^-))$ has a derivative identically equal to 0 if and only if it is equal to a constant.*

If $\mathbb{K}$ has a characteristic $p \neq 0$, then a power series $f(x) \in \mathcal{A}(d(0, r^-))$ has a derivative identically equal to 0 if and only if there exists $g \in \mathcal{A}(d(0, r^-))$ such that $f(x) = (g(x))^p$.

Theorem 18.15 improves Theorems 18.1 and 18.2 concerning derivatives of order $k > 1$.

Theorem 18.15. *Let $a \in \mathbb{K}$, let $R, R', R'' \in \mathbb{R}_+^*$ with $R' < R''$, and let $f \in H(d(0, r))$ (respectively let $f \in H(\mathbb{K} \setminus d(0, R^-))$, respectively let $f \in H(\Gamma(0, R', R'')))$. Then, for every $k \in \mathbb{N}^*$, for every $r < R$ (respectively $r > R$, respectively $r \in]R', R''[$), we have*

$$|f^{(k)}|(r) \leq |k!| \frac{|f|(r)}{r^k}.$$

Proof. Let $f(x) = \sum_{n=0}^{\infty} a_n x^n \in H(d(0, R^-))$. By Theorem 14.6, we have $|f|(r) = \sup_{n \in \mathbb{N}} |a_n| r^n$ and $|f^{(k)}|(r) = \sup_{n \geq k} |\frac{(n!)}{((n-k)!)} a_n| r^{n-1}$. But $\frac{(n!)}{((n-k)!)}$ is an integer multiple of $k!$ because the combination $\binom{n}{k}$ belongs to $\mathbb{N}$. Consequently, $\left|\frac{(n!)}{((n-k)!)}\right| \leq |k!|$, and therefore, we obtain $|f^{(k)}|(r) \leq \sup_{n \geq k} |k!||a_n| r^{n-k}$. The proof is similar when f belongs to $H(\mathbb{K} \setminus d(0, r))$ or to $H(\Gamma(0, r', r''))$. □

Corollary 18.16. *Let $a \in \mathbb{K}$, let $r \in \mathbb{R}_+^*$ and let $f \in H(d(a, r))$. Then,*

$$\|f^k\|_{d(a,r)} \leq |k!| \frac{\|f\|_{d(a,r)}}{r^k}.$$

Theorem 18.17. *Let $f \in H_b(D)$ and let $\rho = \delta(D, (\mathbb{K} \setminus \overline{D}))$. If $\rho > 0$, then f' belongs to $H_b(D)$ and satisfies $\|f'\|_D \leq \frac{1}{\rho} \|f\|_D$.*

Proof. Let $(T_n)_{n \in \mathcal{S}}$ be the sequence of the f-holes and let $D' = \widetilde{D} \setminus \left(\bigcup_{n \in \mathcal{S}} T_n\right)$. By Theorem 15.1, we know that $f \in H_b(D')$ and that

(1) $\|f\|_{D'} = \|f\|_D$.

By Theorem 18.1, f has a derivative f' in D' and we first check that the function f' satisfies $\|f'\|_{D'} \leq \frac{1}{\rho} \|f\|$. Let $a \in D'$. The disk $d(a, \rho^-)$

is obviously included in D because if a point $b \in d(a, \rho^-)$ belonged to $\mathbb{K}\setminus D'$, since D is closed, there would be a disk $d(b, r^-) \subset \mathbb{K}\setminus D'$ with $r < \rho$. Thus, when $x \in d(a, \rho^-)$, $f(x)$ is of the form $\sum_{n=0}^{\infty} a_n(x-a)^n$ and hence the conclusion comes from Theorem 18.1. □

Corollary 18.18. *Let $f \in H(D)$ be such that the set of diameters of the f-holes has a strictly positive lower bound. Then, $f' \in H(D)$.*

Proof. By Theorem 11.5, we know that f is in the form $g + h$ with $h \in R(D)$ and $g \in H_b(\overline{D})$. Obviously, h' belongs to $R(D)$, and by Theorem 18.17, we have $g' \in H_b(\overline{D})$. □

Theorem 18.19. *Let $D \in \text{Alg}$ be closed and open. Let $f \in H_b(D)$, let $(V_n)_{n\in\mathbb{N}^*}$ be the set of f-holes, and let $\sum_{n=0}^{\infty} f_n$ be its Mittag-Leffler series on D defined as $f_0 = \overline{\overline{f_0}}$ and for every $n \in \mathbb{N}^*$, $f_n = \overline{\overline{f_{V_n}}}$. The following three conditions are equivalent:*

(a) *f' belongs to $H(D)$.*
(b) *The series $\sum_{n=0}^{\infty} f_n'$ converges in $H(D)$.*
(c) *The series $\sum_{n=0}^{\infty} f_n'$ converges to f' in $H_b(D)$.*

Proof. We first prove the equivalence between (b) and (c). For each $q \in \mathbb{N}$, the sum $\sum_{n=0}^{q} f_n'$ clearly belongs to $H_b(D)$. Thus, we assume that this series $\sum_{n=0}^{\infty} f_n'$ converges to an element $h \in H(D)$ and we prove that $h = f' \in H_b(D)$. Let $\alpha \in D$. There exists a disk $d(\alpha, r) \subset D$. We show that $h(\alpha) = f'(\alpha)$. For every $\psi \in H(D)$, $\widetilde{\psi}$ denotes the restriction of ψ to $d(\alpha, r)$. Since the sequence $\left(\widetilde{\sum_{j=0}^{n} f_j}\right)$ converges to $\widetilde{f}$, by Theorem 18.17, the sequence of the derivatives $\left(\widetilde{\sum_{j=0}^{n} f_j'}\right)$ does converge to $\widetilde{f}'$, hence it is clearly seen that $f'(\alpha) = h(\alpha)$, and therefore, $f' = h$. We check that f' is bounded in D. The sequence $\|f_n'\|_D$ has limit 0 hence is obviously bounded and therefore its sum f' is bounded in D. Thus, (b) and (c) are equivalent.

Since (c) trivially implies (a), we just have to prove that (a) implies (b). Thus, we suppose (a) to be true and prove (b).

First, we suppose D bounded. For each hole T of D that is either an f-hole or an f'-hole, we denote by $\overline{\overline{f_T}}$ (respectively $\overline{\overline{g_T}}$) the Mittag-Leffler term of f (respectively f'). Let $\mathcal{S}$ be the set of holes

T such that $\overline{\overline{(f_T)'}} \neq \overline{\overline{g_T}}$ and let $\mathcal{T}$ be the set of f-holes such that $\overline{\overline{(f_T)'}} = \overline{\overline{g_T}}$. If we can show that $\mathcal{S} = \emptyset$, then (b) is clearly proven.

Hence, we suppose $\mathcal{S} \neq \emptyset$. All the $\overline{\overline{g_T}}$ are equal to zero except maybe a countable family of them. The series $\sum_{T\in\mathcal{T}} \overline{\overline{g_T}}$ and $\sum_{T\in\mathcal{S}} \overline{\overline{g_T}}$ obviously converge in $H(D)$, and then we have $f' = \sum_{T\in\mathcal{T}} \overline{\overline{(f_T)'}} + \sum_{T\in\mathcal{S}} \overline{\overline{g_T}}$. Since (b) implies (c), the series $\sum_{T\in\mathcal{T}} \overline{\overline{(f_T)'}}$ is clearly equal to the derivative of $\sum_{T\in\mathcal{T}} \overline{\overline{f_T}}$. Let $h = \sum_{T\in\mathcal{S}} \overline{\overline{f_T}} = f - \sum_{T\in\mathcal{T}} \overline{\overline{f_T}}$. Then, $h' = f' - \sum_{T\in\mathcal{T}} \overline{\overline{(f_T)'}} = \sum_{T\in\mathcal{S}} \overline{\overline{g_T}}$. Let $\mathcal{D}$ be the family of diameters of the holes T that belong to $\mathcal{S}$ and let λ be its lower bound. Suppose $\lambda > 0$. By Theorem 18.17, the series $\sum_{T\in\mathcal{S}} \overline{\overline{(f_T)'}}$ converges to h', hence $\sum_{T\in\mathcal{S}} \overline{\overline{(f_T)'}}$ is the Mittag-Leffler series of h' on D, hence $\overline{\overline{(f_T)'}} = \overline{\overline{g_T}}$ for all $T \in \mathcal{S}$ and that contradicts the definition of $\mathcal{S}$. Hence, $\lambda = 0$.

Now, we prove that there exists a hole $V = d(a, r^-) \in \mathcal{S}$ with an annulus $\Gamma(a, r, s)$ such that the set $\mathcal{U}$ of the diameters ρ of the f-holes included in $\Gamma(a, r, s)$ has a strictly positive lower bound. Indeed, suppose such a hole V does not exist. Then, we can easily construct a sequence of f-holes $(T_n)_{n\in\mathbb{N}^*}$ of the form $T_n = d(a_n, r_n^-)$ with (1) $r_n \leq \frac{1}{n}$ and

(2) $|a_{n+1} - a_n| \leq \dfrac{2}{n}$.

For example, assume the sequence has just been constructed up to the rank q, satisfying (1) and (2) for $n \leq q$. Since V does not exist, then in $\Gamma(a_q, r_q, \frac{2}{q})$, we can find an f-hole $T_{q+1} = d(a_{q+1}, r_{q+1}^-)$ with $r_{q+1} < \frac{2}{q+1}$ and then the sequence is clearly constructed by induction by taking any first f-hole $T_1 = d(a_1, r_1^-)$. The sequence $(T_n)_{n\in\mathbb{N}^*}$ clearly converges to a point $w \in D$ and that contradicts the hypothesis "D is closed and open". Hence, we have now proven the existence of the f-hole V with an annulus $\Gamma(a, r, s)$ and a number $\xi > 0$ such that every f-hole $T \subset \Gamma(a, r, s,)$ satisfies

(3) diam $(T) \geq \xi$.

Let $\mathcal{L}$ be this family of f-holes included in $\Gamma(a, r, s)$. Let $l = \sum_{T\in\mathcal{L}} \overline{\overline{f_T}}$. By Theorem 18.17, the series $\sum_{T\in\mathcal{L}} \overline{\overline{(f_T)'}}$ converges to l' in

$H(D)$. Now, let $\psi = h - l - \overline{\overline{f_V}}$. Clearly, ψ belongs to $H(D)$ and no hole T (of D) included in $d(a, s)$ is a ψ-hole. Hence, ψ extends to an element of $H(D \cup d(a, s))$. In $d(a, s), \psi(x)$ is equal to a power series $\phi(x) \in H(d(a, s))$, hence $\phi' \in H(d(a, s))$. Thus, in $D \cap d(a, s)$, $\psi'(x)$ is equal to the series $\phi'(x)$, and then for every hole T of D included in $d(a, s)$, the Mittag-Leffler term of ψ associated with T (with respect to D) is zero.

On the other hand, we have $\psi' = h' - l' - \overline{\overline{(f_V)'}} = \sum_{T \in \mathcal{S}} \overline{\overline{g_T}} - \sum_{T \in \mathcal{L}} \overline{\overline{(f_T)'}} - \overline{\overline{(f_V)'}}$ and then the Mittag-Leffler term of ψ' associated with V (with respect to D) is $\overline{\overline{g_V}} - \overline{\overline{(f_V)'}} \neq 0$. Hence, we have a contradiction with $\psi \in H(d(a, s))$. This finishes proving that (b) is true when D is bounded.

Now, we suppose D unbounded. Since $D \in \text{Alg}$, there exists a disk $d(0, S)$ such that all holes of D are included in this disk. Then for every element $h \in H(D)$, its Mittag-Leffler series in $H(D)$ is the same as in $H(D')$. This is true for both f, f', and therefore (b), which is true in $H(D')$, is obviously true in $H(D)$. That ends the proof of Theorem 18.19. $\square$

When $\mathbb{K}$ has characteristic zero, in most of the cases, we are now able to answer the question "does $f' = 0$ implies $f = ct$?". When D is not infraconnected, it admits an empty annulus $\Lambda = \Gamma(a, r', r'')$, and hence by Proposition 11.15, we know that there exists $w \in H(D)$ such that $w(x) = 1$ whenever $x \in \mathcal{I}(\Lambda)$, while $w(x) = 0$ whenever $x \in \mathcal{E}(\Lambda)$. Thus, the condition "$D$ is infraconnected" is certainly necessary to be able to answer "yes" to the question above.

The two theorems that follow show this condition to be sufficient too, provided D satisfies a little extra condition like to be closed or to belong to Alg.

By Theorems 15.1, 18.19, and 18.8 and Corollary 18.9, we can derive the following:

Corollary 18.20. *Let D be an open closed bounded infraconnected subset of $\mathbb{K}$ and let $(f_n)_{n \in \mathbb{N}}$ be a sequence of $H(D)$ converging uniformly to a function f. Then, the sequence $(f'_n)_{n \in \mathbb{N}}$ converges uniformly to f'.*

Remark. Let D be an open closed infraconnected bounded subset of $\mathbb{K}$, let $\alpha \in D$, and let $(f_n)_{n \in \mathbb{N}}$ be a sequence of $H(D)$

such that the sequence $f_n(\alpha)$ converges in $\mathbb{K}$ and that the sequence $(f'_n)_{n\in\mathbb{N}}$ converges in $H(D)$ to a function $h \in H(D)$. Comparatively to the Archimedean context, we could expect that the sequence $(f_n)_{n\in\mathbb{N}}$ converges in $H(D)$ to a function f such that $f' = h$. Actually, that's wrong. For example, define the sequence $(f_n)_{n\in\mathbb{N}}$ as $f_n(x) = \sum_{k=0}^{n} \frac{x^{p^{2k}}}{p^k}$. Then, $f'_n(x) = 1+\sum_{k=1}^{n} p^k x^{p^{2k-1}}$, and hence, the sequence $(f'_n)_{n\in\mathbb{N}}$ converges in $H(d(0,1))$ to the function $1+\sum_{k=1}^{+\infty} p^k x^{p^{2k-1}}$, whereas $f_n(0) = 1\ \forall n \in \mathbb{N}$. However, the function $\sum_{k=0}^{+\infty} \frac{x^{p^{2k}}}{p^k}$ is unbounded in $d(0,1)$ and hence does not belong to $H(d(0,1))$. That remark does not contradict Theorem 18.10 which only concerned analytic functions in an "open" disk $d(0,r^-)$.

Theorem 18.21. *$\mathbb{K}$ is supposed to have characteristic zero. Let E be an open subset of $\mathbb{K}$ such that $\overline{E}$ also is open. Then, E is infraconnected if and only if for every $f \in H(E)$ such that $f'(x) = 0$ whenever $x \in E$, we have $f = ct$.*

Proof. If E is not infraconnected, it admits at least an empty annulus Λ, and then, by Proposition 11.15, the characteristic function u of $\mathcal{I}(\Lambda)$ belongs to $H(E)$. Hence, there do exist non-constant elements $f \in H(E)$ whose derivative is identically 0. Now, let E be infraconnected and let $f \in H(E)$ satisfy $f'(x) = 0$ whenever $x \in E$. We just have to prove that f is a constant.

Suppose first that E is bounded. By Theorem 11.5, f is in the form $f^* + \overline{f}$ with $f^* \in R_0(\mathbb{K}\setminus(\overline{E}\setminus E))$ and $\overline{f} \in H(\overline{E})$, so $f^{*\prime}(x)+\overline{f}'(x) = 0$ whenever $x \in E$, and therefore, $f^{*\prime} = -\overline{f}'$. Hence, $\overline{f}' \in R_0(\mathbb{K}\setminus(\overline{E}\setminus E)) \cap H(\overline{E})$. Thus, f^* is a rational function that has no pole in $\mathbb{K}$ and then it is a polynomial. Now, as an element of $R_0(\mathbb{K}\setminus(\overline{E}\setminus E))$, it tends to 0 when $|x|$ goes to $+\infty$, hence $f^* = 0$ and therefore, $f^{*\prime}$ is identically 0. Since f^* belongs to $R_0(\mathbb{K}\setminus(\overline{E}\setminus E))$ clearly, $f^* = 0$, and therefore, f belongs to $H(\overline{E})$.

Let $a \in E$ and let $\sum_{n\in\mathcal{S}} h_n$ be the Mittag-Leffler series of f in $\overline{E}$, with $h_0 = \overline{\overline{f_0}}$, and for each $n \in \mathcal{S}$, set $h_n = \overline{\overline{f_{T_n}}}$, for any f-hole T_n. Since $\overline{E}$ is open, we can apply Theorem 18.19 to E and then we have $(h_n)' = 0$ for every $n \in \mathbb{N}$. Since $\overline{E}$ is bounded of diameter r, then by Theorem 16.8, we know that h_0 is a constant. In the same way, by Corollary 18.13, for each $q \in \mathbb{N}$, we know that $h_q = 0$. Hence, f is a constant.

Finally, let E be unbounded. Then, for all $r > 0$, the set $E_r = E \cap d(0,r)$ is such that $\overline{E_r}$ is open, hence f is constant in E_r and therefore in all of E. □

Remark. In particular, Theorem 18.21 applies to open closed sets.

Corollary 18.22. *$\mathbb{K}$ is supposed to have characteristic zero. Let E be open and belong to* Alg. *Then E is infraconnected if and only if for every $f \in H(E)$ such that $f'(x) = 0$ whenever $x \in E$, f is a constant in E.*

Proof. Indeed, since E belongs to Alg, it satisfies Condition (B) in Theorem 17.9: $\overline{D}\setminus D \subset \left(\overset{\circ}{\overline{D}}\right)$. But then, as it is also open, we check that $\overline{E}$ is open. □

We now study thoroughly the question of whether all analytic elements in a set D have derivative in $H(D)$.

Definition. We call *piercing of D* the number $\delta(\overline{D}, \mathbb{K}\setminus D) > 0$ and D is said to be *well pierced* if $\delta(\overline{D}, \mathbb{K}\setminus D) > 0$.

Theorem 18.23. *Let $\overline{D}$ be open. Then, D is well pierced if and only if for every $f \in H(D)$, f' also belongs to $H(D)$.*

Proof. If D is well pierced, by Corollary 18.18, we know that for every $f \in H(D)$, f' belongs to $H(D)$. Now let us suppose D has piercing zero and let $(T_n)_{n\in\mathbb{N}}$ be a sequence of holes $T_n = d(\alpha_n, \rho_n^-)$ such that $\lim_{n\to\infty} \rho_n = 0$. Let λ_n be a sequence in $\mathbb{K}$ such that $\lim_{n\to\infty} \frac{|\lambda_n|}{\rho_n} = 0$, while $\lim_{n\to\infty} \frac{|\lambda_n|}{\rho_n^2} = +\infty$. It is seen that the series $\sum_{n=1}^{\infty} \frac{\lambda_n}{x-\alpha_n}$ converges in $H(D)$ to an element f, while the series $\sum_{n=1}^{\infty} \frac{\lambda_n}{(x-\alpha_n)^2}$ does not. On the other hand, $\sum_{n=1}^{\infty} \frac{\lambda_n}{x-\alpha_n}$ obviously is the Mittag-Leffler series of f. If f' belongs to $H(D)$, by Theorem 18.19, its Mittag-Leffler series must be $\sum_{n=1}^{\infty} \frac{\lambda_n}{(x-\alpha_n)^2}$. Since this series does not converge, this is just impossible. □

Before closing this chapter, we note the following result that may be sometimes helpful in differential equations.

Theorem 18.24. *Let $\overline{D}$ be open. We suppose that both f and f' belong to $H(D)$. For every $\epsilon > 0$, there exists $h \in R(D)$ such that $\|f - h\|_D \le \epsilon$ together with $\|f' - h'\|_D \le \epsilon$.*

Proof. First, we suppose that f belongs to $H_b(\overline{D})$. We have to introduce a notation. Let $g \in H_b(\widetilde{D})$. If D is bounded, $\widetilde{D}$ is a disk $d(a,r)$ and g is of the form $\sum_{m=0}^{\infty} \lambda_m(x-a)^m$. Then, for every $q \in \mathbb{N}$, we put $(g)_q = \sum_{m=0}^{q} \lambda_m(x-a)^m$. If D is unbounded, then g is a constant $\lambda_\circ$, and we put $(g)_q = g$ whenever $q \in \mathbb{N}$. Let $T = d(b, r^-)$ be a hole of D and let $l(x) = \sum_{m=1}^{\infty} \frac{\mu_m}{(x-b)^m}$. For each $q \in \mathbb{N}^*$, we put $(l)_q = \sum_{m=1}^{q} \frac{\mu_m}{(x-b)^m}$. Now, let $\sum_{n=0}^{\infty} f_n$ be the Mittag-Leffler series of f, with $f_0 = \overline{\overline{f_0}}$ and for each $n \in \mathbb{N}^*$, $f_n = \overline{\overline{f_{T_n}}}$, for any f-hole T_n. By Theorem 18.19, the Mittag-Leffler series of f' is $\sum_{n=0}^{\infty} f'_n$, and therefore, there exists an integer $N(\epsilon)$ such that

$$(1) \quad \left\| \sum_{n=0}^{N(\epsilon)} f_n - f \right\|_D \leq \epsilon$$

and

$$(2) \quad \left\| \sum_{n=0}^{N(\epsilon)} f'_n - f' \right\|_D \leq \epsilon.$$

Obviously, we have an integer $Q(\epsilon)$ such that $\|f_n - (f_n)_{Q(\epsilon)}\|_D \leq \epsilon$ whenever $n = 0, \dots, N(\epsilon)$, and then by (1) and (2), it is easily seen that $\|\sum_{n=0}^{N(\epsilon)} (f_n)_{Q(\epsilon)} - f\|_D \leq \epsilon$ and $\|\sum_{n=0}^{\infty} (f'_n)_{Q(\epsilon)} - f'\|_D \leq \epsilon$. By putting $h = \sum_{n=0}^{N(\epsilon)} (f_n)_{Q(\epsilon)}$, we obtain the $h \in R(D)$ we want.

Now, we can consider the general case. By Corollary 11.7, f is of the form $l + \psi$, with $l \in H_b(\overline{D})$ and $\psi \in R(D)$. Hence, $f' = l' + \psi'$. Since f' belongs to $H(D)$, so does ψ'. Hence, by Theorem 18.21, ψ' belongs to $H_b(\overline{D})$. We have just proven that there exists $t \in H_b(\overline{D})$ such that $\|l - t\|_D \leq \epsilon$ and $\|l' - t'\|_D \leq \epsilon$. Hence, we just have to consider $h = t + \psi$, and this ends the proof. □

Theorem 18.25. *Let $\overline{D}$ be open. Suppose $0 \in D$. Let $f \in H(D)$ and let $r \in [\delta(0, D), \mathrm{diam}(D)]$. Then, $|f'|(r) \leq \frac{|f|(r)}{r}$.*

Proof. Suppose first that $|f|(r) \neq 0$. Let $\epsilon > 0$ and let $\eta > 0$ be such that $\frac{|f|+\eta}{r} + \eta < \frac{|f|}{r} + \epsilon$. Now, by Theorem 18.24, we can find $h \in R(D)$ such that $|h'|(r) \leq |f'| + \eta$ and $|h|(r) \leq |f| + \eta$.

By Corollary 18.3, we have $|h'|(r) \leq \frac{|h|(r)}{r}$, hence

$$|f'|(r) \leq |h'|(r) + \eta \leq \frac{|h|(r)}{r} + \eta \leq \frac{|f|(r) + \eta}{r} + \eta \leq \frac{|f|(r)}{r} + \epsilon.$$

Now, suppose $|f|(r) = 0$. Then, by Theorem 18.24, we can find $h \in R(D)$ such that $\max(|h'|(r), |h|(r)) \leq \epsilon$, hence we have again $|f'| \leq \frac{|f|(r)}{r}$. This is true for all $\epsilon > 0$ and hence the claim is proven. □

In the case when $\mathbb{K}$ has a characteristic $p \neq 0$, we have Theorem 18.26.

Theorem 18.26. *Let $\mathbb{K}$ have characteristic $p \neq 0$, let D be closed, and let $f \in H_b(D)$. Then, $f'(x)$ is identically 0 if and only if there exists $g \in H_b(D)$ such that $f = g^p$.*

Proof. Indeed, if there exists $g \in H_b(D)$ such that $f = g^p$, of course, we have $f' = 0$. Now, suppose that $f'(x)$ is identically 0. Let $a \in D$ and let $\sum_{n \in \mathcal{S}} h_n$ be the Mittag-Leffler series of f in D, with $h_0 = \overline{\overline{f_0}}$ and for each $n \in \mathcal{S}$, $h_n = \overline{\overline{f_{T_n}}}$, for any f-hole T_n. By Theorem 18.19, we have $(h_n)' = 0$ for every $n \in \mathbb{N}$.

If D is unbounded, h_0 is a constant, and then we can find $g_0 \in \mathbb{K}$ such that $(g_0)^p = h_0$. If D is bounded of diameter r, then by Corollary 18.13, we can find $g_0 \in H(d(a, r))$ such that $(g_0)^p = h_0$. In the same way, for each $q \in \mathbb{N}$, by Corollary 18.13, we can find $g_q \in H_0(\mathbb{K} \backslash T_q)$ such that $(g_q)^p = h_q$ and then, it is seen that $\lim_{n \to \infty} \|g_n\|_D = 0$ because for each $n \in \mathbb{N}^*$, we have $(\|g_n\|_D)^p = \|h_n\|_D$. So, the series $\left(\sum_{n=0}^{\infty} g_n\right)$ converges in $H_b(D)$ to an element g which clearly satisfies $g^p = f$. This ends the proof of Theorem 18.26. □

Chapter 19

Properties of the Function Ψ for Analytic Elements

Throughout this chapter, D is infraconnected.

The function $\Psi(f, \mu)$ was defined for rational functions in Chapter 4. Here, we generalize that function to analytic elements. Its interest is to transform the multiplicative property of the norm $|\,.\,|$ into an additive property. Overall, Ψ is piecewise affine. Long ago, such a function was first defined in classical works such as the valuation function of an analytic element [1, 35, 44] denoted by $v(f, \mu)$. However, the function $v(f, \mu)$ has the inconvenient of being contravariant: $\mu = -\log(|x|)$ and $v(f, -\log(|x|)) = -\log(|f|(r))$. Here, we change both senses of variation: $\Psi(f, \mu) = -v(f, -\mu)$ [49, 50].

Among applications, we can show that a set E is infraconnected if and only if for all $f \in H(E)$, $f(E)$ is infraconnected and that an analytic element converges along a monotonous filter $\mathcal{F}$ if and only if f' is vanishing along $\mathcal{F}$.

Notations. For every $a \in \widetilde{D}$, we put $\lambda(a) = \log(\delta(a, D))$ if $\delta(a, D) > 0$ and $\lambda(a) = -\infty$ if $\delta(a, D) = 0$. We denote by S the diameter of D, with $S = +\infty$ if D is not bounded.

Let $a \in \widetilde{D}$ and let $\mathcal{F}$ be a circular filter of center a and diameter $r \in [\delta(a, D), S] \cap \mathbb{R}$. By Proposition 3.17, $\mathcal{F}$ is secant with D and then defines an element ${}_D\varphi_{\mathcal{F}}$ of $\mathrm{Mult}(H(D), \mathcal{U}_D)$.

For every $f \in H(D)$ such that ${}_D\varphi_{\mathcal{F}}(f) \neq 0$, we put $\Psi_a(f, \log r) = \log({}_D\varphi_{\mathcal{F}}(f))$. Next, for an $f \in H(D)$ such that ${}_D\varphi_{\mathcal{F}}(f) = 0$, we put $\Psi_a(f, \log r) = -\infty$.

When $a = 0$, for simplicity, we just put $\Psi(f,\mu) = \Psi_0(f,\mu)$. Then, by definition, we have $\Psi(f,\log r) = \log(|f|(r))$.

In the same way, consider an annulus $\Gamma(0,r,t)$ and $f \in \mathcal{A}(\Gamma(0,r,t))$. Then, for any $s \in]r,t[$, f belongs to $H(C(0,s)$, so we can put consider $\Psi(f,\log(s)) = \log(|f|(s))$. If we consider $f \in \mathcal{A}(a,r^-)$, so much the more, we can consider $\Psi_a(f,\ell)$ for each $\ell < \log(r)[$.

Remark. Let $f(x) = \sum_{-\infty}^{+\infty} a_n x^n \in H(C(0,r))$ for some $r > 0$. By Theorem 15.7, we have $\Psi(f,\log r) = \log\big({}_{C(0,r)}\varphi_{0,r}(f)\big) = \log \|f\|_{C(0,r)} = \sup_{n\in\mathbb{Z}} \Psi(a_n) + n\log r$.

Proposition 19.1. *Let $a \in \widetilde{D}$, let $\mu \in [\lambda(a), \log(S)] \cap \mathbb{R}$, and let $f,g \in H(D)$. Then, $\Psi_a(f+g,\mu) \le \max(\Psi_a(f,\mu), \Psi_a(g,\mu))$, and when $\Psi_a(f,\mu) > \Psi_a(g,\mu)$, then $\Psi_a(f+g,\mu) = \Psi_a(f,\mu)$. Moreover, $\Psi_a(fg,\mu) = \Psi_a(f,\mu) + \Psi_a(g,\mu)$.*

Let $r,\ t \in]0,\ +\infty[$ be such that $r \le t$. Let $f \in H(D)$ be such that $\Psi_a(f,\mu)$ is bounded in $[\log(r), \log(t)]$. Then, $\Psi_a(f,\mu)$ is continuous and piecewise affine in $[\log(r), \log(t)]$. Further, there exists $h \in R(D)$ such that $\Psi_a(f,\mu) = \Psi_a(h,\mu)\ \forall \mu \in [\log(r), \log(t)]$.

Inside $D \cap \Gamma(a,r,t)$, the relation $\Psi(f(x)) = \Psi_a(f, \Psi(x-a))$ holds in all classes of all circles $C(a,s)$, except maybe in finitely many classes of finitely many circles $C(a,s)$.

Moreover, if $\Gamma(a,r,t) \subset D$, the function $\Psi_a(f,\mu)$ is convex in $[\log r, \log t]$.

Proof. Without loss of generality, we can assume $a = 0$. The first statements concerning operations and inequalities come directly from those of multiplicative semi-norms ${}_D\varphi$. Now, suppose that $\Psi(f,\mu)$ is bounded in $[\log(r), \log(t)]$, hence there exists $\epsilon > 0$ such that $\Psi(f,\mu) > \log\epsilon\ \forall\mu \in [\log r, \log t]$.

Let $h \in R(D)$ satisfy $\|f-h\|_D < \epsilon$. Particularly, for every circular filter $\mathcal{F}$ secant with D, we have ${}_D\varphi_{\mathcal{F}}(f-h) < \epsilon$, and particularly, ${}_D\varphi_{a,\rho}(f-h) < \epsilon\ \forall \rho \in [r,t]$ i.e. $\Psi(f-h,\mu) < \log(\epsilon) < \Psi(f,\mu)\ \forall\mu \in [\log r, \log t]$. Consequently, $\Psi(f,\mu) = \Psi(h,\mu)\ \forall \mu \in [\log r, \log t]$. Now, by Corollary 4.18, the function $\Psi(h,\mu)$ is continuous, piecewise in $[\log r, \log t]$ and so is $\Psi(f,\mu)$. Moreover, if $\Gamma(a,r,t) \subset D$, the function $\Psi(h,\mu)$ is convex in $[\log r, \log t]$, hence so is $\Psi(f,\mu)$.

By Lemma 4.13, the relation $\Psi(h(x)) = \Psi_a(h, \Psi(x-a))$ holds in all classes of all circles $C(a,s)$, except maybe in finitely many classes of finitely many circles $C(a,s)$. Therefore, the same relation holds for f. □

Proposition 19.2. *Let $a \in D$ and let $f \in H(D)$ satisfy $f(a) \neq 0$. There exists $\mu_\circ \in \mathbb{R}$ such that $\Psi_a(f,\mu) = \Psi(f(a))$ whenever $\mu \leq \mu_\circ$. Let $r \in \mathbb{R}_+^*$, let $\Lambda = C(0,r)$, and let f and $g \in H(\Lambda)$ satisfy $\|f-g\|_\Lambda < \|f\|_\Lambda$. Then, we have $\nu^+(f,\log r) = \nu^+(g,\log r)$, $\nu^-(f,\log r) = \nu^-(g,\log r)$.*

Proof. Indeed, let us take $r > 0$ such that $|f(x) - f(a)| < |f(a)|$ whenever $x \in d(a,r) \cap D$, hence $|f(x)| = |f(a)|$ whenever $x \in d(a,r) \cap D$, and therefore, $\Psi_a(f,\mu) = \Psi(f(a))$ whenever $\mu \leq \log(r)$.

Let $f(x) = \sum_{-\infty}^{+\infty} a_n x^n$ and let $g(x) = \sum_{-\infty}^{+\infty} b_n x^n$. From the hypothesis, we see that $\|f\|_\Lambda = \|g\|_\Lambda$. By Corollary 14.9, we have

(1) $\sup_{n\in\mathbb{Z}} |a_n| r^n = \|f\|_\Lambda = \sup_{n\in\mathbb{Z}} |b_n| r^n$

and $\|f-g\|_\Lambda = \sup_{n\in\mathbb{Z}} |a_n - b_n| r^n$.

Let $s = \nu^-(f,\log r)$ and let $t = \nu^+(f,\log r)$. We see that $|a_s - b_s| r^s \leq \|f-g\|_\Lambda < \|f\|_\Lambda = |a_s| r^s$, hence

(2) $|b_s| = |a_s|$.

In the same way, we have

(3) $|a_t| = |b_t|$.

Now, for every $n < s$ and for every $n > t$, we have $|a_n| r^n < |a_s| r^s = \|f\|_\Lambda$, hence $|b_n| r^n < \|f\|_\Lambda$. Finally, by (1), (2), and (3), we see that $\nu^-(g,\log r) = s$, $\nu^+(g,\log r) = t$. □

By Propositions 19.1, 19.2, and 4.19, we can derive Corollary 19.3.

Corollary 19.3. *Let $f(x) \in H(\Gamma(0,r_1,r_2))$ (respectively $f(x) \in H(\Delta(0,r_1,r_2))$) (with $0 < r_1 < r_2$) and let $\sum_{-\infty}^{+\infty} a_n x^n$ be its Laurent series. The function $\mu \to \Psi(f,\mu)$ is bounded in $]\log r_1, \log r_2[$ (respectively in $[\log(r_1), \log(r_2)]$) and equal to $\sup_{n\in\mathbb{Z}}(\Psi(a_n) + n\mu)$. Next, we have $\Psi(f(x)) \leq \Psi(f,\Psi(x))$ whenever $x \in \Gamma(0,r_1,r_2)$ (respectively whenever $x \in \Delta(0,r_1,r_2)$) and the equality holds in all of $\Gamma(0,r_1,r_2)$ (respectively un all of $\Delta(0,r_1,r_2)$) except in finitely many classes of finitely many circles $C(0,r)$ $(r_1 < r < r_2)$*

(respectively $r_1 \leq r <\leq r_2$*). The right-side derivative (respectively the left side derivative) of the function* $\Psi(f,.)$ *at* μ *is equal to* $\nu^+(f,\mu)$ *(respectively to* $\nu^-(f,\mu)$*). Moreover, if the function in* μ $\Psi(f,\mu)$ *is not derivable at* μ*, then* μ *lies in* $\Psi(\mathbb{K})$*.*

Further, the function $\Psi(f,.)$ *is convex in* $]\log r_1, \log r_2[$ *(respectively in* $[\log r_1, \log r_2]$*). Next, given another* $g \in H(\Gamma(0,r_1,r_2))$ *(respectively* $g \in H(\Delta(0,r_1,r_2))$*), the functions* ν^+ *and* ν^- *satisfy* $\nu^+(fg,\mu) = \nu^+(f,\mu) + \nu^+(g,\mu)$, $\nu^-(fg,\mu) = \nu^-(f,\mu) + \nu^-(g,\mu)$. *Further, the function* $\nu^+(f,.)$ *is continuous on the right and the function* $\nu^-(f,.)$ *is continuous on the left at each point* μ*. They are continuous at* μ *if and only if they are equal.*

Proposition 19.4. *Let* $a \in \widetilde{D}$ *and let* $f \in H(D)$*. If* $f(a) \neq 0$*, there exists* $s > 0$ *such that* $\Psi(f,\mu) = \Psi(f(a))$ $\forall \mu \leq s$*. Let* $b \in D$ *be such that* $|a-b| = r$ *and* $d(b,r^-) \subset D$*. Then we have* $\Psi_b(f,\mu) = \Psi_a(f,\mu)$ $\forall \mu \leq \Psi(b-a)$*.*

Proof. Since $f(a) \neq 0$, the first statement is immediate since $|f(x)|$ is a constant inside a disk of center a. Next, by Lemma 4.14, the relation $\Psi_a(f,\mu) = \Psi_b(f,\mu)$ when $\Psi(a-b) \leq \mu$ is true for every $f \in R(D)$, hence by (2), is obviously generalized to every $f \in H(D)$. □

Proposition 19.5. *Let* $\mu \in \mathbb{R}$ *and let* $f(x) = \sum_{-\infty}^{+\infty} a_n x^n \in H(C(0,\theta^\mu))$*. Then,* $\Psi(f,\mu)$ *is equal to* $\sup_{n\in\mathbb{Z}} \Psi(a_n) + n\mu$ *and we have* $\Psi(f(x)) \leq \Psi(f,\mu)$ *for all* $x \in C(0,\theta^\mu)$*. Moreover, the equality holds in every class except in finitely many classes where* f *admits zeros. Further, if* $\nu^+(f,\mu) = \nu^-(f,\mu)$*, then* $\Psi(f(x)) = \Psi(f,\mu)$ *whenever* $x \in C(0,\theta^\mu)$*.*

If $h \in H(C(0,\theta^\mu))$ *satisfies* $\Psi(f-h,\mu) < \Psi(f,\mu)$*, then* $\nu^+(f,\mu) = \nu^+(h,\mu)$ *and* $\nu^-(f,\mu) = \nu^-(h,\mu)$*.*

Proof. Let $\Lambda = C(0,\theta^\mu)$, let $s = \nu^-(f,\mu)$, and let $t = \nu^+(f,\mu)$. By the remark above, $\Psi(f,\mu)$ is obviously equal to $\sup_{n\in\mathbb{Z}}(\Psi(a_n) + n\mu)$. Let $x \in C(0,\theta^\mu)$. The inequality $\Psi(f(x)) \leq \Psi(f,\mu)$ is true because $\Psi(f,\mu) = \log \|f\|_\Lambda \geq \Psi(f(x))$. Finally, by Proposition 19.1, the equality holds in all the classes except in finitely many. If $\nu^+(f,\mu) = \nu^-(f,\mu)$, then $\Psi(a_s x^s) = \Psi(a_s) + s\mu > \Psi(a_n x^n)$ whenever $n \neq s$, hence $\Psi(f(x)) = \Psi(f,\mu)$.

Now, let $h \in H(\Lambda)$ satisfy $\Psi(f-h,\mu) < \Psi(f,\mu)$ and let $h(x) = \sum_{-\infty}^{+\infty} b_n x^n$. We have $\Psi(a_n - b_n) + n\mu < \Psi(a_s) + s\mu$ whenever

$n \in \mathbb{Z}$, hence $\Psi(b_s) = \Psi(a_s)$, $\Psi(b_t) = \Psi(a_t)$, $\Psi(b_n)+n\mu < \Psi(a_s)+s\mu$ whenever $n < s$ and $n > t$ and $\Psi(b_n) + n\mu \leq \Psi(a_s) + s\mu$ whenever $n \in [s,t]$, hence finally, $\nu^+(h,\mu) = \nu^+(f,\mu)$ and $\nu^-(h,\mu) = \nu^-(f,\mu)$. □

Corollary 19.6. *Let* $f(x) = \sum_{-\infty}^{+\infty} a_n x^n \in \mathcal{A}(\Gamma(0,r_1,r_2))$ *(with* $0 < r_1 < r_2$*). The function* $\mu \to \Psi(f,\mu)$ *defined in* $]\log r_1, \log r_2[$ *is equal to* $\sup_{n\in\mathbb{Z}}(\Psi(a_n)+n\mu)$*. Next, we have* $\Psi(f(x)) \leq \Psi(f,\Psi(x))$ *whenever* $x \in \Gamma(0,r_1,r_2)$ *and the equality holds in all of* $\Gamma(0,r_1,r_2)$ *except in finitely many classes of each circle* $C(0,r)$ $(r_1 < r < r_2)$*. The right-side derivative (respectively the left-side derivative) of the function* $\Psi(f,.)$ *at* μ *is equal to* $\nu^+(f,\mu)$ *(respectively to* $\nu^-(f,\mu)$*). Moreover, if the function in* μ $\Psi(f,\mu)$ *is not derivable at* μ*, then* μ *lies in* $\Psi(\mathbb{K})$*.*

Further, the function $\Psi(f,.)$ *is convex in* $]\log r_1, \log r_2[$*. Next, given another* $g \in \mathcal{A}(\Gamma(0,r_1,r_2))$*, the functions* ν^+ *and* ν^- *satisfy*

$$\nu^+(fg,\mu) = \nu^+(f,\mu) + \nu^+(g,\mu), \ \ \nu^-(fg,\mu) = \nu^-(f,\mu) + \nu^-(g,\mu).$$

Moreover, the function $\nu^+(f,.)$ *is continuous on the right and the function* $\nu^-(f,.)$ *is continuous on the left at each point* μ*. They are continuous at* μ *if and only if they are equal.*

Proof. All statements hold in all annuli $\Gamma(0,r',r'')$ with $r_1 < r' < r'' < r_2$ because the restriction of f to $\Gamma(0,r',r'')$ belongs to $H(\Gamma(0,r',r''))$. □

Proposition 19.7. *Let* $\mu \in \mathbb{R}$ *and let* $f,g \in H(C(0,\theta^\mu))$*. Then,* $\nu^+(fg,\mu) = \nu^+(f,\mu)+\nu^+(g,\mu)$ *and* $\nu^-(fg,\mu) = \nu^-(f,\mu)+\nu^-(g,\mu)$*.*

Proof. By Proposition 19.2, the relations are obvious when f and $g \in R(C(0,\theta^\mu))$ because there is an annulus $\Gamma(0,r_1,r_2) \supset C(0,\theta^\mu)$ such that $f,g \in R(\Gamma(0,r_1,r_2))$. Now, by Corollary 19.6, we may extend them to $H(C(0,\theta^\mu))$ by taking h and $\ell \in R(C(0,\theta^\mu))$ such that $\Psi(f-h,\mu) < \Psi(f,\mu)$ and $\Psi(g-\ell,\mu) < \Psi(g,\mu)$. □

Proposition 19.8. *Let* $r_1, \ r_2 \in \mathbb{R}$ *and let* $f, \ g \in \mathcal{A}(\Gamma(0,r_1,r_2))$ *having no zero in* $\Gamma(0,r_1,r_2)$ *and satisfying* $\nu(f,\mu) \neq \nu(g,\mu)$*,* $\forall \mu \in]\log r_1, \log r_2[$*. Then, both* $\nu^+(f+g,\mu)$ *and* $\nu^-(f+g,\mu)$ *are equal either to* $\nu(f,\mu)$ *or to* $\nu(g,\mu)$*.*

Proof. Let $\mu_j = \log(r_j)$, $j = 1, 2$. Since both f, g have no zero in $\Gamma(0, r_1, r_2)$, $\nu(f, \mu)$ is a constant integer s and $\nu(g, \mu)$ is a constant integer $t \neq s$. Consequently, $\Psi(f, \mu)$ is of the form $a + s\mu$ and $\Psi(g, \mu)$ is of the form $b + t\mu$, therefore the two functions in μ can coincide at most at one point in $[\mu_1, \mu_2]$. So, by Proposition 19.1, we have $\Psi(f + g, \mu) = \max(\Psi(f, \mu), \Psi(g, \mu))$ for all $\mu \in [\log(r_1), \log(r_2)]$ except maybe at all point. But then, by continuity, the equality holds in all $[\log(r_1), \log(r_2)]$.

Let us fix $\mu_0 \in]\mu_1, \mu_2[$. Suppose $\Psi(f + g, \mu) = \Psi(f, \mu)$ in a neighborhood $]\mu_1, \mu_2[$ of μ_0. Then, of course, $\nu(f + g, \mu_0) = \nu(f, \mu_0)$. Suppose now that $\Psi(f + g, \mu) = \Psi(f, \mu)$ in a left neighborhood $]\mu_1, \mu_0]$ of μ_0 and $\Psi(f + g, \mu) = \Psi(g, \mu)$ in a right neighborhood $[\mu_0, \mu_2[$ of μ_0, which implies $\Psi(f, \mu) > \Psi(g, \mu)$ $\forall \mu \in]\mu_1, \mu_0[$ and $\Psi(f, \mu) < \Psi(g, \mu)$ $\forall \mu \in]\mu_0, \mu_2[$. Then, we have $\nu(f + g, \mu) = \nu(f, \mu) \forall \mu \in]\mu_1, \mu_0[$ and $\nu(f + g, \mu) = \nu(g, \mu) \forall \mu \in]\mu_0, \mu_2[$. Consequently, since ν^+ is continuous on the left and ν^- is continuous on the right, we can check that both $\nu^+(f + g, \mu_0)$ and $\nu^-(f + g, \mu_0)$ are equal either to $\nu(f \mu_0)$ or to $\nu(g, \mu_0)$. □

Theorem 19.9. *Let $f \in \mathcal{A}(\mathbb{K} \backslash d(0, R))$. There exists $q \in \mathbb{N}$ such that*

$$\lim_{r \to +\infty} |f|(r) r^q = +\infty.$$

Proof. Let $s \in]R, +\infty[$ be such that $\nu^+(f, \log s) = \nu^-(f, \log s)$ and let $\tau = \nu^+(f, \log s)$. Thus, $\Psi(f, \mu)$ has a derivative at $\log s$ equal to τ. Consequently, since by Proposition 19.1 $\Psi(f, \mu)$ is convex, we have $\Psi(f, \mu) - \Psi(f, \log s) \geq \tau(\mu - \log s)$. Therefore,

$$\lim_{\mu \to +\infty} [\Psi(f, \mu) + (1 - \tau)\mu] = +\infty$$

i.e. $\lim_{r \to +\infty} |f|(r) r^{(1-\tau)} = +\infty$. Finally, we can take $q = \max(0, 1 - \tau)$. □

Chapter 20

Vanishing Along a Monotonous Filter

Throughout this chapter, the set D is supposed to be infraconnected.

From Chapter 7, we know that there exists a spherically complete algebraically closed extension $\widehat{\mathbb{K}}$ of $\mathbb{K}$ whose residue class field is not countable and whose valuation group is equal to $\mathbb{R}$. Given a subset D of $\mathbb{K}$, we denote by $\widehat{D}$ the subset

$$D \cup \left\{ \bigcup_{a \in \overset{\circ}{D}} \widehat{d}(a, \delta(a, (\mathbb{K} \setminus D)) \right\}.$$

The question of whether an analytic element can tend to zero along a monotonous filter is known to be one of the main problems which happen with p-adic analytic functions [35, 36, 44, 50]. Here, we do not describe T-filters. However, we describe sufficient conditions to prevent analytic elements to vanish along a monotonous filter, which is sufficient to study analytic and meromorphic functions inside disks or annuli.

We apply the results to characteristic functions and to the image of an infraconnected set.

Definitions. Let $f \in H(D)$ and let $\mathcal{F}$ be a monotonous filter on D. When $\mathcal{F}$ is decreasing (respectively increasing) of center a and diameter S, f is said to be *strictly vanishing* along $\mathcal{F}$ if $\lim_{\mathcal{F}} f(x) = 0$, and if there exists $S' > S$ (respectively $S' < S$) such that for every $r \in]S, S']$ (respectively $r \in [S', S[$), we have ${}_D\varphi_{a,r}(f) \neq 0$.

When $\mathcal{F}$ is decreasing with no center in $\mathbb{K}$, it admits a canonical basis $(D_n)_{n\in\mathbb{N}}$ with $D_n = d(a_n, r_n) \cap D$ and then f is be said to be *strictly vanishing* along $\mathcal{F}$ if $\lim_{\mathcal{F}} f(x) = 0$ and if there exists $S' > S$ such that ${}_D\varphi_{a_{n+1},r}(f) \neq 0$ whenever $r \in [r_n, S']$, whenever $n \in \mathbb{N}$. Actually, $\mathcal{F}$ admits a center α in $\widehat{\mathbb{K}}$ and then the definition given for decreasing filters with a center in $\mathbb{K}$ also applies and is obviously equivalent.

Lemma 20.1 just translates these definitions by using the function Ψ.

Lemma 20.1. *Let $f \in H(D)$ and let $\mathcal{F}$ be a decreasing (respectively an increasing) filter of center a and diameter S on D. Then, f is strictly vanishing along $\mathcal{F}$ if and only if there exists $S' > S$ (respectively $S' \in]0, S[$) such that $\Psi_a(f, \log S) = -\infty$, $\Psi_a(f, \mu) > -\infty$ whenever $\mu \in]\log S, \log S']$ (respectively $[\log S', \log S[$).*

Let $\mathcal{G}$ be a decreasing filter with no center, of diameter S, and canonical basis $(D_n)_{n\in\mathbb{N}}$, with $D_n = d(a_n, r_n) \cap D$. Then, f is strictly vanishing along $\mathcal{G}$ if and only if there exists $S' > S$ such that $\lim_{n\to\infty} \Psi_{a_n}(f, \log r_n) = -\infty$ and $\Psi_{a_n}(f, \log r) > -\infty$ for $r_n \leq r < S'$, $\forall n \in \mathbb{N}$.

Lemma 20.2. *Let $f \in H(D)$ and let $\mathcal{F}$ be a monotonous filter such that f is strictly vanishing along $\mathcal{F}$. Then, f is properly vanishing along $\mathcal{F}$.*

Proof. Let $S = \text{diam}(\mathcal{F})$. Let (D_n) be a canonical basis of $\mathcal{F}$ and suppose that f is not properly vanishing along $\mathcal{F}$. Since f is vanishing along $\mathcal{F}$, there exists $q \in \mathbb{N}$ such that $f(x) = 0$ whenever $x \in D_q$. But then, we can check that f does not satisfy the definition of an analytic element strictly vanishing along $\mathcal{F}$ because for every multiplicative semi-norm ${}_D\varphi_{a,r}$ whose circular filter is secant to D_q, we have ${}_D\varphi_{a,r}(f) = 0$. In particular, this applies to ${}_D\varphi_{a,r}$, for $r_q < r < S$ (respectively $S < r < r_q$), when $\mathcal{F}$ is increasing (respectively decreasing) of center a, and $D_q = D \cap \Gamma(a, r_q, S)$ (respectively $D_q = D \cap \Gamma(a, S, r_q)$), and this applies to φ_{a_{q+1},r_q}, when $\mathcal{F}$ has no center, whereas $D_q = D \cap d(a_{q+1}, r_q)$. □

Proposition 20.3. *Let $a \in \widetilde{D}$ and $b \in D$ and let $f \in H(D)$ satisfy $f(b) \neq 0$ and ${}_D\varphi_{a,r}(f) = 0$ for some $r \in]0, |a-b|]$. If ${}_D\varphi_{a,|a-b|}(f) = 0$, then f is strictly vanishing along an increasing filter of center b and diameter $S \leq |a-b|$. If ${}_D\varphi_{a,|a-b|}(f) \neq 0$, then f is strictly vanishing along a decreasing filter of center a and diameter $S \in [r, |a-b|\,[$.*

Proof. First, suppose ${}_D\varphi_{a,|a-b|}(f) = 0$, hence we have

$$\Psi_b(f, \log|a-b|) = \Psi_a(f, \log|a-b|) = -\infty.$$

Since $f(b) \neq 0$, by Proposition 19.4, we know that $\lim_{\mu \to -\infty} \Psi_b(f, \mu) = \Psi(f(b))$, hence there exists a unique $\gamma \leq \log|a-b|$ such that $\Psi_b(f, \gamma) = -\infty$ and $\Psi_b(f, \mu) > -\infty$ whenever $\mu \in]\gamma, \log|a-b|[$. Therefore, f is strictly vanishing along the increasing filter of center b and diameter $S = \theta^\gamma$.

Now, suppose ${}_D\varphi_{a,|a-b|}(f) \neq 0$. Since ${}_D\varphi_{a,r}(f) = 0$, we have $\Psi_a(f, \log r) = -\infty$ and $\Psi_a(f, \log|a-b|) > -\infty$, hence there exists a unique $\gamma \in [\log r, \log|a-b|[$ such that $\Psi_a(f, \gamma) = -\infty$ and $\Psi_a(f, \mu) > -\infty$ whenever $\mu \in]\gamma, \log|a-b|]$, so f is strictly vanishing along the decreasing filter of center a and diameter $S = \theta^\gamma$. □

Proposition 20.4. *Let $a \in \widetilde{D}$ and $b \in D$ and let $f \in H(D)$ satisfy $f(b) \neq 0$ and ${}_D\varphi_{a,r}(f) = 0$ for some $r \in \mathbb{R}_+$. Then f is strictly vanishing along a monotonous filter with a center.*

Proof. If $r \leq |a-b|$, the statement comes directly from Proposition 20.3. If $r > |a-b|$, then we have $\Psi_b(f, \log r) = \Psi_a(f, \log r) = -\infty$, whereas $\lim_{\mu \to -\infty} \Psi_b(f, \mu) = \Psi(f(b))$, hence there exists a unique $\gamma \leq \log r$ such that $\Psi_b(f, \gamma) = -\infty$ and $\Psi_b(f, \mu) > -\infty$ whenever $\mu < \gamma$. Thus, f is strictly vanishing along the increasing filter of center b and diameter $S = \theta^\gamma$. □

Proposition 20.5. *Let $\mathcal{F}$ be a monotonous filter on D and let $f \in H(D)$ be strictly vanishing along $\mathcal{F}$. Then, $\mathcal{F}$ is pierced.*

Proof. Suppose that $\mathcal{F}$ is increasing (respectively decreasing) of center a and diameter S and is not pierced. There exists an annulus $\Gamma(a, S, S')$ (respectively $\Gamma(a, S', S)$) included in D such that $\Psi_a(f, \log S) = -\infty$ and $\Psi_a(f, \mu) > -\infty$ whenever $\mu \in [\log S', \log S[$ (respectively $\mu \in]\log S, \log S']$). Hence, by Corollary 19.3, we know that $\Psi_a(f, \mu)$ is bounded in $]\log S', \log S[$ (respectively $]\log S, \log S'[$), a contradiction to the hypothesis.

When $\mathcal{F}$ has no center, we consider a center a of $\mathcal{F}$ in $\widehat{\mathbb{K}}$ and we consider f as an element of $\widehat{H}(\widehat{D})$. Then, the disks $\widehat{d}(a_n, r_n)$ contain no hole of $\widehat{D}$ when n is big enough. Therefore, the restriction of $\widehat{f}$ to $d(a_n, r_n)$ is a power series and therefore we have the same conclusion. □

Corollary 20.6. *Let $a, b \in D$ and let $f \in H(D)$ satisfy $f(b) \neq 0$ and ${}_D\varphi_{a,r}(f) = 0$ for some $r \in \mathbb{R}_+$. Then, f is strictly vanishing along a pierced monotonous filter with a center.*

Proposition 20.7. *Let $a \in \widetilde{D}$. Let $r, r' \in]\log(\delta(a, D)), \log(\text{diam}(D))[$, with $r < r'$. Let $f \in H(D)$ be such that the function $\Psi_a(f, \mu)$ is neither bounded nor identically equal to $-\infty$ in $[\log r, \log r']$. Then there exists a monotonous filter $\mathcal{F}$ of center a and diameter $s \in [r, r']$ such that f is strictly vanishing along $\mathcal{F}$.*

Proof. For convenience we assume $a = 0$. By compacity of $[\log r, \log r']$, there exists $\mu \in [\log r, \log r']$ such that $\Psi(f, \mu) = -\infty$. Since the function $\Psi(f, \mu)$ is not identically $-\infty$ in $[\log r, \log r']$, by continuity, either there exists $\xi, \zeta \in [\log r, \log r']$ with $\xi < \zeta$ such that $\Psi(f, \xi) = -\infty$, $\Psi(f, \mu) > -\infty$ whenever $\mu \in]\xi, \zeta]$ and then f is strictly vanishing along a decreasing filter of center 0 and diameter θ^{ξ}, or there exist $\xi, \zeta \in [\log r, \log r']$ with $\xi < \zeta$ such that $\Psi(f, \zeta) = -\infty$, $\Psi(f, \mu) > -\infty$ whenever $\mu \in [\xi, \zeta[$ and then f is strictly vanishing along an increasing filter of center 0 and diameter θ^{ζ}. This ends the proof. □

Proposition 20.8. *Let $f \in H(D)$ be vanishing along an increasing (respectively a decreasing) filter $\mathcal{F}$ of diameter s. Let a be a center of $\mathcal{F}$ in $\widehat{\mathbb{K}}$ and let $E = \widehat{D} \cup (\widehat{\mathbb{K}} \setminus \widehat{d}(a, s^-))$ (respectively $E = \widehat{D} \cup \widehat{d}(a, s)$). Then, f has continuation to an element F of $\widehat{H}(E)$ such that $F(x) = 0$ whenever $x \in \widehat{\mathbb{K}} \setminus \widehat{d}(a, s^-)$ (respectively $x \in \widehat{d}(a, s)$).*

Proof. We suppose $\mathcal{F}$ increasing. By Theorem 15.9, f has an extension $\widehat{f}$ to $\widehat{D}$. For every $r > 0$, the set of classes of $C(a, r)$ which contain $\widehat{f}$-holes is countable. Since the residue class field of $\widehat{\mathbb{K}}$ is not countable, in $\widehat{C}(a, r)$, there exist classes $\Lambda = d(b, r^-)$ which contain no $\widehat{f}$-holes. Thereby, $\widehat{f}$ has continuation to an infraconnected set D' which contains $\widehat{D}$ and satisfies $\widetilde{D'} = \widehat{\mathbb{K}}$ such that every hole is of the form $\widehat{d}(\alpha, \rho^-)$, with $d(\alpha, \rho^-)$ a hole of D. In this set, we have

(1) $_{D'}\varphi_{a,s}(\widehat{f}) = {}_D\varphi_{\mathcal{F}}(f) = 0.$

First, we suppose that $\mathcal{F}$ is increasing. Let $V = d(a, s^-)$ and let $D'' = D'\setminus V$. Clearly, V is a hole of D''. Then, as an element of $\widehat{H}(D'')$, by Theorem 15.1, (1) implies

(2) $\overline{\overline{(\widehat{f})}}_V = 0.$

Now, by Theorem 15.1, $\widehat{f}$ has a decomposition of the form $g+h$ with $g \in H_0(D' \cup (\widehat{\mathbb{K}}\setminus\widehat{d}(a, s^-)))$ and $h \in H(D' \cup \widehat{d}(a, s^-))$. By (2), it is seen that $_{D'}\varphi_{a,s}(h) = 0$, hence $h = 0$ because h belongs to $\widehat{H}(\widehat{d}(a, s))$. As a consequence, $\widehat{f}$ belongs to $H_0(D' \cup (\widehat{\mathbb{K}}\setminus\widehat{d}(a, s^-)))$, and therefore, in $\mathbb{K}$, f belongs to $H(D \cup (\mathbb{K}\setminus d(a, s^-)))$. Moreover, by (2), we have $\widehat{f}(x) = 0$ whenever $x \in D''$, hence $\widehat{f}(x) = 0$ whenever $x \in \mathbb{K}\setminus d(a, s^-)$.

If $\mathcal{F}$ is decreasing, we can easily perform a symmetric proof. □

Theorem 20.9. *Let $f \in H(D)$ be vanishing along an increasing (respectively a decreasing) filter $\mathcal{F}$ of center a and diameter s. Let $E = D \cup (\mathbb{K}\setminus d(a, s^-))$ (respectively $E = D \cup d(a, s)$). Then, f has continuation to an element of $H(E)$ such that $f(x) = 0$ whenever $x \in \mathbb{K}\setminus d(a, s^-)$ (respectively whenever $x \in d(a, s)$).*

Proof. By Proposition 20.8, f has continuation to an element $\widehat{f} \in \widehat{H}(\widehat{D} \cup (\widehat{\mathbb{K}}\setminus\widehat{d}(a, s^-)))$ (respectively to $H(\widehat{D} \cup \widehat{d}(a, s))$). Therefore, in $\mathbb{K}$, f belongs to $H(D \cup (\mathbb{K}\setminus d(a, s^-)))$. Moreover, we have $\widehat{f}(x) = 0$ whenever $x \in \widehat{\mathbb{K}}\setminus\widehat{d}(a, S^-)$ (respectively $x \in \widehat{d}(a, s)$), hence $f(x) = 0$ whenever $x \in \mathbb{K}\setminus d(a, s^-)$ (respectively $f(x) = 0$ whenever $x \in d(a, s)$). □

Definition. Let $\mathcal{F}$ be a monotonous filter on D and let $f \in H(D)$. Then, f is said to be *collapsing along* $\mathcal{F}$ if there exists $b \in \mathbb{K}$ such that $f - b$ is vanishing along $\mathcal{F}$.

Theorem 20.10. *Let D be open and closed and let $\mathcal{F}$ be a monotonous filter on D. Let $f \in H(D)$ be such that $f' \in H(D)$. Then, f is collapsing along $\mathcal{F}$ if and only if f' is vanishing along $\mathcal{F}$.*

Proof. Without loss of generality, we can suppose that D is bounded because $\mathcal{F}$ is obviously secant with a bounded subset of $\mathbb{K}$.

Suppose first that $\mathcal{F}$ is decreasing, of center a and diameter R. Without loss of generality, we can obviously assume $a = 0$. Suppose f is collapsing along $\mathcal{F}$, of limit ℓ. Since $\lim_{r\to R} \frac{|f-\ell|(r)}{r} = 0$, by Theorem 18.25, we have $\lim_{r\to R}|f'|(r) = 0$, hence f' is vanishing along $\mathcal{F}$.

Conversely, suppose that f' is vanishing along $\mathcal{F}$. Let $(T_i)_{i\in I}$ be the family of holes of D included in $\mathbb{K}\setminus d(0,R)$ and let $(T_i)_{i\in J}$ be the family of holes of D included in $d(0,R)$. Then, the pair $(I,\ J)$ makes a partition of the set of holes of D. Let $D_1 = D\cup d(0,R)$ and let $D_2 = D\cup(\mathbb{K}\setminus d(0,R))$. By Corollary 15.2, we have $H(D) = H(D_1)\oplus H(D_2)$. For every $g \in H(D)$, we set $g = g_1 + g_2$ with $g_k \in H(D_k),\ k = 1,\ 2$. Then, $\|g\|_D = \max(\|g_1\|_{D_1}, \|g_2\|_{D_2})$. By Theorem 18.19, we can check that the decomposition of f' in the form $f' = (f')_1+(f')_2$ is such that $(f')_k = (f_k)'$, $k = 1,\ 2$. When we take numbers $S' > R$ and $S'' < R$, we have $\lim_{S'\to R^+}\|f'\|_{D\cap d(0,S')} = 0 = \lim_{S''\to R^-}\|f'\|_{D\setminus d(0,S'')}$, hence $\lim_{S'\to R^-}\|f_1'\|_{D\cap d(0,S')} = 0 = \lim_{S''\to R^+}\|f_1'\|_{D\setminus d(0,S'')}$ and $\lim_{S'\to R^-}\|f_2'\|_{D\cap d(0,S')} = 0 = \lim_{S''\to R^+}\|f_2'\|_{D\setminus d(0,S'')}$. Consequently, passing to the limit, we get $\varphi_{0,R}(f_1') = \varphi_{0,R}(f_2') = 0$. Therefore, by Corollary 20.6, f_2', which belongs to $H(\mathbb{K}\setminus d(0,R))$ and satisfies $\varphi_{0,R}(f_2') = 0$, is identically zero in $\mathbb{K}\setminus d(0,R)$. Consequently, $f'(x) = f_1'(x)\ \forall x \in D\cap(\mathbb{K}\setminus d(0,R))$. On the other hand, since f_1' belongs to $H(d(0,R))$ and satisfies $\varphi_{0,R}(f_1') = 0$, it is identically zero in $d(0,R)$ and hence $f_1(x)$ is a constant C in $d(0,R)$. Therefore, $\varphi(f_1 - C) = 0$, hence $f_1 - C$ is vanishing along $\mathcal{F}$. But since $f_2(x) = 0\ \forall x \in D\setminus d(0,R)$, actually, $f - C$ is vanishing along $\mathcal{F}$. This finishes showing that f is collapsing along $\mathcal{F}$ when $\mathcal{F}$ is a decreasing filter with a center.

Now, suppose that $\mathcal{F}$ has no center. We can place ourselves in an algebraically closed spherically complete extension of $\mathbb{K}$ and prove the same property for the expansion of f, hence it holds for f. Finally, if $\mathcal{F}$ is increasing, we can make an inversion and prove the same. □

Thanks to monotonous filters, we are now able to complete the study of the characteristic functions.

Theorem 20.11. *Let E be a subset of $\mathbb{K}$ whose interior is not empty. Then, E is not infraconnected if and only if there exists a proper subset B of E whose characteristic function belongs to $H(E)$.*

Proof. If E is not infraconnected, it admits an empty annulus $\Gamma(a,r',r'')$, and then by Proposition 11.15, the characteristic functions of $\mathcal{I}_E(\Gamma(a,r',r''))$ and $\mathcal{E}_E(\Gamma(a,r',r''))$ belong to $H(E)$.

Now, we suppose E to be infraconnected and assume that there is a subset B of E, $B \neq E$ and $B \neq \emptyset$, whose characteristic function u belongs to $H(E)$. Since u, by definition, belongs to $H_b(E)$, it belongs to $H_b(\overline{E})$. And of course, $\overline{E}$ has an interior that is not empty. Hence, without loss of generality, we can assume that E is closed. Let $A = E \backslash B$. Suppose A and B are different from $\emptyset$. Since u is locally constant in all E, at least one of the two subsets A and B has an interior that is not empty. Without loss of generality, we can assume that the interior of A is not empty and hence there exists $a \in A$ and $r > 0$ such that $d(a,r) \subset A$ and $u(x) = 0$ whenever $x \in d(a,r)$ and then we have ${}_D\varphi_{a,r}(u) = 0$. By Proposition 20.3, there exists a monotonous filter $\mathcal{F}$ with center $\alpha \in E$ such that u is strictly vanishing along $\mathcal{F}$, hence by Lemma 20.2, f is properly vanishing along $\mathcal{F}$. But this contradicts the hypothesis "$f(x) = 0$ or 1 for all $x \in E$". This finishes proving Theorem 20.11. □

Corollary 20.12. *Let* $E \in \text{Alg}$. *The algebra* $H(E)$ *has non-trivial idempotents if and only if* E *is not infraconnected.*

Theorem 20.13. *Let* $f \in H(D)$. *Then,* $f(D)$ *is infraconnected.*

Proof. Let $D' = f(D)$ and let us suppose that D' admits an empty annulus $\Gamma(a, r', r'')$. Let $D'' = \overline{D'}$. It is seen that $\Gamma(a, r', r'')$ also is an empty annulus of D''.

Let $A'' = \mathcal{I}_{D''}(\Gamma(a, r', r''))$ and let $B'' = \mathcal{E}_{D''}(\Gamma(a, r', r''))$. Let u be the characteristic function of A''. By Proposition 11.15, we know that u belongs to $H(D'')$. Since D'' is closed and contains $f(D)$, by Corollary 12.3, $u \circ f$ belongs to $H(D)$. Let $A = f^{-1}(A'')$ and $B = f^{-1}(B'')$. Obviously, we have $A \cap B = \emptyset$ and $A \cup B = D$. Now, $u \circ f(x) = 1\ \forall x \in A$, $u \circ f(x) = 0\ \forall x \in B$. But since D is infraconnected, by Theorem 20.11, $H(D)$ contains no characteristic function of any proper subset. This ends the proof of Theorem 20.13. □

Chapter 21

Quasi-Minorated Elements

Throughout this chapter, the set D is supposed to be infraconnected.

The main results given here were published in [35, 44, 50]. According to the definition of quasi-minorated elements, Theorem 21.1 is easy:

Theorem 21.1. *Let $f \in H(D)$. Then, f is not quasi-minorated if and only if there exists a large circular filter $\mathcal{F}$ secant with D such that ${}_D\varphi_{\mathcal{F}}(f) = 0$.*

Proof. By Lemmas 16.7 and 17.12, without loss of generality, we can assume that D is bounded.

Suppose first that there exists a large circular filter $\mathcal{F}$ secant with D such that ${}_D\varphi_{\mathcal{F}}(f) = 0$. Let (a_n) be a monotonous distances sequence thinner than $\mathcal{F}$ such that $\lim_{n\to+\infty} f(a_n) = 0$. Then, f is not quasi-minorated.

Conversely, suppose that f is not quasi-minorated. Then there exists a bounded sequence (a_n) of D such that $\lim_{n\to+\infty} f(a_n) = 0$ and such that one can't extract a sequence converging in $\mathbb{K}$. By Theorem 3.1, we can extract from the sequence (a_n) either a monotonous distances sequence or a constant distances sequence. In both cases, there exists a circular filter $\mathcal{F}$ less thin than this subsequence, and hence, we have $\lim_{\mathcal{F}} f(x) = 0$. □

Theorem 21.2. *Let f be a non-identically zero element of $H(D)$. Then, f is not quasi-minorated if and only if there exists a pierced monotonous filter $\mathcal{F}$ on D such that f is strictly vanishing along $\mathcal{F}$.*

Proof. By Theorem 21.1, if f is vanishing along a monotonous filter; it is not quasi-minorated. Now suppose that f is not quasi-minorated. Since f is not identically zero, we can find a bounded sequence $(a_n)_{n\in\mathbb{N}}$ in D such that $\lim_{n\to\infty} f(a_n) = 0$ such that no subsequence converges in $\mathbb{K}$.

Suppose first we can extract form the sequence $(a_n)_{n\in\mathbb{N}}$ a constant distances sequence $(a_{q(m)})m \in \mathbb{N}$, let $c = a_{q(1)}$, let $r = |a_{q(1)} - a_{q(2)}|$, and let $\phi = \varphi_{b,r}$. Since $\lim_{m\to+\infty} f(a_{q(m)}) = 0$, we have $\varphi_{c,r}(f) = 0$. Since f is not identically zero in D, there exists $b \in D$ such that $f(b) \neq 0$. If $b \in d(0,r)$, then by Proposition 20.3, f is strictly vanishing along an increasing filter of center c, of diameter $\rho \in]0,r]$. If $b \notin d(c,r)$, by Proposition 20.3, if $\varphi_{c,|b-c|}(f) \neq 0$, then f is strictly vanishing along a decreasing filter of center c, of diameter $\rho \in d(c,|b-c|)$. Finally, if $\varphi_{c,|b-c|}(f) = 0$, then f is strictly vanishing along an increasing filter of center b, of diameter $\rho \in d(0,|b-c|)$, and by Proposition 20.5, that filter is pierced.

Suppose now we can't extract from the sequence $(a_n)_{n\in\mathbb{N}}$ a constant distances sequence. Then we can extract from the sequence $(a_n)_{n\in\mathbb{N}}$ a monotonous distances sequence $(a_{q(m)})_{m\in\mathbb{N}}$. There exists a unique monotonous filter $\mathcal{F}$ less thin than the subsequence $(a_{q(m)})_{m\in\mathbb{N}}$. Suppose first that $\mathcal{F}$ has a center c and let $r = \operatorname{diam}(\mathcal{F})$. Then, we have $\varphi_{c,r}(f) = 0$, and hence, the same reasoning shows that f is strictly vanishing along a monotonous filter which by Proposition 20.5 is pierced. Finally, suppose $\mathcal{F}$ is decreasing, with no center. We can find a center in a spherically closed extension and make the same reasoning again. □

Corollary 21.3. *If D has no monotonous pierced filter, every element of $H(D)$ different from zero is quasi-minorated and takes every value finitely many times.*

Definition. A subset D of $\mathbb{K}$ is said to be *analytic* if for every disk $d(a,r)$ included in D and for every $f \in H(D)$, the property $f(x) = 0\ \forall x \in d(a,r)$ implies that f is identically zero in D.

Corollary 21.4. *If D has no monotonous pierced filter, D is analytic. Particularly, if D has finitely many holes, D is analytic. Particularly, if D is infraconnected affinoid, it is analytic.*

Theorem 21.5. *Let f_1, f_2 be quasi-minorated elements of $H(D)$. If f_1f_2 belongs to $H(D)$, then it is also quasi-minorated.*

Proof. Indeed, suppose that $f_1 f_2$ is not quasi-minorated. By Theorem 21.2, there exists a pierced monotonous filter $\mathcal{F}$ on D such that ${}_D\varphi_{\mathcal{F}}(f_1 f_2) = 0$. Hence, either ${}_D\varphi_{\mathcal{F}}(f_1) = 0$ or ${}_D\varphi_{\mathcal{F}}(f_2) = 0$. But, by Theorem 21.1, both options are impossible because both f_1, f_2 are quasi-minorated. Hence, so is $f_1 f_2$. □

Theorem 21.6. *Let $f \in H(D)$. If f is semi-invertible, then it is quasi-minorated.*

Proof. Let f be semi-invertible, of the form $P(x)g(x)$ with g invertible in $H(D)$, and P a polynomial whose zeros $a_1, \dots, a_q$ belong to D. We then suppose f not quasi-minorated. By Theorem 21.5, g is not quasi-minorated either. Hence, there exists a pierced monotonous filter $\mathcal{F}$ on D such that ${}_D\varphi_{\mathcal{F}}(g) = 0$. But, by Lemma 11.3, that contradicts the hypothesis "g invertible in $H(D)$". Hence, f is quasi-minorated. □

Theorem 21.7. *Suppose that $\overline{D}$ is open. Then, an element of $H(D)$ is quasi-minorated if and only if it is quasi-invertible.*

Proof. If f is quasi-invertible, it is semi-invertible, and then, by Theorem 21.6, it is quasi-minorated. Now, assume f to be quasi-minorated. We prove it to be quasi-invertible. As in Theorem 21.1, by Lemmas 16.7 and 17.12, without loss of generality, we can assume that D is bounded. Let $S(D)$ be the set of polynomials whose zeros belong to $\overline{D}\setminus D$. If $D \in \mathrm{Alg}$, then $H(D) = S(D)^{-1}H(\overline{D})$. By Corollary 11.13, there exists $Q \in S(D)$ and $h \in H(\overline{D})$ such that $f = \frac{h}{Q}$. If h is not quasi-minorated, there exists a monotonous filter $\mathcal{F}$ on $\overline{D}$ such that ${}_D\varphi_{\mathcal{F}}(h) = 0$ and then we have ${}_D\varphi_{\mathcal{F}}(f) = 0$: this contradicts the hypothesis "f quasi-minorated" and therefore h is quasi-minorated. Then, by Theorem 16.8, h is quasi-invertible, in the form Pg with P a polynomial whose zeros are interior to $\overline{D}$ and g an element invertible in $H(\overline{D})$. Since $\overline{D}$ is bounded and open, by Corollary 17.10, D belongs to Alg, hence $\frac{g}{Q}$ is invertible in $H(D)$. Let $P = P_1 P_2$ with P_1 (respectively P_2) the polynomial of the zeros of P in $\overset{\circ}{\overline{D}} \cap D$ (respectively in $\overline{D}\setminus D$). Then, $\frac{P_2 g}{Q}$ is invertible in $H(D)$, and then, f is quasi-invertible. □

Theorem 21.8. *If D belongs to* Alg *and has no pierced filter, every element of $H(D)$ different from zero is quasi-invertible.*

Proof. As in Theorem 21.1, by Lemmas 17.12 and 16.7, without loss of generality, we can assume that D is bounded. Now, since D has no pierced filter, by Corollary 21.3, every element of $H(D)$ is quasi-minorated. But since D has no pierced filter, $\overline{D}$ is open, hence by Theorem 21.7, every element of $H(D)$ is quasi-invertible. □

Corollary 21.9. *Let D be closed and let $\mathcal{T}$ be the set of holes of D. If $\{\widetilde{T} | T \in \mathcal{T}\}$ is finite, then every element of $H(D)$ different from zero is quasi-invertible.*

Corollary 21.10. *If D is closed and has finitely many holes, then every element of $H(D)$ different from zero is quasi-invertible.*

Corollary 21.11. *If D is a disk $d(a, r)$ or $d(a, r^-)$, or if D is an annulus $\Gamma(a, r_1, r_2)$ (with $0 < r_1 < r_2$) or $\Delta(a, r_1, r_2)$ (with $0 < r_1 < r_2$) or a circle $C(a, r)$, then every element of $H(D)$ different from zero is quasi-invertible.*

Proof (Corollaries 21.9, 21.10, and 21.11). Indeed, D has no pierced filter, hence the elements different from zero are quasi-minorated and are quasi-invertible because D is open. □

We see that when D belongs to Alg, a quasi-minorated element that has no zero in D actually is invertible in $H(D)$.

Lemma 21.12. *Let $f \in H(\overline{D})$ be quasi-minorated in $H(D)$. Then, f is quasi-minorated in $H(\overline{D})$.*

Proof. Indeed, let $(a_n)_{n\in\mathbb{N}}$ be a bounded sequence in $\overline{D}$ such that $\lim_{n\to\infty} f(a_n) = 0$. There obviously exists a sequence $(b_n)_{n\in\mathbb{N}}$ in D such that $\lim_{n\to\infty} a_n - b_n = 0$ and $\lim_{n\to\infty} f(b_n) = 0$. So, the sequence (b_n) is bounded as the sequence (a_n). Since f is quasi-minorated in $H(D)$, then we can extract a Cauchy sequence $(b_{q(m)})_{m\in\mathbb{N}}$ from the sequence $(b_n)_{n\in\mathbb{N}}$ and then the sequence $(a_{q(m)})_{m\in\mathbb{N}}$ is a Cauchy subsequence of the sequence $(a_n)_{n\in\mathbb{N}}$. Thus, we have proven that f is quasi-minorated in $H(\overline{D})$. □

Theorem 21.13. *Let $D \in$ Alg. Let $f \in H(D)$ be quasi-minorated and have no zero in D. Then, f is invertible in $H(D)$.*

Proof. If D is closed, the statement is given by Theorem 16.9. Consider now the general case. As in Theorem 21.1, by Lemmas 16.7 and 17.12, without loss of generality, we can assume that D is bounded. Let Q be the polynomial of the poles of f in $\overline{D}\setminus D$ and let $h(x) = Q(x)f(x)$. By Theorem 21.5, h is quasi-minorated in $H(D)$. But, by Corollary 21.11, h belongs to $H(\overline{D})$, and therefore, by Lemma 21.12, it is quasi-minorated in $H(\overline{D})$. We prove that h has finitely many zeros in $\overline{D}$. Indeed, we assume that h admits infinitely many zeros in $\overline{D}$. So, we can find a sequence $(a_n)_{n\in\mathbb{N}}$ in $\overline{D}\setminus D$ such that $h(a_n) = 0$ whenever $n \neq m,\ n, m \in \mathbb{N}$. Since h is quasi-minorated in $H(\overline{D})$ and since D is bounded, we can extract a Cauchy subsequence from the sequence $(a_n)_{n\in\mathbb{N}}$. This Cauchy subsequence obviously converges to a point $a \in \overline{D}$ and therefore we have $h(a) = 0$. But, as h has no zero in D, a belongs to $\overline{D}\setminus D$. And now, since D belongs to Alg, then a belongs to $\overset{\circ}{\overline{D}}$. But, by Corollary 16.3, a zero of h which is interior to $\overline{D}$ is isolated in $\overline{D}$ and this contradicts the definition of a. Thus, we have proven that h has finitely many zeros $(b_j)_{1\leq j\leq q}$ in $\overline{D}$, all of them in the set $\overline{D}\setminus D$ which is included in $\overset{\circ}{\overline{D}}$. Then, each zero b_j has a multiplicity order n_j $(1 \leq j \leq q)$. Let $P(x) = \prod_{j=1}^{q}(x - b_j)^{n_j}$ be the polynomial of the zeros of h in $\overset{\circ}{\overline{D}}$. By Corollary 16.6, the function $g(x) = \frac{h(x)}{P(x)}$ belongs to $H(\overline{D})$ and obviously has no zero in $\overline{D}$. As we have already seen when D is closed, g is invertible in $H(\overline{D})$. Now, since both P, Q have all their zeros in $\overline{D}\setminus D$, they are invertible in $R(D)$ and so is $\frac{P}{Q}$. But then $f = \frac{P}{Q}g$ and hence f is invertible in $H(D)$. □

Theorem 21.14. *Let D be closed, bounded, having finitely many holes and let $f \in H(D)$. Then, $f(D)$ is an infraconnected closed and bounded subset of $\mathbb{K}$.*

Proof. Since D has finitely many holes, by Corollary 21.10, f is quasi-minorated. By Theorem 20.13, $f(D)$ is infraconnected. Since f is bounded on a closed bounded subset, $f(D)$ is bounded. Let b belong to the closure of $f(D)$ and let $(a_n)_{n\in\mathbb{N}}$ be a sequence of D such that $\lim_{n\to+\infty} f(a_n) = b$. Since the sequence (a_n) is bounded, there exists a subsequence thinner than a circular filter $\mathcal{F}$ secant with D.

If $\mathcal{F}$ is large, then we have ${}_D\varphi_{\mathcal{F}}(f) = 0$, and hence, by Proposition 21.1, f is not quasi-minorated, a contradiction. Consequently, $\mathcal{F}$ converges to a point a. Since D is closed, a belongs to D. Consequently, $f(a) = b$ and hence b belongs to $f(D)$. Therefore, $f(D)$ is closed. □

Theorem 21.15 shows an example of very simple increasing T-filter, without describing the general theory of T-filters [38, 44, 53].

Theorem 21.15. *Let $(r_n)_{n\in\mathbb{N}}$ be a sequence in $|\mathbb{K}|$ such that $0 < r_n < r_{n+1}$ and $\lim_{n\to+\infty} r_n = R$, and let $(q_n)_{n\in\mathbb{N}}$ be a sequence of $\mathbb{N}$ prime to the characteristic of $\mathbb{K}$ such that $q_n \leq q_{n+1}$ and $\lim_{n\to+\infty}\left(\frac{r_n}{r_{n+1}}\right)^{q_n} = 0$. Let $l \in]0, R[$ and for each $n \in \mathbb{N}$, let $b_n \in C(0,(r_n)^{q_n})$, let $a_{n,1},\dots,a_{n,q_n}$ be the q_nth roots of b_n, and let $E = d(0,R^-)\setminus\left(\bigcup_{n\in\mathbb{N}}(\cup_{j=1}^{q_n} d(a_{n,j},\ l^-))\right)$. Set $f_n(x) = \prod_{k=1}^{n}\prod_{j=1}^{q_k}\left(\frac{1}{1-\left(\frac{x}{a_{k,j}}\right)}\right)$. Then each f_n belongs to $R(E)$ and the sequence $(f_n)_{n\in\mathbb{N}}$ converges in $H(E)$ to an element strictly vanishing along the pierced increasing filter of center 0 and diameter R.*

Proof. Let us first show that the sequence $(f_n)_{n\in\mathbb{N}}$ converges in $H(E)$. We note that each pole $a_{k,j}$ of f_n lies in a hole of diameter l and is unique in that hole and in the class. Moreover, since each q_n is prime to the residue characteristic of $\mathbb{K}$, each pole $a_{k,j}$ of f_n is unique in the class $d(a_{k,j}, r_k^-)$. Consequently, we have $|f_n(x)| \leq \frac{|x||f_n|(|x|)}{l}$ $\forall x \in E$ and hence

$$(1) \qquad |f_n(x)| \leq \frac{R|f_n|(|x|)}{l} \quad \forall x \in E.$$

Let us now show that the sequence $(f_n)_{n\in\mathbb{N}}$ converges in $H(E)$. We first note that each factor $Q_k = \prod_{j=1}^{q_k}\left(\frac{1}{1-\left(\frac{x}{a_{k,j}}\right)}\right)$ satisfies $|Q_k|(r) \leq 1$ $\forall r < R$ because $|Q_k|(r) = 1$ $\forall r \leq r_k$ and $|Q_k|(r) < r$ $\forall r \in]r_k, R[$. Consequently, we have

$$(2) \qquad |f_n|(r) \leq \prod_{k=0}^{n}\prod_{j=1}^{q_k}\left(\frac{r_k}{r}\right)^{q_k} \leq \left(\frac{r_{n-1}}{r_n}\right)^{q_{n-1}} \quad \forall r \geq r_n.$$

On the other hand, when $r < r_n$, we have

$$|1 - Q_{n+1}|(r) = \left|1 - \frac{1}{1 - \left(\frac{x}{b_{n+1}}\right)^{q_{n+1}}}\right|(r) \leq \left(\frac{r_n}{r_{n+1}}\right)^{q_{n+1}},$$

hence

(3)
$$|f_{n+1} - f_n|(r) = |f_n|(r)|1 - Q_{n+1}|(r) < |f_n|(r)\left(\frac{r_n}{r_{n+1}}\right)^{q_{n+1}} \quad \forall r < r_n.$$

By (2) and (3), we can see that $\lim_{n\to+\infty}\big(\sup_{r<R}|f_{n+1}(r) - f_n(r)|\big) = 0$, and hence by (1), we have $\lim_{n\to+\infty}\|f_{n+1} - f_n\|_E = 0$, hence the sequence f_n converges to an element $f \in H(E)$. Moreover, by (1) and (2), we can see that $\lim_{|x|\to R^-,\ x\in E} f(x) = 0$, so f is vanishing along the increasing pierced filter $\mathcal{F}$ of center 0 and diameter R. Further, we may note that $|Q_n|(r) = 1\ \forall r \leq r_n$, hence, when $r \leq r_n$, we have $|f|(r) = |f_s|(r)\ \forall s \geq n$. Consequently, $|f|(r) \neq 0\ \forall r < R$ and hence f is strictly vanishing along $\mathcal{F}$. $\square$

Chapter 22

Zeros of Power Series

Most of classical results on zeros of polynomials are now extended to power series. In particular, power series converging inside a disk satisfy a Schwarz lemma that is even simpler than in $\mathbb{C}$.

Throughout this chapter, r is a strictly positive real number and r', r'' are strictly positive real numbers satisfying $r' < r''$.

Theorem 22.1. *Let $f \in H(C(0,r))$. The number of zeros of f in $C(0,r)$ is equal to $\nu^+(f, \log r) - \nu^-(f, \log r)$ (taking multiplicity into account).*

Proof. This equality was given for a polynomial in Theorem 4.16. First, we prove it when f is an element of $H(C(0,r))$ invertible in $H(C(0,r))$. By Proposition 19.7, we have $\nu^+(\frac{1}{f}, \log r) = -\nu^+(f, \log r)$ and $\nu^-(\frac{1}{f}, \log r) = -\nu^-(f, \log r)$. Since any $h \in H(C(0,r))$ satisfies $\nu^+(h, \log r) \geq \nu^-(h, \log r)$, we see that $\nu^+(f, \log r) = \nu^-(f, \log r)$.

We now consider the general case. By Corollary 21.11, f has a factorization of the form Pg with P a polynomial whose zeros belong to $C(0,r)$ and g invertible in $H(C(0,r))$. Then, $\nu^+(f, \log r) - \nu^-(f, \log r) = \nu^+(P, \log r) - \nu^-(P, \log r) = \deg(P)$, and this just ends the proof of Theorem 22.1. $\square$

Corollary 22.2. *Let $a \in \mathbb{K}$ and $r > 0$. Let $f(x) \in H(C(a, r^-))$. Let $\widehat{\mathbb{K}}$ be a complete algebraically closed extension of $\mathbb{K}$ and let $\widehat{C}(a,r) = \{x \in \widehat{\mathbb{K}} \mid |x - a| = r\}$. Then, the zeros of f in $\widehat{C}(a,r)$ are exactly* *those of* f *in* $C(a,r)$ *(taking multiplicity into account). Similarly, the zeros of* f *in* $\widehat{d}(a,r) = \{x \in \widehat{\mathbb{K}} \mid |x-a| \leq r\}$ *(respectively in* $\widehat{d}(a,r^-) = \{x \in \widehat{\mathbb{K}} \mid |x-a| < r\}$*) are exactly those of* f *in* $d(a,r)$ *(respectively in* $d(a,r^-)$*) (taking multiplicity into account).*

Corollary 22.3. *Let* $f \in H(C(0,r))$ *have* t *zeros in* $C(0,r)$.

Let $q = \nu^+(f,\log r) - \nu^-(f,\log r)$*. Then,* $r = \sqrt[q]{\left|\frac{a_t}{a_{q+t}}\right|}$.

Theorem 22.4. *Let* $\Lambda = C(0,r)$ *and let* $f(x) = \sum_{-\infty}^{+\infty} a_n x^n$ *be a convergent Laurent series in* Λ*, having no zero in* Λ*. Let* $\mu = \log r$*. Then,* $\nu^+(f,\mu) = \nu^-(f,\mu) = q \in \mathbb{Z}$ *and* $|f(x)| = |a_q x^q|$ *whenever* $x \in \Lambda$*. Moreover, if* $q \neq 0$*, then* $f(\Lambda) = C(0,|a_q|r^q)$.

Proof. By Proposition 19.7, we see that $\nu^+(f,\mu) = \nu^-(f,\mu)$. Thus, we have $\Psi(f,\mu) = \Psi(a_q) + q\mu$ whenever $\mu \in I$. Consequently, $f(C(0,r)) \subset C(0,|a_q|r^q)$. Now, suppose $q \neq 0$ and let $s = |a_q|r^q$. Let $b \in C(0,s)$ and let $g = f - b$. So, by definition, $g(x) = \sum_{-\infty}^{-1} a_n x^n + (a_0 - b) + \sum_1^{+\infty} a_n x^n$, and by hypothesis, $|a_0| < |a_q|r^q$, hence $|a_0 - b| = |a_q|r^q$. Consequently, $\nu^-(g,\mu) < \nu^+(g,\mu)$, therefore, by Theorem 22.1, g admits at least one zero in $C(0,r)$ and hence b lies in $f(C(0,r))$. This proves that $f(C(0,r)) = C(0,|a_q|r^q)$. □

Corollary 22.5. *Let* $f(x) = \sum_{-\infty}^{+\infty} a_n x^n$ *be a convergent Laurent series in* $\Lambda = \Gamma(a,R',R'')$ *(respectively in* $\Lambda = \Delta(a,R',R'')$*), having no zero in* $\Gamma(a,R',R'')$*. Then we have* $\nu^+(f,\mu) = \nu^-(f,\mu) = q \in \mathbb{Z}$ *whenever* $\mu \in]\log r', \log r''[$ *(respectively* $\mu \in [\log r', \log r'']$*) and* $|f(x)| = |a_q x^q|$ *whenever* $x \in \Lambda$*. Moreover, if* $q > 0$*, putting* $s' = |a_q|r'^q$*,* $s'' = |a_q|r''^q$*, we have* $f(\Lambda) = \Gamma(0,s',s'')$ *(respectively* $f(\Lambda) = \Delta(0,s',s'')$*), and if* $q < 0$*, putting* $s' = |a_q|r''^q$*,* $s'' = |a_q|r'^q$*, we have* $f(\Lambda) = \Gamma(0,s',s'')$ *(respectively* $f(\Lambda) = \Delta(0,s',s'')$*).*

Corollary 22.6. *Let* $\Lambda = \Gamma(0,r',r'')$ *and let* $f(x)$*,* $g(x)$ *be convergent Laurent series in* Λ*, having no zero in* Λ *such that* $|f|(r) = |g|(r)\ \forall r \in]r',r''[$*. Then,* $\nu(f,\log r) = \nu(g,\log r)\ \forall r \in]r',r''[$.

Proof. By hypothesis, we have $\Psi(f,\mu) = \Psi(g,\mu)\ \forall \mu \in]\log r', \log r''[$. But $\nu(f,\log r) = \frac{d\Psi(f)}{d\mu}(\log r)$ and $\nu(g,\log r) = \frac{d\Psi(g)}{d\mu}(\log r)$, hence $\nu(f,\log r) = \nu(g,\log r)$. □

Theorem 22.7. *Let $f \in H(d(0,r))$. The number of zeros of f in $d(0,r)$ is equal to $\nu^+(f, \log r)$ (taking multiplicity into account).*

Proof. This equality was given for a polynomial in Corollary 4.17. First, we prove it when f is an invertible element of $H(d(0,r))$. By Theorem 22.1, we have $\nu^+(f,\mu) = \nu^-(f,\mu) = 0 \ \forall \mu \leq \log(r)$. Consequently, $\nu^+(f, \log r) = 0$. Consider now the general case. Since, by Theorem 21.11, f is quasi-invertible, it has a factorization of the form $P(x)h(x)$ with P a polynomial whose zeros lie in $d(0,r)$ and h is an invertible element of $H(d(0,r))$. Then, by Corollary 4.17, $\nu^+(P, \log(r))$ is the number of zeros of P (hence of f), and by Proposition 19.7, we have $\nu^+(f, \log(r)) = \nu^+(P, \log(r)) + \nu^+(h, \log(r)) = \nu^+(P, \mu)$ which ends the proof. □

Corollary 22.8. *Let $\Lambda = \Gamma(0, r', r'')$ and let $f(x)$, $g(x)$ be convergent Laurent series in Λ such that $|f|(r) = |g|(r) \ \forall r \in]r', r''[$. Then for each $r \in]r', r''[$, f and g have the same number of zeros in $C(0,r)$ (taking multiplicity into account).*

Theorem 22.9. *Let $f(x) = \sum_{n=0}^{\infty} a_n (x-a)^n \in H(d(a,r))$ and let $s = \sup_{n \geq 1} |a_n| r^n$. Then, $f(d(a,r)) = d(a_0, s)$ and $\Psi_a(f - a_0, \log r) = \log s$.*

Proof. Without loss of generality, we can suppose that $a = 0$. Let $b \in d(a_0, s)$ and consider $f(x) - b = a_0 - b + \sum_{n=1}^{+\infty} a_n x^n$. By hypothesis, $|a_0 - b| \leq s$, hence $\nu^+(f - b, \log(r) = \nu^+(f - a_0, \log(r)) \geq 1$, and hence $f - b$ has at least one zero in $d(0,r)$. Consequently, $d(0,s) \subset f(d(0,r))$. Conversely, when $x \in d(0,r)$, we have $|f(x) - a_0| \leq s$, hence $d(0,s) = f(d(0,r))$. As a consequence, $\Psi_a(f - a_0, \log r) = \log s$. □

Corollary 22.10. *Let $f(x) = \sum_{n=0}^{\infty} a_n (x-a)^n \in \mathcal{A}_b(d(a, r^-))$ (not identically zero) and let $s = \sup_{n \geq 1} |a_n| r^n$. Then, $f(d(a, r^-)) = d(a_0, s^-)$.*

Proof. On the one hand, $f(d(a, r^-))$ is obviously included in $d(a_0, s^-)$. On the other hand, given $b \in d(a_0, s^-)$ and $\rho \in]0, r[$ such that $\sup_{n \geq 1} |a_n| \rho^n \geq |b - a_0|$, by Theorem 22.9, b belongs to $f(d(a, \rho))$ because $f \in H(d(a, \rho))$. □

Corollary 22.11. *Let $f(x) = \sum_{n=0}^{\infty} a_n (x-a)^n \in H(d(a, r^-))$ (not identically zero) and let $s = \sup_{n \geq 1} |a_n| r^n$. Then, $f(d(a, r^-)) = d(a_0, s^-)$.*

Lemma 22.12. *Let $f \in H(d(0,r))$ satisfy $\nu^+(f, \log r) \geq 1$, let $b \in f(d(0,r))$, and let $g(x) = f(x) - b$. Then, we have $\nu^+(g, \log r) = \nu^+(f, \log r)$.*

Proof. Let $f(x) = \sum_{n=0}^{\infty} a_n x^n$ and let $t = \nu^+(f, \log r)$. By hypothesis, we have $|a_t|r^t \geq |a_n|r^n$ whenever $n < t$ and $|a_t|r^t > |a_n|r^n$ whenever $n > t$. Now let $g(x) = \sum_{n=0}^{\infty} b_n x^n$. Hence, $b_0 = a_0 - b, b_n = a_n$ whenever $n \geq 1$. By hypothesis, we have $|b_0| \leq \sup_{n\geq 1} |a_n|r^n$, hence $|b_0| \leq |b_t|r^t$, and finally, $\nu^+(g, \log r) = t$. □

Lemma 22.13. *Let $f \in H(d(a,r))$ have t zeros in $d(a,r)$ with $t \geq 1$ (taking multiplicity into account) and let $b \in f(d(a,r))$. Then, $f - b$ also admits t zeros in $d(a,r)$ (taking multiplicity into account).*

Proof. We assume $a = 0$. By Lemma 22.12, we know that $\nu^+(f, \log r) = t$. Hence, we have $\nu^+(f, \log r) = t$ and $\Psi(f, \log r) = \Psi(a_t) + t \log r$. Next, $\Psi(f(x)) \leq \Psi(f, \log r)$ for all $x \in d(0,r)$, hence $\Psi(b) \leq \Psi(f, \log r)$, and therefore, $\Psi(b) \leq \Psi(a_t) + t \log r$. Hence, $\nu^+(f - b, \log r) = t$. That ends the proof. □

Theorem 22.14. *If an entire function $f \in \mathcal{A}(\mathbb{K})$ is bounded or has no zero, it is a constant.*

Proof. Let $f(x) = \sum_{j=0}^{\infty} a_j x^j$. By Lemma 22.12, the number of zeros of f in any disk $d(0,r)$ is equal to the biggest of the integers l such that $|a_l|r^l = \sup_{j\in\mathbb{N}} |a_j|r^j$ and this is also equal to $\lim_{|x|\to r, |x|\neq r} |f(x)|$. Hence, if f is bounded, or has no zero in $\mathbb{K}$, obviously we have $a_n = 0 \ \forall n > 0$. □

Theorem 22.15. *Let $f(x) \in \mathcal{A}(\mathbb{K})\setminus\mathbb{K}$. Then f admits at least one zero in $\mathbb{K}$. Moreover, if f is not a polynomial, then f has infinitely many zeros in $\mathbb{K}$ and the zeros make a sequence $(\alpha_n)_{n\in\mathbb{N}}$ such that $\lim_{n\to+\infty} |\alpha_n| = +\infty$.*

Proof. Suppose first that f has finitely many zeros in $\mathbb{K}$. Then, by Theorem 14.16, there exists a polynomial P such that f factorizes in the form Pg with $g \in \mathcal{A}(\mathbb{K})$ and $g(x) \neq 0 \forall x \in \mathbb{K}$. Hence, by Theorem 22.14, g is a constant, hence f is a polynomial. Next, by Theorem 22.7, f has finitely many zeros in each disk, hence the sequence $(\alpha_n)_{n\in\mathbb{N}}$ tends to $+\infty$. □

Theorem 22.16. *Let Λ be a disk of the form $d(a, r)$ (respectively $d(a, r^-)$) and let $f \in H(\Lambda)$ have no zero in Λ. Then, $|f(x)|$ is equal to a constant in Λ and f is invertible in $H(\Lambda)$.*

Proof. By Corollary 21.11 and Theorems 21.13, f is invertible in $H(\Lambda)$. We may obviously assume $a = 0$. By Lemma 22.12, we have $\nu^+(f, \mu) = \nu^-(f, \mu) = 0$ whenever $\mu \leq \log r$ (respectively $\mu < \log r$), hence by Proposition 19.7, $\Psi(f, \mu)$ has a derivative equal to 0 and therefore is equal to a constant in $]-\infty, \log r[$. Then, by Corollary 19.6, we have $\Psi(f(x)) = \Psi(f, \Psi(x))$, hence $\Psi(f(x))$ is equal to a constant in Λ. □

Theorem 22.17. *Let Λ be a set in one of the following forms:*

(i) $\Lambda = d(0, r)$.
(ii) $\Lambda = d(0, r^-)$.
(iii) $\Lambda = C(0, r)$.

Let $f \in H(\Lambda)$ (f not identically 0) and let $h \in H(\Lambda)$ satisfy $\|f-h\|_\Lambda < \|f\|_\Lambda$. Then f and h have the same number of zeros in Λ (taking multiplicity into account).

Proof. Regardless of the cases (i), (ii), and (iii), we know that $\|f\|_\Lambda = {}_\Lambda\varphi_{0,r}(f)$ and then we have $\log \|f\|_\Lambda = \Psi(f, \log r)$. Since $\|f - h\|_\Lambda < \|f\|_\Lambda$, then $\Psi((f - h, \log r) < \Psi(f, \log r)$. Hence, by Proposition 19.5, we know that $\nu^+(f, \log r) = \nu^+(h, \log s)$ and $\nu^-(f, \log r) = \nu^-(h, \log r)$. Consequently, f has as many zeros as h in Λ, by Theorem 22.7 if $\Lambda = d(0, r)$ and by Theorem 22.1 if $\Lambda = C(0, r)$.

We now suppose $\Lambda = d(0, r^-)$. By Corollary 21.11, both f, h are quasi-invertible in $H(\Lambda)$. Let $\rho \in]0, r]$ be such that $d(0, \rho)$ contains all zeros of h in Λ. According to the statements (i) and (ii) already proven, we see that f has as many zeros as h in $d(0, \rho)$ (taking multiplicity into account) and has no zero in $C(0, s)$ whenever $s \in]\rho, r[$. This ends the proof. □

Theorem 22.18. *If $f \in H(C(a, r))$, it satisfies $|f(x)| \leq {}_D\varphi_{a,r}(f)$ $\forall x \in C(a, r)$ and the equality $|f(x)| = {}_D\varphi_{a,r}(f)$ holds in all classes except finitely many that are the classes, where f has at least one zero.*

Proof. Let $f \in H(C(a,r))$ and $\Lambda = C(a,r)$. We can find $h \in R(\Lambda)$ such that $\|f - h\|_\lambda < \|f\|_\Lambda$, hence by Theorem 22.17, f has the same number of zeros as h in each class of Λ (taking multiplicity into account). So, in each class of $\Xi = d(b, r^-)$ where f has no zero, $|f(x)|$ is equal to ${}_D\varphi_{b,r}(f)$. But by Lemma 13.4, we have ${}_D\varphi_{b,r}(f) = {}_D\varphi_{a,r}(f)$, which ends the proof. □

Corollary 22.19. *Let $f(x) \in \mathcal{A}(d(0,r^-))$ have infinitely many zeros in $d(0,r^-)$. Then the set of zeros of f in $d(0,r^-)$ is a sequence $(\alpha_n)_{n\in\mathbb{N}}$ such that $\lim_{n\to+\infty} |\alpha_n| = r$.*

Proof. Indeed, by Proposition 14.12, for each $\rho \in]0,r[$, f belongs to $H(d(0,\rho))$ and hence has finitely many zeros in $d(0,\rho)$. □

Theorem 22.20. *Let $f(x) = \sum_{n=0}^{\infty} a_n x^n \in \mathcal{A}(d(0,r^-))$. Then f has finitely many zeros in $d(0,r^-)$ if and only if there exists $q \in \mathbb{N}$ such that $|a_q|r^q \geq \sup_{n\in\mathbb{N}} |a_n|r^n$. Moreover, if t is the smallest of the integers q such that $|a_q|r^q \geq \sup_{n\in\mathbb{N}} |a_n|r^n$, then f has exactly t zeros in $d(0,r^-)$. Further, the three following statements are equivalent:*

(i) *f has no zero in $d(0,r^-)$.*
(ii) *f is invertible in $\mathcal{A}(d(0,r^-))$.*
(iii) *$|f(x)|$ is a non-zero constant.*

Proof. First, we suppose that there exists $q \in \mathbb{N}$ such that $|a_q|r^q \geq \sup_{n\in\mathbb{N}} |a_n|r^n$. Let $t \in \mathbb{N}$ be the smallest of the integers q such that $|a_q|r^q \geq \sup_{n\in\mathbb{N}} |a_n|r^n$. There exists a unique $s \in]0,r[$ such that $|a_t|s^t \geq |a_n|s^n$ for every $n < t$. Then, for all $\rho \in]s,r[$, in $H(d(0,\rho))$, we have $\nu^+(f, \log\rho) = \nu^-(f, \log\rho) = t$, hence f admits exactly t zeros in $d(0,\rho)$, whenever $\rho \in]s,r[$.

Conversely, suppose that f admits exactly t zeros in $d(0,r^-)$. There then there exists $s \in]0,r[$ such that f admits exactly t zeros in $d(0,s)$ and of course in each disk $d(0,\rho)$ for every $\rho \in]s,r[$. Hence, we have $\nu^+(f, \log\rho) = \nu^-(f, \log\rho) = t$ for every $\rho \in]s,r[$. Therefore, we have $|a_t|\rho^t > |a_n|\rho^n$ for every $n \neq t$ and for every $\rho \in]s,r[$. Finally, we see that $|a_t|r^t \geq \sup |a_n|r^n$.

Further, the equivalence between (i) and (ii) is shown at Theorem 14.19. The equivalence of (i) with (iii) comes from the fact that f has no zero if and only if $|a_0| > |a_n|s^n$ $\forall n \in \mathbb{N}$, $\forall s \in]0,r[$. □

Theorem 22.21. *Let $f(x) = \sum_{-\infty}^{+\infty} a_n x^n \in \mathcal{A}(\mathbb{K}\setminus d(0,R))$ be not identically zero, let $R' > R$, and let $q = \nu^+(f, \log(R'))$. The family of zeros of f in $\mathbb{K}\setminus d(0,R'^-)$ either is finite or is a sequence $(\alpha_n)_{n\in\mathbb{N}}$ such that $\lim_{n\to+\infty} |\alpha_n| = +\infty$.*

Proof. By definition, f is of the form $g + h$ with $g(x) = \sum_{-\infty}^{-1} a_n x^n \in \mathcal{A}(\mathbb{K}\setminus d(0,R))$ and $h(x) = \sum_{0}^{+\infty} a_n x^n \in \mathcal{A}(\mathbb{K})$. Particularly, by Theorem 14.6, for every R', $R'' \in]R, +\infty[$ (with $R' < R''$), both g, h belong to $H(\Delta(0,R',R''))$. Then, so does f. Therefore, by Corollary 22.3, f has finitely many zeros in $\Delta(0,R',R'')$ which are on the circles $C(0,r)$ such that $\nu^+(f,\log r) > \nu^-(f,\log r)$. Thus, the family of zeros of f in $\mathbb{K}\setminus d(0,R'^-)$ either is finite or is a sequence $(\alpha_n)_{n\in\mathbb{N}}$ such that $\lim_{n\to+\infty} |\alpha_n| = +\infty$. □

Theorem 22.22. *Let R, $S \in]0, +\infty[$, $R < S$, and let*

$$f(x) = \sum_{-\infty}^{+\infty} a_n x^n \in \mathcal{A}(\mathbb{K}\setminus d(0,R))$$

have infinitely many zeros in $\mathbb{K}\setminus d(0,S))$. Then, for every fixed $t \in \mathbb{N}$, we have $\lim_{r\to+\infty} \frac{|f|(r)}{r^t} = +\infty$.

Proof. By Theorem 22.21, the sequence of zeros $(\alpha_n)_{n\in\mathbb{N}}$ is such that $\lim_{n\to+\infty} |\alpha_n| = +\infty$. Set $r_n = |\alpha_n|$. The sequence $\nu+(f, \log(r_n)$ is strictly increasing and hence there exists $q \in \mathbb{N}$ such that $\nu^+(f, \log(r_q) > t$, therefore $\nu^+(f,\mu) \geq 1\ \forall \mu > \log(r_q)$. Set $g(x) = \frac{f(x)}{x^t}$. Then clearly $|g|(r) = \frac{|f|(r)}{r^t}$ and hence $\nu^+(g,\mu) \geq 1\ \forall \mu \geq \log(r_q)$. Thus, the function (in μ) $\Psi(f,\mu)$ is a convex function, piecewise affine, whose derivative, when it is derivable, is grater than 1 whenever $\mu \geq \log(r_q)$. Consequently, $\lim_{r\to+\infty} |g|(r) = +\infty$, and therefore, $\lim_{r\to+\infty} \frac{|f|(r)}{r^t} = +\infty$. □

Corollary 22.23. *Let $f(x) \in \mathcal{A}(\mathbb{K})\setminus\mathbb{K}[x]$. Then, for every fixed $t \in \mathbb{N}$, we have $\lim_{r\to+\infty} \frac{|f|(r)}{r^t} = +\infty$.*

Theorem 22.24. *Let $r \in \mathbb{R}_+$, let $r_1, r_2 \in]0, R[)$ satisfy $r_1 < r_2$, and let $f \in \mathcal{A}(d(0,R^-))$. If f admits exactly q zeros in $d(0,r_1)$ (taking multiplicity into account) and has no zeros in $\Gamma(0,r_1,r_2)$, then f satisfies*

$$\Psi(f, \log r_2) - \Psi(f, \log r_1) = q(\log r_2 - \log r_1).$$

Proof. By Lemma 22.13, we have $\nu^+(f, \log r_1) = q$, and by Theorem 22.4, we have $\Psi(f, \mu) - \Psi(f, \log r_1) = q(\mu - \log r_1)$ for every $\mu \in [\log r_1, \log r_2[$, hence by continuity, we have $\Psi(f, \log r_2) - \Psi(f, \log r_1) = q(\log r_2 - \log r_1)$. □

Theorem 22.25. *Let $f(x) \in \mathcal{A}(d(a, r^-))$. If f is not bounded, then f has infinitely many zeros in $d(a, r^-)$.*

Proof. Without loss of generality, we can suppose $a = 0$. Suppose f has t zeros in $d(0, r^-)$. Let $d(0, s)$ be a disk containing all zeros of f, with $s < r$. Then, by Theorem 22.22, for all $\rho \in [s, r[$, we have $\Psi(f, \log(\rho)) \leq \Psi(f, \log(s)) + t(\log(\rho) - \log(s)) \leq \Psi(f, \log(s)) + t(\log(r) - \log(s))$. So, $\psi(f, \mu)$ is bounded, hence by Theorem 14.20, f is bounded in $d(0, r^-)$, a contradiction. □

Theorem 22.26. *Let $f \in \mathcal{A}(d(0, R^-))$ and let $r_1, r_2 \in]0, R[$ satisfy $r_1 < r_2$. If f admits exactly q zeros in $d(0, r_1)$ (taking multiplicity into account) and t different zeros α_j, of respective multiplicity order m_j $(1 \leq j \leq t)$ in $\Gamma(0, r_1, r_2)$, then f satisfies*

$$\Psi(f, \log r_2) - \Psi(f, \log r_1) = \sum_{j=1}^{t} m_j(\log(r_2) - \Psi(\alpha_j)) \\ + q(\log r_2 - \log r_1).$$

Proof. Let $C(0, \rho_h)$ $1 \leq h \leq s$ be the circles containing at least one zero of f. For each $h = 1, \ldots, s$, let $\alpha_{n(h)}, \ldots, \alpha_{n(h+1)-1}$ be the zeros of f in $C(0, \rho_h)$. Let $u = \sum_{j=1}^{t} m_j$.

First, by Theorem 22.24, we note that

(1) $\Psi(f, \log(\rho_1)) - \Psi(f, \log(r_1)) = q(\log(\rho_1) - \log(r_1))$

and similarly

(2) $\Psi(f, \log(r_2)) - \Psi(f, \log(\rho_s)) = (q + u)(\log(r_2) - \log(\rho_s))$.

Next, for each $h = 1, \ldots, s - 1$, set $l_h = q + \sum_{j=1}^{n(h+1)-1} m_j$. Then, f has no zero in $\Gamma(0, \rho_h, \rho_{h+1})$ and has l_h zeros in $d(0, \rho_h)$, hence by Theorem 22.24, we have

(3) $\Psi(f, \log(\rho_{h+1})) - \Psi(f, \log(\rho_h)) = l_h(\log(\rho_{h+1}) - \log(\rho_h))$.

Then, by (1), (2), and (3), we can check the conclusion. □

Corollary 22.27. *Let $f(x) \in \mathcal{A}(d(0, R^-))$ be such that $f(0) \neq 0$, let $r \in]0, R[$ and let a_j, $1 \leq j \leq q$ be the zeros of f in $d(0, r)$, of respective multiplicity m_j. Then,*

$$\Psi(f, \log r) = \Psi(f(0)) + \sum_{j=1}^{q} m_j(\log r - \Psi(a_j)).$$

Corollary 22.28. *Let $f(x) = \sum_{n=0}^{\infty} a_n x^n \in \mathcal{A}(d(0, r^-))$ have a set of zeros in $d(0, r^-)$ that consists of a sequence $(\alpha_n)_{n\in\mathbb{N}}$ such that $\alpha_n \neq 0$ $\forall n \in \mathbb{N}$ and where each α_n is of order u_n. Then, f is unbounded if and only if the sequence $(\alpha_n)_{n\in\mathbb{N}}$ satisfies $\prod_{n=0}^{\infty}\left(\frac{|\alpha_n|}{r}\right)^{u_n} = 0$.*

Corollary 22.29. *Let $f(x) = \sum_{n=0}^{\infty} a_n x^n \in \mathcal{A}(d(0, r^-))$ have a set of zeros in $d(0, r^-)$ that consists of a sequence $(\alpha_n)_{n\in\mathbb{N}}$ such that $\alpha_n \neq 0$ $\forall n \in \mathbb{N}$ and where each α_n is of order u_n. Then, $\|f\|_D = |f(0)|\prod_{n=0}^{\infty}\left(\frac{r}{|\alpha_n|}\right)^{u_n}$.*

Corollary 22.30 (Schwarz Lemma). *Let $D = d(a, s)$ and let $f \in H(D)$ have at least (respectively at most) q zeros in $d(a, r)$ with $q > 0$ and $0 < r < s$. Then, we have $\frac{\varphi_{a,s}(f)}{\varphi_{a,r}(f)} \geq \left(\frac{s}{r}\right)^q$* (respectively $\frac{\varphi_{a,s}(f)}{\varphi_{a,r}(f)} \leq \left(\frac{s}{r}\right)^q$).

Corollary 22.31. *Let $f \in \mathcal{A}(\mathbb{K})$. The following two statements are equivalent:*

(i) *f is a polynomial of degree q.*
(ii) *There exists $q \in \mathbb{N}$ such that $\frac{|f|(r)}{r^q}$ has a finite limit when r tends to $+\infty$.*

Corollary 22.32. *Let r, s, $R \in]0, +\infty[$ satisfy $0 < r < s < R$ and let $f \in H((0, R))$. Then,*

$$\begin{aligned}
&\Psi(f, \log(s)) - \Psi(f, \log(r)) \\
&\quad \leq (\Psi(f, \log(R)) - \Psi(f, \log(s)))\left(\frac{\log(s) - \log(r)}{\log(R) - \log(s)}\right).
\end{aligned}$$

Proof. Let q be the total number of zeros of f in $d(0, s)$, each counted with its multiplicity. Then, by Theorem 22.30, we have

$\Psi(f, \log(s)) - \Psi(f, \log(r)) \le q(\log(s) - \log(r))$. On the other hand, $\Psi(f, \log(R)) - \Psi(f, \log(s)) \ge q(\log(R) - \log(s))$. Consequently,

$$q \le \frac{\Psi(f, \log(R)) - \Psi(f, \log(s))}{\log(R) - \log(s)},$$

and hence the proof is over. □

Theorem 22.33. *Let $f(x) = \sum_{n=0}^{\infty} a_n x^n \in \mathcal{A}(\mathbb{K})$ (respectively $f(x) = \sum_{n=0}^{\infty} a_n x^n \in \mathcal{A}(d(0, r^-))$). All zeros of f are of order one and the set of zeros of f is a sequence $(\alpha_n)_{n\in\mathbb{N}}$ such that $|\alpha_n| < |\alpha_{n+1}|$ if and only if the sequence $\left|\frac{a_n}{a_{n+1}}\right|$ is strictly increasing. Moreover, if these properties are satisfied, then the sequence of zeros of f in $\mathbb{K}$ (respectively in $d(0, r^-)$) is a sequence $(\alpha_n)_{n\in\mathbb{N}^*}$ such that $\lim_{n\to+\infty} |\alpha_n| = +\infty$ (respectively $\lim_{n\to+\infty} |\alpha_n| = r$) and $|\alpha_n| = \left|\frac{a_n}{a_{n+1}}\right|$.*

Proof. Suppose first that $f \in \mathcal{A}(d(0, r^-))$. First, we suppose that the set of zeros of f in $d(0, r^-)$ is an increasing distances sequence $(\alpha_n)_{n\in\mathbb{N}^*}$. Then by Corollary 22.19, we know that $\lim_{n\to+\infty} |\alpha_n| = r$. By Corollary 22.3, for each $q \in \mathbb{N}^*$, we have $\nu^+(f, \Psi(\alpha_q)) - \nu^-(f, \Psi(\alpha_q)) = 1$ and $|\alpha_0| = \left|\frac{a_0}{a_1}\right|$. Then by an immediate induction, we deduce that $|\alpha_n| = \left|\frac{a_n}{a_{n+1}}\right|$ for every $n \in \mathbb{N}^*$ and therefore the sequence $\left|\frac{a_n}{a_{n+1}}\right|$ is strictly increasing.

Conversely, we suppose that $\left|\frac{a_n}{a_{n+1}}\right|$ is a strictly increasing sequence. Hence, we have $\left|\frac{a_n}{a_{n+1}}\right| < \left|\frac{a_{n+1}}{a_{n+2}}\right|$ for every $n \in \mathbb{N}$. For each $m \in \mathbb{N}^*$, we put $s_m = \Psi(a_m) - \Psi(a_{m-1})$ and $r_m = \theta^{s_m}$. Clearly, we have $\nu^+(f, s_m) - \nu^-(f, s_m) = 1$ for every $m \in \mathbb{N}^*$ and $\nu^+(f, \mu) = \nu^-(f, \mu)$ for every $\mu \in \left(] - \infty, \log r[\setminus\{s_m |\ m \in \mathbb{N}^*\}\right)$. Hence, by Theorem 22.1, f admits exactly one zero in each circle $C(0, r_m)$ and no other zero in $d(0, r^-)$.

Suppose now that $f \in \mathcal{A}(\mathbb{K})$. The same proof applies with $\lim_{n\to+\infty} |\alpha_n| = +\infty$. □

Corollary 22.34. *Let $f(x) = \sum_{n=0}^{\infty} a_n x^n$ and suppose that the sequence $\left(\frac{|a_n|}{|a_{n+1}|}\right)_{n\in\mathbb{N}}$ is strictly increasing, of limit $+\infty$ (respectively of limit R). Then, f belongs to $\mathcal{A}(\mathbb{K})$ (respectively $\mathcal{A}(d(0, R^-))$). Moreover, putting $r_n = \frac{|a_n|}{|a_{n+1}|}$, $n \in \mathbb{N}$, f admits a unique zero on each circle $C(0, r_n)$ and has no other zero in $\mathbb{K}$ (respectively in $d(0, R)$).*

Proof. Indeed, thanks to the remark at the beginning of Chapter 13, the radius of convergence of f is $+\infty$ (respectively R). Then, the conclusion comes from Theorem 22.33. □

Remark. We can easily construct a sequence $(r_n)_{n\in\mathbb{N}}$ satisfying the hypothesis of Corollary 22.34 and thereby the function h.

It is often uneasy to determine whether a function defined in an infraconnected set is an analytic element. The following example may be useful.

Theorem 22.35. *Let $R > 0$ and let $(a_n)_{n\in\mathbb{N}}$ be a sequence of $d(a, R^-)$ such that $|a_n| < |a_{n+1}|$, $\lim_{n\to+\infty} |a_n| = R$, $\prod_{k=0}^{+\infty} \frac{|a_n|}{R} = 0$ and set $r_n = |a_n|$, $n \in \mathbb{N}$. Let $f \in \mathcal{A}(d(a, R^-))$ admit each a_n as a zero of order 1 and no other zero in $d(a, r_n)$. Let $E = d(a, R^-)\setminus\left(\bigcup_{n\in\mathbb{N}} C(a, r_n)\right)$. Then the function $\frac{1}{f}$ defined in E belongs to $H(E)$.*

Proof. Without loss of generality, we can obviously suppose $a = 0$ and $f(0) = 1$. By Corollary 22.28, f belongs to $\mathcal{A}_u(d(0, R^-))$. For each $n \in \mathbb{N}$, set $r_n = |a_n|$. Let $r \in]0, R[$. If $r \neq r_n$ $\forall n \in \mathbb{N}$, then $|f(x)| = |f|(r)$ $\forall x \in C(0, r)$. Now, let $x \in C(0, r_n)$. If $|x - a_n| = r_n$, we have $|f(x)| = r_n$. And if x belongs to E so that $|x - a_n| < r_n$, then $|f(x)| \geq |f|(r_n)\frac{r_n}{s}$. Consequently,

$$(1) \qquad |f(x)| = |f|(|x|) \quad \forall x \in E.$$

Let $f(x) = \sum_{n=0}^{+\infty} c_n x^n$ and for each $k \in \mathbb{N}$, let $P_k(x) = \sum_{n=0}^{k} c_n x^n$. Of course, the sequence $(P_k)_{k\in\mathbb{N}}$ converges to f uniformly in every disk $d(0, r)$ with $r \in]0, R[$. Moreover, by Theorem 22.34, each zero a_k of f satisfies $|a_k| = \frac{|c_{k-1}|}{|c_k|}$. Consequently, for $n > k$, P_n also admits a unique zero $a_{n,k}$ in $C(0, r_n)$ and has no other zero in E. Therefore, we have $|P_n|(r) = |f|(r)$ $\forall r \leq r_n$. We show that the sequence $(\frac{1}{P_n})_{n\in\mathbb{N}}$ converges to $\frac{1}{f}$ in $H(E)$. Indeed, let us fix $\epsilon > 0$ and let us choose $q \in \mathbb{N}$ such that $|f|(r_q) > \frac{1}{\epsilon}$. Consider now integers $n > q$ such that $|f(x) - P_n(x)| < \epsilon$ $\forall x \in d(0, r_q)$. Then obviously,

$$\frac{1}{f(x)} - \frac{1}{P_n(x)}| < \epsilon \quad \forall x \in E \cap d(0, r_q).$$

On the other hand, given $x \in E\setminus d(0, r_q)$, we have $|P_n|(r) \geq |P_n|(r_q) > \frac{1}{\epsilon}$, hence $\left|\frac{1}{P_n(x)} - \frac{1}{f(x)}\right| < \epsilon$. Therefore, $\left\|\frac{1}{P_n(x)} - \frac{1}{f(x)}\right\|_E \leq \epsilon$. This finishes proving that $\frac{1}{f(x)}$ belongs to $H(E)$. □

Notation. Let $f \in \mathcal{A}(\mathbb{K})$ and let $n \in \mathbb{N}^*$. We denote by $f^{<n>}$ the function $f \circ f \circ f \circ \cdots \circ f$, n times.

Theorem 22.36. *Suppose the residue characteristic p is different from 0. There exists functions $\mathcal{A}(\mathbb{K})$ and points $a \in \mathbb{K}$ such that $\lim_{n\to+\infty} |f^{<n>}(a)| = +\infty$.*

Proof. Let $t(n) = n^2$, $n \in \mathbb{N}$. Let $f(x) = \sum_{n=0}^{+\infty} p^{2t(n)} x^{2n+1}$. For each $n \in \mathbb{N}$, put $r_n = \frac{p^{t(n-1)}}{p^{t(n)}}$. Then, we can check that f admits exactly two zeros (taking multiplicity into account) in the circle $C(0, r_n)$ and has no zeros outside $\bigcup_{n=1}^{\infty} C(0, r_n)$, except $\{0\}$.

Consider now a number $s \in]r_n, r_{n+1}[$ of the form $q\sqrt{p}$, with $q \in \mathbb{Q}$. Inside the disk $d(0, s)$, f admits $2n + 1$ zeros (taking multiplicity into account). Consequently, $|f|(s) = |p^{2t(n)}|s^{2n+1} = \frac{p^{-2t(n)}|q|^{2n+1}p^{-2n}}{\sqrt{p}} = \frac{p^{-2t(n)-2n}|q|^{2n+1}}{\sqrt{p}}$. Thus, we can see that $|f|(s)$ is of the form $q'\sqrt{p}$ with $q' \in \mathbb{Q}$. Moreover, since q belongs to $\mathbb{Q}$, f has no zero in $C(0, s)$, hence $|f(x)| = |f|(s) = \frac{|q'|}{\sqrt{p}}$ $\forall x \in C(0, s)$. Now consider $\frac{|f|(s)}{s} = \frac{|p^{2t(n)}|s^{2n+1}}{s} = p^{-2t(n)} s^{2n}$.

On the other hand, by (1), $r_n = p^{t(n)-t(n-1)}$. Since $r_n < s$, we can derive

$$p^{-2t(n)} s^{2n} > p^{-2t(n)} p^{2n(t(n)-t(n-1))}.$$

Now, $-2t(n) + 2n(t(n) - t(n-1)) = 2\big(-t(n) + n(t(n) - t(n-1))\big) = 2(-n^2 + n(2n-1)) = 4n^2 - 2n$. Consequently, when $s > r_n$, we have $\frac{|f|(s)}{s} > p^{4n^2-2n}$. Now, let us take $s_2 > r_2$, and for each $n \in \mathbb{N}$, $n \geq 2$, let us define by induction $s_{n+1} = |f|(s_n)$. So, we have $\frac{s_{n+1}}{s_n} \geq p^8$, hence $\lim_{n\to+\infty} s_n = +\infty$. But by construction, for all $x \in C(0, s_n)$, we have $|f(x)| = |f|(s_n)$. Consequently, taking $a \in C(0, s_2)$, we have $|f^{<k>}(a)| = s_k$, and hence, $\lim_{n\to+\infty} |f^{<n>}(a)| = +\infty$. □

Chapter 23

Image of a Disk

In this chapter, D is just an open subset of $\mathbb{K}$.

Theorem 23.1. *Let $f(x) = \sum_{n=0}^{\infty} a_n(x-a)^n \in H(d(a,r))$. Then the following statements* (a), (b), (c), (d), and (e) *are equivalent:*

(a) $|a_0| > |a_n|r^n$ *for all* $n > 1$.
(b) $\|f - f(a)\|_{d(a,r)} < |f(a)|$.
(c) f *has no zero in* $d(a,r)$.
(d) $|f(x)|$ *is constant and different from* 0 *in* $d(a,r)$.
(e) f *is invertible in* $H(d(a,r))$.

Proof. First, (a) and (b) are equivalent by Theorem 14.6. Second, (a), (c), and (d) are equivalent by Theorems 22.7 and 22.16. Third, by Lemma 11.3, (d) implies (e), and finally, (e) obviously implies (c). □

Corollary 23.2. *Let $f(x) = \sum_{n=0}^{\infty} a_n(x-a)^n \in \mathcal{A}(d(a,r^-))$ (respectively let $f(x) = \sum_{n=0}^{\infty} a_n(x-a)^n \in H((d(a,r^-)))$. Then statements* (a), (b), (c), (d), and (e) *are equivalent:*

(a) $|a_0| \geq |a_n|r^n$ *for all* $n > 1$.
(b) $|f - f(a)| < |f(a)|\ \forall x \in d(a,r^-)$.
(c) f *has no zero in* $d(a,r^-)$.
(d) $|f(x)|$ *is constant and different from* 0 *in* $d(a,r^-)$.
(e) f *is invertible in* $\mathcal{A}(d(a,r^-))$ *(respectively f is invertible in* $H(d(a,r^-)))$.

Proof. Concerning $\mathcal{A}(d(a,r^-))$, we just have to apply Theorem 23.1 to f in $H(d(a,\rho))$ for every $\rho \in]0,r[$. Concerning $H(d(a,r^-))$, we can use Lemma 11.3 to check that an element of $H(d(0,r^-))$ having no zero is invertible. □

Theorem 23.3. *Let a, $b \in \mathbb{K}$ and $r, s \in \mathbb{R}_+^*$, and let $f \in \mathcal{A}(d(a,r^-))$, $g \in \mathcal{A}(d(b,s^-))$ be such that $f(d(a,r^-)) \subset d(b,s^-)$. Then, $g \circ f$ belongs to $\mathcal{A}(d(a,r^-))$.*

Proof. Without loss of generality, we can clearly assume $a = b = 0$. First, suppose that f has no zero in $d(0,r^-)$. Then, $|f(x)|$ is equal to a constant c in $d(0,r^-)$, with $c < s$. Hence, of course, $f(d(0,r^-))$ is included in $d(0,c)$. Now, let $\rho \in]0,r[$. The restriction of f to $d(0,\rho)$ belongs to $H(d(0,\rho^-))$ and the restriction of g to $d(0,c)$ belongs to $H(d(0,c))$. Hence, the restriction of $g \circ f$ to $d(0,\rho)$ belongs to $H(d(0,\rho))$. Consequently, by Corollary 12.3, $g \circ f$ belongs to $H(d(0,\rho))$. This is true for every $\rho \in]0,r[$, and therefore, this shows that $g \circ f$ belongs to $\mathcal{A}(d(0,r^-))$.

Now, we suppose that f admits at least one zero in $d(0,r^-)$. Hence, there exists $r' \in]0,r[$ such that f has at least one zero in $d(0,r')$. Therefore, by Corollary 22.30, $\|f\|_{d(0,\rho)}$ is strictly increasing in ρ in the interval $[r',r[$. Now, let $\rho \in]0,r[$ and let $\sigma = \|f\|_{d(0,\rho)}$. The restriction of f to $d(0,\rho)$ belongs to $H(d(0,\rho))$, and further, $f(d(0,\rho))$ is included in $d(0,\sigma)$. Since g belongs to $H(d(0,\sigma))$, $g \circ f$ belongs to $H(d(0,\rho))$. As previously, this is true for every $\rho \in]0,r[$, hence $g \circ f$ belongs to $\mathcal{A}(d(0,r^-))$. □

Theorem 23.4 (Dieudonné–Dwork). *Let $f \in \mathcal{A}(d(0,R^-))$ satisfy $f(0) = 1$ and have no zero in $d(0,R^-)$. There exists a sequence $(u_k)_{k\in\mathbb{N}^*}$ in $d(0,R)$ such that $f(x) = \prod_{k=1}^{\infty}(1 - u_k x^k)$ whenever $x \in d(0,R^-)$.*

Proof. Since $f(0) = 1$, we can write $f(x)$ in the form $1 + \sum_{n=1}^{\infty} b_n x^n$. Since f has no zero in $d(0,R^-)$, by Corollary 23.2, we have $|b_n|R^n \leq 1$ for every $n \in \mathbb{N}^*$. Now, suppose that we have already found $u_1, \ldots, u_k$ such that $f(x)$ factorizes in the form

$$(\mathcal{R}_k) \quad \prod_{j=1}^{k}(1 - u_j x^j)\left(1 + x^{k+1}\sum_{n=0}^{\infty}\beta_{n,k+1}x^n\right)$$

$$\text{with } |\beta_{n,k+1}| \leq \frac{1}{R} \quad \text{for every } n \in \mathbb{N}.$$

Actually, we have $(1+\beta_{0,k+1}x^{k+1})\big(1+\sum_{n=1}^{\infty}(-\beta_{0,k+1}x^{k+1})^n\big)=1$, and therefore, we can factorize $\big(1+x^{k+1}\sum_{n=0}^{\infty}\beta_{n,k+1}x^n\big)$ in the form

$$(1+\beta_{0,k+1}x^{k+1})\left(1+\sum_{n=1}^{\infty}(-\beta_{0,k+1}x^{k+1})^n\right)\left(1+x^{k+1}\sum_{n=0}^{\infty}\beta_{n,k+1}x^n\right).$$

Now, consider the function g_{k+1} defined as

$$g_{k+1}(x)=\left(1+\sum_{n=1}^{\infty}(-\beta_{0,k+1}x^{k+1})^n\right)\left(1+x^{k+1}\sum_{n=0}^{\infty}\beta_{n,k+1}x^n\right).$$

In g_{k+1}, it is seen that the term in x^{k+1} is equal to 0, so g_{k+1} is of the form

$$\left(1+x^{k+2}\sum_{n=0}^{\infty}\beta_{n,k+2}x^n\right),\quad \text{with } |\beta_{n,k+2}|\le\frac{1}{R},\quad \text{for every } n\in\mathbb{N}.$$

Now, we just put $u_{k+1}=\beta_{0,k+1}$ and then we have proven $(\mathcal{R}_{k+1})$. Since $(\mathcal{R}_0)$ is trivially satisfied, by induction, we can construct a sequence $(u_k)_{k\in\mathbb{N}^*}$ in $d(0,R^-)$ and a sequence $(g_k)_{k\in\mathbb{N}^*}$ in $\mathcal{A}(d(0,R^-))$ such that for each $k\in\mathbb{N}$, g_k is of the form $1+x^{k+1}\sum_{n=0}^{\infty}\beta_{n,k+1}x^n$ and satisfies $f(x)=\prod_{j=1}^{k}(1-u_jx^j)g_k(x)$. For each $k\in\mathbb{N}^*$, let $f_k(x)=\prod_{j=1}^{k}(1-u_jx^j)$.

It is seen that for each $r\in]0,R[$, we have $\|g_k-1\|_{d(0,r)}\le r^{k+1}$. As a consequence, for every $r\in]0,R[$, the sequence $(f_k)_{k\in\mathbb{N}^*}$ converges to f in $H(d(0,r))$, and therefore, we have $f(x)=\prod_{k=1}^{\infty}(1-u_kx^k)$ for all $x\in d(0,R^-)$. That ends the proof. □

Proposition 23.5. *Let D be a set of the form $\bigcup_{i\in I}d(\alpha_i,r^-)$ with $|\alpha_i-\alpha_j|=r$ whenever $i\neq j$. Let ℓ be fixed in I, let $f\in H(D)$ be such that $f(x)=\sum_{n=0}^{\infty}a_n(x-\alpha_\ell)^n\in H(D)$ whenever $x\in d(\alpha_\ell,r^-)$, and let $s=\sup_{n\ge1}|a_n|r^n$. For every $i\in I$, let $\beta_i=f(\alpha_i)$. Then, we have $f(d(\alpha_i,r^-))=d(\beta_i,s^-)$ for every $i\in I$ and $|\beta_i-\beta_j|\le s$ whenever $i,j\in I$. Moreover, if I is not finite, the equality $|\beta_i-\beta_\ell|=s$ holds for every $i\in I$ but finitely many.*

Proof. We set $a = \alpha_\ell$ and $g = f - \beta_\ell$. By Theorem 22.9, for every $\rho \in]0, r[$, we have $\Psi_a(g, \log \rho) = \sup_{n\geq 1} \Psi(a_n) + n \log \rho$, hence by continuity,

$$\Psi_a(g, \log r) = \sup_{n\geq 1} \Psi(a_n) + n \log r = \log s.$$

Now, let $j \in I$, $j \neq \ell$ and set $b = \alpha_j$. Since $|b - a| = r$, by Proposition 19.4, we have $\Psi_a(g, \log r) = \Psi_b(g, \log r)$, hence $\Psi(g(x)) \leq \Psi_a(g, \log r) = \log s$ for all $x \in d(b, r^-)$. Thus, we see that $|f(x) - \beta_\ell| \leq s$ for all $x \in d(b, r^-)$. Hence, by Corollary 22.11, $f(d(b, r^-))$ is a disk $d(\beta_j, t^-)$ with $t \leq s$. Since α_ℓ and α_j play the same role, in the same way, we show that $s \leq t$, and therefore, $s = t$. Thus, we have proven that $f(d(\alpha_i, r^-)) = d(\beta_i, s^-)$ for every $i \in I$ and $|\beta_i - \beta_j| \leq s$ whenever $i, j \in I$.

Now, we suppose that I is infinite. By Proposition 19.1, the equality $\Psi(g(x)) = \Psi_a(g, \log r) = \log s$ holds in all the classes of $C(a, r)$ but finitely many ones, hence we have $|\beta_j - \beta_\ell| = s$ for every $j \in I$ but finitely many and this ends the proof of Proposition 23.5. □

Theorem 23.6. *Let D be an open analytic subset of $\mathbb{K}$, let $f \in H(D)$, and let $D' = f(D)$. Then, D' is open and satisfies* $\operatorname{codiam}(D') \geq \operatorname{codiam}(D) \inf_{x\in D} |f'(x)|$.

Proof. Let $b \in f(D)$ and let $a \in D$ be such that $f(a) = b$. Since D is open, there exists a disk $d(a, r)$ included in D. Let $f(x) = \sum_{j=0}^{\infty} a_n(x - a)^n$ $\forall x \in d(a, r)$. Since D is an analytic set, f is not identically zero in any disk included in D, hence by Theorem 22.9, $\|f - b\|_{d(a,r)}$ is a number $s = \sup_{n\geq 1} |a_n| r^n$. Then, by Theorem 22.9, $f(d(a, r))$ is the disk $d(b, s)$, hence $d(b, s) \subset D$, which proves that $f(D)$ is open. Particularly, $|f'(a)| = |a_1|$, hence $\|f - b\|_{d(a,r)} \geq r|f'(a)|$, therefore, $\delta(D', (\mathbb{K}\setminus D')) \geq \delta(D, (\mathbb{K}\setminus D)) \inf_{x\in D} |f'(x)|$. □

Theorem 23.7. *Let $f \in H(d(0, r))$, let $t = \nu^+(f, \log r)$, and assume $t \geq 1$. Suppose that f' is not identically zero and let $\alpha_1, \dots, \alpha_q$ be the zeros of f' in $d(0, r)$. For every $b \in f(d(0, r))\setminus f(\{\alpha_1, \dots, \alpha_q\})$, $f - b$ admits exactly t zeros of order* 1 *in $d(0, r)$.*

Proof. Let $b \in f(d(0, r))\setminus f(\{\alpha_1, \dots, \alpha_q\})$. By Theorem 22.7, $f - b$ admits t zeros in $d(0, r)$ (taking multiplicity into account). But for each zero α of $f - b$ (as $\alpha \neq \alpha_j$ whenever $j = 1, \dots, q$), we have $f'(\alpha) \neq 0$, hence the t zeros of $f - b$ are of order one. □

Definition. Let $f \in H(D)$. Then, f is said to be *strictly injective* in D if f is injective and if $f'(x) \neq 0$ whenever $x \in D$.

In the same way, given $a \in \mathbb{K}$ and $r > 0$, an analytic function $f(x) \in \mathcal{A}(d(a, r^-))$ is said to be *strictly injective* in $d(a, r^-)$ if f is injective and if $f'(x) \neq 0$ whenever $x \in D$.

Theorem 23.8. *Let $\mathbb{K}$ have characteristic zero and let $f \in H(D)$ be injective in D. Then, f is strictly injective.*

Proof. Suppose that f is not strictly injective and let $\alpha \in D$ be a zero of f'. Let $d(\alpha, r)$ be a disk included in D. Without loss of generality, we may assume $\alpha = 0$. Hence, in $d(0, r)$, $f(x)$ is equal to a series of the form $a_0 + \sum_{n=q}^{\infty} a_n x^n$ with $q \geq 2$ and $a_q \neq 0$. Therefore, we have $\nu^+(f, \log r) \geq q \geq 2$. Let $t = \nu^+(f, \log r)$. Since $\mathbb{K}$ has characteristic zero, f' is not identically zero and therefore admits finitely many zeros $\alpha_1 = 0, \alpha_2, \ldots, \alpha_s$ in $d(0, r)$. Let $b \in f(d(0, r)) \setminus f(\{\alpha_1, \ldots, \alpha_s\})$. By Theorem 23.7, $f - b$ admits t simple zeros in $d(0, r)$ and this contradicts the hypothesis "f injective in $d(0, r)$". □

Theorem 23.9. *Let $a \in \mathbb{K}$, $r \in \mathbb{R}_+$, let $f(x) = \sum_{n=0}^{\infty} a_n (x - a)^n \in H(d(a, r))$, and let $s = \sup_{n \geq 1} |a_n| r^n$ be > 0. Then the following statements are equivalent:*

(α) $|a_1| > |a_n| r^{n-1}$ *whenever* $n > 1$.
(β) $|f(x) - f(y)| = |x - y||a_1|$ *whenever* $x, y \in d(a, r)$.
(γ) f *is strictly injective in* $d(a, r)$.

Moreover, when conditions (α), (β), and (γ) are satisfied, then we have $s = |a_1| r$ and $|f'(x)| = |a_1|$ whenever $x \in d(a, r)$.

Proof. Without loss of generality, we may obviously assume $a = 0$.

First, we suppose (α) is satisfied and consider

$$f(x) - f(y) = (x - y) \left(a_1 + \sum_{n=2}^{\infty} a_n \left(\sum_{j=0}^{n-1} x^j y^{n-1-j} \right) \right).$$

For every $n \geq 2$, it is seen that $|x^j y^{n-1-j}| \leq r^{n-1}$. Hence, we have $|a_1| > \left| \sum_{n=2}^{\infty} a_n \left(\sum_{j=0}^{n-1} x^j y^{n-1-j} \right) \right|$ and thereby $|f(x) - f(y)| = |a_1|\, |x - y|$. At the same time, we note that (α) implies $|f'(x)| = |a_1|$

whenever $x \in d(a,r)$, while, by Theorem 22.9, we have $s = |a_1|r$. So (α) implies (β).

Second, we suppose (β) is satisfied. Since $s > 0$, by Corollary 22.10, f is not a constant, hence $a_1 \neq 0$. Then, by (β), we have $f(x) \neq f(y)$ whenever $x \neq y$. Moreover, $|f'(x)| = |a_1| \neq 0$, hence (γ) is satisfied.

Third, we suppose (γ) is satisfied. Let $b \in f(d(a,r))$, let $g = f - b$, and let $t = \nu^+(g, \log r)$. If $t \geq 2$, either g admits several different zeros or g admits a zero α of order t. In both cases, we see that g is not strictly injective, hence neither is f. Finally, we have $t = 1$ and hence (α) is satisfied. This ends the proof of Theorem 23.9. □

Theorem 23.9 is easily applied to analytic functions inside a disk.

Corollary 23.10. *Let $f(x) = \sum_{n=0}^{\infty} a_n(x-a)^n \in \mathcal{A}_b(d(a,r^-))$ and suppose that the number $s = \sup_{n\geq 1} |a_n|r^n$ is strictly positive. Then conditions (α), (β), (γ), and (δ) are equivalent:*

(α) $|a_1| \geq |a_n|r^{n-1}$ *whenever* $n > 1$.
(β) $|f(x) - f(y)| = |x-y|\ |a_1|$ *whenever* $x, y \in d(a,r^-)$.
(γ) f *is strictly injective in* $d(a,r^-)$.
(δ) $s = |a_1|r$.

Moreover, when conditions (α), (β), (γ), and (δ) are satisfied, we have $|f'(x)| = |a_1|$ whenever $x \in d(a,r^-)$.

Proof. For every $\rho \in]0,r[$, we apply Theorem 23.9 to $f \in H(d(a,\rho))$. □

Lemma 23.11. *Let $f \in \mathcal{A}(d(\alpha,r^-))$ be injective and such that f' is not identically zero. Then, f is strictly injective.*

Proof. We may obviously assume that $f'(\alpha) \neq 0$ and $\alpha = 0$. Hence, f is of the form $\sum_{n=0}^{\infty} a_n x^n$ with $a_1 \neq 0$. If f' has a zero β, there exists an integer $q > 1$ such that

(1) $|qa_q|\ |\beta|^{q-1} = |a_1|$.

Let $g(x) = f(x) - f(0)$. Then, g is also injective and has a simple zero at 0. But by (1), we have $|a_q|\ |\beta|^q \geq |a_1|\ |\beta|$, hence we have $\nu^+(g,\mu) \geq q$ when μ is close enough to $\log(r)$. Then, by Theorem 22.7, we know that g has at least q zeros in $d(0,r^-)$ and then admits another zero $\gamma \neq 0$, which contradicts the fact that it is injective. Consequently, f' has no zero in $d(\alpha,r^-)$ i.e. f is strictly injective. □

We are now able to study the inverse functions of an analytic element inside a disk $d(a, r)$.

Theorem 23.12. *Let $a \in \mathbb{K}$, $r \in \mathbb{R}_+$, let $f(x) \in H(d(a, r))$ be strictly injective in $d(a, r)$, let $s = |f'(a)|r$, and let $b = f(a)$. The homomorphism Θ from $H(d(b, s))$ into $H(d(a, r))$ defined as $\Theta(h) = h \circ f$ is an isometric isomorphism from $H(d(b, s))$ onto $H(d(a, r))$.*

Proof. Without loss of generality, we may obviously suppose $a = b = f(a) = 0$ and $f'(0) = 1$. We put $E = d(0, r)$. Hence, by Theorem 23.9, in $d(0, r)$, $f(x)$ is equal to a series of the form $x + \sum_{j=2}^{\infty} a_j x^j$ with $\sup_{j \geq 2} |a_j| r^j < r$. Next, by Theorem 22.9, r is equal to s. It is obviously seen that $\|\Theta(h)\|_E = \|h\|_E$ for every $h \in H(E)$. So, we only have to prove that Θ is surjective. Let $\lambda = \inf_{P \in \mathbb{K}[x]} \|\Theta(P) - x\|_E$ and suppose $\lambda > 0$.

Set $\lambda = \frac{\inf_{P \in \mathbb{K}[x]}(\|P(f) - x)\|)}{r}$ and suppose that $\lambda > 0$. By definition, $\lambda < 1$ because $\|f - x\| < r$. Let $P \in \mathbb{K}[x]$ be such that $\|P(f) - x\| < r\lambda^{\frac{2}{3}}$. Then, we can write $x = P(f) + h(x)$ with $h \in H(d(0, r))$ and $h(x) = \sum_{k=0}^{\infty} c_k x^k$ with $|c_n| < r\lambda^{\frac{2}{3}}$ $\forall n \in \mathbb{N}$ and $\lim_{n \to +\infty} c_n = 0$.

Let $q \in \mathbb{N}$ be such that $|c_n < r\lambda^2$ $\forall n > q$ and let $\omega(x) = \sum_{n=0}^{q} c_n x^n$. We first note that for all $n \in \mathbb{N}$, we have

(1) $\|(P(f) + h(x))^n - (P(f) + \omega(x))^n\| \leq \|h - \omega\| \leq r\lambda^{\frac{2}{3}}$.

Now, for each $n \in \mathbb{N}$, set $(P(f) + h(x))^n = (P(f) + \omega(x))^n + \omega_n(x)$. Then by (1), we have

(2) $\|\omega_n\| \leq r\lambda^2$ $\forall n \in \mathbb{N}$.

On the other hand, $\|(P(f) + \omega(x))^n - (P(f))^n| \leq \|\omega\| \leq r\lambda^{\frac{2}{3}}$ $\forall n \in \mathbb{N}$. Consequently, we can write

(3) $\|(P(f) + \omega(x))^n = (P(f))^n + \ell_n(x)$ with $\|\ell_n\| \leq r\lambda^{\frac{2}{3}}$ $\forall n \in \mathbb{N}$.

Now, we have

$$\begin{aligned} x - P(f) = &\sum_{n=0}^{q} c_n\big((P(f) + \omega(x))^n + \omega_n(x)\big) \\ &+ \sum_{n=q+1}^{\infty} c_n\big((P(f) + \omega(x))^n + \omega_n(x)\big), \end{aligned}$$

therefore by (2) and (3), we can write

$$(4)\qquad x - P(f) = \sum_{n=0}^{q} c_n (P(f))^n + \sum_{n=0}^{q} c_n \big(\ell_n(x) + \omega_n(x)\big) + \sum_{n=q+1}^{\infty} c_n \big((P(f) + \omega(x))^n + \omega_n(x)\big)$$

with $\|c_n\big((P(f) + \omega(x))^n + \omega_n(x)\big)\| < r\lambda^2\ \forall n \geq q$ and $\|\ell_n + \omega_n\| < r\lambda^{\frac{2}{3}}\ \forall n = 0, \ldots, q$, hence $\|\sum_{n=0}^{q} c_n(\ell_n(x) + \omega_n(x))\| \leq r\lambda^{\frac{4}{3}}$. Consequently, $\left\|x - P(f) - \sum_{n=0}^{q} c_n(P(f))^n\right\| \leq r\lambda^{\frac{4}{3}}$. Set $Q(x) = P(x) - \sum_{n=0}^{q} c_n(P(x))^n$. Then, the polynomial Q satisfies $\|x - Q(f)\| \leq r\lambda^{\frac{4}{3}}$, a contradiction to the hypothesis $\lambda > 0$. Consequently, $\lambda = 0$, and therefore, x does belong to the closure of $\Theta(H(E))$. But since Θ is isometric, $\Theta(H(E))$ is obviously closed in $H(E)$, and therefore, x belongs to $\Theta(H(E))$. As a consequence, $\Theta(H(E)) = H(E)$. □

Corollary 23.13. *Let $f \in H(d(a, r))$ be strictly injective and let $d(b, s) = f(d(a, r))$. Then, $\overset{-1}{f}$ belongs to $H(d(b, s))$.*

Corollary 23.14. *Let $\mathbb{K}$ have characteristic 0, let $f \in H(a, r))$ be injective, and let $d(b, s) = f(d(a, r))$. Then, $\overset{-1}{f}$ belongs to $H(d(b, s))$.*

Corollary 23.15. *Let $f \in \mathcal{A}(d(0, r^-))$ be strictly injective in $d(0, r^-)$ and let $s = r|f'(0)|$. Then, $\overset{-1}{f}$ belongs to $\mathcal{A}(d(0, s^-))$.*

Proof. Indeed, by Corollary 23.13, $\overset{-1}{f}$ belongs to $H(d(0, u))$ for every $u \in [0, s]$. □

Corollary 23.16. *Let $\mathbb{K}$ have characteristic 0, let $f \in \mathcal{A}(d(0, r^-))$ be injective in $d(0, r^-)$, and let $s = r|f'(0)|$. Then, $\overset{-1}{f}$ belongs to $\mathcal{A}(d(0, s^-))$.*

Definition. An injective analytic function $f \in \mathcal{A}_b(d(a, r^-))$ is said to be *bianalytic* if f^{-1} belongs to $\mathcal{A}(f(d(a, r^-)))$.

We use the following lemma in topology:

Lemma 23.17. *Let E be a topological space and let F and G be subsets dense in E. If F is open, then $F \cap G$ is dense in E.*

Proof. Indeed, let $a \in E$, let V be an open neighborhood of a, and let $u \in V \cap F$. Since F is open, $V \cap F$ is a neighborhood of u. Hence, as G is dense in E, there exists $x \in (V \cap F) \cap G$. Therefore, $V \cap (F \cap G) \neq \emptyset$. □

Lemma 23.18. *Let $\overline{D}$ be open. Let a be a point of $\overline{D} \backslash D$ and let $f \in H(D \cup \{a\})$ be strictly injective in D. There exists an open set E satisfying:*

(i) *$\overline{E}$ is open.*
(ii) *$D \cup \{a\} \subset E \subset \overline{D}$.*
(iii) *f belongs to $H(E)$ and is strictly injective in E.*

Proof. By Theorem 11.5, we know that f is of the form $g+h$ with $g \in H(\overline{D})$ and $h \in R(\mathbb{K} \backslash (\overline{D} \backslash (D \cup \{a\})))$. Since $\overline{D}$ is open, there exists $\sigma > 0$ such that $d(a, \sigma) \subset \overline{D}$. Now, as $h \in R(D \cup \{a\})$, there exists $\tau > 0$ such that $h \in R(D \cup d(a, \tau))$. Let $\rho = \min(\sigma, \tau)$ and let $E = D \cup d(a, \rho)$. It is seen that $\overline{E} = \overline{D} \cup d(a, \rho)$, and therefore, $\overline{E}$ is open. Next, both g, h belong to $H(E)$, hence so does f.

We suppose that f is not injective in E. Let b and $c \in E$ be such that $f(b) = f(c)$ and let $\omega = f(b)$. Let $r \in \mathbb{R}_+^*$ be such that $d(b, r) \cup d(c, r) \subset E$ whereas $d(b, r) \cap d(c, r) = \emptyset$. Then $f(d(b, r))$ is a disk $d(\omega, s)$ whereas $f(d(c, r))$ is a disk $d(\omega, t)$. We may obviously assume $s \leq t$. Let $\Sigma = d(b, r) \cap D$ and let $\Lambda = d(c, r) \cap D$. Obviously, Σ is dense in $d(b, r)$ whereas Λ is dense in $d(c, r)$, hence $f(\Sigma)$ is dense in $d(\omega, s)$ whereas $f(\Lambda)$ is dense in $d(\omega, t)$. Hence, $f(\Lambda) \cap d(\omega, s)$ is dense in $d(\omega, s)$. Since Σ and Λ are open sets in $\mathbb{K}$, by Corollary 22.10, both $f(\Sigma)$ and $f(\Lambda)$ are open sets in $\mathbb{K}$ because f is not a constant in $d(b, r)$ or in $d(c, r)$. Hence, both $f(\Lambda) \cap d(\omega, s)$, $f(\Sigma)$ are dense open subsets of $d(\omega, s)$. Therefore, by Lemma 23.17, we see that $f(\Sigma) \cap f(\Lambda)$ is dense in $d(\omega, s)$ and certainly is not empty. Let $y \in f(\Sigma) \cap f(\Lambda)$ and $\alpha \in \Sigma$, $\beta \in \Lambda$ satisfy $f(\alpha) = f(\beta) = y$. By definition of Σ and Λ, we see that α and $\beta \in D$ whereas $\alpha \neq \beta$. This contradicts the hypothesis "f is injective in D" and finally shows f to be injective in E.

Now, since f is strictly injective in $d(a,r^-)\cap D$, we have $f'(x)\neq 0$ whenever $x\in d(a,r^-)\cap D$, hence by Lemma 23.11, f is strictly injective in $d(a,r^-)$. Finally, we have $f'(x)\neq 0$ whenever $x\in E$, and this ends the proof of Lemma 23.18. □

We remember that Condition (B) was defined in Chapter 17.

Theorem 23.19. *Let $\overline{D}$ be open, let D' satisfy $D\subset D'\subset \overline{D}$, and let $f\in H(D')$ be strictly injective in D. There exists an open set D'' satisfying $D'\subset D''\subset \overline{D}$ such that f belongs to $H(D'')$ and is strictly injective in D''.*

Proof. By Lemma 23.18, for every $a\in D'$, there exists an open set D_a such that $\overline{D_a}$ is open, satisfying Condition (B) such that $D\cup\{a\}\subset D_a\subset \overline{D}$ and such that f belongs to $H(D_a)$ and is strictly injective in D_a. Let $D''=\bigcup_{a\in D'} D_a$. Then, D'' is open and such that $D'\subset D''\subset \overline{D}$. By Theorem 11.5, f has a unique decomposition in the form $g+h$ with $g\in H(\overline{D})$ and $h\in R(\mathbb{K}\setminus(\overline{D}\setminus D'))$. Since $f\in H(D_a)$, obviously, $h\in H(D_a)$. Hence, h has no pole in D_a whenever $a\in D'$, therefore $h\in R(D'')$. Hence, f belongs to $H(D'')$. Moreover, by Lemma 23.18, we have $f'(x)\neq 0$ whenever $x\in D_a$, hence whenever $x\in D''$.

Now, we just have to check that f is injective in D''. Let $a,b\in D'$ satisfy $f(a)=f(b)$. Since D_a satisfies Condition (B), we may apply Lemma 23.18 to D_a and b and then we have an open set E such that $D_a\cup\{b\}\subset E\subset \overline{D}_a=\overline{D}$ and such that f belongs to $H(E)$ and is strictly injective in E. Hence, the hypothesis $f(a)=f(b)$ is impossible, and therefore, f is injective in D'. But the hypothesis made on D' actually is satisfied on D''. Hence, f is injective in D'' and this finishes proving Theorem 23.19. □

Proposition 23.20. *Let $(d(\alpha_i,r_i^-))_{i\in I}$ be a partition of D. Let $h\in H(D)$ be injective in D and let $f\in H(D)$ satisfy*

(i) $|f'(\alpha_i)-h'(\alpha_i)|<|h'(\alpha_i)|$ *whenever* $i\in I$,
(ii) $\|f-h\|_{d(\alpha_i,r_i^-)}<|h'(\alpha_i)|r_i$ *whenever* $i\in I$.

Then f is strictly injective. Furthermore, for all $i\in I$, we have $f(d(\alpha_i,r_i^-))=h(d(\alpha_i,r_i^-))$.

Proof. By (i), h' is not identically zero in $d(\alpha_i,r_i^-)$, hence by Lemma 23.11, h is strictly injective in $d(\alpha_i,r_i^-)$. For every $i\in I$, we put $s_i = |h'(\alpha_i)|r_i$. By Corollary 22.10, we have $h(d(\alpha_i, r_i^-)) = d(h(\alpha_i), s_i^-)$, hence by (ii), it is seen that $f(d(\alpha_i, r_i^-)) \subset d(h(\alpha_i), s_i^-)$. Let $f(d(\alpha_i, r_i^-)) = d(f(\alpha_i), t_i^-)$. Therefore, we have

(1) $t_i \leq s_i$,

while obviously

(2) $t_i \geq |f'(\alpha_i)|r_i$.

But, by (i), we have $|f'(\alpha_i)| = |h'(\alpha_i)|$, hence by (1) and (2), we can derive $t_i = s_i$, hence $t_i = |f'(\alpha_i)|r_i$, and therefore, by Theorem 23.9, f is strictly injective in $d(\alpha_i, r_i^-)$. Suppose that f is not injective in all of D. Then there exists a and $b \in D$ such that $f(a) = f(b)$. Since f is injective in each disk $d(\alpha_i, r_i^-)$, we see that there exist j and $m \in I$ with $j \neq m$ such that $a \in d(\alpha_j, r_j^-)$ and $b \in d(\alpha_m, r_m^-)$. Since we have just proven that $f(d(\alpha_i, r_i^-)) = h(d(\alpha_i, r_i^-))$ for all $i \in I$, we see that there exist $a' \in d(\alpha_j, r_j^-)$ and $b' \in d(\alpha_m, r_m^-)$ such that $h(a') = f(a), h(b') = f(b)$. This clearly contradicts the hypothesis "h is injective in D". Consequently, f is injective in all of D. Therefore, by Lemma 23.11, f is strictly injective. $\square$

Proposition 23.21 shows the set of the strictly injective elements to be open in $H(D)$.

Proposition 23.21. *Let D be such that $\delta(D, \mathbb{K}\setminus D) = \rho > 0$. Let $\lambda \in]0, +\infty[$. Let $h \in H(D)$ be injective and satisfy $|h'(x)| \geq \lambda$ whenever $x \in D$. For every $f \in H(D)$ such that $\|f - h\|_D < \lambda\rho$, f is strictly injective in D and satisfies $f(d(\alpha, \rho^-)) = h(d(\alpha, \rho^-))$ for all $\alpha \in D$.*

Proof. Let $h \in H(D)$ satisfy $\|f - h\|_D < \lambda\rho$. Since the distance from D to $\mathbb{K}\setminus D$ is $\rho > 0$, there exists a partition of D in the form $(d(\alpha_i, \rho^-))_{i \in I}$ and then we have $\|f - h\|_{d(\alpha_i, \rho^-)} < \lambda\rho \leq |h'(\alpha_i)|\rho$. Moreover, given any $g \in H(d(\alpha, \rho^-))$, by Theorem 18.1, we have $\|g'\|_{d(\alpha,\rho^-)} \leq \frac{1}{\rho}\|g\|_{d(\alpha,\rho^-)}$, hence

$$|f'(\alpha_i) - h'(\alpha_i)| \leq \frac{1}{\rho}\|f - h\|_{d(\alpha_i, \rho^-)} < \lambda \leq |h'(\alpha_i)|.$$

So, Conditions (i) and (ii) of Proposition 23.20 are clearly satisfied. $\square$

Remark. As we know, when D has holes, ρ is just the lower bound of the diameters of the holes.

Theorem 23.22. *Let $d(\alpha_i, r_i^-)_{i\in I}$ be a partition of D, let $h \in H(D)$, and let $u \in]0,1[$. Let $\phi \in H(D)$ satisfy $\|\phi\|_D < 1$ and $\|h\phi\|_{d(\alpha_i,r_i^-)} \leq ur_i\|h'\|_{d(\alpha_i,r_i^-)}$ for all $i \in I$. Then, for every $t \in]\max(u, \|\phi\|_D), 1[$ there exists a family $(\beta_i)_{i\in I}$ with $\beta_i \in d(\alpha_i, r_i^-)$ such that*

(i) $|h(\beta_i)\phi'(\beta_i)| \leq t|h'(\beta_i)|$ *whenever* $i \in I$,
(ii) $\|h\phi\|_{d(\alpha_i,r_i^-)} \leq t|h'(\beta_i)|r_i$ *whenever* $i \in I$.

Let h be strictly injective and let $f = h(1+\phi)$. Then, f is strictly injective and satisfies $f(d(\alpha_i, r_i^-)) = h(d(\alpha_i, r_i^-))$ for all $i \in I$.

Proof. Let us fix $i \in I$ and put $\alpha = \alpha_i, r = r_i$. For every $g \in H(d(\alpha, r^-))$, we have $\|g\|_{d(\alpha,r^-)} = \lim_{|x-\alpha|\to r^-} |g(x)|$, hence there clearly exists $\beta \in d(\alpha, r^-)$ such that

(1) $\quad |h'(\beta)| \geq \dfrac{u}{t}\ \|h'\|_{d(\alpha,r^-)}$, hence the hypothesis implies

(2) $\quad \|h\phi\|_{d(\beta,r^-)} \leq t\dfrac{u}{t}\ |h'(\beta)|r.$

Moreover, we know that $\|\phi'\|_{d(\alpha,r^-)} \leq \frac{1}{r}\ \|\phi\|_{d(\alpha,r^-)}$, hence we have

$$|h(\beta)\phi'(\beta)| \leq |h(\beta)|\frac{1}{r}\|\phi\|_{d(\alpha,r^-)} \leq \frac{1}{r}\|h\|_{d(\alpha,r^-)}\|\phi\|_{d(\alpha,r^-)}.$$

As the norm $\|\ .\ \|_{d(a,r^-)}$ is multiplicative, we have $|h(\beta)\phi'(\beta)| \leq \frac{1}{r}\|h\phi\|_{d(\alpha,r^-)}$, hence by the above hypothesis, $|h(\beta)\phi'(\beta)| \leq u\|h'\|_{d(\alpha,r^-)}$, and finally, by (1), we obtain

(3) $\quad |h(\beta)\phi'(\beta)| \leq t|h'(\beta)|.$

Hence, we just have to put $\beta_i = \beta$ and do this for every $i \in I$ in order to obtain (i) and (ii) from (2) and (3). Since $\|\phi\|_D \leq 1$, we may take $t \geq \|\phi\|_D$. We see that $f' - h' = h'\phi + h\phi'$, hence by Condition (i), we obtain $|f'(\beta_i) - h'(\beta_i)| \leq t|h'(\beta_i)|$. Then, as $t < 1$, and as $h'(x) \neq 0$ whenever $x \in D$, we see that $|f'(\beta_i) - h'(\beta_i)| < |h'(\beta_i)|$ whenever $i \in I$. This is just Condition (i) in Proposition 23.20. Moreover, Condition (ii) implies Condition (ii) in Proposition 23.20. Hence, by Proposition 23.20, f is strictly injective and satisfies $f(d(\alpha_i, r_i^-) = h(d(\alpha_i, r_i^-))$ for all $i \in I$. □

Theorem 23.23. *Let D be analytic and let F be the set of the injective elements of $H(D)$. The closure of F in $H(D)$ is equal to $F \cup \mathbb{K}$.*

Proof. Let $\overline{F}$ be the closure of F and let $f \in \overline{F} \setminus \mathbb{K}$. Suppose that f is not injective and let $a, b \in D$ be such that $f(a) = f(b)$. Without loss of generality, we may obviously assume $f(a) = 0$. Now, let $r \in]0, |a-b|[$ be such that $d(a,r) \cup d(b,r)$ is included in D. Suppose that f is not identically zero in D. Since D is an analytic set, the restriction of f to $d(a,r)$ (respectively $d(b,r)$) is not identically zero. Let $h \in F$ satisfy $\|f - h\|_D < \min(\|f\|_{d(a,r)}, \|f\|_{d(b,r)})$. By Theorem 22.16, h admits a zero in $d(a,r)$ and another in $d(b,r)$. But since $r < |a - b|$, these two zeros are different, and therefore, this contradicts the hypothesis "$h \in F$". That ends the proof. □

Chapter 24

Quasi-Invertible Analytic Elements

Throughout this chapter D is supposed to be infraconnected.

Some of the results given here were obtained in [35, 36, 44, 50]. We show that when an ideal of an algebra $H(D)$ contains a quasi-invertible element, this ideal is principal and generated by a polynomial.

Lemma 24.1. *Let T be a hole of D and let $f \in H(D \cup T)$ be invertible in $H(D)$. If f has no zero in T, then f is invertible in $H(D \cup T)$.*

Proof. Let $(f_n)_{n\in\mathbb{N}}$ be a sequence in $R(D \cup T)$ which converges to f in $H(D \cup T)$. Let $T = d(a, r^-)$. Since f has no zero in T, by Theorem 22.15, we have $|f(x)| = |f(a)|$ for all $x \in T$, and therefore, $|f(x)| = {}_D\varphi_{a,r}(f)$ for all $x \in T$. Now, since D is infraconnected, by Corollary 13.2, we have ${}_D\varphi_{a,r}(g) \leq \|g\|_D$ whenever $g \in H(D)$, hence for n big enough,

$$\left|\frac{1}{f(x)} - \frac{1}{f_n(x)}\right| = \frac{|f_n(x) - f(x)|}{|f(x)|^2} \leq \frac{\|f_n - f\|_D}{({}_D\varphi_{a,r}(f))^2}$$

whenever $x \in T$. Hence, we see that the sequence $\frac{1}{f_n}$ converges to $\frac{1}{f}$ in $H(D \cup T)$. □

Theorem 24.2. *Let $D \in$ Alg. If an ideal contains a quasi-invertible element, then it is generated by a polynomial whose zeros belong to $\overline{\overset{\circ}{D}} \cap D$.*

Proof. Let $\mathcal{J}$ be an ideal of $H(D)$ that contains a quasi-invertible element f that is of the form Pg with g invertible in $H(D)$ and $P(x) \in \mathbb{K}[x]$, all the zeros of P lying inside $D \cap \overset{\circ}{\overline{D}}$. Then, P belongs to $\mathcal{J}$. Consequently, the set $\mathcal{J}_0$ of polynomials that belong to $\mathcal{J}$ is not empty and hence is an ideal of $\mathbb{K}[x]$, hence $\mathcal{J}_0$ is of the form $Q(x)\mathbb{K}[x]$ with $Q \in \mathbb{K}[x]$. On the other hand, by hypothesis, f factorizes in $H(D)$ in the form Pg with g invertible in $H(D)$ and $P(x) \in \mathbb{K}[x]$, all the zeros of P lying inside $D \cap \overset{\circ}{\overline{D}}$. Since fg^{-1} belongs to $\mathcal{J}$, obviously P belongs to $\mathcal{J}_0$. Hence, Q divides P and then all zeros of Q lie in $D \cap \overset{\circ}{\overline{D}}$. We show that $\mathcal{J} = QH(D)$.

First, we suppose that D is bounded. Let $\alpha_1, \dots, \alpha_q$ be the zeros of Q and suppose that there exists some $h \in \mathcal{J} \setminus QH(D)$. Since D is bounded, by Theorem 11.12, h is of the form $\frac{\ell}{S}$ with $\ell \in H(\overline{D})$ and S a polynomial whose zeros belong to $\overline{D} \setminus D$. Hence, ℓ belongs to $\mathcal{J}$. Now we can find $r > 0$ such that $d(\alpha_i, r) \subset \overline{D}$, whenever $i = 1, \dots, q$. Let $\Lambda = \bigcup_{i=1}^{q} d(\alpha_i, r)$ and let $D' = D \cup \Lambda$. Since the zeros of Q lie in Λ, there exists $\lambda > 0$ such that $|Q(x)| \geq \lambda$ whenever $x \in D \setminus \Lambda$. Now since $\overline{D}$ is closed and bounded, there exists $b \in \mathbb{K}$ such that $\|b\ell\|_D < \lambda$. We put $\phi = Q + b\ell$. Clearly, outside Λ, we have $|\phi(x)| \geq \lambda$. Next, by Theorem 22.18, in each disk $d(\alpha_i, r)$, ϕ has finitely many zeros, hence in D', ϕ has finitely many zeros, all of them in Λ. Therefore, by Theorem 16.2, it factorizes in the form $V(x)W(x)$ with $W \in H(D'), W(x) \neq 0$ whenever $x \in D'$ and V a polynomial whose zeros belong to Λ. By Theorem 22.16, $|W(x)|$ has a strictly positive lower bound in Λ and another non-zero lower bound in $D' \setminus \Lambda$ because $|V(x)|$ is obviously bounded in D'. Finally, W has a non-zero lower bound in D', therefore it is invertible in $H(D')$. Hence, V belongs to $\mathcal{J}$. But then Q divides V in $\mathbb{K}[x]$. But since $Q + b\ell$ is equal to VW, then Q divides $Q + b\ell$ and hence it divides ℓ and h too. This contradicts the hypothesis $h \in \mathcal{J} \setminus QH(D)$ and finishes proving that Q generates $\mathcal{J}$ when D is bounded.

Now, we suppose D unbounded. We may obviously assume D to have at least one hole $d(a, r^-)$ and without loss of generality, we may assume $a = 0$. Let $\gamma(x) = \frac{1}{x}$ and let $D'' = \gamma(D)$. Then, D'' is a bounded set that belongs to Alg such that $0 \notin D'$. We also have $\gamma = \gamma^{-1}$ and $\gamma(D'') = D$. Let ψ be the mapping from $H(D)$ onto $H(D'')$ defined as $\psi(f) = f \circ \gamma$. Then, ψ is a $\mathbb{K}$-algebra isomorphism from $H(D)$ onto $H(D'')$. Moreover, $\psi(\mathcal{J})$ is an ideal $\mathcal{J}''$ of $H(D'')$. Let $u = \frac{1}{x}$ and $T(u) = Q(x)$. Then we can check that $\mathcal{J}'' = TH(D'')$, which ends the proof. $\square$

Definition and notation. For any integer $n \in \mathbb{N}$, we denote by $\mathcal{Q}_n(D)$ the set of the quasi-invertible elements $f \in H(D)$ that have exactly n zeros, taking multiplicity into account and by $\mathcal{Q}(D)$ the set $\bigcup_{n=0}^{\infty} \mathcal{Q}_n(D)$.

Theorem 24.3 shows that if two analytic elements f, g are close enough, then the zeros of f and g also are respectively close, once correctly ordered. It is known as the convergence of zeros theorem.

Theorem 24.3. *Let D be closed and bounded. Let $n \in \mathbb{N}$, let $f \in \mathcal{Q}_n(D)$, and let $\alpha_1, \ldots, \alpha_n$ be the zeros of f (taking multiplicity into account). For every $\epsilon > 0$, there exists $\eta > 0$ such that for every $h \in H(D)$ satisfying $\|f - h\|_D \leq \eta$, h belongs to $\mathcal{Q}_n(D)$ and the zeros $\beta_1, \ldots, \beta_n$ of h, once correctly ordered, satisfy $|\alpha_i - \beta_i| \leq \epsilon$ $(1 \leq i \leq n)$.*

Proof. Let $f = Pg \in \mathcal{Q}_n(D)$ with g invertible in $H(D)$ and P an n-degree monic polynomial whose zeros are interior to D. Let $\gamma_1, \ldots, \gamma_q$ be the different zeros of P, each γ_j of order s_j (with obviously $\sum_{j=1}^{q} s_j = n$). Let $\xi = \inf_{j \neq \ell} |\gamma_j - \gamma_\ell|$, let $\epsilon \in]0, \xi[$, let $\Lambda_j(\epsilon) = d(\gamma_j, \epsilon)$, and let $\Lambda(\epsilon) = \bigcup_{j=1}^{q} d(\gamma_j, \xi)$. It is easily seen that $|P(x)|$ has a non-zero lower bound in $D \setminus \Lambda(\epsilon)$. Since D is closed and bounded, $|g(x)|$ has a non-zero lower bound in D. Hence, $|f(x)|$ has a lower bound $\lambda > 0$ in $D \setminus \Lambda(\epsilon)$. Let $\eta = \min(\lambda, \min_{1 \leq j \leq q} \|f\|_{\Lambda_j(\epsilon)})$ and let $h \in H(D)$ satisfy

(1) $\quad \|f - h\|_D < \eta.$

Obviously, we have $|f(x)| = |h(x)| \geq \lambda$ whenever $x \in D \setminus \Lambda(\epsilon)$. But then by (1) and by Theorem 22.17, we see that h has exactly s_i zeros like f in $\Lambda_i(\epsilon)$ $(1 \leq i \leq q)$ taking multiplicity into account. Thus, we have already proven the statement when all zeros of f have order 1.

Now, extending this to the general case is just a question of writing. We may assume the α_i to be ordered in such a way that

$$\alpha_1 = \cdots = \alpha_{s_1} = \gamma_1, \quad \alpha_{s_1+1} = \cdots = \alpha_{s_1+s_2} = \gamma_2,$$

$$\alpha_{s_1+\cdots+s_{q-1}+1} = \cdots = \alpha_{s_1+\cdots+s_q} = \gamma_q.$$

Thus, for every $j = 1, \dots, q$ in $\Lambda_j(\epsilon)$, we can check that γ_j is equal to $\alpha_{s_1+\dots+s_{j-1}+k}$ whenever $k = 1, \dots, s_j$. Since f admits s_j zeros in $\Lambda_j(\epsilon)$, as does h, we may denote them by $\beta_{s_1+\dots+s_{j-1}+1}, \dots, \beta_{s_1+\dots+s_j}$ (some of them being eventually equal). So, we obtain $|\alpha_i - \beta_i| \leq \epsilon$ whenever $i = 1, \dots, n$. □

Corollary 24.4. *Let D be closed and bounded. For every $n \in \mathbb{N}$, $\mathcal{Q}_n(D)$ is open in $H(D)$ and so is $\mathcal{Q}(D)$.*

Lemma 24.5. *Let $a \in \overline{D}$ satisfy $a \notin \overset{\circ}{\overline{D}}$. There exists a quasi-minorated element $f \in H_b(D)$ which is not semi-invertible, satisfying $\lim_{\substack{x\to a\\ x\in D}} f(x) = 0$ and $\limsup_{\substack{x\to 0\\ x\in D}} \left|\frac{f(x)}{x}\right| = +\infty$.*

Proof. Without loss of generality, we may obviously assume $a = 0$. Then the Cauchy filter $\mathcal{F}$ of base $\{d(0, r) \cap D \mid r > 0\}$ is pierced. Let $(T_m)_{m\in\mathbb{N}}$ be a sequence of holes of D that runs $\mathcal{F}$ and let $D' = \mathbb{K}\setminus(\bigcup_{m=0}^{\infty} T_m)$. By Corollary 17.5, there exists $f \in H_b(D')$ such that $\lim_{\substack{x\to 0\\ x\in D'}} f(x) = 0$ and such that $\limsup_{\substack{x\to 0\\ x\in D}} \left|\frac{f(x)}{x}\right| = +\infty$. We check that D' has no monotonous pierced filter because its only holes are the T_m. Hence, by Corollary 21.3, f is quasi-minorated. If 0 belongs to D, it is seen that $f(0) = 0$ while $\limsup_{\substack{x\to 0\\ x\in D}} \left|\frac{f(x)}{x}\right| = +\infty$, hence f can't factorize in the form $xg(x)$ with $g \in H(D)$, and therefore, f is not semi-invertible.

Now, suppose $0 \notin D$ and that f semi-invertible. Then it factorizes in the form $P(x)g(x)$ with P the polynomial of the zeros of f in $\overset{\circ}{\overline{D}}$ and g an invertible element in $H(D)$. Since $0 \notin D$, we have $P(0) \neq 0$, hence $\lim_{\substack{x\to 0\\ x\in D}} g(x) = 0$. But since $\lim_{x\to 0} |g(x)| = +\infty$, by Corollary 11.9, $\frac{1}{g(x)}$ admits a pole at 0. Let n be its order. Then, by Corollary 11.9, $\frac{x^n}{g(x)}$ has a finite limit different from zero at 0. But since $\frac{f(x)}{x}$ is unbounded in any set $d(0, r) \cap (D'\setminus\{0\})$, so is $\frac{g(x)}{x}$ and therefore we have $\liminf_{x\to 0}\left|\frac{x^n}{g(x)}\right| = 0$. Hence, g can't be invertible. This finally shows that f is not semi-invertible and finishes the proof of Lemma 24.5. □

Lemma 24.6. *Let D be such that $\widetilde{D}\setminus\overline{D}$ is not bounded. Then there exists a quasi-minorated element $f \in H_b(D)$ satisfying*

(1) $\lim_{\substack{|x|\to\infty\\ x\in D}} f(x) = 0$

and

(2) $\limsup\limits_{\substack{|x|\to\infty \\ x\in D}} |xf(x)| = +\infty.$

Moreover, xf does not belong to $H(D)$.

Proof. Since D has holes, we may obviously assume that 0 belongs to a hole. Let $\gamma(x) = \frac{1}{x}$ and let $D' = \gamma(D)$. Then D' is bounded and 0 belongs to $\overline{D'} \setminus \overset{\circ}{\overline{D'}}$. By Lemma 24.5, there exists a quasi-minorated element $h \in H_b(D')$ satisfying

(3) $\lim\limits_{\substack{x\to 0 \\ x\in D}} f(x) = 0$

and

(4) $\limsup\limits_{\substack{x\to 0 \\ x\in D}} \left|\frac{f(x)}{x}\right| = +\infty.$

Then we set $f = h \circ \gamma$. By (3), f satisfies (1), and by (4), f satisfies (2). Now, by Lemma 16.7, f is quasi-minorated. Finally, we check that xf does not belong to $H(D)$. Indeed, suppose $xf \in H(D)$. By Theorem 11.5, xf is of the form $g(x)+P(x)$, with $g \in H_b(D)$ and $P \in \mathbb{K}[x]$. Let $q = \deg(P)$. Since xf is not bounded, we have $q > 0$. Then, $x^{1-q}f$ has a limit different from 0 when $|x|$ tends to $+\infty$ and this contradicts (1). This ends the proof of Lemma 24.6. □

Theorem 24.7. *If D does not belong to* Alg, *there exist invertible elements $f, g \in H(D)$ such that fg belongs to $H(D)$ but is not semi-invertible.*

Proof. First, we suppose that there exists $a \in (\overline{D}\setminus D)\setminus \overset{\circ}{\overline{D}}$. Without loss of generality, we assume $a = 0$. By Lemma 24.5, there exists $f \in H_b(D)$ such that $\lim_{\substack{x\to 0 \\ x\in D}} f(x) = 0$ while $\frac{f(x)}{x}$ is not bounded in any set $D \cap d(0,r)$ $(r > 0)$. Since $f \in H_b(D)$, we can find $A \in \mathbb{K}$ such that $|A| > \|f\|_D$. Let $g = A + f$. Then g is invertible in $H(D)$. Let $T = d(b, \rho^-)$ be a hole of D and let $F = \frac{x}{(x-b)g}$. Then both $\frac{x}{x-b}$, g^{-1} belong to $H_b(D)$, hence so does F. But by definition, F is the product of two invertible elements of $H(D)$. We also note that F has no zero in D. Next, we note that $\frac{f(x)(x-b)}{x}$ is not bounded in any set $D \cap d(0,r)$ $(r > 0)$, although $\lim_{\substack{x\to 0 \\ x\in D}} f(x)(x-b) = 0$.

Thus, $\frac{f(x)(x-b)}{x}$ cannot admit a pole at 0, and therefore, by Corollary 11.9, $\frac{f(x)(x-b)}{x}$ does not belong to $H(D)$. But then, since $\frac{A(x-b)}{x}$ does belong to $H(D)$, we see that F^{-1} does not belong to $H(D)$. Since F has no zero in D, it is not semi-invertible although both $\frac{x}{x-b}$, g^{-1} are invertible in $H(D)$.

Now, we suppose that $\widetilde{D}\setminus\overline{D}$ is not bounded. Since D has holes, we may obviously assume that 0 belongs to a hole. Let $\gamma(x) = \frac{1}{x}$ and let $D' = \gamma(D)$. Then D' is bounded, and 0 belongs to $\in \overline{D'}\setminus \overset{\circ}{\overline{D'}}$. Hence, as we just saw, there exist invertible elements $h,\ g \in H_b(D')$ such that hg belongs to $H_b(D')$ and has no zero in D' but is not invertible in $H(D')$. Then we put $\tau = h \circ \gamma,\ \psi = g \circ \gamma,\ \phi = (hg) \circ \gamma$. By Theorem 12.7, both $\tau,\ \psi$ are invertible in $H(D)$, and ϕ belongs to $H(D)$ and has no zero in D. Since hg is not invertible in $H(D')$, by Theorem 12.7 again, ϕ is not invertible in $H(D)$. Since it has no zero in D, it is not semi-invertible in $H(D)$ and that finishes the proof of Theorem 24.7. □

Theorem 24.8. *The following three statements are equivalent:*

(i) *D belongs to* Alg *and $\overline{D}$ is open.*
(ii) *$\widetilde{D}\setminus\overline{D}$ is bounded and $\overline{D}$ is open.*
(iii) *The set of the quasi-minorated elements of $H(D)$ is equal to the set of the quasi-invertible elements.*

Proof. By Theorem 17.9, we know that (i) implies (ii). Conversely, suppose (ii) is satisfied. Particularly, $\overline{D}$ is open. Suppose (i) is not satisfied. Then D does not belong to Alg. Since $\widetilde{D}\setminus\overline{D}$ is bounded, there must exist $a \in \overline{D}\setminus D$ that does not belong to $\overset{\circ}{\overline{D}}$, and therefore, this contradicts (ii). Hence, (i) and (ii) are equivalent. Now, since (ii) implies (i), we can apply Theorem 21.7, hence (ii) implies (iii). Finally, it just remains to show that if $\widetilde{D}\setminus\overline{D}$ is not bounded or if $\overline{D}$ is not open, then there exist quasi-minorated elements that are not quasi-invertible.

On one hand, if $\overline{D}$ is not open, by Lemma 24.5, such an element does exist. On the other hand, if $\widetilde{D}\setminus\overline{D}$ is not bounded, then D does not belong to Alg, and therefore, by Theorem 24.7, there exist invertible elements $f,\ g$ in $H(D)$ such that fg is not semi-invertible and hence is not quasi-invertible. But, by Theorem 21.6, both $f,\ g$

are quasi-minorated, and then, by Theorem 21.5, fg also is quasi-minorated. That ends the proof of Theorem 24.8. □

Theorem 24.9. *Let D be closed, let T be a hole of D, and let $f \in H(D \cup T)$ have no zero in T. There exists a bounded closed infraconnected set E such that $T \subset E \subset D \cup T$, $T \neq E$ and such that the restriction of f to E is invertible in $H(E)$.*

Proof. Let $T = d(a, r^-)$ and let $\lambda =_D \varphi_{a,r}(f)$. By Theorem 22.16, we know that $|f(x)| = \lambda$ for all $x \in T$. Moreover, since the restriction of f to T is not identically zero, we have $\lambda > 0$. Let $\ell = \frac{\lambda}{2}$.

First, we suppose $C(a, r) \cap D \neq \emptyset$. Let $b \in D \cap C(a, r)$. Then we have $\lim_{\substack{|x-b|\to r,\ |x-b|<r, \\ x \in D}} |f(x)| =_D \varphi_{b,r}(f) =_D \varphi_{a,r}(f) = \lambda$. Hence, there exists $s \in]0, r[$ such that $|f(x)| \geq \ell$ for every $x \in d(b, r^-) \cap D$. Then, the set $E = T \cup \big(d(b, r^-) \cap D\big)$ is clearly infraconnected, closed, and bounded and we have $|f(x)| \geq \ell$ for all $x \in E$. Hence, the restriction of f to E is invertible in $H(E)$.

Now, we suppose $C(a, r) \cap D = \emptyset$. There exists $s > r$ such that $|f(x)| \geq \ell$ for every $x \in \Gamma(a, r, s) \cap D$. So, we consider the set $E = T \cup \big(\Gamma(a, r, s) \cap D\big)$. It is infraconnected, closed, and bounded and we have $|f(x)| \geq \ell$ for all $x \in E$. Hence the restriction of f to E is invertible in $H(E)$.

In both cases, we can see that T is strictly included in E. That ends the proof of Theorem 24.9. □

We can now briefly examine the ideals of an algebra $H(D)$ when all elements are quasi-invertible.

Theorem 24.10. *Let $D \in$ Alg. If an ideal contains a quasi-invertible element, then it is generated by a polynomial whose zeros belong to $\overline{\overset{\circ}{D}} \cap D$.*

Proof. Let $\mathcal{H}$ be an ideal of $H(D)$ that contains a quasi-invertible element f, and let $\mathcal{H}_0$ be the set of polynomials that belong to $\mathcal{H}$. By hypothesis, f factorizes in $H(D)$ in the form Pg with g invertible in $H(D)$ and $P(x) \in K[x]$, all the zeros of P lying inside $D \cap \overline{\overset{\circ}{D}}$. Since fg^{-1} belongs to $\mathcal{H}$, obviously P belongs to $\mathcal{H}_0$. Hence, T divides P, and then all the zeros of T lie in $D \cap \overline{\overset{\circ}{D}}$. We show that $\mathcal{H} = TH(D)$.

First, we suppose that D is bounded. It is clearly seen that $\mathcal{H}_0$ is an ideal of $\mathbb{K}[x]$, hence there exists $T(x) \in K[x]$ such that $\mathcal{H}_0 = T(x)K[x]$.

Let $\alpha_1, \dots, \alpha_q$ be the zeros of T. Now, we suppose that there exists some $h \in \mathcal{H}\setminus TH(D)$. Since D is bounded, by Theorem 11.12, h is of the form $\frac{\ell}{S}$ with $\ell \in H(\overline{D})$ and S a polynomial whose zeros belong to $\overline{D}\setminus D$. Hence, ℓ belongs to $\mathcal{H}$. Now, we can find $r > 0$ such that $d(\alpha_i, r) \subset \overline{D}$, whenever $i = 1, \dots, q$. Let $\Lambda = \bigcup_{i=1}^{q} d(\alpha_i, r)$ and let $D' = D \cup \Lambda$. Since the zeros of T lie in Λ, there exists $\lambda > 0$ such that $|T(x)| \geq \lambda$ whenever $x \in D\setminus\Lambda$. Now, since $\overline{D}$ is closed and bounded, there exists $b \in K$ such that $\|b\ell\|_D < \lambda$. We put $\phi = T + b\ell$. Clearly, outside Λ, we have $|\phi(x)| \geq \lambda$. Besides, by Theorem 22.18, in each disk $d(\alpha_i, r)$, ϕ has finitely many zeros, hence in D', ϕ has finitely many zeros, all of them in Λ. Hence, it factorizes in the form $Q(x)W(x)$ with $W \in H(D'), W(x) \neq 0$ whenever $x \in D'$ and Q a polynomial whose zeros belong to Λ. By Theorem 22.16, $|W(x)|$ has a strictly positive lower bound in Λ and another non-zero lower bound in $D'\setminus\Lambda$ because $|Q(x)|$ is obviously bounded in D'. Finally, W has a non-zero lower bound in D', therefore it is invertible in $H(D')$. Hence, Q belongs to $\mathcal{H}$. But then T divides Q in $\mathbb{K}[x]$. Since $T + b\ell$ is equal to WQ, then T divides $T + b\ell$, and ℓ, and h too. This contradicts the hypothesis $h \in \mathcal{H}\setminus TH(D)$ and finishes proving that T generates $\mathcal{H}$ when D is bounded.

Now, we suppose D unbounded. We may obviously assume D to have at least one hole $d(a, r^-)$, and without loss of generality, we may assume $a = 0$. Let $\gamma(x) = \frac{1}{x}$, and let $D' = \gamma(D)$. Then D' is a bounded set that belongs to Alg such that $0 \notin D'$. We also have $\gamma = \gamma^{-1}$, and $\gamma(D') = D$. Let ψ be the mapping from $H(D)$ into $H(D')$ defined as $\psi(f) = f \circ \gamma$. Then ψ is a $\mathbb{K}$-algebra isomorphism from $H(D)$ onto $H(D')$. Besides, $\psi(\mathcal{H})$ is an ideal $\mathcal{H}'$ of $H(D')$. Let $P(x) = \prod_{j=1}^{q}(x - a_i)$, for every $j = 1, \dots, q$, let $a'_i = \frac{1}{a_i}$, and let $B(u) = \prod_{j=1}^{q}(u - a'_i)$. Clearly, $\psi(P) = \frac{B(u)}{u^q}$. As $0 \notin D'$, u is invertible in $H(D')$, and B belongs to $\mathcal{H}'$. Hence, $\mathcal{J}'$ is generated by a polynomial whose zeros lie inside $D' \cap \overset{\circ}{\overline{D}'}$. Now, let $W(u) = \prod_{j=1}^{t}(u - c_j)$. Since the c_j lie in D', they are different from 0. For every $j = 1, \dots, t$, let $e_j = \frac{1}{c_j}$, and let $S(x) = \prod_{j=1}^{t}(x - e_j)$. It is seen that for each

$j = 1, \ldots, t$, e_j does belong to $D \cap \overset{\circ}{\overline{D}}$. Now, let $h \in \mathcal{H}$. Then $\psi(h)$ belongs to $\mathcal{H}'$ and is of the form $W(u)G(u)$, with $G \subset H(D')$. Putting $F = \psi^{-1}(G)$, in $H(D)$, we have $h = (-1)^t S(x)\frac{F(x)}{x^t \prod_{i=1}^{t} e_j}$. As $\frac{F(x)}{x^t}$ is an invertible element of $H(D)$, this finishes showing that S generates $\mathcal{H}$. □

Corollary 24.11. *Suppose $D \in$ Alg and all elements are quasi-invertible, except 0. Then $H(D)$ is a principal ring and each ideal is generated by a polynomial whose zeros lie in the opening of D and every maximal ideal of $H(D)$ is of the form $(x-a)H(D)$ with $a \in D$.*

By Corollaries 21.9, 21.10, and 24.11, we can derive again Corollaries 24.12 and 24.13:

Corollary 24.12. *Suppose $D \in$ Alg be closed and let $\mathcal{T}$ be the set of holes of D. If $\{\widetilde{T} | T \in \mathcal{T}\}$ is finite, $H(D)$ is principal, each ideal is generated by a polynomial whose zeros lie in D, and every maximal ideal of $H(D)$ is of the form $(x-a)H(D)$, with $a \in D$.*

Corollary 24.13. *If D is a disk $d(a, r)$ or $d(a, r^-)$, or if D is an annulus $\Gamma(a, r_1, r_2)$ (with $0 < r_1 < r_2$) or $\Delta(a, r_1, r_2)$ (with $0 < r_1 < r_2$) or a circle $C(a, r)$, then $H(D)$ is principal, each ideal is generated by a polynomial whose zeros lie in D and every maximal ideal of $H(D)$ is of the form $(x-b)H(D)$ with $b \in D$.*

Chapter 25

Logarithm and Exponential in a p-adic Field

In this chapter, the field $\mathbb{K}$ is supposed to have characteristic zero.

We define the p-adic logarithm and the p-adic exponential and shortly study them, in connection with the study of the roots of 1 made in Chapter 24. Both functions are also defined in [1]. Here, as in [50], we compute the radius of convergence of the p-adic exponential by using results on injectivity seen in Chapter 23.

Lemma 25.1. *$\mathbb{K}$ is supposed to have residue characteristic $p \neq 0$. Let $r \in]0,1[$, and for each $n \in \mathbb{N}$, let $h_n(x) = (1+x)^{p^n}$. The sequence h_n converges to 1 with respect to the uniform convergence on $d(0,r)$.*

Proof. Without loss of generality, we may assume $|p| = \frac{1}{p}$. Let $E = d(0,r)$, and for each $n \in \mathbb{N}$, let $u_n = p^n$ and let q_n be the integral part of $\frac{n}{2}$. Now, we put $t_n = p^{q_n}$ and we denote by h the identical function on E. Then, $h_n - 1 = \sum_{j=1}^{u_n} \binom{u_n}{j} h^j$. By Lemma 7.1, we have $\left|\binom{u_n}{j}\right| \leq \frac{p^{-n}}{|j|}$, hence

$$(1) \qquad \left|\binom{u_n}{j}\right| \leq j\, p^{-n} \leq t_n\, p^{-n} \leq p^{-\frac{n}{2}} \text{ whenever } j = 1, \ldots, t_n.$$

Next, we have

$h_n - 1 = \displaystyle\sum_{j=1}^{t_n} \binom{u_n}{j} h^j + \sum_{j=t_n+1}^{u_n} \binom{u_n}{j} h^j$. By (1), it is seen that

$$(2)\qquad \left\|\sum_{j=1}^{t_n}\binom{u_n}{j}h^j\right\|_E \le p^{-\frac{n}{2}}$$

while

$$(3)\qquad \left\|\sum_{j=t_n+1}^{u_n}\binom{u_n}{j}h^j\right\|_E \le \|h\|_E^{t_n+1}.$$

Now, by (2) and (3), we see that $\lim_{n\to\infty}\|h_n-1\|_E=0$. □

Definition and notation. As previously defined, for each $q\in\mathbb{N}^*$, we denote by r_q the positive number such that $\log_p(r_q)=-\frac{1}{p^{q-1}(p-1)}$. We denote by $f(x)$ the series $\sum_{n=1}^{\infty}(-1)^{n-1}\frac{x^n}{n}$.

Theorem 25.2. *f has a radius of convergence equal to 1. If the residue characteristic of $\mathbb{K}$ is $p\neq 0$, then f is unbounded in $d(0,1^-)$. If the residue characteristic is zero, then $|f(x)|$ is bounded by 1 in $d(0,1^-)$. The function defined in $d(1,1^-)$ as $Log(x)=f(x-1)$ has a derivative equal to $\frac{1}{x}$ and satisfies $Log(ab)=Log(a)+Log(b)$ whenever $a,\ b\in d(1,1^-)$.*

Proof. It is clearly seen that the radius of f is 1 because $|n|\geq\frac{1}{n}$. As in the Archimedean case, the property $Log(ab)=Log(a)+Log(b)$ comes from the fact that both Log and the function h_a defined as $h_a(x)=Log(ax)$ have the same derivative. The other statements are immediate. □

In Chapter 7, when $\mathbb{K}$ has residue characteristic $p\neq 0$, we have introduced the group W of the p^sth roots of 1, i.e. the set of the $u\in\mathbb{K}$ satisfying $u^{p^s}=1$ for some $s\in\mathbb{N}$.

Theorem 25.3. *$\mathbb{K}$ is supposed to have residue characteristic $p\neq 0$ (respectively 0). All zeros of Log are of order 1. The set of zeros of the function Log is equal to W (respectively 1 is the only zero of Log). The restriction of Log to the disk $d(1,(R_1)^-)$ (respectively $d(1,1^-)$) is injective and is a bijection from $d(1,(R_1)^-)$ onto $d(0,(R_1)^-)$ (respectively from $d(1,1^-)$ onto $d(0,1^-)$).*

Proof. It is obvious that the zeros of Log are of order 1 because the derivative of Log has no zero. First, we suppose $\mathbb{K}$ to have residue characteristic $p \neq 0$. Each root of 1 in $d(1,1^-)$ is a zero of Log. Moreover, by Theorem 7.8, we know that the only roots of 1 in $d(1,1^-)$ are the p^nth roots. Now, we can check that Log admits no zero other than the roots of 1. Indeed, suppose that a is a zero of Log but is not a root of 1, and for each $n \in \mathbb{N}$, let $b_n = a^{p^n}$. Since b_n belongs to $d(1,1^-)$, by Lemma 25.1, we have $\lim_{n\to\infty} b_n = 1$. But obviously $Log(b_n) = 0$ for every $n \in \mathbb{N}$, hence this contradicts the fact that 1 is an isolated zero of Log.

Thus, Log has no zero in the disk $d(1,(R_1)^-)$, except 1, and therefore, by Theorem 22.7, the series $f(x) = \sum_{n=1}^{\infty}(-1)^{n-1}\frac{x^n}{n}$ satisfies $\nu^+(f,\log_p r) = 1$ for every $r \in]0,R_1[$, hence $r > \frac{r^n}{|n|}$ for all $r \in]0,R_1[$, for every $n \in \mathbb{N}^*$. Therefore, by Corollary 23.10, it is injective in $d(0,R_1^-)$. Then, by Corollary 22.10, we see that $Log(d(1,R_1^-)) = d(0,R_1^-)$.

Now, we suppose that $\mathbb{K}$ has residue characteristic zero. Then, the function $f(x) = \sum_{n=1}^{\infty}(-1)^{n-1}\frac{x^n}{n}$ satisfies $\nu^+(f,\log_p r) = 1$ for every $r \in]0,1[$, hence $r > \frac{r^n}{n}$ for all $r \in]0,1[$, for every $n \in \mathbb{N}^*$. Therefore, f has no zero different from 1 in $d(0,1^-)$ and, by Corollary 23.10, is injective in $d(0,1^-)$. Then, by Corollary 22.10, we see that $Log(d(1,1^-)) = d(0,1^-)$. This ends the proof. □

Corollary 25.4. $\mathbb{K}$ *is supposed to have residue characteristic* 0. *There is no root of* 1 *in* $d(1,1^-)$, *except* 1.

Proof. Indeed, any root of 1 should be a zero of Log in $d(1,1^-)$. □

Definition and notation. If $\mathbb{K}$ has residue characteristic $p \neq 0$, we denote by exp the inverse (or reciprocal) function of the restriction of Log to $d(1,R_1^-)$ which obviously is a function defined in $d(0,R_1^-)$, with values in $d(1,R_1^-)$. If $\mathbb{K}$ has residue characteristic 0, we denote by exp the inverse function of Log which is obviously defined in $d(0,1^-)$ and takes values in $d(1,1^-)$.

Theorem 25.5. $\mathbb{K}$ *is supposed to have residue characteristic* $p \neq 0$ *(respectively* $p = 0$*). The function exp belongs to* $\mathcal{A}_b(d(0,R_1^-))$ *(respectively* $\mathcal{A}_b(d(0,1^-))$*), is a bijection from* $d(0,R_1^-)$ *onto* $d(1,R_1^-)$ *(respectively from* $d(0,1^-)$ *onto* $d(1,1^-)$*), and satisfies* $exp(x) = exp'(x) = \sum_{n=0}^{\infty}\frac{x^n}{n!}$ *whenever* $x \in d(0,R_1^-)$ *(respectively* $x \in d(0,1^-)$*). Moreover, the disk of convergence of its series is equal to*

$d(0,R_1^-)$ (respectively $d(0,1^-)$). Further, if $p \neq 0$, then exp does not belong to $H(d(0,R_1^-))$.

Proof. By Corollary 23.15, we know that the function *exp* belongs to $\mathcal{A}_b(d(0,R_1^-))$ (respectively $\mathcal{A}_b(d(0,1^-))$) and is obviously a bijection from $d(0,R_1^-)$ onto $d(1,R_1^-)$ (respectively from $d(0,1^-)$ onto $d(1,1^-)$). As it is the reciprocal of *Log*, it must satisfy $exp(x) = exp'(x)$ for all $x \in d(0,R_1^-)$ (respectively $x \in d(0,1^-)$), and therefore, $exp(x) = \sum_{n=0}^{\infty} \frac{x^n}{n!}$ whenever $x \in d(0,R_1^-)$ (respectively $x \in d(0,1^-)$). Thus, the radius of convergence r is at least R_1 (respectively 1). If the residue characteristic is 0, it is obviously seen that the series cannot converge for $|x| = 1$, hence the disk of convergence is $d(0,1^-)$.

Now, we suppose that the residue characteristic is $p \neq 0$. Suppose that the power series of *exp* converges in $d(0,R_1)$. Then *exp* has continuation to an element of $H(d(0,R_1))$. On the other hand, since $\nu(f,\log_p r) = 1$ for all $r \in]0,R_1[$, we have $\nu^-(f,\log_p R_1) = 1$, and then by Theorem 22.9, $Log(d(1,R_1))$ is equal to $d(0,R_1)$. Hence, we can consider $exp(Log(x))$ in all the disk $d(0,R_1)$. By Corollary 12.3, this is an element of $H(d(1,R_1))$. But this element is equal to the identity in all of $d(1,R_1^-)$ and therefore in all of $d(1,R_1)$. Of course, this contradicts the fact that *Log* is not injective in the circle $C(1,R_1)$. This finishes proving that the disk of convergence of *exp* is just $d(0,R_1^-)$.

Let us show that *exp* does not belong to $H(d(0,R_1^-))$. Indeed, suppose *exp* belongs to $H(d(0,R_1^-))$. Consider the Mittag-Leffler decomposition of *exp* on the infraconnected set $d(0,R_1^-)$. It is of the form $\sum_{n=0}^{\infty} g_n$ with $g_0 \in H(d(0,R_1)$ and $g_n \in H_0(\mathbb{K}\backslash d(a_n,R_n^-))$ with $a_n \in C(0,R_1)$. Set $T_n = d(a_n,R_1^-)$, $n \in \mathbb{N}^*$ and $S = \bigcup_{n=1}^{\infty} T_n$.

Let $\mathcal{K}$ be the residue class field of $\mathbb{K}$. By Theorem 8.4, we can consider a complete algebraically closed extension $\widehat{\mathbb{K}}$ of $\mathbb{K}$ whose residue class field $\widehat{\mathcal{K}}$ is not countable. Thus, we can find a class G of $\widehat{C}(0,R_1)$ that has an empty intersection with S, and then, by Theorem 15.1, *exp* has continuation to an element of $H(d(0,R_1^-) \cup G$. Let $c \in G$. Since $|c| = R_1$, by Theorem 25.3, the function $h(x) = Log(1+x) - c = -(c + \sum_{n=1}^{\infty} \frac{(-x)^n}{n})$ satisfies $\nu^-(h,\log_p(R_1) = 0$, $\nu^+(h,\log_p(R_1) > 1$, hence h admits a zero $a \in \widehat{C}(0,R_1)$. Then, a does not belong to $\mathbb{K}$ because if $a \in \mathbb{K}$, then $\text{Log}(1+a) \in \mathbb{K}$, a contradiction. Now, let ζ be a pth root of 1 different from 1 and let $t = \zeta(1+a)$.

Since $|\zeta - 1| = R_1$, t is of the form $1 + b$ with $b \in \widehat{C}(0, R_1)$. We then have $\text{Log}(1+a) = \text{Log}(1+b)$. Set $E = \widehat{d}(a, R_1^-)$, $F = \widehat{d}(b, R_1^-)$, $D' = d(0, R_1^-) \cup E \cup F$, and $D'' = d(0, R_1^-) \cup G$. Since the image of $d(0, R_1^-)$ by $\text{Log}(1 + x)$ is $d(0, R_1^-)$ and since the derivative of $\text{Log}(1 + x)$ has no zero, by Corollary 22.10, we can check that for each $u \in \widehat{C}(0, R_1)$, the image of $d(u, R_1^-)$ by $\text{Log}(1 + x)$ is $d(\text{Log}(1 + u), R_1^-)$. Consequently, both images of E and F by $\text{Log}(1 + x)$ are equal to G. Now, $\text{Log}(1 + x)$ belongs to $\widehat{H}(D')$, the image of D' by the function $f(x) = \text{Log}(1 + x)$ is D'', and *exp* belongs to $\widehat{H}(D'')$. Consequently, by Corollary 12.3, $\exp \circ \text{Log}(1 + x)$ belongs to $\widehat{H}(D')$ and we have $\exp \circ \text{Log}(1 + x) = 1 + x \ \forall x \in d(0, R_1^-)$. Finally, since D' has no pierced filter, by Proposition 20.5, it is an analytic set. Consequently, the equality $exp \circ \text{Log}(1 + x) = 1 + x \ \forall x \in d(0, R_1^-)$ holds in all D', a contradiction since $\text{Log}(1 + x)$ is not injective in D'. That finishes showing that *exp* does not belong to $H(d(0, R_1^-))$. □

Notation. Henceforth, we put $e^x = exp(x)$.

Theorem 25.6. *Suppose that $p \neq 0$. Let $x \in d(0, R_1^-)$. Then e^x is algebraic over $\mathbb{Q}_p$ if and only if so is x. Let $u \in d(0, 1^-)$. Then $\log(1 + u)$ is algebraic over $\mathbb{Q}_p$ if and only if so is u.*

Proof. By Theorem 14.24, if x is algebraic over $\mathbb{Q}_p$, so is e^x. Similarly, if u is algebraic over $\mathbb{Q}_p$, so is $\log(1 + u)$. Consequently, suppose that e^x is algebraic over $\mathbb{Q}_p$. Then e^x is of the form $1 + t$ with $|t| < 1$, hence $\log(1 + t)$ is algebraic over $\mathbb{Q}_p$. But then, $\log(1 + t) = \log(e^x) = x$, hence x is algebraic over $\mathbb{Q}_p$. Now, more generally, suppose $\log(1 + u)$ is algebraic over $\mathbb{Q}_p$, with $|u| < 1$. Take $q \in \mathbb{N}$ such that $|p^q \log(1 + u)| < R_1$. We have $p^q \log(1 + u) = \log((1 + u)^{p^q})$. Since $|p^q \log(1 + u)| < R_1$, we have $|\log((1 + u)^{p^q})| < R_1$, hence $exp\big(\log((1 + u)^{p^q})\big) = (1 + u)^{p^q}$. Consequently, $(1 + u)^{p^q}$ is algebraic over $\mathbb{Q}_p$ and hence so is u. □

We can show a similar result when $p = 0$:

Theorem 25.7. *Suppose that $p = 0$. Let $x \in d(0, 1^-)$. Then e^x is algebraic over $\mathbb{Q}_p$ if and only if so is x. Let $u \in d(0, 1^-)$. Then $\log(1 + u)$ is algebraic over $\mathbb{Q}_p$ if and only if so is u.*

Chapter 26

Problems on p-adic Exponentials

The author is grateful to Michel Waldschmidt for his advices. On the other hand, most of results first proven in the field $\mathbb{K}$ also hold (with slide changes) in an ultrametric field $\mathcal{K}$ of residue characteristic 0, as, for example, the Levi-Civita field [88].

Notation. Given three functions ϕ, ψ, ζ defined in an interval $J =]a, +\infty[$ (respectively $J =]a, R[$), with values in $[0, +\infty[$, we write $\phi(r) \leq \psi(r) + O(\zeta(r))$ if there exists a constant $b \in \mathbb{R}$ such that $\phi(r) \leq \psi(r) + b\zeta(r)$. We write $\phi(r) = \psi(r) + O(\zeta(r))$ if $|\psi(r) - \phi(r)|$ is bounded by a function of the form $b\zeta(r)$.

Throughout this chapter, the field $\mathbb{K}$ is supposed to have characteristic 0 and residue characteristic $p \neq 0$ and we now denote by $\mathbb{L}$ an algebraically closed complete field with respect to an ultrametric absolute value, with residue characteristic 0.

Next, we denote by Ψ_p the function Ψ on the field $\mathbb{C}_p$ and by Ψ_0 the function Ψ on the field $\mathbb{L}$.

Proposition 26.1 is used in the proof of the next theorems and is proven by induction, similar to (1.4.2) in [92].

Proposition 26.1. *Let $P_1, \ldots, P_q \in \mathbb{K}[X]$ be different from 0 and let $w_1, \ldots, w_q \in \mathbb{K}$ be pairwise distinct. Let $F(x) = \sum_{j=1}^{q} P_j(x)e^{w_j x}$. Then F is not identically zero.*

Corollary 26.2. *Let $b_1, \ldots, b_n \in D_1$ (respectively in D_0). The functions x, $e^{b_1 x}, \ldots, e^{b_n x}$ are algebraically independent over $\mathbb{K}$ (respectively over $\mathbb{L}$) if and only if $b_1, \ldots, b_n$ are $\mathbb{Q}$-linearly independent.*

Theorem 26.3. *Let $\lambda > 0$ and let $F(x) = \sum_{i=1}^{k} f_i(x)e^{\omega_i x}$, where each ω_i satisfies $-\Psi_p(\omega_i) \geq \frac{1}{p-1} + \lambda$, each f_i lies in $\mathbb{C}_p[x]$, and $\deg(f_i) = m_i - 1$. Let $n = \sum_{i=1}^{k} m_i$.*

Then the number N of zeros of F in $d(0,1)$ satisfies $N \leq (n-1)$ $\left(1 + \frac{1}{\lambda(p-1)}\right)$.

Proof. Denoting by $\mathcal{D}$ the derivation $\frac{d}{dx}$, then F satisfies a relation of the form

$$\mathcal{D}^n F = c_1\mathcal{D}^{n-1}F + c_2\mathcal{D}^{n-2}F + \cdots + c_n F, \tag{1}$$

where

$$\prod_{i=1}^{k}(x-\omega_i)^{m_i} = x^n - c_1x^{n-1} - c_2x^{n-2} - \cdots - c_n. \tag{2}$$

Expanding F as a power series, we have

$$F(x) = \sum_{h=0}^{\infty} a_h x^h = \sum_{h=0}^{\infty} \frac{b_h}{h!}x^h, \quad |x| < 1.$$

Moreover, by Proposition 26.1, F is not identically zero. From the differential equation (1), we have the recurrence

$$b_{h+n} = c_1 b_{h+n-1} + \cdots + c_n b_h, \quad h = 0, 1, 2, \ldots. \tag{3}$$

Since $-\Psi_p(\omega_i) \geq \frac{1}{p-1} + \lambda$, the series F converges in $d(0,1)$, hence it lies in $H(d(0,1))$, therefore there exists $t \in \mathbb{N}$ such that $|a_s| = \max_{h\in\mathbb{N}} |a_h|$, and hence without loss of generality, we can suppose that $\max_{1\leq i\leq k} \Psi_p(a_h) = 0$. Let t be the biggest of the $s \in \mathbb{N}$ such that $|a_s| = \max_{h\in\mathbb{N}} |a_h|$. Thus we have $|a_t| = 1$ and $|a_h| < 1\ \forall h > t$. Consequently, by Theorem 22.7, the number of zeros of F in $d(0,1)$ is t.

Let $q = \frac{1}{p-1} + \lambda$, hence $-\Psi_p(\omega_i) \geq q$ $(i = 1, \ldots, k)$. From (2), we can see that $-\Psi_p(c_j) > jq$ $(j = 1,\ 2, \ldots, n)$.

Now, since by assumption we have $\Psi_p(a_h) \leq 0$ and $\Psi_p(a_h) = \Psi_p(b_h) - \Psi_p(h!)$, this implies that $\Psi_p(b_h) \leq 0$ $(h = 0,\ 1, \ldots)$. By (3), we have

$$-\Psi_p(b_n) \geq \min_{1 \leq j \leq n}(-\Psi_p(c_j) - \Psi_p(b_{n-j}) > \min_{1 \leq j \leq n}(qj - \Psi_p(b_{n-j})) \geq q.$$

Then, by (3), with $h = 1,\ 2, \ldots$, we obtain $-\Psi_p(b_{h+n-1}) > hq$, hence

$$\begin{aligned}-\Psi_p(a_{h+n-1}) &> hq + \Psi_p((n-1+h)!) > hq - \frac{(n-1+h)}{p-1} \\ &= h\lambda - \frac{n-1}{p-1}.\end{aligned}$$

Finally, it follows that if $h \geq \frac{(n-1)}{\lambda(p-1)}$, then necessarily $\Psi_p(a_{h+n-1}) < 0$, hence $|a_{h+n-1}| < |a_t|\ \forall h \geq \frac{(n-1)}{\lambda(p-1)}$, $\forall h \geq 1$. Thus, we have $|a_j| < 1 = |a_t|\ \forall j \geq (n-1)(1+\frac{1}{\lambda(p-1)})$, therefore $t < (n-1)\big(1+\frac{1}{\lambda(p-1)}\big)$ and hence the number of zeros of F in $d(0,1)$ is bounded by $(n-1)\big(1+\frac{1}{\lambda(p-1)}\big)$. □

Theorem 26.4. *Let* $F(x) = \sum_{i=1}^{k} f_i(x)e^{\omega_i x}$, *where each* $\omega_i \in \mathbb{L}$ *satisfies* $\Psi_0(\omega_i) < 0$, *each* f_i *lies in* $\mathbb{L}[x]$, *and* $\deg(f_i) = m_i - 1$. *Let* $n = \sum_{i=1}^{k} m_i$.

Then the number N *of zeros of* F *in* $d(0,1)$ *satisfies* $N \leq n - 1$.

Proof. Denoting by $\mathcal{D}$ the derivation $\frac{d}{dx}$, then F satisfies a relation of the form

$$\mathcal{D}^n F = c_1\mathcal{D}^{n-1}F + c_2\mathcal{D}^{n-2}F + \cdots + c_n F, \tag{1}$$

where

$$\prod_{i=1}^{k}(x-\omega_i)^{m_i} = x^n - c_1x^{n-1} - c_2x^{n-2} - \cdots - c_n. \tag{2}$$

Expanding F as a power series, we have

$$F(x) = \sum_{h=0}^{\infty} a_h x^h, \quad |x| < 1.$$

Let $b_h = a_h h!$, $h \in \mathbb{N}$. Then, $F(x) = \sum_{h=0}^{\infty} \frac{b_h}{h!} x^h$. By Proposition 26.1, F is not identically zero. From the differential equation (1), we have the recurrence

$$(3) \qquad b_{h+n} = c_1 b_{h+n-1} + \cdots + c_n b_h, \quad h = 0, 1, 2, \ldots.$$

Since $\Psi_0(\omega_i) < 0 \ \forall i = 1, \ldots, k$, the series F converges in $d(0,1)$, hence it lies in $H(d(0,1))$, therefore there exists $t \in \mathbb{N}$ such that $|a_s| = \max_{h\in\mathbb{N}} |a_h|$, and hence without loss of generality, we can suppose that $\max_{1\leq i\leq k} \Psi_0(a_h) = 0$. Let t be the biggest of the $s \in \mathbb{N}$ such that $|a_s| = \max_{h\in\mathbb{N}} |a_h|$. Thus we have $|a_t| = 1$ and $|a_h| < 1 \ \forall h > t$. Consequently, by Theorem 22.7, the number of zeros of F in $d(0,1)$ is t. Let $\epsilon = \min_{1\leq i\leq k}(-\Psi_p(\omega_i))$. From (2), we can see that $\Psi_0(c_j) \leq j\epsilon$ $(j = 1, 2, \ldots, n)$.

By assumption, we have $\Psi_0(a_h) \leq 0$ and $\Psi_0(a_h) = \Psi_0(b_h) - \Psi_0(h!) = \Psi_0(b_h)$ because $\Psi_0(h!) = 0 \ \forall h$, hence $\Psi_0(b_h) \leq 0$ $(h = 0, 1, \ldots)$. By (3), we have

$$-\Psi_0(b_n) \geq \min_{1\leq j\leq n} (-\Psi_0(c_j) - \Psi_0(b_{n-j})) > \min_{1\leq j\leq n} (\epsilon j - \Psi_0(b_{n-j})) \geq \epsilon.$$

Then, by (3), with $h = 1, 2, \ldots$ we have $-\Psi_0(b_{h+n-1}) > h\epsilon$, hence

$$-\Psi_0(a_{h+n-1}) > h\epsilon + \Psi_0((n-1+h)!) = h\epsilon \quad \forall h \geq 1.$$

Thus, for every $h \geq 1$, we have $-\Psi_0(a_{h+n-1}) > 0$, hence $t \leq n-1$. Consequently, the number of zeros of F in $d(0,1)$ is bounded by $(n-1)$, which ends the proof. □

Hermite–Lindemann's theorem is well known in complex analysis. The same holds in p-adic analysis. We need Siegel's lemma in all the following theorems of this chapter. We choose a particular form of this lemma [92]:

Lemma 26.5 (Siegel). *Let E be a finite extension of $\mathbb{Q}$ of degree q and let $\lambda_{i,j}$ $1 \leq i \leq m$, $1 \leq j \leq n$ be elements of E integral over $\mathbb{Z}$. Let $M = \max(\overline{|\lambda_{i,j}|}\ 1 \leq i \leq m, \mid 1 \leq j \leq n)$ and let $(\mathcal{S})$ be the linear system $\{\sum_{j=1}^{n} \lambda_{i,j} x_j = 0,\ 1 \leq i \leq m\}$. There exists solutions $(x_1, \ldots, x_n)$ of $(\mathcal{S})$ such that $x_j \in \mathbb{Z}\ \forall j = 1, \ldots, n$ and*

$$\log(|x_j|_\infty) \leq \log(M)\frac{qm}{n-qm} + \frac{\log(2)}{2} \quad \forall j = 1, \ldots, n.$$

Lemma 26.6 is necessary in the proof of the following theorems and is easily proved in [92] since its proof implies no change in the field $\mathbb{K}$ since it only concerns algebraic numbers.

Lemma 26.6. *Let $a_1, \dots, a_q \in \mathbb{K}$ be algebraic over $\mathbb{Q}$, let $P(X_1, \dots, X_q) \in \mathbb{Z}[X_1, \dots, x_q]$ be such that $\deg_{X_j}(P) \leq r_j$ $1 \leq j \leq q$, and let $\beta = P(a_1 \cdots a_q)$. Then β is algebraic over $\mathbb{Q}$, $d(a_1)^{r_1} \cdots d(a_q)^{r_q}$ is a multiple of den(β), and we have*

$$s(\beta) \leq \log H(P) + \sum_{j=1}^{q} (r_j s(a_j) + \log(r_j) + 1).$$

The p-adic version of Hermite–Lindemann's theorem was proved by Mahler [73]. Here we give another proof, using specific ultrametric tools.

Notation. We denote by D_0 the disk $d(0, 1^-)$, and if the residue characteristic of $\mathbb{K}$ is $p > 0$, we put $r_1 = p^{\frac{-1}{p-1}}$ and denote by D_1 the disk $d(0, r_1^-)$.

Given a positive real number a, we denote by $[a]$ the biggest integer n such that $n \leq a$.

Remark. In particular, Levi-Civita's fields have residue characteristic 0 [88].

Theorem 26.7. *Let $\alpha \in D_1 \subset \mathbb{K}$ be algebraic. Then e^{α} is transcendental.*

Proof. We suppose that α and e^{α} are algebraic. Let $h = |\alpha|$. Let E be the field $\mathbb{Q}[\alpha, e^{\alpha}]$, let $q = [E : \mathbb{Q}]$, and let w be a common denominator of α and e^{α}. We construct a sequence of polynomials $(P_N(X, Y))_{N \in \mathbb{N}}$ in two variables such that $\deg_X(P_N) = [\frac{N}{\log(N)}]$, $\deg_Y(P_N) = [(\log N)^3]$ and such that the function $F_N(x) = P_N(x, e^x)$ satisfy further, for every $s = 0, \dots, N-1$ and for every $j = 0, \dots,$ $[\log(N)]$,

$$\frac{d^s}{dx^s} F_N(j\alpha) = 0.$$

According to computations in the proof of Hermite Lindemann's theorem in the complex context (Theorem 3.1.1 in [92]), we have

$$(1)\qquad \frac{d^M F_N(\gamma_N)}{dx^M} = \sum_{l=0}^{u_1(N)} \sum_{m=0}^{u_2(N)} b_{l,m,N} \sum_{\sigma=0}^{u_1(N)} \left(\frac{u_1(N)!}{\sigma!(u_1(N)-\sigma)!}\right) \times \left(\frac{l!}{(u_1(N)-\sigma)!}\right) m^{u_1(N)-\sigma}$$

$$j^{u_1(N)-\sigma} \cdot (\alpha)^{u_1(N)-\sigma} \cdot (e^\alpha)^{ju_2(N)}.$$

We put $u_1(N) = \deg_X(P_N)$, $u_2(N) = \deg_Y(P_N)$. We solve the system

$$w^{u_1(N)+u_2(N)} \frac{d^s}{dx^s} F_N(j\alpha) = 0, \;\; 0 \le s \le N-1, \;\; j = 0, \dots, [\log(N)],$$

where the undeterminates are the coefficients $b_{l,m,N}$ of P_N. We then write the system under the form

$$\sum_{l=0}^{u_1(N)} \sum_{m=0}^{u_2(N)} b_{l,m,N} \sum_{\sigma=0}^{\min(s,l)} \left(\frac{s!}{\sigma!(s-\sigma)!}\right)\left(\frac{l!}{(l-\sigma)!}\right) m^{s-\sigma}.j^{l-\sigma},$$

$$(2)\qquad (w\alpha)^{l-\sigma}(we^\alpha)^{jm}.w^{u_1(N)-(l-\sigma)+u_2(N)-jm} = 0.$$

That represents a system of $N[\log(N)]$ equations of at least $N([\log(N)])^2$ undeterminates, with coefficients in E, integral over $\mathbb{Z}$.

According to computations of Hermite–Lindemann's theorem in the complex context (Theorem 3.1.1 in [92]), it appears that in system (2), each factor $\left(\frac{s!}{\sigma!(s-\sigma)!}\right)$, $\left(\frac{l!}{(l-\sigma)!}\right)$, $m^{s-\sigma}$, $j^{l-\sigma}$, $(w\alpha)^{l-\sigma}$, $(we^\alpha)^{jm}$, $w^{u_1(N)-(l-\sigma)+u_2(N)-jm}$ admits a bounding of the form $SN(\log(\log(N))$, when N goes to $+\infty$. On one hand, $w^{u_1(N)+u_2(N)}$ is a common denominator and we have

$$\log(w^{u_1(N)+u_2(N)}) \le \log(\omega)\left(\frac{N}{\log(N)}\right) + \left(\log(N)^3\right),$$

and hence we have a constant $T > 0$ such that

$$(3)\qquad \log(w^{u_1(N)+u_2(N)}) \le \frac{TM}{\log M}.$$

Next, we note that

$$\text{(4)}\qquad \log\left(\frac{u_1(N)!}{\sigma!(u_1(N)-\sigma)!}\right) \leq u_1(N)\log(u_1(N)) \leq \frac{N}{\log(N)}\log\left(\frac{N}{\log(N)}\right) \leq N,$$

and similarly,

$$\text{(5)}\qquad \log\left(\frac{l!}{(u_1(N)-\sigma)!}\right) \leq u_1(N)\log(u_1(N)) \leq N,$$

and

$$\text{(6)}\qquad \log(m^{u_1(N)-\sigma}) \leq \frac{3N}{\log(N)}\log(\log(N)).$$

Now, we check that

$$\log\left(j^{u_1(N)-\sigma}.(|\overline{\alpha}|)^{u_1(N)-\sigma}.(|\overline{e^{\alpha}}|)^{ju_2(N)}\right) \leq N + \frac{N}{\log(N)}\log(|\overline{\alpha}|) + \log(N)(\log(N))^3\log(|\overline{e^{\alpha}}|),$$

and hence there exists a constant $L > 0$ such that

$$\text{(7)}\qquad \log\left(j^{u_1(N)-\sigma}.(|\overline{\alpha}|)^{u_1(N)-\sigma}.(|\overline{e^{\alpha}}|)^{ju_2(N)}\right) \leq LN.$$

Therefore, by (2), (3), (4), (5), (6), and (7), we have a constant $C > 0$ such that each coefficient a of the system satisfies

$$\text{(8)}\qquad s(a) \leq CN(\log(\log(N)).$$

By Siegel's Lemma 26.5 and by (8), there exist integers $b_{l,m,N}$, $0 \leq l \leq u_1(N)$, $0 \leq m \leq u_2(N)$ in $\mathbb{Z}$ such that

$$\text{(9)}\qquad 0 < \max_{l\leq u_1(N),\ m\leq u_2(N)} \log(|b_{l,m,N}|_\infty) \leq \frac{qN\log(N)}{N(\log(N))^2 - qN\log(N)}(CN\log(\log(N))$$

and such that the function

$$F_N(x) = \sum_{l=0}^{u_1(N)} \sum_{m=0}^{u_2(N)} b_{l,m;N} x^l e^{mx} \tag{10}$$

satisfies

$$\frac{d^s}{dx^s} F_N(j\alpha) = 0, \quad 0 \le s \le N-1, \quad j = 0, 1, \dots, [\log(N)].$$

Now, by (9), we can check that there exists a constant $G > 0$ such that

$$\max_{l \le u_1(N),\ m \le u_2(N)} (\log(|b_{l,m,N}|_\infty) \le \frac{GN \log(\log(N))}{\log(N)}. \tag{11}$$

The function F_N we have defined in (10) belongs to $\mathcal{A}(D_1)$ and by Proposition 26.1 is not identically zero, hence at least one of the numbers $\frac{d^s}{dx^s} F_N(0)$ is not null. Let M be the biggest of the integers such that $\frac{d^s}{dx^s} F_N(j\alpha) = 0\ \forall s = 0, \dots, M-1,\ j = 0, 1, 2, \dots, [\log(N)]$. Thus, we have $M \ge N$ and there exists $j_0 \in \{0, 1, \dots, [\log(N)]\}$ such that $\frac{d^M}{dx^M} F_N(j_0\alpha) \ne 0$. We put $\gamma_N = \frac{d^M}{dx^M} F_N(j_0\alpha)$.

Let us now give an upper bound of $s(\gamma_N)$. On one hand, $w^{u_1(N)+u_2(N)}$ is a common denominator, and by (2), we have a constant $T > 0$ such that

$$\log(w^{u_1(N)+u_2(N)}) \le \frac{TM}{\log M}.$$

On the other hand, by (1), we have

$$\begin{aligned} \frac{d^M F_N(\gamma_N)}{dx^M} &= \sum_{l=0}^{u_1(N)} \sum_{m=0}^{u_2(N)} b_{l,m,N} \sum_{\sigma=0}^{u_1(N)} \left(\frac{u_1(N)!}{\sigma!(u_1(N)-\sigma)!} \right) \\ &\quad \times \left(\frac{l!}{(u_1(N)-\sigma)!} \right) m^{u_1(N)-\sigma}. \end{aligned}$$

$$j^{u_1(N)-\sigma} \cdot (\alpha)^{u_1(N)-\sigma} \cdot (e^\alpha)^{ju_2(N)}.$$

Now, by (2), (3), (6), (7), (8), and (10) and taking into account that the number of terms is bounded by $N(\log N)^2$, we can check that there exists a constant B such that

$$s(\gamma_N) \leq BN. \tag{12}$$

Let us now give an upper bound of $|\gamma_N|$. For convenience, we first suppose that $j_0 = 0$, hence $\frac{d^M}{dx^M} F_N(0) \neq 0$. Set $h = |\alpha|$. Then, by Theorem 18.1, we have $|\gamma_N| \leq \frac{|F_N|(h)}{h^M}$. Moreover, we note that F_N admits at least $M[\log(M)]$ zeros in $d(0,h)$, and therefore, by Corollary 22.30, we have $|F_N|(h) \leq \left(\frac{h}{R_1}\right)^{M[\log(M)]}$ because $|F_N|(r) \leq 1\ \forall r < R_1$. Consequently, $|\gamma_N| \leq \frac{h^{M(\log(M-1)}}{(R_1)^{M \log M}}$, and hence

$$\log(|\gamma_N|) \leq M(\log(M) - 1)(\log(h)) - M \log(M)(\log(R_1))).$$

Let $\lambda = \log(h) - \log(R_1)$. Then $\lambda < 0$. And we have $\log(|\gamma_N|) \leq \lambda M \log(M) - M \log(h)$, therefore there exists a constant $A > 0$ such that

$$\log(|\gamma_N|) \leq -AM \log(M). \tag{13}$$

Let us now stop assuming that $j_0 = 0$. Putting $z = x - j\alpha$ and $g(z) = f(x)$, since all points $j\alpha$ belong to $d(0,h)$, it is immediate to go back to the case $j_0 = 0$, which confirms (13) in the general case. But now, by Theorem 9.10, relations (12) and (13) make a contradiction to the relation $-2qs(\gamma_N) \leq \log(|\gamma_N|)$ satisfied by algebraic numbers and shows that γ_N is transcendental. But then, so is e^α. □

In $\mathbb{L}$, we have a similar version:

Theorem 26.8. *Let $\alpha \in \mathbb{L}$ be algebraic such that $|\alpha| < 1$. Then e^α is transcendental over $\mathbb{Q}$.*

Proof. Everything works in $\mathbb{L}$ as in a field of residue characteristic $p \neq 0$ up to Relation (8) in the proof of Theorem 26.7. Here we can replace r_1 by 1 and therefore the conclusion is the same as in Theorem 26.7. □

Theorem 26.9 (Gelfond–Schneider). *Let $\ell \in D_1$, $\ell \neq 0$, and let $b \notin \mathbb{Q}$ belong to $\mathbb{K}$ be such that $b\ell \in D_1$. Then at least one of the three numbers $a = e^\ell$, b, $e^{b\ell}$ is transcendental.*

Proof. A large part of the proof does not involve the topology of the field $\mathbb{K}$ and hence is similar to the proof in the field $\mathbb{C}$ [92] where we can copy many technical relations. We suppose that $a = e^{\ell}$, b and $e^{b\ell}$ are algebraic over $\mathbb{Q}$. Let $E = \mathbb{Q}[e^{\ell},\ b,\ e^{b\ell}]$ and let $\delta = [E : \mathbb{Q}]$ and let d be a common denominator of $b, e^{\ell}, e^{b\ell}$.

Put $S = \max(1, |b|)$, $T \in]S, \frac{R_1}{|\ell|}[$, $\sigma = \log(\frac{T}{S})$, $\tau = \log T$, $\Lambda = d(0, S)$, and $\Delta = d(0, T)$. We consider integers N of the form q^2, with $q \in \mathbb{N}$, and we first show that there exists a non-identically zero polynomial $P_N(X, Y) \in \mathbb{Z}[X, Y]$ such that $\deg_X(P_N) \leq N^{\frac{3}{2}}$, and $\deg_Y(P_N) \leq 2\delta N^{\frac{1}{2}}$ such that the function $F_N(x)$ defined in Δ by $F_N(x) = P_N(x, e^{\ell x})$ satisfy

$$F_N(i + jb) = 0 \quad \forall i = 1, \ldots, N, \quad \forall j = 1, \ldots, N.$$

In order to find P_N, let us write

$$\sum_{h=0}^{N^{\frac{3}{2}}-1} \sum_{k=0}^{2\delta N^{\frac{1}{2}}-1} C_{h,k}(N) X^h Y^k$$

with $C_{h,k}(N) \in \mathbb{Z}$ and consider the system of equations where the $C_{h,k}(N)$ are the undeterminates:

$$d^{(4\delta+1)N^{\frac{3}{2}}} \cdot F_N(i + jb) = 0 \ (1 \leq i \leq N; 1 \leq j \leq N).$$

Thus, we obtain a system of N^2 equations of $2\delta N^2$ undeterminates in $\mathbb{Z}$, with coefficients in E. By Lemma 26.4, these coefficients have size bounded by

$$\begin{aligned} &N^{\frac{3}{2}} \log(N) + N^{\frac{3}{2}}(8\delta + 2)\log(d) + \log(1 + \overline{|b|}) + 2\delta \log(\overline{|e^{\ell + b\ell}|}) \\ &\leq \frac{3}{2} N^{\frac{3}{2}} \log(N). \end{aligned}$$

By Lemma 26.3, we can find in $\mathbb{Z}$ a family of integers not all equal to zero, $(C_{h,k}(N),\ 0 \leq N^{\frac{3}{2}} - 1,\ 0 \leq k \leq 2\delta N^{\frac{1}{2}} - 1)$ satisfying

$$\log\left(\max_{h,k} |C_{h,k}(N)|_\infty\right) \leq 2N^{\frac{3}{2}} \log N \left(\frac{\delta N^2}{2\delta N^2 - \delta N^2}\right) = 2N^{\frac{3}{2}} \log N$$

such that the function F_N defined by $F_N(x) = P_N(x, e^{\ell x})$ satisfies $F_N(i + jb) = 0\ \forall i = 1, \ldots, N,\ j = 1, \ldots, N$.

Now, we can check the function F_N is an analytic element in every disk of the form $d(0, r)$ such that $r|\ell| < R_1$ and hence in $\Delta = d(0, T)$. Since the power of x in the various terms is at most $N^{\frac{3}{2}}$ and since all coefficients are integers, we can check that $\log(|F_N|(T)) \leq \tau N^{\frac{3}{2}}$. On the other hand, since the polynomial P_N is not identically zero, by Proposition 26.1, F_N is not identically zero and then F_N has finitely many zeros in Λ. Particularly, there exists a point of the form $i + jb$ such that $F_N(i + jb) \neq 0$. Consequently, there exists $M \geq N$ such that $F_N(i + jb) = 0\ \forall i \leq M,\ \forall j \leq M$ and there exists a point γ_N of the form $i_0 + j_0 b$ such that $F_N(\gamma_N) \neq 0$ with $M < i_0 \leq M + 1,\ M < j_0 \leq M + 1$. Consequently, the number of zeros of F_N in Λ is at least M^2. Then, by Corollary 22.30, we have $\log(|F_N(\gamma_N)|) \leq \tau N^{\frac{3}{2}} - \sigma M^2$, hence there exists $\lambda > 0$ such that

$$\log(|F_N(\gamma_N)|) \leq -\lambda M^2 \quad \forall N \in \mathbb{N}. \tag{1}$$

By definition, neither σ nor τ depend on N, hence neither does λ.

On the other hand, by Lemma 26.4, we can check that $s\big(F_N(\gamma_N)\big)$ satisfies an inequality of the form $s\big(F_N(\gamma_N)\big) \leq AM^{\frac{3}{2}}\log(M)$ which by (1) contradicts the inequality $-2\delta s\big(F_N(\gamma_N)\big) \leq \log\big(|F_N(\gamma_N)|\big)$ and this ends the proof. □

Theorem 26.10 (Gelfond–Schneider in zero residue characteristic). *Let $\ell \in D_0$, $\ell \neq 0$, and let $b \notin \mathbb{Q}$ belong to $\mathbb{L}$ and be such that $b\ell \in D_0$. Then at least one of the three numbers $a = e^{\ell}$, b, $e^{b\ell}$ is transcendental.*

Proof. The proof is identical to the proof of Theorem 26.9 except that T now belongs to $]S, \frac{1}{|\ell|}[$. □

The six-exponential problem is well known on $\mathbb{C}$ and was solved by Serge Lang [71] and Ramachandra. The problem is the following: *let a_1, a_2, a_3 (respectively b_1, $b_2 \in \mathbb{C}$) be $\mathbb{Q}$-linearly independent. Then at least one of the six numbers $e^{a_i b_j}$ is transcendental.* Next, consider the same problem with only four exponentials: *let a_1, a_2 (respectively b_1, $b_2 \in \mathbb{C}$) be $\mathbb{Q}$-linearly independent.* The question is whether one of the numbers $e^{a_i b_j}$ is transcendental: this is the four exponentials conjecture on $\mathbb{C}$ due to Serge Lang.

The problem, however, has a solution somewhat similar to that of the six exponentials problem, in the particular case when one of the ratios $\frac{a_1}{a_2}$ and $\frac{b_1}{b_2}$ is algebraic.

The same problems make sense on a p-adic field such as $\mathbb{C}_p$ (provided the numbers $a_i b_j$ lie in the disk of convergence of the exponential). Here we give the solution of the six p-adic exponentials problem on the field $\mathbb{C}_p$ and this of the four p-adic exponentials problem when one of the ratios $\frac{a_1}{a_2}$ or $\frac{b_1}{b_2}$ is algebraic. This was described by Jean-Pierre Serre [87].

Theorem 26.11. *Let* $a_1,\ a_2,\ a_3$ *(respectively* $b_1,\ b_2 \in \mathbb{C}_p$*) be* $\mathbb{Q}$*-linearly independent and such that* $\max_{i=1,2,3\ j=1,2} |a_i b_j| < r_1$*. Then at least one of the numbers* $e^{a_i b_j}$ *is transcendental.*

Proof. Assume that all numbers $e^{a_i b_j}$ are algebraic, put $E = \mathbb{Q}[(e^{a_i b_j})_{i=1,2,3,\ j=1,2}]$ and $q = [E : \mathbb{Q}]$. Without loss of generality, we can assume that $a_1 = 1,\ |a_2|,\ |a_3| \leq 1$ and that $\max(|b_1|,\ |b_2|) \leq \frac{1}{p^2}$.

Let $t \in \mathbb{N}^*$ be such that $te^{a_i b_j}$ is integral over $\mathbb{Z}$ for every $i = 1,\ 2,\ 3$ and every $j = 1,\ 2$ and let $B = \log\big(t \max\{\overline{|e^{a_i b_j}|},\ i = 1,\ 2,\ 3,\ j = 1,\ 2\}\big)$. Let $\ell \in \mathbb{N}$ be such that $\ell > 9\sqrt{2}B(q+1)$.

Consider now the linear system of $\ell^2 N^2$ equations with coefficients in E:

$$(\mathcal{S}_N) \qquad \sum_{1\leq m\leq N, 1\leq n\leq N, 1\leq s\leq N} c_{m,n,s,N} e^{(ma_1+na_2+sa_3)(ib_1+jb_2)} = 0,$$

$$1 \leq i \leq \ell N,$$

$1 \leq j \leq \ell N$. We note that the coefficients $e^{(ma_1+na_2+sa_3)(ib_1+jb_2)}$ of $(\mathcal{S}_N)$ satisfy

$$\log\left(\overline{|e^{(ma_1+na_2+sa_3)(ib_1+jb_2)}|}\right) \leq 6B\ell N^2$$

$$1 \leq m \leq N, \quad 1 \leq n \leq N, \quad 1 \leq s \leq N, \quad 1 \leq i \leq \ell N,$$

$1 \leq j \leq \ell N$. Now, by Siegel's lemma 26.5, there exists a family of solutions $(c_{m,n,s,N})_{1\leq m\leq N, 1\leq n\leq N, 1\leq s\leq N}$ in $\mathbb{Z}$ such that

$$(1) \qquad \log|c_{m,n,s,N}|_\infty \leq \frac{\log 2}{2} + 6B\frac{(q\ell^2N^2)\ell N^2}{(N^3 - q\ell^2N^2)}.$$

Let $f_N(x) = \sum_{1\leq m\leq N,\ 1\leq n\leq N,} c_{m,n,s,N} e^{(ma_1+na_2+sa_3)x}$. By Proposition 26.1, f_N is not identically zero. Then, by definition of $(\mathcal{S}_N)$, we have $f_N(ib_1 + jb_2) = 0\ \forall i = 1, \ldots, \ell N,\ j = 1, \ldots, \ell N$, hence f_N admits at least $\ell^2 N^2$ zeros in the disk $d(0, \frac{1}{p^2})$. Let u be a point of the form $ib_1 + jb_2$ with $i,\ j \in \mathbb{N}$ such that $f_N(ib_1 + jb_2) \neq 0$ and such that $i + j$ is minimum and let h be this minimum: say $h = i_0 + j_0$. Thus we can check that when i and j are two positive integers such that $i + j < h$, then $f_N(ib_1 + jb_2) = 0$. Consequently, by construction, we have $h > 2\ell N$ and the number of zeros of f_N in $d(0, \frac{1}{p^2})$ is at least $\frac{(h-1)^2}{2}$. We note that $\|f_N\| \leq 1$ because $|e^x| = 1\ \forall x \in d(0, R_1^-)$ and $c_{m,n} \in \mathbb{Z}\ \forall m, n \in \mathbb{N}$. Consequently, by Corollary 22.30, we have $\log(|f_N|(\frac{1}{p})) \leq -(h-1)N^2$ and therefore

$$\log|f_N(u)| \leq -\frac{(h-1)^2}{2}. \tag{2}$$

Consider now some $c_{m,n,s,N} e^{(ma_1+na_2+sa_3)(ib_1+jb_2)}$ at the point u. By (1), we have $\log(|c_{m,n,s,N}|_\infty \leq \frac{6B(q\ell^2N^2)\ell N^2}{N^3 - q\ell^2N^2} +$ and $\log \overline{|e^{(ma_1+na_2+sa_3)(i_0b_1+j_0b_2)}|} \leq 6BNh$. Consequently, by (1), we can derive

$$\log \overline{|c_{m,n,s,N} e^{(ma_1+na_2+sa_3)(i_0b_1+j_0b_2)}|} \leq 6BNh + \frac{\log 2}{2} + \frac{6Bq\ell^3N^4}{N^3 - q\ell^2N^2},$$

therefore

$$\log(\overline{|f_N(u)|} \leq 6BNh + \frac{\log 2}{2} + \frac{6Bq\ell^3N^4}{N^3 - q\ell^2N^2} + 3\log N. \tag{3}$$

Here we note that the denominator of $F_N(u)$ is bounded by t^{3Nh} because t^{3Nh} is clearly a multiple of the denominator of each term $e^{(ma_1+na_2+sa_3)(ib_1+jb_2)}$ whenever $i + j \leq h$. Therefore, by Corollary 9.12, we can derive

$$\begin{aligned}\log(|f_N(u)|) &\geq -3Nh(q+1)\log(t)\\ &\quad - q\left(6BNh + \frac{\log 2}{2} + \frac{6B\ell^3N^4}{N^3 - q\ell^2N^2} + 3\log(N)\right).\end{aligned}$$

Consequently, by (2) and (3), we obtain

$$\frac{(h-1)^2}{2} \leq (q+1)\left(9BNh + \frac{\log 2}{2} + \frac{6B\ell^3N^4}{N^3 - q\ell^2N^2} + 3\log(N)\right),$$

therefore $\frac{(h-1)^2}{2} \leq (q+1)(9BNh + O(N))$, and hence

$$h - 2 \leq \frac{(h-1)^2}{h} \leq 18B(q+1)N + O(1),$$

and hence $h - 1 \leq 18B(q+1)N + O(1)$. Now, since $\frac{(h-1)^2}{2} \geq \ell^2N^2$, we have $\sqrt{2}\ell N \leq 18B(q+1)N + O(1)$, hence $\ell \leq 9\sqrt{2}B(q+1)$ when N is big enough, a contradiction to the hypothesis on ℓ. □

And similarly, we have the following:

Theorem 26.12. *Let a_1, a_2, a_3 (respectively b_1, $b_2 \in \mathbb{L}$) be $\mathbb{Q}$-linearly independent and such that $\max_{i=1,2,3\ j=1,2} |a_ib_j| < 1$. Then at least one of the numbers $e^{a_ib_j}$ is transcendental over $\mathbb{Q}$.*

As explained above, when reducing to 4 exponentials $e^{a_ib_j}$, $i = 1, 2$, $j = 1, 2$, the transcendence of one of the four numbers is just a conjecture in the general case. Here we give a proof in the particular case when one of the ratios $\frac{a_1}{a_2}$ or $\frac{b_1}{b_2}$ is algebraic.

Theorem 26.13. *Let a_1, a_2 (respectively b_1, $b_2 \in \mathbb{C}_p$) be $\mathbb{Q}$-linearly independent and such that $\max_{i=1,2\ j=1,2} |a_ib_j| < r_1$ and such that $\frac{a_1}{a_2}$ is algebraic. Then at least one of the numbers $e^{a_ib_j}$ is transcendental.*

Proof. Assume that all numbers $e^{a_ib_j}$ are algebraic. Without loss of generality, we can assume that $|b_i| \leq \frac{1}{p^2}$, $i = 1, 2$. Put $a = \frac{a_1}{a_2}$. Let $t \in \mathbb{N}^*$ be such that all the $te^{a_ib_j}$ and ta are integral over $\mathbb{Z}$ for every $i = 1, 2$ and every $j = 1, 2$ and let $B = \log\left(t\max(\overline{|a|}, \max\{\overline{|e^{a_ib_j}|}, i = 1, 2, j = 1, 2\}\right)$. Without loss of generality, we can assume that $a_1 = 1$, hence $a = a_2$. Since a is algebraic, we can put $E = \mathbb{Q}[a, (e^{a_ib_j})_{i=1,2,\ j=1,2}]$ and $q = [E : \mathbb{Q}]$. We can find integers s, $l \in \mathbb{N}$ satisfying

$$(1) \qquad s > (q+1)Bl\left(12 + \frac{qs}{l^2 - qs}\right).$$

Consider now the linear system of sN^2 equations with coefficients in E:

$$(\mathcal{S}_N) \qquad \sum_{1\le m\le lN, 1\le n\le lN} c_{m,n,N}(m+na)^k e^{(m+na)(ib_1+jb_2)} = 0,$$

$$1 \le i \le \mathbb{N},$$

$1 \le j \le N$. We note that the coefficients $(m+na)^k e^{(m+na)(ib_1+jb_2)}$ of $(\mathcal{S}_N)$ satisfy

$$(2) \quad \log\left(\overline{|(m+na)^k e^{(m+na)(ib_1+jb_2)}|}\right) \le 2BlN^2 + k\log(2N+B)$$

$$1 \le m \le lN, \quad 1 \le n \le lN, \quad 1 \le i \le N, \quad 1 \le j \le N, \quad 1 \le k \le s.$$

By Siegel's lemma 26.5 and by (2), there exists a family of solutions $(c_{m,n,N})_{1\le m\le lN, 1\le n\le lN}$ in $\mathbb{Z}$ such that

$$\log|c_{m,n,N}|_\infty \le \frac{\log 2}{2} + (BlN^2 + s(\log 2N + B))\frac{qsN^2}{(l^2N^2 - qsN^2)},$$

therefore

$$(3) \qquad \log|c_{m,n,N}|_\infty \le \frac{qsBlN^4}{l^2N^2 - qsN^2} + O(\log(N)).$$

Now, let $f_N(x) = \sum_{m\le lN, n\le :N} c_{m,n,N}e^{x(m+na)}$ and let h be the smallest integer such that $(f_N)^{(h)}(ib_1+jb_2) \ne 0$ for some pair (i_0, j_0) such that $i_0 \le N$, $j_0 \le N$. By Proposition 26.1, f_N is not identically zero, and by definition, $s < h$. Consequently, the number of zeros of f_N in $d(0, \frac{1}{p^2})$ is at least $(h-1)N^2$, taking multiplicity into account. Set $u = i_0b_1 + j_0b_2$. Consider now some $c_{m,n,N}e^{(m+na)(ib_1+jb_2)}$ at the point u. First, we have

$$\log\overline{|e^{(m+na)(i_0b_1+j_0b_2)}|} \le 4BlN^2.$$

Consequently, by (3), when N is big enough, we can derive

$$\log\overline{|c_{m,n,N}e^{(m+na)(i_0b_1+j_0b_2)}|} \le \frac{qsBlN^4}{l^2N^2 - qsN^2} + 4BlN^2 + O(\log(N)),$$

hence and therefore, by (2),

$$\log \overline{|f_N^{(h)}(u)|} \leq \frac{qsBlN^4}{l^2N^2 - qsN^2} + 4BlN^2 + O(\log(N)).$$

On the other hand, we can check that

$$\log(\text{den}((m + na)^h e^{(m+na)(i_0b_1+j_0b_2)}) \leq 2BlN^2 + hB + O(\log(N)),$$

hence $\log(\text{den}(f_N^{(h)}(u))) \leq 2BlN^2 + sB + O(\log(N))$. Consequently, by Corollary 9.11, we can derive

$$(4) \quad \log(|f_N^{(h)}(u)|) \geq -(q+1)\left(\frac{qsBlN^4}{l^2N^2 - qsN^2} + 4BlN^2 + 8BlN^2 + 2hB + O(\log(N))\right).$$

As in Theorem 26.11, we have $\|f_N\| \leq 1$, hence, by Theorem 18.1, we can derive $\|f_N^{(h)}\| \leq \frac{1}{p^h}$. Next, $c_{m,n,N} \in \mathbb{Z}\ \forall m, n \in \mathbb{N}$, consequently, by Corollary 22.30, we have $\log(|f_N|(\frac{1}{p^2})) \leq -(h-1)N^2$, and by Theorem 18.1, $\log(|f_N^{(h)}|(\frac{1}{p^2})) \leq -(h-1)N^2 + h$. Therefore, by (2) and (4), we obtain

$$(5) \quad (h-1)N^2 - h \leq (q+1)\left(\frac{qsBlN^4}{l^2N^2 - qsN^2} + 12BlN^2 + 2hB\right) + O(\log(N)).$$

Now, by (1), we have $s > (q+1)Bl(12 + \frac{qs}{l^2-qs})$, and since $h > s$, we can see that (5) is impossible when N is big enough, which ends the proof. □

Similarly, we have the following:

Theorem 26.14. *Let a_1, a_2 (respectively b_1, $b_2 \in \mathbb{L}$) be $\mathbb{Q}$-linearly independent and such that $\max_{i=1,2\ j=1,2} |a_ib_j| < 1$ and such that $\frac{a_1}{a_2}$ is algebraic. Then at least one of the numbers $e^{a_ib_j}$ is transcendental over $\mathbb{Q}$.*

Chapter 27

Divisors of Analytic Functions

In this chapter, we define divisors in $\mathbb{K}$ or in a disk $d(a, R^-)$. We then define the divisor of an analytic function and of an ideal.

Definition. We call *a divisor in* $\mathbb{K}$ (respectively *a divisor in a disk* $d(a, R^-)$) a mapping T from $\mathbb{K}$ (respectively from $d(a, R^-)$) to $\mathbb{N}$ whose support is countable and has a finite intersection with each disk $d(a, r)$, $\forall r > 0$ (respectively $\forall r \in]0, R[$). Thus, a divisor on $\mathbb{K}$ (respectively of $d(a, R^-)$) is characterized by a sequence $(a_n, q_n)_{n\in\mathbb{N}}$ with $a_n \in \mathbb{K}$, $\lim_{n\to\infty} |a_n| = \infty$ (respectively $a_n \in d(a, R^-)$, $\lim_{n\to\infty} |a_n - a| = R$), $|a_n| \leq |a_{n+1}|$, and $q_n \in \mathbb{N}^*$ $\forall n \in \mathbb{N}$. So, we frequently denote a divisor by the sequence $(a_n, q_n)_{n\in\mathbb{N}}$ which characterizes it.

The set of divisors on $\mathbb{K}$ (respectively on $d(a, R^-)$) is provided with a natural additive law that makes it a semi-group. It is also provided with a natural order relation: given two divisors T and T', we can set $T \leq T'$ when $T(\alpha) \leq T'(\alpha)$ $\forall \alpha \in d(a, R^-)$. Moreover, if T, T' are two divisors such that $T(\alpha) \geq T'(\alpha)$ $\forall \alpha \in d(0, R^-)$, we can define the divisor $\frac{T}{T'}$.

Given $f \in \mathcal{A}(\mathbb{K})$ (respectively $f \in \mathcal{A}(d(a, R^-))$), we can define *the divisor of* f, denoted by $\mathcal{D}(f)$ on $\mathbb{K}$ (respectively of $d(a, R^-)$), as $\mathcal{D}(f)(\alpha) = 0$ whenever $f(\alpha) \neq 0$ and $\mathcal{D}(f)(\alpha) = s$ when f has a zero of order s at α.

Similarly, given an ideal I of $\mathcal{A}(\mathbb{K})$ (respectively of $\mathcal{A}(d(a, R^-))$), we denote by $\mathcal{D}(I)$ the lower bound of the $\mathcal{D}(f)$ $f \in I$ and $\mathcal{D}(I)$ is called *the divisor of* I.

Conversely, given a divisor E of $\mathbb{K}$ (respectively of $d(a, R^-)$), we denote by $\mathcal{T}(E)$ (respectively by $\mathcal{T}_{a,R}(E)$) the ideal of $\mathcal{A}(\mathbb{K})$ (respectively of $\mathcal{A}(d(a, R^-))$) of the $f \in \mathcal{A}(\mathbb{K})$ (respectively of the $f \in \mathcal{A}(d(a, R^-))$) such that $\mathcal{D}(f) \geq E$.

As we see, given a divisor T on $\mathbb{K}$, there is no problem to construct an entire function whose divisor is T. But given a divisor T on a disk $d(a, R^-)$, if $\mathbb{K}$ is not spherically complete, it is not always possible to find an analytic function (in that disk) whose divisor is T due to Lazard's problem [72].

Finally, given a divisor $T = (a_n, q_n)_{n\in\mathbb{N}}$, we denote by $\overline{T}$ the divisor $(a_n, 1)_{n\in\mathbb{N}}$. Let $T = (a_n, q_n)_{n\in\mathbb{N}}$ be a divisor on $\mathbb{K}$ (respectively of $d(a, R^-)$). For every $r > 0$ (respectively $r \in]0, R[$), we set $|T|(r) = \prod_{|a_j|\leq r} \left(\frac{r}{|a_j|}\right)^{q_j}$. The divisor T on $d(a, R^-)$ is said to be *bounded* if $\lim_{r\to R} |T|(r) < \infty$ and then we put $\|T\| = \lim_{r\to R} |T|(r)$.

The $\mathbb{K}$-algebra $\mathcal{A}(\mathbb{K})$ (respectively $\mathcal{A}(d(a, R^-))$) is provided with the following topology of $\mathbb{K}$-algebra: given $\in \mathcal{A}(\mathbb{K})$ (respectively $f \in \mathcal{A}(d(a, R^-))$), the neighborhoods of f are the sets $\mathcal{W}(f, r, \epsilon) = \{h \in \{\mathcal{A}(\mathbb{K}) \mid |f - h|(r) \leq \epsilon\}$ (respectively $\mathcal{W}(f, r, \epsilon) = \{h \in \mathcal{A}(d(a, R^-)) \mid |f - h|(r) \leq \epsilon\}$, with $0 < r < R$, $\epsilon > 0$).

Remark. Let $f \in \mathcal{A}(d(a, R^-))$ and let $(a_n, q_n)_{n\in\mathbb{N}} = \mathcal{D}(f)$. Then $\omega_{a_n}(f) = q_n \ \forall n \in \mathbb{N}$ and $\omega_\alpha(f) = 0 \ \forall \alpha \in d(a, R^-)\setminus\{a_n \mid n \in \mathbb{N}\}$.

Theorem 27.1 is immediate:

Theorem 27.1. *Let $a \in \mathbb{K}$, $R > 0$. Let f, $g \in \mathcal{A}(\mathbb{K})$ (respectively f, $g \in \mathcal{A}((a, R^-)))$ be such that $\mathcal{D}(f) \geq \mathcal{D}(g)$. Then there exists $h \in \mathcal{A}(\mathbb{K})$ (respectively $h \in \mathcal{A}(d(a, R^-))$) such that $f = gh$.*

Proof. Let $T = \mathcal{D}(g) = (a_n, q_n)_{n\in\mathbb{N}}$. Let us fix $r > 0$ (respectively $r \in]0, R[$), and let $s \in \mathbb{N}$ be such that $|a_n| \leq r \ \forall n \leq s$ and $|a_n| > r \ \forall n > s$. Let $P_r(x) = \prod_{n=0}^{s}(1 - \frac{x}{a_n})^{q_n}$. We can factorize f in the form $P_r\hat{f}$, and similarly, we can factorize g in the form $P_r\hat{g}$, hence $\frac{f}{g} = \frac{\hat{f}}{\hat{g}}$. Since $\hat{g}$ has no zero in $d(0, r)$, it is invertible in $H(d(0, r))$, hence $\frac{f}{g}$ belongs to $H(d(0, r))$. This is true for all $r > 0$ (respectively for all $r \in]0, R[$), and hence $\frac{f}{g}$ belongs to $\mathcal{A}(\mathbb{K})$ (respectively to $\mathcal{A}(d(a, R^-))$). □

Corollary 27.2. *Let $a \in \mathbb{K}$, $R > 0$. Let I be an ideal of $\mathcal{A}(\mathbb{K})$ (respectively an ideal of $\mathcal{A}(d(a, R^-))$) and suppose that there exists*

$g \in I$ such that $\mathcal{D}(g) = \mathcal{D}(I)$. Then $I = g\mathcal{A}(\mathbb{K})$ (respectively $I = g\mathcal{A}(d(a, R^-))$).

As an immediate application of the definitions, by Theorem 22.26, we have Lemma 27.3:

Lemma 27.3. *Let $R \in \mathbb{R}_+^*$ and let $f,\ g \in \mathcal{A}(d(0, R^-)$ be such that $\mathcal{D}(f) \leq \mathcal{D}(g)$. Then, given $r, s \in]0, R[$ such that $r < s$, we have $\Psi(f, \log s) - \Psi(f, \log r) \leq \Psi(g, \log s) - \Psi(g, \log r)$.*

In the whole field $\mathbb{K}$, given a divisor T, it is always possible to find an entire function admitting T for divisor.

Theorem 27.4. *Let $T = (\alpha_n, q_n)_{n\in\mathbb{N}}$ be divisor of $\mathbb{K}$. The infinite product $\prod_{n=1}^{\infty}(1 - \frac{x}{\alpha_n})^{q_n}$ is uniformly convergent in all bounded subsets of $\mathbb{K}$ and defines an entire function $f \in \mathcal{A}(\mathbb{K})$ such that $\mathcal{D}(f) = T$. Moreover, given $g \in \mathcal{A}(\mathbb{K})$ such that $\mathcal{D}(g) = T$, then g is of the form λf.*

Proof. We assume that $|\alpha_n| \leq |\alpha_{n+1}|\ \forall n \in \mathbb{N}$. Let us fix $R > 0$ and set $f_m(x) = \prod_{n=1}^{m}(1 - \frac{x}{\alpha_n})^{q_n}$. Consider $N \in \mathbb{N}$ such that $|\alpha_N| > R$ and $m \geq N$. On the one hand, $|f_m|(R) = |f_N|(R)$. Set $M = |f_N|(R)$. On the other hand, we check that

$$
\begin{aligned}
|f_{m+1}(x) - f_m(x)| &= \left|\prod_{n=1}^{m}\left(1 - \frac{x}{\alpha_n}\right)^{q_n}\right| \left|\left(1 - \frac{x}{\alpha_{m+1}}\right)^{q_{m+1}} - 1\right)\right| \\
&\leq M \left|\sum_{k=1}^{q_{m+1}} (-1)^k \binom{q_{m+1}}{k} \left(\frac{x}{\alpha_{m+1}}\right)^k\right| \\
&\leq M \frac{R}{|\alpha_{m+1}|} \quad \forall x \in d(0, R).
\end{aligned}
$$

Consequently, $|f_{m+1} - f_m|(R) \leq M\frac{R}{|\alpha_{m+1}|}$, which shows that the sequence $(f_m)_{m\in\mathbb{N}}$ is uniformly converging in $d(0, R)$ to an element of $H(d(0, R))$, hence to a power series. This is true for all $R > 0$, hence the limit f defined in $\mathbb{K}$ belongs to $\mathcal{A}(\mathbb{K})$. Now, for each $m \in \mathbb{N}$, let $r_m = |\alpha_m|$. By construction, the zeros of f_m in $d(0, r_m)$ are the α_n with $1 \leq n \leq m$, each with multiplicity q_n. And next, we note that $|(1 - \frac{x}{\alpha_n})^{q_n}| = 1\ \forall n > m,\ \forall x \in d(0, r_m)$. Consequently, the zeros of f in $d(0, r_m)$ are exactly those of f_m. Now, consider $g \in \mathcal{A}(\mathbb{K})$ such that $\mathcal{D}(g) = T$. The function $h \in \mathcal{A}(\mathbb{K})$ such that $f = gh$ has no zero in $\mathbb{K}$ and hence is a constant. □

Corollary 27.5. *For every divisor T on $\mathbb{K}$, there exists $f \in \mathcal{A}(\mathbb{K})$ such that $\mathcal{D}(f) = T$. Moreover, if $f(0) = 1$, f satisfies $|f|(r) = |T|(r)\ \forall r > 0$.*

Corollary 27.6. *Let T be a divisor on $\mathbb{K}$, let $g \in \mathcal{A}(\mathbb{K})$ be such that $\mathcal{D}(g) = T$, and let $f \in \mathcal{A}(\mathbb{K})$ be such that $\mathcal{D}(f) \geq T$. Then there exists $h \in \mathcal{A}(\mathbb{K})$ such that $f = gh$.*

Proof. Indeed, let $E = \frac{\mathcal{D}(f)}{T}$ and let $h \in \mathcal{A}(\mathbb{K})$ be such that $\mathcal{D}(h) = \frac{\mathcal{D}(f)}{T}$. Then $\mathcal{D}(f) = \mathcal{D}(gh)$, hence by Theorem 27.4, $\frac{f}{gh}$ is a constant and we can choose h such that the constant is 1. □

Theorem 27.7. *Let $f \in \mathcal{A}(\mathbb{K})$ have a divisor of the form (a_n, sq_n) with $s \in \mathbb{N}^*$. Then, there exists $g \in \mathcal{A}(\mathbb{K})$ such that $f = g^s$.*

Proof. By Corollary 27.5, there exists $h \in \mathcal{A}(\mathbb{K})$ such that $\mathcal{D}(h) = (a_n, q_n)$. Then, $\frac{f}{h^s}$ has no zero and no pole, and therefore, it is a constant λ. Let $l \in \mathbb{K}$ be such that $l^s = \lambda$ and let $g = lh$. Then, $g^s = f$. □

So, by Theorem 27.4, given a divisor T on $\mathbb{K}$, we can find an entire function whose divisor is just T. It is natural to consider the same problem inside a disk $d(a, r^-)$. Indeed, in $\mathbb{C}$, it is known that a similar problem always admits a solution: in the whole field $\mathbb{C}$ as well as inside an open disk. Actually, in the general context of a complete ultrametric algebraically closed field $\mathbb{K}$, the problem has no solution when $\mathbb{K}$ is not spherically complete.

This problem was first considered by Lazard [72] and we detail the solutions he gave. First, we construct a function f whose divisor is bigger than the given divisor but narrows it.

We deal with the problem by showing that given a sequence $(a_n)_{n\in\mathbb{N}}$ such that $|a_n - a| < R$ for all $n \in \mathbb{N}$ and $\lim_{n\to\infty} |a_n - a| = R$, a sequence of integers $(q_n)_{n\in\mathbb{N}}$ and a number $\epsilon > 0$, there exists an analytic function $f \in \mathcal{A}(d(0, R^-))$ that admits each a_n as a zero of order $t_n \geq q_n$ and such that $|f|(r) \leq (1+\epsilon)\ |T|(r)$. First, we need Lemma 27.8.

Lemma 27.8. *Let $T = (a_n, q_n)_{n\in\mathbb{N}}$ be a divisor on $d(a, R^-)$ and let $f \in \mathcal{A}(d(a, R^-))$ satisfy $f(0) = 1$, $\mathcal{D}(f) \geq T$ and $|f|(r) = |T|(r)\ \forall r \in]0, R[$. Then, $\mathcal{D}(f) = T$.*

Proof. Since $f(0) = 1$, we may write f in the form $\prod_{j=0}^{\infty}(1 - \frac{x}{a_j})^{s_j}$ with $q_j \leq s_j\ \forall j \in \mathbb{N}$. By hypothesis, we have $q_j \leq s_j\ \forall j \in \mathbb{N}$. Suppose that $s_k > q_k$ for some index k and let $r_n = |a_n|$, $n \in \mathbb{N}$. Since $|f|(r_k) = |T|(r_k)$, when $r \in]r_k, r_{k+1}[$, we have $|f|(r) = |f|(r_k)(\frac{r}{r_k})^{s_k}$, $|T|(r) = |f|(r_k)(\frac{r}{r_k})^{q_k}$, and since $|f|(r) = |T|(r)\ \forall r \in]0, R[$, clearly $s_k = q_k$. □

Notation. For each divisor E of $\mathbb{K}$, we denote by $\mathcal{T}(E)$ the set of $f \in \mathcal{A}(\mathbb{K})$ such that $E \leq \mathcal{D}(f)$. Similarly, for each divisor E of $d(a, R^-)$, we denote by $\mathcal{T}_{a,R}(E)$ the set of $f \in \mathcal{A}(d(a, R^-))$ such that $E \leq \mathcal{D}(f)$.

Theorem 27.9. *For every divisor E of $\mathbb{K}$, $\mathcal{T}(E)$ is a closed ideal of $\mathcal{A}(\mathbb{K})$. Moreover, $\mathcal{T}$ is a bijection from the set of divisors of $\mathbb{K}$ onto the set of closed ideals of $\mathcal{A}(\mathbb{K})$. Further, given a closed ideal I of $\mathcal{A}(\mathbb{K})$, then $I = \mathcal{T}(\mathcal{D}(I))$.*

Similarly, we have Theorem 27.10:

Theorem 27.10. *Let $a \in \mathbb{K}$ and take $R > 0$. For every divisor E of $d(a, R^-)$, $\mathcal{T}_{a,R}(E)$ is a closed ideal of $\mathcal{A}(d(a, R^-))$. Moreover, $\mathcal{T}_{a,R}$ is a bijection from the set of divisors of $d(a, R^-)$ onto the set of closed ideals of $\mathcal{A}(d(a, R^-))$. Further, given a closed ideal I of $\mathcal{A}(d(a, R^-))$), then $I = \mathcal{T}_{a,R}(\mathcal{D}(I))$.*

Proof (Theorems 27.9 and 27.10). Let E be a divisor of $\mathbb{K}$ (respectively of $d(a, R^-)$). First, let us check that $\mathcal{T}(E)$ (respectively $\mathcal{T}_{a,R}(E)$) is a closed ideal of $\mathcal{A}(\mathbb{K})$ (respectively of $\mathcal{A}(d(a, R^-))$). Let $E = (a_n, q_n)_{n\in\mathbb{N}}$ and let $(f_m)_{m\in\mathbb{N}}$ be a sequence of elements of $\mathcal{T}(E)$ (respectively of $\mathcal{T}_{a,R}(E)$) converging to a limit f in $\mathcal{A}(\mathbb{K})$ (respectively in $\mathcal{A}(d(a, R^-))$). For every $n \in \mathbb{N}$, each f_m admits a_n as a zero of order at least q_n, hence by Lemma 16.1, so does f. Consequently, f belongs to $\mathcal{T}(E)$ (respectively f belongs to $\mathcal{T}_{a,R}(E)$).

Now, let us show that $\mathcal{T}$ (respectively $\mathcal{T}_{a,R}$) is injective. Let E, F be two distinct divisors of $\mathbb{K}$ (respectively of $d(a, R^-)$). Without loss of generality, we can suppose that E admits a pair (b, s) with $s > 0$ and that F either does not admit any pair (b, m) or admits a pair (b, m) with $m < s$. Let $f \in \mathcal{T}(F)$ (respectively let $f \in \mathcal{T}_{a,R}(F)$) and suppose that $\omega_b(f) \geq s$. Then, by Lemma 16.1, f factorizes in the form $(x - b)^{s-m} g$ with $g \in \mathcal{A}(\mathbb{K})$ (respectively $g \in \mathcal{A}(d(a, R^-)))$ and of course g belongs to $\mathcal{T}(F)$ (respectively to $\mathcal{T}_{a,R}(F)$). But by construction, g does not belong to $\mathcal{T}(E)$ (respectively to $\mathcal{T}_{a,R}(E)$) because $\omega_b(g) < s$. Therefore, $\mathcal{T}(E) \neq \mathcal{T}(F)$ (respectively $\mathcal{T}_{a,R}(E) \neq \mathcal{T}_{a,R}(F)$). So, $\mathcal{T}$ (respectively $\mathcal{T}_{a,R}$) is injective.

Now, let us show that $\mathcal{T}$ (respectively $\mathcal{T}_{a,R}$) is injective. Let E, F be two distinct divisors of $\mathbb{K}$ (respectively of $d(a, R^-)$). By Theorem 27.4, there exists $f \in \mathcal{A}(\mathbb{K})$ (respectively $f \in \mathcal{A}(d(a, R^-)))$ such that $\mathcal{D}(f) = F$, hence $\mathcal{D}(f) \neq E$. Therefore, $\mathcal{D}(f) \notin \mathcal{T}(E)$, and hence $\mathcal{T}(E) \neq \mathcal{T}(F)$. So, $\mathcal{T}$ is injective (respectively $\mathcal{T}_{a,R}(E) \neq \mathcal{T}_{a,R}(F)$. So, $\mathcal{T}_{a,R}$ is injective).

Let us show that it is also surjective. Let I be a closed ideal of $\mathcal{A}(\mathbb{K})$ (respectively of $\mathcal{A}(d(a, R^-)))$ and let $E = \mathcal{D}(I)$. Then E is of the form $(a_n, q_n)_{n \in \mathbb{N}}$ with $|a_n| \leq |a_{n+1}|$ and $\lim_{n \to +\infty} |a_n| = +\infty$ (respectively $\lim_{n \to +\infty} |a_n| = R$), hence there is a unique $s \in \mathbb{N}$ such that $a_n \in d(0, r)\ \forall n \leq s$ and $a_n \notin d(0, r)\ \forall n > s$.

Let $J = \mathcal{T}(E)$ (respectively $J = \mathcal{T}_{a,R}(E)$). Then, of course, $I \subset J$. Let us show that $J \subset I$. Let $f \in J$ and take $r > 0$. Denoting by P_r the polynomial $\prod_{i=0}^{s} (X - a_i)^{q_i}$ by Theorem 24.2, $I \cap H(d(0, r)) = P_r(x) H(d(0, r))$. But now all functions $g \in J \cap H(d(0, r))$ also are of the form $P_r(x) h(x)$ with $h \in H(d(0, r))$. Consequently, in $H(d(0, r))$, we can write f in the form $f = \sum_{j=1}^{m} g_j h_j$ with $g_j \in I$ and $h_j \in H(d(0, r))$. Let $\epsilon > 0$ be fixed.

For each $j = 1, \dots, m$, narrowing each h_j by a polynomial ℓ_j in $H(d(0, r))$, we can find $\ell_j \in \mathbb{K}[x]$ such that $|g_j(h_j - \ell_j)|(r) \leq \epsilon$. Now, let $\phi_r = \sum_{j=1}^{m} g_j \ell_j$. Then ϕ_r belongs to I and satisfies $|\phi_r - f|(r) \leq \epsilon$. This is true for each $r > 0$ and for every $\epsilon > 0$. Consequently, since I is closed, f belongs to I. This finishes proving that $\mathcal{T}$ (respectively $\mathcal{T}_{a,R}$) is surjective. Further, we have proven that $I = \mathcal{T}(\mathcal{D}(I))$ (respectively $I = \mathcal{T}_{a,R}(\mathcal{D}(I))$). □

Theorem 27.11. *Every closed ideal of $\mathcal{A}(\mathbb{K})$ is principal.*

Proof. Indeed, consider a closed ideal I and let $E = \mathcal{D}(I)$. By Theorem 27.9, I is of the form $\mathcal{T}(E)$ with $E = \mathcal{D}(I)$. By Theorem 27.4, there exists $g \in \mathcal{A}(\mathbb{K})$ such that $\mathcal{D}(g) = E$ and of course, g

belongs to I. Hence, $g\mathcal{A}(\mathbb{K}) \subset I$. Now, let $f \in I$. Then, $\mathcal{D}(f) \geq E$, hence by Theorem 27.1, f factorizes in the form gh with $h \in \mathcal{A}(\mathbb{K})$, hence $I = g\mathcal{A}(\mathbb{K})$. □

Theorem 27.12. *Let $r \in |\mathbb{K}^*|$, let $f \in H(C(0,r))$, and let $P \in \mathbb{K}[x]$ have all its zeros in $d(0,r)$. There exists $g \in H(C(0,r))$ and $L \in \mathbb{K}[x]$ unique such that $f = Pg + L$, $\deg(L) < \deg(P)$, $\Psi(R, \log r) \leq \Psi(f, \log r)$, $\Psi(g, \log r) \leq \Psi(f, \log r) - \Psi(P, \log r)$. Moreover, if f belongs to $H(d(0,r))$, then so does g.*

Proof. Since $r \in |K|$, without loss of generality, we may assume that $r = 1$. Similarly, we may also assume that $\Psi(P, 0) = \Psi(f, 0) = 0$, so P is quasi-monic. Thus, the problem now consists of finding $g \in H(C(0,1))$ and $L \in \mathbb{K}[x]$, each unique, satisfying the statements.

Let $f(x) = \sum_{-\infty}^{\infty} a_m x^m$, and for each $n \in \mathbb{N}$, let $f_n(x) = \sum_{m=-n}^{n} a_m x^m$. For each $n \in \mathbb{N}$, set $u_n = \sup_{|j|_\infty > n} |a_j|$. Then, $\lim_{n\to+\infty} a_n = \lim_{n\to-\infty} a_n = 0$. Next, we note that $x^n P$ is quasi-monic like P. By applying Lemma 5.2 to $x^n f_n$, we have a Euclidean division of $x^n f_n$ by $x^n P$ in the form $x^n f_n = x^n P g_n + S_n$ with $S_n \in \mathbb{K}[x]$, $\deg(S_n) < n + \deg(P)$, $\Psi(S_n, 0) \leq 0$, $\Psi(g_n, 0) \leq 0$. Now, by construction, $S_n = x^n(f_n - Pg_n)$, hence S_n is of the form $x^n L_n$, with $L_n \in \mathbb{K}[x]$, $\deg(L_n) < \deg(P)$, and $\Psi(L_n, 0) = \Psi(S_n, 0) \leq 0$. So, $f_n = Pg_n + L_n$.

Consequently, $f_{n+1} - f_n = P(g_{n+1} - g_n) + L_{n+1} - L_n$. By applying again Lemma 5.2 to $x^{n+1}(f_{n+1} - f_n)$, we can check that $\Psi(g_{n+1} - g_n, 0) \leq \log(u_n)$ and $\Psi(L_{n+1} - L_n, 0) \leq \log(u_n)$, hence both sequences $(g_n)_{n\in\mathbb{N}}$, $(L_n)_{n\in\mathbb{N}}$ converge in $H(C(0,1))$, and more precisely, the sequence (L_n) converges to a polynomial L of degree $< \deg(P)$. Moreover, setting $g = \lim_{n\to\infty} g_n$, clearly we have $f = Pg + L$ and $\Psi(g, 0) \leq 0$, $\Psi(L, 0) \leq 0$, which shows the existence of g and L in the first claim.

Now, let us check that they are unique satisfying these relations. Suppose we have $h \in H(C(0,1))$ and $S \in \mathbb{K}[x]$ satisfying the same properties, with particularly $f = Ph + S$, $\deg(S) < \deg(P)$. Then, $P(g - h) = S - L$. Since $\deg(S - L) < \deg(P)$, $S - L$ is an element of $H(C(0,1))$ having strictly less zeros than P in $C(0,1)$, a contradiction, except if $g = h$, hence $L = S$.

Now, assume that f lies in $H(d(0,1))$. Then, by Corollary 21.9, $f - L$ is a quasi-invertible element of $H(d(0,1))$ hence is of the form $Q\phi$ with ϕ invertible in $H(d(0,1))$ and $Q \in \mathbb{K}[x]$, having all its zeros in $d(0,1)$. Hence, $Pg = Q\phi$. This holds in $H(C(0,1))$. But since P has all its zeros in $C(0,1)$, P must divide Q: say $Q = PV$, with $V \in \mathbb{K}[x]$ having all its zeros in $d(0,1)$. So, $Pg = V\phi$, hence $g = V\phi$, and by hypothesis, both V, ϕ lie in $H(d(0,1))$, hence so does g, which completes the proof. □

Definition. Given $r \in |\mathbb{K}^*|$, the division of an element f of $H(C(0,r))$ by a polynomial P having all its zeros in $d(0,r)$, as defined in Theorem 27.12, is called *Euclidean division of f by P in $H(C(0,r))$* or *r-Euclidean division of f by P*.

Lemma 27.13. *Let $(u_n)_{n\in\mathbb{N}}$ be a sequence in $\mathbb{R}_+$ such that $\sum_{n=0}^{\infty} u_n < +\infty$. Let $A = \sum_{n=0}^{\infty} u_n$ and let $B > A$. There exists an increasing sequence $(q_n)_{n\in\mathbb{N}}$ such that $\lim_{n\to\infty} q_n = +\infty$ and such that $\sum_{n=0}^{\infty} q_n u_n \leq B$.*

Proof. Let $E = B - A$. For every $n \in \mathbb{N}$, we denote by s_n the smallest integer such that $\sum_{j=s_n}^{\infty} u_j \leq 4^{-n}E$. Then, for every $j \in \mathbb{N}$ such that $s_n \leq j < s_{n+1}$, we set $q_j = 2^n$. We have $\sum_{j=s_n}^{s_{n+1}-1} q_j u_j \leq 2^{-n}E$, hence $\sum_{j=0}^{s_{n+1}-1} q_j u_j \leq A + E\sum_{k=1}^{n} 2^{-k}$, and finally, $\sum_{j=0}^{\infty} q_j u_j \leq B$. □

Theorem 27.14. *Let $T = (a_n, q_n)_{n\in\mathbb{N}}$ be a divisor on the disk $d(a, R^-)$ with $a_n \neq 0$ $\forall n \in \mathbb{N}$ and let $\epsilon > 0$. There exists $f \in \mathcal{A}(d(a,R^-))$ such that $\mathcal{D}(f) \geq T$, $f(a) = 1$ and $|f|(r) \leq |T|(r)(1+\epsilon)$ $\forall r \in]0,R[$.*

Proof. Without loss of generality, we can assume $a = 0$. The set $\{|a_j| \mid j \in \mathbb{N}\}$ is obviously equal to the image of a strictly increasing sequence of limit R that we denote by $(r_m)_{m\in\mathbb{N}}$. For each $m \in \mathbb{N}$, the set S_m of the a_j lying in $C(0, r_m)$ is of the form $\{a_{h_m}, a_{h_{m+1}}, \dots, a_{k_m}\}$. We set $B = 1 + \epsilon$ and $P_m = \prod_{j=h_m}^{k_m} \left(1 - \frac{x}{a_j}\right)^{q_j}$.

We can construct a polynomial whose divisor is $(a_n, q_n)_{n<t}$. For every $m \in \mathbb{N}$, we set $\mu_m = \log r_m$, and $\lambda = \log(B)$.

Now, by Lemma 27.13, there exists an increasing sequence $(t_n)_{n\in\mathbb{N}}$ in $\mathbb{N}$ such that $\lim_{n\to\infty} t_n = +\infty$ and such that

$$(1) \quad \sum_{j=0}^{\infty} t_j(\mu_{j+1} - \mu_j) \leq \lambda.$$

For every $s \in \mathbb{N}$, we put $\tau(s) = \sum_{j=0}^{s} t_j(\mu_{j+1} - \mu_j)$, and $g_s = \prod_{m=0}^{s} P_m$. We note that $\Psi(g_q, \mu) = \Psi(g_s, \mu)$ whenever $\mu \leq \mu_q$ and $q \leq s$. So, we can define the function ℓ from $]-\infty, \log S[$ into $\mathbb{R}$ as $\ell(\mu) = \lim_{s\to\infty} \Psi(g_s, \mu)$. Then ℓ is an increasing function in μ that satisfies

$$(2) \quad \ell(\mu) = \sum_{j=0}^{k(m)} q_j(\mu - \Psi(a_j)) \text{ whenever } \mu \in [\mu_m, \mu_{m+1}].$$

We construct a sequence $(f_s)_{s\in\mathbb{N}}$ in $\mathbb{K}[x]$ satisfying the following relations $(\alpha_s), (\beta_s), (\gamma_s)$, and (δ_s) for every $s \in \mathbb{N}$ and $(\epsilon_s), (\varphi_s)$ for every $s \in \mathbb{N}^*$:

(α_s) $f_s(0) = 1$.
(β_s) P_j divides f_s for every $j \leq s$.
(γ_s) $\Psi(f_s, \mu_{s+1}) \leq \ell(\mu_{s+1}) + \tau(s)$.
(δ_s) $\Psi(f_s, \mu) \leq \ell(\mu) + \lambda$ whenever $\mu < \log S$.
(ϵ_s) $\Psi(f_s - f_{s-1}, \mu) \leq \ell(\mu) + t_s(\mu - \mu_s) + \lambda$ whenever $\mu \leq \mu_s$.
(φ_s) $\Psi(f_s - f_{s-1}, \mu) \leq \ell(\mu_s) + t_s(\mu - \mu_s) + \tau(s)$ whenever $\mu \in]\mu_s, \log S]$.

We proceed by induction and prove that when (α_s), (β_s), (γ_s), and (δ_s) are satisfied for $s \in \mathbb{N}$, then we can derive (α_{s+1}), (β_{s+1}), (γ_{s+1}), (δ_{s+1}), (ϵ_{s+1}), and (φ_{s+1}). By taking $f_0 = P_0$, we check that $(\alpha_0), (\beta_0), (\gamma_0)$, and (δ_0) are obviously satisfied. We now suppose already constructed f_m satisfying $(\alpha_m), (\beta_m), (\gamma_m)$, and (δ_m) for every $m = 0, \ldots, s$ and $(\epsilon_m), (\varphi_m)$ for every $m = 1, \ldots, s$. We define f_{s+1} satisfying $(\alpha_{s+1}), (\beta_{s+1}), (\gamma_{s+1}), (\delta_{s+1}), (\epsilon_{s+1})$, and (φ_{s+1}). It is seen that each polynomial P_s has all its zeros in $C(0, r_s)$.

Let R_{s+1} be the rest of the Euclidean division of f_s by P_{s+1} in $H(C(0, r_{s+1}))$. Let $Q_{s+1} = x^{t_{s+1}} g_s$. We have

$$(3) \quad \Psi(Q_{s+1}, \mu) = \mu t_{s+1} + \Psi(g_s, \mu) \text{ whenever } \mu \in \mathbb{R}.$$

We note that Q_{s+1} admits no zero in $C(0, r_{s+1})$ and then is invertible in $H(C(0, r_{s+1}))$. As a consequence, according to Theorem 27.12, we can perform the r_{s+1}-Euclidean division of $\frac{R_{s+1}}{Q_{s+1}}$ by P_{s+1} in $H(C(0, r_{s+1}))$. Let V_{s+1} be the rest of this division. Thus, $\frac{R_{s+1}}{Q_{s+1}}$ is of the form $T_{s+1}P_{s+1} + V_{s+1}$ with $T_{s+1} \in H(C(0, r_{s+1}))$ and

$$(4) \quad \Psi(Q_{s+1}V_{s+1}, \mu_{s+1}) \leq \Psi(R_{s+1}, \mu_{s+1}),$$

we have $R_{s+1} = Q_{s+1}(T_{s+1}P_{s+1} + V_{s+1})$. Now we put $f_{s+1} = f_s - Q_{s+1}V_{s+1}$. Of course, f_{s+1} satisfies (α_{s+1}). We check that f_{s+1} satisfies (β_{s+1}). By definition, each P_j divides g_s for every $j = 0, \dots, s$, and hence it divides Q_{s+1}. Next, by (β_s), P_j also divides f_s. Consequently, P_j divides f_{s+1} for every $j = 0, \dots, s$. Moreover, P_{s+1} divides both $f_s - R_{s+1}$ and $Q_{s+1}T_{s+1}P_{s+1}$. Hence it also divides f_{s+1} and thereby (β_{s+1}) is satisfied.

Now, we prove (φ_{s+1}). By (4), R_{s+1} satisfies $\Psi(R_{s+1}, \mu_{s+1}) \leq \Psi(f_s, \mu_{s+1})$. Hence, by Relation (γ_s), we have

(5) $\Psi(R_{s+1}, \mu_{s+1}) \leq \ell(\mu_{s+1}) + \tau(s) = \Psi(g_{s+1}, \mu_{s+1}) + \tau(s)$.

Since $\deg(R_{s+1}) < \deg(P_{s+1}) < \deg(g_{s+1})$ and since all zeros of g_{s+1} lie in $d(0, r_{s+1})$, g_{s+1} has more zeros than R_{s+1} in $d(0, r_{s+1})$, and therefore, by Theorem 22.26, we have

$$\Psi(R_{s+1}, \mu) - \Psi(R_{s+1}, \mu_{s+1}) \leq \Psi(g_{s+1}, \mu) - \Psi(g_{s+1}, \mu_{s+1})$$

whenever $\mu \in]\mu_{s+1}, \log S]$, and therefore, by (5), we obtain

(6) $\Psi(R_{s+1}, \mu) \leq \Psi(g_{s+1}, \mu) + \tau(s)$ whenever $\mu \in]\mu_{s+1}, \log S]$.

Since $\frac{R_{s+1}}{Q_{s+1}} = T_{s+1}P_{s+1} + V_{s+1}$, by Theorem 27.12, we have $\Psi(Q_{s+1}V_{s+1}, \mu_{s+1})) \leq \Psi(R_{s+1}, \mu_{s+1})$, and hence, by (5), $\Psi(Q_{s+1}V_{s+1}, \mu_{s+1}) \leq \ell(\mu_{s+1}) + \tau(s)$. But $\ell(\mu_{s+1}) = \Psi(g_{s+1}, \mu_{s+1})$, hence by (6), we obtain

(7) $\Psi(Q_{s+1}V_{s+1}, \mu_{s+1}) \leq \Psi(g_{s+1}, \mu_{s+1}) + \tau(s)$.

We note that $\deg(Q_{s+1}V_{s+1}) < \deg(g_{s+1}) + t_{s+1}$ and that all zeros of g_{s+1} lie in $d(0, r_{s+1})$. Hence, by Theorem 22.26 and by (3) and (7), we have

(8) $\Psi(Q_{s+1}V_{s+1}, \mu) \leq \Psi(g_{s+1}, \mu) + t_{s+1}(\mu - \mu_{s+1}) + \tau(s)$ for every $\mu \in]\mu_{s+1}, \log S]$.

Actually, by definition of f_{s+1}, we have $\Psi(Q_{s+1}V_{s+1}, \mu) = \Psi(f_{s+1} - f_s, \mu)$ and $\Psi(g_{s+1}, \mu) \leq \ell(\mu)$ for every $\mu \in]\mu_{s+1}, \log S]$, hence by (7), we have proved φ_{s+1}. We deduce (ϵ_{s+1}).

In particular, when $\mu = \mu_{s+2}$, we obtain

$$\Psi(Q_{s+1}V_{s+1}, \mu_{s+2}) \leq \ell(\mu_{s+2}) + t_{s+1}(\mu_{s+2} - \mu_{s+1}) + \tau(s).$$

But we note that $t_{s+1}(\mu_{s+2} - \mu_{s+1}) + \tau(s) = \tau(s+1)$, hence

(9) $\quad \Psi(Q_{s+1}V_{s+1}, \mu_{s+2}) \leq \ell(\mu_{s+2}) + \tau(s+1)$.

And, by (8) and (1), we obtain

(10) $\quad \Psi(f_{s+1} - f_s, \mu) \leq \ell(\mu) + \lambda$ whenever $\mu \in]\mu_{s+1}, \log S]$.

Now we take $\mu \leq \mu_{s+1}$. It is seen that

$$\Psi(Q_{s+1}, \mu) - \Psi(Q_{s+1}, \mu_{s+1}) = \ell(\mu) - \ell(\mu_{s+1}) + t_{s+1}(\mu - \mu_{s+1}).$$

Therefore, we have

$$\begin{aligned}\Psi(Q_{s+1}V_{s+1}, \mu) \leq \Psi(Q_{s+1}V_{s+1}, \mu_{s+1}) + \ell(\mu) - \ell(\mu_{s+1}) \\ + t_{s+1}(\mu - \mu_{s+1}).\end{aligned}$$

But by (8), we have $\Psi(Q_{s+1}V_{s+1}, \mu_{s+1}) \leq \ell(\mu_{s+1}) + \lambda$, hence we obtain $\Psi(Q_{s+1}V_{s+1}, \mu) \leq \ell(\mu) + \lambda + t_{s+1}(\mu - \mu_{s+1})$ whenever $\mu \leq \mu_{s+1}$, and this is (ϵ_{s+1}). In particular, we have $\Psi(f_{s+1} - f_s, \mu) \leq \ell(\mu) + \lambda$ whenever $\mu \leq \mu_{s+1}$, and therefore, by (δ_s), we obtain (δ_{s+1}).

Now, we show (γ_{s+1}). Obviously, we have

(10) $\quad \Psi(f_0, \mu_{s+2}) \leq \ell(\mu_{s+2})$.

Next, by Relations $(\varphi_m)_{1 \leq m \leq s+1}$ for every $m \in \mathbb{N}^*$, we have

$$\Psi(f_m - f_{m-1}, \mu_{s+2}) \leq \ell(\mu_m) + [t_m(\mu_{s+2} - \mu_m) + \tau(m)].$$

But as the sequence $(t_m)_{m \in \mathbb{N}}$ is increasing, it is seen that

$$\tau(m) + t_m(\mu_{s+2} - \mu_m) \leq \sum_{j=0}^{s+1} t_j(\mu_{j+1} - \mu_j) = \tau(s+1).$$

Obviously, $\ell(\mu_{s+2}) \geq \ell(\mu_m)$, hence we obtain $\Psi(f_m - f_{m-1}, \mu_{s+2}) \leq \ell(\mu_{s+2}) + \tau(s+1)$ whenever $m = 1, \ldots, s+1$. Finally, by (10), f_{s+1} satisfies $\Psi(f_{s+1}, \mu_{s+2}) \leq \ell(\mu_{s+2}) + \tau(s+1)$ and this is (γ_{s+1}).

We note that (ϵ_s) and (φ_s) are not used to prove

$$(\alpha_{s+1}),\ (\beta_{s+1}),\ (\gamma_{s+1}),\ (\delta_{s+1}),\ (\epsilon_{s+1}),\ (\varphi_{s+1}).$$

Consequently, (ϵ_1) and (φ_1) are clearly proven by (α_0), (β_0), (γ_0), and (δ_0), and therefore, we are now done with the recurrence. Therefore, we can now construct the sequence $(f_s)_{s\in\mathbb{N}}$ satisfying (α_s), (β_s), (γ_s), (δ_s), (ϵ_s), and (φ_s). By Relations (ϵ_s), the sequence is easily seen to converge in each algebra $H(d(0,u))$ whenever $u \in]0,S[$. Indeed, given $u \in]0,S[$ and $N \in \mathbb{N}$ such that $\mu_N < \log u$, by (ϵ_{s+1}), we have

$$\begin{aligned}\log(\|f_{s+1} - f_s\|_{d(0,u)}) &= \Psi(f_{s+1} - f_s, \log u)\\ &\le \ell(\log u) - t_s(\log u - \mu_{s+1}) + \lambda,\end{aligned}$$

hence $\log(\|f_{s+1} - f_s\|_{d(0,u)}) \le \ell(\log u) - t_s(\log u - \mu_N) + \lambda$, whenever $s > N$. As $\lim_{s\to+\infty} t_s = +\infty$, it is seen that $\lim_{s\to\infty} \|f_{s+1} - f_s\|_{d(0,u)} = 0$.

Let f be the function defined in $d(0,R^-)$ as the limit of the sequence $(f_s)_{s\in\mathbb{N}}$ in each disk $d(0,u)$. Obviously, as an element of $H(d(0,u))$ for every $u \in]0,S[$, f belongs to $\mathcal{A}(d(0,R^-))$. By Relations (α_s), f satisfies $f(0) = 1$.

We check $|f|(r) \le B|T|(r)$. Let $u \in]0,R[$ be such that $\mu_N \le \log u \le \mu_{N+1}$. We have $\log \|f\|_{d(0,u)} = \Psi(f, \log u)$.

When s is big enough, $\Psi(f_s, \log u)$ is clearly equal to $\Psi(f, \log u)$, hence f satisfies $\log \|f\|_{d(0,u)} = \Psi(f_s, \log u) \le \ell(\log u) + \lambda$. Hence, by (2), we obtain $|f|(r) \le B|T|(r)$. Now, we just have to check that every a_j is a zero of f of order $z_j \ge q_j$. Let m be such that $h_m \ge j$. For every $s \ge m$, $(1 - \frac{x}{a_j})^{q_j}$ divides f_s in $H(d(0,u))$ (for every $u \in]0,S]$), hence by Lemma 16.1, $(1 - \frac{x}{a_j})^{q_j}$ divides f in $H(d(0,u))$ and this finishes the proof of Theorem 27.14. □

We can obtain a small improvement of Theorem 27.14:

Theorem 27.15. *Let $T = (a_n, q_n)_{n\in\mathbb{N}}$ be a divisor on the disk $d(a,R^-)$ with $a_n \neq 0$ $\forall n \in \mathbb{N}$, and let $\epsilon > 0$ and $\rho \in]0,R[$. There exists $g \in \mathcal{A}(d(a,R^-))$ such that $\mathcal{D}(g) \ge T$, $g(a) = 1$ and $|g|(r) \le |T|(r)(1+\epsilon)$ $\forall r \in]0,R[$ and $\mathcal{D}(g)(\alpha) = T(\alpha)$ $\forall \alpha \in d(a,\rho)$.*

Proof. By Theorem 27.14, we have a function $f \in \mathcal{A}(d(a, R^-))$ such that $\mathcal{D}(f) \geq T$, $f(a) = 1$ and $|f|(r) \leq |T|(r)(1+\epsilon)\ \forall r \in]0, R[$. Now, we can construct a polynomial $P(x)$ such that $P(0) = 1$ and admitting in $d(a, R^-)$ a divisor $\mathcal{D}(P)$ satisfying $\mathcal{D}(P)(\alpha) = \frac{\mathcal{D}(f)(\alpha)}{T(\alpha)}\ \forall \alpha \in d(a, \rho)$ and $\mathcal{D}(P)(\alpha) = 0\ \forall \alpha \in d(a, R^-) \setminus d(a, \rho)$. Then the function $g = \frac{f}{P}$ satisfies $\mathcal{D}(g)(\alpha) = T(\alpha)\ \forall \alpha \in d(a, \rho)$, $T \leq \mathcal{D}(g) \leq \mathcal{D}(f)$, and hence $|g|(r) \leq |f|(r) \leq |T|(r)(1+\epsilon)\ \forall r \in]0, R[$.

□

Remark. Here we may note that $H(d(0, R^-))$ is much smaller than $\mathcal{A}_b(d(0, R^-))$. Indeed, by Theorem 27.15, there exist functions $f \in \mathcal{A}_b(d(0, R^-))$ having infinitely many zeros in $d(0, R^-)$. But, by Theorem 21.8, any element of $H(d(0, R^-))$ is quasi-invertible and hence has finitely many zeros.

Chapter 28

Michel Lazard's Problem

This chapter is aimed at studying the following problem mentioned in Chapter 26 and first considered by Lazard in a tremendous work [72]. Let T be a divisor on a disk $d(a, R^-)$. Does there exist a function $f \in \mathcal{A}(d(a, R^-))$ such that $\mathcal{D}(f) = T$? The answer depends on whether or not $\mathbb{K}$ is spherically complete.

Theorems 28.1 and 28.4 were first proved in [72]. Proofs are long and much technical. Here we try to give an easier presentation of the proofs which is due to Labib Haddad. More precisely, when $\mathbb{K}$ is spherically complete, Theorem 28.4 shows the following result, as it was done in [72]: let $T = (a_n, q_n)$ be a divisor on $d(a, R^-)$, with $|a_n| \leq |a_{n+1}|$ and $|a_{u(m)}| < |a_{u(m)+1}|$. For each $m \in \mathbb{N}$, let $P_m(x) = \prod_{j=u(m)+1}^{u(m+1)}(x - a_j)$ and let $(Q_m) \in \mathbb{K}[x]$ be such that $|Q_m|(\rho_m) \leq |T|(\rho_m)$. Then there exists a function $f \in \mathcal{A}(d(a, R^-))$ such that P_m divides $f - Q_m$. Hence, in particular, given a divisor T on $d(a, R^-)$, there exists a functions f analytic in $d(a, R^-)$, whose divisor is T.

Theorem 28.1. *Let $\mathbb{K}$ be not spherically complete and let $(D_n)_{n\in\mathbb{N}}$ be a decreasing sequence of disks $d(u_n, \rho_n)$ such that $\bigcap_{n=0}^{\infty} D_n = \emptyset$. Let $R = \frac{1}{\lim_{n\to\infty} \operatorname{diam}(D_n)}$. There exists sequences $(c_n)_{n\in\mathbb{N}}$ of $d(0, R^-)$ such that $\lim_{n\to\infty} |c_n| = R$ and such that no function $f \in \mathcal{A}(d(0, R^-))$ admits for divisor the divisor $T = (c_n, 1)_{n\in\mathbb{N}}$.*

Proof. Without loss of generality, we may obviously assume that $R > 1$, hence $\frac{1}{R} < 1 < R$. Consequently, we may assume that $D_0 \subset d(0, R^-)$. For each $n \in \mathbb{N}$, we set $\rho_n = \operatorname{diam}(D_n)$, hence $\rho_n > \rho_{n+1}$.

For every $n \in \mathbb{N}$, we can take $\alpha_n \in D_n \setminus D_{n+1}$. Let $\beta_n = \alpha_{n+1} - \alpha_n$, $n \in \mathbb{N}$, hence

$$(1) \quad \rho_{n+1} < |\beta_n| \leq \rho_n.$$

Consider the divisor $T = (\frac{1}{\beta_n}, 1)_{n\in\mathbb{N}}$ and suppose that there exists $f \in \mathcal{A}(d(0, R^-))$ whose divisor is exactly T. Without loss of generality, we can assume that $f(x) = \prod_{n\in\mathbb{N}}(1 - \beta_n x)$. Then $f(x)$ is a series of the form $1 + \sum_{n=1}^{\infty} a_n x^n$. We show that $\alpha_1 - a_1 \in D_n \ \forall n \geq 1$.

Let us fix $n \in \mathbb{N}$. We can check that $\alpha_n - \alpha_1 = \sum_{j=1}^{n-1} \beta_j$, hence $\alpha_n - (\alpha_1 - a_1) = a_1 + \sum_{j=1}^{n-1} \beta_j$. Since $\alpha_n \in D_n$, then $\alpha_1 - a_1$ lies in D_n if and only if $|a_1 + \sum_{j=1}^{n-1} \beta_j| \leq \rho_n$.

For all $n \in \mathbb{N}$, we set $t_n = |(\beta_n)^{-1}|$. By (1), the sequence (t_n) is strictly increasing and satisfies

$$(2) \quad |a_n| = \prod_{j=1}^{n} t_j \ \forall n \geq 1.$$

Particularly, $a_n \neq 0 \ \forall n \geq 1$. For each $s \geq 1$, we put $f_s = \sum_{n=0}^{s} a_n x^n$. Then, $\deg(f_s) = s$. By (2), we see that the s zeros of f_s are distinct and are of the form γ_j^s, $1 \leq j \leq s$, with $|\gamma_j^s| = t_j$. Thus, f_s is of the form $\prod_{j=1}^{s}(1 - \beta_{s,j} x)$, with $|\beta_{s,j}| = |\beta_j| = t_j^{-1}$. By identification of coefficients, we obtain $a_1 = -\sum_{j=1}^{s} \beta_{s,j}$. Consequently, when $n \leq s$, we have

$$(3) \quad \sum_{j=1}^{s} \beta_j - \beta_{s,j} = a_1 + \sum_{j=1}^{s} \beta_j.$$

Let us fix $n \in \mathbb{N}^*$. By Theorem 24.3, we can see that for each $j = 1, \ldots, n$, we have $\lim_{s\to\infty} \beta_{s,j} = \beta_j$. Consequently, when s is big enough, we can see that $|\sum_{j=1}^{n-1} \beta_j - \sum_{j=1}^{s} \beta_{s,j}| \leq \rho_n$. Therefore, by (3), we have $|a_1 + \sum_{j=1}^{n-1} \beta_j| \leq \rho_n$.

Consequently, $\alpha_1 - a_1$ lies in D_n. This is true for every $n \in \mathbb{N}$, a contradiction to the hypothesis $\bigcap_{n=0}^{\infty} D_n = \emptyset$. □

In order to prove Theorem 28.4, we must introduce a set of notations that hold throughout this chapter.

Notation. We consider a divisor T on $d(0,R^-)$ of the form $(a_{i,m,},q_{i,m})_{i\leq u_m, m\in\mathbb{N}}$, where the points $a_{i,m}$ lie in the circle $C(0,\rho_m)$ with $0<\rho_m<\rho_{m+1}\ \forall m\in\mathbb{N}$ and $\lim_{m\to+\infty}\rho_m-R$. We denote by $(P_m)_{m\in\mathbb{N}}$ the polynomial $P_m=\prod_{i=1}^{u_m}\left(1-\frac{x}{a_{i,m}}\right)^{q_{i,m}}$ whose zeros by definition belong to the circle $C(0,\rho_m)$, and for each $m\in\mathbb{N}$, we set $d_m=\deg(P_m)$.

We denote by $(Q_m)_{m\in\mathbb{N}}$ a sequence of $\mathbb{K}[x]$ satisfying $|Q_m|(\rho_m)\leq|T|(\rho_m)$, $\deg(Q_m)<d_m\ \forall m\in\mathbb{N}$. We note that $|T|(r)=\prod_{m\geq 1}^{s-1}\left(\frac{r}{\rho_m}\right)^{d_m}$ whenever $r\in[\rho_{s-1},\rho_s]$. Given $q\in\mathbb{N}$, $s\in\mathbb{N}$, $r\in]0,R[$, we set

$$\lambda(q,s,r)=|T|(r)\left(\frac{r}{R}\right)^q\min\left(1,\left(\frac{r}{\rho_s}\right)\right),\text{ i.e.}$$

$$\lambda(q,s,r)=|T|(r)\left(\frac{r}{R}\right)^q\left(\frac{r}{\rho_s}\right)\ \forall r\in]0,\rho_s],$$

$$\lambda(q,s,r)=|T|(r)\left(\frac{r}{R}\right)^q\ \forall r\in[\rho_s,R[.$$

Particularly, we note that $\lambda(q,s,r)\leq|T|(r)\ \forall r<R$. Now, given $n\in\mathbb{N}$, we set $v(q,s,n)=\inf_{r<R}\left(\frac{\lambda(q,s,r)}{r^n}\right)$.

Recall that the Euclidean division in $H(C(0,r))$ is defined in Chapter 27. We denote by $\Lambda(q,s)$ the subset of the $h\in\mathcal{A}(d(0,R^-))$ satisfying the following:

(α) $h(0)=1$.
(β) $|h|(r)\leq|T|(r)\ \forall r<R$.
(γ) P_m divides $h-Q_m$ in $\mathcal{A}(d(0,R^-))\ \forall m=0,\ldots,s$.
(δ) The rest X_m of the Euclidean division of h by P_m in $H(C(0,\rho_m))$ satisfies $|X_m-Q_m|(\rho_m)\leq\lambda(q,m,\rho_m)=|T|(\rho_m)\left(\frac{\rho_m}{R}\right)^q\ \forall m\geq 0$.

Remark. By definition, q and n being fixed, the sequence in s: $(v(q,s,n))_{s\in\mathbb{N}}$ is decreasing.

In the proof of Theorem 28.4, we use Lemmas 28.2 and 28.3. Lemma 28.2 is immediate:

Lemma 28.2. *$\Lambda(q+1,s)\subset\Lambda(q,s)$ and $\Lambda(q,s)$ is a closed subset of $\mathcal{A}(d(0,R^-))$.*

Lemma 28.3. *Let $q\in\mathbb{N}$, $s\in\mathbb{N}^*$, $f\in\Lambda(q,s-1)$. There exists $g\in\Lambda(q,s)$ such that $|g-f|(r)\leq\lambda(q,s,r)\ \forall r<R$.*

Proof. Let $h = \prod_{m=0}^{s-1} P_m$ and $u = x^{q+1}h$. Then P_m and x have no zero in $C(0, \rho_s)$, hence they are invertible in $H(C(0, \rho_s))$ and hence so is u. For each $m \in \mathbb{N}$, we denote by R_m the rest of the Euclidean division of f by P_m in $H(C(0, \rho_m))$. Since $f \in \Lambda(q, s-1)$, by (γ), P_m divides $f - Q_m$ for every $m \leq s-1$, hence $R_m = Q_m \ \forall m = 0, \dots, s-1$.

Consider now the Euclidean division of $(Q_s - R_s)u^{-1}$ by P_s in $H(C(0, \rho_s))$: $(Q_s - R_s)u^{-1} = EP_s + S$, with $\deg(S) < d_s$, $E \in H(C(0, \rho_s))$ and $|S(\rho_s)| \leq |(Q_s - R_s)u^{-1}|(\rho_s)$, hence

(1) $|Su|(\rho_s) \leq |(Q_s - R_s)|(\rho_s)$.

We then take $g = f + Su$, hence $g = f + Shx^{q+1}$. We show that g belongs to $\Lambda(q, s)$ and that $|g - f|(r) \leq \lambda(q, s, r) \ \forall r < R$. We note that g belongs to $\mathcal{A}(d(0, R^-))$ and that $g(0) = f(0) = 1$.

Next, by hypothesis, P_s divides $f - R_s$ and by construction divides $Su + R_s - Q_s$. But $g - Q_s = f + Su - Q_s = f - R_s + Su + R_s - Q_s \in \mathcal{A}(d(0, R^-))$, hence P_s divides $g - Q_s$ in $\mathcal{A}(d(0, R^-))$.

Now, by hypothesis, f lies in $\Lambda(q, s-1)$, hence particularly $|f|(r) \leq |T|(r)$, and we have

(2) $|Q_m - R_m|(\rho_m) \leq |T|(\rho_m)\left(\frac{\rho_m}{R}\right)^q \ \forall m \geq 1.$

Thus, by (1) and (2), we have

(3) $|g - f|(\rho_s) \leq |T|(\rho_s)\left(\frac{\rho_s}{R}\right)^q.$

But since $|S|$ is an increasing function in r, we have $|S|(r) \leq |S|(\rho_s)$ whenever $r \leq \rho_s$. On the other hand, $|h|(r) = |T|(r) \ \forall r \leq \rho_s$. And, as we saw,

$$\begin{aligned} |g - f|(\rho_s) &= |Su|(\rho_s) = |Shx^{q+1}|(\rho_s) = |S|(\rho_s)|T|(\rho_s)(\rho_s)^{q+1} \\ &\leq |T|(\rho_s)\left(\frac{\rho_s}{R}\right)^q. \end{aligned}$$

Consequently, $|S|(r) \leq |S|(\rho_s) \leq \frac{1}{R^q \rho_s} \ \forall r \leq \rho_s$, and hence

$$\begin{aligned} |g - f|(r) &= |Shx^{q+1}|(r) = |S|(r)|h|(r)r^{q+1} \\ &\leq |T|(r)\left(\frac{\rho_s}{R}\right)^q\left(\frac{r}{\rho_s}\right) \forall r \leq \rho_s. \end{aligned} \quad (4)$$

Now, when $r \geq \rho_s$, we have

$$|T|(r) = \prod_{m\geq 1} |P_m|(r) - |h|(r) \prod_{m\geq s} |P_m|(r) \geq |h|(r)|P_s|(r)$$
$$= |h|(r)\Big(\frac{r}{\rho_s}\Big)^{d_s}.$$

Next, Sx^{q+1} is of the form $\lambda \prod_{j=1}^{m}(x - x_j)$ and is a polynomial of degree $m \leq q + d_o$, hence

$$\frac{|x^{q+1}S|(r)}{|x^{q+1}S|(\rho_s)} = \prod_{j=1}^{m} \frac{|x - x_j|(r)}{|x - x_j|(\rho_s)} \leq \Big(\frac{r}{\rho_s}\Big)^{q+d_s}.$$

Thus,

$$|g - f|(r) = |Shx^{q+1}|(r) = |Sx^{q+1}|(r)|h|(r)$$
$$\leq |Sx^{q+1}|(\rho_s)\Big(\frac{r}{\rho_s}\Big)^{q+d_s}\Big(\frac{\rho_s}{r}\Big)^{d_s}|T|(r)$$
$$= |T|(r)\Big(\frac{r}{\rho_s}\Big)^{q}|Sx^{q+1}|(\rho_s) = |T|(r)\Big(\frac{r}{\rho_s}\Big)^{q}(\rho_s)^{q+1}|S|(\rho_s)$$
$$\leq |T|(r)r^q(\rho_s)\frac{1}{R^q\rho_s} = |T|(r)\Big(\frac{r}{R}\Big)^{q}.$$

And finally, with (4), we obtain $|g - f|(r) \leq \lambda(q, s, r)\ \forall r \in]0, R[$. □

Theorem 28.4. *Suppose $\mathbb{K}$ is spherically complete. Assume that $|Q_m|(\rho_m) \leq |T|(\rho_m)\ \forall m \in \mathbb{N}$. Let $R \in]0, +\infty[$. There exists $f \in \mathcal{A}(d(0, R^-))$ satisfying the following:*

(i) $f(0) = 1$.
(ii) $|f|(r) \leq |T|(r)\ \forall r < R$.
(iii) *P_m divides $f - Q_m$ in $\mathcal{A}(d(0, R^-))$.*

Proof. We mean to construct a sequence of functions $(f_q)_{q\in\mathbb{N}}$ which belong to $\mathcal{A}(d(0, R^-))$, converging in $\mathcal{A}(d(0, R^-))$ to a function f satisfying the claim.

We first fix $q \in \mathbb{N}$ and take $f_q \in \Lambda(q, 0)$. Let $f_q(x) = \sum_{n=0}^{\infty} a_n x^n$. We construct a sequence $(g_{q,s})_{s\in\mathbb{N}}$ satisfying $g_{q,s} \in \Lambda(q, s)$ and $|g_{q,s} - g_{q,s-1}|(r) \leq \lambda(q, s, r)$, with $g_{q,0} = f_q$. Suppose already constructed the $g_{q,j}$ for $j = 0, \dots, s-1$. By Lemma 28.3, there exists $h \in \Lambda(q,s)$ such that $|h - g_{q,s-1}|(r) \le \lambda(q,s,r)$. So, we can set $g_{q,s} = h$ and the sequence is then defined by induction for all $s \in \mathbb{N}$.

Now, for each $s \in \mathbb{N}$, we set $g_{q,s}(x) = \sum_{n=0}^{\infty} b_{q,s,n}x^n$. Since by construction the sequence $g_{q,s}$ satisfies

(1) $|g_{q,s} - g_{q,s-1}|(r) \le \lambda(q,s,r) \forall r < R$,

then for each fixed $n \in \mathbb{N}$, the sequence $(b_{q,s,n})_{s\in\mathbb{N}}$ satisfies $|b_{q,s,n} - b_{q,s-1,n}| \le \upsilon(q,s,n)$. Thus, for each fixed $n \in \mathbb{N}$, we consider the sequence of disks $(D_{s,n})_{s\in\mathbb{N}}$ defined as $D_{q,s,n} = d(b_{q,s,n}, \upsilon(q,s,n))$. Since the sequence $(\upsilon(q,s,n))_{s\in\mathbb{N}}$ is decreasing and since $|b_{q,s,n} - b_{q,s-1,n}| \le \upsilon(q,s,n)$, the sequence of disks $(D_{q,s,n})_{s\in\mathbb{N}}$ is decreasing with respect to the inclusion. Consequently, since $\mathbb{K}$ is spherically complete, for each $n \in \mathbb{N}$, there exists $a_{q+1,n} \in \bigcap_{s=0}^{\infty} D_{q,s,n}$. Particularly, since $\upsilon(q,s,n) = 0 \ \forall n \le q$, we note that $b_{q,s,0} = b_{q,s-1,0} = 1$ because $g_{q,s-1} \in \Lambda(q,s-1)$. Consequently, $a_{q+1,0} = 1$.

Now, we show that f_{q+1} belongs to $\Lambda(q+1,0)$. Let $f_{q+1}(x) = \sum_{n=0}^{\infty} a_{q+1,n}x^n$. Since $a_{q+1,0} = 1$, f_{q+1} satisfies Relation (α). Next, by construction, we have $|a_{q+1,n} - b_{q,s-1,n}| \le \upsilon(q,s,n) \ \le \upsilon(q,1,n) \ \forall n \in \mathbb{N}$, hence obviously, $|a_{q+1,n} - a_{q,n}| \le \upsilon(q,1,n) \ \forall n \in \mathbb{N}$. Consequently, $|a_{q+1,n} - a_{q,n}|r^n \le \lambda(q,1,r) \ \forall r < R$, hence

(2) $|f_{q+1} - f_q|(r) \le \lambda(q,1,r) \le |T|(r)\left(\frac{r}{R}\right)^q \ \forall r < R$.

Now since, by hypothesis, $|f_q|(r) \le |T|(r)$, by (2), we can see that $|f_{q+1}|(r) \le |T|(r)$, and therefore, f_{q+1} satisfies Relation (β). Since (γ) is trivial when $s = 0$, it only remains to show that f_{q+1} satisfies (δ).

For each $m \in \mathbb{N}$, let $S_{m,q+1}$ be the rest of the Euclidean division of f_{q+1} by P_m in $H(C(0,\rho_m))$. For each $s \ge m$, since P_m divides $g_{q,s} - Q_m$, the rest of the Euclidean division of $g_{q,s} - f_{q+1}$ by P_m in $H(C(0,\rho_m))$ is equal to $Q_m - S_{m,q+1}$. Consequently, by (1), we have

(3) $|Q_m - S_{m,q+1}|(\rho_m) \le |g_{q,s} - f_{q+1}|(\rho_m) \le \lambda(q,s,\rho_m)$,

and hence

(4) $|Q_m - S_{m,q+1}|(\rho_m) \le |T|(\rho_m)\left(\frac{\rho_m}{R}\right)^q \left(\min\left(1, \frac{\rho_m}{\rho_s}\right)\right) \ \forall s \ge m$.

Now, since $\lim_{s\to\infty}\rho_s = R$, by (4), we have $|Q_m - S_{m,q+1}|(\rho_m) \leq |T|(\rho_m)\big(\frac{\rho_m}{R}\big)^{q+1}$. This finishes showing that (δ) is satisfied by $g_{q+1,s}$, and therefore, $g_{q+1,s}$ belongs to $\Lambda(q+1,0)$. This true for all s, hence by Lemma 28.2, f_{q+1} also belongs to $\Lambda(q+1,0)$.

Thus, we have constructed a sequence $(f_q)_{q\in\mathbb{N}}$ of $\mathcal{A}(d(0,R^-))$ satisfying $f_q \in \Lambda(q,0)\forall q \in \mathbb{N}$. By (2), we can see that the sequence $(f_q)_{q\in\mathbb{N}}$ converges in all $H(d(0,\rho))$, for every $\rho < R$, to a limit f which thereby belongs to $H(d(0,\rho))$ for all $\rho < R$. Consequently, that function f belongs to $\mathcal{A}(d(0,R^-))$. Moreover, since $\Lambda(q,0)$ is closed, f belongs to $\Lambda(q,0)$ for every $q\in\mathbb{N}$. Consequently, by Relation (δ) true for every q, the rest X_m of the Euclidean division of f_q by P_m in $H(C(0,\rho_m))$ satisfies $|X_m - Q_m|(\rho_m) \leq |T|(\rho_m)\big(\frac{\rho_m}{R}\big)^q$ for every $q \in \mathbb{N}$, hence $X_m = Q_m$. So, P_m divides $f - Q_m$ for every $m \in \mathbb{N}$. And by construction, f satisfies (α) and (β), which completes the proof. □

Corollary 28.5. *Suppose $\mathbb{K}$ is spherically complete. Let $(a_n)_{n\in\mathbb{N}}$ be a sequence of $d(0,R^-)$ such that $|a_n| \leq |a_{n+1}|$ $\forall n \in \mathbb{N}$ and $\lim_{n\to+\infty}|a_n| = R$ and let $(b_n)_{n\in\mathbb{N}}$ be a sequence of $\mathbb{K}$. There exists $f \in \mathcal{A}(d(a,R^-))$ such that $f(a_n) = b_n$ $\forall n \in \mathbb{N}$.*

Proof. Indeed, we can define a sequence of integers $(s_n)_{n\in\mathbb{N}}$ such that the divisor $T = (a_n, s_n)_{n\in\mathbb{N}}$ satisfies $|T|(|a_n|) \geq |b_n|$ $\forall n \in \mathbb{N}$. □

Theorem 28.6. *Suppose $\mathbb{K}$ is spherically complete. Let T be a divisor on $d(a,R^-)$. There exists $f \in \mathcal{A}(d(a,R^-))$ such that $\mathcal{D}(f) = T$.*

Proof. Without loss of generality, we may obviously assume $a = 0$. Take $Q_m = 0$ $\forall m \in \mathbb{N}$. By Theorem 28.4, there exists $f \in \mathcal{A}(d(0,R^-))$ such that

(i) $f(0) = 1$,
(ii) $|f|(r) \leq |T|(r)$ $\forall r < R$,
(iii) P_m divides f in $\mathcal{A}(d(0,R^-))$.

By (iii), clearly, $\mathcal{D}(f) \geq T$. Thus, we only have to check that $\mathcal{D}(f) \leq T$. Indeed, for all $s \in \mathbb{N}$, we have

$$|T|(\rho_s) = \prod_{j=1}^{s}\prod_{i=1}^{u_m}\left|\left(1 - \frac{x}{a_{i,m}}\right)^{q_{i,m}}\right|(\rho_s) = \prod_{j=1}^{s}\prod_{i=1}^{u_m}\left(\frac{\rho_s}{\rho_m}\right)^{q_{i,m}}.$$

Now, suppose that $T \neq \mathcal{D}(f)$. Then there exists $\alpha \in d(0, R^-)$ such that $\omega_\alpha(f) > T(\alpha)$. Let s be such that $\rho_s > |\alpha|$. Since $f(0) = 1$, we have

$$|f|(\rho_s) \geq \frac{\rho_s}{|\alpha|} \prod_{j=1}^{s} \prod_{i=1}^{u_m} \left| \left(1 - \frac{x}{a_{i,m}}\right)^{q_{i,m}} \right| (\rho_s) > \prod_{j=1}^{s} \prod_{i=1}^{u_m} \left(\frac{\rho_s}{\rho_m}\right)^{q_{i,m}} = |T|(\rho_s),$$

a contradiction to (iii). □

Similar to $\mathcal{A}(\mathbb{K})$, the algebra $\mathcal{A}(d(a, R^-))$ is provided with the natural topology of uniform convergence on each disk $d(0, r)$ whenever $0 < r < R$. Such a topology makes $\mathcal{A}(d(a, R^-))$ a topological $\mathbb{K}$-algebra.

In Chapter 26, we showed that in $\mathcal{A}(\mathbb{K})$ every closed ideal is principal. Here, following the same methods, provided that $\mathbb{K}$ is spherically complete, we can prove similar results with algebras $\mathcal{A}(d(a, R^-))$:

Theorem 28.7. *Suppose $\mathbb{K}$ is spherically complete. All closed ideals of $\mathcal{A}(d(a, R^-))$ are principal.*

Proof. Let I be a closed ideal of $\mathcal{A}(d(a, R^-))$ and let $E = \mathcal{D}(I)$. By Theorem 27.10, we have $I = \mathcal{T}_{a,R}(E)$. Now, by Theorem 28.6, there exists $g \in \mathcal{A}(\mathbb{K})$ such that $\mathcal{D}(g) = E$ and of course g belongs to $\mathcal{T}_{a,R}(E)$ hence to I. Consequently, $g\mathcal{A}(d(a, R^-)) \subset I$. Conversely, by Corollary 27.2, we have $I = g\mathcal{A}(d(a, R^-))$. □

Chapter 29

Motzkin Factorization and Roots of Analytic Functions

In this chapter, D is a closed infraconnected set and f belongs to $H(D)$.

The idea of factorizing semi-invertible analytic elements into a product of singular factors is a remarkable idea due to Motzkin [75]. This factorization has tight links with the Mittag-Leffler series, as shown in [36].

Lemma 29.1. *Let $T = d(a, r^-)$, with $a \in \mathbb{K}$ and $r > 0$, let $E = \mathbb{K}\backslash T$ and take $b \in T$. Let $g \in H(E)$ be invertible in $H(E)$. Then there exist $\lambda \in \mathbb{K}$, $q \in \mathbb{Z}$ and $h \in H(E)$ invertible in $H(E)$, satisfying $\|h - 1\|_E < 1$, $\lim_{|x|\to+\infty} h(x) = 1$ and $g(x) = \lambda(x-b)^q h(x)$. Moreover, λ, q, and h are respectively unique, satisfying those relations. Further, both λ and q do not depend on b in T and $\frac{g'}{g}$ belongs to $H_0(E)$.*

Proof. Without loss of generality, we may obviously assume $a = 0$. As g is invertible, if g belongs to $H_0(E)$, then $\frac{1}{g}$ does not. So, we may clearly assume that g does not belong to $H_0(E)$. By Theorem 11.5. g is of the form $\widetilde{g} + \widehat{g}$, with $\widetilde{g} \in \mathbb{K}[x]$, $\widetilde{g} \neq 0$, and $\widehat{g} \in H_0(E)$. Let $q = \deg(\widetilde{g})$ and let λ be its coefficient of degree q. Now, we put $h(x) = \frac{g(x)}{\lambda(x-b)^q}$. By definition both λ, q do not depend on b in T. Hence, we may also assume $b = 0$. Clearly, h satisfies $\lim_{|x|\to+\infty} h(x) = 1$. Since $H(E)$ is a $\mathbb{K}$-algebra and since g is invertible, h is invertible in $H(E)$. In particular, we note that h is bounded and admits no zero in E. Now, we check that $\|h - 1\|_E < 1$. Let $s = \frac{1}{r}$, let $A = d(0, s)$

and let $\phi(u) = h(\frac{1}{u})$ whenever $u \in d(0,s)$, $u \neq 0$. Then ϕ belongs to $H(d(0,s)\setminus\{0\})$. But since h is bounded in E, ϕ is bounded in $d(0,s)\setminus\{0\}$. Moreover, the condition $\lim_{|x|\to+\infty} h(x) = 1$ shows that $\lim_{x\to 0}\phi(x) = 1$, hence ϕ belongs to $H(d(0,s))$.

Thus, $\phi(u)$ is of the form $\sum_{n=0}^{\infty} a_n u^n$ with $a_0 = 1$ and hence, by Theorem 22.7, we have $|a_n|s^n < 1\ \forall n > 0$. Let $\epsilon = \sup\{|a_n|s^n | n > 0\}$. Then we have $\|\phi - 1\|_{d(0,s)} = \|h - 1\|_E = \epsilon$. Now, h, q, and λ are easily seen to be unique. Indeed, let $g(x) = \alpha x^t l(x)$ with l invertible in $H(E)$, satisfying $\lim_{|x|\to+\infty} l(x) = 1$. Then we have $1 = x^{q-t}\frac{\lambda h(x)}{\alpha l(x)}$. Consequently, considering the limit when $|x|$ tends to $+\infty$, we have $q = t$, $\lambda = \alpha$, and therefore, $h = l$. Finally, we check that $\frac{g'}{g}$ belongs to $H_0(E)$. Indeed, $\frac{g'}{g} = \frac{q}{x-b} + \frac{h'}{h}$. Obviously, $\frac{q}{x-b}$ belongs to $H_0(E)$. Since $\lim_{|x|\to\infty} |h(x)| = 1$, it is seen that $h(x)$ is of the form $1 + \sum_{n=1}^{\infty} \frac{a_n}{x^n}$ with $\lim_{n\to\infty} |\frac{a_n}{s^n}| = 0$, and therefore, h' is an element of $H(E)$ such that $\lim_{|x|\to\infty} |h'(x)| = 0$. As a consequence, $\frac{h'}{h}$ belongs to $H_0(E)$. Hence, so does $\frac{g'}{g}$ and this ends the proof of Lemma 29.1. □

Definition. Let $E = \mathbb{K}\setminus d(a,r^-)$ with $a \in \mathbb{K}$ and $r > 0$. Let $f \in H(E)$ be invertible in $H(E)$ and let $\lambda(x-a)^q h(x)$ be the factorization given in Lemma 29.1. The integer q will be named *the index of f associated to $d(a,r^-)$* and will be denoted by $mo(f, d(a,r^-))$. If $\lambda = 1$, the element f will be called *a pure factor associated to $d(a,r^-)$*. Let $\mathcal{G}^T$ be the group of invertible elements of $H(\mathbb{K}\setminus T)$.

The following Corollary is then immediate.

Corollary 29.2. *Let $T = d(a,r^-)$. The set of pure factors associated to T is a sub-multiplicative group of the group $\mathcal{G}^T$. Further, every element of $\mathcal{G}^T$ is of the form λh with h a pure factor associated to T and $\lambda \in \mathbb{K}^*$.*

Lemma 29.3. *Let $T = d(a,r^-)$, let $E = \mathbb{K}\setminus T$ with $a \in \mathbb{K}$ and let f be a pure factor associated to T such that $\|f - 1\|_E < 1$. Then $mo(f,T) = 0$.*

Proof. Without loss of generality, we can assume $a = 0$. Let $q = mo(f,T)$. By Lemma 29.1, there exists a unique element h invertible in $H(E)$ such that $f = x^q h$ and $\lim_{|x|\to+\infty} |h(x)| = 1$. Therefore, by Theorem 14.6, $h(x)$ is of the form $1 + \sum_{n=1}^{+\infty} \frac{a_n}{x^n}$, hence

$\|h\|_E < 1$. So, if $q > 0$, then f is unbounded, a contradiction. Next, by Corollary 29.2, $\frac{1}{f}$ is a pure factor satisfying again the hypothesis of Corollary 29.2, hence the hypothesis $q < 0$ gets to a contradiction again. □

Definition. Let f belong to $H(D)$. Let T be a hole of D and let h be a pure factor associated to T. If $\frac{f}{h}$ belongs to $H(D \cup T)$ and has no zero inside T, h is called the *Motzkin factor of f in the hole T.*

Theorem 29.4. *Let T be a hole of D and let f have a Motzkin factor h in T. Then h is unique. Further, if T is not an f-hole, h is the polynomial of the zeros of f inside T. Moreover, if E is another infraconnected set included in D admitting T as a hole and if g denotes the restriction of f to E, then g admits a Motzkin factor in the hole T as an element of $H(E)$ and this Motzkin factor is equal to h.*

Proof. Let f have another Motzkin factor l in T, let $F = \frac{f}{h}$ and let $G = \frac{f}{l}$. Since G has no zeros inside T, by Theorem 24.9, there exists a closed bounded infraconnected set D' satisfying $T \subset D' \subset (D \cup T)$, $T \neq D'$, such that G is invertible in $H(D')$. Hence, in $H(D')$, we have $\frac{F}{G} = \frac{l}{h}$, and hence, $\frac{l}{h}$ belongs to $H(D')$. Since $T \neq D'$, it is seen that $D' \cap (\mathbb{K} \setminus T)$ is an infraconnected closed bounded set included in D that admits T as a hole. Moreover, we have $D' \cup (\mathbb{K} \setminus T) = \mathbb{K}$. Therefore, by Theorem 15.10, we see that $\frac{l}{h}$ belongs to $H(\mathbb{K})$ and is hence a polynomial P. Since $\frac{F}{G}$ belongs to $H(D')$ and has no zeros in T, it is seen that $mo(h, T) = mo(l, T)$, so we have $\lim_{|x| \to \infty} \frac{h(x)}{l(x)} = 1$. Hence, $P = 1$ and this proves that h is unique.

Now, we assume that T is not an f-hole, hence f belongs to $H(D \cup T)$. Let Q be the polynomial of the zeros of f inside T. Then by Corollary 16.6, $\frac{f}{Q}$ belongs to $H(D \cup T)$ and has no zeros inside T. Since its Motzkin factor h is unique, we have $h = Q$. The last statement about g is obvious because $\frac{g}{h}$ clearly belongs to $H(E \cup T)$ and has no zero inside T. This ends the proof of Theorem 29.4. □

Definition and notation: We will call *the f-supersequence of D* the sequence of the holes $(T_n)_{n \in I}$ such that either T_n is an f-hole or f belongs to $H(D \cup T)$ and has at least one zero inside T_n. If f admits a Motzkin factor h in a hole T, it will be denoted by f^T and $mo(h, T)$ will be called *the Motzkin index of f in T*. For every hole which does not belong to the f-supersequence, we put $f^T = 1$.

Lemma 29.5 is immediate.

Lemma 29.5. *Let $D \in \text{Alg}$, let T be a hole of D and let $f,\ g \in H(D)$ admitting Motzkin factors in T. Then $(fg)^T = f^T g^T$ and $mo(fg,T) = mo(f,T) + mo(g,T)$. Moreover, if f is invertible in $H(D)$, then $(f^{-1})^T = (f^T)^{-1}$ and $mo(f^{-1},T) = -mo(f,T)$.*

Lemma 29.6. *Let $f \in H(D)$ and let $(T_n)_{n\in\mathbb{N}}$ be the f-supersequence. Suppose that for each $n \in \mathbb{N}$, f admits Motzkin factors in T_n. Then there exists $N \in \mathbb{N}$ such that $mo(f^{T_n},T_n) = 0$ whenever $n > N$. Moreover, if $D \in \text{Alg}$, the product $\left(\prod_{n=1}^{t} f^{T_n}\right)\left(\prod_{n=t+1}^{\infty} f^{T_n}\right)$ does not depend on t whenever $t \geq N$.*

Proof. Indeed, there exists $N \in \mathbb{N}$ such that we have $\|f^{T_n} - 1\|_D < 1$ whenever $n \geq N$ and therefore, by Corollary 29.2, $mo(f^{T_n},T_n) = 0$. Now, in $H_b(D)$, we have $\left(\prod_{n=N+1}^{\infty} f^{T_n}\right) = \left(\prod_{n=N+1}^{t} f^{T_n}\right)\left(\prod_{n=t+1}^{\infty} f^{T_n}\right)$.

But then, if D belongs to Alg, we have $\left(\prod_{n=1}^{t} f^{T_n}\right)\left(\prod_{n=t+1}^{\infty} f^{T_n}\right) = \left(\prod_{n=1}^{N} f^{T_n}\right)\left(\prod_{n=N+1}^{t} f^{T_n}\right)\left(\prod_{n=t+1}^{\infty} f^{T_n}\right) = \left(\prod_{n=1}^{N} f^{T_n}\right)\left(\prod_{n=N+1}^{\infty} f^{T_n}\right)$. □

Definition. Let $(T_n)_{n\in I}$ be the f-supersequence of D with I a subset of $\mathbb{N}$ which is either finite or equal to $\mathbb{N}$.

If I is finite, f is said to have *a finite Motzkin factorization* if it factorizes in $H(D)$ in the form $\left(f^0 \prod_{n\in I} f^{T_n}\right)$ with f^0 an element of $H(\widetilde{D})$ whose zeros belong to D and for each $n \in \mathbb{N}$, f^{T_n} is a Motzkin factor in T_n.

If I is infinite and equal to $\mathbb{N}$, f is said to have *an infinite Motzkin factorization* if it admits a sequence of Motzkin factors f^{T_n} satisfying $\lim_{n\to\infty} f^{T_n} - 1 = 0$ such that f factorizes in $H(D)$ in the form $\left(f^0 \prod_{n=1}^{t} f^{T_n}\right)\left(\prod_{n=t+1}^{\infty} f^{T_n}\right)$, with f^0 an element of $H(\widetilde{D})$ whose zeros belong to D. In both cases, f^0 will be called *the principal factor of f.*

Corollary 29.7. *Let D be bounded and let f have an infinite Motzkin factorization with a f-supersequence $(T_n)_{n\in\mathbb{N}}$. Then we have $f = f^0\left(\prod_{n=1}^{\infty} f^{T_n}\right)$.*

Corollary 29.8. *Let f have an infinite Motzkin factorization with a f-supersequence $(T_n)_{n\in\mathbb{N}}$ such that $mo(f^{T_n},T_n) = 0$ for all $n > 0$. Then we have $f = f^0\left(\prod_{n=1}^{\infty} f^{T_n}\right)$.*

Remark 1. Let $f \in H(D)$ be unbounded and have Motzkin factorization of the form $\left(f^0 \prod_{n=1}^{N} f^{T_n}\right)\left(\prod_{n=N+1}^{\infty} f^{T_n}\right)$. One cannot claim that the product $\left(f^0 \prod_{n=1}^{\infty} f^{T_n}\right)$ converges in $H(D)$ even if D is closed and belongs to Alg. Indeed, let $r \in]0,1[$, let $(a_n)_{n\in\mathbb{N}}$ be a sequence in $d(0,1)$ such that $|a_n - a_m| = 1$ whenever $n \neq m$ and $a_1 = 0$. For every $n \in \mathbb{N}^*$, we put $T_n = d(a_n, r^-)$ and $E = \mathbb{K}\setminus\left(\bigcup_{n=1}^{\infty} T_n\right)$. The holes of E are T_n. Let $(\lambda_n)_{n\geq 2}$ be a sequence in $d(0,r^-)$ such that $\lim_{n\to\infty} \lambda_n = 0$. For every $n \geq 2$, we put $g_n = 1 + \frac{\lambda_n}{x-a_n}$. The sequence $(g_n)_{n\geq 2}$ is seen to satisfy $\|g_n - 1\|_E \leq \frac{|\lambda_n|}{\rho} < 1$, and therefore, we have $\lim_{n\to\infty} \|g_n - 1\|_E = 0$. Hence, the product $h = \prod_{n=2}^{\infty} g_n$ obviously converges in $H(E)$.

Since E clearly belongs to Alg, we see that $x^2 h$ belongs to $H(E)$ and is invertible in $H(E)$. Now, f clearly has Motzkin factorization with $f^{T_n} = g_n$ for every $n \geq 2$, $f^{T_1} = x^2$ and $f^0 = 1$. However, we will check that the sequence $(f_n)_{n\in\mathbb{N}^*}$ defined by $f_n = x^2\prod_{j=2}^{n} g_j$ does not converge in $H(E)$. Indeed, we have $f_{n+1}(x) - f_n(x) = x^2\left(\prod_{j=2}^{n} g_j(x)\right)(g_{n+1}(x) - 1)$. For every $x \in \mathbb{K}\setminus d(0,1)$, we have $\left|x^2\prod_{j=2}^{n} g_j(x)\right| = |x^2|$ and $|g_{n+1}(x) - 1| = \left|\frac{\lambda_{n+1}}{x}\right|$, hence $|f_{n+1}(x) - f_n(x)| = |x||\lambda_{n+1}|$. Thus, $f_{n+1} - f_n$ is not bounded in $H(E)$, and therefore, the sequence $(f_n)_{n\in\mathbb{N}^*}$ does not converge in $H(E)$. According to Theorem 4 in [75], the product $\prod_{n=1}^{\infty} f_n$ should converge to $x^2 h$ in $H(E)$. Here, we see that this is not true in the general case. Actually, the proof given in [75] only shows the simple convergence of the sequence (f_n) and the uniform convergence on bounded subsets of D.

By Lemma 29.5, Lemma 29.9 is immediate.

Lemma 29.9. *Let $D \in$ Alg, let f, $g \in H(D)$ have Motzkin factorization. Then so does fg. Moreover, we have $(fg)^0 = f^0 g^0$. Further, if f is invertible, f^{-1} also has Motzkin factorization and it satisfies $(f^{-1})^0 = (f^0)^{-1}$.*

Corollary 29.10 ((Boussaf) [13]). *Let $D \in$ Alg, let f have an infinite Motzkin factorization of the form $\left(f^0 \prod_{n=1}^{t} f^{T_n}\right)$ $\left(\prod_{n=t+1}^{\infty} f^{T_n}\right)$. Let $N \in \mathbb{N}$ be such that $mo(f^{T_n}, T_n) = 0$ for all $n > N$. Then we have*

$$f = f^0\left(\prod_{n=1}^{N} f^{T_n}\right)\left(\prod_{n=N+1}^{\infty} f^{T_n}\right).$$

Proposition 29.11. *Let $f \in H(D)$ satisfy $\|f-1\|_D < 1$ and have Motzkin factorization of the form $f^0\Big(\prod_{n=1}^{\infty} f^{T_n}\Big)$ with $\big(T_n\big)_{n\in\mathbb{N}^*}$ the f-supersequence of D. Then for each $n \geq 1$, we have $mo(f^{T_n}, T_n) = 0$.*

Proof. For every $n \in \mathbb{N}^*$, we put $q_n = mo(f^{T_n}, T_n)$. By Lemma 29.6, we may assume the $(T_n)_{n\in\mathbb{N}^*}$ ranged in such a way that $q_n \neq 0$ for $n \leq N$ while $q_n = 0$ whenever $n > N$. When $n \leq N$, f^{T_n} is of the form $(x-\alpha_n)^{q_n}(1+\omega_n)$ with $\omega_n \in H_0(\mathbb{K}\setminus T_n)$, $\| \omega_n \|_{\mathbb{K}\setminus T_n} < 1$ and $\alpha_n \in T_n$. When $n > N$, f^{T_n} is just in the form $(1+\omega_n)$ with $\omega_n \in H_0(\mathbb{K}\setminus T_n)$ and $\|\omega_n\|_{\mathbb{K}\setminus T_n} < 1$. On the other hand, since f has no zero in D, obviously f^0 has no zero in D, and therefore, it has no zero in $\widetilde{D}$. Hence, by Theorem 22.7, f^0 is of the form $A(1+\omega_0(x))$ with $\omega_0 \in H(\widetilde{D})$, $\|\omega_0\|_D < 1$. Let $h(x) = A\prod_{n=1}^{N}(x-\alpha_n)^{q_n}$. We see that f factorizes in the form $h\Big(\prod_{n=0}^{\infty}(1+\omega_n)\Big)$. Since $\|\omega_n\|_D < 1$ for every $n \in \mathbb{N}$ and since $\lim_{n\to\infty}\omega_n = 0$, it is seen that h satisfies (1) $\|h-1\|_D < 1$ as does f. Let us suppose $q_1 \neq 0$. We may obviously assume $\alpha_1 = 0$. Let $T_1 = d(0, r^-)$. Thus, in T_1, h admits 0 as a zero of order q_1 if $q_1 > 0$ (respectively a pole of order $-q_1$ if $q_1 < 0$) and has neither any zero nor any pole different from 0. Anyway, when $x \in T_1$, we have

(2) $|h(x)| = B|x^{q_1}|$ with $B = |A|\prod_{n=2}^{N}|\alpha_n|^{q_n}$.

We will show that (2) contradicts (1) except if $q_1 = 0$.

Suppose $q_1 > 0$. In T_1, $h(x)$ is a series of the form $\sum_{n=q_1}^{+\infty} c_n x^n$ and $h-1 = (\sum_{n=q_1}^{+\infty} c_n x^n) - 1$, hence $1 \leq \|h-1\|_{T_1} \leq \|h-1\|_D$, which contradicts (1).

Now, suppose $q_1 < 0$. By definition, h is obviously invertible in $R(D)$. Hence, we put $F = \frac{1}{h}$ and we see that F satisfies $\|F-1\|_D < 1$ and admits 0 as a unique zero in T_1 while it has no pole in T_1. Hence, the same process lets us get to the same contradiction and finishes showing that $q_1 = 0$, and similarly, $q_n = 0$ for every $n \geq 1$. $\square$

Proposition 29.12. *Let $f \in H(D)$ be invertible in $H(D)$ and have Motzkin factorization and let $a \in D$. Then f satisfies $\left\|\frac{f}{f(a)} - 1\right\|_D < 1$ if and only if for every hole T of the f-supersequence of D, we have $mo(f, T) = 0$.*

Proof. Without loss of generality, we may obviously assume $f(a)=1$. By Proposition 29.11, we already know that if f satisfies $\|f-1\|_D < 1$, then for every hole of the f-supersequence, we have $mo(f,T) = 0$. Now, we suppose that for every hole T of the f-supersequence, we have $mo(f,T) = 0$ and we will prove that $\|f-1\|_D < 1$. Indeed, by Lemma 29.3, for each hole of the f-supersequence, we have $\|f^T - 1\|_D < 1$. Moreover, since f is invertible, f^0 must also be invertible, hence by Theorem 23.1, it is of the form $(1+\psi(x))$, with $\|\psi\|_D < 1$. Then $\|f-1\|_D < 1$. □

We will show that all semi-invertible elements have Motzkin factorization step after step, and first, we consider rational functions.

Proposition 29.13. *Let $f \in R(D)$. Then f admits the Motzkin factorization.*

Proof. The f-supersequence is obviously finite. Let $T_1, \dots, T_s$ be this f-supersequence. We can obviously factorize f in a unique way in the form $\prod_{j=1}^{s}\left(\frac{h_j}{l_j}\right)$, whereas for each $j = 1, \dots, s$, both h_j and l_j are monic polynomials whose zeros lie in T_j. Thus, we can check that $\frac{h_j}{l_j}$ is the Motzkin factor f^{T_j} of f in the hole T_j. Therefore, putting $f^0 = \frac{f}{\prod_{j=1}^{s}\left(\frac{h_j}{l_j}\right)}$, we have the Motzkin factorization: $f = f^0 \prod_{j=1}^{s} f^{T_j}$. □

Proposition 29.14. *Let $\phi \in H(D)$ satisfy $\|\phi - 1\|_D < 1$. Then ϕ admits the Moztkin factorization $\phi^0\left(\prod_{n=1}^{\infty}\phi^{T_n}\right)$ with $(T_n)_{n\in\mathbb{N}^*}$ the f-supersequence. For every ϕ-hole T, we have $\|\phi^T - 1\|_D = \|\overline{\overline{\phi_T}}\|_D$. Moreover, ϕ^0 satisfies $\|\phi^0 - 1\|_D = \|\overline{\overline{\phi_0}} - 1\|_D$.*

Proof. First, we suppose $\phi \in R(D)$. Then by Proposition 29.13, ϕ admits the Motzkin factorization. Now, by Proposition 29.12, for each $n > 0$, we have $mo(\phi, T_n) = 0$, and therefore, ϕ^{T_n} is of the form $1+\omega_n$ with $\|\omega_n\|_D < 1$ whenever $n > 0$, while $\phi^0 = 1 + \omega_0$ with $\|\omega_0\|_D < 1$. Hence, we see that $\overline{\overline{\phi_{T_n}}} = \overline{\overline{\left(\omega_n \prod_{\substack{j\neq n\\ j\in\mathbb{N}}}(1+\omega_j)\right)_{T_n}}}$. Clearly, $\prod_{\substack{j\neq n\\ j\in\mathbb{N}}}(1+\omega_j)$ is of the form $1+\phi_n$ with $\|\phi_n\|_D < 1$, hence $\|(\omega_n\phi_n)\|_D < \|\omega_n\|_D$, and we obtain (1) $\|\overline{\overline{(\omega_n\phi_n)_{T_n}}}\|_D \leq \|\omega_n\phi_n\|_D < \|\omega_n\|_D$.

But ω_n is clearly equal to $\overline{\overline{(\phi^{T_n})}}_{T_n}$ and then we have (2) $\|\overline{\overline{(\omega_n)}}_{T_n}\|_D = \|\omega_n\|_D > \|\omega_n\phi_n\|_D \geq \|\overline{\overline{(\omega_n\phi_n)}}_{T_n}\|_D$.
Moreover, $\overline{\overline{(\omega_n + \omega_n\phi_n)}}_{T_n} = \overline{\overline{(\omega_n)}}_{T_n} + \overline{\overline{(\omega_n\phi_n)}}_{T_n}$, hence by (1) and (2) we have $\|\overline{\overline{(\omega_n(1+\phi_n))}}_{T_n}\|_D = \|\omega_n\|_D$, and finally,

(3) $\|\overline{\overline{\phi}}_{T_n}\|_D = \|\overline{\overline{(\omega_n(1+\phi_n))}}_{T_n}\|_D = \|\omega_n\|_D = \|\phi^{T_n} - 1\|_D$. In the same way, we put $\prod_{n=1}^{\infty}(1+\omega_n) = 1+\psi$ with

(4) $\|\psi\|_D < 1$.

It is seen that ψ belongs to $H_0(\mathbb{K}\setminus(\bigcup_{n=1}^{\infty} T_n))$. Hence, Theorem 15.1, when applied to ψ, shows that

(5) $\overline{\overline{\psi}}_0 = 0$.

Next, we have $\phi = (1+\omega_0)(1+\psi) = 1+\omega_0+\psi+\omega_0\psi$, hence $\overline{\overline{\phi}}_0 = 1 + \overline{\overline{(\omega_0)}}_0 + \overline{\overline{\phi}}_0 + \overline{\overline{(\omega_0\psi)}}_0$. By definition, $\omega_0 \in H(\widetilde{D})$, hence $\omega_0 = \overline{\overline{(\omega_0)}}_0$, and then by (5), we have $\overline{\overline{\phi}}_0 = 1+\omega_0+\overline{\overline{(\omega_0\psi)}}_0$. But by (4), it is seen that $\|\overline{\overline{(\omega_0\psi)}}_0\|_D < \|\omega_0\|_D$, and hence finally, we obtain $\|\overline{\overline{\phi}}_0 - 1\|_D = \|\omega_0\|_D = \|\phi^0 - 1\|_D$. Thus, we have proven the inequalities satisfied by the ϕ^{T_n} and by ϕ^0 when ϕ belongs to $R(D)$.

Now, we consider the general case when $\phi \in H(D)$. Let $(f_m)_{m\in\mathbb{N}}$ be a sequence in $R(D)$ such that $\lim_{m\to\infty} \|\phi - f_m\|_D = 0$. Let $\varepsilon \in]0,1[$ and let $N \in \mathbb{N}$ be such that $\|f_m - \phi\|_D \leq \varepsilon$ whenever $m \geq N$. Let T be a hole of the ϕ-supersequence. We will show that the sequence $((f_m)^T)_{m\in\mathbb{N}}$ converges in $H(D)$ and that this convergence is uniform with respect to the ϕ-supersequence. We fix $m \geq N$. It is seen that $\|f_m - 1\|_D < 1$ and then by Lemma 29.1 and Proposition 29.12, we have $\|(f_m)^T - 1\|_D < 1$, and in particular, $\|(f_m)^T\|_D = 1$. Moreover, we remember that in $H(\mathbb{K}\setminus T)$, the norm $\|\ .\ \|_D$ is multiplicative and actually equal to ${}_D\varphi_T$. Now, let $s \geq N$. We have $\|(f_m)^T - (f_s)^T\|_D = \|\frac{(f_m)^T}{(f_s)^T} - 1\|_D$. But by Lemma 29.5, we have $\frac{(f_m)^T}{(f_s)^T} = \left(\frac{f_m}{f_s}\right)^T$ and then by (3), in $R(D)$, we have $\left\|\left(\frac{f_m}{f_s}\right)^T - 1\right\|_D = \left\|\overline{\overline{\left(\frac{f_m}{f_s}\right)}}_T\right\|_D$. Finally, by Theorem 15.1, we obtain

(6) $\|(f_m)^T - (f_s)^T\|_D \leq \varepsilon$. Relation (6) does not depend on the hole T and it shows that, for each fixed $n \in \mathbb{N}^*$, the sequence $((f_m)^{T_n})_{m\in\mathbb{N}}$ is

a Cauchy sequence which converges in $H(\mathbb{K}\setminus T_n)$ to an element whose index is equal to 0 and this convergence is uniform with respect to n. For each $n \in \mathbb{N}^*$, we put $\phi_n = \lim_{m\to\infty}(f_m)^{T_n}$. Then it is seen that $\prod_{n=1}^{\infty} \phi_n = \lim_{m\to\infty} \prod_{n=1}^{\infty}(f_m)^{T_n}$. As a consequence, the sequence $(f_m)^0$ is also convergent in $H(D)$ and actually in $H_b(\widetilde{D})$. Let ϕ_0 be its limit. Then we have this factorization: $\phi = \prod_{n=0}^{\infty} \phi_n$. We recognize the Motzkin factorization for ϕ. Obviously, for each fixed $n > 0$, the equality satisfied by the $(f_m)^{T_n}$ holds for ϕ^{T_n} and shows that $\|\overline{\overline{(\phi)_{T_n}}}\|_D = \|(\phi)^{T_n} - 1\|_D$. In the same way, the equality satisfied by the $(f_m)^0$ shows that $\|\overline{\overline{(\phi)_0}} - 1\|_D = \|(\phi)^0 - 1\|_D$. This ends the proof of Proposition 29.14. □

Theorem 29.15 is given in [36, 42, 50].

Theorem 29.15. *Let $a \in D$. Let $\phi \in H_b(D)$ be such that $|\phi(a)| \neq 0$. The following statements* (i), (ii), and (iii) *are equivalent:*

(i) $\|\phi - \phi(a)\|_D < |\phi(a)|$.
(ii) *For every hole T, we have $\|\overline{\overline{\phi_T}}\|_D < |\phi(a)|$ and $\|\overline{\overline{\phi_0}} - \overline{\overline{\phi_0}}(a)\|_D < |\phi(a)|$.*
(iii) *ϕ is invertible, admits a Motzkin factorization and for every hole T, ϕ^T satisfies $\|\phi^T - 1\|_D < 1$ and ϕ^0 satisfies $\|\phi^0 - \phi^0(a)\|_D < |\phi(a)|$.*

Further, if statements (i)–(iii) *are satisfied, then we have the following:*

(u) *$m(\phi, T) = 0$ for every hole T.*
(v) *$\|\overline{\overline{\phi_T}}\|_D = \|\phi^T - 1\|_D|\phi(a)|$ for every hole T.*
(w) $\|\phi^0 - \phi^0(a)\|_D = \|\overline{\overline{\phi_0}} - \overline{\overline{\phi_0}}(a)\|_D$.

Proof. Without loss of generality, we may obviously assume $|\phi(a)| = 1$ and

(1) $|\phi(a) - 1| < 1$. Let $(T_m)_{m\in I}$ be the ϕ-supersequence of D. We note that when (i) is satisfied, ϕ is obviously invertible.

First, we suppose (i) is satisfied and will show that so is (ii). By Theorem 15.1, we have

(2) $\|\overline{\overline{(\phi - \phi(a))_{T_m}}}\|_D \le \|\phi - \phi(a)\|_D$.

But it is seen that $\overline{\overline{(\phi - \phi(a))_{T_m}}} = \overline{\overline{\phi_{T_m}}}$. Hence, by (2), we have

(3) $\|\overline{\overline{\phi}}_{T_m}\|_D \leq \|\phi - \phi(a)\|_D < 1$, whenever $m \in I$.

In the same way, $\overline{\overline{(\phi - \phi(a))}}_0 = \overline{\overline{\phi}}_0 - \phi(a)$, and then by Theorem 15.1, we have

(4) $\|\overline{\overline{\phi}}_0 - \phi(a)\|_D < 1$.

Besides by (3), we see that $\|\sum_{m=1}^{\infty} \overline{\overline{\phi}}_{T_m}\|_D < 1$, hence $\|\phi - \overline{\overline{\phi}}_0\|_D = \|\sum_{m=1}^{\infty} \overline{\overline{\phi}}_{T_m}\|_D < 1$, and therefore, $|\phi(a) - \overline{\overline{\phi}}_0(a)| < 1$, hence by (4), we see that

(5) $\|\overline{\overline{\phi}}_0 - \overline{\overline{\phi}}_0(a)\|_D < 1$.

Finally by (3) and (5), statement (ii) is clearly proven.

Now, we will show that each of the statements (ii) and (iii) separately imply (i). We suppose that (ii) is satisfied. Hence, we have

(6) $\| \sum_{m\in I} \overline{\overline{\phi}}_{T_m} \|_D < 1$.

If D is bounded, by statement (ii), and by (6), we obtain (i). Now, let D be not bounded. Then $\overline{\overline{\phi}}_0$ is a constant λ. Hence, ϕ is in the form $\lambda + \sum_{m=1}^{\infty} \overline{\overline{\phi}}_{T_m}$ with $\|\overline{\overline{\phi}}_{T_m}\|_D < 1$ whenever $m \geq 1$, hence

(7) $\|\sum_{m=1}^{\infty} \overline{\overline{\phi}}_{T_m}\|_D < 1$.

Now, we have $\phi - \phi(a) = \sum_{m=0}^{\infty} \overline{\overline{\phi}}_{T_m} - \overline{\overline{\phi}}_{T_m}(a) = \sum_{m=1}^{\infty} (\overline{\overline{\phi}}_{T_m} - \overline{\overline{\phi}}_{T_m}(a))$. By (7), we see that $\|\sum_{m=1}^{\infty}(\overline{\overline{\phi}}_{T_m} - \overline{\overline{\phi}}_{T_m}(a))\|_D < 1$, hence finally (i) $\|\phi - \phi(a)\|_D < 1$.

We now suppose (iii) is satisfied. Hence, we have (8) $\|\phi^{T_m} - 1\|_D < 1$ for all $m \in I$.

If D is bounded, we have $\|\phi^0 - \phi^0(a)\|_D < 1$, hence by (8), we directly have (i). If D is not bounded, then ϕ^0 is a constant B such that $\phi(a) = B\prod_{m\in I} \phi^{T_m}(a)$, hence by (1) and (8), we see that $|B-1| < 1$, and hence by (8), we obtain (i) again.

Thus, (i) is also implied by (ii) as by (iii). Obviously, by (1), (i) implies $\|\phi - 1\|_D < 1$, and therefore, we may apply Proposition 29.14. Next, we suppose that either (ii) or (iii) is satisfied. Hence, so is (i) and so are (u) and (v) by Proposition 29.14.

Finally, we will show (w), and at the same time, we will finish proving the equivalence between (ii) and (iii). Let $\psi = (\overline{\overline{\phi_0}}(a))^{-1}\phi$. We may apply Proposition 29.14 to ψ and we have

(9) $\|\overline{\overline{\psi}}_0 - 1\|_D = \|\psi^0 - 1\|_D$.

But we have

(10) $\|\psi^0 - 1\|_D \geq \|\psi^0 - \psi^0(a)\|_D = \|\phi^0 - \phi^0(a)\|_D$.

(11) $\|\overline{\overline{\phi}}_0 - \overline{\overline{\phi}}_0(a)\|_D = \|\overline{\overline{\psi}}_0 - \overline{\overline{\psi}}_0(a)\|_D = \|\overline{\overline{\psi}}_0 - 1\|_D$.

Hence by (9), (10), (11) we obtain

(12) $\|\phi^0 - \phi^0(a)\|_D \leq \|\overline{\overline{\phi}}_0 - \overline{\overline{\phi}}_0(a)\|_D$.

Now, let $\gamma = \phi^0(a)$ and let $\chi = \gamma^{-1}\phi$. By (1) and (7), we see that $|\gamma - 1| < 1$, hence we may apply Proposition 29.14 to χ and we have

(13) $\|\chi^0 - 1\|_D = \|\overline{\overline{\phi}}_0 - \overline{\overline{\phi}}_0(a)\|_D$

while $\|\overline{\overline{\phi}}_0 - \overline{\overline{\phi}}_0(a)\|_D = \|\overline{\overline{\chi}}_0 - \overline{\overline{\chi}}_0(a)\|_D \leq \|\overline{\overline{\chi}}_0 - 1\|_D$ and $\|\chi^0 - 1\|_D = \|\chi^0 - \chi^0(a)\|_D = \|\phi^0 - \phi^0(a)\|_D$. Hence, by (13), we see that $\|\overline{\overline{\phi}}_0 - \overline{\overline{\phi}}_0(a)\|_D \leq \|\phi^0 - \phi^0(a)\|_D$, and therefore, by (12), we obtain (w). This finishes proving the equivalence between (ii) and (iii), and ends the proof of Theorem 29.15. □

Remark 2. If D is not bounded, as ϕ is bounded, both ϕ^0 and $\overline{\overline{\phi_0}}$ are constant, and therefore, the statements $\|\overline{\overline{\phi_0}} - \overline{\overline{\phi_0}}(a)\|_D < |\phi(a)|$ and $\|\phi^0 - \phi^0(a)\|_D < |\phi(a)|$ are automatically satisfied. Statement (ii) is then equivalent to

(ii′) *For every hole T, we have* $\|\overline{\overline{\phi_T}}\|_D < |\phi(a)|$ and statement (iii) is equivalent to

(iii′) *ϕ is invertible and for every hole T, ϕ^T satisfies $\|\phi^T - 1\|_D < 1$.*

Theorem 29.16. *Let $D \in$ Alg. Then f has the Motzkin factorization if and only if it is semi-invertible.*

Proof. Without loss of generality, we may assume the f-supersequence to be infinite. We denote it by $(T_n)_{n\in\mathbb{N}^*}$. Let f have the Motzkin factorization

$$f^0\left(\prod_{n=1}^{t} f^{T_n}\right)\left(\prod_{n=t+1}^{\infty} f^{T_n}\right),$$

where the product $\left(\prod_{n=t+1}^{\infty} f^{T_n}\right)$ converges in $H(D)$. By definition, f^0 is semi-invertible in $H(\widetilde{D})$ and hence in $H(D)$. Moreover,

$\left(\prod_{n=1}^{t} f^{T_n}\right)\left(\prod_{n=t+1}^{\infty} f^{T_n}\right)$ is clearly invertible in $H(D)$. So, f is semi-invertible.

Now, we suppose f to be semi-invertible and will show it to have the Motzkin factorization. By Lemma 29.9, we may clearly suppose that f is invertible without loss of generality.
First, we suppose that there exists M in $\mathbb{R}_+^*$ satisfying

(1) $M \leq |f(x)|$ whenever $x \in D$.

Let $h \in R(D)$ satisfy

(2) $\|f - h\|_D < \frac{M}{2}$,

and let $h^0 \prod_{n=1}^{N} h^{T_n}$ be the Motzkin factorization of h. For every $n = 1, ...N$, let $q_n = mo(h, T_n)$, let $a_n \in T_n$, let $h_n = (x - a_n)^{-q_n} h^{T_n}$ and let $h_0 = h^0$. Let $u(x) = \prod_{n=1}^{N}(x - a_n)^{q_n}$ and let $l(x) = h^0 \prod_{n=1}^{N} h_n$. By (1) and (2), it is seen that h has no zero in D. Let $a \in D$. Then, by Theorem 23.1, h^0 satisfies $\|h^0 - h^0(a)\|_D < |h^0(a)|$, and of course for every $n > 0$, h_n satisfy $\|h_n - 1\|_D < 1$. Hence , we have

(3) $\|l - l(a)\|_D < |l(a)|$. Let $b = |l(a)|$.

In particular, we have $|l(x)| = b$ whenever $x \in D$. Moreover, we note that we have

(4) $\frac{M}{|l(a)|} \leq |u(x)|$.

Let $F = \frac{f}{u}$. Then F does belong to $H_b(D)$. By (3) and (4), we check that $|F(x) - l(x)| < \frac{b}{2}$, and therefore, by (3) again, we have $|F(x)| = b$ and $\|F - F(a)\|_D \leq \frac{|F(a)|}{2}$. Now, we can apply Theorem 29.15 to F and then F has Motzkin factorization $F^0 \prod_{n=1}^{\infty} F^{T_n}$, with $mo(F, T_n) = 0$ whenever $n > 0$. As a consequence f also has the Motzkin factorization, $\left(f^0 \prod_{n=1}^{N} f^{T_n}\right)\left(\prod_{n=N+1}^{\infty} f^{T_n}\right)$ with $f^0 = F^0$ and for each $n = 1, \ldots, N$, $f^{T_n} = (x - a_n)^{q_n} F^{T_n}$ and finally for each $n > N$, $f^{T_n} = F^{T_n}$.

Now, we suppose that $\inf\{|f(x)| \mid x \in D\} = 0$. Since D is closed and since f is invertible, we see that D is unbounded and that the element $G = \frac{1}{f}$ is not bounded in D. Hence, by Corollary 11.7, there exists $q \in \mathbb{N}^*$ such that $\frac{x^{-q}}{f}$ has a non-zero limit when $|x|$ tends to $+\infty$, $(x \in D)$. Then it is easily seen that there exists $c > 0$ such that $|G(x)| \geq c$ for all $x \in D$. Indeed, on one hand, there exists r such that $|G(x)| \geq 1$ for all $x \in D \setminus d(0, r)$, and on the other hand, f is bounded in $D \cap d(0, r)$, hence we can find $c \in]0, 1[$ such that

$|G(x)| \geq c$ whenever $x \in D\backslash d(0,r)$. Thus, G admits the Motzkin factorization, and then by Lemma 29.9, so does f. This ends the proof of the theorem. □

Remark 3. If a closed set B does not belong to Alg, there are counter-examples of invertible elements F which admit certain Motzkin factor F^T such that $\frac{F}{F^T}$ does not belong to $H(B)$ (and obviously does not belong to $H(B \cup T)$). Indeed, don't let B belong to Alg. Since hypothesis B is closed, we know that $\widetilde{B}\backslash\overline{B}$ is not bounded. Hence, by Lemma 29.6, there exists a quasi-minorated element $f \in H_b(B)$ satisfying

(1) $\lim_{\substack{|x|\to\infty \\ x\in B}} f(x) = 0,$

such that xf does not belong to $H(B)$. Since $f \in H_b(B)$, we can take it such that $\|f\|_B < 1$. Without loss of generality, we may assume that 0 belongs to a hole of B. Let $T = d(a, r^-)$ be another hole of B and let $F = \frac{x(1+f)}{(x-a)}$. Then it is seen that F belongs to $H_b(B)$ and is invertible in $H_b(B)$ because both $\frac{x}{(x-a)}$, $1 + f$ are invertible in $H_b(B)$. Hence, F admits the Motzkin factorization. In particular, we see that $F^T = \frac{1}{(x-a)}$. However, we check that $(x-a)F$ does not belong to $H(B)$ because $(x-a)F = x(1+f)$ and by hypothesis, xf does not belong to $H(B)$.

In the same way, let $G = \frac{1}{F}$. Since F is invertible in $H_b(B)$, so is G. But then we see that $\frac{1}{x-a}G$ does belong to $H_b(B)$ and has no zero in B, but obviously, its inverse does not belong to $H(B)$. Therefore, $\frac{1}{x-a}G$ is not semi-invertible in $H(B)$. Thus, there exist invertible elements h and g in $H(B)$ such that hg is not semi-invertible, although it belongs to $H(B)$. This contradicts Theorem 1 in [75], which states that $\frac{f}{f^T}$ extends to an element of $H(D \cup T)$.

Theorem 29.17 ((Boussaf) [13]). *Let D belong to* Alg *and let $T = d(a, r^-)$ be a hole of D. Then f admits a Motzkin factor in the hole T if and only if ${}_D\varphi_{a,r}(f) \neq 0$.*

Proof. On the one hand, we suppose that f admits a Motzkin factor in the hole T. Let $f = gf^T$. Since g belongs to $H(D \cup T)$ and has no zero in T, of course, by Theorem 22.16, we have ${}_D\varphi_{a,r}(g) \neq 0$. Next, as an invertible element of $H(D)$, it is seen that ${}_D\varphi_{a,r}(f^T) \neq 0$. Hence, ${}_D\varphi_{a,r}(f) \neq 0$.

On the other hand, we suppose ${}_D\varphi_{a,r}(f) \neq 0$. Let $\mathcal{F}$ be the circular filter of center a of diameter r and let $M = {}_D\varphi_{a,r}(f)$. There do exist $a_1, \dots, a_q \in d(a,r)$ and s, t satisfying $s < r < t$, such that $|f(x)| \geq M$ whenever $x \in D \bigcap (\bigcap_{j=1}^{q} \Gamma(a_j, s, t))$. Let $F = D \bigcap (\bigcap_{j=1}^{q} \Gamma(a_j, s, t))$. Then T is clearly a hole of F. Next, the restriction g of f to F is invertible in $H(F)$, and therefore, by Theorem 29.16, it admits a Motzkin factor g^T in the hole T. But then, $\frac{f}{g^T}$ belongs to $H(D)$ and $H(F \cup T)$. Let $E = F \cup T$. Clearly, a hole of $D \cap E$ is either a hole of D included in E or a hole of E. Hence, D and E are infraconnected sets that satisfy the hypothesis of Theorem 15.10 and then $\frac{f}{g^T}$ belongs to $H(D \cup F) = H(D \cup T)$. Finally, as g^T is the Motzkin factor of g in T, $\frac{g}{g^T}$ has no zero inside T. This ends the proof. □

Theorem 29.18. *Let $D \in \mathrm{Alg}$ and let $\mathcal{G}$ be the multiplicative group of the invertible elements in $H(D)$. Let $\mathcal{T}$ be the set of the holes of D. Let $\mathcal{G}^0$ be the subgroup of the elements invertible in $H(\widetilde{D})$. Let $\mathcal{H} = \mathcal{G}^0 \prod_{T \in \mathcal{T}} \mathcal{G}^T$. The product $\mathcal{H}$ is a direct product and is dense in $\mathcal{G}$.*

Proof. The product is direct because for each element, the Motzkin factorization is unique. Thus, $\mathcal{H}$ is the set of the invertible elements whose Motzkin factorization is finite. Since every element of $\mathcal{G}$ has the Motzkin factorization, it obviously belongs to the closure of $\mathcal{H}$. □

Thanks to the Motzkin factorization, the question on whether the nth root of an analytic element is an analytic element appears to be linked to the number of zeros of each Motzkin factor.

Theorem 29.19. *Let D be a closed bounded infraconnected set and let $f \in H(D)$ be semi-invertible. Let T be a hole of D. We assume f^s to have continuation to an element of $H(D \cup T)$ for some $s \in \mathbb{N}^*$. Then the number of the zeros of f^s inside T is a multiple of s (taking mutiplicities into account). Moreover, if f does not belong to $H(D \cup T)$, then the number of the zeros of f^s inside T is different from 0.*

Proof. By Lemma 29.9, we have $(f^s)^T = (f^T)^s$. But as f^s belongs to $H(D \cup T)$, by Theorem 29.4, $(f^s)^T$ is the polynomial of the zeros of f^s inside T. Let $Q = (f^s)^T$. Then we have $\deg(Q) = mo((f^s), T) = smo(f, T)$. So, s divides $\deg(Q)$. Now, assume $\deg(Q) = 0$. We have $Q = 1$, $mo(f, T) = 0$, and therefore, $(f^T)^s = 1$ and $\lim_{|x|\to\infty} f^T(x) = 1$. Thus, f^T is just the constant 1, and therefore, f belongs to $H(D \cup T)$, which ends the proof of Theorem 29.19. □

In particular, Theorem 29.19 applies to open disks.

Theorem 29.20. *Let $r \in |\mathbb{K}|$ and let $f \in H(d(0, r^-))\setminus H(d(0, r))$ satisfy $f^s \in H(d(0, r))$. Then f has continuation to an analytic element in a set D of the form $d(0, r)\setminus\left(\bigcup_{i=1}^{t} d(a_i, r^-)\right)$ with $|a_i| = r = |a_i - a_j|$ whenever $i \neq j$, such that for each $i = 1, \ldots, t$, the number of zeros of f^s in $d(a_i, r^-)$ is a multiple of s different from 0.*

Proof. The Mittag-Leffler series of f in $d(0, r^-)$ is of the form $\sum_{n=0}^{\infty} f_n$ with $f_0 = \overline{\overline{f_0}} \in H(d(0, r))$, and for every $n > 0$, $f_n = \overline{\overline{f_{T_n}}}$, with $T_n = d(a_n, r^-)$ and $|a_n - a_j| = |a_n| = r$ whenever $n \neq j$. Since $f \notin H(d(0, r))$, at least one of the f_n is different from 0. Let l be an integer such that $f_l \neq 0$. Now, since f^s belongs to $H(d(0, r))$, by Theorem 29.19, f^s has a number of zeros inside T_l which is different from 0 and a multiple of s. Since any element of $H(d(0, r))$ has finitely many zeros in $d(0, r)$, we see that there are finitely many integers l such that $f_l \neq 0$. Let I be the finite set of the $l \in \mathbb{N}^*$ such that $f_l \neq 0$ and let $D = d(0, r)\setminus\left(\bigcup_{l\in I} d(a_n, r^-)\right)$. Then by definition, f belongs to $H(D)$ and for every $l \in I$, the number of zeros of f^s in T_n is different from 0 and a multiple of s. This ends the proof of Theorem 29.20. □

Corollary 29.21. *Let f be a power series whose radius of convergence is r though f does not belong to $H(d(0, r))$. If for some $s \in \mathbb{N}^*$, f^s has a radius of convergence r' strictly superior to r and if f^s has strictly less than s zeros inside $C(0, r)$ (taking multiplicities into account), then f does not belong to $H(d(0, r^-))$.*

Proof. Since $r' > r$, obviously we have $s > 1$. We assume that f belongs to $H(d(0, r^-))$, and therefore, r must belong to $|\mathbb{K}|$. Since $r' > r$, f^s belongs to $H(d(0, r))$, and then by Theorem 29.20, its number of zeros inside $C(0, r)$ is different from 0 and is a multiple of s, which contradicts the hypothesis. Hence, finally, f does not belong to $H(d(0, r^-))$. □

We have now got to recall the definition of the function $\sqrt[q]{u}$ when $u \in d(1, 1^-)$.

Notation. Henceforth, we suppose that $\mathbb{K}$ has characteristic zero and residue characteristic p. Let $q \in \mathbb{N}^*$. Let $(1+x)^q = 1+qx+\sum_{j=2}^{q} b_j x^j$. So, $|q| = 1$ and $|b_j| \leq 1$ whenever $j = 2, \ldots, q$. Recall that r_k was defined in Chapter 7.

Theorem 29.22. *Let $q \in \mathbb{N}^*$. If q is prime to p, the mapping $g_q(x) = (1+x)^q$ is injective in $d(1, 1^-)$ and maps $d(1, 1^-)$ onto $d(1, 1^-)$. If $p \neq 0$ and if $q = p$, the mapping $g_p(x) = (1+x)^p$ is injective in $d(1, r_1^-)$ and maps $d(1, (r_1)^-)$ onto $d(1, (r_2)^-)$.*

Proof. Suppose first q prime to p. Since $|q| = 1 \geq |b_j| \; \forall j \geq 2$, by Corollary 22.10 and Corollary 23.14, the mapping g_q defines a bijection from $d(0, 1^-)$ onto $d(1, 1^-)$.

Suppose now $p \neq 0$ and take $q = p$. By Theorems 25.3 and 25.5, inside $d(0, (r_1)^-)$, we can write $g_p(x) = \exp(p\text{Log}(1+x))$. This way, we note that when $x \in d(0, (r_1)^-)$, g_p is injective and that we have $|\text{Log}(1+x)| = |x|$ and $|\exp(x) - 1| = |x|$, hence

$$|(1+x)^p - 1| = |\exp(p\text{Log}(1+x)) - 1| = \frac{x}{p}.$$

Consequently, the image of $d(0, r_1^-)$ by g_p is the disk $d(1, (\frac{r_1}{p})^-) = d(1, (r_2)^-)$. □

Notation. Suppose $q \in \mathbb{N}^*$ prime to p. The mapping η_q defined in $d(1, 1^-)$ by $\eta_q(u) = u^q$ is a bijection from $d(1, 1^-)$ onto $d(1, 1^-)$. We denote by $\sqrt[q]{u}$ the inverse mapping from $d(1, 1^-)$ onto $d(1, 1^-)$ and we put $\phi_q(x) = \sqrt[q]{1+x}$ whenever $x \in d(0, 1^-)$.

Suppose now $p \neq 0$ and take $q = p$. The mapping η_p defined in $d(1, (r_1)^-)$ by $\eta_p(u) = u^p$ is a bijection from $d(1, (r_1)^-)$ onto $d(1, (r_2)^-)$. So, we can denote by $\sqrt[p]{u}$ that inverse mapping from $d(1, (r_2)^-)$ onto $d(1, (r_1)^-)$ and we put $\phi_p(x) = \sqrt[p]{1+x}$ whenever $x \in d(0, (r_2)^-)$.

Theorem 29.23. *Let $q \in \mathbb{N}^*$. If q is prime to p, ϕ_q belongs to $\mathcal{A}_b(d(0, 1^-))$ but does not belong to $H(d(0, 1^-))$. Next, suppose now $p > 0$. Then ϕ_p belongs to $\mathcal{A}_b(d(0, r_1^-))$ but does not belong to $H(d(0, r_1^-))$.*

Proof. Suppose first that q is prime to p. By construction and by Corollary 23.15, ϕ_q belongs to $\mathcal{A}_b(d(0, 1^-))$. Suppose that ϕ_q belongs to $H(d(0, 1))$. Then it must satisfy $(\phi_q(-1))^q = 0$, hence $\phi_q(-1) = 0$, and therefore, $\phi_q(x)^q$ admits a zero of order q at -1. But this contradicts the identity $(\phi_q(x))^q = 1 + x$, and therefore, $\phi_q(x)$ does not belong to $H(d(0,1))$. Finally, since $(\phi_q' x))^q$ has a unique zero in $C(0,1)$, by Corollary 29.21, we see that ϕ_q does not belong to $H(d(0,1^-))$.

Suppose now that $p > 0$. By Theorem 29.22, the function $f(x) = (1+x)^p$ is strictly injective inside $d(0, r_1^-)$ and maps $d(0, r_1^-)$ onto itself. So, by Corollary 23.15, it admits an inverse mapping ϕ_p defined inside $d(0, r_1^-)$ that belongs to $\mathcal{A}_b(d(0, r_1^-))$.

Let us show that ϕ_p does not belong to $H(d(0, r_1^-))$. Indeed, suppose ϕ_p belongs to $H(d(0, r_1^-))$. Consider the Mittag-Leffler decomposition of ϕ_p on the infraconnected set $d(0, r_1^-)$. It is of the form $\sum_{n=0}^{\infty} g_n$ with $g_0 \in H(d(0, r_1)$ and $g_n \in H_0(\mathbb{K} \setminus d(a_n, r_n^-))$ with $a_n \in C(0, r_1)$. Now, by Theorem 8.4, we can consider a complete algebraically closed extension $\widehat{\mathbb{K}}$ of $\mathbb{K}$ whose residue class field is not countable.

Let χ be the residue class field of $\mathbb{K}$. Thus, we can find a class G of $\widehat{C}(0, r_1)$ that has an empty intersection with $\bigcup_{n=1}^{\infty} d(a_n, r_1^-)$ and then ϕ_p has continuation to an element of $H(d(0, r_1^-) \cup G$. Let $c \in G$. Since $|c| = r_1$, the function $h(x) = (1+x)^p - c$ satisfies $\nu^-(h, \log(r_1) = 0$, $\nu^+(h, \log(r_1) > 1$, hence h admits a zero $a \in \widehat{C}(0, r_1)$. Then of course a does not belong to $\mathbb{K}$. Now, let ζ be a pth root of 1 different from 1 and let $t = \zeta(1+a)$. By Corollary 7.7, we have $|\zeta - 1| = r_1$, hence t is of the form $1 + b$ with $b \in \widehat{C}(0, r_1)$. We then have $(1+a)^p = (1+b)^p$. Set $E = \widehat{d}(a, r_1^-)$, $F = \widehat{d}(b, r_1^-)$, $D' = d(0, r_1^-) \cup E \cup F$ and $D'' = d(0, r_1^-) \cup G$. Since the image of $d(0, r_1^-)$ by the function f is $d(0, r_1^-)$, we can check that for each $u \in \widehat{C}(0, r_1)$, the image of $d(u, r_1^-)$ by f is $d(1, r_1^-)$. Consequently, both images of E and F by f are equal to G. Now, since f belongs to $\widehat{H}(D')$, the image of D' by f is D'' and ϕ_p belongs to $\widehat{H}(D'')$. Consequently, by Corollary 12.3, $\phi_p \circ f$ belongs to $\widehat{H}(D')$ and we have $\phi_p \circ f(x) = 1 + x \ \forall x \in d(0, r_1^-)$. But since D' has no pierced filter, by Corollary 21.4, it is an analytic set. Consequently, the equality $\phi_p \circ f(x) = 1 + x \ \forall x \in d(0, r_1^-)$ holds in all D', a contradiction since f is not injective in D'. This finally shows that ϕ_p does not belong to $H(d(0, r_1^-))$. □

Remark 4. Let q be prime to p. Since ϕ_q belongs to $\mathcal{A}_b(d(0,1^-))$, obviously ϕ_q belongs to $H(d(0,r))$, whenever $r \in]0,1[$. Now, let E

be a closed bounded set in $\mathbb{K}$ and let $h \in H(E)$ satisfy $\|h\|_E < 1$. Then by Corollary 12.3, $\phi_q \circ h$ belongs to $H(E)$. In other words, if $g \in H(E)$ and if $\|g-1\|_E < 1$, then $\sqrt[q]{g}$ also belongs to $H(E)$.

Remark 5. Theorems 29.19 and 29.20 couldn't be significantly improved as this example shows. Let q be an integer prime to p, let $a, b \in d(1, 1^-)$, with $a \neq b$, and let $P(x) = (a-x)^{q-1}(b-x)$. It is easily seen that $|P(x)-1| < 1$ whenever $x \in d(0, 1^-)$ and then we can consider $f(x) = \sqrt[q]{P(x)}$. We will show that $f \in H(d(0,1^-)) \setminus H(d(0,1))$. Indeed, we have

$$\begin{aligned}
f(x) &= \sqrt[q]{(a-x)^q \left(\frac{b-x}{a-x}\right)} = (a-x)\sqrt[q]{1 + \left(\frac{b-a}{a-x}\right)} \\
&= (a-x)\sum_{n=0}^{\infty} \binom{\frac{1}{q}}{n} \left(\frac{b-a}{a-x}\right)^n = a - x + \frac{1}{q}(b-a) \\
&\quad + \sum_{j=1}^{\infty} \binom{\frac{1}{q}}{j+1} \frac{(b-a)^{j+1}}{(a-x)^j}.
\end{aligned}$$

This is just a Mittag-Leffler series of the form $f_0 + f_1 \in H(d(0,1^-))$, with

$$f_0 = -x + a + \frac{1}{q}(b-a) \in H(d(0,1)),$$

$$f_1 = -\sum_{j=1}^{\infty} \binom{\frac{1}{q}}{j+1} \frac{(a-b)^{j+1}}{(x-a)^j} \in H_0(\mathbb{K} \setminus d(1,1^-)).$$

Thus, we see that f belongs to $H(d(0,1^-))$, and more precisely, $f \in H(\mathbb{K} \setminus d(1,1^-))$, but $f \notin H(d(0,1))$. Actually, f^q has exactly q zeros in $d(1,1^-)$.

Chapter 30

Order of Growth for Entire Functions

Here we mean to introduce and study the notion of order of growth of an entire function on $\mathbb{K}$ in relation with the distribution of zeros in disks and in relation with the question whether an entire function can be divided by its derivative inside the algebra of entire functions. First, results were published in [16, 18, 50]. The notion of order of growth was defined by Valiron [90].

Definition and notation. Let $f \in \mathcal{A}(\mathbb{K})$. Similar to the definition known on complex entire functions, $\limsup_{r\to+\infty} \frac{\log(\log(|f|(r)))}{\log(r)}$ is called *the order of growth of f* or *the growth order of f* in brief and is denoted by $\rho(f)$. We say that f has *finite order* if $\rho(f) < +\infty$. In this chapter and in Chapters 30, 31, 32, and 33 we denote by Log the Neperian logarithm function and by e the number such that $\text{Log}(e) = 1$. Next, the function Ψ is now defined by $\Psi(x) = \text{Log}(|x|)$.

Theorem 30.1 is easily proven:

Theorem 30.1. *Let f, $g \in \mathcal{A}(\mathbb{K})$. Then*

(i) *if $c(|f|(r))^{\alpha} \geq |g|(r)$ with α and $c > 0$, when r is big enough, then $\rho(f) \geq \rho(g)$,*

(ii) *$\rho(f+g) \leq \max(\rho(f), \rho(g))$ and if $\rho(g) < \rho(f)$, then $\rho(f+g) = \rho(f)$,*

(iii) *$\rho(fg) = \max(\rho(f), \rho(g))$.*

Proof. Suppose $c(|f|(r))^\alpha \geq |g|(r)$ with α and $c > 0$ when r is big enough. Then we check that $\big(\limsup_{r\to+\infty} \frac{\mathrm{Log}(\mathrm{Log}(c(|f|(r))^\alpha))}{\mathrm{Log}(r)}\big) \geq \limsup_{r\to+\infty} \frac{\mathrm{Log}(\mathrm{Log}(|g|(r)))}{\mathrm{Log}(r)}$, hence $\rho(f) \geq \rho(g)$.

Next, we have $|f+g|(r) \leq \max(|f|(r), |g|(r))$, hence

$$\frac{\mathrm{Log}(\mathrm{Log}(|f+g|(r)))}{\mathrm{Log}(r)} \leq \max\left(\frac{\mathrm{Log}(\mathrm{Log}(|f|(r)))}{\mathrm{Log}(r)}, \frac{\mathrm{Log}(\mathrm{Log}(|g|(r)))}{\mathrm{Log}(r)}\right),$$

and hence $\rho(f+g) \leq \max(\rho(f), \rho(g))$.

Now, suppose $\rho(f) > \rho(g)$. There exists a sequence $(r_n)_{n\in\mathbb{N}}$ of limit $+\infty$ such that $\lim_{n\to+\infty} \frac{\mathrm{Log}((\mathrm{Log}(|f|(r_n)}{\mathrm{Log}(r_n)} = \rho(f)$ and $|f|(r_n) > |g|(r_n)$, hence $|f+g|(r_n) = |f|(r_n)$ and hence $\rho(f+g) \geq \rho(f)$. But we have proved that $\rho(f+g) \leq \max(\rho(f), \rho(g)) = \rho(f)$.

Let us now show that $\rho(fg) = \max(\rho(f), \rho(g))$. It is clear that $\rho(fg) \geq \max(\rho(f), \rho(g))$, since $|fg|(r) = |f|(r)|g|(r)$. Now, let $t = \max(\rho(f), \rho(g)$. Then there exists a function ω defined in $\mathbb{R}_+$, of limit 0 at ∞, such that $\frac{\mathrm{Log}(\mathrm{Log}(|f|(r)))}{\mathrm{Log}(r)} \leq t + \omega(r)$ and $\frac{\mathrm{Log}(\mathrm{Log}(|g|(r)))}{\mathrm{Log}(r)} \leq t + \omega(r)$. Hence, we have

$$\mathrm{Log}(|f|(r)) \leq r^{t+\omega(r)},\ \mathrm{Log}(|g|(r)) \leq r^{t+\omega(r)},$$

hence

$$\mathrm{Log}(|f|(r)) + \mathrm{Log}(|g|(r)) \leq 2r^{t+\omega(r)},$$

therefore

$$\mathrm{Log}(\mathrm{Log}(|f|(r)) + \mathrm{Log}(g|(r))) \leq \mathrm{Log}(2) + (t+\omega(r))\mathrm{Log}(r),$$

hence,

$$\mathrm{Log}(\mathrm{Log}(|f|(r)) + \mathrm{Log}(g|(r))) \leq \mathrm{Log}(2) + (t+\omega(r))\mathrm{Log}(r),$$

hence

$$\frac{\mathrm{Log}(\mathrm{Log}(|f|(r).|g|(r)))}{\mathrm{Log}(r)} \leq \frac{\mathrm{Log}(2)}{\mathrm{Log}(r)} + t + \omega(r),$$

and hence

$$\limsup_{r\to+\infty} \frac{\mathrm{Log}(\mathrm{Log}(|f|(r).|g|(r)))}{\mathrm{Log}(r)} \le t.$$

Consequently, $\rho(fg) \le \max(\rho(f), \rho(g))$, which ends the proof. □

Corollary 30.2. *Let $f,\ g \in \mathcal{A}(\mathbb{K})$. Then, $\rho(f^n) = \rho(f)\ \forall n \in \mathbb{N}^*$.*

Notation. Given a number $t \ge 0$, we denote by $\mathcal{A}(\mathbb{K}, t)$ the $\mathbb{K}$-algebra of entire functions of order inferior or equal to t and we put $\mathcal{A}^*(\mathbb{K}) = \bigcup_{t>0} \mathcal{A}(\mathbb{K}, t)$.

Corollary 30.3. *For any $t \ge 0$, $\mathcal{A}(\mathbb{K}, t)$ is a $\mathbb{K}$-subalgebra of $\mathcal{A}(\mathbb{K})$.*

Corollary 30.4. *Consider the differential equation*

$$(\mathcal{E})\ f^{(n)} + a_{n-1}(x) f^{(n-1)}(x) + \cdots + a_0(x) f(x) = 0$$

with $a_j \in \mathcal{A}^*(D)$, $j = 0, \ldots, n-1$ and $\rho(a_j) < \rho(a_0)\ \forall j = 1, \ldots, n-1$. Then every non-trivial solution f of $(\mathcal{E})$ satisfies $\rho(f) \ge \rho(a_0)$.

Theorem 30.5. *Let $f \in \mathcal{A}(\mathbb{K})$ and let $P \in \mathbb{K}[x]$. Then, $\rho(P \circ f) = \rho(f)$ and $\rho(f \circ P) = \deg(P)\rho(f)$.*

Proof. Let $n = \deg(P)$. For r big enough, we have

$$\begin{aligned}\mathrm{Log}(\mathrm{Log}(|f|(r))) &\le \mathrm{Log}(\mathrm{Log}(|P \circ f|(r))) \le \mathrm{Log}((n+1)\mathrm{Log}(|f|(r))) \\ &= \mathrm{Log}(n+1) + \mathrm{Log}(\mathrm{Log}(|f|(r))).\end{aligned}$$

Consequently,

$$\begin{aligned}\limsup_{r\to+\infty} \left(\frac{\mathrm{Log}(\mathrm{Log}(|f|(r)))}{\mathrm{Log}(r)}\right) &\le \limsup_{r\to+\infty} \left(\frac{\mathrm{Log}(\mathrm{Log}(|P \circ f|(r)))}{\mathrm{Log}(r)}\right) \\ &\le \limsup_{r\to+\infty} \left(\frac{\mathrm{Log}(n+1) + \mathrm{Log}(\mathrm{Log}(|f|(r)))}{\mathrm{Log}(r)}\right),\end{aligned}$$

and therefore, $\rho(P \circ f) = \rho(f)$.

Next, for r big enough, we have

$$\frac{\mathrm{Log}(\mathrm{Log}(|f|(r)))}{\mathrm{Log}(r)} \le \frac{\mathrm{Log}(\mathrm{Log}(|f \circ P|(r)))}{\mathrm{Log}(r)}$$
$$= \left(\frac{\mathrm{Log}(\mathrm{Log}(|f \circ P|(r))}{\mathrm{Log}(|P|(r))}\right)\left(\frac{\mathrm{Log}(|P|(r))}{\mathrm{Log}(r)}\right).$$

Now,

$$\limsup_{r\to+\infty}\left(\frac{\mathrm{Log}(\mathrm{Log}(|f \circ P|(r))}{\mathrm{Log}(|P|(r))}\right) = \limsup_{r\to+\infty}\left(\frac{\mathrm{Log}(\mathrm{Log}(|f|(r))}{\mathrm{Log}(r)}\right)$$

because the function h defined in $[0,+\infty[$ as $h(r) = |P|(r)$ is obviously an increasing continuous bijection from $[0,+\infty[$ onto $[|P(0)|,+\infty[$. On the other hand, it is obviously seen that $\limsup_{r\to+\infty}\left(\frac{\mathrm{Log}(|P|(r))}{\mathrm{Log}(r)}\right) = n$. Consequently,

$$\limsup_{r\to+\infty}\left(\frac{\mathrm{Log}(\mathrm{Log}(|f \circ P|(r))}{\mathrm{Log}(|P|(r))}\right) = n\limsup_{r\to+\infty}\left(\frac{\mathrm{Log}(\mathrm{Log}(|f|(r))}{\mathrm{Log}(r)}\right),$$

and hence $\rho(f \circ P) = n\rho(f)$. □

Theorem 30.6. *Let $f, g \in \mathcal{A}(\mathbb{K})$ be transcendental. Then, $\rho(f \circ g) \ge \max(\rho(f), \rho(g))$. If $\rho(f) \ne 0$, then $\rho(f \circ g) = +\infty$.*

Proof. Let $f(x) = \sum_{j=0}^{\infty} a_n x^n$ and $g(x) = \sum_{j=0}^{\infty} b_n x^n$. Since g is transcendental, for every $n \in \mathbb{N}$, there exists r_n such that $s(r_n, g) \ge n$. Then, $|g|(r) \ge |b_n|r^n \ \forall r \ge r_n$, and hence, by Theorem 30.5, we have

$$\rho(f \circ g) \ge n\rho(f). \tag{1}$$

Therefore, $\rho(f \circ g) \ge \rho(f)$.

Now, let $k \in \mathbb{N}$ be such that $a_k \ne 0$ and let s_0 be such that $s(s_0, f) \ge k$. Then, $|f|(r) \ge |a_k|r^k \ \forall r \ge s_0$, hence $|f \circ g|(r) \ge |a_k|(|g|(r))^k \ \forall r \ge s_0$, hence by Theorems 30.1 and 30.5, we have $\rho(f \circ g) \ge \rho(g)$. Next, Relation (1) is true for every $n \in \mathbb{N}$. Suppose now that $\rho(f) \ne 0$. Then, by (1), we have $\rho(f \circ g) = +\infty$. □

Notation. Let $f \in \mathcal{A}(d(0, R^-))$. For each $r \in]0, R[$, we denote by $s(r, f)$ the number of zeros of f in $d(0, r)$, taking multiplicity into account.

Theorem 30.7. *Let $f \in \mathcal{A}(\mathbb{K})$ be not identically zero and such that for some $t \geq 0$, $\limsup_{r\to+\infty} \frac{s(r,f)}{r^t}$ is finite. Then, $\rho(f) \leq t$.*

Proof. Set $\limsup_{r\to+\infty} \frac{s(r,f)}{r^t} = b \in [0, +\infty[$. Let us fix $\epsilon > 0$. We can find $R > 0$ such that $\frac{s(r,f)}{r^t} \leq b + \epsilon \ \forall r \geq R$, and hence, by Corollary 22.30, we have $\frac{|f|(r)}{|f|(R)} \leq \left(\frac{r}{R}\right)^{s(r,f)} \leq \left(\frac{r}{R}\right)^{r^t(b+\epsilon))}$. Therefore, putting $M = |f|(R)$, we have

$$\mathrm{Log}(|f|(r)) \leq \mathrm{Log}(M) + r^t(b+\epsilon)(\mathrm{Log}(r) - \mathrm{Log}(R)).$$

Now, when $u > 2$, $v > 2$, we know that $\mathrm{Log}(u+v) \leq \mathrm{Log}(u) + \mathrm{Log}(v)$. Applying that inequality with $u = M$ and $v = r^t(b+\epsilon)$ $(\mathrm{Log}(r) - \mathrm{Log}(R))$ when $r^t(b+\epsilon)(\mathrm{Log}(r) - \mathrm{Log}(R)) > 2$, since $\mathrm{Log}R \geq 0$, yields

$$\begin{aligned}
&\mathrm{Log}(\mathrm{Log}(|f|(r))) \\
&\quad \leq \mathrm{Log}(\mathrm{Log}(M)) + t\mathrm{Log}(r) + \mathrm{Log}(b+\epsilon) + \mathrm{Log}(\mathrm{Log}(r) - \mathrm{Log}R)) \\
&\quad \leq \mathrm{Log}(\mathrm{Log}(M)) + t\mathrm{Log}(r) + \mathrm{Log}(b+\epsilon) + \mathrm{Log}(\mathrm{Log}(r)).
\end{aligned}$$

Consequently,

$$\begin{aligned}
&\frac{\mathrm{Log}(\mathrm{Log}(|f|(r)))}{\mathrm{Log}(r)} \\
&\quad \leq \frac{\mathrm{Log}(\mathrm{Log}(M)) + t\mathrm{Log}(r) + \mathrm{Log}(b+\epsilon) + \mathrm{Log}(\mathrm{Log}(r))}{\mathrm{Log}(r)},
\end{aligned}$$

and hence we can check that

$$\limsup_{r\to+\infty} \frac{\mathrm{Log}(\mathrm{Log}(|f|(r)))}{\mathrm{Log}(r)} \leq t.$$

□

Theorem 30.8. *Let $f \in \mathcal{A}(\mathbb{K})$ be not identically zero. If there exists $q \geq 0$ such that*

$$\limsup_{r\to+\infty} \left(\frac{s(r,f)}{r^q}\right) < +\infty,$$

then $\rho(f)$ is the lowest bound of the set of $q \in [0, +\infty[$ such that

$$\limsup_{r\to+\infty} \left(\frac{s(r,f)}{r^q}\right) = 0.$$

Moreover, if $\limsup_{r\to+\infty}\left(\frac{s(r,f)}{r^t}\right)$ *is a number* $b \in]0,+\infty[$, *then* $\rho(f) = t$.

If there exists no q *such that* $\limsup_{r\to+\infty}\left(\frac{s(r,f)}{r^q}\right) < +\infty$, *then* $\rho(f) = +\infty$.

Proof. The proof holds in two statements. First, we prove that given $f \in \mathcal{A}(\mathbb{K})$ non-constant and such that for some $t \geq 0$, $\limsup_{r\to+\infty} \frac{s(r,f)}{r^t}$ is finite, then $\rho(f) \leq t$.

Set $\limsup_{r\to+\infty}\left(\frac{s(r,f)}{r^t}\right) = b \in [0,+\infty[$. Let us fix $\epsilon > 0$. We can find $R > 1$ such that $|f|(R) > e^2$ and $\frac{s(r,f)}{r^t} \leq b+\epsilon\ \forall r \geq R$, and hence, by Corollary 22.30, we have $\frac{|f|(r)}{|f|(R)} \leq \left(\frac{r}{R}\right)^{s(r,f)} \leq \left(\frac{r}{R}\right)^{r^t(b+\epsilon))}$. Therefore, since $R > 1$, we have

$$\mathrm{Log}(|f|(r)) \leq \mathrm{Log}(|f|(R)) + r^t(b+\epsilon)(\mathrm{Log}(r)).$$

Now, when $u > 2$, $v > 2$, we check that $\mathrm{Log}(u+v) \leq \mathrm{Log}(u) + \mathrm{Log}(v)$. Applying that inequality with $u = \mathrm{Log}(|f|(R))$ and $v = r^t(b+\epsilon)(\mathrm{Log}(r))$ when $r^t(b+\epsilon)(\mathrm{Log}(r)) > 2$ yields

$$\begin{aligned}\mathrm{Log}(\mathrm{Log}(|f|(r))) &\leq \mathrm{Log}(\mathrm{Log}(|f|(R))) + t\mathrm{Log}(r)\\ &\quad + \mathrm{Log}(b+\epsilon) + \mathrm{Log}(\mathrm{Log}(r)).\end{aligned}$$

Consequently,

$$\begin{aligned}&\frac{\mathrm{Log}(\mathrm{Log}(|f|(r)))}{\mathrm{Log}(r)}\\ &\quad \leq \frac{\mathrm{Log}(\mathrm{Log}(|f|(R))) + t\mathrm{Log}(r) + \mathrm{Log}(b+\epsilon) + \mathrm{Log}(\mathrm{Log}(r))}{\mathrm{Log}(r)},\end{aligned}$$

and hence we can check that

$$\limsup_{r\to+\infty} \frac{\mathrm{Log}(\mathrm{Log}(|f|(r)))}{\mathrm{Log}(r)} \leq t$$

which proves the first claim.

Second, we prove that given $f \in \mathcal{A}(\mathbb{K})$ not identically zero and such that for some $t \geq 0$, we have $\limsup_{r\to+\infty} \frac{s(r,f)}{r^t} > 0$, then $\rho(f) \geq t$.

By hypotheses, there exists a sequence $(r_n)_{n\in\mathbb{N}}$ such that $\lim_{n\to+\infty} r_n = +\infty$ and such that $\lim_{n\to+\infty} \frac{s(r_n,f)}{r_n^t} > 0$. Thus, there exists $b > 0$ such that $\lim_{n\to+\infty} \frac{s(r_n,f)}{r_n^t} \geq b$. We can assume that $|f|(r_0) \geq 1$, hence by Corollary 22.27, $|f|(r_n) \geq 1\ \forall n$. Let $\lambda \in]1, +\infty[$. By Corollary 22.30, we have

$$\frac{|f|(\lambda r_n)}{|f|(r_n)} \geq (\lambda)^{s(r_n,f)} \geq (\lambda)^{[b(r_n)^t]},$$

hence

$$\mathrm{Log}(|f|(\lambda r_n) \geq \mathrm{Log}(|f|(r_n)) + b(r_n)^t \mathrm{Log}(\lambda).$$

Since $|f|(r_n) \geq 1$, we have $\mathrm{Log}(\mathrm{Log}(|f|(\lambda r_n))) \geq \mathrm{Log}(b\mathrm{Log}(\lambda)) + t\mathrm{Log}(r_n)$, therefore

$$\frac{\mathrm{Log}(\mathrm{Log}(|f|(\lambda r_n))}{\mathrm{Log}(r_n)} \geq t + \frac{\mathrm{Log}(b\mathrm{Log}(\lambda))}{\mathrm{Log}(r_n)} \quad \forall n \in \mathbb{N},$$

and hence

$$\limsup_{r\to+\infty} \frac{\mathrm{Log}(\mathrm{Log}(|f|(r)))}{\mathrm{Log}(r)} \geq t$$

which ends the proof of the second claim. □

Definition and notation. Let $t \in [0, +\infty[$ and let $f \in \mathcal{A}(\mathbb{K})$ be of order t. We set $\psi(f) = \limsup_{r\to+\infty} \frac{s(r,f)}{r^t}$ and call $\psi(f)$ the *cotype of growth* of f, or just the *cotype* of f in brief and we put $\widetilde{\psi}(f) = \liminf_{r\to+\infty} \frac{s(r,f)}{r^t}$.

Lemma 30.9. *Let $f,\ g \in \mathcal{A}(\mathbb{K})$. Then,* $\max\big(s(r,f), s(r,g)\big) \leq s(r, fg) = s(r,f) + s(r,g)$.

Proof. Let $r' > r$ be such that both $s(u,f)$ and $s(u,g)$ are constant in $[r,r']$. Then, when $\mu = \mathrm{Log}(u)$, by Corollary 19.3, we have

$$s(u,f) = \frac{d\Psi(f,\mu)}{d\mu}, \quad s(u,g) = \frac{d\Psi(g,\mu)}{d\mu} \quad \forall \mu \in]\mathrm{Log}(r), \mathrm{Log}(r')[.$$

But since $\Psi(fg,\mu) = \Psi(f,\mu) + \Psi(g,\mu)$, the inequalities

$$\max\big(s(r,f), s(r,g)\big) \le s(r,fg) = s(r,f) + s(r,g)$$

are clear. □

Theorem 30.10. *Let $f,\ g \in \mathcal{A}^*(\mathbb{K})$. Then $\psi(fg) \le \psi(f) + \psi(g)$. Moreover, if $\rho(f) \ge \rho(g)$, then $\psi(f) \le \psi(fg)$. If $\rho(f) = \rho(g)$, then $\max\big(\psi(f), \psi(g)\big) \le \psi(fg)$.*

Proof. Set $q = \rho(f)$, $t = \rho(g)$ and suppose $q \ge t$. By Theorem 30.1, we have $\rho(f.g) = \rho(f) = q$. By Lemma 30.9, for each $r > 0$, we have $\max\big(s(r,f), s(r,g)\big) \le s(r,f.g) = s(r,f) + s(r,g)$. Consequently,

$$\begin{aligned}\limsup_{r\to+\infty} \frac{s(r,f.g)}{r^q} &\le \limsup_{r\to+\infty} \frac{s(r,f)}{r^q} + \limsup_{r\to+\infty} \frac{s(r,g)}{r^q}\\ &\le \limsup_{r\to+\infty} \frac{s(r,f)}{r^q} + \limsup_{r\to+\infty} \frac{s(r,g)}{r^t}\\ &= \psi(f) + \psi(g).\end{aligned}$$

Moreover, assuming again $q \ge t$, then

$$\psi(f) = \limsup_{r\to+\infty} \frac{s(r,f)}{r^q} \le \limsup_{r\to+\infty} \frac{s(r,fg)}{r^q} = \limsup_{r\to+\infty} \frac{s(r,fg)}{r^{\rho(fg)}} = \psi(fg).$$

Consequently, if $\rho(f) = \rho(g)$, then $\max\big(\psi(f), \psi(g)\big) \le \psi(fg)$. □

Theorem 30.11. *Let $f(x) = \sum_{n=0}^{+\infty} a_n x^n \in \mathcal{A}(\mathbb{K})$. Then,*

$$\rho(f) = \limsup_{n\to+\infty} \left(\frac{n\mathrm{Log}(n)}{-\mathrm{Log}|a_n|}\right).$$

Proof. We follow a similar way as this of [85] when $\rho(f) < +\infty$.

Let $t = \rho(f)$ and suppose first that $t < +\infty$. Let $\alpha = \limsup_{n\to+\infty} \frac{n\mathrm{Log}(n)}{-\mathrm{Log}|a_n|}$. Take $q > t$. For all $n \in \mathbb{N}$, we have $|a_n|r^n \le |f|(r)$, and therefore, $|a_n|r^n \le e^{(r^q)}$ and hence $|a_n| \le r^{-n}e^{(r^q)}$ i.e.

$$\mathrm{Log}|a_n| \le r^q - n\mathrm{Log}(r), \tag{1}$$

when r is big enough.

Now, choose $r = \left(\frac{n}{s}\right)^{\frac{1}{q}}$. So, we have $\mathrm{Log}|a_n| \leq \frac{n}{q} - \frac{n}{q}\mathrm{Log}(\frac{n}{q})$, i.e.

$$-\mathrm{Log}(|a_n|) \geq -\frac{n}{q} + \frac{n}{q}\mathrm{Log}\left(\frac{n}{q}\right).$$

Consequently, when n is big enough, we have

$$\frac{n\mathrm{Log}(n)}{(-\mathrm{Log}|a_n|)} \leq \frac{n\mathrm{Log}(n)}{\frac{n}{q}\mathrm{Log}(\frac{n}{q}) - \frac{n}{q}} \leq q + O(1).$$

Therefore, we have $\alpha \leq q$ and since this is true for each $q > t$, that shows that $\alpha \leq t$.

Now, take $\beta > \alpha$ so that $\frac{n\mathrm{Log}(n)}{(-\mathrm{Log}|a_n|)} < \beta$ for n big enough. Then, when n is big enough, we have $n\mathrm{Log}(n) \leq \beta(-\mathrm{Log}|a_n|)$, hence $n^{\frac{n}{\beta}} \leq \frac{1}{|a_n|}$, and hence $|a_n| \leq \frac{1}{n^{\frac{n}{\beta}}}$. Consequently, $|a_n|r^n \leq \frac{r^n}{n^{\frac{n}{\beta}}}$. Now, for r big enough, $|f|(r) = \sup_{n\in\mathbb{N}} |a_n|r^n \leq \sup_{n\in\mathbb{N}} \frac{r^n}{n^{\frac{n}{\beta}}}$.

Putting $\varphi(n) = \frac{n}{\beta}$ and $R = \frac{r}{\beta}$, we have

$$|f|(r^{\frac{1}{\beta}}) \leq \sup_{n\in\mathbb{N}} \frac{r^{\varphi(n)}}{n^{\varphi(n)}} \leq \sup_{x>0} \frac{R^x}{x^x}.$$

Now, we check that the maximum on $[0, +\infty[$ of the function $g(x) = \frac{R^x}{x^x}$ is reached when $x = \frac{R}{e}$ and hence is $e^{\frac{R}{e}} = e^{\frac{r}{\beta e}}$. Therefore, we have $|f|(r^{\frac{1}{\beta}}) \leq e^{\frac{r}{\beta e}}$. Putting now $u = r^{\frac{1}{\beta}}$, we can derive $|f|(u) \leq e^{\frac{u^\beta}{\beta e}}$, hence

$$\mathrm{Log}(\mathrm{Log}(|f|(u))) \leq \beta\mathrm{Log}(u) - \mathrm{Log}(e\beta).$$

Consequently,

$$\limsup_{r\to+\infty} \frac{\mathrm{Log}(\mathrm{Log}(|f|(r)))}{\mathrm{Log}(r)} \leq \beta.$$

So, we have $t \leq \beta$ and since this is true for all $\beta > \alpha$, we have proven that $t \leq \alpha$, which ends the proof when $t < +\infty$.

Suppose now that $t = +\infty$ and suppose that $\limsup_{n\to+\infty} \frac{n\mathrm{Log}(n)}{(-\mathrm{Log}|a_n|)} < +\infty$. Let us take $q \in \mathbb{N}$ such that

$$\frac{n\mathrm{Log}(n)}{(-\mathrm{Log}|a_n|)} < q \quad \forall n \in \mathbb{N}. \tag{2}$$

By Theorem 30.8, we have $\limsup_{r\to+\infty} \frac{s(r,f)}{r^q} = +\infty$. So, we can take a sequence $(r_m)_{m\in\mathbb{N}}$ such that

$$\lim_{m\to+\infty} \frac{s(r_m, f)}{(r_m)^q} = +\infty. \tag{3}$$

For simplicity, set $u_m = s(r_m, f)$, $m \in \mathbb{N}$. By (2), for m big enough, we have

$$u_m\mathrm{Log}(u_m) < q(-\mathrm{Log}(|a_{u_m}|) = q\mathrm{Log}\left(\frac{1}{|a_{u_m}|}\right),$$

hence

$$\frac{1}{(u_m)^{u_m}} > |a_{u_m}|^q,$$

therefore

$$|a_{u_m}|^q (r_m)^{qu_m} < \frac{(r_m)^{qu_m}}{(u_m)^{u_m}}$$

i.e.

$$(|f|(r_m))^s < \left(\frac{(r_m)^s}{u_m}\right)^{u_m}.$$

But by Corollary 22.27, we have $\lim_{r\to+\infty} |f|(r_m) = +\infty$, hence $(r_m)^q > u_m$ when m is big enough, and therefore, $\limsup_{m\to+\infty} \frac{s(r_m,f)}{(r_m)^q} \leq 1$, a contradiction to (3). Consequently, (2) is impossible, and therefore,

$$\limsup_{n\to+\infty} \left(\frac{n\mathrm{Log}(n)}{-\mathrm{Log}|a_n|}\right) = +\infty = \rho(f).$$

□

Remark. Of course, polynomials have a growth order equal to 0. On $\mathbb{K}$ as on $\mathbb{C}$, we can easily construct transcendental entire functions of order 0 or of order ∞.

Example 1. Suppose that for each $r > 0$ we have $s(r, f) \in [r^t \text{Log} r, r^t \text{Log} r + 1]$. Then of course, for every $s > t$, we have $\limsup_{r\to+\infty} \frac{s(r,f)}{r^s} = 0$ and $\limsup_{r\to+\infty} \frac{s(r,f)}{r^t} = +\infty$, so there exists no $t > 0$ such that $\frac{s(r,f)}{r^t}$ have non-zero superior limit $b < +\infty$. Consequently, $\rho(f) = +\infty$.

Example 2. Let $(a_n)_{n\in\mathbb{N}}$ be a sequence in $\mathbb{K}$ such that $-\text{Log}|a_n| \in [n(\text{Log}(n))^2, n(\text{Log}(n))^2 + 1]$. Then clearly, $\lim_{n\to+\infty} \frac{\text{Log}|a_n|}{n} = -\infty$, hence the function $\sum_{n=0}^{\infty} a_n x^n$ has radius of convergence equal to $+\infty$. On the other hand, $\lim_{n\to+\infty} \frac{n\text{Log}(n)}{-\text{Log}|a_n|} = 0$, hence $\rho(f) = 0$.

Example 3. Let $(a_n)_{n\in\mathbb{N}}$ be a sequence in $\mathbb{K}$ such that $-\text{Log}|a_n| \in [n\sqrt{\text{Log}(n)}, n\sqrt{\text{Log}(n)} + 1]$. Then, $\lim_{n\to+\infty} \frac{\text{Log}|a_n|}{n} = -\infty$ again, and hence the function $\sum_{n=0}^{\infty} a_n x^n$ has radius of convergence equal to $+\infty$. On the other hand,

$$\lim_{n\to+\infty} \left(\frac{n\text{Log}(n)}{-\text{Log}|a_n|} \right) = +\infty \text{ hence } \rho(f) = +\infty.$$

Theorem 30.12. *Let $f \in \mathcal{A}^*(\mathbb{K})$. Then,*

$$\rho(f) = \inf \left\{ q \in]0, +\infty[\ \middle| \ \lim_{r\to+\infty} \frac{\text{Log}(|f|(r))}{r^q} = 0 \right\}.$$

Proof. Indeed, let $M = \inf\{q \in]0, +\infty[\mid \lim_{r\to+\infty} \frac{\log(|f|(r))}{r^q} = 0\}$. First, we prove that $\rho(f) \le M$. Let q be such that $\lim_{r\to+\infty} \frac{\text{Log}(|f|(r))}{r^q} = 0$. Let us fix $\epsilon > 0$. For r big enough, we have $\frac{\text{Log}(|f|(r))}{r^q} \le \epsilon$, hence $\text{Log}(|f|(r)) \le \epsilon r^q$, therefore $\text{Log}(\text{Log}(|f|(r))) \le \text{Log}\epsilon + q\text{Log}(r)$, hence $\frac{\text{Log}(\text{Log}(|f|(r)))}{\text{Log}(r)} \le q + \frac{\epsilon}{\text{Log}(r)}$. This is true for every $\epsilon > 0$, therefore $\limsup_{r\to+\infty} \frac{\text{Log}(\text{Log}(|f|(r)))}{\text{Log}(r)} \le q$ i.e. $\rho(f) \le q$, and hence, $\rho(f) \le M$.

On the other hand, we note that

$$M = \sup \left\{ q \in]0, +\infty[\ \middle| \ \limsup_{r\to+\infty} \frac{\text{Log}(|f|(r))}{r^q} > 0 \right\}.$$

Now, suppose that for some $q > 0$, we have $\limsup_{r\to+\infty} \frac{\text{Log}(|f|(r))}{r^q} = b > 0$. Let us fix $\epsilon \in]0, b[$. There exists a sequence $(r_n)_{n\in\mathbb{N}}$ such that, when n is big enough, we have $b - \epsilon \leq \frac{\text{Log}(|f|(r_n))}{(r_n)^q} \leq b + \epsilon$, hence $q\text{Log}(r_n) + \text{Log}(b - \epsilon) < \text{Log}(\text{Log}(|f|(r_n))) < q\text{Log}(r_n) + \text{Log}(b - \epsilon)$, therefore

$$q + \frac{\text{Log}(b-\epsilon)}{\text{Log}(r_n)} < \frac{\text{Log}(\text{Log}(|f|(r_n)}{\text{Log}(r_n)} < q + \frac{\text{Log}(b+\epsilon)}{\text{Log}(r_n)}.$$

Consequently, $\lim_{n\to+\infty} \frac{\text{Log}(\text{Log}(|f|(r_n))}{\text{Log}(r_n)} = q$, and therefore, $\rho(f) \geq q$, hence $\rho(f) \geq M$. Finally, $\rho(f) = M$. □

Chapter 31

Type of Growth for Entire Functions

Definition and notation. In complex analysis, the *type of growth* is defined for an entire function having a finite order of growth t as $\sigma(f) = \limsup_{r\to+\infty} \frac{\text{Log}(M_f(r))}{r^t}$, with $t < +\infty$ [90]. Of course, the same notion may be defined for $f \in \mathcal{A}(\mathbb{K})$. Here, as in Chapters 30, we denote by Log the Neperian logarithm and by e the number such that $\text{Log}(e) = 1$. Then, given $f \in \mathcal{A}^*(\mathbb{K})$ of order t, we set $\sigma(f) = \limsup_{r\to+\infty} \frac{\text{Log}(|f|(r))}{r^t}$. Moreover, we put $\widetilde{\sigma}(f) = \liminf_{r\to+\infty} \frac{\text{Log}(|f|r))}{r^t}$.

A function $f \in \mathcal{A}(\mathbb{K})$ is said to be *clean* if $\rho(f) < +\infty$ and $\sigma(f) = \widetilde{\sigma}(f)$ and f is said to be *regular* if $\rho(f) = \widetilde{\rho}(f)$.

Theorem 31.1. *Let f, $g \in \mathcal{A}^*(\mathbb{K})$. Then, $\sigma(fg) \le \sigma(f) + \sigma(g)$. If $\rho(f) \ge \rho(g)$, then $\sigma(f) \le \sigma(fg)$, and if $\rho(f) > \rho(g)$, then $\sigma(fg) = \sigma(f)$. If f is clean and such that $\rho(f) > \rho(g)$, then fg is clean. If f and g are clean and if $\rho(f) = \rho(g)$, then fg is clean and $\sigma(fg) = \sigma(f) + \sigma(g)$ and $\psi(fg) = \psi(f) + \psi(g)$. If $\rho(f) = \rho(g)$, then* $\max(\sigma(f), \sigma(g)) \le \sigma(fg)$. *If $\rho(f) = \rho(g)$ and $\sigma(f) > \sigma(g)$, then $\rho(f+g) = \rho(f)$ and $\sigma(f+g) = \sigma(f)$. If $\rho(f+g) = \rho(f) \ge \rho(g)$, then $\sigma(f+g) \le \max(\sigma(f), \sigma(g))$.*

Proof. Let $s = \rho(f) \ge t = \rho(g)$. Then, $\rho(fg) = s$, hence

$$\sigma(fg) = \limsup_{r\to+\infty} \frac{\text{Log}(|fg|(r))}{r^s} = \limsup_{r\to+\infty} \frac{\text{Log}(|f|(r).|g|(r))}{r^s}.$$

Now, if $s \geq t$, then

$$\sigma(fg) = \limsup_{r\to+\infty} \left(\frac{\text{Log}(|f|(r))}{r^s} + \frac{\text{Log}(|g|(r))}{r^s} \right)$$

$$\leq \limsup_{r\to+\infty} \frac{\text{Log}(|f|(r))}{r^s} + \limsup_{r\to+\infty} \frac{\text{Log}(|g|(r))}{r^t} = \sigma(f) + \sigma(g).$$

Then we note that when $t < s$, we have

$$\limsup_{r\to+\infty} \frac{\text{Log}(|g|(r))}{r^s} = 0,$$

hence

$$\limsup_{r\to+\infty} \frac{\text{Log}(|fg|(r))}{r^s} = \limsup_{r\to+\infty} \frac{\text{Log}(|f|(r))}{r^s} = \sigma(f).$$

Particularly, if f is clean, we have limits instead of limitsup as long as f is concerned. Consequently, if $t < s$, then fg is clean.

Now, suppose that f and g are clean and that $s = t$. Then,

$$\limsup_{r\to+\infty} \frac{\text{Log}(|f|(r)) + \text{Log}(|g|(r))}{r^t} = \lim_{r\to+\infty} \frac{\text{Log}(|f|(r)) + \text{Log}(|g|(r))}{r^t}$$

$$= \sigma(f) + \sigma(g).$$

Thus, fg is clean. Moreover, since f and g are clean, we can see that $\sigma(fg) = \sigma(f) + \sigma(g)$ and $\psi(fg) = \psi(f) + \psi(g)$.

Now, suppose again $s > t$. Then, by Theorem 30.1, $\rho(f+g) = \rho(f) = s$. Consequently,

$$\sigma(f+g) = \limsup_{r\to+\infty} \left(\frac{\text{Log}|f+g|(r)}{r^s} \right)$$

$$\leq \limsup_{r\to+\infty} \left(\frac{\max(\text{Log}|f|(r), \text{Log}|g|(r))}{r^s} \right)$$

$$= \max\left(\limsup_{r\to+\infty} \left(\frac{\text{Log}|f|(r)}{r^s} \right), \limsup_{r\to+\infty} \left(\frac{\text{Log}|g|(r)}{r^s} \right) \right)$$

$$\leq \max\left(\limsup_{r\to+\infty} \left(\frac{\text{Log}|f|(r)}{r^s} \right), \limsup_{r\to+\infty} \left(\frac{\text{Log}|g|(r)}{r^t} \right) \right)$$

$$= \max\left(\sigma(f), \sigma(g)\right).$$

Now, suppose $\rho(f) = \rho(g) = s$. Then,

$$\max\left(\limsup_{r\to+\infty}\left(\frac{\mathrm{Log}(|f|(r))}{r^s}\right), \limsup_{r\to+\infty}\left(\frac{\mathrm{Log}(|g|(r))}{r^s}\right)\right)$$
$$\leq \limsup_{r\to+\infty}\left(\frac{\mathrm{Log}(|f.g|(r))}{r^s}\right)$$

because the two both $|f|(r)$ and $|g|(r)$ tend to $+\infty$ with r. Consequently, $\sigma(fg) \geq \max(\sigma(f), \sigma(g))$.

Now, suppose again that $\rho(f) = \rho(g)$ and suppose $\sigma(f) > \sigma(g)$. Let $s = \rho(f)$, $b = \sigma(f)$. Then, $b > 0$. Let $(r_n)_{n\in\mathbb{N}}$ be a sequence such that $\lim_{n\to+\infty} r_n = +\infty$ and $\lim_{n\to+\infty} \frac{\mathrm{Log}(|f|(r_n))}{(r_n)^s} = b$. Since $\sigma(g) < \sigma(f)$, we note that when n is big enough we have $|g|(r_n) < |f|(r_n)$. Consequently, when n is big enough, we have $|f+g|(r_n) = |f|(r_n)$, and hence

$$(1) \qquad \lim_{n\to+\infty} \frac{\mathrm{Log}(|f+g|(r_n))}{(r_n)^s} = b.$$

Now, by definition of σ, we have $\sigma(f+g) \geq \lim_{n\to+\infty} \frac{\mathrm{Log}(|f+g|(r_n))}{(r_n)^{\rho(f+g)}}$. By Theorem 30.1, we have $\rho(f+g) \leq s$, hence

$$\sigma(f+g) \geq \lim_{n\to+\infty} \frac{\mathrm{Log}(|f+g|(r_n))}{(r_n)^{\rho(f+g)}} \geq \lim_{n\to+\infty} \frac{\mathrm{Log}(|f+g|(r_n))}{(r_n)^s}$$
$$= \lim_{n\to+\infty} \frac{\mathrm{Log}(|f|(r_n))}{(r_n)^s} = \sigma(f),$$

therefore by (1), $\sigma(f+g) \geq \sigma(f)$.

Suppose that $\sigma(f+g) > \sigma(f)$. Putting $h = f+g$, we have $f = h - g$ with $\sigma(g) < \sigma(h)$, hence $\sigma(h-g) \geq \sigma(h)$ i.e. $\sigma(f) > \sigma(f+g)$, a contradiction. Consequently, $\sigma(f+g) = \sigma(f)$. Thus, $\limsup_{r\to+\infty} \frac{\mathrm{Log}(|f+g|(r))}{r^s} = b > 0$. But then, $\limsup_{r\to+\infty} \frac{\mathrm{Log}(|f+g|(r))}{r^m} = 0\ \forall m > s$. Therefore, by Theorem 30.12, we have $\rho(f+g) = \rho(f)$.

Finally, suppose now that $\rho(f+g) = \rho(f) \geq \rho(g)$. Then,

$$\begin{aligned}\sigma(f+g) &= \limsup_{r\to+\infty} \frac{\mathrm{Log}(|f+g|(r))}{r^s}\\ &\leq \max\left(\limsup_{r\to+\infty}\frac{\mathrm{Log}(|f|(r))}{r^s}, \limsup_{r\to+\infty}\frac{\mathrm{Log}(|g|(r))}{r^s}\right)\\ &\leq \max\left(\limsup_{r\to+\infty}\frac{\mathrm{Log}(|f|(r))}{r^s}, \limsup_{r\to+\infty}\frac{\mathrm{Log}(|g|(r))}{r^t}\right)\\ &= \max(\sigma(f),\sigma(g)).\end{aligned}$$

The last statement derives from the previous ones and from Theorem 30.1(iii). □

Corollary 31.2. *Let $f,\ g \in \mathcal{A}(\mathbb{K})$ be such that $\rho(f) \neq \rho(g)$. Then, $\sigma(f+g) \leq \max(\sigma(f),\sigma(g))$. The set of clean functions $f \in \mathcal{A}(\mathbb{K})$ is a multiplicative semi-group. The set $(\mathcal{C}(t),.)$ of clean functions $f \in \mathcal{A}(\mathbb{K})$ of order t is a submultiplicative semi-group and σ and ψ are semi-group morphisms from $(\mathcal{C}(t),.)$ into $(\mathbb{R}^+,+)$.*

Proof. Indeed, assuming that $\rho(f) > \rho(g)$, we have $\rho(f+g) = \rho(f)$ and hence the conclusion comes from the last statement of Theorem 31.1. □

Now we show that $\sigma(f)$ may be computed by the same formula as on $\mathbb{C}$.

Theorem 31.3 [16]. *Let $f(x) = \sum_{n=0}^{\infty} b_n x^n \in \mathcal{A}(\mathbb{K})$ be such that $0 < \rho(f) < +\infty$. Then, $\sigma(f)\rho(f)e = \limsup_{n\to+\infty}\left(n\sqrt[n]{|b_n|^{\rho(f)}}\right)$.*

Proof. Let $t = \rho(f)$. First, let us show that $et\sigma(f) \geq \limsup_{n\to+\infty} n|b_n|^{\frac{t}{n}}$. We follow a similar way as in [86]. Let $u = \sigma(f)$ and let us take $w > \sigma(f)$. For r big enough, we have $\mathrm{Log}(|f|(r)) \leq wr^t$, hence for all $n \in \mathbb{N}$, we can derive

$$|b_n| \leq \frac{|f|(r)}{r^n} \leq \frac{e^{wr^t}}{r^n}. \tag{1}$$

Now, let us take r such that the derivative of the logarithm of the function $\frac{e^{wr^t}}{r^n}$ vanishes: we have $wtr^{t-1} - \frac{n}{r} = 0$. So, we can choose

$r_n = \left(\frac{n}{wt}\right)^{\frac{1}{t}}$ and we can check that

$$|b_n| \leq \frac{e^{\frac{n}{t}}}{\left(\frac{n}{wt}\right)^{\frac{n}{t}}} = \left(\frac{ewt}{n}\right)^{\frac{n}{t}}.$$

Consequently, we have $n|b_n|^{\frac{t}{n}} \leq ewt$, therefore $\limsup_{n\to+\infty} n|b_n|^{\frac{t}{n}} \leq etw$. This is true for all $w > \sigma(f)$, and hence $\limsup_{n\to+\infty} n|b_n|^{\frac{t}{n}} \leq et\sigma(f)$.

Now, let us show the reverse inequality. Take $c > \frac{1}{et}$ $\limsup_{n\to+\infty} n|b_n|^{\frac{t}{n}}$. When n is big enough, we have $|b_n| \leq \left(\frac{ect}{n}\right)^{\frac{n}{t}}$, hence $|b_n|r^n \leq \left(\frac{ect}{n}\right)^{\frac{n}{t}} r^n$, and consequently, $|f|(r) \leq \sup_{n\geq 1}$ $\left(\frac{ect}{n}\right)^{\frac{n}{t}} r^n$. Therefore, $|f|(r) \leq \sup_{x>1} \frac{(ect)^{\frac{x}{t}} r^x}{x^{\frac{x}{t}}}$. Now, set $y = \frac{x}{t}$ and $R = ecr$. Then,

$$|f|(r^{\frac{1}{t}}) \leq \sup_{y>0} \frac{(ect)^y r^y}{(ty)^y} = \sup_{y>0} \left(\frac{ecr}{y}\right)^y = \sup_{y>0} \frac{R^y}{y^y} = e^{\frac{R}{e}} = e^{cr}.$$

Thus, we have $|f|(r) \leq e^{cr^t}$, and hence $\limsup_{r\to+\infty} \frac{\operatorname{Log}(|f|(r))}{r^t} \leq c$. Therefore, $\sigma(f) \leq c$, which ends the proof. □

In the proof of Theorem 31.5, we use the following trivial lemma:

Lemma 31.4. *Let g, h be the real functions defined in $]0, +\infty[$ as $g(x) = \frac{e^{tx}-1}{x}$ and $h(x) = \frac{1-e^{-tx}}{x}$ with $t > 0$. Then,*

(i) $\inf\{|g(x)| \mid x > 0\} = t$,
(ii) $\sup\{|h(x)| \mid x > 0\} = t$.

Notations. Let $f \in \mathcal{A}(\mathbb{K})$ and let $(a_n)_{n\in\mathbb{N}}$ be the sequence of zeros of f with $|a_n| \leq |a_{n+1}|$, $n \in \mathbb{N}$ and for each $n \in \mathbb{N}$, let w_n be the multiplicity order of a_n. For every $r > 0$, let $k(r)$ be the integer such that $|a_n| \leq r\ \forall n \leq k(r)$ and $|a_n| > r\ \forall n > k(r)$. Then, by Corollary 22.27, $\operatorname{Log}(|f|(r))$ is of the form $\sum_{n=0}^{k(r)} w_n(\operatorname{Log}(r) - \operatorname{Log}(|a_n|))$. We put $\sigma(f, r) = \frac{\sum_{n=0}^{k(r)} w_n(\operatorname{Log}(r)-\operatorname{Log}(|a_n|))}{r^{\rho(f)}}$. In the same way, for any $r > 0$ and $n \in \mathbb{N}$, $\psi(f, r) = \frac{s(r,f)}{r^{\rho(f)}}$.

Theorem 31.5. *Let $f \in \mathcal{A}(\mathbb{K})$ be such that $0 < \rho(f) < +\infty$. Then, $\sigma(f) < +\infty$ if and only if $\psi(f) < +\infty$. Suppose that these hypotheses* *are satisfied. Then,*

$$\rho(f)\sigma(f) \le \psi(f) \le \rho(f)\Big(e\sigma(f) - \widetilde{\sigma}(f)\Big)$$

and

$$\rho(f)\Big(\widetilde{\sigma}(f) - \frac{\sigma(f)}{e}\Big) \le \widetilde{\psi}(f) \le \rho(f)\widetilde{\sigma}(f).$$

Further, the hypotheses $\sigma(f) = \widetilde{\sigma}(f)$ and $\psi(f) = \widetilde{\psi}(f)$ are equivalent, and if they are satisfied, then $\psi(f) = \rho(f)\sigma(f)$.

Proof. Without loss of generality, we can assume that $f(0) = 1$. Let $u = \rho(f)$. Let $(a_n)_{n\in\mathbb{N}}$ be the sequence of zeros of f with $|a_n| \le |a_{n+1}|$, $n \in \mathbb{N}$, and for each $n \in \mathbb{N}$, let w_n be the multiplicity order of a_n. For every $r > 0$, let $k(r)$ be the integer such that $|a_n| \le r\ \forall n \le k(r)$ and $|a_n| > r\ \forall n > k(r)$. We put $c_n = |a_n|$, $n \in \mathbb{N}$.

We first show the inequality $\rho(f)\sigma(f) \le \psi(f)$. By the definition of $\sigma(f, r)$, we can derive

$$\begin{aligned}\sigma(f,r) = &\sum_{n=0}^{k(re^{-\alpha})} \frac{w_n\big(\mathrm{Log}(r) - \mathrm{Log}(re^{-\alpha})\big)}{r^u}\\ &+ \sum_{n=0}^{k(re^{-\alpha})} \frac{w_n\big(\mathrm{Log}(re^{-\alpha}) - \mathrm{Log}(c_n)\big)}{r^u}\\ &+ \sum_{k(re^{-\alpha})<n\le k(r)} \frac{w_n(\mathrm{Log}(re^{-\alpha}) - \mathrm{Log}(c_n))}{r^u},\end{aligned}$$

hence

$$\begin{aligned}\sigma(f,r) \le &\sum_{n=0}^{k(re^{-\alpha})} \frac{w_n\big(\mathrm{Log}(r) - \mathrm{Log}(re^{-\alpha})\big)}{r^u}\\ &+ \sum_{n=0}^{k(re^{-\alpha})} \frac{w_n\big(\mathrm{Log}(re^{-\alpha}) - \mathrm{Log}(c_n)\big)}{r^u} + \alpha \sum_{k(re^{-\alpha})<n\le k(r)} \frac{w_n}{r^u}\end{aligned}$$

because $\mathrm{Log}(r) - \mathrm{Log}(c_n) \leq \alpha\ \forall n \in [k(re^{-\alpha}), k(r)] \cap \mathbb{N}$. Consequently,

$$\sigma(f,r) \leq \alpha \sum_{n=0}^{k(re^{-\alpha})} \frac{w_n}{r^u} + \sum_{n=0}^{k(re^{-\alpha})} \frac{w_n\big(\mathrm{Log}(re^{-\alpha}) - \mathrm{Log}(c_n)\big)}{r^u} + \alpha \sum_{k(re^{-\alpha})<n\leq k(r)} \frac{w_n}{r^u},$$

therefore

$$\sigma(f,r) \leq \alpha \sum_{n=0}^{k(r)} \frac{w_n}{r^u} + \sum_{n=0}^{k(re^{-\alpha})} \frac{w_n\big(\mathrm{Log}(re^{-\alpha}) - \mathrm{Log}(c_n)\big)}{r^u},$$

hence

$$\sigma(f,r) \leq e^{-u\alpha} \sum_{n=0}^{k(re^{-\alpha})} \frac{w_n(\mathrm{Log}(re^{-\alpha}) - \mathrm{Log}(c_n))}{(re^{-\alpha})^u} + \alpha \sum_{0\leq n\leq k(r)} \frac{w_n}{r^u}.$$

Thus, we have

$$\sigma(f,r) \leq e^{-u\alpha}\sigma(f,re^{-\alpha}) + \alpha\psi(f,r). \tag{1}$$

Suppose first that $\sigma(f) < +\infty$. We check that we can pass to superior limits on both sides, so we obtain $\sigma(f) \leq e^{-u\alpha}\sigma(f)+\alpha\psi(f)$, therefore $\sigma(f)\frac{(1-e^{-u\alpha})}{\alpha} \leq \psi(f)$. That holds for every $\alpha > 0$, hence, by Lemma 31.4 (ii), we can derive

$$\psi(f) \geq \rho(f)\sigma(f). \tag{2}$$

Now, by (1), we have

$$\sigma(f,r) - e^{-u\alpha}\sigma(f,re^{-\alpha}) \leq \alpha\psi((r,f),$$

hence passing to inferior limits on both sides, we deduce

$$\frac{\widetilde{\sigma}(f) - e^{-u\alpha}\sigma(f)}{\alpha} \leq \widetilde{\psi}(f),$$

hence

$$\frac{u(\widetilde{\sigma}(f) - e^{-u\alpha}\sigma(f))}{u\alpha} \leq \widetilde{\psi}(f),$$

therefore when $\alpha u = 1$, we obtain

(3) $$\rho(f)\left(\widetilde{\sigma}(f) - \frac{\sigma(f)}{e}\right) \leq \widetilde{\psi}(f).$$

We now show the inequality

$$\psi(f) \leq \rho(f)(e\sigma(f) - \widetilde{\sigma}(f)).$$

Let us fix $\alpha > 0$. We can write

$$\begin{aligned}\sigma(f,r) = &\sum_{n=0}^{k(re^{-\alpha})} \frac{w_n(\mathrm{Log}(r) - \mathrm{Log}(re^{-\alpha}))}{r^u}\\ &+ \sum_{j=0}^{k(re^{-\alpha})} \frac{w_j(\mathrm{Log}(re^{-\alpha}) - \mathrm{Log}(c_n))}{r^u}\\ &+ \sum_{k(re^{-\alpha})<j\leq k(r)} \frac{w_j(\mathrm{Log}(r) - \mathrm{Log}(c_j))}{r^u},\end{aligned}$$

hence

$$\sigma(f,r) \geq \alpha \sum_{n=0}^{k(re^{-\alpha})} \frac{w_n}{r^u} + \sum_{j=0}^{k(re^{-\alpha})} \frac{w_j(\mathrm{Log}(re^{-\alpha}) - \mathrm{Log}(c_n))}{r^u},$$

hence
(4)

$$\begin{aligned}\sigma(f,r) \geq \alpha e^{-u\alpha} &\sum_{n=0}^{k(re^{-\alpha})} \frac{w_n}{(re^{-\alpha})^u}\\ + e^{-u\alpha} &\sum_{j=0}^{k(re^{-\alpha})} \frac{w_n(\mathrm{Log}(re^{-\alpha}) - \mathrm{Log}(c_n))}{(re^{-\alpha})^u}\end{aligned}$$

and hence

(5) $$\sigma(f,r) \geq \alpha e^{-u\alpha}\psi(f,re^{-\alpha}) + e^{-u\alpha}\sigma(f,re^{-\alpha}).$$

Therefore, we can deduce

$$\alpha e^{-u\alpha}\psi(f) \leq \limsup_{r\to+\infty}\left(\sigma(f,r) - e^{-u\alpha}\sigma(f,re^{-\alpha})\right),$$

and therefore

$$\alpha e^{-u\alpha}\psi(f) \le \sigma(f) - e^{-u\alpha}\widetilde{\sigma}(f). \tag{6}$$

That holds for every $\alpha > 0$ and hence, when $u\alpha = 1$, by (6), we obtain

$$\psi(f) \le \rho(f)\big(e\sigma(f) - \widetilde{\sigma}(f)\big) \tag{7}$$

which is the left-hand inequality of the general conclusion.

Moreover, we see that if $\psi(f) = +\infty$, then $\sigma(f) = +\infty$.

Now, on (4), we can also take the inferior limit on both sides and we deduce

$$\widetilde{\sigma}(f) \ge \alpha e^{-u\alpha}\widetilde{\psi}(f) + e^{-u\alpha}\widetilde{\sigma}(f),$$

therefore

$$\alpha e^{-u\alpha}\widetilde{\psi}(f) \le \widetilde{\sigma}(f)(1 - e^{-u\alpha}).$$

Then when $u\alpha$ tends to 0, we have

$$\widetilde{\psi}(f) \le \rho(f)\widetilde{\sigma}(f). \tag{8}$$

Now, suppose that $\sigma(f) = +\infty$. We can find an increasing sequence r_n of limit $+\infty$ such that $\sigma(f, r_n) = \sup\{\sigma(f, r) \mid r \le r_n\}$, $n \in \mathbb{N}$. Consider (1) when $\min(\alpha, u\alpha) > 1$. Then, $\sigma(f, r_n e^{-\alpha}) < \sigma(f, r)_n$, hence of course,

$$(1 - e^{-u\alpha})\sigma(f, r_n e^{-\alpha}) \le (1 - e^{-u\alpha})\sigma(f, r_n),$$

and hence $\sigma(f, r_n) - e^{-u\alpha}\sigma(f, r_n e^{-\alpha}) \ge (1 - e^{-u\alpha})\sigma(f, r_n)$, therefore $(1 - e^{-u\alpha})\sigma(f, r_n) \le \psi(f, r_n)$, which proves that $\psi(f) = +\infty$.

Thus, $\sigma(f) < +\infty$ is equivalent to $\psi(f) < +\infty$. Consequently, Relations (2), (3), (5), (7), and (8) still apply and hence hold as soon as $\sigma(f) < +\infty$ or $\psi(f) < +\infty$.

Now, suppose that $\widetilde{\psi}(f) = \psi(f)$. Then we have

$$\rho(f)\sigma(f) \le \psi(f) \le \rho(f)\widetilde{\sigma}(f),$$

therefore $\sigma(f) = \widetilde{\sigma}(f)$ since $\rho(f) > 0$.

Conversely, suppose that $\sigma(f) = \widetilde{\sigma}(f)$. Then, by (6), we have $\psi(f) \leq \sigma(f)\Big(\frac{e^{u\alpha}-1}{\alpha}\Big)$. That holds for every $\alpha > 0$, and then, $\psi(f) \leq u\sigma(f)$, i.e. $\psi(f) \leq \rho(f)\sigma(f)$, hence by (2), we have $\psi(f) = \rho(f)\sigma(f)$.

But now, by (1), we see that $\alpha\psi(f,r) \geq \sigma(f,r) - e^{-u\alpha}\sigma(f,re^{-\alpha})$, hence, passing to the inferior limit, $\alpha\widetilde{\psi}(f) \geq \sigma(f)(1 - e^{-u\alpha})\ \forall\alpha > 0$, therefore $\widetilde{\psi}(f) \geq u\sigma(f)$. But we just showed that $\psi(f) = u\sigma(f)$, hence $\widetilde{\psi}(f) = \psi(f)$.

Thus, we have shown that the hypotheses $\widetilde{\psi}(f) = \psi(f)$ and $\widetilde{\sigma}(f) = \sigma(f)$ are equivalent.

Now, suppose again that $\psi(f) = \widetilde{\psi}(f)$. Then, by the inequalities already proved, on the one hand, we have $\psi(f) \geq \rho(f)\sigma(f)$, and on the other hand, we have $\psi(f) \leq \rho(f)\sigma(f)$. Consequently, $\psi(f) = \rho(f)\sigma(f)$. □

By Theorem 31.3, we obtain Corollary 31.6:

Corollary 31.6. *Let* $f(x) = \sum_{n=0}^{\infty} b_n x^n \in \mathcal{A}(\mathbb{K})$ *be such that* $0 < \rho(f) < +\infty$. *Then,*

$$\rho(f)\sigma(f) \leq \psi(f) \leq \limsup_{n\to+\infty}\left(n\sqrt[n]{|b_n|^{\rho(f)}}\right) - \rho(f)\widetilde{\sigma}(f)$$

and $e\sigma(f) = \dfrac{\limsup_{n\to+\infty}\left(n\sqrt[n]{|b_n|^{\rho(f)}}\right)}{\rho(f)}$.

Theorem 31.7. *Let* $f(x) \in \mathcal{A}(\mathbb{K})$ *be such that* $\widetilde{\sigma}(f) > 0$. *Then,* f *is regular.*

Proof. Let $a = \widetilde{\sigma}(f)$, hence $a = \liminf_{r\to,+\infty} \frac{\text{Log}(|f|(r))}{r^{\rho(f)}}$, then there exists $R > 0$ such that $\frac{Log(|f|(r))}{r^{\rho(f)}} \geq \frac{a}{2}\ \forall r > R$, hence $\text{Log}(|f|(r)) > \frac{a}{2}r^{\rho(f)}\ \forall r > R$ and hence

$$\frac{\text{Log}(\text{Log}(|f|(r))}{\text{Log}(r)} \geq \frac{\text{Log}(a) - \text{Log}(2)}{\text{Log}(r)} + \rho(f) \quad \forall r \geq R.$$

Thus, $\widetilde{\rho}(f) \geq \rho(f)$, which ends the proof. □

Theorem 31.8. *Let* $f(x) = \sum_{n=0}^{\infty} b_n x^n \in \mathcal{A}(\mathbb{K})$ *be clean. Then,*

$$e\psi(f) = \limsup_{n\to+\infty}\left(n\sqrt[n]{|b_n|^{\rho(f)}}\right)$$

and f *is regular.*

Proof. The first statement is an immediate consequence of Corollary 31.6. Now, suppose that f is clean. Let $u = \rho(f)$ and $s = \sigma(f)$. Then, we have

$$\lim_{r\to+\infty} \frac{\mathrm{Log}(|f|(r))}{r^u} = s,$$

hence $\lim_{r\to+\infty}(\mathrm{Log}(\mathrm{Log}(|f|(r))) - u\mathrm{Log}(r)) = \mathrm{Log}(s)$, therefore

$$\lim_{r\to+\infty} \frac{\mathrm{Log}(\mathrm{Log}(|f|(r)))}{\mathrm{Log}(r)} \qquad u = 0$$

i.e.

$$\rho(f) = \lim_{r\to+\infty} \frac{\mathrm{Log}(\mathrm{Log}(|f|(r)))}{\mathrm{Log}(r)},$$

hence f is regular. □

Remark. In [32], the authors claimed that if f is a regular complex entire function, then $\sigma(f) = \widetilde{\sigma}(f)$. We see in the following that such a statement does not hold for functions in $\mathcal{A}(\mathbb{K})$ and that the proof in [4] is put in doubt.

Chapter 32

Growth of the Derivative of an Entire Function

Similar to the situation in complex entire functions, here we see that the order and the type of the derivative of an entire function f are respectively equal to those of f. As in Chapters 30 and 31, we denote by Log the Neperian logarithm and by e the number such that $\text{Log}(e) = 1$.

Throughout this chapter, $\mathbb{K}$ is supposed to have characteristic 0.

Theorem 32.1. *Let $f \in \mathcal{A}(\mathbb{K})$ be not identically zero. Then, $\rho(f) = \rho(f')$.*

Proof. Let $f(x) = \sum_{n=0}^{\infty} a_n x^n$. By Theorem 30.11, we have $\rho(f') = \limsup_{n\to+\infty} \left(\frac{n\text{Log}(n)}{-\text{Log}(|(n+1)a_{n+1}|)}\right)$. But since $\frac{1}{n} \leq |n| \leq 1$, we have

$$\text{Log}(|a_{n+1}|) - \text{Log}(n+1) \leq \text{Log}(|(n+1)a_{n+1}|) \leq \text{Log}(|a_{n+1}|),$$

hence

$$-\frac{\text{Log}(|a_{n+1}|)}{n\text{Log}(n)} \leq -\frac{\text{Log}(|(n+1)a_{n+1}|)}{n\text{Log}(n)} \leq -\frac{\text{Log}(|a_{n+1}|) - \text{Log}(n+1)}{n\text{Log}(n)},$$

hence

$$\begin{aligned}\liminf_{n\to+\infty}\left(-\frac{\mathrm{Log}(|a_{n+1}|)}{n\mathrm{Log}(n)}\right) &\le \liminf_{n\to+\infty}\left(-\frac{\mathrm{Log}(|(n+1)a_{n+1}|)}{n\mathrm{Log}(n)}\right)\\ &\le \liminf_{n\to+\infty}\left(-\frac{\mathrm{Log}(|a_{n+1}|)-\mathrm{Log}(n+1)}{n\mathrm{Log}(n)}\right).\end{aligned}$$

But since

$$\lim_{n\to+\infty}\frac{\mathrm{Log}(n+1)}{n\mathrm{Log}(n)}=0,$$

we have

$$\liminf_{n\to+\infty}\left(-\frac{\mathrm{Log}(|a_{n+1}|)}{n\mathrm{Log}(n)}\right)=\liminf_{n\to+\infty}\left(-\frac{\mathrm{Log}(|a_{n+1}|)-\mathrm{Log}(n+1)}{n\mathrm{Log}(n)}\right),$$

therefore

$$\liminf_{n\to+\infty}\left(\frac{-\mathrm{Log}(|a_{n+1}|)}{n\mathrm{Log}(n)}\right)=\liminf_{n\to+\infty}\left(-\frac{\mathrm{Log}((n+1)|a_{n+1}|)}{n\mathrm{Log}(n)}\right).$$

But since all quantities are positive, we can derive

$$\limsup_{n\to+\infty}\frac{n\mathrm{Log}(n)}{-\mathrm{Log}(|a_{n+1}|)}=\limsup_{n\to+\infty}\left(\frac{n\mathrm{Log}(n)}{-\mathrm{Log}(|(n+1)a_{n+1}|)}\right),$$

therefore

$$\limsup_{n\to+\infty}\frac{n\mathrm{Log}(n)}{-\mathrm{Log}(|a_{n+1}|)}=\limsup_{n\to+\infty}\left(\frac{(n+1)\mathrm{Log}(n+1)}{-\mathrm{Log}(|(n+1)a_{n+1}|)}\right)=\rho(f),$$

and hence $\rho(f')=\rho(f)$. □

Corollary 32.2. *Consider the differential equation*

$$(\mathcal{E})\ f^{(n)}+a_{n-1}(x)f^{(n-1)}(x)+\cdots+a_0(x)f(x)=0$$

with $a_j\in\mathcal{A}^*(\mathbb{K})$, $j=0,\dots,n-1$, *and* $\rho(a_j)<\rho(a_0)\ \forall j=1,\dots,n-1$. *Then, every non-trivial solution* f *of* $(\mathcal{E})$ *satisfies* $\rho(f)\ge\rho(a_0)$.

Corollary 32.3. *The derivation on $\mathcal{A}(\mathbb{K})$ restricted to the algebra $\mathcal{A}(\mathbb{K}, t)$ (respectively to $\mathcal{A}^*(\mathbb{K})$) provides that algebra with a derivation.*

In complex analysis, it is known that if an entire function f has order $t < +\infty$, then f and f' have the same type. We check that it is the same here.

Theorem 32.4. *Let $f \in \mathcal{A}(\mathbb{K})$ of order $t \in]0, +\infty[$. Then, $\sigma(f) = \sigma(f')$.*

Proof. By Theorem 31.3, we have

$$\begin{aligned}
e\rho(f')\sigma(f') &= \limsup_{n\to+\infty} \left(n\left(|n+1||a_{n+1}|\right)^{\frac{t}{n}}\right)\\
&= \limsup_{n\to+\infty} \left(\left((n+1)\left(|n+1||a_{n+1}|\right)^{\frac{t}{n}}\right)^{\frac{n}{n+1}}\left(\frac{n}{n+1}\right)\right)\\
&= \limsup_{n\to+\infty} \left((n+1)\left(|n+1||a_{n+1}|\right)^{\frac{t}{n+1}}\right) = e\rho(f)\sigma(f).
\end{aligned}$$

But since $\rho(f) = \rho(f')$ and since $\rho(f) \neq 0$, we can see that $\sigma(f') = \sigma(f)$. □

Theorem 32.4 shows a way to compare the growth of an entire function f to this of its derivative. Of course, we know that the inequality $|f'|(r) \leq |f|(r)$ holds always. But we don't have an inequality in the other side. However, thanks to Theorem 32.4, we can get this corollary:

Corollary 32.5. *Let $f \in \mathcal{A}^*(\mathbb{K})$ be not identically zero of order $t \in]0, +\infty[$. Given $\epsilon > 0$, there exists a sequence of intervals $[r'_n, r''_n]$, with $\lim_{n\to+\infty} r'_n = +\infty$, such that $|f'|(r) \geq |f|(r)e^{-(\epsilon r^t)}$ $\forall r \in \bigcup_{n\in\mathbb{N}}[r'_n, r''_n]$.*

By Theorems 31.5 and 32.4, we can now derive Corollary 32.6:

Corollary 32.6. *Let $f \in \mathcal{A}^*(\mathbb{K})$ be not identically zero of order $t \in]0, +\infty[$. If f is clean, then $\psi(f) \leq \psi(f')$, and if f' is clean, then $\psi(f) \geq \psi(f')$.*

Corollary 32.7. *Let $f \in \mathcal{A}^*(\mathbb{K})$ be not identically zero of order $t \in]0, +\infty[$. If f and f' are clean, then $\psi(f) = \psi(f')$.*

Remark. More generally, if $\psi(f) = \rho(f)\sigma(f)$ and $\psi(f') = \rho(f')\sigma(f')$, then $\psi(f') = \psi(f)$.

By Theorem 31.5 and Corollary 32.7, the following conjecture is natural and actually would just be an immediate consequence of the conjecture following Corollary 31.6.

Conjecture. *For every $f \in \mathcal{A}(\mathbb{K})$ such that $0 < \rho(f) < +\infty$, we have $\psi(f') = \psi(f)$.*

Applying the above remark, in residue characteristic 0, we have Theorem 32.8:

Theorem 32.8. *Suppose that $\mathbb{K}$ has residue characteristic 0. Then for every $f \in \mathcal{A}(\mathbb{K})$ such that $0 < \rho(f) < +\infty$, we have $\psi(f') = \psi(f)$.*

Proof. Indeed, we can check that for every $r > 0$, $s(r, f') = s(r, f) - 1$, hence $\psi(f') = \psi(f)$. □

Of course, polynomials have a growth order 0. On $\mathbb{K}$ as on $\mathbb{C}$, we can easily construct transcendental entire functions of order 0 or of order ∞.

Example 1. Let $(a_n)_{n\in\mathbb{N}}$ be a sequence in $\mathbb{K}$ such that $-\text{Log}|a_n| \in [n(\text{Log}(n))^2, n(\text{Log}(n))^2 + 1]$. Then clearly, $\lim_{n\to+\infty} \frac{\text{Log}|a_n|}{n} = -\infty$, hence the function $\sum_{n=0}^{\infty} a_n x^n$ has a radius of convergence equal to $+\infty$. On the other hand, $\lim_{n\to+\infty} \frac{n\text{Log}(n)}{-\text{Log}|a_n|} = 0$ hence $\rho(f) = 0$.

Example 2. Let $(a_n)_{n\in\mathbb{N}}$ be a sequence in $\mathbb{K}$ such that $-\text{Log}|a_n| \in [n\sqrt{\text{Log}(n)}, n\sqrt{\text{Log}(n)}+1]$. Then, $\lim_{n\to+\infty} \frac{\text{Log}|a_n|}{n} = -\infty$ again, and hence the function $\sum_{n=0}^{\infty} a_n x^n$ has radius of convergence equal to $+\infty$. On the other hand, $\lim_{n\to+\infty} \left(\frac{n\text{Log}(n)}{-\text{Log}|a_n|}\right) = +\infty$ hence $\rho(f) = +\infty$.

Similarly, comparing the number of zeros of f' to this of f inside a disk is very uneasy. Now, we can give some precisions thanks to Theorems 30.9 and 32.1.

Theorem 32.9. *Let f, $g \in \mathcal{A}(\mathbb{K})$ be transcendental and of order $t \in]0, +\infty[$. Then for every $\epsilon > 0$,*

$$\limsup_{r\to+\infty} \left(\frac{r^\epsilon s(r, g)}{s(r, f)}\right) = +\infty.$$

Proof. Suppose first $t = 0$. The proof then is almost trivial. Indeed, for all $\epsilon > 0$, we have $\lim_{r\to+\infty} \frac{s(r,f)}{r^\epsilon} = 0$, hence $\lim_{r\to+\infty} \frac{r^\epsilon}{s(r,f)} = +\infty$, therefore $\lim_{r\to+\infty} \frac{r^\epsilon s(r,g)}{s(r,f)} = +\infty$.

Now, suppose $t > 0$. By Theorem 30.8, there exists $\lambda > 0$ such that

$$s(r,f) \leq \lambda r^t \quad \forall r > 1. \tag{1}$$

Let us fix $s \in]0,t[$. By hypothesis, $\rho(g) = \rho(f)$, and hence by Theorem 30.8, we have $\limsup_{r\to+\infty} \frac{s(r,g)}{r^s} = +\infty$, so there exists an increasing sequence $(r_n)_{n\in\mathbb{N}}$ of $\mathbb{R}_+$ such that $\lim_{n\to+\infty} r_n = +\infty$ and $\frac{s(r_n,g)}{(r_n)^s} \geq n$. Therefore, by (1), we have

$$\frac{\lambda(r_n)^t s(r_n,g)}{(r_n)^s s(r_n,f)} > \frac{s(r_n,g)}{(r_n)^s} > n,$$

and hence

$$\lambda \lim_{n\to+\infty} \left(\frac{(r_n)^{t-s} s(r_n,g)}{s(r_n,f)} \right) = +\infty.$$

Consequently,

$$\limsup_{r\to+\infty} \left(\frac{(r)^{t-s} s(r,g)}{s(r,f)} \right) = +\infty. \tag{2}$$

Now, since that holds for all $s \in]0,t[$, the statement comes from (2). □

Finally, by Theorem 32.1, we can derive Corollary 32.10:

Corollary 32.10. *Let $f \in \mathcal{A}(\mathbb{K})$ be transcendental and of order $t \in]0,+\infty[$. Then for every $\epsilon > 0$, we have $\limsup_{r\to+\infty} \left(\frac{r^\epsilon s(r,f')}{s(r,f)}\right) = +\infty$ and $\limsup_{r\to+\infty} \left(\frac{r^\epsilon s(r,f)}{s(r,f')}\right) = +\infty$.*

The conjecture $\psi(f') = \psi(f)$ in the general context then seems natural, as suggested in [16, 18], but a proof does not seem easy. However, by Theorems 31.5, 32.1, and 32.4, we can write Corollary 32.11:

Corollary 32.11. *Suppose that $\mathbb{K}$ is of characteristic 0. Let $f \in \mathcal{A}(\mathbb{K})$ be such that $0 < \rho(f) < +\infty$. Then,*

$$|\psi(f) - \psi(f')|_\infty \leq \rho(f)[(e-1)\sigma(f) - \widetilde{\sigma}(f)].$$

Chapter 33

Growth of an Analytic Function in an Open Disk

In Chapters 30, 31, and 32, we defined the order of growth and the type of growth for entire functions in $\mathbb{K}$ in a similar way as it is done for complex entire functions and we also defined a cotype of growth strongly linked to the order and the type: in most of the cases, the cotype is the product of the order of growth by the type of growth.

Here we consider analytic functions in an "open" disk $d(a, R^-)$ that we denote by E throughout this chapter.

Notations and definitions. Let $f = \sum_{n=0}^{\infty} a_n x^n \in \mathcal{A}(E)$. In order to define a growth order similarly as it was done in the algebra of entire functions in $\mathbb{K}$, we can define in $\mathcal{A}(E)$ a growth order in the following way: given $r \in]0, R[$, as it was done in complex analysis, given an unbounded function $f \in \mathcal{A}(E)$, when r is close enough to R, we put $\rho(f, r) = \frac{\mathrm{Log}(\mathrm{Log}(|f|(r)))}{-\mathrm{Log}(R-r)}$ and $\rho(f) = \limsup_{r \to R^-} \rho(f, r)$, hence $\rho(f) = \limsup_{r \to R^-} \frac{\mathrm{Log}(\mathrm{Log}(|f|(r)))}{-\mathrm{Log}(R-r)}$. Then $\rho(f)$ is called *the order of growth of f* [20].

On the other hand, for every $r \in]0, R[$, if the set of the $u > 0$ such that $\lim_{r \to R^-} s(r, f)(R - r)^u = 0$ is empty, we put $\theta(f) = +\infty$. Else, we then denote by $\theta(f)$ the lowest bound of the $u > 0$ such that $\lim_{r \to R^-} s(r, f)(R - r)^u = 0$. Similarly, if the set of $u > 0$ such that $\lim_{r \to R^-} \mathrm{Log}(|f|(r))(R - r)^u = 0$ is empty, we put $\lambda(f) = +\infty$. Else, we denote by $\lambda(f)$ the lowest bound of $u > 0$ such that $\lim_{r \to R^-} \mathrm{Log}(|f|(r))(R - r)^u = 0$. And if $0 < \rho(f) < +\infty$, we put $\sigma(f, r) = \mathrm{Log}(|f|(r))(R - r)^{\rho(f)}$, $\sigma(f) = \limsup_{r \to R^-} \sigma(f, r)$,

$\psi(f,r) = s(r,f)(R-r)^{\rho(f)}$, and $\psi(f) = \limsup_{r\to R^-} \psi(f,r)$. We call $\sigma(f)$ *the type of growth of* f and $\psi(f)$ *the cotype of growth of* f.

Let us recall that, as far as ultrametric entire functions are concerned, the order of growth is equal to the lowest bound of $s > 0$ such that $\lim_{r\to+\infty} \frac{\operatorname{Log}(|f|(r))}{r^s} = 0$ and to the lowest bound of $s > 0$ such that $\lim_{r\to+\infty} \frac{s(r,f)}{r^s} = 0$. Here we try to prove similar results. This chapter is aimed at showing relations between these expressions $\rho(f), \sigma(f), \psi(f)$.

Notation. We denote by $\mathcal{A}^*(E)$ the set of unbounded functions $f \in \mathcal{A}(E)$ such that $0 < \rho(f) < +\infty$.

Theorems 33.1 and 33.3 are easy and don't need any proof:

Theorem 33.1. *Let f, $g \in \mathcal{A}^*(E)$. Then, $\rho(f+g) \leq \max(\rho(f), \rho(g))$ and $\rho(fg) = \max(\rho(f), \rho(g))$.*

Corollary 33.2. *Let f, $g \in \mathcal{A}^*(E)$. Then, $\rho(f^n) = \rho(f)$ $\forall n \in \mathbb{N}^0$. If $\rho(f) > \rho(g)$, then $\rho(f+g) = \rho(f)$.*

Theorem 33.3. *Let $f \in \mathcal{A}^*(E)$ and let $P \in \mathbb{K}[x]$ be non-constant. Then, $\rho(P \circ f) = \rho(f)$.*

Theorem 33.4. *Let f, $g \in \mathcal{A}^*(E)$. Then, $\psi(fg) \leq \psi(f) + \psi(g)$. Moreover, if $\rho(f) = \rho(g)$, then $\max(\psi(f), \psi(g)) \leq \psi(fg)$.*

Proof. Set $\rho(f) = u$, $\rho(g) = t$. Without loss of generality, we can assume $u \geq t$. By Theorem 30.1, we have $\rho(f.g) = \rho(f) = u$. Now, for each $r > 0$, we have $s(f.g,r) = s(r,f) + s(g,r)$, hence

$$\begin{aligned}\psi(fg) &= \limsup_{r\to R^-}(s(r,f) + s(g,r))(R-r)^s \\ &\leq \limsup_{r\to R^-} s(r,f)(R-r)^s + \limsup_{r\to R^-} s(g,r)(R-r)^t,\end{aligned}$$

hence $\psi(fg) \leq \psi(f) + \psi(g)$. Now, suppose $u = t$. Then,

$$\begin{aligned}\psi(fg) &= \limsup_{r\to R^-}(s(r,f) + q(g,r))(R-r)^s \\ &\geq \limsup_{r\to R^-} \max(s(r,f), s(g,r))(R-r)^s \\ &= \max(\psi(f), \psi(g)).\end{aligned}$$

□

Remark 1. Let $f \in \mathcal{A}^*(E)$. If $u > \theta(f)$, then by definition, $\lim_{r\to R^-} s(r,f)(R-r)^u = 0$. But if $u < \theta(f)$, then $\limsup_{r\to R^-} s(r,f)(R-r)^u - +\infty$ because if $\limsup_{r\to R^-} s(r,f)(R-r)^u < +\infty$, we can find $u' \in]s,\theta(f)[$ and then we can check that $\lim_{r\to R^-} s(r,f)(R-r)^{u'} = 0$, a contradiction.

Thanks to the classical inequality $|f'|(r) \leq \frac{|f|(r)}{r}$, Theorem 33.5 is then immediate:

Theorem 33.5. *Suppose $\mathbb{K}$ has characteristic 0. Let $f \in \mathcal{A}^*(E)$. Then, $\rho(f') \leq \rho(f)$.*

Remark 2. In a field of characteristic $p \neq 0$, certain analytic functions have a null derivative. This is why we must suppose that $\mathbb{K}$ has characteristic 0 in all statement involving derivatives.

In complex analysis, many estimates were given concerning the growth order of solutions of linear differential equations. Here, by Corollary 33.2 and Theorem 33.5, we can immediately obtain Corollary 33.6:

Corollary 33.6. *Suppose $\mathbb{K}$ has characteristic 0. Consider the differential equation*

$$(\mathcal{E})\ f^{(n)} + a_{n-1}(x)f^{(n-1)}(x) + \cdots + a_0(x)f(x) = 0$$

with $a_j \in \mathcal{A}^*(E)$, $j = 0,\ldots,n-1$ and $\rho(a_j) < \rho(a_0)\ \forall j = 1,\ldots,n-1$. Then every non-trivial solution f of $(\mathcal{E})$ satisfies $\rho(f) \geq \rho(a_0)$.

Theorem 33.7. *Suppose $\mathbb{K}$ has residue characteristic 0. Then for every $f \in \mathcal{A}^*(D)$, we have $\rho(f') = \rho(f)$, $\theta(f') = \theta(f)$, $\sigma(f') = \sigma(f)$ and $\psi(f') = \psi(f)$.*

Remark 3. Theorem 33.7 does not hold in residue characteristic $p > 0$ because there exist functions $f \in \mathcal{A}^*(D)$ such that $\rho(f) > 0$ and that f' is bounded, as shown in the following example with $R = 1$: $g(x) = \sum_{m=0}^{\infty} \frac{x^{p^m}}{p^m}$. We can see that $g'(x) = \sum_{n=0}^{\infty} x^{p^m-1}$, hence g' is bounded, and therefore $\rho(g') = 0$. However, consider the sequence $(r_m)_{m\in\mathbb{N}}$ defined as $r_m = 1 - \frac{1}{p^m}$. We can check that $|g|(r_m) \geq p^m(r_m)^{p^m}$, hence

$$\mathrm{Log}(|g|(r_m)) \geq m + p^m\mathrm{Log}(r_m) = m + p^m\mathrm{Log}\left(1 - \frac{1}{p^m}\right).$$

When m is big enough, we have $\text{Log}\left(1-\frac{1}{p^m}\right) \geq \frac{-2}{p^m}$, hence

$$\text{Log}\big(|g|(r_m)\big) \geq m - p^m\left(\frac{2}{p^m}\right) = m-2.$$

Therefore, when m is big enough, we have

$$\frac{\text{Log}\big(\text{Log}(|g|(r_m))\big)}{-\text{Log}(r_m)} \geq \frac{\text{Log}(m-2)}{-\text{Log}(1-\frac{1}{p^m})} > \frac{\text{Log}(m-2)}{\frac{2}{p^m}} = \frac{p^m}{2}\text{Log}(m-2).$$

Thus, we have $\rho(g) = +\infty$.

Remark 4. Theorem 33.7 applies, for instance, to the complex Levi-Civita field whose residue characteristic is 0 [88].

Theorem 33.8. *Let $f \in \mathcal{A}^*(E)$. Then, $\lambda(f) = \rho(f)$.*

Proof. First, we prove that $\rho(f) \leq \lambda(f)$. Obviously, we can assume that $\lambda(f) < +\infty$. Let u be such that $\lim_{r\to R^-} \text{Log}(|f|(r))(R-r)^u = 0$. Let us fix $\epsilon > 0$. For r close enough to R, we have $\text{Log}(|f|(r))(R-r)^u \leq \epsilon$, hence $\text{Log}(|f|(r)) \leq \frac{\epsilon}{(R-r)^u}$, therefore $\text{Log}(\text{Log}(|f|(r))) \leq \text{Log}\epsilon - u\text{Log}(R-r)$, hence

$$\frac{\text{Log}(\text{Log}(|f|(r)))}{(-\text{Log}(R-r))} \leq \frac{\text{Log}(\epsilon)}{(-\text{Log}(R-r))} + u,$$

and hence

$$\limsup_{r\to R^-} \frac{\text{Log}(\text{Log}(|f|(r)))}{(-\text{Log}(R-r))} \leq u$$

i.e. $\rho(f) \leq u$. This is true for every s such that $\lim_{r\to R^-} \text{Log}(|f|(r))$ $(R-r)^u = 0$ and hence $\rho(f) \leq \lambda(f)$.

On the other hand, we note that, by definition of $\lambda(f)$, either $\lambda(f) = 0$ and then $\lambda(f) \leq \rho(f)$, or

$$\lambda(f) = \sup\{u \in]0,+\infty[\mid \limsup_{r\to R^-} \text{Log}(|f|(r))(R-r)^u > 0\}.$$

Thus, suppose that $\lambda(f) > 0$. Let us take $u \in]0, \lambda(f)[$. We have a number $b > 0$ such that

$$\limsup_{r\to R^-}(\text{Log}(|f|(r)(R-r)^u) \geq b > 0.$$

Let us fix $\epsilon \in]0, b[$. There exists a sequence $(r_n)_{n\in\mathbb{N}}$ in $]0, R[$ such that $\lim_{n\to+\infty} r_n = R$ and such that, when n is big enough, we have

$b - \epsilon \leq \text{Log}(|f|(r_n))(R - r_n)^u$, hence $-u\text{Log}(R - r_n) + \text{Log}(b - \epsilon) <$ $\text{Log}(\text{Log}(|f|(r_n)))$, therefore

$$u + \frac{\text{Log}(b - \epsilon)}{(-\text{Log}(R - r_n))} \leq \frac{\text{Log}(\text{Log}(|f|(r_n))}{(-\text{Log}(R - r_n))}.$$

Consequently, $\limsup_{n\to+\infty} \frac{\text{Log}(\text{Log}(|f|(r_n))}{(-\text{Log}(R-r_n))} \geq u$, therefore $\rho(f) \geq u$. But this holds for every $u < \lambda(f)$. Thus, $\rho(f) \geq \lambda(f)$, and finally, $\rho(f) = \lambda(f)$. □

Theorem 33.9. *Let $f,\ g \in \mathcal{A}^*(E)$. Then, $\sigma(fg) \leq \sigma(f) + \sigma(g)$. If $\rho(f) \geq \rho(g)$, then $\sigma(f) \leq \sigma(fg)$. If $\rho(f) = \rho(g)$, then $\max(\sigma(f), \sigma(g)) \leq \sigma(fg)$.*

If $\rho(f) = \rho(g)$ and $\sigma(f) > \sigma(g)$, then $\sigma(f+g) \geq \sigma(f)$. If $\rho(f+g) = \rho(f) \geq \rho(g)$, then $\sigma(f+g) \leq \max(\sigma(f), \sigma(g))$.

Proof. Let $u = \rho(f)$, $t = \rho(g)$ and suppose $u \geq t$. When r is close enough to R, we have $\max(\text{Log}(|f|(r)), \text{Log}(|g|(r)) \leq \text{Log}(|f.g|(r)) = \text{Log}(|f|(r)) + \text{Log}(|g|(r))$, and by Theorem 1, we have $\rho(fg) = u$. Therefore,

$$\sigma(fg) = \limsup_{r\to R^-} \left(\text{Log}(|f.g|(r))(R - r)^u\right)$$

$$\leq \limsup_{r\to R^-} \left(\text{Log}(|f|(r))(R - r)^u\right) + \limsup_{r\to R^-} \left(\text{Log}(|g|(r))(R - r)^t\right)$$

$$= \sigma(f) + \sigma(g).$$

On the other hand,

$$\sigma(f) = \limsup_{r\to R^-} \text{Log}(|f|(r))(R - r)^u \leq \limsup_{r\to+R^-} (\text{Log}(|fg|(r))(R - r)^u.$$

But $\rho(fg) = u$, hence $\sigma(f) \leq \sigma(fg)$. Particularly, if $\rho(f) = \rho(g)$, then $\max(\sigma(f), \sigma(g)) \leq \sigma(fg)$.

Now, suppose again that $\rho(f) = \rho(g) = u$ and suppose $\sigma(f) > \sigma(g)$. Let $u = \rho(f)$, $b = \sigma(f)$. Then, $b > 0$. Let $(r_n)_{n\in\mathbb{N}}$ be a sequence such that $\lim_{n\to+\infty} r_n = R$ and $\lim_{n\to+\infty}(\text{Log}(|f|(r_n))(R - r_n)^u) = b$. Since $\sigma(g) < \sigma(f)$, we note that when n is big enough, we have $|g|(r_n) < |f|(r_n)$. Consequently, when n is big enough, we have $|f+g|(r_n) = |f|(r_n)$, and hence

$$(1) \qquad \lim_{n\to+\infty}(\mathrm{Log}(|f+g|(r_n)))(R-r_n)^u) = b.$$

By definition of σ, we have $\sigma(f+g) \geq \lim_{n\to+\infty}(\mathrm{Log}(|f+g|(r_n)))(R-r_n)^{\rho(f+g)}$. By Theorem 1, we have $\rho(f+g) \leq u$, hence

$$\begin{aligned}\sigma(f+g) &\geq \lim_{n\to+\infty}(\mathrm{Log}(|f+g|(r_n)))(R-r_n)^{\rho(f+g)}\\ &\geq \lim_{n\to+\infty}(\mathrm{Log}(|f+g|(r_n)))(R-r_n)^{u}\\ &= \lim_{n\to+\infty}\mathrm{Log}(|f|(r_n))(R-r_n)^{u} = \sigma(f),\end{aligned}$$

therefore by (1), $\sigma(f+g) \geq \sigma(f)$.

Finally, suppose now that $\rho(f+g) = \rho(f) \geq \rho(g)$. Let $u = \rho(f)$ and $t = \rho(g)$. Then,

$$\sigma(f+g) = \limsup_{r\to R^-}(\mathrm{Log}(|f+g|(r)))(R-r)^u$$

$$\leq \max\left(\limsup_{r\to R^-}(\mathrm{Log}(|f|(r)))(R-r)^u, \limsup_{r\to R^-}(\mathrm{Log}(|g|(r)))(R-r)^u\right)$$

$$\leq \max\left(\limsup_{r\to R^-}(\mathrm{Log}(|f|(r)))(R-r)^u, \limsup_{r\to R^-}(\mathrm{Log}(|g|(r)))(R-r)^t\right)$$

$$= \max(\sigma(f), \sigma(g))$$

which ends the proof. □

Corollary 33.10. *Let $f,\ g \in \mathcal{A}^*(E)$ be such that $\rho(f) \neq \rho(g)$. Then,*

$$\sigma(f+g) \leq \max(\sigma(f), \sigma(g)).$$

Lemma 33.11. *Let $a \in [1, +\infty[$ and $b \in [0, +\infty[$. Then, $\mathrm{Log}(a+b) \leq \mathrm{Log}(a) + \mathrm{Log}(b+1)$.*

Proof. Indeed, since $a \geq 1$, we have $\mathrm{Log}(a+b) \leq \mathrm{Log}(a(b+1)) = \mathrm{Log}(a) + \mathrm{Log}(b+1)$. □

Theorem 33.12. *Let $f \in \mathcal{A}^*(E)$. Then, $\theta(f) - 1 \leq \rho(f) \leq \theta(f)$. Moreover, if $0 < \psi(f) < \infty$, then $\rho(f) = \theta(f)$.*

Proof. We denote by $| \, . \, |_\infty$ the Archimedean absolute value of $\mathbb{R}$. Let us first choose $u > \theta(f)$. Then, $\lim_{r\to R^-} s(r,f)(R-r)^u = 0$. Now, since $\lim_{r\to R^-} |f|(r) - +\infty$, we can take $\ell \in]0, R[$ such that $|f|(\ell) > 1$. Then we can take $b > 0$ such that

$$s(r,f) \leq b(R-r)^{-u} \quad \forall r \in [\ell, R[.$$

Now, taking $r \in [\ell, R[$, by Theorem 22.26, we have

$$\mathrm{Log}(|f|(r)) < \mathrm{Log}(|f|(\ell))) + s(r,f)\left(\mathrm{Log}\left(\frac{r}{\ell}\right)\right)$$

which leads to

$$\mathrm{Log}(|f|(r)) \leq \mathrm{Log}(|f|(\ell))) + b(R-r)^{-u}\left(\mathrm{Log}\left(\frac{r}{\ell}\right)\right),$$

hence

$$\mathrm{Log}(\mathrm{Log}(|f|(r))) \leq \mathrm{Log}\left(\mathrm{Log}(|f|(\ell))) + b(R-r)^{-u}\left(\mathrm{Log}\left(\frac{r}{\ell}\right)\right)\right),$$

therefore, by Lemma 33.11, we can derive
(1)

$$\mathrm{Log}(\mathrm{Log}(|f|(r))) \leq \mathrm{Log}(\mathrm{Log}(|f|(\ell)))+\mathrm{Log}\Big(b(R-r)^{-u}\left(\mathrm{Log}\left(\frac{r}{\ell}\right)\right)+1\Big).$$

Now, since $u > 0$, there obviously exists $h \in [\ell, R[$ such that $b(R-r)^{-u} \geq 1 \ \forall r \in [h, R^-[$, therefore by Lemma 33.11 again,

$$\begin{aligned}\mathrm{Log}(\mathrm{Log}(|f|(r))) \leq{}& \mathrm{Log}(\mathrm{Log}(|f|(\ell))) + \mathrm{Log}\Big(b(R-r)^{-u}(\mathrm{Log}\left(\frac{r}{\ell}\right)\Big)\\ &+ \mathrm{Log}(1+1)\end{aligned}$$

i.e.
(2)

$$\begin{aligned}\mathrm{Log}(\mathrm{Log}(|f|(r))) \leq{}& \mathrm{Log}(\mathrm{Log}(|f|(\ell))) + \mathrm{Log}(b) - s\mathrm{Log}(R-r)\\ &+ \mathrm{Log}((\mathrm{Log}(\tfrac{r}{\ell})) + \mathrm{Log}(2)\end{aligned}$$

Consequently, by (2), we obtain

$$\begin{aligned}\frac{\mathrm{Log}(\mathrm{Log}(|f|(r)))}{-\mathrm{Log}(R-r)} \leq{}& \frac{\mathrm{Log}(\mathrm{Log}(|f|(\ell)))}{-\mathrm{Log}(R-r)} + \frac{\mathrm{Log}(b)}{-\mathrm{Log}(R-r)}\\ &+ u + \frac{\mathrm{Log}(\mathrm{Log}(\frac{r}{\ell})) + \mathrm{Log}(2)}{-\mathrm{Log}(R-r)}.\end{aligned}$$

We can check that

$$\lim_{r\to R^-}\frac{\text{Log}(\text{Log}(|f|(\ell)))+\text{Log}(b)}{-\text{Log}(R-r)}=\lim_{r\to R^-}\frac{\text{Log}(\text{Log}(\frac{r}{\ell})+\text{Log}(2)}{-\text{Log}(R-r)}=0,$$

and hence $\limsup_{r\to R^-}\frac{\text{Log}(\text{Log}(|f|(r)))}{-\text{Log}(R-r)}\leq u$. Consequently, choosing $\epsilon>0$, there exists $u\in[\ell,1[$ such that $\frac{\text{Log}(\text{Log}(|f|(r)))}{-\text{Log}(R-r)}\leq u+\epsilon\ \forall r\in[u',R[$, and hence $\rho(f)\leq u+\epsilon$. But since that holds for every $u>\theta(f)$ and for every $\epsilon>0$, we have $\rho(f)\leq u$ and hence $\rho(f)\leq\theta(f)$.

Let us now show that $\rho(f)\geq\theta(f)-1$. By Theorem 22.26, we have

$$\text{(3)}\quad \begin{aligned}\text{Log}(|f|(r))-\text{Log}(|f|(\tfrac{r^2}{R}))&\geq s(\tfrac{r^2}{R},f)(\text{Log}(r)-\text{Log}(\tfrac{r^2}{R}))\\&=(\tfrac{r^2}{R},f)(\text{Log}(R)-\text{Log}(r)).\end{aligned}$$

Consider now a number $s<\theta(f)$ and a sequence $(r_n)_{n\in\mathbb{N}}$ of $]0,R[$ such that $\lim_{n\to+\infty}r_n=R$ and such that $\limsup_{n\to+\infty}s(r_n,f)$ $(R-r_n)^s\geq b>0$. Then, by (3), we have

$$\text{Log}(|f|(r_n))\geq\frac{b(\text{Log}(R)-\text{Log}(r_n))}{\left(R-\frac{r_n^2}{R}\right)^s}.$$

Consequently,

$$\begin{aligned}\text{Log}(\text{Log}(|f|(r_n)))\geq{}&\text{Log}(b)+\text{Log}(\text{Log}(R)-\text{Log}(r_n)))\\&-s\big(\text{Log}(R-r_n)+\text{Log}(R+r_n)\big)+2u\text{Log}(R),\end{aligned}$$

and therefore

$$\begin{aligned}\frac{\text{Log}(\text{Log}(|f|(r_n)))}{-\text{Log}(R-r_n)}\geq{}&\frac{\text{Log}(b)}{-\text{Log}(R-r_n)}+\frac{\text{Log}(\text{Log}(R)-\text{Log}(r_n))}{-\text{Log}(R-r_n)}\\&+u\left(1+\frac{\text{Log}(R+r_n)+2\text{Log}(R)}{-\text{Log}(R-r_n)}\right).\end{aligned}$$

Clearly,

$$\lim_{n\to+\infty}\left(\frac{\text{Log}(b)}{\text{Log}(R-r_n)}\right)=\lim_{n\to+\infty}\frac{\text{Log}(R+r_n)+2\text{Log}(R)}{\text{Log}(R-r_n)}=0,$$

and by elementary reasonings, we can check that

$$\lim_{t\to R^-} \frac{\text{Log}(\text{Log}(R) - \text{Log}(t))}{\text{Log}(R-t)} = 1,$$

therefore

$$\lim_{n\to+\infty} \frac{\text{Log}(\text{Log}(R) - \text{Log}(r_n))}{\text{Log}(R-r_n)} = 1.$$

Consequently,

$$\limsup_{n\to+\infty} \frac{\text{Log}(\text{Log}(|f|(r_n)))}{-\text{Log}(R-r_n)} \geq u-1,$$

and therefore

$$\limsup_{r\to R^-} \frac{\text{Log}(\text{Log}(|f|(r)))}{-\text{Log}(R-r)} \geq u-1.$$

That holds for every $u < \theta(f)$ and shows that if $\theta(f) < +\infty$, then $\rho(f) \geq \theta(f) - 1$. Next, if $\theta(f) = +\infty$, then we would have $\rho(f) = +\infty$, which is excluded by hypothesis since $f \in \mathcal{A}^*(E)$. Consequently, the inequality $\rho(f) \geq \theta(f) - 1$ is established.

Let us now show that $\rho(f) \geq \theta(f)$ when $\psi(f) < +\infty$. Suppose $\theta(f) > \rho(f)$ and let $u \in]\rho(f), \theta(f)[$. Then, by Remark 1, we have $\limsup_{r\to R^-} s(r,f)(R-r)^s = +\infty$, but then $\limsup_{r\to R^-} s(r,f)$ $(R-r)^{\rho(f)} = +\infty$, i.e. $\psi(f) = +\infty$, a contradiction. Therefore, $\theta(f) \leq \rho(f)$, and hence whenever $\psi(f) < +\infty$, we have $\theta(f) = \rho(f)$. □

Theorem 33.12 obviously suggests the following conjecture:

Conjecture. *Let $f \in \mathcal{A}^*(E)$. Then, $\rho(f) = \theta(f)$.*

Theorem 33.13 is much different from the relations concerning ρ, σ, ψ obtained for entire functions.

Theorem 33.13. *Let $f \in \mathcal{A}^*(E)$ be such that $\psi(f) < +\infty$. Then, $\sigma(f) = 0$.*

Proof. Without loss of generality, we can assume that $f(0) \neq 0$. Let us fix $\epsilon > 0$ and let R' be such that $\text{Log}(R) - \text{Log}(R') = \epsilon$. Let $(a_n)_{n\in\mathbb{N}}$ be the sequence of zeros of f, for each $n \in \mathbb{N}$, let w_n be the order of a_n, and let $r_n = |a_n|$. Now, let u be the biggest integer n such that $r_n < R'$ and for each $r > 0$, let $m(r)$ be the biggest integer n such that $r_n \leq r$.

Let $A_u = \sum_{n=0}^{u} w_n$ and let $B_u = \text{Log}(|f(0)|) + \sum_{n=0}^{u} w_n(\text{Log}(R') - \text{Log}(r_n))$. Let us take $r \in]R', R[$. Now, we can write

$$\frac{\sigma(f,r)}{\psi(f,r)} = \frac{B_u + \sum_{n=u+1}^{m(r)} w_n(\text{Log}(r) - \text{Log}(r_n))}{A_u + \sum_{n=u+1}^{m(r)} w_n}.$$

But by hypothesis, $\text{Log}(r) - \text{Log}(r_n) \leq \epsilon\ \forall n \geq u$, hence

$$\sigma(f,r)\text{over}\psi(f,r) \leq \frac{B_u + \epsilon \sum_{n=u+1}^{m(r)} w_n}{A_u + \sum_{n=u+1}^{m(r)} w_n}.$$

Let us put $\phi(r) = \sum_{n=u+1}^{m(r)} w_n$. Thus,

$$\frac{\sigma(f,r)}{\psi(f,r)} \leq \frac{B_u + \epsilon\phi(r)}{A_u + \phi(r)}.$$

But since f belongs to $\mathcal{A}^*(D)$, it has infinitely many zeros in D, hence $\phi(r)$ is an increasing unbounded function tending to $+\infty$ when r tends to R. Consequently, it is obvious that

$$\lim_{r\to R} \sigma(f,r)\text{over}\psi(f),r = 0.$$

Therefore, if $\limsup_{r\to R^-} \psi(f,r) < +\infty$, then $\sigma(f) = 0$. □

Chapter 34

The Set $\text{Mult}(\mathcal{A}_b(d(0, R^-))$

Throughout this chapter, we fix $R \in]0, +\infty[$, we denote by D the disk $d(0, R^-)$ and by A the $\mathbb{K}$-algebra $\mathcal{A}_b(d(0, R^-))$.

In this chapter, we examine the continuous multiplicative norms and semi-norms on A. Following a work by Araujo [6], we show that there exist continuous multiplicative semi-norms whose kernel is a prime closed ideal that is neither null nor a maximal ideal. Theorem 34.1 is easily checked.

Theorem 34.1. *Each circular filter of diameter $r \in]0, R[$ secant with the disk $d(0, R^-)$ defines a continuous multiplicative norm on A. The norm $\| \cdot \|_D$ is a multiplicative norm on A. Each ultrafilter $\mathcal{U}$ on D defines an element $\varphi_{\mathcal{U}} \in \text{Mult}(A, \| \cdot \|)$ as $\varphi_{\mathcal{U}}(f) = \lim_{\mathcal{U}} |f(x)|$.*

Proof. Let $\mathcal{F}$ be a circular filter of diameter $r \in]0, R[$, secant with D. There exists a disk D' of diameter $r' \in [r, R[$ such that $\mathcal{F}$ is secant with D'. And then, by Proposition 14.12, A is included in $H(D')$. Consequently, $\varphi_{0,r'}$ belongs to $\text{Mult}(H(D'), \| \cdot \|_{D'})$ and therefore belongs to $\text{Mult}(A, \| \cdot \|)$. On the other hand, by Theorem 14.20, $\| \cdot \|_D$ is a multiplicative norm defined on A. $\square$

Thus, given an ultrafilter $\mathcal{U}$ on D, the function $|f(x)|$ from D to $[0, \|f\|]$ has a limit $\varphi_{\mathcal{U}}(f)$ which clearly defines an element of $\text{Mult}(A, \| \cdot \|)$.

Definition. Two filters $\mathcal{F}$ and $\mathcal{G}$ on a subset S of $\mathbb{K}$ are said to be *contiguous* if for every $F \in \mathcal{F}$ and $G \in \mathcal{G}$ we have $\delta(F, G) = 0$.

Proposition 34.2 is immediate due to the uniform continuity of functions in A:

Proposition 34.2. *Let $\mathcal{U}$ and $\mathcal{V}$ be two contiguous ultrafilters on D. Then, $\varphi_{\mathcal{U}} = \varphi_{\mathcal{V}}$.*

Proof. Let $f \in A$, and let $l = \varphi_{\mathcal{U}}|(f)$ and $m = \varphi_{\mathcal{V}}(f)$. Let ϵ be > 0. Since f is uniformly continuous, we can find $\eta > 0$ such that $|f(x) - f(y)| \leq \epsilon \ \forall x, y \in D$ such that $|x - y| \leq \eta$. Let us take $F \in \mathcal{U}$ and $G \in \mathcal{V}$ such that $|f(x) - l|_\infty \leq \epsilon \ \forall x \in F$ and $|f(x)| - m|_\infty \leq \epsilon \ \forall y \in G$. But since $\mathcal{U}$ and $\mathcal{V}$ are contiguous, there exists $x \in F$ and $y \in G$ such that $|x - y| \leq \eta$, hence $|l - m|_\infty \leq 3\epsilon$. That holds for every $\epsilon > 0$, hence $l = m$. □

Remark. It seems very hard to know whether two ultrafilters $\mathcal{U}$ and $\mathcal{V}$ on D such that $\varphi_{\mathcal{U}} = \varphi_{\mathcal{V}}$ are contiguous.

Definition. Similar to the definition for analytic elements, an element $f \in A$ is said to be *quasi-invertible* if it factorizes in the form Pg with $P \in \mathbb{K}[x]$ and g an invertible element of A.

By Corollary 14.16 and Theorem 14.19, we have the following obvious lemma:

Lemma 34.3. *An element $f \in A$ is quasi-invertible if and only if it has finitely many zeros.*

Notation. Let $\mathcal{Y}$ be the filter on $d(0, R^-)$ admitting for basis the family of sets of the form $d(0, R^-)\setminus \bigcup_{n=1}^{\infty}(\bigcup_{j=1}^{q_n} d(a_{n,j}, r_n^-))$ with $|a_{j,n}| = r_n < r_{n+1} < R \ \forall n \in \mathbb{N}$ and $\lim_{n\to\infty} r_n = R$.

Theorem 34.4. *For every $f \in A$, $\|f\|_D = \lim_{\mathcal{Y}} |f(x)|$.*

Proof. Let $f \in A$, let $(C(0, r_n))_{n\in\mathbb{N}}$ be the sequence of circles of center 0 containing zeros of f and for every $n \in \mathbb{N}$, and let $a_{n,1}, \ldots, a_{n,q_n}$ be the zeros of f in $C(0, r_n)$. Then we can see that $|f(x)| = |f|(|x|) \ \forall x \notin d(0, R^-)\setminus \bigcup_{n=1}^{\infty}(\bigcup_{j=1}^{q_n} d(a_{n,j}, r_n^-))$. Next, by Theorem 14.20, we have $\|f\|_D = \lim_{r\to R} |f|(r)$ which ends the proof. □

Theorem 34.5. *Let $\phi \in \mathrm{Mult}(A, \| \,.\, \|_D)$ and assume that its restriction $\varphi_{\mathcal{F}}$ to $H(D)$ is not $\| \,.\, \|_D$. Then, $\phi(f) = \lim_{\mathcal{F}} |f(x)| \ \forall f \in A$.*

Proof. By hypothesis, $\mathcal{F}$ is a circular filter of diameter $l < R$. Suppose first $l = 0$. Then $\mathcal{F}$ is the filter of neighborhoods of a point $a \in D$. Let $r \in]0, R[$. Then f belongs to $H(d(a,r))$ and we show that $\phi(h) \le \varphi_{\mathcal{F}}(h)\ \forall h \in A$.

Suppose first f is invertible. Since f has no zero in D, by Theorem 22.16, $|f(x)|$ is a constant $b > 0$. Consequently, $\|f\|_D = b = \varphi_{\mathcal{F}}(f)$. Suppose $\phi(f) \neq b$. Then $\phi(f) < b$ because $b = \|f\|_D$. Now consider $h = \frac{1}{f}$. Since $\|\ .\ \|_D$ is multiplicative, we see that $\phi(h) > \|h\|_D$, a contradiction. Consequently, $\phi(f) = \varphi_{\mathcal{F}}(f)$.

Suppose now f is quasi-invertible. Then f is of the form Pg with $P \in \mathbb{K}[x]$ and g invertible in A. Then, $\phi(f) = \phi(P)\phi(g) = \varphi_{\mathcal{F}}(P)\varphi_{\mathcal{F}}(g) = \varphi_{\mathcal{F}}(f)$.

We now suppose that f is not quasi-invertible. By Lemma 34.3, f has a sequence of zeros $(a_n)_{n\in\mathbb{N}}$ in D, each having a multiplicity order u_n. By Corollary 22.19, we have $\lim_{n\to+\infty} |a_n| = R$, so we can assume $|a_n| \le |a_{n+1}|\ \forall n \in \mathbb{N}$. Let $t = \phi(f)$ and $s = \lim_{\mathcal{F}} |f(x)|$.

First, we show that $t \le s$.

Suppose first that $\mathcal{F}$ is secant with a disk $d(a,r)$ which contains none of the a_n. By Corollary 22.29, we have $\frac{\|f\|_{d(a,r)}}{\|f\|_D} = \prod_{n=0}^{\infty}\left(\frac{|a_n-a|}{R}\right)^{u_n}$, hence inside the disk $d(a,r)$, $|f(x)|$ is a constant equal to $\|f\|_D\prod_{n=0}^{\infty}\left(\frac{|a_n-a|}{R}\right)^{u_n}$, and therefore,

$$s = \|f\|_D \prod_{n=0}^{\infty}\left(\frac{|a_n - a|}{R}\right)^{u_n}. \tag{1}$$

For each $q \in \mathbb{N}$, let $f_q = \frac{f}{\prod_{n=0}^{q}(x-a_n)^{u_n}}$ and let $l_q = \sum_{k=0}^{q} u_k$. So, clearly, $\|f_q\|_D = \frac{\|f\|_D}{R^{l_q}}\ \forall q \in \mathbb{N}$.

Now, since $\phi(P) = \varphi_{\mathcal{F}}(P)\ \forall P \in \mathbb{K}[x]$, we have $\phi(\prod_{n=0}^{q}(x-a_n)^{u_n}) = \prod_{n=0}^{q}|a_n - a|^{u_n}$, hence $\phi(f_q) = \frac{t}{\prod_{n=0}^{q}|a_n-a|^{u_n}}$. But since $\phi(f_q) \le \|f_q\|_D$, that yields

$$\frac{t}{\prod_{n=0}^{q}|a_n - a|^{u_n}} \le \|f_q\|_D = \frac{\|f\|_D}{R^{l_q}} \quad \forall q \in \mathbb{N},$$

hence

$$\frac{tR^{l_q}}{\prod_{n=0}^{q}|a_n - a|^{u_n}} \le \|f\|_D \quad \forall q \in \mathbb{N}.$$

Since this is true for every $q \in \mathbb{N}$, we can derive

$$t\prod_{n=0}^{\infty}\Big(\frac{R^{u_n}}{|a_n-a|^{u_n}}\Big) \leq \|f\|_D,$$

hence by (1), $\frac{t\|f\|_D}{s} \leq \|f\|_D$ and therefore $t \leq s$.

Now, consider the case when there exists no disk $d(a,r)$ such that none of the a_n lie in $d(a,r)$ and such that $\mathcal{F}$ is secant with that disk. Since $\lim_{n\to\infty}|a_n| = R$, $\mathcal{F}$ is a filter admitting a center α. Let ρ be its diameter: of course $\rho < R$ because $\varphi_{\mathcal{F}}$ is not $\|\ .\ \|_D$. Consequently, $d(\alpha,\rho)$ contains finitely many zeros of f. Suppose $|a_j-\alpha| < \rho$ whenever $j = 0,\dots,q$. Let $h = \frac{f}{\prod_{j=0}^{q}(x-a_j)^{u_j}}$. Then, $\varphi_{\mathcal{F}}(h) = \frac{s}{\prod_{j=0}^{q}|a_j-\alpha|^{u_j}}$ and $\phi(h) = \frac{t}{\prod_{j=0}^{q}|a_j-\alpha|^{u_j}}$. Thus, we are led to the same problem with h. Setting $s' = \frac{s}{\prod_{j=0}^{q}|a_j-\alpha|^{u_j}}$, $t' = \frac{t}{\prod_{j=0}^{q}|a_j-\alpha|^{u_j}}$, we have $t' \leq s'$, hence $t \leq s$ in all cases and therefore we have proven

(2) $$\phi(h) \leq \varphi_{\mathcal{F}}(h) \quad \forall h \in A.$$

Suppose now that for some $f \in A$, we have $\phi(f) < \varphi_{\mathcal{F}}(f)$. Let $r \in]l,R[$. The filter $\mathcal{F}$ is then secant with a unique disk $d(a,r)$. Without loss of generality, we can suppose that 0 belongs to $d(a,r)$ and then we can take $a = 0$. Let $f(x) = \sum_{n=0}^{\infty} b_n x^n$. For every $q \in \mathbb{N}$, let $g_q(x) = \sum_{n=0}^{q} b_n x^n$. We note that when q is big enough we have $\varphi_{\mathcal{F}}(g_q) = \sup_{n\in\mathbb{N}}|b_n|r^n$. Set $w = \sup_{n\in\mathbb{N}}|b_n|r^n$. Now $\varphi_{\mathcal{F}}(f-g_q) \leq \sup_{n>q}|b_n|r^n$, therefore $\lim_{q\to+\infty}\varphi_{\mathcal{F}}(f-g_q) = 0$, and hence, by (2),

(3) $$\lim_{q\to+\infty}\phi(f-g_q) = 0.$$

So, we can take q such that $\varphi_{\mathcal{F}}(f-g_q) < \varphi_{\mathcal{F}}(f)$, and hence, by (2), $\phi(f-g_q) < \varphi_{\mathcal{F}}(f)$. But since g_q is a polynomial, we have $\phi(g_q) = \varphi_{\mathcal{F}}(g_q)$, hence $\phi(g_q) > \phi(f)$. Consequently, $\phi(f-g_q) = \phi(g_q) = w$ when q is big enough, a contradiction to (3). □

By Corollary 22.29, we can derive Lemma 34.6:

Lemma 34.6. *Let $f \in A$ be not quasi-invertible such that $f(0) = 1$ and let $(a_n)_{n\in\mathbb{N}}$ be the sequence of zeros with respective multiplicity q_n. Then the series $\sum_{n=0}^{\infty} q_n \log\Big(\frac{R}{|a_n|}\Big)$ converges to $\log\|f\|_D$.*

Lemma 34.7. *Let $\phi \in \text{Mult}(A, \|\ .\ \|_D)$ satisfy $\phi(P) = \|P\|_D\ \forall P \in \mathbb{K}[x]$. Every quasi-invertible element $f \in A$ also satisfies $\phi(f) = \|f\|_D$.*

Proof. First, suppose $f \in A$, invertible in A. Then, $1 = \phi(f)\phi(f^{-1})$. But $\phi(f) \leq \|f\|_D$, $\phi(f^{-1}) \leq \|f^{-1}\|_D$, hence both inequalities must be equalities. Now, take $f = Pg \in A$ which is quasi-invertible, with $P \in \mathbb{K}[x]$ and $g \in A$, invertible in A. Then, $\phi(f) = \phi(P)\phi(g) = \|P\|_D\|g\|_D = \|Pg\|_D = \|f\|_D$. □

Theorem 34.8. *Let $\phi \in \text{Mult}(A, \|\ .\ \|_D)$ be a norm and have a restriction to $\mathbb{K}[x]$ equal to $\|\ .\ \|_D$. Then, $\phi(f) = \|f\|_D\ \forall f \in A$.*

Proof. Suppose the claim is wrong. Suppose that there exists $f \in A$ such that $\phi(f) \neq \|f\|_D$. By Lemma 10.5, we have $\phi(f) < \|f\|_D$. Actually, without loss of generality, we can choose $f \in A$ such that $\phi(f) < 1 \leq \|f\|_D$ and that $f(0) \neq 0$. By Lemma 34.7, f is not quasi-invertible, hence f has a sequence of zeros $(a_n)_{n\in\mathbb{N}}$ in D, with $|a_n| \leq |a_{n+1}|$ and $\lim_{n\to+\infty} |a_n| = R$. For each $n \in \mathbb{N}$, let q_n be the multiplicity order of a_n. By Lemma 34.6, we know that $\sum_{n=0}^{\infty} q_n \log\left(\frac{R}{|a_n|}\right) < +\infty$. Consequently, by Lemma 34.6, there exists a sequence t_n of strictly positive integers satisfying
$t_n \leq t_{n+1},\ n \in \mathbb{N}$,
$\lim_{n\to\infty} t_n = +\infty$,
$\sum_{n=0}^{\infty} t_n q_n \log\left(\frac{R}{|a_n|}\right) < +\infty$.
By Theorem 27.14, there exists a function $g \in \mathcal{A}(d(0,R^-))$ admitting each a_n as a zero of order $s_n \geq t_n q_n$ such that $g(0) = f(0)$ and $|g|(|a_n|) \leq 2\left|\prod_{k=0}^{n}\left(\frac{a_n}{a_k}\right)^{t_n q_n}\right|\ \forall n \in \mathbb{N}$. Consequently, $\lim_{n\to+\infty} |g|(|a_n|) < +\infty$, and hence g belongs to A.

Now, for each $n \in \mathbb{N}$ and for each $k = 0, \ldots, n$, set $u_{n,k} = \max(0, t_n q_k - s_k)$, and let $P_n(x) = \prod_{k=0}^{n}(\frac{x}{a_k} - 1)^{u_{n,k}}$. Since the sequence t_n is increasing, we can check that for each fixed $n \in \mathbb{N}$, each zero a_k of f^{t_n} is a zero of $P_n g$ of order $\geq t_n q_k$: indeed, this is obvious, by definition of g, for all $k \geq n$ and when $k < n$, the order of a_k as a zero of $P_n g$ is at least $u_{n,k} + s_k = \max(s_k, t_n q_k)$. Consequently, by Theorem 27.1, in the ring A, we can write $P_n g$ in the form $f^{t_n}\sigma_n$, with $\sigma_n \in A$.

We have $\|\sigma_n\|_D\|f^{t_n}\|_D = \|P_n\|_D\|g\|_D$, hence, since $\|f\|_D \geq 1$, we can see that

$$(1) \qquad \|\sigma_n\|_D \leq \|P_n\|_D\|g\|_D.$$

On the other hand, since the restriction of ϕ to $\mathbb{K}[x]$ is $\| \,.\, \|_D$, we have $\phi(P_n) = \|P_n\|_D$, hence $\phi(P_n g) = \phi(P_n)\phi(g) = \|P_n\|_D\phi(g)$, and therefore, by (1), we have

$$\|P_n\|_D\phi(g) = \phi(P_n g) = \phi(f^{t_n}\sigma_n) = (\phi(f))^{t_n}\phi(\sigma_n) \leq (\phi(f))^{t_n}\|\sigma_n\|_D$$
$$\leq (\phi(f))^{t_n}\|P_n\|_D\|g\|_D,$$

and hence $\phi(g) \leq (\phi(f))^{t_n}\|g\|_D$. Now, since $\phi(f) < 1$, we can see that $g = 0$, a contradiction. That ends the proof of Theorem 34.8. □

Let us now look at the maximal spectrum of A.

Theorem 34.9. *Let $\mathcal{M}$ be an ideal of A. Then $\mathcal{M}$ is a maximal ideal of codimension 1 if and only if it is of the form $(x - a)A$ with $a \in D$.*

Proof. Given $a \in D$, the ideal $(x - a)A$ is obviously a maximal ideal of codimension 1. because the mapping χ_a from A to $\mathbb{K}$ defined as $\chi_a(f) = f(a)$ maps A onto $\mathbb{K}$.

Now, let $\mathcal{M}$ be a maximal ideal of codimension 1 and let τ be the $\mathbb{K}$-algebra homomorphism from A onto $\mathbb{K}$ admitting $\mathcal{M}$ for kernel. Let $a = \tau(x)$. Since $\tau(x - a) = \tau(x) - a$ is not invertible, we have $|a| < 1$ because if $|a| \geq 1$, then by Theorem 22.20, $x - a$ is invertible in A. Thus, a belongs to D. By Corollary 10.10, we know that τ is continuous, hence $\tau(f) = f(a)\ \forall f \in A$. Consequently, $Ker(\tau) = (x - a)A$. □

Notation. We denote by $\mathrm{Mult}_1(A, \| \,.\, \|)$ the set of multiplicative semi-norms of A whose kernel is a maximal ideal of codimension 1.

By Theorems 34.5 and 34.9, we have the following corollary:

Corollary 34.10. *Let $\phi \in \mathrm{Mult}(A, \| \,.\, \|)\setminus\mathrm{Mult}_1(A, \| \,.\, \|)$. If ϕ is not a norm, its restriction to $\mathbb{K}[x]$ is $\| \,.\, \|_D$.*

Remark. A admits maximal ideals of infinite codimension. The following theorem shows such examples.

Theorem 34.11. *Let $(a_n)_{n\in\mathbb{N}}$ be a sequence of D such that $\lim_{n\to+\infty} |a_n| = 1$ and $\prod_{n\in\mathbb{N}} |a_n| > 0$ and let $\mathcal{I}$ be the ideal of the $f \in A$ such that $\lim_{n\to+\infty} f(a_n) = 0$. Then $\mathcal{I}$ is not null and is included in a maximal ideal of infinite codimension.*

Proof. Let T be the divisor of D defined by the sequence $(a_n)_{n\in\mathbb{N}}$. By Theorem 27.14, there exists an analytic function f in D such that $f(a_n) = 0\ \forall n \in \mathbb{N}$ and such that $|f|(r) \leq |T|(r) + 1\ \forall r \in]0, R[$. But since $\prod_{n\in\mathbb{N}} |a_n| > 0$, we can check that $|T|(r)$ is bounded in $]0, R[$ because

$$\lim_{r\to R} |T||(r) = \frac{1}{\prod_{n\in\mathbb{N}} |a_n|}.$$

Consequently, f is bounded in D and hence $f \in A$. Therefore, $\mathcal{I}$ is not null. But clearly, it is not included in any maximal ideal of the form $(x - a)A$. Consequently, it is included in a maximal ideal of infinite codimension. $\square$

Chapter 35

The Corona Problem on $\mathcal{A}_b(d(0,1^-))$

The Corona conjecture, stated by Kakutani in 1941 in the field $\mathbb{C}$, was solved by Carlesson in 1962 [28]. Consider the open unit disk O in $\mathbb{C}$ and the Banach algebra B of bounded holomorphic functions in O. Each point of O obviously defines a maximal ideal of B. On the other hand, all maximal ideals are of codimension 1 and the Gelfand transform defines a topology on the maximal spectrum. The question was whether the set of maximal ideals defined by points of O was dense inside the whole spectrum of maximal ideals, with respect to the Gelfand topology.

Consider now the situation in our field $\mathbb{K}$. Let us keep the notation introduced in Chapter 34 but put $D = d(0,1^-)$ and $A = \mathcal{A}_b(d(0,1^-))$. The maximal ideals of A of codimension 1 are defined by the points of D, but looking for a Gelfand topology on the maximal spectrum makes no sense because of maximal ideals of infinite codimension. However, the only link existing between the various kind of maximal ideals comes from continuous multiplicative semi-norms. For simplicity, we denote $\| \, . \, \|$ the Gauss norm on A. Here, we have the topology of pointwise convergence on $\text{Mult}(A, \| \, . \, \|)$ and we can ask whether the set of continuous multiplicative semi-norms whose kernel is a maximal ideal of codimension 1 is dense in the whole set of continuous multiplicative semi-norms whose kernel is a maximal ideal, or even whether it is dense in the whole set of all continuous multiplicative semi-norms. Both questions are very hard. Here we solve the first with the use of the ultrametric holomorphic functional calculus [40, 45, 49, 53]. A partial answer is given in [48] when the field is spherically complete. We have generalized it by using Banachic processes.

Notation. Recall that we denote by U the disk $d(0,1)$. Let $\mathcal{H}$ be the family of ideals J of $U[x]$ such that $J \cap U \neq \{0\}$, and, given an integer $s \in \mathbb{N}^*$, let $\mathcal{H}_s$ be the set of $J \in \mathcal{H}$ generated by s elements. For every ideal $J \in \mathcal{H}$, we put $t(J) = \sup\{|x| \mid x \in J \cap U\}$ and $\ell(J) = \inf\{\sup_{f\in J} |f(x)| \mid x \in V\}$ and we denote by $u(J)$ the number such that $t(J) = \ell(J)^{u(J)}$. Finally, we put $m(s) = \sup\{u(J) \mid J \in \mathcal{J}_s\}$. Henceforth, given $f_1, \dots, f_s \in H(U)$ such that $\|f_i\| < 1\ \forall i = 1, \dots, s$, we set $w(f_1, \dots, f_s) = \inf\{\max_{1\leq i\leq s} |f_i(x)| \mid x \in U\}$. Moreover, given $f_1, \dots, f_s \in A$, we set $\lambda(f_1, \dots, f_s) = \inf\{\max_{1\leq i\leq s} |f_i(x)| \mid x \in D\}$.

Proposition 35.2 is also called Corona Statement in dimension 1. However, on a non-archimedean field, it is not at all proven to be equivalent to a property of density for maximal Proposition 35.1 is proved in [91] and is indispensable for further results.

Proposition 35.1. *Let J be a finitely generated ideal of $U[x]$ such that $J \cap U \neq \{0\}$. Then, $m(s) = 2\ \forall s \geq 2$.*

Proposition 35.2. *Let $s \in \mathbb{N}^*$. For any $f_1, \dots, f_s \in A$ satisfying $\|f_i\| < 1$ $(1 \leq i \leq s)$ and $\lambda(f_1, \dots, f_s) > 0$, there exist $g_1, \dots, g_s \in A$ satisfying $\sum_{i=1}^{s} f_i g_i = 1$ and $\|g_i\| < \lambda(f_1, \dots, f_s))^{-2}$.*

Proof. As the first step, we prove that for any $f_1, \dots, f_s \in H(U)$ satisfying $\|f_i\| < 1\ \forall i = 1, \dots, s$ and $w(f_1, \dots, f_s) > 0$, there exist $g_1, \dots, g_s \in H(d(0,R))$ satisfying $\sum_{i=1}^{s} f_i g_i = 1$ and $\|g_i\| < (w(f_1, \dots, f_s))^{-2}\ \forall i = 1, \dots, s$.

Since $\mathbb{K}[x]$ is dense in $H(U)$, we can find polynomials $P_1, \dots, P_s \in \mathbb{K}[x]$ such that $\|P_i - f_i\| \leq (w(f_1, \dots, f_s))^2\ \forall i = 1, \dots, s$. Since $w(f_1, \dots, f_s) > 0$, by Corollary 24.13, there exists no maximal ideal $\mathcal{M}$ of $H(U)$ containing the ideal generated by $f_1, \dots, f_s$. Consequently, there exist $g_1, \dots, g_s \in H(U)$ such that $\sum_{i=1}^{s} g_i f_i = 1$. Therefore, we can define $h_1, \dots, h_s \in H(U)$ such that $\|h_i\| \leq 1\ \forall\ i = 1, \dots, s$ and such that $\sum_{i=1}^{s} h_i f_i$ is an element $P_0 \in U$. Let I be the ideal of $U[x]$ generated by $P_0, P_1, \dots, P_s$. Since $I \cap U \neq \{0\}$, we have $0 < t(I) < \ell(I)$. Consequently, by Proposition 35.1, we can find $Q_0, Q_1, \dots, Q_s \in U[x]$ such that $\sum_{i=0}^{s} Q_i P_i$ is an element $a \in U$

satisfying $|a| \geq (w(f_1,\ldots,f_s))^{m(s)} = (w(f_1,\ldots,f_s))^2$. Then,

$$\sum_{i=1}^{s} a^{-1}f_i + a^{-1}Q_0\sum_{i=1}^{s} h_i f_i = 1 + \sum_{i=1}^{s} a^{-1}Q_i(f_i - P_i).$$

By construction, we have $\|\sum_{i=1}^{s} a^{-1}Q_i(f_i - P_i)\| < 1$, and hence $1+\sum_{i=1}^{s} a^{-1}Q_i(f_i - P_i)$ is a unit u in $H(U)$. So, $\sum_{i=1}^{s} u^{-1}\big(a^{-1}(Q_i + Q_0h_i)\big)f_i = 1$, and hence for every $i = 1,\ldots,s$, we have $\|u^{-1}a^{-1}(Q_i + Q_0h_i)\| \leq (w(f_1,\ldots,f_s))^{-2}$ $\forall i = 1,\ldots,s$. Thus, putting $g_i = u^{-1}\big(a^{-1}(Q_i+Q_0h_i)\big)$, $i = 1,\ldots,s$, we have proven that for any $f_1,\ldots,f_s \in H(U)$ satisfying $\|f_i\| < 1$ $\forall i = 1,\ldots,s$ and $w(f_1,\ldots,f_s) > 0$, there exist $g_1,\ldots,g_s \in H(U)$ satisfying $\sum_{i=1}^{s} f_ig_i = 1$ and $\|g_i\| < (w(f_1,\ldots,f_s))^{-2}$ $\forall i = 1,\ldots,s$.

Now, let us prove the conclusion of Proposition 35.2. So, we take $f_1,\ldots,f_s \in A$. Let $(u_n)_{n\in\mathbb{N}}$ be a sequence in $\mathbb{K}$ such that $0 < |u_n| < |u_{n+1}| < \cdots < 1$ and $\lim_{n\to+\infty}|u_n|^n = 1$. For each $i = 1,\ldots,s$, set $f_{i,n}(x) = f_i(u_nx)$. Since each u_n belongs to D, each $f_{i,n}$ is a power series of radius $r > 1$ and hence belongs to $H(U)$. Then by the claim we have just proven, for each $n \in \mathbb{N}$, there are $g_{i,n} \in H(U)$ such that $\sum_{i=1}^{s} g_{i,n}f_{i,n} = 1$ and $\|g_{i,n}\| < (w(f_{1,n},\ldots,f_{s,n}))^{-2}$. Now, each $g_{i,n}$ is a power series $\sum_{k=0}^{+\infty} g_{i,n,k}x^k$. Let $h_{i,n}$ be the power series $\sum_{k=0}^{2n} g_{i,n,k}(u_n)^{-k}x^k$. Then we have $\|h_{i,n}\| < |u_n|^{-2n}(w(f_1,\ldots,f_s))^{-2}$ and hence $\sum_{i=1}^{s} h_{i,n}f_i$ is of the form $1 + x^nt_n(x)$ with $t_n \in H(U)$.

We get to a conclusion thanks to a Banach process. Let $\mathcal{E}$ be the Banach space of bounded sequences in $\mathbb{K}$ provided with the classic norm $\|(a_n)_{n\in\mathbb{N}}\|_0 = \sup_{n\in\mathbb{N}}|a_n|$ and let $\mathcal{C}$ be the closed sub-space of converging sequences. For every sequence $(a_n)_{n\in\mathbb{N}} \in \mathcal{C}$, we put $\mathcal{L}((a_n)) = \lim_{n\to+\infty} a_n$.

Suppose first that $\mathbb{K}$ is spherically complete, there exists a linear map of norm 1 from $\mathcal{E}$ to $\mathbb{K}$ expanding $\mathcal{L}$ to $\mathcal{E}$. Let $\widetilde{\mathcal{L}}$ be this continuation of $\mathcal{L}$. Set $h_{i,n} = \sum_{k=0}^{+\infty} x^k$ and let $l_{i,k}$ be the sequence $(h_{i,n,k})_{1\leq n}$. For each pair (i,k), we can now define $g_{i,k} = \widetilde{\mathcal{L}}(l_{i,k})$ and put $g_i = \sum_{k=0}^{+\infty} g_{i,k}x^k$. Since $\widetilde{\mathcal{L}}$ is of norm 1, each sequence $g_{i,k}$ is bounded and hence g_i belongs to A. Moreover, by construction, g_i satisfies $\|g_i\| < (\lambda(f_1,\ldots,f_s))^{-2}$ $\forall i = 1,\ldots,s$. Now, we have

$$\sum_{i=1}^{s} g_if_i = \sum_{k=0}^{+\infty}\left(\sum_{i=1}^{s}(\sum_{j=0}^{k}\widetilde{\mathcal{L}}(l_{i,j})f_{i,k-j})\right)x^k \tag{1}$$

and for each fixed $k \in \mathbb{N}$,

$$\sum_{i=1}^{s}\left(\sum_{j=0}^{k}\widetilde{\mathcal{L}}(l_{i,j})f_{i,k-j}\right) = \widetilde{\mathcal{L}}\left(\sum_{i=1}^{s}\left(\sum_{j=0}^{k} f_{i,j}h_{i,n,k-j}\right)\right).$$

Consequently, since

$$\sum_{i=1}^{s} h_{i,n}f_i = 1 + x^n t_n(x),$$

we can check that

$$\lim_{n\to+\infty}\left(\sum_{i=1}^{s}\left(\sum_{j=0}^{k} f_{i,j}h_{i,n,k-j}\right)\right) = 1 \text{ whenever } k = 0$$

and

$$\lim_{n\to+\infty}\left(\sum_{i=1}^{s}\left(\sum_{j=0}^{k} f_{i,j}h_{i,n,k-j}\right)\right) = 0 \text{ whenever } k \neq 0.$$

Consequently, since $\widetilde{\mathcal{L}}$ extends to $\mathcal{E}$ the limit on $\mathcal{C}$, we have

$$\widetilde{\mathcal{L}}\left(\sum_{i=1}^{s}\left(\sum_{j=0}^{k} f_{i,j}h_{i,n,k-j}\right)\right) = 1 \text{ whenever } k = 0$$

and

$$\widetilde{\mathcal{L}}\left(\sum_{i=1}^{s}\left(\sum_{j=0}^{k} f_{i,j}h_{i,n,k-j}\right)\right) = 0 \text{ whenever } k \neq 0.$$

Therefore, by (1), we obtain $\sum_{i=1}^{s} g_i f_i = 1$.

Consider now the general case when $\mathbb{K}$ is no longer supposed to be spherically complete. Consider a spherically complete algebraically closed extension $\widehat{\mathbb{K}}$ of $\mathbb{K}$. Set $B = \lambda(f_1, \dots, f_s))^{-2}$. Then by what forgoes, there exist $h_1, \dots, h_s \in \widehat{A}$ such that $\sum_{j=1}^{s} h_j f_i = 1$ and $\max_{1\leq j\leq s} \|h_j\| < B$.

Inside $\widehat{A}$, let E be the closed subspace of the $\mathbb{K}$-Banach space generated by 1 and all coefficients of all the h_j. Let us take $\epsilon > 0$

such that $(1+\epsilon)\max_{1\le j\le s}\|h_j\| \le B$. Since E is a $\mathbb{K}$-Banach space of countable type, there exists a $\mathbb{K}$-linear map ℓ from E to $\mathbb{K}$ satisfying $\ell(1) = 1$ with a norm $\|\ell\|$ satisfying $\|\ell\| \le 1+\epsilon$. Let T be the closed subspace of $\widehat{A}$ consisting of the power series with coefficients in E. Then T is a A-module and then we have an extension $\mathcal{L}$ of ℓ from T to A defined as $\mathcal{L}\left(\sum_{k=0}^{\infty} e_k x^k\right) = \sum_{k=0}^{\infty} \ell(e_k)x^k$ which is A-linear and its norm satisfies $\|\mathcal{L}\| \le 1+\epsilon$. Now, putting $g_j = \mathcal{L}(h_j),\ 1 \le j \le s$, we have $\sum_{i=1}^{s} f_i g_i = 1$ and $\max_{1\le j\le s}\|g_j\| < B$. □

Notation. Let $\mathcal{I}$ be an ideal of A. For each $f \in A$ and for each $\epsilon > 0$, we set $\mathcal{D}(f,\epsilon) = \{x \in D \mid |f(x)| \le \epsilon\}$.

Given $\phi \in \text{Mult}(A, \|\ .\ \|)$, $f, \dots, f_q \in A$ and $\epsilon > 0$, we denote by $\mathcal{V}(\phi, f_1, \dots, f_q, \epsilon)$ the set $\{\psi \in \text{Mult}(A, \|\ .\ \|)\}$ such that $|\phi(f_j) - \psi(f_j)|_\infty \le \epsilon, j = 1, \dots, q$.

Corollary 35.3. *Let $\mathcal{I}$ be a proper ideal of A. The family $(\mathcal{D}(f,\epsilon),\ f \in \mathcal{I},\ \epsilon > 0)$ generates a filter $\mathcal{F}$ on D such that $\mathcal{I} \subset \mathcal{J}(\mathcal{F})$.*

Proof. Let $(\mathcal{D}(f_j,\epsilon_j),\ 1 \le j \le n$ be such that $\bigcap_{j=1}^{n} \mathcal{D}(f,\epsilon_j) = \emptyset$. Let $\epsilon = \min_{1\le j\le n}(\epsilon_j)$ and let $F_j = D\setminus\mathcal{D}(f_j,\epsilon),\ 1 \le j \le n$. Then, $\bigcup_{j=1}^{n} F_j = D$, hence by Theorem 35.2, $1 \in \mathcal{I}$, a contradiction. □

By Theorem 34.9 and Corollary 35.3, we can derive Corollary 35.4:

Corollary 35.4. *Let $\mathcal{M}$ be a maximal ideal of A. Then there exists an ultrafilter $\mathcal{U}$ on D such that $\mathcal{M} = \mathcal{J}(\mathcal{U})$. Moreover, if $\mathcal{U}$ converges, $\mathcal{M}$ is of codimension 1. If $\mathcal{U}$ does not converge, then $\mathcal{M}$ is of infinite codimension.*

Proof. Indeed, let $\mathcal{M}$ be a maximal ideal of A. The family $(\mathcal{D}(f,\epsilon),\ f \in \mathcal{U},\ \epsilon > 0)$ generates a filter $\mathcal{F}$ on D such that $\mathcal{M} \subset \mathcal{J}(\mathcal{U})$. But since $\mathcal{M}$ is a maximal ideal, then $\mathcal{M} = \mathcal{J}(\mathcal{U})$. The conclusion on the codimension comes from Theorem 34.9. □

Conversely, now we see that certain ultrafilters define maximal ideals. First, we need a few basic results.

Definition and notation. We denote by $\mathcal{W}$ the filter on D admitting for basis the annuli $\Gamma(0,r,1)$. A sequence $(a_n)_{n\in\mathbb{N}}$ is called *a regular sequence* if $\inf_{j\in\mathbb{N}} \prod_{\substack{n\in\mathbb{N}\\ n\ne j}} |a_n - a_j| > 0$.

An ultrafilter $\mathcal{U}$ is said to be *regular* if it is thinner than a regular sequence.

We denote by $B(\mathbb{N}, \mathbb{K})$ the $\mathbb{K}$-algebra of bounded sequences of $\mathbb{K}$. Let $S = (a_n)_{n\in N}$ be a sequence in D thinner than $\mathcal{W}$. We denote by T_S the mapping from A into $B(\mathbb{N}, \mathbb{K})$ which associates with each $f \in A$ the sequence $(f(a_n)_{n\in\mathbb{N}})$, and we denote by $\Sigma(S)$ the set of ultrafilters thinner than S and by $\mathcal{I}(S)$ the ideal of the $f \in A$ such that $f(a_n) = 0 \ \forall n \in \mathbb{N}$.

From Corollary 4.6 of [91], we have Theorem 35.5 which is purely technical:

Theorem 35.5. *Let S be a sequence in D thinner than $\mathcal{W}$. Then T_S is surjective on $B(\mathbb{N}, \mathbb{K})$ if and only if S is regular.*

Notations. Let S be a regular sequence. Since T_S is surjective, there exists a $\mathbb{K}$-algebra isomorphism Λ_S from $\frac{A}{Ker(T_S)}$ onto $B(N, K)$, where $Ker(T_S) = \mathcal{I}(S)$.

For every ultrafilter $\mathcal{G}$ on $\mathbb{N}$, we denote by $\Theta(\mathcal{G})$ the ideal of $B(\mathbb{N}, \mathbb{K})$ consisting of sequences $(a_n)_{n\in\mathbb{N}}$ such that $\lim_{\mathcal{G}} a_n = 0$.

Theorem 35.6 is also classical:

Theorem 35.6. *Θ is a bijection from the set of ultrafilters on $\mathbb{N}$ onto $Max(B(\mathbb{N}, \mathbb{K}))$. The restriction of Θ to the subset of non-principal ultrafilters on $\mathbb{N}$ is a bijection from this set onto the set of non-principal maximal ideals of $B(\mathbb{N}, \mathbb{K})$. Moreover, a maximal ideal of $B(\mathbb{N}, \mathbb{K})$ is principal if and only it is of codimension 1.*

Theorem 35.7. *Let S be a regular sequence and let $\mathcal{M}$ be a maximal ideal of A. The following two statements are equivalent:*

(i) $\mathcal{I}(S) \subset \mathcal{M}$.
(ii) *There exists an ultrafilter $\mathcal{U}$ thinner than S such that $\mathcal{M} = \mathcal{J}(\mathcal{U})$.*
Moreover, the mapping G which associates with each ultrafilter $\mathcal{U}$ thinner than S the ideal $\mathcal{J}(\mathcal{U})$ is a bijection from $\Sigma(S)$ onto the set of maximal ideals of A containing $\mathcal{I}(S)$.

Proof. Obviously, (ii) implies (i). Thus, suppose (i) true. Let $S = (a_n)_{n\in\mathbb{N}}$. By Theorem 35.6, the isomorphism Λ_S makes a bijection G from the set of maximal ideals of A containing $\mathcal{I}(S)$ to the set of maximal ideals of $B(\mathbb{N}, \mathbb{K})$, and more precisely, it makes a bijection from the set of maximal ideals of A of infinite codimension containing $\mathcal{I}(S)$ to the set of maximal ideals of $B(\mathbb{N},\mathbb{K})$ of infinite codimension which actually are the non-principal maximal ideals of $B(\mathbb{N},\mathbb{K})$. Let $\mathcal{N} = G(M)$. By Theorem 35.5, there exists an ultrafilter $\mathcal{U}$ on $\mathbb{N}$ such that $\mathcal{N}$ is the ideal of the bounded sequences tending to zero along $\mathcal{U}$. Now, let Ξ be the natural bijection from the set of non-principal ultrafilters of $\mathbb{N}$ onto the set of ultrafilters thinner than S and let $V = \Xi(\mathcal{U})$. Then, $\mathcal{N} = \Lambda_S(M)$, hence $\mathcal{M} = \mathcal{I}(S)$. Moreover, in this way, we can see that $G \circ \Xi^{-1}$ is a bijection from $\Sigma(S)$ onto the set of maximal ideals of A containing $\mathcal{I}(S)$.

Corollary 35.8. *If $\mathcal{U}$ is a regular ultrafilter on D, $\mathcal{J}(\mathcal{U})$ is a maximal ideal of A.*

Proposition 35.9. *Let $(B, \| \,.\, \|)$ be a unital commutative ultrametric $\mathbb{K}$-Banach algebra. Suppose there exist $\ell \in B$, ϕ, $\psi \in \text{Mult}(B, \| \,.\, \|)$ such that $\psi(\ell) < \phi(\ell)$, $sp(\ell) \cap \Gamma(0, \psi(\ell), \phi(\ell)) = \emptyset$ and there exists $\epsilon \in]0, \phi(\ell) - \psi(\ell)[$ satisfying further $\|(\ell - a)^{-1}\| \leq M \ \forall a \in \Gamma(0, \psi(\ell), \phi(\ell) - \epsilon)$. Then there exists $g \in B$ such that $\psi(g) = 1$, $\phi(g) = 0$.*

Proof. By Theorem 10.6, we have $\| \ell \|_{si} = \sup\{\theta(\ell) \mid \theta \in \text{Mult}(B, \| \,.\, \|)\}$.

Let $s = \psi(\ell)$, $t = \phi(\ell) - \epsilon$. Consider now the construction made in Theorem 21.15 by taking $s = r_0$, $Q = \|\ell\|$, $R = t - \epsilon$, and $l = \frac{1}{M}$ and consider the sequence $(a_{n,j})_{n\in\mathbb{N}, 1\leq j\leq q_n}$ and the set $E = d(0, Q^-) \setminus \Big(\bigcup_{n\in\mathbb{N}} (\cup_{j=1}^{q_n} d(a_{n,j},\ l^-)) \Big)$ defined in Theorem 21.15. Then, in $H(E)$, we have

$$(1) \qquad \left\| \frac{1}{x-b} \right\|_E \leq l \ \forall b \in \bigcup_{n\in\mathbb{N}} (\cup_{j=1}^{q_n} d(a_{n,j},\ l^-)).$$

There exists a natural homomorphism σ from $R(E)$ into B such that $\sigma(x) = \ell$. And since $Q = \|\ell\|$ and $\|(\ell - b)^{-1}\| \leq M \ \forall b \in \Gamma(0, s, t)$, by Theorem 15.1 and by (1), σ is clearly continuous with respect to the norms $\| \,.\, \|_E$ of $R(E)$ and $\| \,.\, \|$ of B. Consequently, σ has continuation to a continuous homomorphism from $H(E)$ to B.

Now, let $\psi' = \psi \circ \sigma$, $\phi' = \phi \circ \sigma$. Then both ϕ', ψ' belong to $\text{Mult}(H(E), \| \,.\, \|)$ and satisfy $\psi'(x) = s$, $\phi'(x) = t - \epsilon$. So, ψ' is of the form $\varphi_{\mathcal{F}}$ with $\mathcal{F}$ a circular on E secant with $C(0, s)$ and ϕ' is of the form $\varphi_{\mathcal{G}}$ with $\mathcal{G}$ a circular on E secant with $C(0, t)$.

Consider now the function f constructed in Theorem 21.15 which, by construction, belongs to $H(E)$ and has no zero and no pole in $d(0, s^-)$. Consequently, $|f(x)| = |f(0)| = 1\ \forall x \in d(0, s^-)$. Moreover, as showed in Theorem 21.15, we have $\lim_{\mathcal{G}} f(x) = 0$, hence $\phi'(f) = 0$. Putting $g = \sigma(f)$, we have $\psi(g) = \psi'(f) = 1$ and $\phi(g) = \phi'(f) = 0$, which ends the proof. □

Definition and notation. We denote by $\mathcal{W}$ the filter on D admitting for basis the annuli $\Gamma(0, r, R)$, $r \in]0, R[$. Given an ideal $\mathcal{I}$ of A, we denote by $\mathcal{G}_I$ the filter generated by the sets $\mathcal{D}(f, \epsilon)$, $f \in I, \epsilon > 0$. By definition, $\mathcal{G}_I$ is minimal, with respect to the relation of thinness, among the filters $\mathcal{H}$ such that $\lim_{\mathcal{H}} f(x) = 0\ \forall f \in I$. A filter $\mathcal{F}$ on D is said to be *coroner* if it is thinner than $\mathcal{W}$.

By Theorem 10.9, each maximal ideal of a Banach $\mathbb{K}$-algebra is the kernel of at least one element of $\mathrm{Mult}(A, \|\ .\ \|)$. A maximal ideal $\mathcal{M}$ of A is said to be *univalent* if there exists only one $\phi \in \mathrm{Mult}_1(A, \|\ .\ \|)$ such that $Ker(\phi) = \mathcal{M}$ and the algebra A is said to be *multibijective* if every maximal ideal of A is univalent.

Theorem 35.10. *Let $\mathcal{M}$ be a non-principal maximal ideal of A and let $\mathcal{U}$ be an ultrafilter thinner than $\mathcal{G}_{\mathcal{M}}$. Then $\varphi_{\mathcal{U}}$ belongs to the closure of* $\mathrm{Mult}_1(A, \|\ .\ \|)$ *in* $\mathrm{Mult}_m(A, \|\ .\ \|)$.

Proof. Let V be neighborhood of $\varphi_{\mathcal{U}}$ in $\mathrm{Mult}(A, \|\ .\ \|)$. It contains some set of the form $\mathcal{V}(\varphi_{\mathcal{U}}, f_1, \ldots, f_q, \epsilon)$. For each $j = 1, \ldots, q$, there exists $E_j \in \mathcal{U}$ such that

$$\Big|\ |f_j(x)| - \varphi_{\mathcal{U}}(f_j)\ \Big|_\infty \leq \epsilon \quad \forall x \in E_j.$$

Let $E = \bigcap_{j=1}^q E_j$. Then,

$$\Big|\ |f_j(x)| - \varphi_{\mathcal{U}}(f_j)\ \Big|_\infty \leq \epsilon \quad \forall x \in E, \forall j = 1, \ldots, q.$$

Consequently, φ_a belongs to $\mathcal{V}(\varphi_{\mathcal{U}}, f_1, \ldots, f_q, \epsilon)$ for all $a \in E$. □

Corollary 35.11. *Let $\mathcal{M}$ be a univalent non-principal maximal ideal of A and let $\phi \in \mathrm{Mult}_m(A, \|\ .\ \|)$ satisfy $Ker(\phi) = \mathcal{M}$. Then ϕ is of the form $\phi(f) = \lim_{\mathcal{U}} |f(x)|$ with $\mathcal{U}$ a coroner ultrafilter and such that $\mathcal{J}(\mathcal{U}) = \mathcal{M}$. Moreover, ϕ belongs to the closure of* $\mathrm{Mult}_1(A, \|\ .\ \|)$ *in* $\mathrm{Mult}_m(A, \|\ .\ \|)$.

Proposition 35.12. *Let $\mathcal{U}$ be a coroner ultrafilter on D, let $f \in A \setminus \mathcal{J}(\mathcal{U})$ be non-invertible in A, such that $\|f\| \leq 1$, and let $g \in A$, $h \in \mathcal{J}(\mathcal{U})$ be such that $fg = 1 + h$. Let $\tau = \varphi_{\mathcal{U}}(f)$, let $\epsilon \in]0, \tau[$, and let $\Lambda = \{x \in D \mid \left| |f(x)g(x)| - 1 \right|_\infty < \epsilon, \left| |f(x)| - \tau \right|_\infty < \epsilon\}$.*

Suppose that there exist a function $\widetilde{h} \in A$ admitting for zeros in D the zeros of h in $D \setminus \Lambda$ and a function $\overline{h} \in A$ admitting for zeros the zeros of h in Λ, each counting multiplicities, so that $h = \overline{h}\widetilde{h}$. Then $|\widetilde{h}(x)|$ has a strictly positive lower bound in Λ and $\overline{h}$ belongs to $\mathcal{J}(\mathcal{U})$.

Moreover, there exists $\omega \in]0, \tau[$ such that $\omega \leq \inf\{\max(|f(x)|, |\overline{h}(x)|) \mid x \in D\}$. Further, for every $a \in d(0, (\tau - \epsilon))$, we have $\omega \leq \inf\{\max(|f(x) - a|, |\overline{h}(x)|) \mid x \in D\}$.

Proof. Let $u \in \Lambda$ and let s be the distance of u from $\mathbb{K} \setminus \Lambda$. So, the disk $d(u, s^-)$ is included in Λ, hence fg has no zero inside this disk. Consequently, $|f(x)g(x)|$ is a constant b in $d(u, s^-)$. Consider the family F_u of radii of circles $C(u, r)$, containing at least one zero of fg. By Theorem 22.20, F_u has no cluster point different from 1. Consequently, there exists $\rho \geq s$ such that fg admits at least one zero in $C(u, \rho)$ and admits no zero in $d(u, \rho^-)$. And then $|f(x)g(x)|$ is a constant c in $d(u, \rho^-)$. But then, at u, we see that $b = c$, and therefore, $d(u, \rho^-)$ is included in Λ. Hence, $\rho = s$, and therefore, fg admits at least one zero α in $C(u, s)$. Thus, at α, we have $h(\alpha) = -1$. Therefore, in the disk $d(\alpha, s^-)$, we can check that $\varphi_{\alpha,s}(h) \geq 1$. But by Lemma 13.4, $\varphi_{\alpha,s}(h) = \varphi_{u,s}(h)$, hence $\varphi_{u,s}(h) \geq 1$.

Now,

$$\frac{\|h\|}{\varphi_{u,s}(h)} = \frac{\|\widetilde{h}\|}{\varphi_{u,s}(\widetilde{h})} \frac{\|\overline{h}\|}{\varphi_{u,s}(\overline{h})} \geq \frac{\|\widetilde{h}\|}{\varphi_{u,s}(\widetilde{h})}.$$

Therefore, since $\varphi_{u,s}(h) \geq 1$, we obtain

$$\frac{\|\widetilde{h}\|}{\varphi_{u,s}(\widetilde{h})} \leq \|h\|. \tag{1}$$

But since by definition $d(u, s^-)$ is included in Λ, $\widetilde{h}$ has no zero in this disk, hence $|\widetilde{h}(x)|$ is constant and equal to $\varphi_{u,s}(\widetilde{h})$. Consequently, by

(1), we obtain $\frac{\|\widetilde{h}\|}{|\widetilde{h}(u)|} \leq \|h\|$, and therefore,

$$|\widetilde{h}(u)| \geq \frac{\|\widetilde{h}\|}{\|h\|} \quad \forall u \in \Lambda.$$

This shows that $\widetilde{h}$ does not belong to $\mathcal{J}(\mathcal{U})$, hence, by Theorem 35.1, $\varphi_{\mathcal{U}}(\widetilde{h}) \neq 0$. Consequently, $\varphi_{\mathcal{U}}(\overline{h}) = 0$.

Now, by hypothesis, we have $fg - \overline{h}\widetilde{h} = 1$. Since both g, $\widetilde{h}$ belong to A and therefore are bounded in D, it is obvious that $\inf\{\max(|f(x)|, |\overline{h}(x)|) \mid x \in D\} > 0$. So, we may obviously choose $\omega \in]0, \tau - \epsilon[$ such that

$$\text{(2)} \qquad \omega \leq \inf\{\max(|f(x)|, |\overline{h}(x)|) \mid x \in D\}.$$

Let us now show that for every $a \in d(0, (\tau - \epsilon))$, we have $\omega \leq \inf\{\max(|f(x) - a|, |\overline{h}(x)|) \mid x \in D\}$.

Let $\Lambda' = \{x \in D \mid |f(x)| \geq \tau - \epsilon\}$ and let $a \in d(0, (\tau - \epsilon)^-)$. When β lies in Λ', we have $|f(\beta)| > |a|$, hence by (2), $\max(|f(\beta) - a|, |\overline{h}(\beta)|) \geq \omega$ because by(2), either $\omega \leq |\overline{h}(\beta)|$, or $\omega \leq |f(\beta)| = |f(\beta) - a|$.

Now, let β lie in $D\setminus\Lambda'$ and let t be the distance from β to Λ'. Since $D\setminus\Lambda'$ is open, t is > 0. Consider $\varphi_{\beta,t}(f)$. Either there exists $\ell \in \Lambda'$ such that $|\beta - \ell| = t$ and then $\varphi_{\beta,t}(f) \geq |f(\ell)|$ or there exists a sequence $(x_n)_{n\in\mathbb{N}} \in \Lambda'$ such that $\lim_{n\to+\infty} |\beta - x_n| = t$. Now, let β lie in $D\setminus\Lambda'$ and let t be the distance from β to Λ'. Since $D\setminus\Lambda'$ is open, t is > 0. Consider $\varphi_{\beta,t}(f)$. Either there exists $\mu \in \Lambda'$ such that $|\beta - \mu| = t$ and then $\varphi_{\beta,t}(f) \geq |f(\mu)| \geq \tau - \epsilon$ or there exists a sequence $(x_n)_{n\in\mathbb{N}} \in \Lambda'$ such that $\lim_{n\to+\infty} |\beta - x_n| = t$ and $|x_n - \beta| > t$. Suppose that we are in the second case: there exists a sequence $(x_n)_{n\in\mathbb{N}} \in \Lambda'$ such that $\lim_{n\to+\infty} |\beta - x_n| = t$ and $|x_n - \beta| > t$. Then the sequence is thinner than the circular filter of center β and diameter t, hence

$$\lim_{n\to+\infty} |f(x_n)| = \varphi_{\beta,t}(f),$$

hence $\varphi_{\beta,t}(f) \geq \tau - \epsilon$ again. If f has no zero in $d(\beta, t^-)$, then $|f(x)|$ is a constant in that disk, hence of course $\varphi_{\beta,t}(f) < \tau - \epsilon$, a contradiction. Consequently, f must have a zero γ in $d(\beta, t^-)$. Therefore, due to (2), we have $|\overline{h}(\gamma)| \geq \omega$. But since, by definition, $\Lambda \subset \Lambda'$, the zeros of $\overline{h}$ belong to Λ'. And since $d(\beta, t^-) \cap \Lambda' = \emptyset$ actually $\overline{h}$ has no zero in $d(\beta,t^-)$. Consequently, $|\overline{h}(x)|$ is constant in $d(\beta,t^-)$, and hence $|\overline{h}(\beta)| \geq \omega$, which completes the proof. □

The following basic lemma is easily checked:

Lemma 35.13. *Let S be a set and let E be a subset. Let $\mathcal{F}$ be an ultrafilter on E. Then the filter $\widetilde{\mathcal{F}}$ on S with basis $\mathcal{F}$ is an ultrafilter inducing on E the ultrafilter $\mathcal{F}$.*

Corollary 35.14. *Let S be a set and let E be subset of S. Let $\mathcal{F}$ be an ultrafilter on E and let $\widehat{\mathcal{F}} = \mathcal{G}$ be the ultrafilter on S having $\mathcal{F}$ as a basis of filter. Let f be a function defined on S with values in a compact topological space T. Then, $\lim_{\mathcal{G}} f(x) = \lim_{\mathcal{F}} f(x)$.*

Proof. Suppose that f admits distinct limits on $\mathcal{F}$ and $\mathcal{G}$. Then $\mathcal{F}$ is a basis of a filter on S that is not secant with $\mathcal{G}$, a contradiction since $\mathcal{F}$ is the ultrafilter induced by $\mathcal{G}$ on E. □

Proposition 35.15. *Let $\mathcal{M}$ be a non-principal maximal ideal of A and let $\mathcal{U}$ be an ultrafilter on D such that $\mathcal{M} = \mathcal{J}(\mathcal{U})$. Let $f \in A\backslash\mathcal{M}$ satisfy $\|f\| < 1$, let $\tau = \varphi_{\mathcal{U}}(f)$, and let $\epsilon \in]0,\tau[$. There exists $c > 0$ such that, for every $a \in d(0,\tau-\epsilon)$, there exists $g_a \in A$ satisfying $(f-a)g_a - 1 \in \mathcal{M}$ and $\|g_a\| \leq c$.*

Proof. Suppose first that f is invertible in A. By Theorem 22.20, $|f(x)|$ is a constant and hence is equal to τ. Therefore, $|f(x)-a| = \tau\ \forall a \in d(0,\tau-\epsilon)$. Consequently, $f-a$ is invertible and its inverse g_a satisfies $\|g_a\| = \tau^{-1}$. Thus, we only have to show the claim when f is not invertible.

Since f does not belong to $\mathcal{M}$, we can find $g \in A$ and $h \in \mathcal{M}$ such that $fg = 1 + h$ with $h \in \mathcal{M}$.

Let $\widehat{\mathbb{K}}$ be an algebraically closed spherically complete extension of $\mathbb{K}$, and let $\widehat{D}$ be the disk $\{x \in \widehat{\mathbb{K}} \mid |x| < 1\}$. Let $\widehat{A}$ be the algebra of bounded power series converging in $\widehat{D}$ with coefficients in $\widehat{\mathbb{K}}$.

$\mathcal{U}$ makes a basis of a filter $\widehat{\mathcal{U}}$ on $\widehat{D}$, and by definition, $\mathcal{U}$ is the the filter induced by $\widehat{\mathcal{U}}$ on D. By Lemma 35.13, $\widehat{\mathcal{U}}$ is an ultrafilter on $\widehat{D}$.

Consider now f as an element of $\widehat{A}$. Then $\widehat{\mathcal{U}}$ defines an element ψ of $\text{Mult}(\widehat{A}, \|\ .\ \|)$ as $\psi(\ell) = \lim_{\widehat{\mathcal{U}}} |\ell(x)|, \forall \ell \in \widehat{A}$. Consequently, by

Corollary 35.14, τ is equal to $\lim_{\mathcal{U}} |f(x)|$. Let

$$\Lambda = \{x \in \widehat{D} \mid \mid |f(x)g(x)| - 1|_\infty < \epsilon, \mid |f(x)| - \tau|_\infty < \epsilon\}.$$

By Proposition 28.6, we can factorize h in the form $\widetilde{h}\overline{h}$, where $\widetilde{h} \in \widehat{A}$ is a function admitting for zeros in $\widehat{D}$ the zeros of h in $\widehat{D}\backslash\Lambda$ and $\overline{h} \in \widehat{A}$ is a function admitting for zeros the zeros of h in Λ, each counting multiplicities. Moreover, we can choose $\overline{h}$ so that $\|\overline{h}\| < 1$.

Now, in the field $\widehat{\mathbb{K}}$, by Proposition 35.12, there exists $\omega > 0$ such that for every $a \in \widehat{d}(0, (\tau - \epsilon))$, we have $\omega \leq \inf\{\max(|f(x) - a|, |\overline{h}(x)|) \mid x \in \widehat{D}\}$. This implies that $\inf\{\max(|f(x) - a|, |\overline{h}(x)|) \,|x \in D\} \geq \omega \ \forall a \in \widehat{d}(0, \tau - \epsilon)$. We note that $\|f - a\| < 1$ for every $a \in \widehat{d}(0, \tau - \epsilon)$, so we may apply Theorem 35.1 and obtain a bound b only depending on f and h and functions ℓ_a, $h_a \in \widehat{A}$ such that $(f - a)\ell_a + \overline{h}h_a = 1$, with

$$\text{(1)} \qquad \|\ell_a\| < b, \ \|h_a\| < b \quad \forall a \in \widehat{d}(0, \tau - \epsilon).$$

By hypothesis, we have $\lim_{\mathcal{U}} h(x) = 0$. Hence, by Corollary 35.14, on $\widehat{D}$ we have $\lim_{\widehat{\mathcal{U}}} h(x) = 0$. Then, by Proposition 35.12, we have $\lim_{\widehat{\mathcal{U}}} \overline{h}(x) = 0$, hence, on D,

$$\text{(2)} \qquad \lim_{\mathcal{U}} \overline{h}h_a(x) = 0 \quad \forall a \in d(0, \tau - \epsilon).$$

Now, let us fix $a \in d(0, \tau - \epsilon)$. Let G be the closed $\mathbb{K}$-vector subspace of $\widehat{\mathbb{K}}$ (considered as a $\mathbb{K}$-Banach space), linearly generated over $\mathbb{K}$ by 1 and all coefficients of ℓ_a. Take $\eta > 0$ such that $(1 + \eta)\max(\|\ell_a\|, \|h_a\|) \leq b$. We note that G is a $\mathbb{K}$-Banach space of countable type, hence there exists a $\mathbb{K}$-linear mapping Ξ from G to $\mathbb{K}$ of norm $\leq 1 + \eta$ such that $\Xi(1) = 1$ [89]. Let F be the closed $\mathbb{K}$-vector subspace of $\widehat{A}$ consisting of all power series with coefficients in E. Then F is a A-module and Ξ has continuation to a A-linear mapping $\widehat{\Xi}$ from F to A defined as $\widehat{\Xi}(\sum_{n=0}^{\infty} b_n x^n) = \sum_{n=0}^{\infty} \Xi(b_n) x^n$. This mapping $\widehat{\Xi}$ has a norm bounded by $1+\eta$. Set $g_a = \widehat{\Xi}(\ell_a)$. Then, by (1), we have

$$\text{(3)} \qquad \|g_a\| \leq b(1 + \eta) \quad \forall a \in d(0, \tau - \epsilon).$$

On the other hand, by construction, for every $z \in G$, we have $|\widehat{\Xi}(z)| \leq |z|(1+\eta)$: that holds particularly for elements of $G \cap D$. Now, since $(f-a)(l_a) - \overline{h}h_a = 1$, for all $x \in D$, we have $l_a(x) \in G$, $f(x) - a \in K$ and hence $\overline{h}h_a(x)$ belongs to G. Therefore, the inequality applies and shows that $|\widehat{\Xi}(\overline{h}h_a)(x)| \leq |(\overline{h}h_a)(x)|(1+\eta)$, hence by (2), we can derive $\lim_{\mathcal{U}} \widehat{\Xi}(\overline{h}h_a)(x) = 0 \ \forall a \in d(0, \tau - \epsilon)$. But since $\widehat{\Xi}$ is a A-module linear mapping, we have $\widehat{\Xi}((f-a)h_a - 1) = (f-a)$ $g_a - 1$. Consequently, $\lim_{\mathcal{U}} |(f(x)-a)g_a(x) - 1| = 0 \ \forall a \in d(0, \tau-\epsilon)$ and hence $(f-a)g_a - 1$ belongs to $\mathcal{J}(\mathcal{U})$. Putting $c = b(1+\eta)$, by (3), we are done. □

Theorem 35.16. *A is multibijective.*

Proof. Suppose A is not multibijective and let $\mathcal{M}$ be a maximal ideal which is not univalent. Let F be the quotient field $\frac{A}{\mathcal{M}}$, let θ be the canonical surjection from A onto F, and let $\| \ . \ \|_q$ be the $\mathbb{K}$-Banach algebra quotient norm of F. By Corollary 35.11, there exists an ultrafilter $\mathcal{U}$ on D such that $\mathcal{M} = \mathcal{J}(\mathcal{U})$. Thus, there exists $\psi \in \text{Mult}(A, \| \ . \ \|)$ such that $Ker(\psi) = \mathcal{M}$ and $\psi \neq \varphi_{\mathcal{U}}$. Consequently, there exists $f \in A$ such that $\psi(f) \neq \varphi_{\mathcal{U}}(f)$, with $\psi(f) \neq 0$, $\varphi_{\mathcal{U}}(f) \neq 0$. We check that we may also assume $\psi(f) < \varphi_{\mathcal{U}}(f)$. Indeed, suppose $\psi(f) > \varphi_{\mathcal{U}}(f)$. Let $g \in A$ be such that $\theta(g) = \theta(f)^{-1}$. Then we can see that $\psi(g) = \psi(f)^{-1}$, $\varphi_{\mathcal{U}}(g) = (\varphi_{\mathcal{U}}(f))^{-1}$, therefore $\psi(g) < \varphi_{\mathcal{U}}(g)$. Thus, we may assume $\psi(f) < \varphi_{\mathcal{U}}(f)$ without loss of generality. Similarly, we may obviously assume that $\|f\| < 1$. By Lemma 35.7, we know that f is not invertible.

By construction, $\varphi_{\mathcal{U}}$ factorizes in the form $\phi_1 \circ \theta$ and similarly, ψ factorizes in the form $\phi_2 \circ \theta$ with $\phi_1, \ \phi_2 \in \text{Mult}(F, \| \ . \ \|_q)$. So, on F we have $\phi_1(\theta(f)) > \phi_2(\theta(f))$.

Let $\tau = \varphi_{\mathcal{U}}(f)$ and let $\epsilon \in]0, \tau[$. By Proposition 35.15, there exists $c > 0$ such that, for every $a \in d(0, \tau - \epsilon)$, there exists $g_a \in A$ satisfying $(f-a)g_a - 1 \in \mathcal{M}$ and $\|g_a\| \leq c$. Now, $\theta(g_a) = (\theta(f-a))^{-1}$. Thus, $\|(\theta(f-a))^{-1}\|_q \leq c \ \forall a \in d(0, \tau - \epsilon)$. Therefore, by applying Proposition 35.9 to the $\mathbb{K}$-Banach algebra F, we can see that there exists $y \in F$ such that $\phi_1(y) = 1, \phi_2(y) = 0$. Therefore, taking $g \in A$ such that $\theta(g) = y$, we get $\varphi_{\mathcal{U}}(g) = 0, \ \psi(g) = 1$, a contradiction to the hypothesis $Ker(\varphi_{\mathcal{U}}) = Ker(\psi)$. This finishes showing that A is multibijective. □

By Corollary 35.11, we have the following two corollaries:

Corollary 35.17. *For every $\phi \in \text{Mult}_m(A, \| \,.\, \|)\setminus\text{Mult}_1(A, \| \,.\, \|)$, there exists a coroner ultrafilter $\mathcal{U}$ such that $\phi(f) = \lim_{\mathcal{U}} |f(x)|$ $\forall f \in A$.*

Corollary 35.18. *$\text{Mult}_1(A, \| \,.\, \|)$ is dense in $\text{Mult}_m(A, \| \,.\, \|)$.*

Another interesting question was whether certain elements of $\text{Mult}(A, \| \,.\, \|)$ may have a kernel that is neither null nor a maximal ideal. The question was solved by Jesus Araujo thanks to this nice example [6].

Theorem 35.19 (Araujo, [6]). *Let $h(x) = \sum_{n=0}^{\infty} a_n x^n$ and suppose that the sequence $\left(\frac{|a_n|}{|a_{n+1}|}\right)_{n\in\mathbb{N}}$ is strictly increasing of limit 1. Then, h belongs to A. Moreover, putting $r_n = \frac{|a_n|}{|a_{n+1}|}$, $n \in \mathbb{N}$, h admits a unique zero on each circle $C(0, r_n)$ and has no other zero in D.*

Let $\mathcal{N}$ be an ultrafilter on $\mathbb{N}$, and for every $f \in A$, let $\phi(f, n) = \|f\|_{d(\alpha_n, r)}$. Let $\varphi_r(f) = \lim_{\mathcal{N}} \phi(f, n)$.

Then φ_r belongs to $\text{Mult}(A, \| \,.\, \|)$ and $Ker(\varphi_r)$ is neither null nor a maximal ideal of A. Moreover, $Ker(\varphi_r)$ does not depend on $r \in]0, 1[$.

However, each so defined semi-norm φ_r belongs to the closure of $\text{Mult}_1(A, \| \,.\, \|)$ in $\text{Mult}(A, \| \,.\, \|)$.

Proof. Let $\mathcal{M}$ be the ideal of the $f \in A$ such that $\lim_{\mathcal{N}} |f(\alpha_n)| = 0$. Of course, h belongs to $\mathcal{M}$ and $Ker(\varphi_r)$ is strictly included in $\mathcal{M}$. Indeed, since h admits a unique zero in the disk $d(\alpha_n, r)$, it satisfies $\|h\|_{d(\alpha_n,,r)} = |h|(r_n)\frac{r_n}{r}$ and therefore $\lim_{\mathcal{N}} \|h\|_{d(\alpha_n,r)}) = \frac{R}{r}$, which proves that h does not belong to $Ker(\varphi_r)$.

On the other hand, we prove that $Ker(\varphi_r)$ is not null. Let $(q_n)_{n\in\mathbb{N}^*}$ be a sequence of positive integers satisfying $q_n \leq q_{n+1}$ $\forall n \in \mathbb{N}^*$, $\lim_{n\to+\infty} q_n = +\infty$ and such that the series $\sum_{n=1}^{+\infty} q_n \log(\frac{R}{r_n})$ converges: we can easily find the sequence (q_n) since $\lim_{n\to+\infty} r_n = R$. Now, consider the divisor $(\alpha_n, q_n)_{n\in\mathbb{N}}$ of the disk $d(0, R^-)$. By Theorem 27.14, there exists $g \in \mathcal{A}(d(0, R^-))$ admitting each α_n as a zero of order $t_n \geq q_n$ and such that $|g|(r_n) \leq |T|(r_n) + 1$ $\forall n \in \mathbb{N}^*$. Consequently, g is bounded in $d(0, R^-)$ and hence belongs to A. Next, for every $n \in \mathbb{N}^*$, by Corollary 22.30, we have

$$\|g\|_{d(\alpha_n,r)} \leq |g|(r_n)\Big(\frac{r}{r_n}\Big)^{t_n} \leq \|g\|\Big(\frac{r}{r_n}\Big)^{q_n}.$$

Since the sequence $(q_n)_{n\in\mathbb{N}^*}$ tends to $+\infty$ and the sequence (r_n) is increasing, we have $\lim_{n\to+\infty} \|g\|_{d(\alpha_n,r)} = 0$, which proves that g belongs to $Ker(\varphi_r)$. Let $f \in Ker(\varphi_r)$ and let $s \in]0, R[$. If $s < r$, it is obvious that f belongs to $Ker(\varphi_s)$. Now, suppose $s > r$. Consider an element L of $\mathcal{N}$ such that $\inf_{n\in L} \|f\|_{d(\alpha_n,r)} = 0$. We prove that $\inf_{n\in L} \|f\|_{d(\alpha_n,s)} = 0$. For each $n \in \mathbb{N}^*$, by Corollary 22.32, we have

(1) $$\log(\|f\|_{d(\alpha,s)}) - \log(\|f\|_{d(\alpha,r)})$$

$$\leq \Big(\log(\|f\|_{d(\alpha,r_n)}) - \log(\|f\|_{d(\alpha,s)})\Big)\left(\frac{\log(s) - \log(r)}{\log(r_n) - \log(s)}\right).$$

Suppose that the sequence $\|f\|_{d(\alpha_n,s)}$ does not tend to 0. There exists a sequence $(u_m)_{m\in\mathbb{N}}$ of $\mathbb{N}^*$ such that $\|f\|_{d(\alpha_{u_m},s)} > b\ \forall m \in \mathbb{N}$ with $b > 0$. But then we get to a contradiction with (1). Consequently, $\inf_{n\in L} \|f\|_{d(\alpha_n,s)} = 0$ and therefore f belongs to $Ker(\varphi_s)$, which proves that $Ker(\varphi_s) = Ker(\varphi_r)$ and finishes the proof of Theorem 35.19. □

Question. The characterization of all multiplicative semi-norms is not yet complete: does $\phi \in \text{Mult}(A, \| \,.\, \|)$ exist different from those we have studied?

Chapter 36

Meromorphic Functions in $\mathbb{K}$

In this chapter, we define and examine the basic properties of meromorphic functions: relations with poles of analytic elements, absolute values on fields of meromorphic functions defined by circular filters, value of the derivative on a circular filter, development in a Laurent series in an annulus, and existence of primitives [19].

Definition and notation. We denote by $\mathcal{M}(\mathbb{K})$ the field of fractions of $\mathcal{A}(\mathbb{K})$. The elements of $\mathcal{M}(\mathbb{K})$ are called *meromorphic functions in* $\mathbb{K}$.

In the same way, given $a \in \mathbb{K}$ and $r > 0$, we denote by $\mathcal{M}(d(a, r^-))$ (respectively $\mathcal{M}_b(d(a, r^-))$, respectively $\mathcal{M}_u(d(a, r^-)))$ the field of fractions of $\mathcal{A}(d(a, r^-))$ (respectively the field of fractions of $\mathcal{A}_b(d(a, r^-))$, respectively the set $\mathcal{M}(d(a, r^-))\backslash\mathcal{M}_b(d(a, r^-)))$.

The topology defined on $\mathcal{A}(\mathbb{K})$ (respectively $\mathcal{A}(d(a, R^-)))$ has expansion to $\mathcal{M}(\mathbb{K})$ (respectively to $\mathcal{M}(d(a, R^-)))$. The neighborhoods of a function $f \in \mathcal{M}(\mathbb{K})$ (respectively $\mathcal{M}(d(a, R^-)))$ are the sets $\mathcal{W}(f, r, \epsilon) = \{h \in \{\mathcal{M}(\mathbb{K}) \mid |f - h|(r) \leq \epsilon\}$, with $r > 0$ (respectively $\mathcal{W}(f, r, \epsilon) = \{h \in \{\mathcal{M}(d(a, R^-)) \mid |f - h|(r) \leq \epsilon\}$, with $r > 0$).

Let $b \in \mathbb{K}$ (respectively $b \in d(a, R^-)$) and let $r \in \mathbb{R}_+^*$ (respectively $r \in]0, R[$). The absolute value $\varphi_{b,r}$ defined on $\mathcal{A}(\mathbb{K})$ (respectively on $\mathcal{A}(d(a, R^-)))$ has an immediate continuation to $\mathcal{M}(\mathbb{K})$ (respectively to $\mathcal{M}(d(a, R^-)))$ that we denote again by $\varphi_{b,r}$. In the same way, $\varphi_{0,r}$ is denoted by $| \,.\, |(r)$ on $\mathcal{M}(\mathbb{K})$ and on $\mathcal{M}(d(0, R^-))$. Similarly, the function $\Psi(\ .\mu)$ defined on $\mathcal{A}(\mathbb{K})$ and on $\mathcal{A}(d(0,R^-))$ has an immediate continuation to $\mathcal{M}(\mathbb{K})$ and to $\mathcal{M}(d(0,R^-))$ as $\Psi(\frac{h}{l},\mu) = \Psi(h,\mu) - \Psi(l,\mu)$, with $h, l \in \mathcal{A}(\mathbb{K})$ (respectively $h, l \in \mathcal{A}(d(0,R^-))$).

Let $f = \frac{h}{l} \in \mathcal{M}(\mathbb{K})$ (respectively $f = \frac{h}{l} \in \mathcal{M}(d(a,R^-))$). For each $\alpha \in \mathbb{K}$ (respectively $\alpha \in d(a,R^-)$), the number $\omega_\alpha(h) - \omega_\alpha(l)$ does not depend on the functions h, l chosen to make $f = \frac{h}{l}$. Thus, we can generalize the notation by setting $\omega_\alpha(f) = \omega_\alpha(h) - \omega_\alpha(l)$. If $\omega_\alpha(f)$ is an integer $q > 0$, α is called *a zero of f of order q*. If $\omega_\alpha(f)$ is an integer $q < 0$, α is called *a pole of f of order $-q$*. If $\omega_\alpha(f) \geq 0$, f will be said to be *holomorphic* at α.

Similarly, as for $\mathcal{A}(\mathbb{K})$, given $f \in \mathcal{M}(\mathbb{K})$ (respectively $f \in \mathcal{M}(d(a,R^-))$), we can define the divisor $\mathcal{D}(f)$ on $\mathbb{K}$ (respectively of $d(a,R^-)$) as $\mathcal{D}(f)(\alpha) = 0$ whenever $f(\alpha) \neq 0$ and $\mathcal{D}(f)(\alpha) = s$ when f has a zero of order s at α.

Lemma 36.1. *Let $r, R \in \mathbb{R}_+^*$ with $0 < r < R$ and let $f \in \mathcal{M}(d(a,R^-))$. Then f has finitely many poles $a_1, \dots, a_q$ in $d(a,r^-)$. Let $E = d(a,r)\setminus\{a_1,\dots,a_q\}$. Then f belongs to $H(E)$. If there exists $s \in \mathbb{N}^*$ such that f^s is a constant, then so is f.*

Proof. Without loss of generality, we can assume $a = 0$. Let $f = \frac{h}{l}$ with h, $l \in \mathcal{A}(d(0,R))$. Since l belongs to $\mathcal{A}(d(0,R))$, by Corollary 22.19, l has finitely many zeros in $d(0,r)$, hence f has finitely many poles $a_1, \dots, a_q$ in $d(0,r)$. Suppose first f is of the form $\frac{1}{l}$ with $l \in \mathcal{A}(d(0,R^-))$. By Corollary 14.16, l factorizes in the form $P(x)u(x)$ with $P \in \mathbb{K}[x]$ a polynomial whose zeros in $d(a,r)$ are $a_1, \dots, a_q$ and $u \in \mathcal{A}(d(0,R^-))$ is invertible in $H(d(0,r))$. On the other hand, $\frac{1}{P(x)}$ obviously belongs to $R(E)$. And by Proposition 17.3, E belongs to Alg. Consequently, l is invertible in $H(E)$. Consider now the general case $f = \frac{h}{l}$ with h, $l \in \mathcal{A}(d(0,R))$. Then both h, $\frac{1}{l}$ belong to $H(E)$), hence by Proposition 17.3, so does f.

Suppose now that f^s is a constant. Since $\mathbb{K}$ is algebraically closed and since $\mathcal{M}(d(a,R^-))$ is a field extension of $\mathbb{K}$, f belongs to $\mathbb{K}$. □

Corollary 36.2. *Let $f \in \mathcal{M}(d(a,R^-))$, let $r \in]0,R[$, let $\alpha_j, 1 \leq j \leq q$ be the poles of f in $d(a,r)$, let $\rho \in]0, \min_{i\neq j} |\alpha_i - \alpha_j|[$, and for each $j = 1,\dots,q$, let $\rho_j \in]0,\rho[$ and let $T_j = d(\alpha_j,\rho_j^-)$. Let $D = d(a,r)\setminus\left(\bigcup_{j=1}^q T_j\right)$. Then f belongs to $H(D)$.*

Lemma 36.3. *Let $f \in \mathcal{M}(\mathbb{K})$. There exists $h \in \mathcal{A}(\mathbb{K})$ such that $\mathcal{D}(h) = \mathcal{D}(f)$ and then the function $l = \frac{h}{f}$ belongs to $\mathcal{A}(\mathbb{K})$. Then $\mathcal{D}(\frac{1}{f}) = \mathcal{D}(l)$ and we can write f in the form $\frac{h}{l}$ with $h,\ l \in \mathcal{A}(\mathbb{K})$, having no common zero.*

Proof. Indeed, by Theorem 27.4, there exists $h \in \mathcal{A}(\mathbb{K})$ such that $\mathcal{D}(h) = \mathcal{D}(f)$ and hence conclusion follows. □

Remark. Let $f \in \mathcal{M}(d(a, R^-))$, let $r \in]0, R[$, and let α_j, $1 \leq j \leq n$ be the poles of f in $d(a, r)$, of respective order q_j. By Lemma 36.1, f belongs to $H(d(a, r) \setminus \{\alpha_1, \dots, \alpha_n\})$. Now, according to the definition of poles for analytic elements (see Chapter 11), f also admits each α_j as a pole of order q_j, considered as an element of $H(d(a, r) \setminus \{\alpha_1, \dots, \alpha_n\})$.

By Theorem 27.4, we have already seen that if $f \in \mathcal{M}(\mathbb{K})$ has no zero and no pole in $\mathbb{K}$, then it is a constant. Here we can generalize that with functions inside a disk.

Theorem 36.4. *Let $f \in \mathcal{M}(\mathbb{K})$ (respectively $f \in \mathcal{M}(d(a, R^-))$) have no pole in $\mathbb{K}$ (respectively in $d(a, R^-)$). Then f belongs to $\mathcal{A}(\mathbb{K})$ (respectively to $\mathcal{A}(d(a, R^-))$).*

Proof. Suppose $f \in \mathcal{M}(\mathbb{K})$ has no pole in $\mathbb{K}$. By Lemma 36.3, we can write f in the form $\frac{h}{l}$ with $\mathcal{D}(f) = \mathcal{D}(h)$. Since f has no pole in $\mathbb{K}$, l has no zero and hence is a constant, which ends the proof when f belongs to $\mathcal{M}(\mathbb{K})$.

Suppose now that f belongs to $\mathcal{M}(d(a, R^-))$ and has no pole in $d(a, R^-)$. By Proposition 14.12, it is sufficient to show that for each $\rho \in]0, R[$, f belongs to $H(d(a, \rho))$. Let $f = \frac{h}{l}$, with $h,\ l \in \mathcal{A}(d(a, R^-))$. By Proposition 14.12, both $h,\ l$ belong to $H(a, \rho))$.

By hypothesis, each zero α of l is a zero of h such that $\omega_\alpha(h) \geq \omega_\alpha(l)$. Let P be the polynomial admitting for zeros the zeros of l inside $d(a, \rho)$ with the same multiplicity and no other zero. Then P divides h and l in $\mathcal{A}(d(a, R^-))$, say $h = P\phi$, $l = P\psi$. So, ψ is a power series that has no zero in $d(a, \rho)$, hence by Theorem 16.9, it is an invertible element of $H(d(a, \rho))$, which ends the proof. □

Corollary 36.5. *Let $f,\ g \in \mathcal{A}(\mathbb{K})$ (respectively $f,\ g \in \mathcal{A}(d(a, R^-))$) be such that $\mathcal{D}(g) \leq \mathcal{D}(f)$. There exists $h \in \mathcal{A}(\mathbb{K})$ (respectively $h \in \mathcal{A}(d(a, R^-))$) such that $f = gh$.*

Proof. Indeed, $\frac{f}{g}$ belongs to $\mathcal{M}(\mathbb{K})$ (respectively to $\mathcal{M}(d(a,R^-))$) and has no pole. □

Corollary 36.6. *Let $f \in \mathcal{M}(\mathbb{K})$ (respectively $f \in \mathcal{M}(d(a,R^-))$) have no zero and no pole in $\mathbb{K}$ (respectively in $d(a,R^-)$). Then it is a constant (respectively an invertible element of $\mathcal{A}_b(d(a,R^-))$).*

Corollary 36.7. *Let $f,\ g \in \mathcal{A}(\mathbb{K})$ (respectively $f,\ g \in \mathcal{A}(d(a,R^-))$) satisfy $\mathcal{D}(f) = \mathcal{D}(g)$. Then $\frac{f}{g}$ belongs to $\mathbb{K}$ (respectively is invertible in $\mathcal{A}_b(d(a,R^-))$).*

Corollary 36.8. *Let $f \in \mathcal{A}(\mathbb{K})$ be such that $\mathcal{D}(f) = (a_n, q_n)_{n\in\mathbb{N}}$ with $a_n \neq 0\ \forall n \in \mathbb{N}$, $\lim_{n\to+\infty} |a_n| = +\infty$. Then $f(x)$ is of the form $\lambda\prod_{n=0}^{\infty}\left(1-\frac{x}{a_n}\right)^{q_n}$ with $\lambda \in \mathbb{K}$.*

Theorem 36.9. *Let f be a function defined in $\mathbb{K}$ such that for every $r > 0$, the restriction of f to $d(0,r^-)$ belongs to $\mathcal{M}(d(0,r^-))$. Then, $f \in \mathcal{M}(\mathbb{K})$.*

Proof. Let $(a_n, q_n)_{n\in\mathbb{N}}$ be the sequence of the poles of f in $\mathbb{K}$ of respective order q_n, with $|a_n| \leq |)a_{n+1}|$. Since the sequence $(|a_n|)_{n\in\mathbb{N}}$ tends to $+\infty$, the product $\prod_{n=1}^{\infty}\left(1-\frac{x}{a_n}\right)^{q_n}$ converges in ql $\mathbb{K}$ and the function $g(x) = \prod_{n=1}^{\infty}\left(1-\frac{x}{a_n}\right)^{q_n}$ belongs to $\mathcal{A}(d(0,r^-))$ for every $r > 0$, hence it is a power series converging in all $\mathbb{K}$, therefore it lies in $\mathcal{A}(\mathbb{K})$. Consequently, f lies in $\mathcal{M}(\mathbb{K})$. □

By Theorem 28.6, Lemma 36.9 and Corollary 36.10 are immediate:

Lemma 36.10. *Let $\mathbb{K}$ be spherically complete, let $a \in \mathbb{K}$, $r \in \mathbb{R}_+^*$, and let $\mathcal{B}$, $\mathcal{C}$ be divisors on $d(a,R^-)$. There exists $f \in \mathcal{M}(d(a,R^-))$ such that $\mathcal{D}(f) = \mathcal{B}$ and $\mathcal{D}(\frac{1}{f}) = \mathcal{C}$.*

Corollary 36.11. *Let $\mathbb{K}$ be spherically complete, let $a \in \mathbb{K}$, $r \in \mathbb{R}_+^*$, and let $f \in \mathcal{M}(d(a,R^-))$. There exist $g,\ h \in \mathcal{A}(d(a,R^-))$, having no common zero, such that $f = \frac{g}{h}$.*

Remark. If $\mathbb{K}$ is not spherically complete, in the general case, as shown in Theorem 28.1, we cannot find an analytic function $h \in \mathcal{A}(d(a,R^-))$ such that $\mathcal{D}(h) = \mathcal{D}(f)$. Consequently, in a field such as $\mathbb{C}_p$, we can't write f in the form $f = \frac{h}{l}$ with $h,\ l \in \mathcal{A}(d(a, R^-))$, having no common zero (this gap was forgotten in several works).

However, by Theorem 8.4, we can take an algebraically closed spherically complete extension $\widehat{\mathbb{K}}$ of $\mathbb{K}$ and consider f as an analytic function on the disk $\widehat{d}(a, R^-)$ in the field $\widehat{\mathbb{K}}$: then f may be written in the form $f = \frac{\widehat{h}}{\widehat{l}}$ with $\widehat{h},\ \widehat{l} \in \mathcal{A}(\widehat{d}(a, R^-))$, with $\widehat{h},\ \widehat{l}$ having no common zero.

Theorem 36.12. *Let* $r \in \mathbb{R}_+^*$, *let* $f(x) = \mathcal{M}(\mathbb{K})$ *(respectively* $f \in \mathcal{M}(d(a, r^-))$*), let* S *be the set of zeros and poles of* f *in* $\mathbb{K}$ *(respectively in* $d(a, r^-)$*), let* t *be the g.c.d. of* $\{\omega_\alpha(f) \mid \alpha \in S\}$*, and let* $n \in \mathbb{N}^*$. *If there exists* $g \in \mathcal{M}(\mathbb{K})$ *(respectively* $g \in \mathcal{M}(d(a, r^-))$*) such that* $g^q = f$*, then* q *divides* t*. Conversely, if* q *divides* t*, then there exists* $g \in \mathcal{M}(\mathbb{K})$ *such that* $g^q = f$ *(respectively if* p *is prime to* q *and if* q *divides* t *then there exists* $g \in \mathcal{M}(d(a, R^-))$ *such that* $g^q = f$*).*

Proof. If there exists $g \in \mathcal{M}(d(a, r^-))$ (respectively $g \in \mathcal{M}(\mathbb{K})$) such that $g^q = f$, then of course, $\omega_\alpha(g)$ divides $\omega_\alpha(f)$ for every $\alpha \in S$ and hence it divides t. Now suppose q divides t and set $t = lq$. For each $\alpha \in S$, $\omega_\alpha(f)$ is of the form $ts_\alpha = qls_\alpha$.

Suppose first $f \in \mathcal{M}(\mathbb{K})$. By Lemma 36.3, in $\mathcal{M}(\mathbb{K})$, there exists $g \in \mathcal{M}(\mathbb{K})$ admitting each zero α of f as a zero of order ls_α and each pole α of f as a pole of order $-ls_\alpha$. Then $\frac{f}{g^q}$ has no zero and no pole in $\mathbb{K}$, hence, by Corollary 36.6, it is a constant λ. Let υ be a $q-th$ root of λ. Then, $f = (\upsilon g)^q$.

Now, suppose that q is prime to p and suppose $f \in \mathcal{M}(d(a, R^-))$. Suppose first that $\mathbb{K}$ is spherically complete. By Lemma 36.10 there exists $g \in \mathcal{M}(d(a, R^-))$ admitting each zero α of f as a zero of order ls_α and each pole α of f as a pole of order $-ls_\alpha$. Then $\frac{f}{g^q}$ has no zero and no pole in $d(a, R^-)$, hence, it belongs to $\mathcal{A}(d(a, r^-))$. But since it has no zero, by Corollary 23.2, it satisfies $|h(x) - h(a)| < |h(a)|\ \forall x \in d(a, r^-)$. Let $\psi(x) = \frac{h(x)}{h(a)}$. Then we have $|\psi(x) - 1| < 1\ \forall x \in d(a, r^-)$ and then, since q is prime to p, by Theorem 29.23, we can apply the function $\sqrt[q]{.}$ to $\psi(x)$ in order to get a function $\sqrt[q]{\psi(x)} \in \mathcal{A}(d(a, r^-))$. Now, let υ be a qth root of $h(a)$. We have $f(x) = h(a)\psi(x)(g(x))^q = \left(\upsilon \sqrt[n]{\psi(x)} g(x)\right)^q$ which ends the proof when $\mathbb{K}$ is spherically complete.

Consider now the general case, when $\mathbb{K}$ is no longer supposed to be spherically complete. Let $\widehat{\mathbb{K}}$ be a spherically complete algebraically closed extension of $\mathbb{K}$. The function f has continuation to a function $\widehat{f}$ which belongs to $\mathcal{A}(\widehat{d}(\alpha, R^-))$ and hence there exists a function $g \in \mathcal{A}(\widehat{d}(\alpha, R^-))$ such that $g^q = f$. Then, by Lemma 14.5, g is a power series that has all coefficients in $\mathbb{K}$ and hence belongs to $\mathcal{A}(d(a, R^-))$. □

Corollary 36.13. *Let $f(x) \in \mathcal{M}(\mathbb{K})$, let S be the set of zeros and poles of f in $\mathbb{K}$, and let t be the* g.c.d. *of $\{\omega_\alpha(f) \mid \alpha \in S\}$. Then t is the greatest of the integers n such that there exists $g \in \mathcal{M}(\mathbb{K})$ satisfying $g^n = f$.*

Theorem 36.14. *Let $f \in \mathcal{M}(\mathbb{K})$ (respectively $f \in \mathcal{M}(d(a, R^-))$, respectively $f \in \mathcal{M}(D)$) be constant inside a disk included in $\mathbb{K}$ (respectively in $d(a, R^-)$, respectively in D). Then f is constant in $\mathbb{K}$ (respectively in $d(a, R^-)$, respectively in D).*

Proof. For a non-identically zero meromorphic function, the zeros and the poles of f are isolated. Consequently, if $f(x)$ is equal to a constant inside a disk, it is constant in the set of definition. □

Definition. Given $f \in \mathcal{M}(\mathbb{K})$ (respectively $f \in \mathcal{M}(d(a, R^-))$, respectively $f \in \mathcal{M}(D)$), we call *divisor of the poles of f on $\mathbb{K}$ (respectively on $d(a, R^-)$)* the divisor of $\frac{1}{f}$ on $\mathbb{K}$ (respectively on $d(a, R^-)$, respectively on D).

Lemma 36.15. *Let $f \in \mathcal{M}(\mathbb{K})\backslash\mathcal{A}(\mathbb{K})$ (respectively $f \in \mathcal{M}(d(0, R^-))\backslash\mathcal{A}(d(0, R^-))$) and suppose 0 is not a pole of f. Let r be the minimal distance of the poles of f to 0. Then f belongs to $\mathcal{A}(d(0, r^-))$ and its radius of convergence is r.*

Proof. Consider the divisor T of the poles of f on $d(0, R^-)$. If $f \in \mathcal{M}(\mathbb{K})\backslash\mathcal{A}(\mathbb{K})$, there is no problem to write f in the form $\frac{h}{l}$ with h, $l \in \mathcal{A}(\mathbb{K})$, where l has no zero in $d(0, r^-)$. Consequently, by Corollary 23.2, the restriction of l to $d(0, r^-)$ is invertible in $\mathcal{A}(d(0, r^-))$. Therefore, $\frac{h}{l}$ belongs to $\mathcal{A}(d(0, r^-))$ and hence its radius of convergence is $\geq r$. Conversely, since f has a pole in $C(0, r)$, it is not equal to a power series in x in $d(0, r)$ and hence, the radius of convergence is r.

Now, suppose $f \in \mathcal{M}(d(0, R^-))\backslash\mathcal{A}(d(0, R^-))$. By Theorem 27.15, we can find a function $l \in \mathcal{A}(d(0, R^-))$ such that $\mathcal{D}(l) \geq T$ and such that none of the zeros of l lie in $d(0, r^-)$. Next, we set $h = fl$ and see that h has no pole in $d(0, R^-)$. So, in both cases, we have made f in the form $\frac{h}{l}$ with $h,\ l \in \mathcal{A}(\mathbb{K})$, where l has no zero in $d(0, r^-)$. The proof is then similar to the case $f \in \mathcal{A}(\mathbb{K})$. □

Corollary 36.16. *Let $f \in \mathcal{M}(\mathbb{K})\setminus\mathcal{A}(\mathbb{K})$ (respectively $f \in \mathcal{M}(d(0, R^-))\setminus\mathcal{A}(d(0, R^-))$). If 0 is not a pole, $f(x)$ has a development in a power series whose radius of convergence is the minimal distance of poles of f to 0. If 0 is a pole of order q of f, then $f(x)$ has a development in a Laurent series $\sum_{k=-q}^{\infty} a_k x^k$ with $a_{-q} \neq 0$ and the radius of convergence of the series $\sum_{k=0}^{\infty} a_k x^k$ is equal to the minimal distance of non-zero poles of f to 0.*

Theorem 36.17. *Let $f \in \mathcal{M}(d(0, R^-))$ have no pole in an annulus $\Gamma(0, r, s)$ with $s < R$. Then $f(x)$ is equal to a Laurent series $\sum_{-\infty}^{+\infty} a_n x^n$ converging in all $\Gamma(0, r, s)$. For each $\mu \in [\log r, \log s]$, if f has q zeros and t poles in $d(0, \theta^{\mu})$ taking multiplicity into account, one has $\nu^+(f, \mu) = q - t$, and if f has q' zeros and t' poles in $d(0, (\theta^{\mu})^-)$, one has $\nu^-(f, \mu) = q' - t'$. Then the functions in ρ: $\nu^+(f, \log \rho)$, $\nu^-(f, \log \rho)$ and $|f|(\rho)$ are increasing. Let $k = \nu^+(f, \log r)$. Then $|f|(\rho) \geq |a_k|\rho^k\ \forall \rho \in [r, s]$.*

Proof. Since f belongs to $\mathcal{M}(d(0, R^-))$ and since $s < R$, f has finitely many zeros and poles in $d(0, s)$, hence we can write it $\frac{h}{l}$ with $h,\ l \in \mathcal{A}(d(0, R^-))$ having no common zero in $d(0, s)$. Since f has no pole in $\Gamma(0, r, s)$, l has no zero in $\Gamma(0, r, s)$. Let $\rho = \theta^{\mu}$. By Corollary 19.3, we have $\nu^+(f, \mu) = \nu^+(h, \mu) - \nu^+(l, \mu) = q - t$. Similarly, in $d(0, \rho^-)$, we have $\nu^-(f, \mu) = \nu^-(h, \mu) - \nu^-(l, \mu) = q' - t'$.

We can write $f(x)$ in the form $\frac{h(x)}{Q(x)}$ with $h \in \mathcal{A}(d(0, s^-))$ and $Q \in \mathbb{K}[x]$. Then Q has no zero in $\Gamma(0, r, s)$ and hence $\nu^+(Q, \mu)$ is constant in $[\log r, \log s[$. On the other hand, $\nu^+(h, \mu)$ is increasing hence so is $\nu^+(f, \mu)$. Consequently, the function $|f|(\rho)$ is increasing. Therefore, $|f|(\rho) \geq |a_k|\rho^k\ \forall \rho \in [r, s]$. □

Corollary 36.18. *Let $f \in \mathcal{M}(d(0, R^-))$ have no pole in $\Gamma(0, r, s)$, with $0 < r < s < R$ and let $q = \nu^-(f, \log s)$, $k = \nu^+(f, \log r)$. Then,*

$$\left(\frac{s}{r}\right)^k \leq \frac{|f|(s)}{|f|(r)} \leq \left(\frac{s}{r}\right)^q.$$

Corollary 36.19. *Let $f \in \mathcal{M}(\mathbb{K})\setminus\mathbb{K}(x)$ have finitely many poles. For every $q \in \mathbb{N}$, f satisfies $\lim_{r\to\infty} \frac{|f|(r)}{r^q} = +\infty$.*

Proof. Let $f = \frac{h}{Q}$ with Q a monic polynomial and $h \in \mathcal{A}(\mathbb{K})$. Since $f \notin \mathbb{K}(x)$, h does not lie in $\mathbb{K}[x]$ hence it has infinitely many zeros and therefore infinitely many terms $a_n \neq 0$ when $n > 0$. □

We also need the following lemma in the future:

Lemma 36.20. *Let $f \in \mathcal{M}(\mathbb{K})$ be transcendental and have finitely many poles and let P be a polynomial. There exists $s > 0$ such that $|f + P|(r) = |f|(r)\ \forall r \geq s$ and then f has the same number of zeros as $f + P$ in $d(0, r)$.*

Proof. Let $R \in \mathbb{R}_+^*$ be such that all poles of f and all zeros of P lie in $d(0, R)$, let q be the number of poles of f and let $t = \deg(P)$. Then $f(x)$ is of the form $\frac{g(x)}{\prod_{j=1}^{q}(x-b_j)}$ with $g(x)$ of the form $\sum_{n=0}^{\infty} a_n x^n$. Now, when $r > R$, we have $|f|(r) = \frac{|g|(r)}{r^q}$ and $|P|(r)| = \frac{|P|(R)}{R^t} r^t$. Consequently, by Theorem 22.22, $|f|(r)$ gets bigger than $|P|(r)$ when r is big enough and hence there exists $s > R$ such that $|P|(r) < |f|(r)\ \forall r > s$. But then, by Corollary 22.23, we have $\nu(f, \log r) = \nu(f + P, \log r)\ \forall r > s$ and hence f and $f + P$ have the same number of zeros in $d(0, r)$. □

Definitions. Let $f \in \mathcal{M}(\mathbb{K})$ (respectively $f \in \mathcal{M}_u(d(a, R^-))$) and let $b \in \mathbb{K}$. Then b is said to be *an exceptional value for f* or a Picard value if $f - b$ has no zero in $\mathbb{K}$ (respectively in $d(a, R^-)$) and b is said to be *a pseudo-exceptional value for f* if $\lim_{r\to\infty} |f - b|(r) = 0$ (respectively $\lim_{r\to R^-} |f - b|(r) = 0$). Moreover, if $f \in \mathcal{M}(\mathbb{K})\setminus\mathbb{K}(x)$ (respectively if $f \in \mathcal{M}_u(d(a, R^-))$), b is said to be *a quasi-exceptional value for f* if $f - b$ has finitely many zeros in $\mathbb{K}$ (respectively in $d(a, R^-)$) [49, 52, 81].

Theorem 36.21. *Let $f \in \mathcal{M}(\mathbb{K})\setminus\mathbb{K}$ (respectively $f \in \mathcal{M}_u(d(a, R^-))$). If b is an exceptional value for f, then it is a pseudo-exceptional value for f. Let $f \in \mathcal{M}(\mathbb{K})\setminus\mathbb{K}(x)$ (respectively $f \in \mathcal{M}_u(d(a, R^-))$). If b is a quasi-exceptional value for f, then it is a pseudo-exceptional value for f.*

Proof. Without loss of generality, we may assume that $a = b = 0$. Suppose first that $f \in \mathcal{M}(\mathbb{K})\setminus\mathbb{K}$ and that 0 is an exceptional value for f. So, $\frac{1}{f}$ has no pole in $\mathbb{K}$ (respectively in $d(0, R^-)$), hence it is a function $h \in \mathcal{A}(\mathbb{K})\setminus\mathbb{K}$ (respectively $h \in \mathcal{A}_u(d(0, R^-))$) so that $f = \frac{1}{h}$.

Then, by Corollary 14.18 (respectively by Theorem 14.20), we have $\lim_{r\to+\infty}|h|(r) = +\infty$ (respectively $\lim_{r\to R^-}|h|(r) = +\infty$).

Suppose now that $f \in \mathcal{M}(\mathbb{K})\backslash\mathbb{K}(x)$ and that 0 is a quasi-exceptional value for f. Then f is of the form $\frac{P(x)}{h(x)}$ with $P \in \mathbb{K}[x]$ and $h \in \mathcal{A}(\mathbb{K})\backslash\mathbb{K}(x)$. By Corollary 14.7, we have $\lim_{r\to+\infty}\frac{|P|(r)}{|h|(r)} = 0$, which proves the claim when $f \in \mathcal{M}(\mathbb{K})\backslash\mathbb{K}(x)$.

Next, suppose that $f \in \mathcal{M}_u(d(0, R^-))$ admits 0 as a quasi-exceptional value. Then f is of the form $\frac{P(x)}{h(x)}$ with $P \in \mathbb{K}[x]$ and $h \in \mathcal{A}_u(d(0, R^-))$. So P is bounded in $d(0, R^-)$, hence of course $\lim_{r\to+\infty}\frac{|P|(r)}{|h|(r)} = 0$, which ends the proof. □

Theorem 36.22. *Let $f \in \mathcal{M}(\mathbb{K})\backslash\mathbb{K}$ (respectively $f \in \mathcal{M}_u(d(a, R^-))$). Then f admits at most one pseudo-exceptional value. Moreover, if $f \in \mathcal{A}(\mathbb{K})\backslash\mathbb{K}$ (respectively $f \in \mathcal{A}(d(a, R^-))$), then f has no pseudo-exceptional value.*

Proof. Suppose that b is a pseudo-exceptional value for f. Without loss of generality, we may assume that $a = b = 0$. Let $t \in \mathbb{K}^*$. Since $\lim_{r\to+\infty}|f|(r) = 0$ (respectively $\lim_{r\to R^-}|f|(r) = 0$), it is obvious that $\lim_{r\to+\infty}|f - t|(r) = |t|$ (respectively $\lim_{r\to R^-}|f - t|(r) = |t|$), so t is not a pseudo-exceptional value for f.

Now, suppose $f \in \mathcal{A}(\mathbb{K})\backslash\mathbb{K}$. Since $\lim_{r\to+\infty}|f|(r) = +\infty$, of course, 0 is not a pseudo-exceptional value of f. Finally, suppose $\mathcal{A}(d(0, R^-)))$. Then if $f \in \mathcal{A}_u(d(0, R^-))$, we have $\lim_{r\to R}|f|(r) = +\infty$, hence 0 is not a pseudo-exceptional value of f. And if $f \in \mathcal{A}_b(d(0, R^-))$, we have $\lim_{r\to R}|f|(r) = \|f\|_{d(0,R^-)}$ which is not 0, hence 0 is not a pseudo-exceptional value of f either. □

Corollary 36.23. *Let $f \in \mathcal{M}(\mathbb{K})\backslash\mathbb{K}$ (respectively $f \in \mathcal{M}_u(d(a, R^-))$). Then f admits at most one exceptional value. Moreover, if $f \in \mathcal{M}(\mathbb{K})\backslash\mathbb{K}(x)$ (respectively $f \in \mathcal{M}_u(d(a, R^-))$), then f admits at most one quasi-exceptional value. Further, if $f \in \mathcal{A}(\mathbb{K})\backslash\mathbb{K}$ (respectively if $f \in \mathcal{A}_u(d(a, R^-))$), then f admits no exceptional value. And if $f \in \mathcal{A}(\mathbb{K})\backslash\mathbb{K}[x]$ (respectively if $f \in \mathcal{A}_u(d(a, R^-))$), then f admits no quasi-exceptional value.*

Chapter 37

Residues of Meromorphic Functions

Throughout this chapter, D is infraconnected, T is a hole of D, and V is a disk of the form $d(a,r)$ or $d(a,r^-)$, included in $\widetilde{D}$, such that $\widetilde{V} \cap D \neq \emptyset$.

Definition and notation. Let $f \in \mathcal{M}(\mathbb{K})$ (respectively $f \in \mathcal{M}(d(0,R^-)))$ have a pole α of order q and let $f(x) = \sum_{k=-q}^{-1} a_k(x-\alpha)^k + h(x)$ with $a_{-q} \neq 0$ and $h \in \mathcal{M}(\mathbb{K})$ (respectively $f \in \mathcal{M}(d(0,R^-)))$ and h holomorphic at α. Accordingly to previous notations for analytic elements in Chapters 11 and 15, the coefficient a_{-1} is called *residue of f at α* and denoted by $\text{res}(f,\alpha)$.

We can now compare residues on a hole defined for analytic elements and residues at a point, we just defined for a meromorphic function:

Theorem 37.1. *Let $a \in \mathbb{K}$, let $R \in \mathbb{R}_+^*$, let $f \in \mathcal{M}(d(a,R^-))$, and let $r \in]0,R[$. Let α_j, $1 \leq j \leq q$ be the poles of f in $d(a,r)$, let $\rho \in]0, \min_{i\neq j} |\alpha_i - \alpha_j|[$ and for each $j = 1,\dots,q$, let $\rho_j \in]0,\rho[$, and let $T_j = d(\alpha_j,\rho_j^-)$. Let $D = d(\alpha,r)\setminus\left(\bigcup_{j=1}^q T_j\right)$. Then f belongs to $H(D)$ and* $\text{res}(f,\alpha_j) = \text{res}(f,T_j)$, $j = 1,\dots,q$.

Proof. By Corollary 36.2, f belongs to $H(D)$. On the other hand, assuming that α_j is a pole of order s_j, by Corollary 36.16, $f(x)$ has a development at α_j in a Laurent series $\sum_{m=-s_j}^{\infty} b_{m,j}(x-\alpha_j)^m$. Consequently, by Theorem 15.1, the Mittag-Leffler term of f on T_j with respect to the infraconnected set D is $\sum_{m=-s_j}^{-1} b_{m,j}(x-\alpha_j)^m$. Then $\text{res}(f,T_j) = b_{-1,j} = \text{res}(f,\alpha_j)$, which ends the proof. □

Corollary 37.2. *Let $f \in H_b(D)$ be meromorphic in $T = d(b, r^-)$ and admit only one pole b inside T. Let q be the multiplicity order of b. Then the Mittag-Leffler term of f associated with T is of the form $\sum_{j=1}^{q} \frac{a_j}{(x-b)^j}$, with $a_q \neq 0$ and also is of the form $\frac{P}{(x-a_j)^q}$, where P is a polynomial of degree $s < q$. Moreover, it does not depend on r when r tends to 0.*

Definition. Let $f \in \mathcal{M}(d(a, R^-))$, let b be a pole of order t of f, let $r > 0$ be such that $d(b, r^-)$ contains no pole of f other than b, and let $\frac{P(x)}{(x-b)^t}$ be the Mittag-Leffler term of f associated with $d(b, r^-)$. Then $\frac{P(x)}{(x-b)^t}$ is called *the singular part of f at b*.

Thus, when we consider the Laurent development of f around b, it is of the form $\sum_{-t}^{\infty} \frac{a_n}{(x-b)^n}$ and the singular part of f at the pole b is $\sum_{-t}^{-1} \frac{a_n}{(x-b)^n}$.

An element $f \in H(D)$ is said to be *meromorphic in V* if there exist finitely many points $(a_i)_{(1 \leq i \leq n)}$ in V such that f has continuation to an element of $H((D \cup V) \setminus \{a_i |\ 1 \leq i \leq n\})$.

Let f be meromorphic in V and belong to $H((D \cup V) \setminus \{a_i |\ 1 \leq i \leq n\})$. For each $i = 1, \ldots, n$, if $f \notin H((D \cup V) \setminus \{a_h\ |h \neq i\})$, then by Corollary 11.9 and Theorem 11.10, a_i is a pole of f as an element of $H((D \cup V) \setminus \{a_j | 1 \leq j \leq n\})$. Let q_i be its order. Then a_i is called *a pole of f of order q_i in V*. The polynomial $P(x) = \prod_{i=1}^{n} (x - a_i)^{q_i}$ is called *the polynomial of the poles of f in V*.

Lemma 37.3. *Let D be bounded or belong to* Alg *and let $f \in H(D)$. If f is meromorphic in T, the polynomial of its poles P in T satisfies $Pf \in H(D \cup T)$.*

Proof. Indeed, let $D' = (D \cup T) \setminus \{a_1, \ldots, a_n\}$. If D is bounded, then so is D', and therefore by Theorem 11.4, Pf belongs to $H(D')$. But by construction Pf is bounded at each point a_i, and therefore by Corollary 11.6, Pf belongs to $H(D \cup T)$.

Now, suppose $D \in$ Alg. Then by Theorem 17.9, D' belongs to Alg and therefore Pf belongs to $H(D')$, so we have the same conclusion. □

Lemma 37.4. *Let D be bounded (respectively let $D \in$ Alg) and let f be invertible in $H(D)$. Then f is meromorphic in T if and only if so is $\frac{1}{f}$.*

Proof. First, we suppose f meromorphic in T. Let P be the polynomial of its poles in T and let $g(x) = f(x)P(x)$. Since D is bounded (respectively belongs to Alg) by Lemma 37.3, g belongs to $H(D \cup T)$. Let Q be the polynomial of the zeros of g in T. Since f has no zero in D, Q actually is the polynomial of the zeros of g in $D \cup T$ and then g is of the form $Q(x)h(x)$ with h an element of $H(D \cup T)$ that has no zero in T. Hence, we have $\frac{1}{h} = \frac{1}{f}\frac{Q}{P}$. If $D \in \mathrm{Alg}$, $\frac{1}{h}$ obviously belongs to $H(D)$. If D is bounded, we have $\frac{Q}{P} \in R_b(D)$, and then by Theorem 11.4, $\frac{1}{h}$ belongs to $H(D)$. Thus, $\frac{1}{h}$ belongs to $H(D)$ anyway. Now, by Lemma 24.1, h is invertible in $H(D \cup T)$. Hence, f factorizes in the form $\frac{P}{Q}h$ and then $\frac{1}{f} = \frac{1}{Q}\frac{P}{h}$. But $\frac{P}{h}$ belongs to $H(D \cup T)$ and therefore $\frac{1}{f}$ is meromorphic in T and admits Q as the polynomial of its poles in T. We may obviously apply the same reasoning to $\frac{1}{f}$ and this shows the converse. □

Theorem 37.5. *Let $F \in H(D)$ be meromorphic in T and satisfy $\|F - 1\|_D < 1$. Then F has as many poles as many zeros in T.*

Proof. Without loss of generality, we may obviously assume that D is bounded because the hypothesis remains true in any set $D \cap d(0, R)$. We may also assume that $T = d(0, r^-)$. Let P (respectively Q) be the polynomial of the zeros (respectively the poles) of F in T. Then by Lemma 37.3, F factorizes in the form $\frac{f}{Q}$ with $f \in H(D \cup T)$. Now, the zeros of f inside T are just those of F, hence f factorizes in the form Pg with $g \in H(D \cup T)$, g having no zero in T. Let $h = \frac{P}{Q}$. Since g has no zero in T, by Theorem 22.14, $|g(x)|$ is equal to a constant inside T. Let $s \in]0, r[$ be such that all zeros of P and of Q lie in $d(0, s)$. Then obviously F belongs to $H(\Gamma(0, s, r))$. Now, by hypothesis, there exists $\lambda > 0$ such that $\Psi(F, \mu) \leq \lambda$ for all $\mu \geq \log r$. Hence by continuity, there exists $\lambda' > 0$ and s' in $]s, r[$ such that $\Psi(F, \mu) \leq \lambda'$ for all $\mu \geq \log s'$. Thus, there exists $b \in]0, 1[$ and t in $]s', r[$ such that $|F(x) - 1| \leq b$ for all $x \in \Gamma(0, t, r)$ and then, $|h(x)|$ is constant in $\Gamma(0, t, r)$. Hence, we have $\frac{d}{d\mu}\Psi(h, \mu) = 0$ for all $\mu \in [\log t, \log r]$. Since h has neither any zero nor any pole in $C(0, t)$, by Corollary 4.17, h has as many zeros as many poles in $d(0, t)$ and therefore in $d(0, r)$ and this ends the proof. □

It is useful to consider again elements meromorphic at a point.

Lemma 37.6. *Let $a \in \overline{D}$ and let $f \in H(D)$ be meromorphic but not holomorphic at a. Then a is a pole of f.*

Proof. By hypothesis, there exists $r > 0$ such that f belongs to $H\big(D \cup \big(d(a,r)\setminus\{a\}\big)\big)$. Suppose that a is not a pole of f. Then by Theorem 36.4, f belongs to $H(D \cup d(a,r))$ and then f is holomorphic at a. □

Corollary 37.7. *Let $a \in \overline{D}$ and let $f \in H(D)$. Then f is meromorphic at a and admits a as a pole of order q if and only if there exists a disk $d(a,r)$ included in $\overline{D}$ and an element $h \in H(d(a,r))$ such that $f(x)(x-a)^q = h(x)$ whenever $x \in d(a,r)\setminus\{a\}$ and $h(a) \neq 0$.*

Corollary 37.8. *Let D satisfy Condition (B) and let $f \in H(D)$. For every $a \in \overline{D}$, f is meromorphic at a. For every $a \in D$, f is holomorphic at a.*

Remark. Let $a \in \overline{D} \setminus \overset{\circ}{D}$ and let f admit a as a pole of order q. This does not imply that f is meromorphic at a. Indeed, by [44], we know that there exist infraconnected sets E with a point $a \in E \setminus \overset{\circ}{E}$ and elements $h \in H(E)$ such that $\lim_{|x-a|\to 0} h(x) = 0$ and such that $\limsup_{|x-a|\to 0}\left|\frac{h(x)}{x-a}\right| = +\infty$. Let $E' = E\setminus\{a\}$ and let $g = \frac{1+h}{x-a}$. It is easily seen that a is a pole of order 1 for g. But h does not belong to any space $H(d(a,r))$, whenever $r > 0$ (because if it did, it should factorize in $H(d(a,r))$ in the form $(x-a)\ell(x)$, with $\ell \in H(d(a,r))$). Thus, it is seen that $(x-a)g$ does not belong to $H(E)$ and therefore g is not meromorphic at a.

Concerning the derivation, Theorem 37.9 is easy and follows the classical rules:

Theorem 37.9. *Let $f \in \mathcal{M}(\mathbb{K})$ (respectively $f \in \mathcal{M}(d(a,R^-))$). For each $\alpha \in \mathbb{K}$ (respectively $\alpha \in d(a,R^-)$) such that f is holomorphic at α, f has a derivative $f'(\alpha)$ at α. Further, given a point $\beta \in \mathbb{K}$ (respectively $\beta \in d(a,R^-)$) and the Laurent development of f at β: $\sum_{k=-q}^{\infty} a_k(x-\beta)^k$ with $a_{-q} \neq 0$, the development of f' at β is $\sum_{k=-q}^{0} ka_k(x-\beta)^{k-1} + \sum_{k=1}^{\infty} ka_k(x-\beta)^{k-1}$.*

Proof. Suppose first f is holomorphic at α. By Theorem 36.4, $f(x)$ is equal to a power series $\sum_{k=0}^{\infty} a_k(x-\alpha)^k$ converging inside a disk $d(\alpha,r^-)$, where r is the minimal distance from α to the various poles.

Then by Theorem 18.1, we know that f has a derivative whose development is obtained by deriving term by term.

Suppose now that β is a pole of order q and let $r \in]0, R[$ be strictly inferior to the minimal distance from β to the other poles (with just $r < R$ if β is the unique pole of f). By Lemma 36.1, for every $\rho \in]0, r[$, f belongs to $H(d(\beta, r) \backslash d(\beta, \rho^-))$ and the Laurent development of f at β is its Mittag-Leffler development as an element of $H(d(\beta, r) \backslash d(\beta, \rho^-))$: the Mittag-Leffler term associated with the hole $d(\beta, \rho^-)$ is $\sum_{k=-q}^{-1} a_k(x-\beta)^k$ with $a_{-q} \neq 0$ and the term associated with $d(\beta, r)$ is $\sum_{k=0}^{\infty} a_k(x-\beta)^k$ with $a_{-q} \neq 0$. Consequently, by Theorem 18.19, the derivative has a Mittag-Leffler development at β: $\sum_{k=-q}^{-1} ka_k(x-\beta)^{k-1} + \sum_{k=1}^{\infty} ka_k(x-\beta)^{k-1}$. This is true for all $r \in]0, R[$ strictly inferior to the minimal distance from β to the other poles and for every $\rho \in]0, r[$, which ends the proof. □

Theorem 37.10 is an improvement of the classical upper bound f' in function of f. That is due to Bézivin [10].

Theorem 37.10. *For each $n \in \mathbb{N}$ and for all $r \in]0, R[$, we have*

$$|f^{(n)}|(r) \leq |n!| \frac{|f|(r)}{r^n}.$$

Moreover, given $r \in]0, R[$ such that $\nu^+(f, \log r) = \nu^-(f, \log r)$, if the residue characteristic p does not divide $\nu(f, \log r)$, then $\nu(f', \log r) = \nu(f, \log r) - 1$ and

$$|f'|(r) = \frac{|f|(r)}{r}.$$

Proof. When $f \in \mathcal{A}(d(0, R^-))$, this is shown at Corollary 18.16. Now, consider the general case and set $f = \frac{U}{V}$ with $U, V \in \mathcal{A}(d(0, R^-))$. The stated inequality is trivial when $q = 1$. So, we assume it holds for $q \leq n-1$ and consider $f^{(n)}$. Writing $U = V\left(\frac{U}{V}\right)$, by Leibnitz theorem, we have

$$U^{(n)} = \sum_{q=0}^{n} \binom{n}{q} V^{(n-q)} \left(\frac{U}{V}\right)^{(q)},$$

and hence

$$V\left(\frac{U}{V}\right)^{(n)} = U^{(n)} - \sum_{q=0}^{n-1} \binom{n}{q} V^{(n-q)} \left(\frac{U}{V}\right)^{(q)}.$$

Now, by Corollary 18.16, we have

$$(1) \qquad |U^{(n)}|(R) \leq |n!|\frac{|U|(R)}{R^n}$$

and for each $q \leq n-1$, we have $|V^{(n-q)}|(R) \leq |(n-q)!|\frac{|V|(R)}{R^{n-q}}$ and $\left|\left(\frac{U}{V}\right)^{(q)}\right|(R) \leq |q!|\frac{|U|(R)}{|V|(R)R^q}$. Consequently,

$$\left|\left(\frac{U}{V}\right)^{(q)}\right|(R)\left|\left(\frac{U}{V}\right)^{(n-q)}\right|(R) \leq |((n-q)!)q!|\frac{|U|(R)}{|V|(R)R^n}$$

and then we can derive

$$(2) \qquad \left|\binom{n}{q}\right|\left|\left(\frac{U}{V}\right)^{(q)}\right|(R)\left|\left(\frac{U}{V}\right)^{(n-q)}\right|(R) \leq |n!|\frac{|U|(R)}{|V|(R)R^n}.$$

So, by (1) and (2), the first conclusion holds for $q = n$.

Suppose now that $\nu^+(f, \log r) = \nu^-(f, \log r)$ and that the residue characteristic of $\mathbb{K}$ does not divide $\nu(f, \log r)$. Without loss of generality, we may assume that f has no pole in $C(0,r)$ because all conclusions hold by continuity. In $C(0,r)$, $f(x)$ is equal to a power series $\sum_{-\infty}^{+\infty} a_n x^n$. Set $q = \nu(f, -\log(r))$. Then, $|f|(r) = |a_q|r^q$ and $|f'|(r) = |q||a_q|r^{q-1} = |a_q|r^{q-1}$, which ends the proof. □

It seems obvious that the condition for a meromorphic function to admit primitives is that all residues are null. This is stated by Theorem 37.11 but the proof is not this immediate.

Let us remark that the topology of uniform convergence in all disks of $\mathbb{K}$ (respectively of all disk included inside an open disk $d(a, R^-)$) is obviously defined on the algebra $\mathcal{M}(\mathbb{K})$ (respectively $\mathcal{M}(d(a, R^-))$) and that $\mathcal{M}(\mathbb{K})$ (respectively $\mathcal{M}(d(a, R^-))$) is complete for that topology.

Theorem 37.11 is a Mittag-Leffler theorem similar to this known in complex analysis.

Theorem 37.11. *Let $(a_m, q_m)_{m\in\mathbb{N}}$ be a divisor of $\mathbb{K}$ and for every $n \in \mathbb{N}$, and let $Q_m \in \mathbb{K}[x]$ be of degree $< q_m$, prime with $(x - a_m)$ such that $\|Q_m\| \leq 1$ $\forall m$. There exists $f \in \mathcal{M}(\mathbb{K})$ admitting for poles each a_m of order q_m and no other pole and such that its singular part is $\frac{Q_m}{(x-a_m)^{q_m}}$.*

Proof. Since $\deg(Q_m) < q_m$, we can check that the series $\sum_{n=0}^{\infty} \frac{Q_m(x)}{(x-a_m)^{q_m}}$ converges for all $x \notin \{a_0, \dots, a_m..\}$ because when $|x - a_m| \geq |a_m| \geq 1$, then $\left|\frac{Q_m(x)}{(x-a_m)^{q_m}}\right| \leq \|Q_m\|$. The series defines a meromorphic function f in each disk $d(0, r^-)$, and hence, by Theorem 36.9, it is a meromorphic function in $\mathbb{K}$. Then for each $m \in \mathbb{N}$, the function $f(x) - \frac{Q_m}{(x-a_m)^{q_m}}$ is holomorphic at a_m, hence $\frac{Q_m}{(x-a_m)^{q_m}}$ is its singular part at a_m. Moreover, since Q_m is prime with $x - a_m$, then a_m is a pole of order q_m. □

Now, we can give an easy proof Theorem 37.12:

Theorem 37.12. *$\mathbb{K}$ is supposed to have characteristic 0. A function $f \in \mathcal{M}(\mathbb{K})$ (respectively $f \in \mathcal{M}(d(a, R^-))$, $a \in \mathbb{K}$, $R > 0$) admits primitives in $\mathcal{M}(\mathbb{K})$ (respectively in $\mathcal{M}(d(a, R^-))$) if and only if all residues of f are null.*

Proof. Let a be a pole of f. According to the Laurent series of f at a, if f admits primitives, then f has no residue different from zero at a because the function $\frac{1}{x-a}$ has no primitive in $\mathcal{M}(d(a, r))$ (whenever $r > 0$).

Now let $(a_m)_{m\in\mathbb{N}}$ be the sequence of poles of f, each of respective order q_m and suppose that $\mathrm{res}(f, a_m) = 0\ \forall m \in \mathbb{N}$. Then the singular part of f at a_m is of the form $\frac{Q_m(x-a_m))}{(x-a_m)^{q_m}}$ with $q_m \geq 2$ and $Q_m(x)$ is a polynomial of degree $\leq q_m - 2$. Consequently, the singular part l_m of f at a_m admits a primitive L_m of the form $\frac{P_m(x-a_m)}{(x-a_m)^{q_m-1}}$ with $\deg(P_m) \leq q_m - 2$. Suppose first that f has finitely many poles $a_1, \dots, a_q$. Then $f(x)$ is of the form $g(x) + \sum_{m=1}^{q} l_m(x)$, where g is a power series converging in $\mathbb{K}$ and hence g admits a primitive G vanishing at 0, which is also a power series converging in $\mathbb{K}$. Then f admits a primitive $F(x) = G(x) + \sum_{m=1}^{q} L_m(x)$.

Consider now the general case and take a disk $d(0, R)$. Let $a_1, \dots, a_q(R)$ be the poles of f in $d(0, R)$. Then in the same way, i $d(0, R)$ we can write $f(x) = g_R(x) + \sum_{m=1}^{q(R)} l_m(x)$, where g_R is a power series converging in $d(0, R)$. Then, g_R has a primitive G_R converging in $d(0, R)$, and hence in $d(0, R)$, $f(x)$ admits primitive S $G_R + \sum_{m=1}^{q(R)} L_m(x) + c$, where C is a constant. Let F_R be the primitive that vanishes at 0.

This holds for all $R > 0$, hence when $R < R'$, the decomposition in $d(0, R)$ is the restriction of that in $d(0, R')$. Thus, f admits in each disk $d(0, R)$ a primitive F_R vanishing at 0 such that $F_R(x) = F_{R'}(x)\ \forall x \in d(0, R)$. Consequently, we can define a function F which is a primitive of f in $\mathbb{K}$, and by Theorem 36.9, it is meromorphic in all $\mathbb{K}$. □

Corollary 37.13. *The field* $\mathbb{K}$ *is supposed to have characteristic* 0. *Let* $f \in \mathcal{M}((\mathbb{K})$ *(respectively* $f \in \mathcal{M}(d(a, R^-))$, $a \in \mathbb{K}$, $R > 0$). *Then* f' *belongs to* $\mathbb{K}(x)$ *if and only if so does* f.

Proof. If f belongs to $\mathbb{K}(x)$, of course, so does f'. Now, suppose f' belongs to $\mathbb{K}(x)$. We can write it in the form $\sum_{j=1}^{q} \frac{b_j}{(x-a_j)^{q_j}}$. And by Theorem 37.11, we have $q_j \geq 2\ \forall j = 1, \dots, q$. Consequently, since $\mathbb{K}$ has characteristic 0, $f(x)$ is of the form $-\sum_{j=1}^{q} \frac{b_j}{q_j(x-a_j)^{q_j-1}} + c$ with $c \in \mathbb{K}$ and hence f belongs to $\mathbb{K}(x)$. □

Chapter 38

Bezout Algebras of Analytic Functions

Let us recall that the ring of analytic functions on a region of the complex number field is well known to be a Bezout ring. The two fundamental theorems necessary for a proof are the Weierstrass factorization theorem and the Mittag-Leffler theorem.

According to results of [72], it appears that in several hypotheses, rings of analytic functions on complete ultrametric algebraically closed fields are Bezout rings. However, that interesting property is not stated. Moreover, it derives from a Mittag-Leffler theorem referred in general topology whose justification is not relevant. Here we plan to give proofs of all these properties, using results on quasi-invertible analytic elements and on a Mittag-Leffler theorem for meromorphic functions similar to this of complex analysis but is quite different from Krasner's Mittag-Leffler theorem for analytic elements on an infraconnected subset of $\mathbb{K}$.

In Chapter 26, we showed that in $\mathcal{A}(\mathbb{K})$ every closed ideal is principal. Here, following the same methods, provided that $\mathbb{K}$ is spherically complete, we can now prove similar results with algebras $\mathcal{A}(d(a, R^-))$:

Theorem 38.1. *Suppose $\mathbb{K}$ is spherically complete. Let B be a divisor on $d(a, R^-)$. There exists $f \in \mathcal{A}(d(a, R^-))$ such that $\mathcal{D}(f) = B$.*

Proof. Without loss of generality, we may obviously assume $a = 0$. Take $Q_m = 0 \ \forall m \in \mathbb{N}$. By Theorem 28.4, there exists $f \in \mathcal{A}(d(0, R^-))$ such that

(i) $f(0) = 1$,
(ii) $|f|(r) \leq |B|(r) \ \forall r < R$,
(iii) P_m divides f in $\mathcal{A}(d(0, R^-))$.

By (iii), clearly $\mathcal{D}(f) \geq B$. Thus, we only have to check that $\mathcal{D}(f) \leq B$. Indeed, for all $s \in \mathbb{N}$, we have

$$|B|(\rho_s) = \prod_{j=1}^{s} \prod_{i=1}^{u_m} \left|\left(1 - \frac{x}{a_{i,m}}\right)^{q_{i,m}}\right|(\rho_s) = \prod_{j=1}^{s} \prod_{i=1}^{u_m} \left(\frac{\rho_s}{\rho_m}\right)^{q_{i,m}}.$$

Now, suppose that $B \neq \mathcal{D}(f)$. Then there exists $\alpha \in d(0, R^-)$ such that $\omega_\alpha(f) > B(\alpha)$. Let s be such that $\rho_s > |\alpha|$. Since $f(0) = 1$, we have

$$|f|(\rho_s) \geq \frac{\rho_s}{|\alpha|} \prod_{j=1}^{s} \prod_{i=1}^{u_m} \left|\left(1 - \frac{x}{a_{i,m}}\right)^{q_{i,m}}\right|(\rho_s) > \prod_{j=1}^{s} \prod_{i=1}^{u_m} \left(\frac{\rho_s}{\rho_m}\right)^{q_{i,m}} = |T|(\rho_s),$$

a contradiction to (iii). □

By Theorems 27.10 and 38.1, we have Corollary 38.2:

Corollary 38.2. *Suppose $\mathbb{K}$ is spherically complete. Let B be a divisor on $d(a, R^-)$. Then $\mathcal{T}_{a,R}(B)$ is principal* [34].

Corollary 38.3. *Suppose $\mathbb{K}$ is spherically complete. Then all closed ideals of $\mathcal{A}(d(a, R^-))$ are principal* [34].

Proof. Let I be a closed ideal of $\mathcal{A}(d(a, R^-))$ and let $E = \mathcal{D}(I)$. By Theorem 27.10, we have $I = \mathcal{T}_{a,R}(E)$. Now, by Corollary 38.2, $\mathcal{T}_{a,R}(E)$ is principal. □

Lemma 38.4. *Let E be a divisor of $\mathbb{K}$ (respectively a divisor of $d(a, R^-)$, $a \in \mathbb{K}$, $R > 0$) and for each $r > 0$ (respectively $r \in]0, R[$), let $g_r \in H(C(0, r))$. There exists $g \in \mathcal{A}(\mathbb{K})$ (respectively $g \in \mathcal{A}(d(a, R^-))$), not depending on r, such that $\mathcal{D}(g - g_r) \geq E_r$.*

Proof. Let $f \in \mathcal{A}(\mathbb{K})$ (respectively $f \in \mathcal{A}(d(a, R^-))$) be such that $\mathcal{D}(f) \geq E$. By Theorem 37.11, there exists $F \in \mathcal{M}(\mathbb{K})$ whose principal parts at the poles located in $C(0, r)$ are respectively the same as those of $g_r f^{-1}$ for each $r > 0$. Then fF belongs to $\mathcal{A}(\mathbb{K})$ (respectively to $\mathcal{A}(d(a, R^-))$). Putting $g = fF$, we can see that $\mathcal{D}(g - g_r) \geq E_r$, which ends the proof. □

Theorem 38.5. *Every ideal of finite type of $\mathcal{A}(\mathbb{K})$ is closed and is of the form $\mathcal{T}(E)$ with E a divisor of $\mathbb{K}$.*

Proof. Let I be an ideal of finite type of $\mathcal{A}(\mathbb{K})$ generated by $f_1, \dots, f_q$ and let $E = \mathcal{D}(I)$. Then, $I = \mathcal{T}(E)$, and by Theorem 27.9, $\mathcal{T}(E)$ is closed, hence the closure of I is $\mathcal{T}(E)$. Consequently, we can see that $E = \min(\mathcal{D}(f_1), \dots, \mathcal{D}(f_q))$. Let us fix $r > 0$. In $H(C(0,r))$, there exist $g_{1,r}, \dots, g_{q,r} \in H(C(0,r))$ such that $g = \sum_{j=1}^{q} g_{j,r} f_j$. For each $j = 2, \dots, q$, let $f_{j,r}$ be the polynomial of the zeros of f_j in $C(0,r)$. By Lemma 38.4, there exists $g_j \in \mathcal{A}(\mathbb{K})$, not depending on r, such that $g_{j,r} - g_j$ be divisible in $H(C(0,r))$ by $\mathcal{D}(f_{1,r})$. Now, set $h = g - \sum_{j=2}^{q} g_j f_j$. We have $\mathcal{D}(h) \geq \mathcal{D}(f_1)$, hence h factorizes in the form $g_1 f_1$ with $g_1 \in \mathcal{A}(\mathbb{K})$ (respectively $g_1 \in \mathcal{A}(d(a, R^-))$) and then

$$g = h + \sum_{j=2}^{q} g_j f_j = \sum_{j=1}^{q} g_j f_j.$$

□

Similarly, when $\mathbb{K}$ is spherically complete, we have Theorem 38.6:

Theorem 38.6. *Suppose $\mathbb{K}$ is spherically complete. Given $a \in \mathbb{K}$ and $R > 0$, every ideal of finite type of $\mathcal{A}(d(a, R^-))$ is closed and is of the form $\mathcal{T}_{a,R}(E)$ with E a divisor of $d(a, R^-)$ [34].*

Proof. Let I be an ideal of finite type of $\mathcal{A}(d(a, R^-))$ generated by $f_1, \dots, f_q$ and let $E = \mathcal{D}(I)$. Then, $I = \mathcal{T}_{a,R}(E)$, and by Theorem 27.10, $\mathcal{T}_{a,R}(E)$ is closed, hence the closure of I is $\mathcal{T}_{a,R}(E)$. Consequently, we can see that $E = \min(\mathcal{D}(f_1), \dots, \mathcal{D}(f_q))$. Let us fix $r > 0$. In $H(C(0,r))$, there exist $g_{1,r}, \dots, g_{q,r} \in H(C(0,r))$ such that $g = \sum_{j=1}^{q} g_{j,r} f_j$. For each $j = 2, \dots, q$, let $f_{j,r}$ be the polynomial of the zeros of f_j in $C(0,r)$. By Lemma 38.4, there exists $g_j \in \mathcal{A}(d(a, R^-))$, not depending on r, such that $g_{j,r} - g_j$ is divisible in $H(C(0,r))$ by $\mathcal{D}(f_{1,r})$. Now, set $h = g - \sum_{j=2}^{q} g_j f_j$. We have $\mathcal{D}(h) \geq \mathcal{D}(f_1)$, hence h factorizes in the form $g_1 f_1$ with $g_1 \in \mathcal{A}(d(a, R^-))$ and then

$$g = h + \sum_{j=2}^{q} g_j f_j = \sum_{j=1}^{q} g_j f_j.$$

□

By Theorem 38.5, we have Corollary 38.7:

Corollary 38.7. $\mathcal{A}(\mathbb{K})$ *is a Bezout ring* [34].

Proof. Indeed, consider an ideal of finite type I. By Theorem 38.5, it is closed, and hence, by Theorem 27.11, it is principal. □

And by Theorem 38.6, we have Corollary 38.8:

Corollary 38.8. *Let* $a \in \mathbb{K}$ *and let* $R > 0$. *If* $\mathbb{K}$ *is spherically complete,* $\mathcal{A}(d(a, R^-))$ *is a Bezout ring.*

Proof. Indeed, consider an ideal of finite type I. By Theorem 38.7, it is of the form $\mathcal{T}_{a,R}(E)$ with E a divisor of $d(a, R^-)$ and it is closed. But then, by Corollary 38.3, it is principal. □

Chapter 39

Meromorphic Functions Out of a Hole

Notation. We fix $R > 0$ and denote by I the interval $[R, +\infty[$. Throughout this chapter, we denote by S the disk $d(0, R^-)$ and put $D = \mathbb{K} \setminus S$.

We denote by $H_0(D)$ the $\mathbb{K}$-subvector space of the $f \in H(D)$ such that $\lim_{|x| \to +\infty} f(x) = 0$.

By classical properties of analytic elements, we know that given a circle $C(a, R)$ and an element f of $H(C(a, R))$ i.e. a Laurent series $f(x) = \sum_{-\infty}^{+\infty} c_n (x - a)^n$ converging whenever $|x| = r$, then $|f(x)|$ is equal to $\sup_{n \in \mathbb{Z}} |c_n| r^n$ in all classes of the circle $C(a, r)$ except maybe in finitely many. When $a = 0$, we put $|f|(r) = \sup_{n \in \mathbb{Z}} |c_n| r^n$. Then, $|f|(r)$ is a multiplicative norm on $H(C(0, r))$.

We denote by $\mathcal{A}(D)$ the $\mathbb{K}$-algebra of Laurent series converging in D and by $\mathcal{A}^c(D)$ the set of $f \in \mathcal{A}(D)$ having infinitely many zeros in D. Similarly, we denote by $\mathcal{M}(D)$ the field of fractions of $\mathcal{A}(D)$ that we call *field of meromorphic functions in* D and we denote by $\mathcal{M}^c(D)$ the set of functions $f \in \mathcal{M}(D)$ which have infinitely many zeros or poles in D. We denote by $\mathcal{M}_0(D)$ the set of functions $f \in \mathcal{M}(D)$ which have finitely many zeros and poles in D.

Similarly, as we did in $\mathbb{K}$ and inside a disk, here we define a pseudo-exceptional value and a quasi-exceptional value in D. Given a meromorphic function $f \in \mathcal{M}(D)$, a value $b \in \mathbb{K}$ is called *a pseudo-exceptional value for* f if $\lim_{|x| \to +\infty} f(x) = b$ it is called *a quasi-exceptional value for* f if $f - b$ has finitely many zeros in D, and it is called *an exceptional value for* f if has no zero in in D.

Proposition 39.1. *Let $f \in \mathcal{M}^c(D)$ and let b be a quasi-exceptional value. Then b is a pseudo-exceptional value.*

Proof. Without loss of generality, we may assume that $b = 0$. Therefore, we can write f in the form $\frac{P}{h}$ with P a polynomial whose zeros lie in D and $h \in \mathcal{A}^c(D)$. On the other hand, $h(x)$ is a Laurent series $\sum_{-\infty}^{+\infty} a_n x^n$ converging in all D, having infinitely many zeros, hence infinitely many coefficients a_n with $n > 0$ are different from zero, therefore one sees that $\lim_{|x| \to +\infty} \frac{|P|(r)|}{|h|(r)} = 0$, and hence $\lim_{|x| \to +\infty} \frac{|P(x)|)|}{|h(x)|} = 0$. □

Proposition 39.2. *Let $f \in \mathcal{M}(D)$. If f has infinitely many zeros in D (respectively infinitely many poles in D), the set of zeros (respectively the set of poles) is a sequence $(\alpha_n)_{n \in \mathbb{N}}$ such that $\lim_{n \to +\infty} |\alpha_n| = +\infty$.*

Proof. Suppose first $f \in \mathcal{A}(D)$. For each $L > R$, f belongs to $H(\Delta(0, R, L))$, and by Corollary 21.11, it is quasi-invertible in $H(\Delta(0, R, L))$, hence it has finitely many zeros in $\Delta(0, R, L)$, for every $L > R$. Consequently, if f has infinitely many zeros in D, the zeros form a sequence $(\alpha_n)_{n \in \mathbb{N}}$ such that $\lim_{n \to +\infty} |\alpha_n| = +\infty$. Suppose now $f \in \mathcal{M}(D)$. Then f is of the form $\frac{g}{h}$ with $g,\ h \in f \in \mathcal{A}(D)$. If f has infinitely many zeros, so does g, and each zero of f is a zero of g, hence the zeros of f form a sequence $(\alpha_n)_{n \in \mathbb{N}}$ such that $\lim_{n \to +\infty} |\alpha_n| = +\infty$. Similarly, if f has infinitely many poles, the h has infinitely many zeros, and each pole of f is a zero of h, hence the poles of f form a sequence $(\beta_n)_{n \in \mathbb{N}}$ such that $\lim_{n \to +\infty} |\beta_n| = +\infty$. □

Theorem 39.3. *Let $f \in \mathcal{M}(D)$ have no zero and no pole in D. Then $f(x)$ is of the form $\sum_{-\infty}^{q} a_n x^n$ with $|a_q| r^q > |a_n| r^n\ \forall n < q,\ \forall r \geq R$, and $|f(x)| = |a_q| r^q\ \forall x \in D$.*

Proof. For every $r \geq R$, f belong to $H(C(0, r))$, and by Theorem 22.1, we have $\nu^+(f, \log(r)) = \nu^-(f, \log(r))$. Consequently, by continuity, $\nu(f, \log(r)$ is a constant q in $\log(R), +\infty[$. It is then clear that $|f(x)| = |a_q| r^q\ \forall x \in D$. □

Theorem 39.4. *Let $f \in \mathcal{M}(D)$ have at least infinitely many zeros or infinitely many poles in D. Then f admits at most one pseudo-exceptional value.*

Proof. Suppose that $f \in \mathcal{M}^c(D)$ has two distinct pseudo-exceptional values a and b. Without loss of generality, we can assume that $a = 0$, and hence f is of the form $\frac{\phi}{\psi}$ with ϕ and $\psi \in \mathcal{A}(D)$, ψ admitting infinitely many zeros and satisfying $\lim_{r\to+\infty} \frac{\phi|(r)}{|\psi|(r)} = 0$. Then, $f - b - \frac{\phi - b\psi}{\psi}$. But when r is big enough, we have $|\psi - b\psi(r)| = |b\psi|(r)$, therefore $\phi - b\psi$ does not admit 0 as a pseudo-exceptional value, a contradiction. □

Corollary 39.5. *Let $f \in \mathcal{M}(D)$ have at least infinitely many zeros or infinitely many poles in D. Then f admits at most one exceptional value.*

Definition. Let $f \in H(D)$ have no zero in D, $f(x) = \sum_{-\infty}^{q} a_n x^n$ with $|a_q|R^q > |a_n|R^n \ \forall n < q$ and $a_q = 1$. Then f is a Motzkin factor associated with S and the integer q is called *the Motzkin index of f* and is denoted by $mo(f, S)$.

Theorem 39.6. *Let $f \in \mathcal{M}(D)$. We can write f in a unique way in the form $f^S f^0$ with $f^S \in H(\mathbb{K}\backslash D)$ a Motzkin factor associated with S and $f^0 \in \mathcal{M}(\mathbb{K})$, having no zero and no pole in S.*

Proof. Suppose first $f \in \mathcal{A}(D)$ and take $V > R$. Then as a quasi-invertible element of $H(\Delta(0, R, V))$, by Theorem 29.16, f admits a factorization in the form $f^S f^0$, where f^S is a Motzkin factor and f^0 belongs to $H(d(0, V))$ and has no zero in S. Moreover, by Lemma 29.6, f^S does not depend on V. Consequently, since f^S is obviously invertible in $\mathcal{A}(D)$, we can factorize $f \in \mathcal{A}(D)$ in the form $f^S f^0$, where f^0 belongs to $\mathcal{A}(\mathbb{K})$ and has no zero in S.

Consider now the general case: $f = \frac{g}{h}$ with $g,\ h \in \mathcal{A}(D)$. Then we can write $g = g^S g^0,\ h = h^S h^0$, hence $f = \left(\frac{g^S}{h^S}\right)\left(\frac{g^0}{h^0}\right)$. Then we can check that this is the factorization announced in the statement: $f^S = \frac{g^S}{h^S}$ and $f^0 = \frac{g^0}{h^0}$.

Lemma 39.7 is immediate:

Lemma 39.7. *The set of Motzkin factors associated with S makes a multiplicative group. Let $f,\ g \in \mathcal{M}(D)$. Then,* $(fg)^S = (f^S)(g^S)$, $\left(\frac{1}{f}\right)^S = \frac{1}{f^S}$, $(fg)^0 = (f^0)(g^0)$, $\left(\frac{1}{f}\right)^0 = \frac{1}{f^0}$, *and* $mo(fg, S) = mo(f, S) + mo(g, S)$, $mo(\frac{1}{f}, S) = -mo(f, S)$.

Chapter 40

Shilov Boundary for Algebras $H(D)$

Given a holomorphic function f in an open bounded connected subset D of $\mathbb{C}$, $|f(z)|$ reaches its maximum on the boundary of D. Consider now a closed bounded infraconnected subset D of $\mathbb{K}$ and $f \in H(D)$. We show that the supremum of the $\phi(f)$ when ϕ runs in $\text{Mult}(H(D), \| \cdot \|_D)$ is reached on the Shilov boundary that we characterize. We show that the set of circular filters is provided with a tree structure and that the diameter is an increasing function with values in $\mathbb{R}$, defining distances associated with this structure. The first remarks on that tree structure are due to Motzkin [74], and it was thoroughly examined in [14] and in [49]. Here the structure is helpful to determine the Shilov boundary for algebras $H(D)$.

Throughout this chapter and in the following, according to Theorem 14.1, we identify $\Phi(D)$ with $\text{Mult}(H(D), \| \cdot \|_D)$.

Definition and notation. Given $a \in \mathbb{K}$ and $r > 0$, we denote by $\mathcal{F}_{a,r}$ the circular filter of center a and diameter r. Given a large circular filter $\mathcal{F}$, the set of its centers is denoted by $\mathcal{Q}(\mathcal{F})$. Given two circular filters $\mathcal{F}$ and $\mathcal{G}$, $\mathcal{F}$ is said *to surround* $\mathcal{G}$ if either $\mathcal{G}$ is secant with $\mathcal{Q}(\mathcal{F})$ or if $\mathcal{F} = \mathcal{G}$. Then, $\mathcal{F}$ is said *to strictly surround* $\mathcal{G}$ if $\mathcal{F}$ surrounds $\mathcal{G}$ and is different from $\mathcal{G}$.

Similarly, a circular filter $\mathcal{F}$ is said *to surround a monotonous filter* $\mathcal{G}$ if it surrounds the circular filter associated with $\mathcal{G}$. A monotonous filter $\mathcal{F}$ is said *to surround a circular filter* $\mathcal{G}$ if its associated circular filter surrounds $\mathcal{G}$ and $\mathcal{F}$ is said *to surround a monotonous filter* $\mathcal{G}$ if the circular filter associated with $\mathcal{F}$ surrounds the circular filter associated with $\mathcal{G}$.

We denote by $\preceq$ the relation on the set of circular filters defined as $\mathcal{F} \preceq \mathcal{G}$ if $\mathcal{G}$ surrounds $\mathcal{F}$ and by $\prec$ the relation defined as $\mathcal{F} \prec \mathcal{G}$ if $\mathcal{F} \preceq \mathcal{G}$ and $\mathcal{F} \neq \mathcal{G}$.

Given an order relation $\mathcal{R}$ defined on a set E, we call *strict-order associated with* $\mathcal{R}$ the relation $\overline{\mathcal{R}}$ defined as $a\overline{\mathcal{R}}b$ if $a\mathcal{R}b$ and $a \neq b$.

Let D be an infraconnected subset of $\mathbb{K}$. If D is bounded, of diameter R, the circular filter of center $a \in D$ and diameter R is called *the peripheral* of D.

A circular filter $\mathcal{F} \in \Phi(\mathbb{K})$ is said to be *D-bordering* if it is secant with both D and $\mathbb{K}\setminus D$. Moreover, it is said to be *strictly D-bordering* if it is the peripheral of D or the peripheral of a hole of D.

Remark. If a circular filter $\mathcal{F} \in \Phi(\mathbb{K})$ is strictly D-bordering it has a center.

Lemma 40.1. *Let D be a closed infraconnected subset of $\mathbb{K}$. Every circular filter which is strictly D-bordering is D-bordering. Further, if D is infraconnected affinoid, then a circular filter is D-bordering if and only if it is strictly D-bordering.*

Proof. Let $\mathcal{F}$ be the peripheral of a hole $T = d(a, r^-)$ of D. All annuli $\Gamma(b, r', r'')$ with $b \in d(a, r)$ and $r' < r < r''$ have a non-empty intersection with T and D, which shows that $\mathcal{F}$ is D-bordering. And we have a similar situation with the peripheral of D if D is bounded.

Now, if D is infraconnected affinoid, the set of holes is finite. If a circular filter $\mathcal{F} \in \Phi(D)$ is not strictly D-bordering, then it admits annuli having an empty intersection with any hole and with $\mathbb{K}\setminus\widetilde{D}$, hence it is not D-bordering. □

Lemmas 40.2 and 40.3 are immediate:

Lemma 40.2. *Let $\mathcal{F}, \mathcal{G}$ be two circular filters such that $\mathcal{Q}(\mathcal{F}) \neq \emptyset$ and $\mathcal{Q}(\mathcal{G}) \neq \emptyset$. Then, $\mathcal{F} \preceq \mathcal{G}$ if and only if $\mathcal{Q}(\mathcal{F}) \subset \mathcal{Q}(\mathcal{G})$.*

Lemma 40.3. *Let $\mathcal{F}, \mathcal{G}$ be two circular filters such that $\mathcal{F} \preceq \mathcal{G}$. Then,* $\operatorname{diam}(\mathcal{F}) \leq \operatorname{diam}(\mathcal{G})$. *Moreover, $\mathcal{F} = \mathcal{G}$ if and only if* $\operatorname{diam}(\mathcal{F}) = \operatorname{diam}(\mathcal{G})$.

Theorem 40.4. *The relation $\preceq$ is an order relation on $\Phi(\mathbb{K})$ and $\prec$ is the strict-order associated with this order relation.*

Proof. Indeed, the relation is reflexive by definition. Let $\mathcal{F}, \mathcal{G}, \mathcal{H}$ be three circular filters. Suppose that $\mathcal{F} \preceq \mathcal{G}$ and $\mathcal{G} \preceq \mathcal{F}$. If $\mathcal{F} \neq \mathcal{G}$, then by definition $\mathcal{Q}(\mathcal{F}) \neq \emptyset$, $\mathcal{Q}(\mathcal{G}) \neq \emptyset$, $\mathcal{F}$ is secant with $\mathcal{Q}(\mathcal{G})$, and $\mathcal{G}$ is secant with $\mathcal{Q}(\mathcal{F})$. Consequently, $\mathcal{Q}(\mathcal{F}) = \mathcal{Q}(\mathcal{G})$, and therefore, by Lemma 40.3, we have $\mathcal{F} = \mathcal{G}$. Finally, suppose that $\mathcal{F} \preceq \mathcal{G}$ and $\mathcal{G} \preceq \mathcal{H}$, with $\mathcal{F} \neq \mathcal{G} \neq \mathcal{H}$. Then, $\mathcal{F}$ is secant with $\mathcal{Q}(\mathcal{G})$ and with $\mathcal{Q}(\mathcal{H})$, and therefore, $\mathcal{F} \preceq \mathcal{H}$. □

Corollary 40.5. *Let D be infraconnected. The mapping* diam *is strictly increasing from $\Phi(D)$, ordered by $\preceq$, to $\mathbb{R}_+$.*

Theorem 40.6. *Let D be infraconnected and let $\mathcal{F} \in \Phi(D)$. Then $\mathcal{F}$ is a minimal element in $\Phi(D)$ if and only if*

(i) *either it is punctual,*
(ii) *or it has no center,*
(iii) *or $\mathcal{Q}(\mathcal{F}) \cap D = \emptyset$.*

Proof. On one hand, it is easily seen that if $\mathcal{F}$ is punctual, or has no center, then it is minimal in $\Phi(\mathbb{K})$ and therefore in $\Phi(D)$. Suppose now that $\mathcal{Q}(\mathcal{F}) \cap D = \emptyset$. Suppose that $\mathcal{G} \in \Phi(D)$ satisfies $\mathcal{G} \prec \mathcal{F}$. Then $\mathcal{G}$ is secant with $\mathcal{Q}(\mathcal{F})$ and has a diameter strictly inferior to the diameter r of $\mathcal{F}$, hence there exists a disk $d(b, s) \in \mathcal{G}$, with $s < r$. Consequently, $d(b, s) \subset \mathcal{Q}(\mathcal{F})$. Since $d(b, s) \cap D = \emptyset$, we see that $\mathcal{G}$ is not secant with D.

Now, let $\mathcal{F} \in \Phi(D)$ be minimal in $\Phi(D)$. Suppose that $\mathcal{F}$ is not punctual and has a center and let $r = \text{diam}(\mathcal{F})$. If $\mathcal{Q}(\mathcal{F}) \cap D \neq \emptyset$, then for any $a \in \mathcal{Q}(\mathcal{F}) \cap D$, of course, $\mathcal{F}$ surrounds the filter of neighborhoods of a, a contradiction to the hypothesis "$\mathcal{F}$ minimal in $\Phi(D)$". Thus, $\mathcal{Q}(\mathcal{F}) \cap D = \emptyset$. If $\mathcal{Q}(\mathcal{F}) = \emptyset$, then $\mathcal{F}$ has no center. Else, $\mathcal{Q}(\mathcal{F}) \cap D = \emptyset$, which ends the proof. □

Lemma 40.7. *Let $\mathcal{F}, \mathcal{G}$ be two circular filters such that $\mathcal{Q}(\mathcal{F}) \cap \mathcal{Q}(\mathcal{G}) \neq \emptyset$. Then $\mathcal{F}$ and $\mathcal{G}$ are comparable with respect to the relation $\preceq$.*

Proof. Since $\mathcal{Q}(\mathcal{F}) \cap \mathcal{Q}(\mathcal{G}) \neq \emptyset$ and since both $\mathcal{Q}(\mathcal{F})$, $\mathcal{Q}(\mathcal{G})$ are disks, we can suppose, for instance, $\mathcal{Q}(\mathcal{F}) \subset \mathcal{Q}(\mathcal{G})$. Then $\mathcal{F}$ is secant with $\mathcal{Q}(\mathcal{G})$, hence $\mathcal{F} \preceq \mathcal{G}$. □

Theorem 40.8 is a very typical theorem concerning circular filters:

Theorem 40.8. *Let $\mathcal{F}$, $\mathcal{G}$ be two circular filters surrounding a certain circular filter $\mathcal{H}$. Then $\mathcal{F}$ and $\mathcal{G}$ are comparable with respect to the relation $\preceq$.*

Proof. If $\mathcal{F} = \mathcal{H}$, then $\mathcal{G}$ just surrounds $\mathcal{F}$. Now, suppose that both $\mathcal{F}$, $\mathcal{G}$ strictly surround $\mathcal{H}$. Then $\mathcal{H}$ is secant with both $\mathcal{Q}(\mathcal{F})$ and $\mathcal{Q}(\mathcal{G})$, hence $\mathcal{Q}(\mathcal{F}) \cap \mathcal{Q}(\mathcal{G}) \neq \emptyset$. Therefore, either $\mathcal{Q}(\mathcal{F}) \subset \mathcal{Q}(\mathcal{G})$ or $\mathcal{Q}(\mathcal{G}) \subset \mathcal{Q}(\mathcal{F})$, hence by Lemma 40.7, $\mathcal{F}$ and $\mathcal{G}$ are comparable. □

Theorem 40.9 is also very important:

Theorem 40.9. *Let $\mathcal{F} \in \Phi(\mathbb{K})$ and let $s > \mathrm{diam}(\mathcal{F})$. There exists a unique disk $d(a, s)$ such that $\mathcal{F}$ is secant with it and there exists a unique circular filter on $\mathbb{K}$, of diameter s, surrounding $\mathcal{F}$.*

Proof. Let $r = \mathrm{diam}(\mathcal{F})$. We can check that there exists a unique disk $d(b, s)$ such that $\mathcal{F}$ is secant with this disk. Indeed, if $\mathcal{F}$ has a center a, this disk is just $d(a, s)$. Then $\mathcal{F}$ is surrounded by the filter $\mathcal{G}$ of center a and diameter s.

Now, if $\mathcal{F}$ has no center, it admits as one of its elements a disk of $d(b, s')$ with $r < s' < s$ and hence $\mathcal{F}$ is secant with $d(b, s)$. Consider now another disk $d(c, s)$ such that $\mathcal{F}$ is secant with it. Then we have $d(b, s') \cap d(c, s) \neq \emptyset$, hence $d(b, s') \subset d(c, s)$, and hence $d(b, s) = d(c, s)$. So, we have shown again that $\mathcal{F}$ is surrounded by the filter $\mathcal{G}$ of center a and diameter s.

Conversely, let $\mathcal{H}$ be another circular filter of diameter s, surrounding $\mathcal{F}$. Then, by Theorem 40.8, $\mathcal{G}$ and $\mathcal{H}$ are comparable. But they have the same diameter, hence by Lemma 40.3, they are equal. □

By Proposition 3.14, we obtain Theorem 40.10.

Theorem 40.10. *Let D be infraconnected, let $\mathcal{F}$ be a circular filter on $\mathbb{K}$ secant with D, and let $r \in]\mathrm{diam}(\mathcal{F}), \mathrm{diam}(D)[$. The unique circular filter of diameter r surrounding $\mathcal{F}$ is secant with D.*

Proof. By Theorem 40.9, there exists a unique disk $d(a, r)$ such that $\mathcal{F}$ is secant with it. By Proposition 3.14, the circular filter $\mathcal{G}$ of center a and diameter r is secant with D and obviously surrounds $\mathcal{F}$.

This filter is unique by Theorem 40.9. Now, there exists a $\mathcal{G}$-affinoid G containing annuli $\Gamma(a, s, r)$ and $\Gamma(a, r, t)$ such that $G \cap D = \emptyset$. On the other hand, by definition, D being infraconnected admits points in $\Gamma(a, r, t)$ and also in $\Gamma(a, s, r)$, a contradiction. This shows that $\mathcal{G}$ is secant with D. □

Proposition 40.11. *Let $\mathcal{F}$, $\mathcal{G}$ be two circular filters on $\mathbb{K}$ which are not comparable for the relation $\preceq$. There exist disks $F \in \mathcal{F}$ and $G \in \mathcal{G}$ such that $F \cap G = \emptyset$. Moreover, given disks $F' = d(a, r') \in \mathcal{F}$, $G' = d(b, s') \in \mathcal{G}$ such that $F' \cap G' = \emptyset$, we have $\delta(F, G) = \delta(F', G') >$ $\max(\mathrm{diam}(\mathcal{F}), \mathrm{diam}(\mathcal{G}))$.*

Proof. Suppose first that both $\mathcal{F}$, $\mathcal{G}$ have no center. Then $\mathcal{F}$ (respectively $\mathcal{G}$) admits a canonical basis $\mathcal{D}$ (respectively $\mathcal{E}$) consisting of a decreasing sequence of disks. But since the two filters are not secant, we can obviously find $F \in \mathcal{D}$ and $G \in \mathcal{E}$ such that $F \cap G = \emptyset$.

Suppose now that $\mathcal{F}$ has centers and let $d(a, r) = \mathcal{Q}(\mathcal{F})$. If all disks $G \in \mathcal{G}$ contain $d(a, r)$, then $\mathcal{G}$ surrounds $\mathcal{F}$, a contradiction to the hypothesis. Hence, there exists a disk $G \in \mathcal{G}$ which does not contain $d(a, r)$. Let $d(b, s) \in \mathcal{G}$ be a disk such that $d(b, s) \cap d(a, r) = \emptyset$. Then we have $|a - b| > r$, therefore we can take $r' \in]r, s[$ and we have $d(a, r) \cap d(b, r') = \emptyset$.

Consider now disks $F' = d(a, r') \in \mathcal{F}$, $G' = d(b, s') \in \mathcal{G}$ such that $F' \cap G' = \emptyset$. Let $R = |a - b|$. Since $F' \cap G' = \emptyset$, we have $\delta(F', G') = R > \max(r', s')$ and $\mathcal{F}$ is secant with $d(a, R^-)$, $\mathcal{G}$ is secant with $d(b, R^-)$. Therefore, F is included in $d(a, R^-)$ and G is included in $d(b, R^-)$. Consequently, $\delta(F, G) = R$ and $\mathrm{diam}(\mathcal{F}) < \mathrm{diam}(F) < R$, $\mathrm{diam}(\mathcal{G}) < \mathrm{diam}(G) < R$. That ends the proof. □

Notation. Let $\mathcal{F}$, $\mathcal{G}$ be two circular filters which are not comparable for $\preceq$. By Proposition 40.11, we note that given disks $F \in \mathcal{F}$ and $G \in \mathcal{G}$ such that $F \cap G = \emptyset$, $\delta(F, G)$ does not depend on the choice of the disks $F \in \mathcal{F}$ and $G \in \mathcal{G}$, so we can put $\lambda(\mathcal{F}, \mathcal{G}) = \delta(F, G)$ with F, G disks such that $F \in \mathcal{F}, G \in \mathcal{G}$, $F \cap G = \emptyset$.

Theorem 40.12. *Let D be infraconnected and let $\mathcal{F}, \mathcal{G} \in \Phi(D)$. There exists* $\sup(\mathcal{F}, \mathcal{G}) \in \Phi(D)$ *and it is the unique circular filter of diameter $\lambda(\mathcal{F}, \mathcal{G})$ which surrounds both $\mathcal{F}$ and $\mathcal{G}$.*

Proof. The claim is trivial when the two filters are comparable for $\preceq$. So we assume they are not. Let $l = \lambda(\mathcal{F}, \mathcal{G})$. By Theorem 40.9, there exists a unique circular filter $\mathcal{S} \in \Phi(\mathbb{K})$ of diameter l surrounding $\mathcal{F}$. Then $\mathcal{S}$ has centers and $\mathcal{Q}(\mathcal{S})$ is a disk $d(a, l)$. By Proposition 40.11, there are disks included in $d(a, l)$ which belong to $\mathcal{G}$ and therefore are secant with $d(a, l)$, hence $\mathcal{S}$ surrounds $\mathcal{G}$. We check that $\mathcal{S}$ is the smallest element of the set of filters on $\mathbb{K}$ surrounding both $\mathcal{F}$ and $\mathcal{G}$. Indeed, let $\mathcal{H}$ be a circular filter surrounding $\mathcal{F}$ and $\mathcal{G}$. Then both $\mathcal{F}$ and $\mathcal{G}$ are secant with $\mathcal{Q}(\mathcal{H})$. Let $d(b, s) = \mathcal{Q}(\mathcal{H})$. Consider disks $F = d(\alpha, \rho) \in \mathcal{F}$ and $G = d(\beta, \sigma) \in \mathcal{G}$ such that $F \cap G = \emptyset$. Since $F \cap d(b, s) \neq \emptyset$ and $G \cap d(b, s) \neq \emptyset$ and since $F \cap G = \emptyset$, it is easily seen that both F, G are included in $d(b, s)$, hence $s \geq l$. But since both $\mathcal{S}$, $\mathcal{H}$ surround $\mathcal{F}$, by Theorem 40.8, they are comparable for $\preceq$. Then, since $l \leq s$, $\mathcal{H}$ surrounds $\mathcal{S}$ and therefore $\mathcal{S}$ is the smallest element of the set of filters on $\mathbb{K}$ surrounding both $\mathcal{F}$ and $\mathcal{G}$.

Now, suppose that both $\mathcal{F}$, $\mathcal{G}$ are secant with D. Then, $d(\alpha, \rho) \cap D \neq \emptyset$ and $d(\beta, \sigma) \cap D \neq \emptyset$. Since $\mathcal{S}$ is the circular filter of center α and diameter $l = |\alpha - \beta|$, it is secant with D because D is infraconnected. □

Notation. Let $\mathcal{F}$, $\mathcal{G}$ be two circular filters and let $\mathcal{S} = \sup(\mathcal{F}, \mathcal{G})$. We set $\delta(\mathcal{F}, \mathcal{G}) = \text{diam}(\mathcal{S}) - \frac{\text{diam}(\mathcal{F}) + \text{diam}(\mathcal{G})}{2}$.

Remark. Particularly, if $\mathcal{F} \preceq \mathcal{G}$, we have $\delta(\mathcal{F}, \mathcal{G}) = \text{diam}(\mathcal{G}) - \text{diam}(\mathcal{F})$. Moreover, given a, $b \in K$, the filter of neighborhoods $\mathcal{F}$ of a and the filter of neighborhoods $\mathcal{G}$ of b satisfy $\delta(\mathcal{F}, \mathcal{G}) = |a - b|$.

Lemma 40.13 is immediate:

Lemma 40.13. *Let $\mathcal{F}, \mathcal{G}$ be circular filters. Suppose there are infraconnected subsets $F \in \mathcal{F}$, $G \in \mathcal{G}$ such that $F \cap G = \emptyset$. Then, $\delta(F, G) = \text{diam}(\sup(\mathcal{F}, \mathcal{G}))$, $2\delta(\mathcal{F}, \mathcal{G}) = 2\delta(F, G) - \text{diam}(\mathcal{F}) - \text{diam}(\mathcal{G})$, and $|\text{diam}(\mathcal{F}) - \text{diam}(\mathcal{G})|_\infty \leq \delta(\mathcal{F}, \mathcal{G})$. If $\mathcal{F} \preceq \mathcal{G}$, we have $\delta(\mathcal{F}, \mathcal{G}) = \text{diam}(\mathcal{G}) - \text{diam}(\mathcal{F})$.*

Proof. Let $\mathcal{S} = \sup(\mathcal{F}, \mathcal{G})$. Since both F and G are infraconnected, we have $\delta(F, G) = \text{diam}(\mathcal{S})$ hence $2\delta(F, G) = \delta(\mathcal{F}, \mathcal{S}) + \delta(\mathcal{G}, \mathcal{S}) = 2\text{diam}(\mathcal{S}) - \text{diam}(\mathcal{F}) - \text{diam}(\mathcal{G})$. The other claims are immediate. □

Theorem 40.14. *δ is a distance on $\Phi(\mathbb{K})$.*

Proof. We only have to show the triangular inequality. Let $\mathcal{F}, \mathcal{G}, \mathcal{H} \in \Phi(\mathbb{K})$, with $\text{diam}(\mathcal{F}) = r$, $\text{diam}(\mathcal{G}) = s$, $\text{diam}(\mathcal{H}) = t$. Let

$\mathcal{X} = \sup(\mathcal{F}, \mathcal{G})$, $\mathcal{Y} = \sup(\mathcal{G}, \mathcal{H})$, $\operatorname{diam}(\mathcal{X}) = M$, $\operatorname{diam}(\mathcal{Y}) = N$. Suppose, for instance, $M \leq N$. We note that both $\mathcal{X}, \mathcal{Y}$ surround $\mathcal{G}$. Consequently, by Theorem 40.8, they are comparable. And since $M \leq N$, we have $\mathcal{X} \preceq \mathcal{Y}$. Consequently, $\mathcal{F} \preceq \mathcal{Y}$, and hence, $\sup(\mathcal{F}, \mathcal{H})$ has a diameter $V \leq N$. Then, $2\delta(\mathcal{F}, \mathcal{H}) = 2V - r - t \leq N - r - t$.

On the other hand, $2\delta(\mathcal{F}, \mathcal{G}) + 2\delta(\mathcal{G}, \mathcal{H}) = 2M - r - s + 2N - s - t$. But since $s \leq M$, we have $2\delta(\mathcal{F}, \mathcal{G}) + 2\delta(\mathcal{G}, \mathcal{H}) \geq 2N - r - t \geq 2V - r - t = 2\delta(\mathcal{F}, \mathcal{H})$. □

Remark. δ is not a non-Archimedean distance on $\Phi(K)$. Indeed, consider three circular filters $\mathcal{F}_{a,r}$, $\mathcal{F}_{a,s}$, $\mathcal{F}_{a,t}$ with $r < s < t$. Then, $\delta(\mathcal{F}_{a,r}, \mathcal{F}_{a,t}) = t - r = \delta(\mathcal{F}_{a,r}, \mathcal{F}_{a,s}) + \delta(\mathcal{F}_{a,s}, \mathcal{F}_{a,t})$.

Definition. Henceforth, we call *δ-topology* the topology defined by the distance δ.

Theorem 40.15. *Every bounded monotonous sequence of $\Phi(\mathbb{K})$ with respect to the order $\preceq$ has a limit with respect to the δ-topology.*

Proof. Let $(\mathcal{F}_n)_{n\in\mathbb{N}}$ be a bounded monotonous sequence of circular filters and for each $n \in \mathbb{N}$, let $r_n = \operatorname{diam}(\mathcal{F}_n)$. Without loss of generality, we can obviously assume that the sequence is strictly monotonous. Clearly, for each $n \in \mathbb{N}^*$, each filter $\mathcal{F}_n$ has a center. Let $l = \lim_{n\to\infty} r_n$. Suppose first that it is a decreasing sequence. For each $n \in \mathbb{N}$, we can find a center $a_n \notin d(a_{n+1}, r_{n+1})$. Thus, the sequence $(a_n)_{n\in\mathbb{N}}$ is such that the sequence $(|a_{n+1} - a_n|)_{n\in\mathbb{N}}$ is strictly decreasing. Hence, by Proposition 3.18, there exists a unique circular filter $\mathcal{F}$ less thin than the sequence $(a_n)_{n\in\mathbb{N}}$ of radius l. And then, we check that the sequence $(\mathcal{F}_n)_{n\in\mathbb{N}}$ converges to $\mathcal{F}$ with respect to the δ-topology. If the sequence is increasing, it is easily seen that the sequence converges to the circular filter of center a_1 and diameter l. □

Theorem 40.16. *$\Phi(\mathbb{K})$ is complete with respect to the δ-topology.*

Proof. Let $(\mathcal{F}_n)_{n\in\mathbb{N}}$ be a Cauchy sequence with respect to the δ-topology. Without loss of generality, we can suppose that each $\mathcal{F}_n$ has a center. Indeed, for each $n \in \mathbb{N}$, we can take a circular filter $\mathcal{F}'_n$ with a center such that $\delta(\mathcal{F}'_n, \mathcal{F}_n) < \frac{1}{n}$ and then we check that the sequence $(\mathcal{F}'_n)_{n\in\mathbb{N}}$ is a Cauchy sequence again. Then, if it converges to a limit $\mathcal{G}$, so does the sequence $(\mathcal{F}_n)_{n\in\mathbb{N}}$.

So, we suppose that for each $n \in \mathbb{N}$, $\mathcal{F}_n$ has a center a_n. Let $r_n = \operatorname{diam}(\mathcal{F}_n)$. For every $m, n \in \mathbb{N}$, let $\mathcal{S}_{m,n}$ denote $\sup(\mathcal{F}_m, \mathcal{F}_n)$ and let

$s_{m,n} = \text{diam}(\mathcal{S}_{m,n})$. Next, for each $n \in \mathbb{N}$, let $s_n = \sup_{m \geq n, t \geq n,}(s_{m,t})$. So, by definition, the sequence (s_n) is decreasing. We show that the sequence $d(a_n, s_n)_{n\in\mathbb{N}}$ is decreasing with respect to the inclusion. Indeed, by definition, for every $m \geq n$, we have $\mathcal{Q}(\mathcal{S}_{n,m}) \subset d(a_n, s_n)$, therefore $\mathcal{Q}(\mathcal{F}_{n+1}) \subset d(a_n, s_n)$, hence $a_{n+1} \in d(a_n, s_n)$, and hence $d(a_{n+1}, s_{n+1}) \subset d(a_n, s_n)$.

Now, for every $n \in \mathbb{N}$, let $\mathcal{G}_n$ be the circular filter of center a_n and diameter s_n. The sequence $(\mathcal{G}_n)_{n\in\mathbb{N}}$ is then decreasing with respect to the order $\preceq$, and therefore, by Theorem 40.15, has a limit $\mathcal{G}$ with respect to the δ-topologies. Let $s = \lim_{n\to\infty} s_n$. We prove that $\lim_{n\to+\infty} \delta(\mathcal{F}_n, \mathcal{G}_n) = 0$.

Let $\epsilon > 0$. By definition of the sequence s_n, we can find a rank $t(\epsilon)$ such that

(1) $s_n - s \leq \frac{\epsilon}{4}\ \forall n \geq t(\epsilon)$.

And since the sequence $(\mathcal{F}_n)$ is a Cauchy sequence, there exists $k(\epsilon) \in \mathbb{N}$ such that $\delta(\mathcal{F}_m, \mathcal{F}_q) \leq \frac{\epsilon}{4}\ \forall m,\ q \geq k(\epsilon)$. We can choose $k(\epsilon) \geq t(\epsilon)$. By Lemma 40.13, we can derive

(2) $|r_n - r_m|_\infty \leq \frac{\epsilon}{4}\ \ \forall m, n \geq k(\epsilon)$.

On the other hand, since $\mathcal{G}_n$ surrounds $\mathcal{F}_n$, we have

(3) $\delta(\mathcal{G}_n, \mathcal{F}_n) = s_n - r_n$.

Now, let $q > m \geq n \geq t(\epsilon)$. By definition of the distance δ, we have $\delta(\mathcal{F}_m, \mathcal{S}_{m,q}) \leq \delta(\mathcal{F}_m, \mathcal{F}_q) \leq \frac{\epsilon}{4}$, hence

(4) $s_{m,q} - r_m \leq \frac{\epsilon}{4}$.

Now, we can take m, q bigger than n and such that $s_n - s_{m,q} \leq \frac{\epsilon}{4}$, therefore $s_m - s_{m,q} \leq \frac{\epsilon}{4}$ because $s_m \leq s_n$. Consequently,

$$\begin{aligned}\delta(\mathcal{G}_n, \mathcal{F}_n) &= s_n - r_n = (s_n - s_m) + (s_m - s_{m,q}) \\ &\quad + (s_{m,q} - r_m) + (r_m - r_n) \\ &\leq (s_n - s_m) + (s_m - s_{m,q}) + (s_{m,q} - r_m) + |r_m - r_n|_\infty,\end{aligned}$$

and hence, by (1), (2), (3), and (4), we obtain $\delta(\mathcal{G}_n, \mathcal{F}_n) \leq \epsilon\ \forall n \geq k(\epsilon)$. Thus, $\lim_{n\to\infty} \delta(\mathcal{F}_n, \mathcal{G}) = 0$, and therefore, the Cauchy sequence $\mathcal{F}_n$ has limit $\mathcal{G}$. $\square$

Remark. In [85], another distance $\widetilde{\delta}$ is defined on the set of large circular filters, as $\widetilde{\delta}(\mathcal{F},\mathcal{G}) = 2\log(\text{diam}(\sup(\mathcal{F},\mathcal{G})) - \log(\text{diam}(\mathcal{F})) - \log(\text{diam}(\mathcal{G}))$. This last distance $\widetilde{\delta}$ defines the same topology as δ on the set of large circular filters but does not admit the same Cauchy sequences and does not expand the usual distance on the field $\mathbb{K}$ defined by the absolute value. As a consequence, the set of large circular filters appears to be complete with respect to $\widetilde{\delta}$ and therefore a sequence converging to a point $a \in \mathbb{K}$ in the usual sense, actually has no limit for $\widetilde{\delta}$. We use $\widetilde{\delta}$ in Chapter 40.

Theorem 40.17. *Let B be a totally ordered subset of $\Phi(\mathbb{K})$. Then B admits an infimum $\mathcal{T}$ with respect to the order $\preceq$. Further, a subset A of $\Phi(\mathbb{K})$ admits a supremum with respect to the order $\preceq$ if and only if it is bounded with respect to the δ-topology.*

Proof. Let B be a totally ordered subset of $\Phi(\mathbb{K})$ and set $r = \inf\{\text{diam}(\mathcal{G}) \mid \mathcal{G} \in B\}$. By Lemma 40.13, we can see that all sequences $(\mathcal{G}_n)_{n\in\mathbb{N}}$ of B satisfying $\lim_{n\to\infty} \text{diam}(\mathcal{G}_n) = r$ are Cauchy sequences and admit the same limit $\mathcal{T}$. Therefore, $\mathcal{T}$ is clearly the infimum of B with respect to $\preceq$. Now, let A be a subset of $\Phi(\mathbb{K})$. If it admits a supremum with respect to the order $\preceq$, it is obviously bounded. Conversely, assume that A is bounded. We can clearly find a disk $d(b,s)$ such that all elements of A are secant with $d(b,s)$. Consequently, the circular filter $\mathcal{F}$ of center b and diameter s is such that $\mathcal{Q}(\mathcal{F})$ contains all elements of A. Now, by Theorem 40.8, the set A^* of circular filters surrounding all elements of A is totally ordered and therefore admits an infimum $\mathcal{S}$ and we have $\mathcal{G} \preceq \mathcal{S}$ for every $\mathcal{G} \in A$. Consequently, $\mathcal{S}$ is the supremum of A. $\square$

Remark. By Lemma 3.15, it is easily seen that there exist circular filters without countable bases if and only if the residue class field of $\mathbb{K}$ is not countable.

Theorem 40.18. *Let D be a closed bounded subset of $\mathbb{K}$ and let $\varphi_{\mathcal{F}} \in \text{Mult}(H(D), \|\cdot\|_D)$. There exists a basis of neighborhoods of $\varphi_{\mathcal{F}}$ in $\text{Mult}(H(D), \|\cdot\|_D)$, with respect to the topology of pointwise convergence, consisting of the family of sets of the form $\text{Mult}(H(E\cap D), \|\cdot\|_{E\cap D})$, where E is a $\mathcal{F}$-affinoid.*

Proof. Let $f_1, \ldots, f_q \in H(D)$ and consider a neighborhood W of $\varphi_{\mathcal{F}}$ of the form $\{\psi \in \text{Mult}(H(D), \|\cdot\|_D) \mid |\varphi_{\mathcal{F}}(f_j) - \psi(f_j)|_\infty \leq \epsilon\}$: we show that it contains a neighborhood of the form $\text{Mult}(H(E \cap D), \| \,.\, \|_{E\cap D})$, where E is a $\mathcal{F}$-affinoid. Indeed, by Lemma 3.15 and Theorem 14.1, there exist infraconnected affinoid sets $E_j \in \mathcal{F}$, $(1 \leq j \leq q)$ such that $\| \,|f_j(x)| - \varphi_{\mathcal{F}}(f_j)\|_\infty \leq \epsilon\ \forall x \in E_j\ (1 \leq j \leq q)$. Then, by Proposition 2.16, $\bigcap_{j=1}^{q} E_j$ is an infraconnected affinoid set E which belongs to $\mathcal{F}$. So, E is a $\mathcal{F}$-affinoid such that $\| \,|f_j(x)| - \varphi_{\mathcal{F}}(f_j)\|_\infty \leq \epsilon\ \forall x \in E_j\ (1 \leq j \leq q)$, therefore $\| \,|\psi(f_j)| - \varphi_{\mathcal{F}}(f_j)\|_\infty \leq \epsilon\ \forall x \in E_j\ (1 \leq j \leq q)$, $\forall \psi \in \text{Mult}(H(E \cap D), \| \,.\, \|_{E\cap D})$. Then, $\text{Mult}(H(E), \| \,.\, \|_E)$ is a neighborhood of $\varphi_{\mathcal{F}}$ in $\text{Mult}(\mathbb{K}(x))$ and $\text{Mult}(H(E \cap D), \| \,.\, \|_{E\cap D})$ is a neighborhood of $\varphi_{\mathcal{F}}$ in $\text{Mult}(H(D), \| \,.\, \|_D)$. □

Corollary 40.19. *Let $\varphi_{\mathcal{F}} \in \text{Mult}(\mathbb{K}[x])$. There exists a basis of neighborhoods of $\varphi_{\mathcal{F}}$ in $\text{Mult}(\mathbb{K}[x])$, with respect to the topology of pointwise convergence, consisting of the family of sets of the form $\text{Mult}(R(E), \| \,.\, \|_E)$, where E is a $\mathcal{F}$-affinoid.*

Theorem 40.20. *Let D be a closed bounded set. The topology of pointwise convergence is weaker than the δ-topology on $\text{Mult}(H(D), \| \,.\, \|_D)$.*

Proof. Let $\varphi_{\mathcal{F}} \in \text{Mult}(H(D), \| \,.\, \|_D)$ and let $r = \text{diam}(\mathcal{F})$. Since the $\mathcal{F}$-affinoids form a basis of the filter $\mathcal{F}$, it is sufficient to show that given an $\mathcal{F}$-affinoid E there exists $\epsilon > 0$ such that if $\delta(\mathcal{F}, \mathcal{G}) \leq \epsilon$, then $\mathcal{F}$ is secant with E.

Suppose first that $\mathcal{F}$ has no center. There exists a disk F included in E, of diameter $s > r$ with $s - r \leq \frac{\epsilon}{2}$, and hence for every circular filter $\mathcal{G}$ secant with F, of diameter $t \in]r, s]$, we have $\delta(\mathcal{F}, \mathcal{G}) \leq 2(s-r)$. Consequently, fixing $\epsilon \leq 2(s-r)$, we can see that all $\mathcal{G}$ secant with F satisfy $\delta(\mathcal{F}, \mathcal{G}) \leq \epsilon$, hence F contains a neighborhood of $\mathcal{F}$.

Suppose now that $\mathcal{F}$ has a center a. Let l be the minimum of the diameters of holes of E inside $\mathcal{Q}(\mathcal{F})$ and let t be an annulus $\Gamma(a, r, t)$ containing no hole of E. Let us now fix $\epsilon < \frac{\min(t-r, r-l)}{2}$. Take $\mathcal{G}$ secant with E such that $\delta(\mathcal{F}, \mathcal{G}) \leq \epsilon$.

Suppose first that $\mathcal{G}$ is secant with $\mathcal{Q}(\mathcal{F})$ and then, for instance, is secant with $d(a, r^-)$. Then we can see that $\mathcal{G}$ is secant with $\Gamma(a, l, r)$ and hence is secant with E. Suppose now that $\mathcal{G}$ is not secant with $\mathcal{Q}(\mathcal{F})$ and hence is secant with $\mathbb{K} \backslash d(a, r)$. Since $\epsilon < \frac{\min(t-r, r-l)}{2}$, we can see that $\mathcal{G}$ is secant with $\Gamma(a, r, t) \subset E$.

Consequently, we have found a neighborhood V of $\varphi_{\mathcal{F}}$ with respect to the δ-topology such that if $\delta(\mathcal{F}, \mathcal{G}) \leq \epsilon$, then $\mathcal{G}$ is secant with E.

That finishes proving that every neighborhood of $\mathcal{F}$ with respect to the pointwise topology contains a neighborhood with respect to the δ-topology. $\square$

Remark. The two topologies do not coincide in the general case. Suppose that a set D has an increasing filter $\mathcal{F}$ of center a and diameter R. Let $(a_n)_{n\in\mathbb{N}}$ be a sequence of D such that $|a_n - a| < |a_{n+1} - a|$ and $\lim_{n\to+\infty} |a_n - a| = R$.

The sequence $(\varphi_{a_n})_{n\in\mathbb{N}}$ converges to $\varphi_{\mathcal{F}}$ with respect to the pointwise topology but does not converge to any limit with respect to the δ-topology.

Notation. We denote by $\Sigma(D)$ the set of D-bordering circular filters and by $\Sigma_0(D)$ the set of strictly D-bordering circular filters.

Remark. Since the topology of pointwise convergence is weaker than the δ-topology, the boundary of $\text{Mult}(R(D), \|\ .\ \|_D)$ inside $\text{Mult}(\mathbb{K}[x])$, with respect to the δ-topology, is obviously included in the boundary of $\text{Mult}(R(D), \|\ .\ \|_D)$ inside $\text{Mult}(\mathbb{K}[x])$, with respect to the topology of pointwise convergence. The equality does not hold in the general case.

Lemma 40.21. *Let D be a closed bounded infraconnected subset of $\mathbb{K}$ and let $\mathcal{F}$ be a D-bordering but not strictly D-bordering circular filter. For every disk $M \in \mathcal{F}$, $\mathcal{F}$ is secant with $M\setminus D$ but not secant with any hole of D. Moreover, $\Sigma_0(D)$ is dense in $\Sigma(D)$ with respect to the topology of pointwise convergence.*

Proof. Let $\varphi_{\mathcal{F}} \in \Sigma(D)\setminus\Sigma_0(D)$. Then every element of $\mathcal{F}$ contains holes of D and hence is secant with $M\setminus D$. Suppose that $\mathcal{F}$ is secant with a hole $T = d(b, \rho^-)$ of D. Since $\mathcal{F}$ is not the peripheral of T, it is secant with a disk $d(c, s) \subset d(b, \rho^-)$ with $s < \rho$, and of course, it has elements B of diameter $l \in]s, \rho[$ that contain no hole of D, a contradiction. So, $\mathcal{F}$ is not secant with any hole of D. Now, take $h_1, \dots, h_n \in H(D)$ and $\epsilon > 0$. Since $\mathcal{F}$ admits a basis of infraconnected affinoid sets, we can find an infraconnected affinoid set $E \in \mathcal{F}$ such that $|\ |h_j(x)| - \varphi_{\mathcal{F}}(h_j)|_\infty < \epsilon\ \forall x \in E \cap D, \forall j = 1, \dots, n$. Then there exists a hole T of D included in E. Consequently, $|\varphi_T(h_j) - \varphi_{\mathcal{F}}(h_j)|_\infty < \epsilon\ \forall j = 1, \dots, n$. This shows that $\Sigma_0(D)$ is dense in $\Sigma(D)$. $\square$

Corollary 40.22. *Let D be a closed infraconnected subset of $\mathbb{K}$ such that $\Sigma_0(D)$ is finite. Then, $\Sigma(D) = \Sigma_0(D)$.*

Remark. When $\Sigma_0(D)$ is not finite, it is not necessarily dense in $\Sigma(D)$ with respect to the δ-topology. Consider a sequence $(a_n)_{n\in\mathbb{N}}$ with $|a_n| < |a_{n+1}|$ and $\lim_{n\to+\infty} |a_n| = 1$, let $r \in]0, |a_1|$, and let

$$D = d(0,2)\Bigg\backslash\left(\bigcup_{n=1}^{\infty} d(a_n, r^-)\right).$$

Then $\Sigma_0(D)$ consists of the set of circular filters $\mathcal{F}_{a_n,r}$ and $\mathcal{F}_{0,2}$. But $\Sigma(D)$ also contains $\mathcal{F}_{0,1}$. Then the sequence $(\mathcal{F}_{a_n,r})_{n\in\mathbb{N}}$ does converge to $\mathcal{F}_{0,1}$ with respect to the pointwise convergence but is not a Cauchy sequence with respect to the δ-topology, and more precisely, $\delta(\mathcal{F}_{0,1}, \mathcal{F}_{a_n,r}) = 1 - r$.

Theorem 40.23 ([15, 49, 50]). *Let D be a closed bounded subset of $\mathbb{K}$. Then $\Sigma(D)$ is equal to the boundary of* $\text{Mult}(H(D, \| \,.\, \|_D)$ *inside* $\text{Mult}(\mathbb{K}[x])$ *with respect to the topology of pointwise convergence.*

Proof. First, we show that $\Sigma(D)$ is included in the boundary of $\text{Mult}(H(D, \| \,.\, \|_D)$ inside $\text{Mult}(\mathbb{K}[x])$. Since $\Sigma_0(D)$ is dense inside $\Sigma(D)$ and since the boundary of $\text{Mult}(H(D, \| \,.\, \|_D)$ inside $\text{Mult}(\mathbb{K}[x])$ is obviously closed, it is sufficient to show that $\Sigma_0(D)$ is included in that boundary. Let $\widetilde{D} = d(a, r)$ and let $\varphi_{\mathcal{F}} \in \Sigma_0(D)$. Let F be the subset of $\text{Mult}(\mathbb{K}[x])\backslash\text{Mult}(H(D), \| \,.\, \|_D)$ consisting of all $\varphi_{\mathcal{G}}$ such that $\mathcal{G}$ is secant with $\mathbb{K}\backslash d(a, s)$ for some $s > r$. If $\mathcal{F}$ is the peripheral of $d(a, r)$, it is obvious that $\varphi_{\mathcal{F}}$ belongs to the closure of F and hence belongs to the boundary of $\text{Mult}(H(D), \| \,.\, \|_D)$.

Consequently, we are led to assume that $\mathcal{F}$ is the peripheral of certain hole $d(b_j, r_j^-)$ of D and then symmetrically, it is also obvious that $\varphi_{\mathcal{F}}$ belongs to the closure of $\text{Mult}(\mathbb{K}[x])\backslash\text{Mult}(H(D), \| \,.\, \|_D)$ inside $\text{Mult}(\mathbb{K}[x])$.

Now, let us show that the boundary of $\text{Mult}(H(D, \| \,.\, \|_D)$ inside $\text{Mult}(\mathbb{K}[x])$ is included in $\Sigma(D)$. Indeed, let $\varphi_{\mathcal{F}} \in \text{Mult}(H(D), \|.\, \|_D)$ belong to the closure of $\text{Mult}(\mathbb{K}[x])\backslash\text{Mult}(H(D), \|.\, \|_D)$ in $\text{Mult}(\mathbb{K}[x])$. Since $\mathcal{F}$ is secant with D, we only need to show that it is also secant with $\mathbb{K}\backslash D$. Let $t = \text{diam}(\mathcal{F})$. Suppose it is not secant with $\mathbb{K}\backslash D$. Then we can find an $\mathcal{F}$-affinoid F of the form $d(b, u)\backslash\left(\bigcup_{j=1}^{q} d(c_j, \rho)\right)$ that is included in D, with $t < u < r$, $\rho < t$

and $|c_i - c_j| = t \ \forall i \neq j$. Now, let us take $s' \in]\rho, t[$ and $s'' \in]t, u[$ and set

$$E = d(b, s'') \setminus \left(\bigcup_{j=1}^{q} d(c_j, s'^-) \right).$$

Let $h(x) = \prod_{j=1}^{q}(x - c_j)$. On one hand, we can check that when $x \in \mathbb{K} \setminus D$, either $|x - b| \geq u$, and then $|h(x)| \geq u^q$, or $|x - c_k| \leq \rho$ for some k and then $|h(x)| \leq \rho(\prod_{j \neq k} |c_j - c_k|) = \rho t^{q-1}$.

On the other hand, when x belongs to E, then we can check that $\rho t^{q-1} < s' t^{q-1} \leq |h(x)| \leq s''^q < u^q$, which contradicts the hypothesis that $\varphi_{\mathcal{F}}$ belongs to the closure of $\mathrm{Mult}(\mathbb{K}[x]) \setminus \mathrm{Mult}(H(D), \| \,.\, \|_D)$ in $\mathrm{Mult}(\mathbb{K}[x])$. Consequently, $\mathcal{F}$ is secant with $\mathbb{K} \setminus D$ and hence $\varphi_{\mathcal{F}}$ belongs to $\Sigma(D)$. This finishes showing that the boundary of $\mathrm{Mult}(H(D), \| \,.\, \|_D)$ is included in $\Sigma(D)$ and hence is equal to $\Sigma(D)$. □

Definition. A subset S of $\Phi(D)$ is called *a Shilov set* if for every $f \in H(D)$, there exists $\mathcal{F} \in S$ such that $\varphi_{\mathcal{F}}(f) = \|f\|_D$. A Shilov set that is closed with respect to the topology of pointwise convergence is called *Shilov boundary for $H(D)$* if it is the smallest closed Shilov set.

Theorem 40.24. *Let D be a closed bounded infraconnected subset of $\mathbb{K}$. Then $\Sigma(D)$ is equal to the Shilov boundary for $H(D)$.*

Proof. First, we note that $\Sigma(D)$ is closed inside $\mathrm{Mult}(\mathbb{K}[x])$ because, by Theorem 40.23, it is the boundary of $\mathrm{Mult}(H(D), \| \,.\, \|_D)$. By the Mittag-Leffler Theorem 16.1, it is easily seen that $\Sigma_0(D)$ is a Shilov set for $H(D)$.

Now, we show that $\Sigma(D)$ is the smallest closed Shilov set. Suppose it is not the smallest. Then, there exists another closed Shilov set S which does not contain $\Sigma(D)$. Since S is closed and since $\Sigma_0(D)$ is dense in $\Sigma(D)$, there exists $\psi \in \Sigma_0(D) \setminus S$. Let $\mathcal{G}$ be the strictly D-bordering filter such that $\psi = \varphi_{\mathcal{G}}$. Thus, ψ is of the form $\varphi_{b,s}$.

Since S is closed, there exists an infraconnected affinoid set E of the form $d(b, s'') \setminus \bigcup_{j=1}^{q} d(c_j, s')$, with $s' < s < s''$ such that $\mathrm{Mult}(H(E), \| \,.\, \|_E) \cap S = \emptyset$. This means that for every $\varphi_{\mathcal{F}} \in S$, $\mathcal{F}$ is not secant with E. Let $r = \mathrm{diam}(D)$ and suppose first that $s = r$. Consider $g(x) = \prod_{j=1}^{q}(x - c_j)$. Then, $\varphi_{\mathcal{G}}(h) = r^q$, and for every $\varphi_{\mathcal{F}} \in S$, we can check that $\varphi_{\mathcal{F}}(g) \leq s'^q$, a contradiction to the hypothesis "S is a Shilov set".

Consequently, we have $s < r$, and therefore, since $\varphi_{\mathcal{G}} \in \Sigma_0(D)$, the disk $d(b, s^-)$ is a hole of D. Let $h(x) = \frac{\prod_{j=1}^{q}(x-c_j)}{(x-b)^{q+1}}$. Then, $\varphi_{\mathcal{G}}(h) = \frac{1}{s}$, and for every $\varphi_{\mathcal{F}} \in S$, we can check that $\varphi_{\mathcal{F}}(h) = \max(\frac{s'}{s^2}, \frac{1}{s''}) < \frac{1}{s}$, a contradiction to the hypothesis "S is a Shilov set". Thus, we have proven that any closed Shilov set must contain $\Sigma(D)$ and this finishes proving that $\Sigma(D)$ is the Shilov boundary for $H(D)$. □

Corollary 40.25. *Let D be an infraconnected affinoid subset of $\mathbb{K}$. The Shilov boundary for $H(D)$ is equal to $\Sigma_0(D)$.*

Corollary 40.26. *Let D be a closed bounded infraconnected subset of $\mathbb{K}$. The following three sets are equal:*

(i) $\Sigma(D)$,
(ii) *the boundary of* $\text{Mult}(H(D), \| \cdot \|_D)$ *inside* $\text{Mult}(\mathbb{K}[x])$ *with respect to the topology of pointwise convergence,*
(iii) *the Shilov boundary for* $(H(D), \| \cdot \|_D)$ *with respect to the topology of pointwise convergence.*

More precisely, we can state Corollary 40.27:

Corollary 40.27. *Let D be a closed bounded infraconnected subset of $\mathbb{K}$ having finitely many D-bordering filters. The following four sets are equal:*

(i) $\Sigma(D)$,
(ii) *the boundary of* $\text{Mult}(H(D), \| \cdot \|_D)$ *inside* $\text{Mult}(\mathbb{K}[x])$ *with respect to the topology of pointwise convergence,*
(iii) *the boundary of* $\text{Mult}(H(D), \| \cdot \|_D)$ *inside* $\text{Mult}(\mathbb{K}[x])$ *with respect to the δ-topology,*
(iv) *the Shilov boundary for* $(H(D), \| \cdot \|_D)$ *with respect to the topology of pointwise convergence.*

Proof. If D has finitely many D-bordering filters, the boundary of $\text{Mult}(H(D), \| \cdot \|_D)$ inside $\text{Mult}(\mathbb{K}[x])$ is finite for both topologies and hence is reduced to $\Sigma(D) = \Sigma_0(D)$. □

Remark. In particular, Corollary 40.27 applies to $H(D)$ when D is an infraconnected affinoid set and then $H(D)$ is also called Krasner–Tate algebra [37].

Chapter 41

Mappings from $\Phi(D)$ to the Tree $\Phi(\mathbb{K})$

We show that every element $f \in H(D)$ has continuation to a mapping f_* from $\Phi(D))$ to $\Phi(\mathbb{K})$. Given a circular filter $\mathcal{F} \in \Phi(D)$, the mapping that associates with each $f \in H(D)$ the circular filter $f_*(\mathcal{F})$ is uniformly continuous with respect to the norm of $H(D)$ and the metric δ on $\Phi(\mathbb{K})$ [492].

Every analytic element on a closed bounded set D has expansion to a function from $\mathrm{Mult}(H(D), \|\cdot\|_D)$ to $\mathrm{Mult}(\mathbb{K}[x])$ which is continuous with respect to both topologies on $\mathrm{Mult}(\mathbb{K}[x])$. Results apply to meromorphic functions in $\mathbb{K}$ because, in particular, they define open continuous functions in $\mathrm{Mult}(\mathbb{K}[x])$ with respect to the topology of pointwise convergence. Such a mapping is increasing with respect to the order of the tree $\mathrm{Mult}(\mathbb{K}[x])$ if and only if it is an entire function.

Notation. Throughout this chapter, we denote by D an infraconnected affinoid subset of $\mathbb{K}$.

Lemma 41.1 is easy:

Lemma 41.1. *Let $\mathcal{F}$, $\mathcal{G} \in \Phi(\mathbb{K})$ be secant with a disk $d(a, r)$. Then, $\delta(\mathcal{F}, \mathcal{G}) \leq r$.*

Proof. Since $\mathcal{F}$, $\mathcal{G} \in \Phi(\mathbb{K})$ are secant with $d(a, r)$, their diameters are $\leq r$. On the other hand, $\sup(\mathcal{F}, \mathcal{G})$ is also secant with $d(a, r)$ and hence has a diameter $\leq r$. Consequently, $\delta(\mathcal{F}, \mathcal{G}) \leq r$. □

Theorem 41.2. *Let the annuli $\Gamma(a, r', r)$ and $\Gamma(a, r, r'')$ be included in D and let $h \in H(D)$. Then $h(\mathcal{F}_{a,r})$ is a basis of a filter thinner than a circular filter $\mathcal{H}$ of center $h(a)$ and diameter $\varphi_{a,r}(h - h(a))$.*

Proof. Since D is affinoid, up to a change of variable, we can assume that $d(a, r^-) \subset D$, hence $h \in \mathcal{A}(d(a, r^-))$. Then, putting $s = \varphi_{a,r}(h - h(a))$, we can see that, given $\rho < r$, $h(\Gamma(a, \rho, r))$ is an annulus of the form $\Gamma(h(a), \psi(\rho), s)$ with $\psi(\rho) < s$ and that $\lim_{\rho \to r} \psi(\rho) = s$. Consequently, $h(\mathcal{F}_{a,r})$ is a basis of a filter $\mathcal{C}$ secant with the increasing filter of center $h(a)$ and diameter $s = \varphi_{a,r}(h - h(a))$ and hence is secant with the circular filter $\mathcal{H}$ of center $h(a)$ and diameter s.

On the other hand, we have $\varphi_{a,r}(g \circ f) = \lim_{\mathcal{C}} |g(u)| \ \forall g \in \mathbb{K}(u)$. Hence, for every $g \in \mathbb{K}(u)$, $|g(u)|$ admits a limit along $\mathcal{C}$. Therefore, by Theorem 4.22, $\mathcal{C}$ is thinner than a circular filter $\mathcal{H}'$ and hence $\mathcal{H}'$ is a circular filter secant with $\mathcal{H}$. But then it is equal to $\mathcal{H}$, which ends the proof. □

Theorem 41.3. *Let $h \in H(D)$ and let $\mathcal{F}$ be a circular filter with no center secant with D. Then $h(\mathcal{F})$ is a basis of a filter $\mathcal{F}'$ thinner than a circular filter $\mathcal{H}$ with no center.*

Proof. Let $(D_n)_{n \in \mathbb{N}}$ be a canonical basis of $\mathcal{F}$. Since $\bigcap_{n \in \mathbb{N}} D_n = \emptyset$, without loss of generality, we can assume that there exists $q \in \mathbb{N}^*$ such that $D_q \subset D$. Consequently, by Theorem 22.9, each $h(D_n)$ is a disk and hence the sequence $h(D_n)$ is a decreasing sequence of disks. Suppose now the intersection of the sequence $(h(D_n))_{n \in \mathbb{N}}$ is not empty and let $b \in \bigcap_{n \in \mathbb{N}} h(D_n)$. For each $n \in \mathbb{N}$, there exists $a_n \in D_n$ such that $h(a_n) = b$. Since the equation $h(x) = b$ has finitely many solutions inside a disk, there exists $\alpha \in \bigcap_{n \in \mathbb{N}} h(D_n)$ such that $h(\alpha) = b$, a contradiction. This finishes showing that $h(\mathcal{F})$ is a basis of a circular filter with no center. □

Notation. Let $h \in H(D)$ and let $\mathcal{F}$ be a circular filter secant with D. We denote by $h_*(\mathcal{F})$ the circular filter less thin than the filter admitting $h(\mathcal{F})$ for basis.

Corollary 41.4. *Let $\mathcal{F}$ be a large circular filter on D and let $h \in H(D)$. Then $h_*(\mathcal{F})$ has a center if and only if so does $\mathcal{F}$. Moreover, given $f \in H(h(D))$, then $\varphi_{h_*(\mathcal{F})}(f) = \varphi_{\mathcal{F}}(f \circ h)$.*

Proof. We only have to check that $\varphi_{h_*(\mathcal{F})}(f) = \varphi_{\mathcal{F}}(f \circ h)\ \forall f \in H(h(D))$. Indeed,

$$\varphi_{h_*(\mathcal{F})}(f) = \lim_{h_*(\mathcal{F})}(f(x)) = \lim_{\mathcal{F}}(h(f(x))) = \varphi_{\mathcal{F}}(h \circ f)\ \forall f \in H(h(D)).$$

□

Corollary 41.5. *Suppose $\Gamma(a, r, s) \subset D$. Let $h(x) = \sum_{-\infty}^{+\infty} a_n (x - a)^n \in H(D)$. Assume that $\nu(h, \log u) = m \neq 0\ \forall u \in]r, s[$. Let $\mathcal{F}$ (respectively $\mathcal{G}$) be the circular filter of center a and diameter r (respectively s). Then, $h_*(\mathcal{F})$ (respectively $h_*(\mathcal{G})$) is the circular filter of center 0 and diameter $|h|(r) = |a_m|r^m$ (respectively $|h|(s) = |a_m|s^m$). Moreover, $h_*(\mathcal{F})$ and $h_*(\mathcal{G})$) are circular filters comparable with respect to the order $\preceq$.*

Proof. By Lemma 41.2, $h(\mathcal{F})$ (respectively $h(\mathcal{G})$) is a basis of a filter $\mathcal{F}'$ (respectively $\mathcal{G}'$) thinner than the circular filter of center a and diameter $\varphi_{a,r}(h - h(a))$ (respectively $\varphi_{a,s}(h - h(a))$) i.e. $|a_m|r^m$ (respectively $|a_m|s^m$). The last statement is obvious. □

Definition. Let F be a field. Given $h = \frac{P(x)}{Q(x)} \in F(x)$ with $P,\ Q \in F[x]$ relatively prime, we call *degree of* h the number $\deg(h) = \max(\deg(P), \deg(Q))$ and we call *absolute degree of* h the number $|\deg(P) - \deg(Q)|_\infty$.

Theorem 41.6. *Let $\mathcal{F} \in \Phi(D)$ and let $f,\ h \in H(D)$ be such that $\varphi_{\mathcal{F}}(f - h) < \operatorname{diam}(h_*(\mathcal{F}))$. Then, $h_*(\mathcal{F}) = f_*(\mathcal{F})$. Further, if E is a closed bounded set containing $f(D)$, then $\varphi_{f_*(\mathcal{F})} \in \operatorname{Mult}(H(E), \|\ .\ \|_E)$, and for all $g \in H(E)$, it satisfies $\varphi_{f_*(\mathcal{F})}(g) = \varphi_{\mathcal{F}}(g \circ f)\ \forall g \in H(E)$.*

Proof. Suppose first that $\mathcal{F}$ has no center. By Theorem 41.3, we know that $h(\mathcal{F})$ is a basis of $h_*(\mathcal{F})$. Then, given a decreasing sequence of disks $(D_n)_{n\in\mathbb{N}}$ making a basis of $\mathcal{F}$, we have $D_n \subset D$ when n is big enough and then we can check that $h(D_n) = f(D_n)$, which shows the conclusion.

Suppose now that $\mathcal{F}$ has a center a and let r be its diameter. Since D is infraconnected affinoid, either it contains an annulus $\Gamma(a, r', r)$ or it contains an annulus $\Gamma(a, r, r'')$ where f has no zero. In such an annulus, $f(x)$ is equal to a Laurent series $\sum_{-\infty}^{+\infty} a_n(x - a)^n$. Now, we have $\varphi_{\mathcal{F}}(f - h) < \operatorname{diam}(h_*(\mathcal{F}))$. Then, $\varphi_{\mathcal{F}}(f - a_0) = \varphi_{a,r}(f - a_0) = \varphi_{a,r}(h - a_0) = \operatorname{diam}(h_*(\mathcal{F}))$.

When r' or r'' are close enough to r, if D contains the annulus $\Gamma(a, r', r)$, we can check that the image of $\Gamma(a, r', r)$ by $f - a_0$ is equal to this by $h - a_0$, and similarly, if D contains the annulus $\Gamma(a, r, r'')$, the image of $\Gamma(a, r, r'')$ by $f - a_0$ is equal to this by $h - a_0$. This holds for every such annuli in D secant with $\mathcal{F}$ and shows that $h(\mathcal{F})$ and $f(\mathcal{F})$ are bases of a same filter. Consequently, $f(\mathcal{F} \cap D)$ generates the same filter as $h(\mathcal{F} \cap D)$. Then, by Lemma 41.2, there exists a unique circular filter $\mathcal{H} = h_*(\mathcal{F})$, less thin than the filter admitting for basis $f(\mathcal{F})$. Moreover, by Lemma 41.2, $\mathcal{H}$ admits a_0 as a center and its diameter, by construction, is $\varphi_{a,r}(h - a_0)$. Then, by Corollary 41.5, we check that $\text{diam}(f_*(\mathcal{F}))$ belongs to $|\mathbb{K}|$ if and only if so does $\text{diam}(\mathcal{F})$.

Now, for every $P \in \mathbb{K}[x]$, we set $\psi(P) = \varphi_{\mathcal{F}}(P \circ f)$. Then ψ belongs to $\text{Mult}(\mathbb{K}[x])$ and therefore is of the form $\varphi_{\mathcal{G}}$. Hence, by Theorem 4.10, $\mathcal{G} = \mathcal{H}$ and the last statement is immediate. □

Theorem 41.7. *Let $\mathcal{F} \in \Phi(D)$ and let f, $h \in H(D)$. Then, $\delta(f_*(\mathcal{F}), h_*(\mathcal{F})) \leq \|f - h\|_D$.*

Proof. Let $\mathcal{F}' = f_*(\mathcal{F})$ and let $\mathcal{F}'' = h_*(\mathcal{F})$. Let $r = \|f - h\|_D$. Suppose first $\text{diam}(\mathcal{F}') > r$. Then, by Theorem 41.6, we have $\mathcal{F}' = \mathcal{F}''$. Similarly, if $\text{diam}(\mathcal{F}'') > r$, we have $\mathcal{F}' = \mathcal{F}''$.

Suppose now $\text{diam}(\mathcal{F}') \leq r$ and $\text{diam}(\mathcal{F}'') \leq r$. Suppose first that $\mathcal{F}$ has no center. Then, by Corollary 41.4, both $\mathcal{F}'$ and $\mathcal{F}''$ have no center. Set $s' = \text{diam}(\mathcal{F}')$ and $s'' = \text{diam}(\mathcal{F}'')$. Let $(D_n)_{n \in \mathbb{N}}$ be a decreasing sequence of disks making a basis of $\mathcal{F}$ and let $E'_n = f(D_n)$, $E''_n = h(D_n)$, $n \in \mathbb{N}$. Then the sequence $(E'_n)_{n \in \mathbb{N}}$ is a basis of $\mathcal{F}'$ and the sequence $(E''_n)_{n \in \mathbb{N}}$ is a basis of $\mathcal{F}''$. But then, $\delta(E'_n, E''_n) \leq r \ \forall n \in \mathbb{N}$ and hence $E'_n = E''_n \ \forall n \in \mathbb{N}$ because $\text{diam}(E'_n) \geq \text{diam}(\mathcal{F}')$ and $\text{diam}(E''_n) \geq \text{diam}(\mathcal{F}'')$. Consequently, $\mathcal{F}' = \mathcal{F}''$.

Suppose now that $\mathcal{F}$ has a center. Since D is affinoid, without loss of generality, we can assume that $\mathcal{F}$ is the filter of center 0 and diameter 1 and that $d(0, 1) \subset D$. Then, $f(d(0, 1))$ is a disk $d(a, s)$ and $h(d(0, 1))$ is a disk $d(b, t)$ and we have $|a - b| \leq r$, with $a = f(0)$ and $b = h(0)$. Now, by Theorem 41.2, $\mathcal{F}'$ is secant with $d(a, s)$ and $\mathcal{F}''$ is secant with $d(b, t)$ and we have $s = \text{diam}(\mathcal{F}')$, $t = \text{diam}(\mathcal{F}'')$. Thus, both $\mathcal{F}'$, $\mathcal{F}''$ are secant with the disk $d(a, r) = d(b, r)$, and hence $\delta(\mathcal{F}', \mathcal{F}'') \leq \|f - h\|_D$. □

Corollary 41.8. *For every $f \in H(D)$, f_* is the uniform limit of a sequence $(h_n)_*$ with $h_n \in R(D)$, with respect to the distance δ on $\Phi(D)$.*

Definition and notation. Let D be a subset of $\mathbb{K}$ and let u be > 0. We denote by $\Phi(D, u)$ the set of $\mathcal{F} \in \Phi(D)$ such that $\text{diam}(\mathcal{F}) < u$. We call *segment of* $\Phi(D)$ a connected totally ordered subset of $\Phi(D)$ with respect to the δ-topology.

Lemma 41.9. *Let $\mathcal{F}$, $\mathcal{G}$ be circular filters such that $\text{diam}(\mathcal{F}) < r$, $\delta(\mathcal{F}, \mathcal{G}) \leq r$. Then $\text{diam}(\mathcal{G}) < 2r$ and both $\mathcal{F}$, $\mathcal{G}$ are secant with a same disk $d(a, 2r)$.*

Proof. If $\mathcal{G} \preceq \mathcal{F}$, then $\text{diam}(\mathcal{G}) < r$ and there is nothing to show. Suppose now that $\mathcal{F} \preceq \mathcal{G}$. Then $\mathcal{G}$ has a center a. Since $\delta(\mathcal{F}, \mathcal{G}) \leq r$, we have $\text{diam}(\mathcal{G}) < s + r \leq 2r$, hence both $\mathcal{F}$, $\mathcal{G}$ are secant with $d(a, 2r)$.

Suppose now that $\mathcal{F}$ and $\mathcal{G}$ are not comparable with respect to $\preceq$. Let $\mathcal{S} = \sup(\mathcal{F}, \mathcal{G})$, and let $s = \text{diam}(\mathcal{S})$. We have $\delta(\mathcal{F}, \mathcal{S}) < \delta(\mathcal{F}, \mathcal{G}) \leq r$, hence $\text{diam}(\mathcal{S}) \leq \text{diam}(\mathcal{F}) + \delta(\mathcal{F}, \mathcal{G}) < 2r$. Consequently, both $\mathcal{F}$ and $\mathcal{G}$ are secant with the disk $\mathcal{Q}(\mathcal{S})$ of diameter $s < 2r$, and therefore, $\text{diam}(\mathcal{G}) < 2r$. □

Lemma 41.10. *Let $\sigma = \text{codiam}(D)$, let $f \in H(D)$, and let ϵ be > 0. There exists $u \in]0, \sigma[$ such that the restriction of f_* to the set $\Phi(D, u)$ satisfies the following property: $\delta(f_*(\mathcal{F}), f_*(\mathcal{G})) \leq \epsilon \ \forall \mathcal{F}, \mathcal{G} \in \Phi(D, u)$ such that $\delta(\mathcal{F}, \mathcal{G}) < u$.*

Proof. Since f is uniformly continuous, there exists $t \in]0, \sigma[$ such that $|f(x) - f(y)| \leq \epsilon \ \forall x, \ y \in D$ such that $|x - y| \leq t$. Let $u = \frac{t}{2}$ and let $\mathcal{F}$, $\mathcal{G} \in \Phi(D, u)$ be such that $\delta(\mathcal{F}, \mathcal{G}) < u$. Then $\mathcal{F}$ is secant with a disk $d(a, u)$, and since $\delta(\mathcal{F}, \mathcal{G}) < u$, by Lemma 41.9, $\mathcal{G}$ is secant with $d(a, t)$. Consequently, $\delta(f_*(\mathcal{F})$ and $f_*(\mathcal{G}))$ are secant with a same disk of diameter ϵ, and therefore, by Lemma 41.1 again, $\delta(f_*(\mathcal{F}), f_*(\mathcal{G})) \leq \epsilon$. □

Lemma 41.11. *Let $\sigma = \text{codiam}(D)$, let $h \in R(D)$, and let $u \in]0, \sigma[$ be fixed. Let w be the absolute degree of h. Let $\mathcal{F}_{a,r}, \mathcal{F}_{a,s}$ belong to $\Phi(D)$ with $u \leq r < s$. Then,*

$$\delta(h_*(\mathcal{F}_{a,r}), h_*(\mathcal{F}_{a,s})) \leq \|h\|_D \left(\frac{s^w - r^w}{u^w} \right).$$

Proof. Without loss of generality, we can assume $a = 0$. Suppose first that h has no pole in $\Gamma(0, r, s)$. Then $h(x)$ is equal to a Laurent series $\sum_{-\infty}^{+\infty} a_n x^n$ converging in that annulus. Without loss of generality, we can suppose that $a_0 = 0$. Let $k = \nu^+(h, \log r)$ and $q = \nu^-(h, \log s)$. Suppose first $|h|(s) > |h|(r)$. Then, by Corollary 36.18, we have $1 < \frac{|h|(s)}{|h|(r)} \le \left(\frac{s}{r}\right)^q$. Now, suppose $|h|(s) < |h|(r)$. Then we have $1 < \frac{|h|(r)}{|h|(s)} \le \left(\frac{s}{r}\right)^{-q}$. On the other hand, by Theorem 36.17, we have $|q| \le w$. Consequently, when $|h|(s) > |h|(r)$, we obtain

$$0 < |h|(s) - |h|(r) \le |h|(r)\Big(\left(\frac{s}{r}\right)^w - 1\Big) \le \|f\|_D\Big(\left(\frac{s}{r}\right)^w - 1\Big),$$

and when $|h|(s) < |h|(r)$, we obtain

$$0 < |h|(r) - |h|(s) \le |h|(s)\Big(\left(\frac{s}{r}\right)^w - 1\Big) \le \|f\|_D\Big(\left(\frac{s}{r}\right)^w - 1\Big).$$

So, in both cases, we have

$$0 < \Big|\ |h|(s) - |h|(r)\Big|_\infty \le \|f\|_D\Big(\left(\frac{s}{r}\right)^w - 1\Big) \le \|f\|_D\Big(\frac{s^w - r^w}{u^w}\Big),$$

therefore

$$\delta(h_*(\mathcal{F}_{0,r}), h_*(\mathcal{F}_{0,s})) \le \|h\|_D\Big(\frac{s^w - r^w}{u^w}\Big).$$

Consider now the general case when h may admit poles in $\Gamma(0, r, s)$. Let $C(0, r_1), \dots, C(0, r_k)$ be the circles included in $\Gamma(0, r, s)$, containing at least one pole of h, with $r < r_1 < \cdots < r_k < s$. Set $r_0 = r$ and $r_{k+1} = s$. Thus, we have $\mathcal{F}_{0,r} \prec \mathcal{F}_{0,r_1} \prec \cdots \prec \mathcal{F}_{0,r_k} \prec \mathcal{F}_{0,s}$ and hence $\delta(\mathcal{F}_{0,r}, \mathcal{F}_{0,s}) = \sum_{j=0}^{k} \delta(\mathcal{F}_{0,r_j}, \mathcal{F}_{0,r_{j+1}})$. As we just showed, $\delta(h_*(\mathcal{F}_{0,r_j}), h_*(\mathcal{F}_{0,r_{j+1}})) \le \|h\|_D\Big(\frac{r_{j+1}^w - r_j^w}{u^w}\Big)$, $j = 0, \dots, k$. Therefore,

$$\delta(h_*(\mathcal{F}_{0,r}), h_*(\mathcal{F}_{0,s})) \le \|h\|_D \sum_{j=0}^{k} \Big(\frac{r_{j+1}^w - r_j^w}{u^w}\Big) = \|h\|_D\Big(\frac{s^w - r^w}{u^w}\Big).$$

□

Theorem 41.12. *Let $\sigma = \mathrm{codiam}(D)$, let $h \in R(D)$, let $u \in]0, \sigma[$ be fixed, and let w be the absolute degree of h. Let $\mathcal{F}, \mathcal{G} \in \Phi(D)$ with $u \le \mathrm{diam}(\mathcal{F}) = r < \mathrm{diam}(\mathcal{G}) = s$ and $\mathcal{F} \prec \mathcal{G}$. Then,*

$$\delta(h_*(\mathcal{F}), h_*(\mathcal{G})) \le \|h\|_D\Big(\frac{s^w - r^w}{u^w}\Big).$$

Proof. By Lemma 41.10, we only have to check the conclusion when $\mathcal{F}$ has no center. Then $\mathcal{F}$ has a basis consisting of a decreasing sequence of disks $(D_n)_{n\in\mathbb{N}}$ and these disks are included in D when n is big enough. The sequence $(h(D_n))_{n\in\mathbb{N}}$ is then a basis of $h_*(\mathcal{F})$. For each $n \in \mathbb{N}$, there is a unique circular filter $\mathcal{F}_n \in \Phi(D)$ such that $\mathcal{Q}(\mathcal{F}_n) = D_n$. Setting $r_n = \text{diam}(D_n) = \text{diam}(\mathcal{F}_n)$, we have $\lim_{n\to\infty}(r_n) = r$, and we can check that each filter $h_*(\mathcal{F}_n)$ has a center and, by Lemma 41.11, satisfies

$$\delta(h_*(\mathcal{F}_n), h_*(\mathcal{G})) \leq \|h\|_D\Big(\frac{s^w - r_n^w}{u^w}\Big).$$

Now, when n tends to $+\infty$, $\delta(h_*(\mathcal{F}_n), h_*(\mathcal{F}))$ tends to 0 and $\|h\|_D\Big(\frac{s^w - r^w}{u^w}\Big)$, which ends the proof. □

Theorem 41.13. *Let* $\text{diam}(D) = R$ *and let* $f \in H(D)$*. The mapping* f_* *from* $\Phi(D)$ *to* $\Phi(\mathbb{K})$ *is uniformly continuous with respect to the* δ*-topology on both sets.*

Proof. By Corollary 41.8, f_* is a uniform limit of a sequence $((h_*)_n)_{n\in\mathbb{N}}$ with $h_n \in R(D)$, so it is sufficient to prove the statement when $h \in R(D)$, what we assume. For convenience, we can also suppose that $D \subset d(0, R)$ with $R < 1$. Let $\sigma = \text{codiam}(D)$, let $h \in R(D)$, and let ϵ be > 0. By Lemma 41.10, there exists $u \in]0, \sigma[$ such that the restriction of h_* to the set $\Phi(D, u)$ satisfies the following property: $\delta(h_*(\mathcal{F}), h_*(\mathcal{G})) \leq \epsilon \ \forall \mathcal{F}, \mathcal{G} \in \Phi(D, u)$ such that $\delta(\mathcal{F}, \mathcal{G}) \leq u$. So, we fix u in that way. Let $\mathcal{F}$, $\mathcal{G} \in \Phi(D)$ be such that

$$\delta(\mathcal{F}, \mathcal{G}) \leq \min\left(\frac{\epsilon u^w}{2^{w+1}\|f\|_D w R^2}, \frac{u}{2}\right).$$

Suppose first that $\text{diam}(\mathcal{F}) < \frac{u}{2}$. Since $\delta(\mathcal{F}, \mathcal{G}) \leq \frac{u}{2}$, by Lemma 41.9, we can check that $\mathcal{F}, \mathcal{G}$ are secant with a same disk $d(a, u)$, and hence by Lemma 41.10, $\delta(h_*(\mathcal{F}), h_*(\mathcal{G})) \leq \epsilon$.

Now, suppose that $\text{diam}(\mathcal{F}) \geq \frac{u}{2}$, $\text{diam}(\mathcal{G}) \geq \frac{u}{2}$. Let $\mathcal{S} = \sup(\mathcal{F}, \mathcal{G})$, let $r = \text{diam}(\mathcal{F})$, and let $t = \text{diam}(\mathcal{S})$. Then, $\delta(f_*(\mathcal{F}), f_*(\mathcal{G})) \leq \delta(f_*(\mathcal{F}), f_*(\mathcal{S})) + \delta(f_*(\mathcal{S}), f_*(\mathcal{G}))$. Since $\mathcal{F} \prec \mathcal{S}$, by Theorem 41.12, we have

$$\delta(f_*(\mathcal{F}), f_*(\mathcal{S})) \le \|f\|_D \left(\frac{t^w - r^w}{\left(\frac{u}{2}\right)^w}\right).$$

But now, since $r < t < 1$, we note that $t^w - r^w \le (t-r)wR^2 = \delta(\mathcal{S}, \mathcal{F})wR^2$, hence

$$\delta(f_*(\mathcal{F}), f_*(\mathcal{S})) \le \|f\|_D \left(\frac{(t-r)wR^2}{\left(\frac{u}{2}\right)^w}\right) = \|f\|_D \left(\frac{\delta(\mathcal{F}, \mathcal{S})wR^2}{\left(\frac{u}{2}\right)^w}\right) \le \frac{\epsilon}{2}.$$

Similarly, we have $\delta(f_*(\mathcal{G}), f_*(\mathcal{S})) \le \frac{\epsilon}{2}$. Therefore, $\delta(f_*(\mathcal{F}), f_*(\mathcal{G})) \le \epsilon\ \forall \mathcal{F}, \mathcal{G} \in \Phi(D)$ such that $\delta(\mathcal{F}, \mathcal{G}) \le \min(\frac{\epsilon u^w}{2^{w+1}\|f\|_D}, \frac{u}{2})$. Consequently, f_* is uniformly continuous. □

Notation. Given $f \in \mathcal{M}(K)$, we denote by $\Pi(f)$ the set of poles of f.

Lemma 41.14. *Let $f \in \mathcal{M}(\mathbb{K})$ and suppose that $\Pi(f) \cap D = \emptyset$. Then, $f \in H(D)$.*

Proof. We can write f in the form $\frac{g}{h}$, where g, $h \in \mathcal{A}(\mathbb{K})$ have no common zeros and $h(x) \neq 0\ \forall x \in D$, hence h is invertible in $H(D)$. □

Theorem 41.15. *Let $f \in H(D)$ and let $B = f(D)$. Then, $f_*(\Phi(D)) = \Phi(B)$.*

Proof. First, let $\mathcal{F} \in \Phi(D)$. The circular filter $f_*(\mathcal{F})$ is clearly secant with B and hence belongs to $\Phi(B)$.

Now, let $\mathcal{G} \in \Phi(B)$. Let $(b_n)_{n\in\mathbb{N}}$ be a sequence of B thinner than $\mathcal{G}$ and let $(a_n)_{n\in\mathbb{N}}$ be a sequence of D such that $f(a_n) = b_n\ \forall n \in \mathbb{N}$. Since the sequence $(a_n)_{n\in\mathbb{N}}$ is bounded, by Corollary 3.20, we can extract a subsequence $(a_{\zeta(m)})_{m\in\mathbb{N}}$ which is thinner than a circular filter $\mathcal{H}$. Then, $\mathcal{H} \in \Phi(D)$. And then the sequence $(b_{\zeta(m)})_{m\in\mathbb{N}}$ is thinner than $f_*(\mathcal{H})$, hence $f_*(\mathcal{H}) = \mathcal{G}$. □

Lemma 41.16. *Let $f \in H(D)$ and $\mathcal{F}_{a,r} \in \Phi(D)$. Then $f_*(\mathcal{F}_{a,r})$ is of the form $\mathcal{F}_{b,s}$. Moreover, s belongs to $|\mathbb{K}|$ if and only if so does r. Let $\mathcal{F}_{e,u} \in \Phi^\circ(f(D))$ with $u \in |\mathbb{K}|$. There exists $\mathcal{F}_{c,t} \in \Phi^\circ(D)$ such that $f_*(\mathcal{F}_{c,t}) = \mathcal{F}_{e,u}$, with $t \in |\mathbb{K}|$.*

Proof. By Corollary 41.5, it is obvious that $f_*(\mathcal{F}_{a,r})$ is a circular filter with a center b and a diameter s. Moreover, by Corollary 41.5 again, s belongs to $|\mathbb{K}|$ if and only if so does r. Now, consider $\mathcal{F}_{e,u}$. By Lemma 41.15, there exists $\mathcal{F}_{c,t} \in \Phi(D)$ such that $f_*(\mathcal{F}_{c,t}) = \mathcal{F}_{e,u}$ and then, as we just saw, t belongs to $|\mathbb{K}|$ if and only if so does u. □

Theorem 41.17. *Let $f \in \mathcal{M}(\mathbb{K})$ (respectively $f \in \mathcal{M}(d(a, R^-))$). Then f_* is increasing from $\Phi(\mathbb{K})\setminus\Pi(f)$ to $\Phi(\mathbb{K})$ if and only if $f \in \mathcal{A}(\mathbb{K})$ (respectively $f \in \mathcal{A}(d(a, R^-))$).*

Proof. Let $f \in \mathcal{A}(\mathbb{K})$ (respectively $f \in \mathcal{A}(d(a, R^-))$). Let $\mathcal{F}$, $\mathcal{G} \in \Phi(\mathbb{K})$ be such that $\mathcal{F} \prec \mathcal{G}$. Let $s = \text{diam}(\mathcal{G})$ and $r = \text{diam}(\mathcal{F})$, hence $r < s$. Suppose first that $\mathcal{F}$ has a center a. Without loss of generality, we can assume $a = 0$. Let $f(x) = \sum_{n=0}^{\infty} a_n x^n$, and let $k = \nu(f, \log r)$, $q = \nu(f, \log s)$. Then, by Theorem 41.6, we have $f_*(\mathcal{F}) = |a_k|r^k$, $f_*(\mathcal{G}) = |a_q|r^q$, and of course, $|a_k|r^k = |f|(r)$, $|a_q|r^q = |f|(s)$, hence $|f|(r) < |f|(s)$. Let $\rho = |f|(r)$, $\sigma = |f|(s)$. So we have $f_*(\mathcal{F}) = \mathcal{F}_{0,\rho}$, $f_*(\mathcal{G}) = \mathcal{F}_{0,\sigma}$, hence $f_*(\mathcal{F}) \prec f_*(\mathcal{G})$. Now, suppose that $\mathcal{F}$ has no center. Then $\mathcal{F}$ admits a center in a spherically complete extension $\widehat{\mathbb{K}}$ of $\mathbb{K}$ and f has an expansion $\widehat{f}$ in $\widehat{\mathbb{K}}$, hence the relation $f_*(\mathcal{F}) \prec f_*(\mathcal{G})$ holds in $\widehat{\mathbb{K}}$, so it holds in $\mathbb{K}$.

Now, consider $f \in \mathcal{M}(\mathbb{K})\setminus\mathcal{A}(\mathbb{K})$ (respectively $f \in \mathcal{M}(d(a, R^-))\setminus \mathcal{A}(d(a, R^-))$) and let a be a pole of f. Let $d(a, l)$ be a disk such that f admits neither zeros nor poles different from a inside $d(a, l)$. Without loss of generality, we can assume $a = 0$. Let r, $s \in]0, l[$ be such that $r < s$. Set $\omega_0(f) = -q$. Inside $d(0, l)$, $|f|(r)$ is of the form $|a|r^{-q}$. Let $\rho = |f|(r)$, $\sigma = |f|(s)$. Then $f_*(\mathcal{F}_{0,r}) = \mathcal{F}_{0,\rho}$, $f_*(\mathcal{F}_{0,t}) = \mathcal{F}_{0,\sigma}$, hence of course $\mathcal{F}_{0,\sigma} \prec \mathcal{F}_{0,\rho}$, which shows that f_* is not increasing. □

Chapter 42

Injective Analytic Elements

In this chapter, D is an infraconnected affinoid subset of $\mathbb{K}$.

This chapter is aimed at characterizing the injective meromorphic functions in a subset that is a chained union of infraconnected affinoid subsets. The relation satisfied by such injective functions recalls the one obtained in [58] by Yvette Perrin. The equality

$$|f(x) - f(y)|^2 = |x - y|^2 |f'(x) f'(y)| \quad \forall x, y \in D$$

was shown for various kinds of injective analytic functions. In Chapter 49 of [44], it was suggested that all injective analytic elements should satisfy that by making the conjecture that an injective analytic element should be the product of a major Moebius function by a train of minor Moebius functions.

In [80], Rivera Letelier stated the above equality for all injective meromorphic functions. Here we give a complete proof of this equality by using tools of circular filters and the Shilov boundary.

Notation. Recall that we denote by U the unit disk $d(0,1)$ and by $\mathcal{K}$ the residue class field of $\mathbb{K}$. We use the hyperbolic distance $\widetilde{\delta}$ and defined on $\Phi^{\circ}(\mathbb{K})$ as follows: consider two circular filters $\mathcal{F}$, $\mathcal{G} \in \Phi^{\circ}(\mathbb{K})$, and let $\mathcal{S} = \sup(\mathcal{F}, \mathcal{G})$. We set $\widetilde{\delta}(\mathcal{F}, \mathcal{G}) = 2\log(\mathrm{diam}(\mathcal{S})) - \log(\mathrm{diam}(\mathcal{F})) - \log(\mathrm{diam}(\mathcal{G}))$.

Thus, Lemma 42.1 is immediate:

Lemma 42.1. *Let a, $b \in \mathbb{K}$, r, $s \in \mathbb{R}_{+}^{*}$. Then,*

$$\widetilde{\delta}(\mathcal{F}_{a,r}, \mathcal{F}_{b,s}) = 2\max(\log r, \log s, \log|a - b|) - \log r - \log s.$$

Lemma 42.2. *$\widetilde{\delta}$ is a distance on $\Phi^\circ(\mathbb{K})$ that defines the same topology as δ on $\Phi^\circ(\mathbb{K})$.*

Proof. We only have to show that $\widetilde{\delta}$ satisfies the triangular inequality. The proof is similar to this of Theorem 40.14, by replacing each diameter by its logarithm, but it also takes into account the concavity of the logarithm. Next, as the logarithm function is locally bicontinuous, it is easily seen that each neighborhood with respect to δ is a neighborhood with respect to $\widetilde{\delta}$ and vice versa. □

Next, we need several lemmas.

Lemma 42.3. *Let $\mathcal{F}_{a,r} \in \Phi^\circ(\mathbb{K})$ and let $f(x) = \frac{1}{x}$. Then,*
if $|a| \leq r$, then $f_*(\mathcal{F}_{a,r}) = \mathcal{F}_{0,\frac{1}{r}}$,
if $|a| > r$, then $f_*(\mathcal{F}_{a,r}) = \mathcal{F}_{\frac{1}{a},\frac{r}{|a|^2}}$.

Proof. Suppose first $|a| \leq r$. The annuli $\Gamma(0, r-\epsilon, r+\epsilon)$ belong to $\mathcal{F}_{a,r}$ and their images by f are annuli of the form $\Gamma(0, \frac{1}{r+\epsilon}, \frac{1}{r-\epsilon})$. But the unique circular filter admitting all these annuli for elements is clearly $\mathcal{F}_{0,\frac{1}{r}}$.

Suppose now that $|a| > r$. The annuli $\Gamma(a, r-\epsilon, r+\epsilon)$ belong to $\mathcal{F}_{a,r}$ and their images are annuli of the form $\Gamma(\frac{1}{a}, \frac{r-\epsilon}{|a|^2}, \frac{r+\epsilon}{|a|^2})$. But the unique circular filter admitting these annuli for elements is clearly $\mathcal{F}_{\frac{1}{a},\frac{r}{|a|^2}}$, which ends the proof. □

Lemma 42.4. *Let $f \in H(D)$ be a Moebius function. Then f_* is isometric in $\Phi^\circ(D)$ with respect to the distance $\widetilde{\delta}$.*

Proof. f is equal to a composition of particular Moebius functions of three types:

(i) $h_1(x) = x + c,\ c \in \mathbb{K}$,
(ii) $h_2(x) = \lambda x,\ \lambda \in \mathbb{K}$,
(iii) $h_3(x) = \dfrac{1}{x}$.

Consequently, it is sufficient to prove that each h_j is isometric. h_1 is obviously isometric with respect to $\widetilde{\delta}$. Let us check that so is h_2. For simplifying notation, put $g = f_2$ and $l = h_3$. Consider two circular filters $\mathcal{F}_{a,r}$, $\mathcal{F}_{b,s}$ with $r \leq s$. It is immediately seen that

$g_*(\mathcal{F}_{a,r}) = \mathcal{F}_{\lambda a,|\lambda|r}$ and $g_*(\mathcal{F}_{b,s}) = \mathcal{F}_{\lambda b,|\lambda|s}$. Then,

$$\begin{aligned}
&\widetilde{\delta}(g_*(\mathcal{F}_{a,r}), g_*(\mathcal{F}_{b,s})\\
&\quad = 2\max(\log(|\lambda|r), \log(|\lambda|s), \log(|\lambda a - \lambda b|)) - \log(|\lambda a|) - \log|\lambda s|\\
&\quad = 2\max(\log r, \log s, \log(|a-b|)) - \log r - \log s = \widetilde{\delta}(\mathcal{F}_{a,r}, \mathcal{F}_{b,s}).
\end{aligned}$$

We now consider $l = h_3$ and check that l preserves the distance in each of the following six cases:

(1) Suppose first $a \neq 0, b \neq 0, |a| \leq |b|, |a-b| = |b|, r < |a|, s < |b|$. By Lemma 42.3, we have $l_*(\mathcal{F}_{a,r}) = \mathcal{F}_{\frac{1}{a},\frac{r}{|a|^2}}$, $l_*(\mathcal{F}_{b,s}) = \mathcal{F}_{\frac{1}{b},\frac{s}{|b|^2}}$, hence

$$\begin{aligned}
\widetilde{\delta}(l_*(\mathcal{F}_{a,r}), l_*(\mathcal{F}_{b,s})) &= 2\max\left(\log\frac{r}{|a|^2}, \log\frac{s}{|b|^2}, \log\left(\left|\frac{1}{a} - \frac{1}{b}\right|\right)\right)\\
&\quad - \log\frac{r}{|a|^2} - \log\frac{s}{|b|^2}.
\end{aligned}$$

Since $\frac{1}{|a|} \geq \frac{1}{|b|}$ and $\frac{r}{|a|^2} < \frac{1}{|a|}$, $\frac{s}{|b|^2} < \frac{1}{|b|}$, we get

$$\begin{aligned}
\widetilde{\delta}(l_*(\mathcal{F}_{a,r}), l_*(\mathcal{F}_{b,s})) &= 2\log\frac{1}{|a|} - \log\frac{r}{|a|^2} - \log\frac{s}{|b|^2}\\
&= \log|a-b| - \log r - \log s = \widetilde{\delta}(\mathcal{F}_{a,r}, \mathcal{F}_{b,s}).
\end{aligned}$$

(2) Suppose $|a| = |b|$, $|a-b| < |b|$, $r < |a-b|$, $s < |a-b|$. By Lemma 42.3, we have again $l_*(\mathcal{F}_{a,r}) = \mathcal{F}_{\frac{1}{a},\frac{r}{|a|^2}}$, $l_*(\mathcal{F}_{b,s}) = \mathcal{F}_{\frac{1}{b},\frac{s}{|b|^2}}$, hence

$$\begin{aligned}
\widetilde{\delta}(l_*(\mathcal{F}_{a,r}), l_*(\mathcal{F}_{b,s})) &= 2\max\left(\log\frac{r}{|a|^2}, \log\frac{s}{|b|^2}, \log\left(\left|\frac{1}{a} - \frac{1}{b}\right|\right)\right)\\
&\quad - \log\frac{r}{|a|^2} - \log\frac{s}{|b|^2}.
\end{aligned}$$

Now, $\log(|\frac{1}{a} - \frac{1}{b}|) = \log\frac{|b-a|}{|a|^2} > \max(\log\frac{r}{|a|^2}, \log\frac{s}{|b|^2})$, so

$$\begin{aligned}
\widetilde{\delta}(l_*(\mathcal{F}_{a,r}), l_*(\mathcal{F}_{b,s})) &= 2\log\left(\frac{|b-a|}{|a|^2}\right) - \log\frac{r}{|a|^2} - \log\frac{s}{|b|^2}\\
&= 2\log|b-a| - \log r - \log s = \widetilde{\delta}(\mathcal{F}_{a,r}, \mathcal{F}_{b,s}).
\end{aligned}$$

(3) Suppose $a = b \neq 0$, $r < s < |b|$. Then, $l_*(\mathcal{F}_{a,r}) = \mathcal{F}_{\frac{1}{a},\frac{r}{|a|^2}}$, $l_*(\mathcal{F}_{a,s}) = \mathcal{F}_{\frac{1}{a},\frac{s}{|a|^2}}$, hence

$$\begin{aligned}\widetilde{\delta}(l_*(\mathcal{F}_{a,r}), l_*(\mathcal{F}_{a,s})) &= 2\max\left(\log\frac{r}{|a|^2}, \log\frac{s}{|a|^2}\right) - \log\frac{r}{|a|^2} - \log\frac{s}{|a|^2}\\ &= \log r - \log s = \widetilde{\delta}(\mathcal{F}_{a,r}, \mathcal{F}_{b,s}).\end{aligned}$$

(4) Suppose that $a = 0, b \neq 0$, $r < |b|, s < |b|$. Then, $l_*(\mathcal{F}_{0,r}) = \mathcal{F}_{0,\frac{1}{r}}$, $l_*(\mathcal{F}_{b,s}) = \mathcal{F}_{\frac{1}{b},\frac{s}{|b|^2}}$, hence

$$\begin{aligned}&\widetilde{\delta}(l_*(\mathcal{F}_{a,r}), l_*(\mathcal{F}_{b,s}))\\ &= 2\max\left(\log\frac{1}{r}, \log\frac{s}{|b|^2}, \log\left(\left|\frac{1}{b}\right|\right)\right) - \log\frac{1}{r} - \log\frac{s}{|b|^2}\\ &= 2\log\frac{1}{r} - \log\frac{1}{r} - \log\frac{s}{|b|^2} = 2\log|b| - \log r - \log s\\ &= \widetilde{\delta}(\mathcal{F}_{0,r}, \mathcal{F}_{b,s}).\end{aligned}$$

(5) Suppose finally $a = b = 0$, $r < s$. Then, $l_*(\mathcal{F}_{0,r}) = \mathcal{F}_{0,\frac{1}{r}}$, $l_*(\mathcal{F}_{0,s}) = \mathcal{F}_{0,\frac{1}{s}}$, so

$$\begin{aligned}\widetilde{\delta}(l_*(\mathcal{F}_{0,r}), l_*(\mathcal{F}_{0,s})) &= \widetilde{\delta}(\mathcal{F}_{0,\frac{1}{r}}, \mathcal{F}_{0,\frac{1}{s}})\\ &= 2\max\left(\log\frac{1}{r}, \log\frac{1}{s}\right) - \log\frac{1}{r} - \log\frac{1}{s}\\ &= \log s - \log r = \widetilde{\delta}(\mathcal{F}_{0,r}, \mathcal{F}_{0,s}).\end{aligned}$$

(6) Suppose now that $b = 0$, $r < |a|$, $s \geq |a|$. As in the previous case, we have $l_*(\mathcal{F}_{0,s}) = \mathcal{F}_{0,\frac{1}{s}}$, $l_*(\mathcal{F}_{a,r}) = \mathcal{F}_{\frac{1}{|a|},\frac{r}{|a|^2}}$, hence

$$\begin{aligned}&\widetilde{\delta}(l_*(\mathcal{F}_{a,r}), l_*(\mathcal{F}_{b,s}))\\ &= 2\max\left(\log\frac{r}{|a|^2}, \log\left(\left|\frac{1}{|a|}\right|\right)\right) - \log\frac{1}{s} - \log\frac{r}{|a|^2}\\ &= -2\log|a| + \log s - \log r + 2\log|a| = \log s - \log r\\ &= \widetilde{\delta}(\mathcal{F}_{0,r}, \mathcal{F}_{b,s}).\end{aligned}$$

Now, we check that one of those six cases occurs:

(A) Suppose that $\mathcal{F}_{a,r} \prec \mathcal{F}_{b,s}$. Then we can assume $a = b$. If $a = b = 0$, we are in case (5). If 0 is not a center of both filters, we are in case (3). If $a = 0$ but b is not a center of $\mathcal{F}_{b,s}$, we are in case (6). There is no other possibility when $\mathcal{F}_{a,r} \prec \mathcal{G}_{b,s}$.

(B) Suppose now that $\mathcal{F}_{a,r}$ and $\mathcal{G}_{b,s}$ are not comparable for the relation $\preceq$. If $a = 0$ and $b \neq 0$, we have $r < |b|$ and $s < |b|$, so we are in case (4). Suppose now $a \neq 0,\ b \neq 0$. Then we are in case (1) or in case (2). That finishes the proof of the lemma. □

In order to prove Lemma 42.7, we have to recall a classical lemma in algebra:

Lemma 42.5. *Let F be an algebraically closed field and let f, g be rational functions such that $g \circ f$ is the identity on F. Then both f and g are Moebius functions.*

We also need Lemma 42.6:

Lemma 42.6. *Let $h(x) = \frac{ax+b}{cx+d}$ with $a,\ b,\ c,\ d \in U$ and $|ad-bc| = 1$ and let $\mathcal{G} = \mathcal{F}_{0,1}$. Then, $h_*(\mathcal{G}) = \mathcal{G}$.*

Proof. Let $\overline{h}$ be the residue class of h in $\mathcal{K}(u)$. The Moebius function $\overline{h}$ is injective and maps all elements of $\mathcal{K}$ to $\mathcal{K}$ except at most one. Consider now a $\mathcal{G}$-affinoid D. Since $\mathcal{G} = \mathcal{F}_{0,1}$, all classes of U are included in D except at most finitely many, and therefore, D is sent by h onto a set containing all classes of U except a finite number of them. Therefore, $h_*(\mathcal{G})$ is less thin than the filter $\mathcal{T}$ generated by the sets consisting of all classes of U except finitely many. And clearly, $\mathcal{G}$ is the unique circular filter thinner than $\mathcal{T}$. Hence, $h_*(\mathcal{G}) = \mathcal{G}$. □

Lemma 42.7. *Suppose $\widetilde{D} = U$. Let $f \in H(D)$ be such that $f_*(\mathcal{F}_{0,1}) = \mathcal{F}_{0,1}$. Then $|f(x)| = 1$ holds in all classes of U except maybe finitely many and there exists $h \in R(D)$ such that*

(i) $\|f - h\|_D < 1$,

(ii) $h = \dfrac{P}{Q}$ *with* $P,\ Q \in \mathbb{K}[x],\ \|P\| = 1,\ \|Q\| = 1$,

(iii) $h_*(\mathcal{F}_{0,1}) = \mathcal{F}_{0,1}$.

Moreover, given any function $h \in R(D)$ such that $\|f - h\|_D < 1$, then h satisfies (ii) *and* (iii).

Proof. We know that $|f(x)| = |f|(1)$ holds in all classes of U except maybe a finite number of them. But since $f_*(\mathcal{F}_{0,1}) = \mathcal{F}_{0,1}$, we have $|f|(1) = 1$. Consider now $h \in \mathbb{K}(x)$ satisfying $\|f - h\|_D < 1$. By Theorem 41.6, we have $h_*(\mathcal{F}_{0,1}) = f_*(\mathcal{F}_{0,1}) = \mathcal{F}_{0,1}$. Then we can write h in the form $\frac{P}{Q}$ with with $P,\ Q \in \mathbb{K}[x]$, $\|P\| = 1$, $\|Q\| = 1$. □

Notation. Let $\mathcal{F}_{0,1} \in \Phi(D)$ and let $f \in H(D)$ be such that $f_*(\mathcal{F}_{0,1}) = \mathcal{F}_{0,1}$. Let $h \in R(D)$ satisfy $\|f - h\|_D < 1$.

Lemma 42.8. *Suppose $\widetilde{D} = U$ and let $\mathcal{F}_{0,1} \in \Phi(D)$, let $f \in H(D)$ be bianalytic and assume that $f_*(\mathcal{F}_{0,1}) = \mathcal{F}_{0,1}$. There exists a neighborhood V of $\mathcal{F}_{0,1}$ with respect to the metric $\widetilde{\delta}$ such that the restriction of f_* to V is isometric. Moreover, the image $\overline{f}$ of f in the residue class field $\mathcal{K}(t)$ is a Moebius function.*

Proof. Let us set again $\mathcal{G} = \mathcal{F}_{0,1}$. Since f is bianalytic, the inverse function $g = \overset{-1}{f}$ belongs to $H(f(D))$. We define a neighborhood V of $\mathcal{F}_{0,1}$ such that f_* is isometric in V with respect to the distance $\widetilde{\delta}$. Consider the image $\overline{f}$ of f in the residue class field $\mathcal{K}(u)$. Since $\varphi_{0,1}(f) = 1$, $|f(x)|$ is equal to 1 in all classes of U except maybe finitely many. Let B be the union of the classes of U included in D. Then B is of the form $U\setminus\Big(\cup_{j=1}^{q} d(a_i, 1^-)\Big)$ with $|a_j| \leq 1$, $|a_i - a_j| = 1\ \forall i,\ j \leq q$. By definition, f is injective in B. Since $f_*(\mathcal{G}) = \mathcal{G}$, for each class Λ of U included in D, by Corollary 36.11, $f(\Lambda)$ is a class of U, and hence, $\overline{f}$ takes different values in different classes of U. Now, similarly, the function g maps all classes of U but finitely many onto classes of U. Moreover, when f takes values in a class Λ of $C(0,1)$, then g is defined in Λ. So, similarly, $\overline{g}$ is not a constant and satisfies $\overline{g} \circ \overline{f}(u) = u$ for all elements $u \in \mathcal{K}$ but finitely many. Consequently, $\overline{g} \circ \overline{f}$ is the identity in $\mathcal{K}(u)$, and hence by Lemma 42.5, $\overline{f}$ is a Moebius function. Then we can find a Moebius function $h(x) = \frac{ax+b}{cx+d}$ such that $\overline{h} = \overline{f}$, which implies $a,\ b,\ c,\ d \in U$ and $|ad - bc| = 1$.

Therefore, $f(x)$ is of the form $h(x) + \tau(x)$ with $\tau \in H(D)$ satisfying $\varphi_{0,1}(\tau) < 1$. Then, by Lemma 42.6, we have $h_*(\mathcal{G}) = \mathcal{G}$. Now, since $\varphi_{0,1}(\tau) < 1$, by Theorem 41.6, we have $h_*(\mathcal{G}) = f_*(\mathcal{G}) = \mathcal{G}$. Moreover, by Theorem 41.12, both functions h_* and f_* are continuous with respect to the δ-topology, so there exists a neighborhood

V of $\mathcal{G}$ in $\Phi^\circ(D)$) such that $\operatorname{diam}(\tau_*(\mathcal{H})) < \operatorname{diam}(h_*(\mathcal{H}))\ \forall \mathcal{H} \in V$, therefore $(h+\tau)_*$ and h_* give the same image of $\mathcal{H}$ for every $\mathcal{H} \in V$. But by Lemma 42.4, h_* is isometric, hence f_* is isometric in V. $\square$

Corollary 42.9. *Let $f \in H(D)$ be bianalytic and let $\mathcal{F}_{a,r} \in \Phi^\circ(D)$ with $r \in |\mathbb{K}|$. There exists a neighborhood V of $\mathcal{F}_{a,r}$ with respect to the metric $\widetilde{\delta}$ such that the restriction of f_* to V is isometric.*

Proof. Since $r \in |\mathbb{K}|$, without loss of generality, we can suppose that $\mathcal{F}_{a,r} = \mathcal{F}_{0,1}$ and consider the restriction of f to $D \cap U$. Then we can apply Lemma 42.8 to this restriction. $\square$

Lemma 42.10. *Let I and J be closed bounded intervals in $\mathbb{R}_+^*$ with a homeomorphism ϕ from I onto J and assume that for each $a \in I$, there exists a neighborhood V of a such that $|\log a - \log b|_\infty = |\log(\phi(a)) - \log(\phi(b))|_\infty\ \forall a,\ b \in V$. Then, $|\log a - \log b|_\infty = |\log(\phi(a)) - \log(\phi(b))|_\infty\ \forall a,\ b \in I$.*

Proof. By elementary properties of real functions in an interval, since ϕ is a homeomorphism from I onto J, it is strictly monotonous. Without loss of generality, we can assume that it is increasing. Now, let $a, b \in I$ with $a < b$. By compacity, we can find finitely many open intervals V_k, $k = 1, \dots, q$ such that $[a,b] = \bigcup_{k=1}^q V_k$ and $|\log x - \log y|_\infty = |\log(\phi(x) - \log(\phi(y)|_\infty\ \forall x,\ y \in V_k,\ \forall k = 1, \dots, q-1$. Set $V_k =]a_k, a_{k+1}[$. Without loss of generality, we can assume that $a < a_k < a_{k+1} < a_q < b$. Then,

$$\begin{aligned}
&|\log(\phi(a)) - \log(\phi(b))|_\infty = \log(\phi(b)) - \log(\phi(a)) \\
&\quad = \log(\phi(a_1)) - \log(\phi(a)) + \sum_{k=1}^{q-1} \log(\phi(a_{k+1})) - \log(\phi(a_k)) \\
&\qquad + \log(\phi(b)) - \log(\phi(a_q)) \\
&\quad = \log(a_1) - \log(a) + \sum_{k=1}^{q-1} \log(a_{k+1}) - \log(a_k) + \log(b) - \log(a_q) \\
&\quad = \log(b) - \log(a),
\end{aligned}$$

which ends the proof. $\square$

Definition. Let $\mathcal{F}_{a,r}$, $\mathcal{F}_{b,s}$ be two circular filters on $\mathbb{K}$. If $\mathcal{F}_{a,r} \prec \mathcal{F}_{b,s}$, we call *shortest path from* $\mathcal{F}_{a,r}$ *to* $\mathcal{F}_{b,s}$ the path γ defined by the following function γ from $[r, s]$ to $\Phi^\circ(\mathbb{K})$: $\gamma(u) = \mathcal{F}_{a,u}$.

Now, if $\mathcal{F}_{a,r}$ and $\mathcal{F}_{b,s}$ are not comparable, we set $\mathcal{S} = \sup(\mathcal{F}_{a,r}, \mathcal{F}_{b,s})$, with $R = \operatorname{diam}(\mathcal{S})$, and we call *shortest path from* $\mathcal{F}_{a,r}$ *to* $\mathcal{F}_{b,s}$ the path γ defined by the following injective function from $[r, 2R - s]$ to $\Phi^\circ(\mathbb{K})$: $\gamma(u) = \mathcal{F}_{a,u}$, $\forall u \in [r, R]$ and $\gamma(u) = \mathcal{F}_{b,2R-u}$, $\forall u \in [R, 2R - s]$.

We prove Proposition 42.12 in two steps. Lemma 42.11 is immediate and just comes from the definition of $\widetilde{\delta}$ and lets us define the distance $\widetilde{d}$ on intervals in $\mathbb{R}$:

Lemma 42.11. *Let $\mathcal{F}_{a,r}$, $\mathcal{F}_{a,s}$ be two circular filters in $\mathbb{K}$ such that $\mathcal{F}_{a,r} \prec \mathcal{F}_{a,s}$. Let γ be the shortest path defined in the interval $I = [r, s]$ with values in $\Phi^\circ(\mathbb{K})$, as $\gamma(u) = \mathcal{F}_{a,u}$ $\forall u \in [r, s]$. On I, let $\widetilde{d}$ be defined by $\widetilde{d}(t, u) = |\log u - \log t|_\infty$. Then $\widetilde{d}$ is a distance on I and γ is an isometric homeomorphism from I to $\gamma(I)$ when I is equipped with the distance $\widetilde{d}$ and $\gamma(I)$ is equipped with the distance $\widetilde{\delta}$.*

Lemma 42.11 is generalized in Proposition 42.12:

Proposition 42.12. *Let $\mathcal{F}_{a,r}$, $\mathcal{F}_{b,s}$ be two circular filters in $\mathbb{K}$ and let $\mathcal{F}_{a,R} = \mathcal{F}_{b,R} = \sup(\mathcal{F}_{a,r}, \mathcal{F}_{b,s})$. Let γ be the shortest path from $\mathcal{F}_{a,r}$ to $\mathcal{F}_{b,s}$ defined in the interval $I = [r, 2R - s]$. Then γ is an isometric homeomorphism from I to $\gamma(I)$ when I is equipped with the distance $\widetilde{d}$ and $\gamma(I)$ is equipped with the distance $\widetilde{\delta}$.*

Proof. If $\mathcal{F}_{a,r} \prec \mathcal{F}_{b,s}$ or if $\mathcal{F}_{b,s} \prec \mathcal{F}_{a,r}$, then the claim is given by Lemma 42.11. Now, assume that $\mathcal{F}_{a,r}$ and $\mathcal{F}_{b,s}$ are not comparable. Then the restriction of γ to both sections $[r, R]$, $[R, 2R - s]$ is isometric. So, we only have to check that given $u \in [r, R]$ and $w \in [R, 2R - s]$, we have $\widetilde{\delta}(\mathcal{F}_{a,u}, \mathcal{F}_{b,w}) = \widetilde{d}(u, w)$. But actually we check that

$$\begin{aligned}\widetilde{d}(u, w) &= |\log(u) - \log(w)|_\infty \\ &= |(\log(R) - \log(u)) + (\log(w) - \log(R))|_\infty \\ &= |\log(R) - \log(u)|_\infty + |\log(w) - \log(R)|_\infty \\ &= \widetilde{d}(u, R) + \widetilde{d}(R, w),\end{aligned}$$

and similarly, $\widetilde{\delta}(\mathcal{F}_{a,u}, \mathcal{F}_{b,w}) = \widetilde{\delta}(\mathcal{F}_{a,u}, \mathcal{F}_{a,R}) + \widetilde{\delta}(\mathcal{F}_{b,R}, \mathcal{F}_{b,w})$, which ends the proof. □

Theorem 42.13. *Let $f \in H(D)$ be injective. Then f_* is injective on $\Phi(D)$.*

Proof. Consider first two circular filters with centers, secant with D: $\mathcal{F}_{a,r}$, $\mathcal{F}_{b,s}$, with r, $s \in |\mathbb{K}|$ and assume that $f_*(\mathcal{F}_{a,r}) = f_*(\mathcal{F}_{b,s})$. We prove that $\mathcal{F}_{a,r} = \mathcal{F}_{b,s}$. By Theorem 41.6, we can assume that f belongs to $R(D)$ and we know that $f_*(\mathcal{F}_{a,r})$ is a circular filter $\mathcal{F}_{c,t}$ such that t lies in $|\mathbb{K}|$. Let $E = f(C(a,r) \cap D)$. Since D is affinoid, $C(a,r) \cap D$ contains all classes of $C(a,r)$ except maybe finitely many. Let us show that $f(D \cap C(a,r))$ contains all classes of $C(c,t)$ except maybe finitely many. Without loss of generality, we can assume that $a = c = 0$ and that $r = t = 1$. Thus, we have $f_*(\mathcal{F}_{0,1}) = \mathcal{F}_{0,1}$. Then by Lemma 42.8, the residue class $\overline{f}$ of f in the residue field $\mathcal{K}(X)$ is a Moebius function. Let $u_1, \ldots, u_m$ be the classes of $d(0,1)$ that are not included in D. Then, since $\overline{f}$ is a Moebius function, $\overline{f}(\mathcal{K} \setminus \{u_1, \ldots, u_m\})$ is a set of the form $(\mathcal{K} \setminus \{w_1, \ldots, w_m\})$ with $w_1, \ldots, w_m$ classes of $C(0,1)$, hence $f(D \cap C(0,1))$ contains all classes of $C(0,1)$ except finitely many, what we wanted to show.

Thus, we have proven that $f(C(a,r) \cap D)$ contains all classes of $C(c,t)$ except maybe finitely many. But then, similarly, all classes of $C(c,t)$ but finitely many are included in $f(D \cap C(b,s))$. Consequently, there exists a class of $C(c,t)$ included in both $f(D \cap C(a,r))$ and $f(D \cap C(b,s))$, a contradiction to the hypothesis of injectivity on f, except if $\mathcal{F}_{a,r} = \mathcal{F}_{b,s}$.

Consider now general case and let $\mathcal{G}$, $\mathcal{H} \in \Phi(D)$ be such that $f_*(\mathcal{G}) = f_*(\mathcal{H})$, assuming that $\mathcal{G}$, for instance, has no center. Then, by Corollary 41.4, $f_*(\mathcal{G})$ also has no center and neither has $\mathcal{H}$. Since $\mathcal{G}$ and $\mathcal{H}$ are two distinct circular filters with no center, there exist disks $d(a,r)$ and $d(b,s)$ such that $d(a,r) \cap d(b,s) = \emptyset$ and such that $\mathcal{G}$ is secant with $d(a,r)$ and $\mathcal{H}$ is secant with $d(b,s)$. Let $(D_n)_{n \in \mathbb{N}}$ be a basis of $\mathcal{G}$ consisting in a decreasing sequence of disks included in $d(a,r)$ and let $(E_n)_{n \in \mathbb{N}}$ be a basis of $\mathcal{H}$ consisting in a decreasing sequence of disks included in $d(b,s)$. Since f is injective, we have $f(D_m) \cap f(E_n) = \emptyset \ \forall m,\ n \in \mathbb{N}$, and hence $f_*(\mathcal{G}) \neq f_*(\mathcal{H})$, which ends the proof. □

Theorem 42.14. *Let $f \in H(D)$ be bianalytic. Then the function f_* is isometric with respect to the distance $\widetilde{\delta}$.*

Proof. Since the set of circular filters with a center is dense in the whole set $\Phi^\circ(D)$ with respect to the δ-topology and therefore with respect to the topology defined by $\widetilde{\delta}$, it is sufficient to prove that for any two circular filters $\mathcal{F}_{a,r}, \mathcal{F}_{b,s} \in \Phi^\circ(D)$, we have $\widetilde{\delta}(f_*(\mathcal{F}_{a,r}), f_*(\mathcal{F}_{b,s})) = \widetilde{\delta}(\mathcal{F}_{a,r}, \mathcal{F}_{b,s})$.

Let $\mathcal{F}_{a,r}$, $\mathcal{F}_{b,s} \in \Phi^\circ(D)$. Let $\mathcal{F}_{a,R} = \mathcal{F}_{b,R} = \sup(\mathcal{F}_{a,r}, \mathcal{F}_{b,s})$. Let $\mathcal{F}_{\alpha,\rho} = f_*(\mathcal{F}_{a,r})$ and $\mathcal{F}_{\beta,\upsilon} = f_*(\mathcal{F}_{b,s})$ and let $\mathcal{F}_{\alpha,\upsilon} = \sup(\mathcal{F}_{\alpha,\rho}, \mathcal{F}_{\beta,\tau})$.

Consider the shortest path γ defined in the interval $I = [r, 2R-s]$, with values in $\Phi^\circ(D)$. Similarly, consider the shortest path σ defined in the interval $J = [\rho, 2\upsilon - \tau]$, with values in $\Phi^\circ(f(D))$ such that $\sigma(\rho) = \mathcal{F}_{\alpha,\rho}$, $\sigma(2\upsilon - \tau) = \mathcal{F}_{\beta,\tau}$.

By Theorem 42.13, f_* is injective, hence the mapping $h = \sigma^{-1} \circ f_* \circ \gamma$ is injective and hence is a bijection from I onto J. Moreover, since both σ, ϕ are homeomorphism, h is continuous. Consequently, it is a continuous bijection from I onto J with respect to the metric $\widetilde{d}$. But the topology defined by the metric $\widetilde{d}$ on a bounded interval of $\mathbb{R}$ is clearly the same as the usual topology. Consequently, h is a continuous bijection from I onto J with respect to the usual topology of $\mathbb{R}$ on both I and J. Therefore, h, as an injective continuous function, is a strictly monotonous function. Without loss of generality, we can assume that h is strictly increasing.

Now, by Corollary 42.9, for every $u \in I \cap |\mathbb{K}|$, there exists a neighborhood $\mathcal{V}(u)$ such that h is isometric in $\mathcal{V}(u)$ with respect to $\widetilde{d}$. This means that in such an interval $\mathcal{V}(u)$, we have $|\log(h(y)) - \log(h(x))|_\infty = \log y - \log x \ \forall x,\ y \in \mathcal{V}(u),\ x < y$.

But since $|\mathbb{K}|$ is dense in $\mathbb{R}_+$, by compacity, we can find finitely many intervals $[a_j, a_{j+1}]$, with $a_j \in |\mathbb{K}|$, $j = 0, \dots, q-1$, such that $I = [a_0, a_q]$ and that h is isometric in each interval $]a_j, a_{j+1}[$. Actually, by continuity, h is then isometric in each interval $[a_j, a_{j+1}]$. So, in each interval $[a_j, a_{j+1}]$, we have $\frac{h(y)}{h(x)} = \frac{y}{x}$, and then, by continuity, we can check that the equality $\frac{h(y)}{h(x)} = \frac{y}{x}$ holds in all I. For instance, let $a \in [a_k, a_{k+1}]$ and let $b \in [a_{k+1}, a_{k+2}]$. Then,

$$\frac{h(b)}{h(a)} = \Big(\frac{h(b)}{h(a_{k+1})}\Big)\Big(\frac{h(a_{k+1})}{h(a)}\Big) = \Big(\frac{b}{a_k+1}\Big)\Big(\frac{a_{k+1}}{a}\Big) = \frac{b}{a}.$$

So, by induction, we can extend the relation to the whole interval I. And similarly, if h is decreasing, then we have

$$\frac{h(b)}{h(a)} = \frac{a}{b} \quad \forall a,\ b \in I,\ a < b.$$

Consequently, h is isometric.

Now, by Proposition 42.12, both γ, σ are isometric homeomorphisms from I (respectively J) to their respective images, when I (respectively J) is provided with $\widetilde{d}$ and $\Phi^\circ(D)$ is provided with $\widetilde{\delta}$. Consequently, f_* is isometric with respect to $\widetilde{\delta}$. □

Notation. We denote by $\mathcal{P}$ the peripheral of D.

Theorem 42.15. *Suppose $\widetilde{D} = U$ and let $f \in H(D)$ be bianalytic and such that $\widetilde{f(D)} = U$ and $f_*(\mathcal{P}) = \mathcal{P}$. Then f_* preserves the order $\preceq$ and the distance δ on circular filters of $\Phi(D)$. Moreover, given $\mathcal{G}, \mathcal{G}' \in \Phi(D)$, f_* satisfies $f_*(\sup(\mathcal{G}, \mathcal{G}')) = \sup(f_*(\mathcal{G}), f_*(\mathcal{G}'))$.*

Proof. Since the set of centered circular filters is dense inside $\Phi(D)$, it is sufficient to prove the statement when both $\mathcal{G}, \mathcal{G}'$ are centered and large. So we assume $\mathcal{G} = \mathcal{F}_{a,r}$, $\mathcal{G}' = \mathcal{F}_{a',r'}$ with $rr' > 0$.

Let us first show that f_* preserves the order and hence take $a' = a$, $r < r'$. Set $f_*(\mathcal{F}_{a,r}) = \mathcal{F}_{b,s}$, $f_*(\mathcal{F}_{a,r'}) = \mathcal{F}_{b',s'}$. Since $f_*(\mathcal{P}) = \mathcal{P}$ and since $\mathcal{P}$ also is the peripheral of $f(D)$, by Theorem 42.14, we have $\widetilde{\delta}(\mathcal{F}_{a,r}, \mathcal{P}) = -\log r$, $\widetilde{\delta}(\mathcal{F}_{b,s}, \mathcal{P}) = -\log s$, hence $r = s$, and similarly, $r' = s'$. Consequently, $\widetilde{\delta}(\mathcal{F}_{b,s}, \mathcal{F}_{b',r'}) = 2\max(r, r', \log|b'-b|) - r - r'$. But by Theorem 42.14, $\widetilde{\delta}(\mathcal{F}_{b,r}, \mathcal{F}_{b',r'}) = \widetilde{\delta}(\mathcal{F}_{a,r}, \mathcal{F}_{a,r'}) = \log r' - \log r$. Thus, $\log(|b'-b|) \le r'$, hence $\mathcal{F}_{b',r'} = \mathcal{F}_{b,r'}$, and therefore, $f_*(\mathcal{F}_{a,r}) \prec f_*(\mathcal{F}_{a,r'})$, i.e. f_* preserves the order.

Consequently, we can derive that f_* preserves the distance δ on the set $\{\mathcal{F}_{a,\rho} \mid \rho \ge r\}$. Indeed, we have $\delta(\mathcal{F}_{a,\rho}, \mathcal{F}_{a,r}) = \rho - r$, but we have seen that $f_*(\mathcal{F}_{a,\rho})$, $f_*(\mathcal{F}_{a,r})$ are of the form $\mathcal{F}_{b,\rho}, \mathcal{F}_{b,r}$ respectively, hence $\delta(f(\mathcal{F}_{a,\rho}), f(\mathcal{F}_{a,r})) = \rho - r = \delta(\mathcal{F}_{a,\rho}, \mathcal{F}_{a,r})$. Particularly, since f_* preserves $\mathcal{P}$, it is obvious that f_* also preserves the diameters of circular filters.

Now, we prove that f_* preserves the supremum. We consider $\mathcal{F}_{a,r}$, $\mathcal{F}_{a',r'}$ and set $\mathcal{M} = \sup(\mathcal{F}_{a,r}, \mathcal{F}_{a',r'})$ and $\mathcal{N} = \sup(f_*(\mathcal{F}_{a',r}), f_*(\mathcal{F}_{a,r'}))$. The statement is trivial when $a = a'$. So, we can assume that the two filters are not comparable. Therefore, $\text{diam}(\mathcal{M}) = |a - a'| > \max(r, r')$ and $\widetilde{\delta}(\mathcal{F}_{a,r}, \mathcal{F}_{a,r'}) = 2\log|a - a'| - \log r - \log r'$. Moreover, since f_* preserves the diameter, we have $\text{diam}(f_*(\mathcal{M})) = \text{diam}(\mathcal{M}) = |a - a'|$. On the other hand, we have $f_*(\mathcal{F}_{a,r}) = \mathcal{F}_{b,r}$, $f_*(\mathcal{F}_{a',r'}) = \mathcal{F}_{b',r'}$. Since f_* preserves the order, we have $\mathcal{F}_{b,r} \prec f_*(\mathcal{M})$, $\mathcal{F}_{b,r'} \prec f_*(\mathcal{M})$, hence $\mathcal{N} \preceq f_*(\mathcal{M})$. On the other hand, since f_* preserves the distance $\widetilde{\delta}$, we have

$$\begin{aligned}&2\max(\log r, \log r', \log|a - a'|) - \log r - \log r'\\&\quad = 2\max(\log r, \log r', \log|b - b'|) - \log r - \log r',\end{aligned}$$

hence $\log|a - a'| = \max(\log r, \log r', \log|b - b'|)$, and hence $|b - b'| = |a - a'|$. Consequently, $\text{diam}(\mathcal{N}) = |a - a'| = \text{diam}(f_*(\mathcal{M}))$, which proves the equality, and therefore, $f_*(\mathcal{M}) = \mathcal{N}$. Consequently, $f_*(\sup(\mathcal{G}, \mathcal{G}')) = \sup(f_*(\mathcal{G}), f_*(\mathcal{G}'))$. Moreover, since f_* preserves the distance δ on a totally ordered subset of $\Phi(D)$ and since $f_*(\mathcal{P}) = \mathcal{P}$, we can check that $\delta(\mathcal{M}, \mathcal{P}) = \delta(\mathcal{N}, \mathcal{P})$, and hence $\text{diam}(\mathcal{N}) = \text{diam}(\mathcal{M})$.

Now, let us show that f_* preserves the distance δ in general. Let $R = \text{diam}(\mathcal{M}) = \text{diam}(\mathcal{N})$. We have $\sup(f_*(\mathcal{F}_{a,r}), f_*(\mathcal{F}_{a',r'})) = \mathcal{F}_{b,R} = \mathcal{F}_{b',R}$. Then, $f_*(\mathcal{F}_{a,r}) = \mathcal{F}_{b,r}$, $f_*(\mathcal{F}_{a',r'}) = \mathcal{F}_{b',r'}$ and $f_*(\mathcal{F}_{a,R}) = \mathcal{F}_{b,R} = \mathcal{F}_{b',R}$. Consequently, $\delta(\mathcal{F}_{a,r}, \mathcal{F}_{a',r'}) = 2R - r - r'$, and hence $\delta(f_*(\mathcal{F}_{a,r}), f_*(\mathcal{F}_{a',r'}) = \delta(\mathcal{F}_{b,r}, \mathcal{F}_{b',r'}) = 2R - r - r'$, which ends the proof. □

Theorem 42.16. *Let $f \in H(D)$ be bianalytic. Then, $f_*(\Sigma(D)) \subset \Sigma_0(f(D))$.*

Proof. Let $\mathcal{P}$ be the peripheral of D, let $R = \text{diam}(\mathcal{P})$, and let $E = f(D)$. Then $f_*(\mathcal{P})$ is a circular filter $\mathcal{F}_{a,r}$. Suppose $\mathcal{F}_{a,r}$ does not belong to $\Sigma_0(E)$. By definition of $\Sigma_0(E)$, there exists no annulus $\Gamma(a, r', r)$ with empty intersection with E, and similarly, there is no annulus $\Gamma(a, r, r'')$ with empty intersection with E. Now, since D is infraconnected, by Theorem 21.12, so is E. Therefore, by Proposition 3.17, we can find circular filters $\mathcal{F}_{a,r'}$, $\mathcal{F}_{a,r''}$ secant with E, with $r' < r < r''$ and r', $r'' \in |\mathbb{K}|$. Consequently, by Theorem 41.6 and Lemma 41.16, there exist circular filters $\mathcal{F}_{\alpha',\rho'}$, $\mathcal{F}_{\alpha'',\rho''}$ secant with D such that $f_*(\mathcal{F}_{\alpha',\rho'}) = \mathcal{F}_{a,r'}$ and $f_*(\mathcal{F}_{\alpha'',\rho''}) = \mathcal{F}_{a,r''}$ (with ρ', $\rho'' \in |\mathbb{K}|$). Next, since f is injective, by Theorem 42.13, f_* is injective on the set of circular filters $\mathcal{F}_{\alpha,\rho} \in \Phi(D)$, hence both $\mathcal{F}_{\alpha',\rho'}$, $\mathcal{F}_{\alpha'',\rho''}$ are different from $\mathcal{P}$, therefore, $\rho' < R$, $\rho'' < R$. Let $R' = \max(\rho', \rho'')$. Since D is infraconnected, there exists a path γ' from $[0, 1]$ to $\Phi(D)$ such that $\gamma(0) = \mathcal{F}_{\alpha',\rho'}$, $\gamma(1) = \mathcal{P}$, and similarly, there exists a path γ'' from $[1, 2]$ to $\Phi(D)$ such that $\gamma(1) = \mathcal{P}$, $\gamma(1) = \mathcal{F}_{\alpha'',\rho''}$. Thus, we can derive a path γ from $[0, 2]$ to $\Phi(D)$ such that $\gamma(0) = \mathcal{F}_{\alpha',\rho'}, \gamma(2) = \mathcal{F}_{\alpha'',\rho''}$. Then $f_* \circ \gamma$ defines a path from $[0, 2]$ to $\Phi(E)$ such that $f_* \circ \gamma(0) = \mathcal{F}_{a,r'}$ and $f_* \circ \gamma(2) = \mathcal{F}_{a,r''}$. Let $s \in]R', R]$.

The function in t: $\operatorname{diam}(f_* \circ \gamma(t))$ is obviously continuous in t, and therefore, by intermediate values theorem, there exists $\tau' \in]0, 1[$ and $\tau'' \in]1, 2[$ such that $\operatorname{diam}(f_* \circ \gamma(\tau')) = \operatorname{diam}(f_* \circ \gamma(\tau''))$, a contradiction to the injectivity of f_*. This proves that one of the two annuli $\Gamma(a, r', r)$ and $\Gamma(a, r, r'')$ has an empty intersection with E, and therefore, $\mathcal{F}_{a,r}$ belongs to $\Sigma_0(E)$.

By inversion, we can generalize this proof to any other circular filter of $\Sigma(D)$. Consider a circular filter $\mathcal{F} \in \Sigma(D)$ that is not $\mathcal{P}$. Since D is infraconnected affinoid, by Corollary 40.25, it is the peripheral of a hole $T = d(b, s^-)$ of D. Without loss of generality, we can assume that $b = 0$. Let $u(x) = \frac{1}{x}$ and let $B = u(D)$. Obviously, $u_*(\mathcal{F})$ is the peripheral $\mathcal{P}'$ of B and $\Sigma(f(D)) = \Sigma(f \circ u(B))$. So, $(f \circ u)_*(\mathcal{P}')$ belongs to $\Sigma_0(f \circ u(B))$. □

Remark. Actually, we know that given an infraconnected affinoid set D, $H(D)$ is a Krasner–Tate algebra [37], so for every $f \in H(D)$, $f(D)$ also is infraconnected affinoid, hence $\Sigma(f(D)) = \Sigma_0(f(D))$. On the other hand, if the field $\mathbb{K}$ has characteristic zero, an injective analytic element of the Krasner–Tate algebra $H(D)$ is a universal generator [39], which proves that the inverse function belongs to $H(f(D))$ and then f_* is a bijection from $\Sigma(D)$ onto $\Sigma(f(D))$. And if $\mathbb{K}$ has characteristic $p \neq 0$, an injective function is a p^qth root of a universal generator [39], which leads to the same result.

Theorem 42.17. *Let $f \in H(D)$ be bianalytic. Then,*

$$|f(x) - f(y)|^2 = |x - y|^2 |f'(x) f'(y)| \quad \forall x, y \in D.$$

Proof. Since D is infraconnected affinoid, we can assume that $\widetilde{D} = U$. Let $E = f(D)$.

Suppose first that $f_*(\mathcal{P})$ is the peripheral of E. Without loss of generality, we can assume that $0 \in E \cap D$. Let $R = \operatorname{diam}(E)$. By Lemma 41.16, R belongs to $|\mathbb{K}|$, so we can take $\lambda \in \mathbb{K}$ such that

$|\lambda| = \frac{1}{R}$. Let $g = \lambda f$. Then, $g_*(\mathcal{P}) = \mathcal{P}$, hence we can apply Theorem 42.15 to g: g_* preserves the distance δ on D, so we have $|g(x)-g(y)| = |x-y|\ \forall x, y \in D$.

We can deduce that $|g'(x)|$ is identically equal to 1 in all D. Indeed, let us fix y in D. Put $a = y$ and $b = g(y)$. Then of course, $\left|\lim_{x\to a}\left(\frac{g(x)-g(a)}{x-a}\right)\right| = 1$, hence $|g'(x)|| = 1\ \forall x \in D$. Therefore, we can write

$$|g(x) - g(y)|^2 = |x - y|^2 |g'(x)g'(y)| \quad \forall x, y \in D.$$

Suppose now that $f_*(\mathcal{P})$ is not the peripheral of $f(D)$. By Theorem 42.16, $f_*(\mathcal{P})$ is the peripheral of a hole $T = d(a, r^-)$ of E. Without loss of generality, we can suppose that $0 \in T$. By Lemma 41.16, diam(T) belongs to $|\mathbb{K}|$, so we can find $\lambda \in \mathbb{K}$ such that $|\lambda| =$ diam(T). Consider now the function $h(x) = \frac{\lambda}{f(x)}$. Clearly, $h_*(\mathcal{P}) = \mathcal{P}$. Consequently, we have $|h(x)-h(y)|^2 = |x-y|^2|h'(x)h'(y)|\ \forall x, y \in D$. This means

$$\left|\frac{\lambda(f(x) - f(y))}{f(x)f(y)}\right|^2 = |x - y|^2 \left|\left(\frac{\lambda f'(x)}{f(x)^2}\right)\left(\frac{\lambda f'(y)}{f(y)^2}\right)\right| \quad \forall x, y \in D,$$

and hence $|f(x) - f(y)|^2 = |x - y|^2 |f'(x)f'(y)|\ \forall x, y \in D$ again. □

Remark. In [58], Yvette Perrin showed that in a Hederic field, an injective meromorphic function is a Moebius function, i.e. it satisfies

$$(f(x) - f(y))^2 = (x - y)^2 f'(x)f'(y) \quad \forall x, y \in D.$$

Here we obtain a formula that looks like that one.

In Corollary 23.13, we saw that a strictly injective analytic element on a disk $d(a, r)$ is bianalytic. This is much more general and concerns particularly analytic elements on affinoid sets, but the theory of Krasner–Tate algebras is then necessary to give a clear explanation of this [37, 39]. In order to generalize Theorem 42.17, we must recall those statements.

Theorem 42.18. *Let D be an infraconnected affinoid subset of $\mathbb{K}$. Then, $H(D)$ id a Krasner–Tate algebra.*

Theorem 42.19. *Let D be an infraconnected affinoid subset of $\mathbb{K}$ and let $f \in H(D)$. Then, $f(D)$ is an affinoid subset of $\mathbb{K}$.*

Now, when $\mathbb{K}$ is of characteristic 0, we can add a few precisions.

Theorem 42.20. *Let $\mathbb{K}$ have characteristic 0, let D be an infraconnected affinoid subset of $\mathbb{K}$, and let $f \in H(D)$ be injective. For every $b \in sp(f)$, either $f - b$ is invertible in $H(D)$, of $(f - b)H(D)$ is a maximal ideal of $H(D)$. Moreover, f^{-1} belongs to $H(f(D)$.*

By Corollary 36.2, we can also derive Corollary 42.21:

In a consequence of Theorems 42.17 and 42.20, we obtain the following theorem:

Theorem 42.21. *Let $\mathbb{K}$ have characteristic 0, let D be an infraconnected affinoid subset of $\mathbb{K}$, and let $f \in H(D)$ be injective. Then,*

$$|f(x) - f(y)|^2 = |x - y|^2 |f'(x)f'(y)| \quad \forall x, y \in D.$$

By Corollary 36.2, we can also derive Corollary 42.20:

Corollary 42.22. *Let $\mathbb{K}$ have characteristic 0. Let $a \in \mathbb{K}$, $r \in \mathbb{R}_+^*$ and let $f \in \mathcal{M}(d(a, R^-))$ and let D be an affinoid infraconnected subset of $d(a, R)$. Then f is injective in D if and only if f' has no zero in D and f satisfies*

$$|f(x) - f(y)|^2 = |x - y|^2 |f'(x)f'(y)| \quad \forall x, y \in D.$$

Corollary 42.23. *Let $\mathbb{K}$ have characteristic 0. Let E be a chained union of infraconnected affinoid subsets $(D_j)_{j \in J}$ of $\mathbb{K}$ and let f be a function from E to $\mathbb{K}$ whose restriction to each D_j belongs to $H(D_j)$ and is injective in E. Then for every x, $y \in E$, we have*

$$|f(x) - f(y)|^2 = |x - y|^2 |f'(x)f'(y)|.$$

Proof. Indeed, given x, $y \in E$, there exists a finite chained family of infraconnected subsets $D_{j(1)}, \dots, D_{j(q)}$ whose union E' contains x, y. By Proposition 2.13, E' is infraconnected, and by Proposition 2.16, it is affinoid. Moreover, by Theorem 15.12, f belongs to $H(E')$, so the conclusion follows from Theorem 42.17. □

Examples.

(1) Let $(a_n)_{n \in \mathbb{N}}$ be a sequence of $\mathbb{K}$ such that $\lim_{n \to \infty} |a_n| = 1$, let $(r_n)_{n \in \mathbb{N}}$ be a sequence of $|\mathbb{K}|$ such that

$0 < r_n < |a_n|$, and let $D = \{x \in d(0, 1^-) \setminus \bigcup_{n=1}^{\infty} d(a_n, r_n^-)$. Let $f \in H(D)$ be injective.

For every $r \in |K| \cap]0, 1[$, the set $D_r = D \cap d(0, r)$ is clearly affinoid, hence the restriction of f to D_r is bianalytic, and hence, we have

$$|f(x) - f(y)|^2 = |x - y|^2 |f'(x) f'(y)| \quad \forall x, y \in D_r.$$

This holds for every $r < 1$, and hence, this holds for all $x, y \in D$.

(2) Let f be a meromorphic function in $\mathbb{K}$ which is injective in the annulus $\Gamma(0, r, s)$. Given s', $r'|K|$ such that $rleqR' < s'|eqs$, then $\Gamma(0, r', s')$ is an affinoid, and hence,

$$|f(x) - f(y)|^2 = |x - y|^2 |f'(x) f'(y)| \quad \forall x, y \in \Gamma(0, r', s').$$

This holds for every r', s', and hence,

$$|f(x) - f(y)|^2 = |x - y|^2 |f'(x) f'(y)| \quad \forall x, y \in \Gamma(0, r, s).$$

Proof. This is given by Theorem 41.5 for Example 1 and Theorem 41.7 for Example 2. □

Chapter 43

Counting Functions and Nevanlinna Theory

The Nevanlinna theory was made by Rolf Nevanlinna on complex functions [79] and widely used by many specialists of complex functions, particularly Walter Hyman [65]. It consists of defining counting functions of zeros and poles of a meromorphic function f and giving an upper bound for multiple zeros and poles of various functions $f - b$, $b \in \mathbb{C}$.

A similar theory for functions in a p-adic field was constructed and correctly proved by Boutabaa [21] in the field $\mathbb{K}$, after some previous work by Ha Huy Khoai [63]. In [23], the theory was extended to functions in $\mathcal{M}(d(0, R^-))$ by taking into account Lazard's problem [72].

Notations. Recall that given three functions ϕ, ψ, ζ defined in an interval $J =]a, +\infty[$ (respectively $J =]a, R[$), with values in $[0, +\infty[$, we write $\phi(r) \leq \psi(r) + O(\zeta(r))$ if there exists a constant $b \in \mathbb{R}$ such that $\phi(r) \leq \psi(r) + b\zeta(r)$. We write $\phi(r) = \psi(r) + O(\zeta(r))$ if $|\psi(r) - \phi(r)|$ is bounded by a function of the form $b\zeta(r)$.

Similarly, we write $\phi(r) \leq \psi(r) + o(\zeta(r))$ if there exists a function h from $J =]a, +\infty[$ (respectively from $J =]a, R[$) to $\mathbb{R}$ such that $\lim_{r \to +\infty} \frac{h(r)}{\zeta(r)} = 0$ (respectively $\lim_{r \to R} \frac{h(r)}{\zeta(r)} = 0$) and such that $\phi(r) \leq \psi(r) + h(r)$. And we write $\phi(r) = \psi(r) + o(\zeta(r))$ if there exists a function h from $J =]a, +\infty[$ (respectively from $J =]a, R[$) to $\mathbb{R}$ such that $\lim_{r \to +\infty} \frac{h(r)}{\zeta(r)} = 0$ (respectively $\lim_{r \to R} \frac{h(r)}{\zeta(r)} = 0$) and such that $\phi(r) = \psi(r) + h(r)$.

Throughout this chapter, we denote by I the interval $[t, +\infty[$ and by J an interval of the form $[t, R[$ with $t > 0$.

We have to introduce the counting function of zeros and poles of a meromorphic function $f \in \mathcal{M}(\mathbb{K})$ or $f \in \mathcal{M}(d(0, R^-))$ counting or not multiplicity. Here we choose a presentation that avoids assuming that all functions we consider admit no zero and no pole at the origin.

Definitions. We denote by $Z(r, f)$ the counting function of zeros of f in $d(0, r)$ in the following way.

Let (a_n), $1 \leq n \leq \upsilon(r)$, be the finite sequence of zeros of f such that $0 < |a_n| \leq r$ of respective order s_n.

We set $Z(r, f) = \max(\omega_0(f), 0) \log r + \sum_{n=1}^{\upsilon(r)} s_n(\log r - \log |a_n|)$, and so, $Z(r, f)$ is called *the counting function of zeros of f in $d(0, r)$, counting multiplicity.*

In order to define the counting function of zeros of f without multiplicity, we put $\overline{\omega_0}(f) = 0$ if $\omega_0(f) \leq 0$ and $\overline{\omega_0}(f) = 1$ if $\omega_0(f) \geq 1$.

Now, we denote by $\overline{Z}(r, f)$ the counting function of zeros of f without multiplicity:

$\overline{Z}(r, f) = \overline{\omega_0}(f) \log r + \sum_{n=1}^{\upsilon(r)} (\log r - \log |a_n|)$, and so, $\overline{Z}(r, f)$ is called *the counting function of zeros of f in $d(0, r)$ ignoring multiplicity.*

In the same way, considering the finite sequence (b_n), $1 \leq n \leq \tau(r)$, of poles of f such that $0 < |b_n| \leq r$, with respective multiplicity order t_n, we put $N(r, f) = \max(-\omega_0(f), 0) \log r + \sum_{n=1}^{\tau(r)} t_n(\log r - \log |b_n|)$ and then $N(r, f)$ is called *the counting function of the poles of f, counting multiplicity.*

Next, in order to define the counting function of poles of f without multiplicity, we put $\overline{\overline{\omega_0}}(f) = 0$ if $\omega_0(f) \geq 0$ and $\overline{\overline{\omega_0}}(f) = 1$ if $\omega_0(f) \leq -1$ and we set $\overline{N}(r, f) = \overline{\overline{\omega_0}}(f) \log r + \sum_{n=1}^{\tau(r)} (\log r - \log |b_n|)$ and then $\overline{N}(r, f)$ is called *the counting function of the poles of f, ignoring multiplicity.*

Now we can define the Nevanlinna function $T(r, f)$ in I or J as $T(r, f) = \max(Z(r, f), N(r, f))$ and the function $T(r, f)$ is called *characteristic function of f or Nevanlinna function of f.*

Finally, if S is a subset of $\mathbb{K}$, we denote by $Z_0^S(r, f')$ the counting function of zeros of f', excluding those which are zeros of $f - a$ for any $a \in S$.

Remark. If we change the origin, the functions Z, N, T are not changed, up to an additive constant.

By Corollary 21.2, Lemma 43.1 is easy:

Lemma 43.1. *Let $\widehat{\mathbb{K}}$ be a complete algebraically closed extension of $\mathbb{K}$ whose absolute value extends that of $\mathbb{K}$ and let $f \in \mathcal{M}(\mathbb{K})$ (respectively let $f \in \mathcal{M}(d(0, R^-))$). Let $\widehat{d}(0, R) = \{x \in \widehat{\mathbb{K}} \mid |x| < R\}$. The meromorphic function $\widehat{f}$ defined by f in $\widehat{d}(0, R)$ has the same Nevanlinna functions as f.*

In a p-adic field such as $\mathbb{K}$, the first main theorem is almost immediate and is an immediate consequence of Corollary 22.27.

Theorem 43.2. *Let $f \in \mathcal{M}(\mathbb{K})$ (respectively $f \in \mathcal{M}(d(0, R^-))$) have no zero and no pole at 0. Then $\log(|f|(r)) = \Psi(f, \log r) = \log(|f(0)|) + Z(r, f) - N(r, f)$ and $\log^+(|f|(r)) \leq \log(|f(0)|) + T(r, f) - N(r, f)$.*

Proof. We can write $f(x) = \frac{h}{l}$ with h, $l \in \mathcal{A}(\mathbb{K})$ (respectively h, $l \in \mathcal{A}(d(0, R^-))$) such that $l(0)h(0) \neq 0$. By Corollary 22.27, we have $\log(|h|(r)) = \log(|f(0)|) + Z(r, h)$, $\log(|l|(r)) = \log(|l(0)|) + N(r, l)$, so the conclusion is obvious. □

Theorem 43.3 is now immediate:

Theorem 43.3. *Let f, $g \in \mathcal{M}(\mathbb{K})$ (respectively f, $g \in \mathcal{M}(d(0, R^-))$). Then, $Z(r, fg) \leq Z(r, f) + Z(r, g)$, $N(r, fg) \leq N(r, f) + N(r, g)$, $T(r, fg) \leq T(r, f) + T(r, g)$, $T(r, f + g) \leq T(r, f) + T(r, g) + O(1)$, $T(r, cf) = T(r, f)$ $\forall c \in \mathbb{K}^*$, $T(r, \frac{1}{f}) = T(r, f))$, $T(r, \frac{f}{g}) \leq T(r, f)) + T(r, g)$.*

Suppose now $f, g \in \mathcal{A}(\mathbb{K})$ (respectively f, $g \in \mathcal{A}(d(0, R^-))$). Then, $Z(r, fg) = Z(r, f) + Z(r, g)$, $T(r, f) = Z(r, f))$, $T(r, fg) = T(r, f) + T(r, g) + O(1)$, and $T(r, f + g) \leq \max(T(r, f), T(r, g))$. Moreover, if $\lim_{r\to+\infty} T(r, f) - T(r, g) = +\infty$, then $T(r, f + g) = T(r, f)$ when r is big enough.

Lemma 43.4. *Let $\alpha_1, \dots, \alpha_n \in \mathbb{K}$ be pairwise distinct, let $P(u) = \prod_{i=1}^n (u - \alpha_i)$, and let $f \in \mathcal{M}(d(0, R^-))$. Then, $Z(r, P(f)) = \sum_{i=1}^n Z(r, f - \alpha_i)$ and $\overline{Z}(r, P(f)) = \sum_{i=1}^n \overline{Z}(r, f - \alpha_i)$.*

Lemma 43.5. *Let $f \in \mathcal{M}(\mathbb{K})$. Then f belongs to $\mathbb{K}(x)$ if and only if $T(r, f) = O(\log r)$.*

Proof. If f belongs to $\mathbb{K}(x)$, one can write it $\frac{P(x)}{Q(x)}$ with P, $Q \in \mathbb{K}[x]$ having no common zeros, hence $Z(r,f) = Z(r,P)$ and $N(r,f) = Z(r,Q)$, and hence $T(r,f) = O(\log r)$.

Now suppose that $f \notin \mathbb{K}(x)$. Suppose, for instance, that f has infinitely many zeros (a_n) of respective order q_n. Then let us fix $s \in \mathbb{N}$ and let $t \in \mathbb{N}$ be $> s+1$. For r big enough, we have $Z(r,f) > \sum_{n=1}^{t}(\log r - \Psi(a_n)) > s\log r$, hence $Z(r,f)$ is not $O(\log r)$. Similarly, if f has infinitely many poles, we get to the same conclusion. □

Applying Lemma 43.1 and Theorem 37.10 to $\frac{f'}{f}$, up to a change of origin, we can derive Corollary 43.6:

Corollary 43.6. *Let $f \in \mathcal{M}(\mathbb{K})$ (respectively $f \in \mathcal{M}(d(0,R^-))$). Then,*

$$Z\left(r,\frac{f'}{f}\right) - N\left(r,\frac{f'}{f}\right) \leq -\log r + O(1).$$

Theorem 43.7. *Let $f \in \mathcal{A}(\mathbb{K})$ (respectively $f \in \mathcal{A}(d(0,R^-))$) and let $b \in \mathbb{K}$. Then, $Z(r,f) = Z(r,f-b) + O(1)$ $r \in I$ (respectively $r \in J$).*

Proof. Let $f(x) = \sum_{n=0}^{\infty} a_n x^n$ and let $\rho \in \mathbb{R}_+^*$ (respectively $\rho \in]0,R[$) be such that $\nu^+(f,\log\rho) > 0$ and $\nu^+(f-b,\log\rho) > 0$. Then we have $\nu^+(f,\mu) = \nu^+(f-b,\mu)$ $\forall \mu > \log\rho$ (respectively $\forall \mu \in]\log\rho, \log R[$). Consequently, on each circle $C(0,r)$ such that $r > \rho$ (respectively $r \in]\rho, R[$), f and $f-b$ have the same number of zeros, taking multiplicity into account. Let (a_n) be the sequence of zeros of f, with respective multiplicity q_n, with $|a_n| \leq |a_{n+1}|$, $n \in \mathbb{N}^*$ and $|a_n| > \rho$ if and only if $n \geq t$.

Similarly, let (b_n) be the sequence of zeros of $f-b$, with respective multiplicity s_n, with $|b_n| \leq |b_{n+1}|, n \in \mathbb{N}^*$ and $|b_n| > \rho$ if and only if $n \geq u$. Since f and $f-b$ have the same number of zeros in $d(0,\rho)$, we also have

(1) $\sum_{n=1}^{t} q_n = \sum_{n=1}^{u} s_n$.

Consequently, for all $r > \rho$ (respectively $r \in]\rho, R[$), we have

$$\sum_{\substack{n\geq t,\\ |a_n|\leq r}} q_n(\log r - \Psi(a_n)) = \sum_{\substack{n\geq u\\ |b_n|\leq r}} s_n(\log r - \Psi(b_n)).$$

Now, suppose that both $f(0)$, $f(0)-b$ are not 0. Then,

$$Z(r,f) = \sum_{|a_n|\le r} q_n(\log r - \Psi(a_n)),$$

$$Z(r,f-b) = \sum_{|b_n|\le r} s_n(\log r - \Psi(b_n)).$$

Therefore, $Z(r,f) - Z(r,f-b)$ is reduced to

$$\sum_{\substack{|a_n|\le r,\\ |a_n|\le\rho}} (\log r - \Psi(a_n)) - \sum_{\substack{|b_n|\le r,\\ |b_n|\le\rho}} (\log r - \Psi(b_n))$$

$$= \sum_{n=1}^{t} q_n(\log r - \Psi(a_n)) - \sum_{n=1}^{u} s_n(\log r - \Psi(b_n))$$

that this is a constant by (1), for $r > \rho$ (respectively for $r \in]0,R[$).

And now, suppose that 0 is a zero of order q_1 of f. Then,

$$Z(r,f) = q_1 \log r + \sum_{n\ge 2, |a_n|\le r} q_n(\log r - \Psi(a_n)),$$

and therefore, $Z(r,f) - Z(r,f-b)$ is reduced to

$$q_1 \log r + \sum_{\substack{n\ge 2,\\ |a_n|\le\rho}} q_n(\log r - \Psi(a_n)) - \sum_{|b_n|\le\rho} s_n(\log r - \Psi(b_n))$$

$$= q_1 \log r + \sum_{n=2}^{t} q_n(\log r - \Psi(a_n)) - \sum_{n=1}^{u} s_n(\log r - \Psi(b_n)),$$

and we check that this is a constant again thanks to (1).

Similarly, if $f(0) = b$, then f and $f-b$ playing the same role, we have the same conclusion. □

Theorem 43.8 (First Main Theorem). *Let* $f,\ g \in \mathcal{M}(\mathbb{K})$ *(respectively let* $f,\ g \in \mathcal{M}(d(0,R^-)))$*. Then* $T(r,f+b) = T(r,f) + O(1)$ *and* $\log^+|f|(r) \le T(r,f) - N(r,f) + O(1)$.

Let h *be a Moebius function. Then,* $T(r,f) = T(r,h\circ f) + O(1)$. *Let* $P(X) \in \mathbb{K}[X]$*. Then,* $T(r,P(f)) = \deg(P)T(r,f) + O(1)$ *and* $T(r,f'P(f) \ge T(r,P(f))$.

Suppose now f and g have finitely many poles. Then, $Z(r, fg) = Z(r, f) + Z(r, g)$, $T(r, f) = Z(r, f)) + O(1)$, $T(r, fg) = T(r, f) + T(r, g) + O(1)$, and $T(r, f + g) \leq \max(T(r, f), T(r, g)) + O(|og(r))$. Moreover, if $\lim_{r\to+\infty} T(r, f) - T(r, g) = +\infty$, then $T(r, f + g) = T(r, f)$ when r is big enough.

Proof. $T(r, f + b) \leq T(r, f) + O(1) \leq T(r, f + b) + O(1)$, hence $T(r, f + b) = T(r, f) + O(1)$ $\forall b \in \mathbb{K}$. Now, consider $T(r, f + g)$ when f, g have finitely many poles. We have $T(r, f + g) = Z(r, f + g) + O(\log r) \leq \max(Z(r, f), Z(r, g)) + O(\log(r))$.

Let $h(X) = \frac{aX+b}{cX+d}$ be a Moebius function and let $g(x) = h \circ f(x)$. We can write $h(X) = \frac{a}{c} + \frac{\lambda}{cX+d}$ with $\lambda = d(1 - \frac{a}{c})$. Then,

$$\begin{aligned} T(r, g) &= T\left(r, \frac{\lambda}{cf(x) + d}\right) + O(1) = T(r, cf(x) + d) + O(1) \\ &= T(r, f(x)) + O(1). \end{aligned}$$

Now, let $P(X) = \prod_{k=1}^{q}(X - a_k) \in \mathbb{K}[x]$ be a polynomial of degree q and let $F(x) = P(f(x))$. Then, $T(r, f - a_k) = T(r, f) + O(1)$ $\forall k = 1, \ldots, q$, and hence, $T(r, F) = qT(r, f) + O(1)$. Moreover, zeros of $P(f)$ are not poles of f' and poles of f' are poles of f and hence are not zeros of $P(f)$. Consequently, $N(r, f'P(f) = N(r, P(f)) + N(r, f') = N(r, P(f)) + N(r, f) + \overline{N}(r, f)$, $Z(r, f'P(f)) = Z(r, P(f) + Z(r, f')$. Therefore, $T(r, f'P(f) \geq T(r, P(f))$.

It now only remains to prove that $T(r, P(f)) = qT(r, f) + O(1)$. Let $P(X) = \prod_{j=1}^{q}(X - a_j)$. It is immediate to check that $Z(r, P(f)) = \sum_{j=1}^{q} Z(r, f - a_j) = qZ(r, f) + O(1)$ and that $N(r, P(f)) = qN(r, f)$, therefore $T(r, P(f)) = qT(r, f) + O(1)$. □

Theorem 43.9. *Let $f \in \mathcal{M}(\mathbb{K})$ (respectively $f \in \mathcal{M}(d(0, R^-))$). There exists ϕ, $\psi \in \mathcal{A}(\mathbb{K})$ (respectively ϕ, $\psi \in \mathcal{A}(d(0, R^-))$) such that $f = \frac{\phi}{\psi}$ and $\max(T(r, \phi), T(r, \psi)) \leq T(r, f) + O(1)$, $r \in I$ (respectively ($r \in J$)).*

Proof. Let $V_1 = \mathcal{D}(f)$ and let $V_2 = \mathcal{D}(\frac{1}{f})$. Suppose first $f \in \mathcal{M}(\mathbb{K})$. By Theorem 27.4, there exists ϕ, $\psi \in \mathcal{A}(\mathbb{K})$ such that $\mathcal{D}(f) = \mathcal{D}(\phi)$, $\mathcal{D}(\frac{1}{f}) = \mathcal{D}(\psi)$. Consequently, $Z(r, f) = Z(r, \phi), N(r, f) = Z(r, \psi)$ and the claim is immediate. Now, suppose $f \in \mathcal{M}(d(0, R^-))$. By Theorem 27.14, there exists $\phi \in \mathcal{A}(\mathbb{K})$ such that $\mathcal{D}(f) \leq \mathcal{D}(\phi)$ and such that $|\mathcal{D}(\phi)|(r) \leq |V_1|(r)+1,\ r \in J$, hence $Z(r,\phi) \leq Z(r,f)+1$, $r \in J$. Let $\psi = \frac{\phi}{f}$. Then ψ lies in $\mathcal{A}(d(0,R^-))$ because $\mathcal{D}(f) \leq \mathcal{D}(\phi)$. And $\mathcal{D}(\psi) = \mathcal{D}(\frac{1}{f}) + \mathcal{D}(\phi) - \mathcal{D}(f)$. Consequently, $|\mathcal{D}(\psi)|(r) \leq |\mathcal{D}(\frac{1}{f})|(r)+1$. But $T(r,\phi) = Z(r,\phi)+O(1) = \log(|\mathcal{D}(\phi)|(r)) + o(1)$ and $T(r,\psi) = Z(r,\psi)+O(1) = \log(|\mathcal{D}(\psi)|(r)) + O(1)$. Therefore, $\max(T(r,\phi), T(r,\psi)) \leq \max(Z(r,f), N(r,f)) + O(1),\ r \in I$ (respectively $(r \in J)$). □

Theorem 43.10. *Let $f \in \mathcal{M}(d(0,R^-))$. Then f belongs to $\mathcal{M}_b(d(0, R^-))$ if and only if $T(r,f)$ is bounded in $[0,R[$.*

Proof. Suppose first $f \in \mathcal{A}(d(0,R^-))$. Without loss of generality, we can obviously suppose that $f(0) \neq 0$. By Theorem 43.2, we have $\log|f|(r) = \log(|f(0)|) + Z(r,f)$. And $|f|(r) = \sup\{|f(x)| \mid x \in d(0,r^-)\}$, so the claim is clear. Now, consider the general case. Suppose $T(r,f)$ is not bounded, so either $Z(r,f)$ or $N(r,f)$ is not bounded. Let $f = \frac{\phi}{\psi}$ with $\phi,\ \psi \in \mathcal{A}(d(0,R^-))$. If $Z(r,f)$ is not bounded, then $\phi \notin \mathcal{A}_b(d(0,R^-))$. If $N(r,f)$ is not bounded, then $\psi \notin \mathcal{A}_b(d(0,R^-))$. Thus, f cannot be put in the form $\frac{\phi}{\psi}$ with ϕ, $\psi \in \mathcal{A}_b(d(0,R^-))$, and therefore, $f \notin \mathcal{A}_b(d(0,R^-))$.

Conversely, if $f \in \mathcal{A}_b(d(0,R^-))$, then it is of the form $\frac{\phi}{\psi}$ with ϕ, $\psi \in \mathcal{A}_b(d(0,R^-))$, hence both $Z(r,\phi), Z(r,\psi)$ are bounded. But since $Z(r,f) \leq Z(r,\phi)$ and $N(r,f) \leq Z(r,\psi)$, $T(r,f)$ is clearly bounded in $[0,R[$. □

Corollary 43.11. *Let $f \in \mathcal{M}_u(d(a,R^-))$, and let $h \in \mathcal{M}_b(d(a, R^-))$, $h \neq 0$. Then, fh belongs to $\mathcal{M}_u(d(a,R^-))$.*

By Theorems 43.8 and 43.10, we can also derive Corollary 43.12.

Corollary 43.12. *Let $f \in \mathcal{M}(d(a,R^-))$ and let $P \in \mathbb{K}[x]$. Then $P(f)$ belongs to $\mathcal{M}_b(d(a,R^-))$ if and only if so does f.*

Theorem 43.13. *Let $\mathbb{K}$ be of characteristic 0. Let $f \in \mathcal{M}(\mathbb{K})$ (respectively $f \in \mathcal{M}(d(0,R^-))$). Then $Z(r,f') - N(r,f') \leq Z(r,f) - N(r,f) - \log r + O(1),\ r \in I$ (respectively $r \in J$). Moreover, $N(r,f^{(k)}) = N(r,f) + k\overline{N}(r,f) + O(1),\ r \in I$ and $Z(r,f^{(k)}) \leq Z(r,f) + k\overline{N}(r,f) - k\log r + O(1),\ r \in I$ (respectively $r \in J$).*

Proof. Without loss of generality, we can assume that f, $f', \dots, f^{(k)}$ have no zero and no pole at 0.

The first statement is immediate and just comes from this basic property: if α is a pole of f of order q, then it is a pole of $f^{(k)}$ of order $q+k$. Next, by Theorem 43.2, we have $Z(r,f) - N(r,f) = \Psi(f,\log r) - \log(|f(0)|)$ and $Z(r,f') - N(r,f') = \Psi(f',\log r) - \log(|f'(0)|)$. But $\Psi(f',\log r) \le \Psi(f,\log r) - \log r$, hence we obtain $Z(r,f') \le N(r,f') - N(r,f) + Z(r,f) - \log(r) + O(1)$. Actually, $N(r,f') - N(r,f) = \overline{N}(r,f)$, hence $N(r,f^{(k)}) = N(r,f) + k\overline{N}(r,f)$. Next, $Z(r,f') \le Z(r,f) + \overline{N}(r,f) - \log(r) + O(1)$. Now, suppose that the second statement has been proved for $k \le t$. Thus, we have $Z(r,f^{(t+1)}) \le Z(r,f^{(t)}) + \overline{N}(r,f^{(t)}) - \log(r) = O(1)$. But as we just noted, $\overline{N}(r,f^{(t)}) = \overline{N}(r,f)$, hence $Z(r,f^{(t+1)}) \le Z(r,f) + (t+1)$ $\overline{N}(r,f) - (t+1)\log(r) + O(1)$. □

Lemma 43.14 is now classical and easily checked:

Lemma 43.14. *Let $\alpha_1,\dots,\alpha_q \in \mathbb{K}$ be pairwise distinct, let $S = \{\alpha_1,\dots,\alpha_q\}$, and let $P(x) = \prod_{j=1}^{q}(x-\alpha_j)$. Let $f \in \mathcal{M}(\mathbb{K})$ (respectively $f \in \mathcal{M}(d(0,R^-))$). Then,*

$$\sum_{j=1}^{n} Z(r,f-\alpha_j) = Z(r,P(f)), \quad \sum_{j=1}^{n} \overline{Z}(r,f-\alpha_j) = \overline{Z}(r,P(f))\ \forall r \in I$$

(respectively $\forall r \in J$). Moreover, assuming that $\mathbb{K}$ is of characteristic 0, we have

$$Z(r,P(f)) - \overline{Z}(r,P(f)) \le Z(r,f') \le Z(r,f) + \overline{N}(r,f) - \log(r) \\ + O(1)\ \forall r \in I \quad (\textit{respectively } \forall r \in J).$$

Corollary 43.15. *We assume that $\mathbb{K}$ is of characteristic 0. Let $f \in \mathcal{M}(\mathbb{K})$ (respectively $f \in \mathcal{M}(d(0,R^-))$). Then, $T(r,f^{(k)}) \le (k+1)T(r,f) + O(1)$ $(r \in I)$ (respectively $r \in J$).*

Theorem 43.16. *We assume that $\mathbb{K}$ is of characteristic 0. Let $f \in \mathcal{M}(\mathbb{K})$ (respectively $f \in \mathcal{M}(d(0,R^-))$). Then, $T(r,f) - Z(r,f) \le T(r,f') - Z(r,f') + O(1)$. Further, given $\alpha \in \mathcal{M}(d(0,R^-))$, we have $T(r,\alpha f) - Z(r,\alpha f) \le T(r,f) - Z(r,f) + T(r,\alpha)$.*

Proof. By Theorem 43.13, the first statement is immediate. Let us check the last one. On one hand, $T(r,f) - Z(r,f) =$

$\max(Z(r,f),N(r,f)) - Z(r,f) = \max(0, N(r,f) - Z(r,f))$ $r < R$. On the other hand,

$$\begin{aligned} T(r,f') - Z(r,f') &= \max(Z(r,f'),N(r,f')) - Z(r,f') \\ &= \max(0, N(r,f') - Z(r,f')) \\ &= \max(0, N(r,f) + \overline{N}(r,f) - Z(r,f'))\ r < R. \end{aligned}$$

But by Theorem 43.13, $-Z(r,f') \geq -Z(r,f) - \overline{Z}(r,f) + \log(r) + O(1)$ $r < R$, hence $T(r,f') - Z(r,f') \geq \max(0, N(r,f) - Z(r,f) + \log(r)) + O(1) \geq T(r,f) - Z(r,f) + O(1)$, $r < R$.

Now, let $\alpha \in \mathcal{M}(d(0,R^-))$. Suppose $N(r,\alpha f) \geq Z(r,\alpha f)$, $r < R$. Then, $T(r,\alpha f) - Z(r,\alpha f) = N(r,\alpha f) - Z(r,\alpha f)$ $r < R$. We can write α in the form $\frac{\beta(x)}{\lambda(x)}$ with β, $\lambda \in H(d(0,r))$, β, λ having no common zero. Next, we can write λ in the form $\lambda_1\lambda_2$ were each zero of λ_1 is not a zero of f and each zero of λ_2 is a zero of f. Then we can check that $N(r,\alpha f) = N(r,f) + Z(r,\lambda_1)$ and $Z(r,\alpha f \geq Z(r,f) - Z(r,\lambda_2)$. Consequently, $N(r,\alpha f) - Z(r,\alpha f) \leq N(r,f) + Z(r,\lambda_1) - (Z(r,f) - Z(r,\lambda_2)) = N(r,f) - Z(r,f) + Z(r,\lambda_1) + Z(r,\lambda_2) = N(r,f) - Z(r,f) + Z(r,\lambda) \leq N(r,f) - Z(r,f) + T(r,\lambda)$ $r < R$.

Suppose now that $N(r,\alpha f) \leq Z(r,\alpha f)$. We can do a symmetric reasoning with the zeros of β. □

Lemma 43.17 is an immediate consequence of Corollary 22.27 and Theorem 37.10:

Lemma 43.17. *Let $f \in \mathcal{M}(\mathbb{K})$ (respectively $f \in \mathcal{M}(d(0,R^-))$) and let $G = \frac{f'}{f}$. Then, G satisfies $Z(r,G) \leq N(r,G) - \log r + O(1)$ $r \in I$ (respectively ($r \in J$)).*

Proof. Without loss of generality, we can assume that 0 is neither a pole of f nor a zero for ff'. By Theorem 37.10, G satisfies $\Psi(G,\log r) \leq -\log r$. On the other hand, by Theorem 43.2, we have $\Psi(G,\log r) = \log|G(0)| + Z(r,G) - N(r,G)$. Consequently, we obtain $\log|G(0)| + Z(r,G) - N(r,G) \leq -\log r + O(1)$, which proves the claim. □

Corollary 43.18. *Let $f \in \mathcal{M}(\mathbb{K})$ (respectively $f \in \mathcal{M}(d(0,R^-))$) and let $G = \frac{f'}{f}$. Then, $T(r,G) = N(r,G) + O(1)$.*

Proof. By Lemma 43.17, we have $T(r,G)) \leq N(r,G) + O(1)$. But as all meromorphic functions, G satisfies $T(r,G) \geq N(r,G)$. □

Chapter 44

A Non-Clean Entire Function

In Chapter 31, we noted the following question:

Let $f(x) = \sum_{n=0}^{\infty} b_n x^n \in \mathcal{A}(\mathbb{K})$ *be such that* $0 < \rho(f) < +\infty$*. Do we have*

$$\rho(f)\sigma(f) = \psi(f) = \frac{\limsup_{n\to+\infty}\left(n\sqrt[n]{|b_n|^{\rho(f)}}\right)}{e}?$$

Actually, the answer is no in the general case. Using the counting function of an entire function, we construct a counter-example. In order to prove that, we need two basic lemmas.

Lemma 44.1. *Let* f_1*,* f_2 *be two functions from* $\mathbb{R}_+$ *to* $\mathbb{R}_+$ *such that* $\lim_{x\to+\infty} f_1(x) = \lim_{x\to+\infty} f_2(x) = +\infty$ *and*

$$\limsup_{x\to+\infty}\frac{f_1(x)}{f_2(x)} = b \in \mathbb{R}_+, \quad \liminf_{x\to+\infty}\frac{f_1(x)}{f_2(x)} = a > 0.$$

Then,

$$\lim_{x\to+\infty}\frac{\mathrm{Log}(f_1(x))}{\mathrm{Log}(f_2(x))} = 1.$$

Lemma 44.2. *Let* α*,* $\beta \in \mathbb{R}_+$ *and let* $g(x) = e^{-x}(\alpha x - \beta)$*. Then* g' *has a unique zero at* $1 + \frac{\beta}{\alpha}$ *and* $g(1 + \frac{\beta}{\alpha}) = \alpha e^{-(1+\frac{\beta}{\alpha})}$*. Moreover,* g *is increasing in* $[0, 1 + \frac{\beta}{\alpha}]$ *and is decreasing in* $[1 + \frac{\beta}{\alpha}, +\infty[$ *and tends to* 0 *when* x *tends to* $+\infty$*.*

Theorem 44.3. *Suppose that $\mathbb{K}$ is of characteristic 0. There exist regular non-clean functions $f \in \mathcal{A}(\mathbb{K})$ such that $\psi(f) > \rho(f)\sigma(f)$.*

Proof. We begin the definition of positive increasing sequences $(r_m)_{m\in\mathbb{N}}$, $(\alpha_m)_{m\in\mathbb{N}}$ in $\mathbb{N}^*$, $(\beta_m)_{m\in\mathbb{N}}$, where $r_0 = 1$, $r_{2m} \in |\mathbb{K}|$, $2r_{2m} \leq \alpha_m < 2r_{2m}+1$, We put $q_m = \alpha_m - \alpha_{m-1}$, $\nu_m = \mathrm{Log}(r_m)$, $\beta_0 = 0$, and $\beta_m = \beta_{m-1} + q_m(Log(r_{2m}))$.

In $[0, +\infty[$, we define $g_k(\nu) = e^{-\nu}(\alpha_k\nu - \beta_k)$ up to the rank m and suppose that the function g_k satisfies $1 \leq g_k(\nu_{2k}) \leq 1 + \frac{1}{4k^2}$ and $1 \leq g_k(\nu_{2k+2}) \leq 1 + \frac{1}{4(k+1)^2}$ $\forall k = 1, \dots, m-1$ and $g_{k-1}(\nu_{2k}) = g_k(\nu_{2k})$.

By Lemma 44.2, g_k is increasing in $[\frac{\beta_k}{\alpha_k}, 1 + \frac{\beta_k}{\alpha_k}]$ from 0 to a maximum equal to $\alpha_k e^{1+\frac{\beta_k}{\alpha_k}}$ and is decreasing to 0 when ν tends to $+\infty$. Hence, g_k takes the value 1 at a unique point λ_{2k} in $[\frac{\beta_k}{\alpha_k}, 1+\frac{\beta_k}{\alpha_k}]$ and at a unique point $\lambda_{2k+2} \in [1+\frac{\beta_k}{\alpha_k}, +\infty[$. We then have $g_k(\lambda_{2k}) = e^{-\lambda_{2k}}(\alpha_k\lambda_{2k} - \beta_k)$ and $g_k(\lambda_{2k+2}) = e^{-\lambda_{2k+2}}(\alpha_k\lambda_{2k+2} - \beta_m)$, hence $\lambda_{2k} = \frac{e^{\lambda_{2k}} + \beta_k}{\alpha_k}$ and $\lambda_{2k+2} = \frac{e^{\lambda_{2k+2}} + \beta_k}{\alpha_k} > \nu_{2k+1}$ and we can take the value $r_{2k+2} \in |\mathbb{K}|$ close enough to $e^{\lambda_{2k+2}}$ such that, putting $r_{2k+2} = e^{\nu_{2k+2}}$, we then have

$$1 \leq g_k(\nu_{2k+2}) = 1 + x_k \leq 1 + \frac{1}{4(k+1)^2}$$

and

$$1 \leq g_k(\nu_{2k}) = 1 + y_k \leq 1 + \frac{1}{4(k)^2}. \tag{1}$$

We note that $r_{2k+2} > r_{2k+1}$, hence $\nu_{2k+2} > \nu_{2k+1}$. Next, the function g_{k+1} is defined in the same way in $[\nu_{2k+2}, \nu_{2k+4}]$ as $g_{k+1}(\nu) = e^{-\nu}(\alpha_{k+1}\nu - \beta_{k+1})$. And we can check that $g_{k+1}(\nu_{2k+2}) = g_k(\nu_{2k+2})$.

Then by Lemma 44.2, g_m has a maximum at

$$\nu_{2m+1} = 1 + \frac{\beta_m}{\alpha_m}, \tag{2}$$

and g_{m+1} has a maximum at $\nu_{2m+3} = 1 + \frac{\beta_{m+1}}{\alpha_{m+1}}$ and $g_{m+1}(\nu_{2m+3}) = \alpha_{m+1}e^{\frac{\beta_{m+1}}{\alpha_{m+1}}} > 1$, hence $\nu_{2m+3} > \nu_{2m+2}$. Consequently, the sequence $(r_n)_{n\in\mathbb{N}}$ and is strictly increasing. This way, the sequences are now defined for all $m \in \mathbb{N}$. Recall that $q_m = \alpha_m - \alpha_{m-1}$. We put $\Theta_m = \nu_{2m+1} - \nu_{2m}$. Then, $\nu_{2m} = \frac{\beta_m + e^{\nu_{2m}}(1+x_m)}{\alpha_m}$, and hence by (1) and (2), we obtain

$$(3)\ \ \Theta_m = 1 - \frac{e^{\nu_{2m}}(1+y_m)}{\alpha_m} = 1 - \frac{r_{2m}(1+y_m)}{\alpha_m} = 1 - \frac{r_{2m}(1+y_m)}{2r_{2m}+\eta_m},$$

where $(\eta_m)_{m\in\mathbb{N}}$ is a positive sequences bounded by 1 and the sequence (y_m), by (1), satisfies $0 \le y_m \le \frac{1}{4(m)^2}$. Then,

$$(4) \qquad \Theta_m \ge \frac{1}{2} - \frac{1}{8(m)^2} > \frac{15}{32}.$$

We can now define by induction the sequences (r_m), (ν_m), (g_m), (y_m), (Θ_m) and then $\lim_{n\to+\infty} r_n = +\infty$. Consequently, by (3) and (4),

$$(5) \qquad \lim_{m\to+\infty} \Theta_m = \frac{1}{2}.$$

We now obtain

$$g_m(\nu_{2m+1}) = e^{-\nu_{2m+1}}(\alpha_m \nu_{2m+1} - \beta_m) = \alpha_m e^{-(1+\frac{\beta_m}{\alpha_m})},$$

and hence, by (2),

$$(6) \qquad g_m(\nu_{2m+1}) = \frac{(2r_{2m}+\eta_m)}{r_{2m+1}} = 2e^{-\Theta_m} + \zeta_m,$$

where $(\zeta_m)_{m\in\mathbb{N}}$ is a positive sequence of limit 0, since $\lim_{m\to+\infty} r_{2m+1} = +\infty$.

We can now define a function g in $[0, +\infty[$ as $g(\nu) = g_m(\nu)$ when $\nu \in [\text{Log}(r_{2m}), \text{Log}(r_{2m+2})]$.

So, by (5), we have

$$(7) \qquad \lim_{m\to+\infty} g(\nu_{2m+1}) = \frac{2}{\sqrt{e}}.$$

Thus, we can check that

$$(8) \qquad \limsup_{\nu\to+\infty} g(\nu) = \limsup_{m\to+\infty} g(\nu_{2m+1}) < 2.$$

Now, by Theorem 27.4, we can consider the entire function f admitting q_m zeros on each circle $C(0, r_{2m})$ and no other zero. Let $(a_{j,m})_{(1\le j\le q_m)}$ be the zeros of f on the circle $C(0, r_{2m})$.

Then, when $2m \le r < r_{2m+2}$, the counting functions of zeros of f (counting multiplicity) is of the form

$$Z(f,r) = \sum_{k=1}^{m}\sum_{j=1}^{q_k}(a_{k,j}(\text{Log}(r) - \text{Log}(r_{2k})))$$
$$= \sum_{k=1}^{m} q_k(\text{Log}(r) - \text{Log}(r_{2k}))),$$

and hence, putting $\alpha_m = \sum_{k=1}^{m} q_k$ and $\beta_m = \sum_{k=1}^{m} q_k r_{2k}$, the function g appears as the quotient of the counting function of zeros of f (counting multiplicity) by e^{ν} when we put $\nu = \text{Log}(r)$. So, we have $\frac{\text{Log}(|f|(r))}{r} = g(\nu)$ whenever $\text{Log}(r) = \nu \in [\nu_{2m}, \nu_{2m+2}[$, and therefore by (1), we can see that

$$\liminf_{r\to+\infty} \frac{\text{Log}(|f|(r))}{r} = 1$$

and

$$\limsup_{r\to+\infty} \frac{\text{Log}(|f|(r))}{r} = \lim_{m\to+\infty} |g(\nu_{2m+1})| = \frac{2}{\sqrt{e}}. \tag{9}$$

Moreover, by (9) and Lemma 44.1, we have

$$\lim_{r\to+\infty} \frac{\text{Log}(\text{Log}(|f|(r)))}{\text{Log}(r)} = 1,$$

hence $\rho(f) = 1$.

Further, since $\limsup_{r\to+\infty} \frac{\text{Log}(|f|(r))}{r} = \frac{2}{\sqrt{e}}$ and $\liminf_{r\to+\infty} \frac{\text{Log}(|f|(r))}{r} = 1$, we can see that $\sigma(f) = \frac{2}{\sqrt{e}}$ and $\widetilde{\sigma}(f) = 1$. Thus, f is not clean though it is regular.

More precisely, by construction, for every $r \in [r_{2m}, r_{2m+2}[$, we have $\psi(f,r) = \frac{2r_{2m}+\eta_m}{r}$, and hence $\psi(f, r_{2m})$ is of the form $2 + y_m$, where $(y_m)_{m\in\mathbb{N}}$ is a sequence of limit 0. Therefore, $\psi(f) \ge 2$, while $\rho(f) = 1$. This shows that f does not satisfy the relation $\psi(f) = \rho(f)\sigma(f)$, and hence, this is not always satisfied when a function f is not clean. □

Remark 1. Of course, by Theorem 31.5, we know that the function f built in the proof of Theorem 44.3 satisfies $\psi(f) > \widetilde{\psi}(f)$. But we can directly verify this: on one hand, $\psi(f) = 2$, and on the other hand, we can see that $\psi(f, r_{2m+1}) = \frac{\alpha_m}{r_{2m+1}}$, and hence by (5), $\widetilde{\psi}(f) \leq \frac{2}{\sqrt{e}}$.

Next, f must satisfy Theorem 31.5: $\rho(f)\sigma(f) \leq \psi(f) \leq \rho(f)(e\sigma(f) - \widetilde{\sigma}(f))$. Let us check. We have seen that $\psi(f) = 2$, $\rho(f) = 1$, $\sigma(f) = \frac{2}{\sqrt{e}}$, $\widetilde{\sigma}(f) = 1$. Then, $\rho(f)(e\sigma(f) - \widetilde{\sigma}(f)) = 2\sqrt{e} - 1 > 2$. That is okay.

Remark 2. By Corollary 31.7, a clean entire function such that $\sigma(f) > 0$ is regular. The converse is not true, as shown in Theorem 44.3.

In complex analysis, given an entire function f, we put $M(f, r) = \sup\{|f(z)|_\infty,\ |z|_\infty = r\}$, where $|\ .\ |_\infty$ is the archimedean modulus on $\mathbb{C}$. In [4], the authors claimed that if a complex entire function f satisfies

$$\limsup_{r \to +\infty} \frac{\mathrm{Log}(\mathrm{Log}(M(f, r)))}{\mathrm{Log}(r)} = \liminf_{r \to +\infty} \frac{\mathrm{Log}(\mathrm{Log}(M(f, r)))}{\mathrm{Log}(r)},$$

then

$$\limsup_{r \to +\infty} \frac{\mathrm{Log}(M(f, r))}{r^\rho} = \liminf_{r \to +\infty} \frac{\mathrm{Log}(M(f, r))}{r^\rho},$$

where $\rho = \lim_{r \to +\infty} \frac{\mathrm{Log}(\mathrm{Log}(M(f,r)))}{\mathrm{Log}(r)}$. In the field $\mathbb{K}$, we just checked that such a theorem does not hold. Actually, the proof of [4] is put in doubt by the following argument held in Lemma 2 of [4]:

since

$$\int_{r_0}^{+\infty} \frac{\exp(\mathrm{Log}(M(r, f))}{(\exp(r^\lambda))^{t-\varepsilon+1}} dr = +\infty,$$

"then"

$$\liminf_{r \to +\infty} \frac{\exp(\mathrm{Log}(M(r, f))}{(\exp(r^\lambda))^{t-\varepsilon}} = +\infty.$$

Suppose, for example, that in $[r_0, +\infty[$, $M(r, f)$ is equivalent to $\frac{\exp(r\lambda(t-\varepsilon+1))}{r}$.

Then, $\frac{\exp(r\lambda(t-\varepsilon+1))}{\exp(r\lambda(t-\varepsilon))}$ is equivalent to $\frac{1}{r}$, and hence

$$\int_{r_0}^{+\infty} \frac{\exp(\mathrm{Log}(M(r,f))}{(\exp(r^{\lambda}))^{t-\varepsilon+1}} dr = +\infty,$$

but

$$\liminf_{r\to+\infty} \frac{\exp(\mathrm{Log}(M(r,f))}{(\exp(r^{\lambda}))^{t-\varepsilon}} = 0.$$

Chapter 45

Nevanlinna Theory in $\mathbb{K}$ and Inside a Disk

We can now prove the second main theorem under different forms. Lemma 45.1 is essential and directly leads to the theorems.

Lemma 45.1. *Let $f \in \mathcal{M}(\mathbb{K})$ (respectively $f \in \mathcal{M}_u(d(0, R^-))$). Suppose that there exists $\xi \in \mathbb{K}$ (respectively $\xi \in \mathcal{M}_b(d(0, R^-))$) and a sequence of intervals $I_n = [u_n, v_n]$ such that $u_n < v_n < u_{n+1}$, $\lim_{n\to+\infty} u_n = +\infty$ (respectively $\lim_{n\to+\infty} u_n = R$) and*

$$\lim_{n\to+\infty} \Big(\inf_{r\in I_n} T(r, f) - Z(r, f - \xi) \Big) = +\infty$$

(respectively $\lim_{n\to+\infty} (\inf_{r\in I_n} T(r, f) - Z(r, f - \xi)) = +\infty$).

Let $\tau \in \mathbb{K}$ (respectively let $\tau \in \mathcal{M}_b(d(0, R^-))$), $\tau \neq \xi$. Then, $Z(r, f - \tau) = T(r, f) + O(1)$ $\forall r \in I_n$, when n is big enough.

Proof. We know that the Nevanlinna functions of a meromorphic function f are the same in $\mathbb{K}$ and in an algebraically closed complete extension of $\mathbb{K}$ whose absolute value extends that of $\mathbb{K}$. Consequently, without loss of generality, we can suppose that $\mathbb{K}$ is spherically complete because we know that such a field does admit a spherically complete algebraically closed extension whose absolute value expands that of $\mathbb{K}$. If f belongs to $\mathcal{M}(\mathbb{K})$, we can obviously set it in the form $\frac{g}{h}$, where g, h belong to $\mathcal{A}(\mathbb{K})$ and have no common zero. Next, since $\mathbb{K}$ is supposed to be spherically complete, if f belongs to $\mathcal{M}(d(0, R^-))$, we can also set it in the form $\frac{g}{h}$, where g, h

belong to $\mathcal{A}(d(0, R^-))$ and have no common zero. Consequently, we have $T(r, f) = \max(Z(r, g), Z(r, h))$.

When ξ is a constant, we can obviously suppose that $\xi = 0$. Suppose now $\xi \in \mathcal{M}_u(d(0, R^-))$. Then, $f - \xi$ also belongs to $\mathcal{M}_u(d(0, R^-))$ and $\tau - \xi$ belongs to $\mathcal{M}_b(d(0, R^-))$. Consequently, in both cases, we can assume $\xi = 0$ to prove the claim. Next, up to a change of origin, we can also assume that none of the functions we consider have a pole or a zero at the origin.

Now, we have $\lim_{n\to+\infty}\Big(\inf_{r\in I_n} T(r, f) - Z(r, f)\Big) = +\infty$ i.e.

$$(1) \qquad \lim_{n\to+\infty}\Big(\inf_{r\in I_n}(Z(r, h) - Z(r, g))\Big) = +\infty.$$

Particularly, we note that $T(r, f) = Z(r, h) + O(1)$ whenever $r \in I_n$ when n is big enough.

Consider now $Z(r, f - \tau) = Z(r, g - \tau h)$. But by (1), we can see that $|g|(r) < |\tau|h|(r)\ \forall r \in I_n$ when n is big enough, hence $Z(r, g - \tau h) = Z(r, \tau h)\ \forall r \in I_n$ when n is big enough, hence $Z(r, \tau h) = Z(r, h)\ \forall r \in I_n$ when n is big enough. Therefore, $Z(r, f - \tau) = Z(r, h) + O(1) = T(r, f) + O(1)$, $\forall r \in I_n$ when n is big enough. So the claim is proven when τ is a constant.

Suppose now that $f \in \mathcal{M}(d(0, R^-))$ and $\tau \in \mathcal{M}_b(d(0, R^-))$. By Theorem 36.11, we can write τ in the form $\frac{\phi}{\psi}$, where $\phi, \psi \in \mathcal{A}_b((d(0, R^-))$ have no common zero. Consider $Z(r, f - \tau h) = Z(r, \frac{\psi g - \phi h}{\psi h})$. Since g and h have no common zero and since both ϕ, ψ are bounded, we have $Z(r, \frac{\psi g - \phi h}{\psi h}) = Z(r, \psi g - \phi h) + O(1)$. By (1), in I_n, we have $|\psi g|(r) < |\phi h|(r)$ when n is big enough, and since $|\,.\,|(r)$ is an absolute value, $|\psi g - \phi h|(r) = |\phi h|(r)$ in I_n when n is big enough. Therefore, we have $Z(r, \psi g - \phi h) = Z(r, \phi h) = Z(r, h) + O(1)$ in I_n when n is big enough. Consequently, $Z(r, f - \tau) = Z(r, h) + O(1) = T(r, h) + O(1) = T(r, f) + O(1)\ \forall r \in I_n$ when n is big enough. That finishes proving Lemma 45.1. □

Theorem 45.2. *Let $f \in \mathcal{M}(\mathbb{K})$ and let $a_1, \ldots, a_q \in \mathbb{K}$ be distinct. Then,*

$$(q-1)T(r, f) \le \max_{1\le k\le q}\left(\sum_{j=1, j\ne k}^{q} Z(r, f - a_j)\right) + O(1).$$

Corollary 45.3. *Let $f \in \mathcal{M}(\mathbb{K})$ and let $a_1, \ldots, a_q \in \mathbb{K}$ be distinct. Then, $(q-1)T(r,f) \leq \sum_{j=1}^{q} Z(r, f-a_j) + O(1)$.*

Theorem 45.4. *Let $f \in \mathcal{M}(d(0,R^-))$ and let $\tau_1, \ldots, \tau_q \in \mathcal{M}_b(d(0,R^-))$ be distinct. Then,*

$$(q-1)T(r,f) \leq \max_{1 \leq k \leq q} \left(\sum_{j=1, j\neq k}^{q} Z(r, f-\tau_j) \right) + O(1).$$

Corollary 45.5. *Let $f \in \mathcal{M}(d(0,R^-))$ and let $\tau_1, \ldots, \tau_q \in \mathcal{M}_b(d(0,R^-))$ be distinct. Then, $(q-1)T(r,f) \leq \sum_{j=1}^{q} Z(r, f-\tau_j) + O(1)$.*

Proof (Proof of Theorems 45.2 and 45.4). Suppose Theorems 45.2 (respectively Theorem 45.4) is wrong. In order to make a unique proof for the two theorems, in Theorem 45.2, we set $\tau_j = a_j$. Thus, there exists $f \in \mathcal{M}(\mathbb{K})$ (respectively $f \in \mathcal{M}(d(0,R^-))$) and $\tau_1, \ldots, \tau_q \in \mathbb{K}$ (respectively $\tau_1, \ldots, \tau_q \in \mathcal{M}_b(d(0,R^-))$) such that $(q-1)T(r,f) - \max_{1\leq k\leq q}\left(\sum_{j=1,j\neq k}^{q} Z(r,f-\tau_j)\right)$ admits no superior bound in $]0,+\infty[$. So, there exists a sequence of intervals $J_s = [w_s, y_s]$ such that $w_s < y_s < w_{s+1}$, $\lim_{s\to+\infty} w_s = +\infty$ (respectively $\lim_{s\to+\infty} w_s = R$) and two distinct indices $m \leq q$ and $t \leq q$ such that

$$\lim_{s\to+\infty} \inf_{r\in J_s} \big(T(r,f) - Z(r,f-\tau_m)\big) = +\infty$$

and

$$\lim_{s\to+\infty} \inf_{r\in J_s} \big(T(r,f) - Z(r,f-\tau_t)\big) = +\infty.$$

But by Lemma 45.1, this is impossible. This ends the proof of Theorems 45.2 and 45.4. □

Remark. Theorem 45.2 does not hold in complex analysis. Indeed, let f be a meromorphic function in $\mathbb{C}$ omitting two values a and b, such as $f(x) = \frac{e^x}{e^x-1}$. Then, $Z(r,f-a) + Z(r,f-b) = 0$.

Theorem 45.6. *Let $\mathbb{K}$ be of characteristic 0. Let $\alpha_1, \ldots, \alpha_q \in \mathbb{K}$, with $q \geq 2$, let $S = \{\alpha_1, \ldots, \alpha_q\}$, and let $f \in \mathcal{M}(\mathbb{K})$ (respectively*

$f \in \mathcal{M}(d(0, R^-)))$. Then, $(q-1)T(r,f) \leq \sum_{j=1}^{q} \overline{Z}(r, f-\alpha_j) + Z(r,f') - Z_0^S(r,f') + O(1) \forall r \in I$ (respectively $\forall r \in J$).

Moreover, if f belongs to $f \in \mathcal{A}(\mathbb{K})$ (respectively $\mathcal{A}(d(0, R^-)))$, then $qT(r,f) \leq \sum_{j=1}^{q} \overline{Z}(r, f-\alpha_j) + Z(r,f') - Z_0^S(r,f') + O(1) \forall r \in I$ (respectively $\forall r \in J$).

Theorem 45.7 (Second Main Theorem) Boutabaa. *Let $\mathbb{K}$ be of characteristic 0. Let $\alpha_1, \dots, \alpha_q \in \mathbb{K}$, with $q \geq 2$, let $S = \{\alpha_1, \dots, \alpha_q\}$, and let $f \in \mathcal{M}(\mathbb{K})$ (respectively $f \in \mathcal{M}_u(d(0, R^-)))$. Then,*

$$(q-1)T(r,f) \leq \sum_{j=1}^{q} \overline{Z}(r, f-\alpha_j) + \overline{N}(r,f) - Z_0^S(r,f') - \log r + O(1)$$

$$\forall r \in I \text{ (respectively } \forall r \in J).$$

Proof (Proof of Theorems 45.6 and 45.7). By Theorem 45.2 (respectively 45.4), there exists a constant $B > 0$, and for each $r > 0$ (respectively for each $r \in]0, R[$), there exists $k(r) \in \mathbb{N}$, $k(r) \leq q$, such that $(q-1)T(r,f) \leq \sum_{j=1, j\neq k(r)}^{q} Z(r, f-a_j) + B$ i.e. $(q-1)T(r,f) \leq \sum_{j=1}^{q} Z(r, f-a_j) - Z(r, f-a_{k(r)}) + B$. Now, $\sum_{j=1}^{q} Z(r, f-a_j) = \sum_{j=1}^{q} \overline{Z}(r, f-a_j) + Z(r,f') - Z_0^S(r,f') - \log r$. Consequently, $(q-1)T(r,f) \leq \sum_{j=1}^{q} \overline{Z}(r, f-a_j) + Z(r,f') - Z_0^S(r,f') - Z(r, f-a_{k(r)}) + B$, and this proves the first claim of Theorem 45.6. Particularly, if $f \in \mathcal{A}(\mathbb{K})$ (respectively if $f \in \mathcal{A}(d(0, R^-)))$, then we have $Z(r, f-a_j) = T(r, f-a_j) = T(r,f) + O(1)$ $\forall j = 1, \dots, q$, hence $Z(r, f-a_{k(r)}) = T(r,f) + O(1)$, and therefore, $qT(r,f) \leq \sum_{j=1}^{q} \overline{Z}(r, f-a_j) + Z(r,f') - Z_0^S(r,f') + O(1)$, which ends the proof of Theorem 45.6.

Henceforth, by Theorem 43.13, there exists a constant $c_j > 0$ such that $Z(r,f') \leq Z(r, f-a_j) = N(r, f-a_j) - \log r + c_j$. Let $c = \max(c_1, \dots, c_q)$. Then, $Z(r,f') - Z_0^S(r,f') - Z(r, f-a_{k(r)} \leq \overline{N}(r, f-a_{k(r)}) + c - \log r = \overline{N}(r,f) + c - \log r$. Consequently,

$$\sum_{j=1}^{q} Z(r, f-a_j) = \sum_{j=1}^{q} \overline{Z}(r, f-a_j) + \overline{N}(r,f) - \log r + O(1).$$

That finishes the proof of Theorem 45.7. □

Remark. In Theorem 45.7, in the hypothesis $f \in \mathcal{M}(d(0, R^-))$, the term $-\log r$ has no veritable meaning since r is bounded.

Corollary 45.8. *Let $\mathbb{K}$ be of characteristic 0. Let $\alpha_1, \dots, \alpha_q \in \mathbb{K}$, with $q \geq 2$, let $S = \{\alpha_1, \dots, \alpha_q\}$, and let $f \in \mathcal{M}(\mathbb{K})$ (respectively $f \in \mathcal{M}(d(0, R^-))$). Then,*

$$\sum_{j=1}^{q} \left(Z(r, f - \alpha_j) - \overline{Z}(r, f - \alpha_j)\right) \leq T(r, f) + \overline{N}(r, f) - Z_0^S(r, f')$$
$$- \log r + O(1) \quad \forall r \in I \text{ (respectively } \forall r \in J).$$

We can check that given $f \in \mathcal{A}(\mathbb{K})$ such that $0 < \rho(f) < +\infty$ and $0 < \sigma(f)$, we have $\sigma(f, r) = \frac{T(r,f)}{r^{\rho(f)}}$. Consequently, Corollary 45.9 is just a corollary of Corollary 45.8.

Using $\rho(f)$, we can derive a new form of the main Nevanlinna theorem for entire functions:

Corollary 45.9. *Let $\mathbb{K}$ be of characteristic 0. Let $f \in \mathcal{A}(\mathbb{K})$ be such that $0 < \rho(f) < +\infty$ and let $a_1, \dots, a_q \in \mathbb{K}$. Then,*

$$(q-1)\sigma(f) \leq \limsup_{r \to +\infty} \left(\frac{1}{r^{\rho(f)}} \sum_{i=1}^{q} \overline{Z}(r, f - a_i) \right).$$

Proof. We have $\sigma(f, r) = \frac{\mathrm{Log}(|f|(r))}{r^{\rho(f)}}$, hence by Theorem 45.7,

$$r^{\rho(f)}(q-1)\sigma(f, r) \leq \sum_{i=1}^{q} \overline{Z}(r, f - a_i) - \mathrm{Log}(r) + O(1).$$

The conclusion is then obvious. □

Corollary 45.10. *Let $\mathbb{K}$ be of characteristic 0. Let $f \in \mathcal{A}(\mathbb{K})$ be clean. Let $a_1, \dots, a_q \in \mathbb{K}$. Then,*

$$(q-1)\sigma(f) \leq \liminf_{r \to +\infty} \left(\frac{1}{r^{\rho(f)}} \sum_{i=1}^{q} \overline{Z}(r, f - a_i) \right)$$

and

$$(q-1)\psi(f) \leq \liminf_{r \to +\infty} \left(\frac{\rho(f)}{r^{\rho(f)}} \sum_{i=1}^{q} \overline{Z}(r, f - a_i) \right).$$

Chapter 46

Nevanlinna Theory Out of a Hole

Throughout Chapters 46–57, the field $\mathbb{K}$ is supposed to be of characteristic 0.

Now, we mean to construct a Nevanlinna theory for meromorphic functions in the complement of an open disk thanks to the use of specific properties of the analytic elements on infraconnected subsets of $\mathbb{K}$ already examined.

Given $f \in \mathcal{M}(D)$, for $r > R$, here we denote by $Z_R(r, f)$ the counting function of zeros of f between R and r, i.e. if $\alpha_1, \ldots, \alpha_m$ are the distinct zeros of f in $\Delta(0, R, r)$, with respective multiplicity u_j, $1 \leq j \leq m$, then $Z_R(r, f) = \sum_{j=1}^{m} u_j(\log(r) - \log(|\alpha_j|))$. Similarly, we denote by $N_R(r, f)$ the counting function of poles of f between R and r, i.e. if $\beta_1, \ldots, \beta_n$ are the distinct poles of f in $\Delta(0, R, r)$, with respective multiplicity v_j, $1 \leq j \leq m$, then $N_R(r, f) = \sum_{j=1}^{n} v_j(\log(r) - \log(|\beta_j|))$. Finally, we put $T_R(r, f) = \max\big(Z_R(r, f), N_R(r, f)\big)$.

Next, we denote by $\overline{Z}_R(r, f)$ the counting function of zeros without counting multiplicity: if $\alpha_1, \ldots, \alpha_m$ are the distinct zeros of f in $\Delta(0, R, r)$, then we put $\overline{Z}_R(r, f) = \sum_{j=1}^{m} \log(r) - \log(|\alpha_j|)$.

Similarly, we denote by $\overline{N}_R(r, f)$ the counting function of poles without counting multiplicity: if $\beta_1, \ldots, \beta_n$ are the distinct poles of f in $\Delta(0, R, r)$, then we put $\overline{N}_R(r, f) = \sum_{j=1}^{n} \log(r) - \log(|\beta_j|)$.

Finally, putting $W = \{a_1, \ldots, a_q\}$, we denote by $Z_R^W(r, f')$ the counting function of zeros of f' on points x where $f(x) \notin W$.

Given two functions defined in an interval $I = [b, +\infty[$, we write $\phi(r) = \psi(r) + O(\log(r))$ (respectively $\phi(r) \leq \psi(r) + O(\log(r))$)

if there exists a constant $B > 0$ such that $|\phi(r) - \psi(r)|_\infty \leq B\log(r),\ r \in I$ (respectively $\phi(r) - \psi(r) \leq B\log(r),\ r \in I$).

We write $\phi(r) = o(\psi(r)),\ r \in I$, if $\lim_{r\to+\infty} \frac{\phi(r)}{\psi(r)} = 0$.

Theorem 46.1. *Let $f \in \mathcal{M}(D)$. Then,* $\log(|f|(r)) - \log(|f|(R)) = Z_R(r,f) - N_R(r,f) + mo(f,S)(\log r - \log R)$ $(r \in I)$.

Proof. By Theorem 39.6, we have $f = f^S f^0$. Since f^S has no zero and no pole in D, by Theorem 39.3, it satisfies $|f^S|(r)) = r^{mo(f,S)}$ $\forall r \in I$, hence $\log(|f^S|(r)) - \log(|f^S|(R)) = mo(f,S)(\log r - \log R)$ $(r \in I)$. Next, since f^0 has no zero and no pole in S, we have $\log(|f^0|(r)) - \log(|f^0|(R)) = Z_R(r,f^0) - N_R(r,f^0)$ $(r \in I)$, therefore the statement is clear. □

Corollary 46.2. *Let $f \in \mathcal{M}(D)$. Then $T_R(r,f)$ is identically zero if and only if f is a Motzkin factor.*

Corollary 46.3. *Let $f,\ g \in \mathcal{A}(D)$ satisfy* $\log(|f|(r)) \leq \log(|g|(r))$ $\forall r \geq R$ $(r \in I)$. *Then, $Z_R(r,f) \leq Z_R(r,g) + (mo(g,S) - mo(f,S))(\log(r) - \log(R)),\ (r \in I)$.*

Theorem 46.4. *Let $f \in \mathcal{A}(D)$. Then, $Z_R(r,f') \leq Z_R(r,f) + O(\log(r))$ $(r \in I)$.*

Proof. Indeed, by Theorem 18.2, we have $|f'|(r) \leq \frac{|f|(r)}{r}$. Therefore, the conclusion comes from Theorem 46.1. □

We can now characterize the set $\mathcal{M}^c(D)$:

Theorem 46.5. *Let $f \in \mathcal{M}(D)$. The three following statements are equivalent:*

(i) $\displaystyle\lim_{r\to+\infty} \frac{T_R(r,f)}{\log(r)} = +\infty$ $(r \in I)$,

(ii) $\displaystyle\frac{T_R(r,f)}{\log(r)}$ *is unbounded,*

(iii) *f belongs to $\mathcal{M}^c(D)$.*

Proof. Consider an increasing sequence $(u_n)_{n\in\mathbb{N}}$ in $\mathbb{R}_+$ such that $\lim_{n\to+\infty} u_n = +\infty$ and let $(k_n)_{n\in\mathbb{N}}$ be a sequence of $\mathbb{N}^*$.

Clearly, we have

$$\lim_{r\to+\infty}\frac{\sum_{u_n\le r}k_n(\log(r)-\log(u_n))}{\log(r)}=+\infty.$$

Consequently, if a function $f\in\mathcal{M}^c(D)$ has infinitely many zeros (respectively infinitely many poles in D), then $\lim_{n\to+\infty}\frac{Z_R(r,f)}{\log(r)}=+\infty$ (respectively $\lim_{n\to+\infty}\frac{N_R(r,f)}{\log(r)}=+\infty$), hence in both cases, $\lim_{n\to+\infty}\frac{T_R(r,f)}{\log(r)}=+\infty$. Conversely, if f has finitely many zeros and finitely many poles in D, then we check that $\lim_{n\to+\infty}\frac{T_R(r,f)}{\log(r)}<+\infty$. Thus, the equivalence of the three statements is clear. □

Theorem 46.6. *The set $\mathcal{M}_0(D)=\mathcal{M}(D)\backslash\mathcal{M}^c(D)$ is a subfield of $\mathcal{M}(D)$. Every $f\in\mathcal{M}^c(D)$ is transcendental over $\mathcal{M}_0(D)$.*

Proof. By Theorem 46.5, $\mathcal{M}_0(D)$ is a subfield of $\mathcal{M}(D)$. Now, consider a polynomial $P(Y)=\sum_{j=0}^{n}a_jY^j\in\mathcal{M}_0(D)[Y]$ with $a_n=1$. Let $f\in\mathcal{M}^c(D)$ and suppose that $P(f)=0$. Then, $f^n=-\sum_{j=0}^{n-1}a_jf^j$. Set $\Xi=\sum_{j=0}^{n-1}a_jf^j$ and $f=f^0\frac{g}{h}$ with $g,\ h\in\mathcal{A}(D)$ having no zero in S. Then, $\Xi=\frac{\sum_{j=0}^{n-1}a_jg^jh^{n-1-j}}{h^{n-1}}$.

Since $\sum_{j=0}^{n-1}a_jg^jh^{n-1-j}$ belongs to $\mathcal{A}(D)$, by Theorem 46.7, we have

$$T_R\left(r,\sum_{j=0}^{n-1}a_jg^jh^{n-1-j}\right)\le(n-1)T_R(r,f)+O(\log(r)),\ \ (r\in I),$$

and of course, $T_R(r,h^{n-1})\le(n-1)T_R(r,f),\ (r\in I)$. Consequently

$$T_R(r,\Xi)\le(n-1)T_R(r,f)+O(\log(r),\ \ (r\in I).$$

But on the other hand, by Theorem 46.7, $T_R(r,f^n)=nT_R(r,f)$. Therefore, we should have $nT_R(r,f)\le(n-1)T_R(r,f)+O(\log(r),\ (r\in I)$, which is impossible by Theorem 46.6 because f belongs to $\mathcal{M}^c(D)$. Consequently, the equality $P(f)=0$ is impossible, which proves that f is transcendental over $\mathcal{M}_0(D)$. □

Operations on $\mathcal{M}(D)$ work almost like for meromorphic functions in the whole field.

Theorem (First Main Theorem) 46.7. *Let $f,\ g \in \mathcal{M}(D)$. Then for every $b \in \mathbb{K}$, we have $T_R(r, f+b) = T_R(r, f) + O(\log(r))$ $(r \in I)$, $T_R(r, f.g) \leq T_R(r, f) + T_R(r, g) + O(\log(r))$ $(r \in I)$, $T_R(r, \frac{1}{f}) = T_R(r, f)$, $T_R(r, f+g) \leq T_R(r, f) + T_R(r, g) + O(\log(r))$ $(r \in I)$, and $T_R(r, f^n) = nT_R(r, f)$.*

Let h be a Moebius function. Then, $T_R(r, h \circ f) = T_R(r, f) + O(\log(r))$ $(r \in I)$.

Moreover, if both f and g belong to $\mathcal{A}(D)$, then

$$T_R(r, f+g) \leq \max(T_R(r, f), T_R(r, g)) + O(\log(r)) \ (r \in I)$$

and $T_R(r, fg) = T_R(r, f) + T_R(r, g)$, $(r \in I)$. Particularly, if $f \in \mathcal{A}^c(D)$, then $T_R(r, f+b) = T_R(r, f) + O(1)$ $(r \in I)$. Given a polynomial $P(X) \in \mathbb{K}[X]$ of degree q, then $T_R(r, P \circ f) = qT_R(r, f) + O(\log(r))$.

Proof. Suppose first $f,\ g \in \mathcal{A}(D)$. It is immediate to check that $T_R(r, fg) = Z_R(fg) = Z_R(f) + Z_R(r, g) = T_R(r, f) + T_R(r, g)$, that $T_R(r, f^n) = nT_R(r, f)$, and that $T_R(r, \frac{1}{f}) = T_R(r, f)$.

Then, $T_R(r, f+g) = Z_R(r, f+g) = \log(|f+g|(r)) - mo(f+g, S)(\log(r) - \log(R))$, $(r \in I)$. But $\log(|f+g|(r)) \leq \max\big(\log(|f|(r)), \log(|g|(r))\big)$, hence

$$\begin{aligned}&Z_R(r, f+g) \leq \max\big(Z_R(r, f) + mo(f, S)(\log(r) - \log(R)),\\ &Z_R(r, g) + mo(g, S)(\log(r) - \log(R))\big),\end{aligned}$$

and hence $T_R(r, f+g) \leq \max\big(T_R(r, f), T_R(r, g)\big) + O(\log(r))$.

Particularly, given $b \in \mathbb{K}$, we have $T_R(r, f+b) \leq T_R(r, f) + O(\log(r)) \leq T_R(r, f) + O(\log(r))$, hence $T_R(r, f+b) = T_R(r, f) + O(\log(r))$.

Now, given a polynomial of degree q, we have $Z_R(r, P \circ f) = qZ_R(r, f) + O(\log(r))$ and $N_R(r, P \circ f) = qN_R(r, f)$, hence $T_R(r, P \circ f) = qT_R(r, f)$.

Now, suppose $f \in \mathcal{A}^c(D)$. Then $f(x)$ is a Laurent series $\sum_{-\infty}^{+\infty} a_n x^n$ convergent in all D such that $\lim_{r\to+\infty} |f|(r) = +\infty$. Let $b \in \mathbb{K}$ and take V be such that $|f|(r) > |b|\ \forall r \geq V$. Then for every $r > V$, $|f|(r)$ is of the form $|a_k|r^k$ with $k > 0$, $|a_n|r^n < |a_k|r^k\ \forall n > k$ and the number of zeros of f in $\Delta(0, R, r)$ is $k - mo(f, S)$. Next, $f - b$ is of the form $\sum_{-\infty}^{+\infty} c_n x^n$ with $c_n = a_n\ \forall n \neq 0$ and $c_0 = a_0 - b$. Consequently, $f - b$ has the same number of zeros in $\Delta(0, R, r)$ and in each circle $C(0, r)$ for $r > V$. Therefore, $T_R(r, f) = T_R(r, f - b)$ when r is big enough.

Next, consider the general case: $f,\ g \in \mathcal{M}(D)$. First, it is immediate to check that $T_R(r, fg) \leq T_R(r, f) + T_R(r, g)$. Similarly, for $T_R(r, \frac{1}{f})$. By definition, we have $Z_R(r, \frac{1}{f}) = N_R(r, f)$ and $N_R(r, \frac{1}{f}) = Z_R(r, f)$, hence $T_R(r, \frac{1}{f}) = T_R(r, f)$.

Now, consider $T_R(r, f + g)$ in the general case: $f,\ g \in \mathcal{M}(D)$. By Theorem 39.6, we can write

$$f + g = f^S \left(\frac{f_1^0}{f_2^0} \right) + g^S (g_1^0,\ g_2^0),$$

hence

$$T_R(r, f + g) = T_R \left(r, \frac{f^S f_1^0 g_2^0 + g^S g_1^0 f_2^0}{f_2 g_2} \right)$$

with $f_1^0,\ f_2^0,\ g_1^0,\ g_2^0 \in \mathcal{A}(\mathbb{K})$, having no zero in T and $f^S,\ g^S$ Motzkin factors associated with S. Then, $Z_R(r, f^S f_1^0 g_2^0) = Z_R(r, f_1^0) + Z_R(r, g_2^0), Z_R(r, g^S g_1^0 f_2^0) = Z_R(r, g_1^0) + Z_R(r, f_2^0)$, hence, by what we just saw, $Z_R(r, f^S f_1^0 g_2^0 + g^S g_1^0 f_2^0) \leq \max\left(T_R(r, f), T_R(r, g)\right) + O(\log(r))$. And obviously, $Z_R(r, f_2 g_2) \leq T_R(r, f) + T_R(r, g)$. So we obtain in the general case $T_R(r, f + g) \leq T_R(f) + T_R(r, g) + O(\log(r))$.

Finally, consider a Moebius function h. Then, $h \circ f(x)$ is of the form $C + \frac{\epsilon}{\alpha f(x) + \beta}$, and thereby, $T_R(r, h \circ f) = T_R(r, f) + O(\log(r))$. □

Theorem 46.8. *Every $f \in \mathcal{M}^c(D)$ is transcendental over $\mathcal{M}_0(D)$.*

Proof. Consider a polynomial $P(Y) = \sum_{j=0}^{n} a_j Y^j \in \mathcal{M}_0(D)[Y]$ with $a_n = 1$. Let $f \in \mathcal{M}^c(D)$ and suppose that $P(f) = 0$. Then, $f^n = -\sum_{j=0}^{n-1} a_j f^j$. Set $\Xi = \sum_{j=0}^{n-1} a_j f^j$ and $f = f^0 \frac{g}{h}$ with $g,\ h \in \mathcal{A}(D)$ having no zero in S. Then, $\Xi = \frac{\sum_{j=0}^{n-1} a_j g^j h^{n-1-j}}{h^{n-1}}$.

Since $\sum_{j=0}^{n-1} a_j g^j h^{n-1-j}$ belongs to $\mathcal{A}(D)$, by Theorem 46.7, we have

$$T_R \left(r, \sum_{j=0}^{n-1} a_j g^j h^{n-1-j} \right) \leq (n-1) T_R(r, f) + O(\log(r)),\ \ (r \in I),$$

and of course, $T_R(r, h^{n-1}) \leq (n-1) T_R(r, f),\ (r \in I)$. Consequently,

$$T_R(r, \Xi) \leq (n-1) T_R(r, f) + O(\log(r),\ \ (r \in I).$$

But on the other hand, by Theorem 46.7, $T_R(r, f^n) = nT_R(r, f)$. Therefore, we should have $nT_R(r, f) \leq (n-1)T_R(r, f) + O(\log(r)$, $(r \in I)$, which is impossible by Theorem 46.6 because f belongs to $\mathcal{M}^c(D)$. Consequently, the equality $P(f) = 0$ is impossible, which proves that f is transcendental over $\mathcal{M}_0(D)$. □

Theorem 46.9. *Let $f \in \mathcal{M}(D)$. Then, $N_R(r, f^{(k)}) = N_R(r, f) + k\overline{N}_R(r, f)$, $(r \in I)$ and $Z_R(r, f^{(k)}) \leq Z_R(r, f) + k\overline{N}_R(r, f) + O(\log(r))$, $(r \in I)$.*

Proof. The inequality $N_R(r, f^{(k)}) = N_R(r, f) + k\overline{N}_R(r, f) + O(1)$, $r \in I$, is obvious. Next, consider f in the form $\frac{g}{h}$ with g, $h \in \mathcal{A}(\mathbb{K})$. Recall that we can write h in the form $\overline{h}\widetilde{h}$ with $\overline{h}$ and $\widetilde{h}$ in $\mathcal{A}(\mathbb{K})$, each zero of $\overline{h}$ being of order one and all zeros of h being a zero of $\overline{h}$. So, h' is of the form $\widetilde{h}\widehat{h}$, where $\widehat{h}$ belong to $\mathcal{A}(\mathbb{K})$ and none of the zeros of $\widehat{h}$ is a zero of h. Then, f' is of the form $\frac{g'\overline{h}-g\widehat{h}}{h\overline{h}}$. So, $Z_R(r, f') \leq Z_R(r, g'\overline{h} - g\widehat{h})$, and hence, by Theorem 46.7,

$$Z_R(r, f') \leq \max(Z_R(r, g'\overline{h}), Z_R(r, g\widehat{h}). \tag{1}$$

On one hand, by Theorem 46.5, $Z_R(r, g') \leq Z_R(r, g) + O(\log r)$, and by Corollary 46.4, we have $Z_R(r, g') \leq Z_R(r, f) + O(\log(r))$. Obviously, $Z_R(r, \overline{h}) \leq \overline{Z}_R(r, h) = \overline{N}_R(r, \ell) = \overline{N}_R(r, f)$, hence $Z_R(r, g'h) \leq Z_R(r, f) + \overline{N}_R(r, f)$.

Now, let us estimate $Z_R(r, \widehat{h})$. Since $\log(|h'|(r)) \leq \log(|h|(r)) - \log r$, we have $Z_R(r, h') \leq Z_R(r, h) + O(\log(r))$. But since $h' = \widetilde{h}\widehat{h}$, we have $Z_R(r, \widehat{h}) = Z_R(r, h') - Z_R(r, \widetilde{h}) \leq Z_R(r, h) - Z_R(r, \widetilde{h}) + O(\log(r)) = Z_R(r, \overline{h}) + O(\log(r)) = \overline{N}_R(r, f) + O(\log(r))$. Consequently,

$$\begin{aligned} Z_R(r, g\widehat{h}) &\leq Z_R(r, g) + \overline{N}_R(r, f) + O(\log(r)) \\ &= Z_R(r, f) + \overline{N}_R(r, f) + O(\log(r)). \end{aligned}$$

Thus, by (1), we have proven the claim when $k = 1$ and then it is immediately derived by induction on k. □

Similar to Lemmas 43.17 and 46.9, now we have Lemma 46.10 and Corollary 46.11:

Lemma 46.10. *Let $f \in \mathcal{M}(D)$ and let $G = \frac{f'}{f}$. Then, G satisfies $Z_R(r, G) \leq N_R(r, G) - \log r + O(1)$ $(r \geq R)$.*

Corollary 46.11. *Let $f \in \mathcal{M}(D)$ and let $G = \frac{f'}{f}$. Then, $T_R(r, G) - N_R(r, G) = 0 + O(1)$.*

Lemma 46.12 is necessary in the proof of Theorem 46.13.

Lemma 46.12. *Let $f \in \mathcal{M}(D)$. Suppose that there exists $\xi \in \mathbb{K}$ and a sequence of intervals $J_n = [u_n, v_n]$ such that $u_n < v_n < u_{n+1}$, $\lim_{n\to+\infty} u_n = +\infty$, and $\lim_{n\to+\infty} \left[\inf_{r\in J_n} \frac{T_R(r,f)-Z_R(r,f-\xi)}{\log(r)}\right] = +\infty$.*

Let $\tau \in \mathbb{K}$ $\tau \neq \xi$. Then, $Z_R(r, f-\tau) = T_R(r, f) + O(\log(r)))$ $\forall r \in J_n$ when n is big enough.

Proof. Without loss of generality, we can obviously suppose that $\xi = 0$. By Theorem 39.6, f is of the form $f^S f^0$ and f^0 is of the form $\frac{g}{h}$ with $g,\ h \in \mathcal{A}(D)$, having no zero in S. Set $w = f^S$. Thus, we have

$$\lim_{n\to+\infty}\left[\inf_{r\in J_n} \frac{Z_R(r,h) - Z_R(r,g)}{\log(r)}\right] = +\infty.$$

Consequently, by Theorem 46.1,

$$\lim_{n\to+\infty}\left[\inf_{r\in J_n} \frac{\log(|h|(r) - \log(|g|(r)}{\log(r)}\right] = +\infty. \tag{1}$$

Consider now $f - \tau$. We have $f - \tau = \frac{wg-\tau h}{h}$, hence

$$\log(|f|(r)) = \log\left(|wg - \tau h|(r) - \log(|h|(r)\right).$$

But by (1), we have $\log(|\tau h|(r)) > \log(|wg|(r))$ because $\log(|w|(r) = O(\log(r))$, therefore $\log\left(|wg - \tau h|(r)\right) = \log(|\tau h|(r))$ $\forall r \in J_n$ when n is big enough, and hence

$$\lim_{n\to+\infty}\left[\sup_{r\in J_n} \frac{\log(|\tau h - wg|(r) - \log(|h|(r)}{\log(r)}\right] = 0. \tag{2}$$

Consequently, by (2) and by Theorem 46.1,

$$\lim_{n\to+\infty}\left[\sup_{r\in J_n} \frac{Z_R(r,\tau h - wg) - Z_R(r,h)}{\log(r)}\right] = 0$$

i.e.

$$\lim_{n\to+\infty}\left[\sup_{r\in J_n} \frac{Z_R(r,f-\tau) - T_R(r,f)}{\log(r)}\right] = 0,$$

which proves the claim. □

The Nevanlinna second main theorem is based on the following theorem:

Theorem 46.13. *Let $f \in \mathcal{M}(D)$ and let $a_1, \dots, a_q \in \mathbb{K}$ be distinct. Then,*

$$(q-1)T_R(r,f) \leq \max_{1\leq k\leq q}\left(\sum_{j=1, j\neq k}^{q} Z_R(r, f-a_j)\right) + O(\log(r)) \ (r \in I).$$

Proof. Suppose Theorem 46.13 is wrong. Thus, there exists $f \in \mathcal{M}(D)$ and $a_1, \dots, a_q \in \mathbb{K}$ such that $(q-1)T_R(r,f) - \max_{1\leq k\leq q}\left(\sum_{j=1, j\neq k}^{q} Z_R(r, f-a_j)\right)$ admits no superior bound in $]0, +\infty[$. So, there exists a sequence of intervals $J_s = [w_s, y_s]$ such that $w_s < y_s < w_{s+1}$, $\lim_{s\to+\infty} w_s = +\infty$ and two distinct indices $m \leq q$ and $t \leq q$ such that

$$\lim_{s\to+\infty}\left[\inf_{r\in J_s} \frac{(T_R(r,f) - Z_R(r, f-a_m))}{\log(r)}\right] = +\infty$$

and

$$\lim_{s\to+\infty}\left[\inf_{r\in J_s} \frac{(T_R(r,f) - Z_R(r, f-a_t))}{\log(r)}\right] = +\infty.$$

But by Lemma 46.12, that is impossible. □

We can now state and prove the second main theorem for $\mathcal{M}(D)$.

Theorem 46.14. *Let $\mathbb{K}$ be of characteristic 0. Let $f \in \mathcal{M}(D)$, let $\alpha_1, \dots, \alpha_q \in \mathbb{K}$, with $q \geq 2$, and let $W = \{\alpha_1, \dots, \alpha_q\}$. Then,*

$$\begin{aligned}(q-1)T_R(r,f) \leq \sum_{j=1}^{q} \overline{Z}_R(r, f-\alpha_j) + Z_R(r, f') \\ - Z_R^W(r, f') + O(\log(r)) \quad (r \in I).\end{aligned}$$

Moreover, if f belongs to $\mathcal{A}(D)$, then

$$\begin{aligned}qT_R(r,f) \leq \sum_{j=1}^{q} \overline{Z}_R(r, f-\alpha_j) + Z_R(r, f') \\ - Z_R^W(r, f') + O(\log(r)) \quad (r \in I).\end{aligned}$$

Theorem 46.15 (Second Main Theorem). *Let $\mathbb{K}$ be of characteristic 0. Let $f \in \mathcal{M}(D)$, let $\alpha_1, \ldots, \alpha_q \in \mathbb{K}$, $q \geq 2$, and let $W = \{\alpha_1, \ldots, \alpha_q\}$. Then,*

$$(q-1)T_R(r, f) \leq \sum_{j=1}^{q} \overline{Z}_R(r, f-\alpha_j) + \overline{N}_R(r, f) - Z_R^W(r, f') + O(\log(r)) \quad (r \in I).$$

Proof (Proof of Theorems 46.14 and 46.15). By Theorem 46.13, there exists a constant $B > 0$, and for each $r > R$, there exists $k(r) \in \mathbb{N}$, $k(r) \leq q$, such that

$$(q-1)T_R(r, f) \leq \sum_{j=1, j\neq k(r)}^{q} Z_R(r, f-a_j) + B\log(r).$$

Now, since $\mathbb{K}$ is of characteristic 0, we have

$$\sum_{j=1}^{q} Z_R(r, f-a_j) = \sum_{j=1}^{q} \overline{Z}_R(r, f-a_j) + Z_R(r, f') - Z_R^W(r, f') + B\log(r).$$

Consequently,

(9)
$$(q-1)T_R(r, f) \leq \sum_{j=1}^{q} \overline{Z}_R(r, f-a_j) + Z_R(r, f') - Z_R^W(r, f') - Z_R(r, f-a_{k(r)}) + O(\log(r))$$

and this proves the first claim of Theorem 46.14.

Particularly, if $f \in \mathcal{A}(D)$, then we have $Z_R(r, f-a_j) = T_R(r, f-a_j) = T_R(r, f) + O(\log(r))$ $\forall j = 1, \ldots, q$, hence $Z_R(r, f-a_{k(r)}) = T_R(r, f) + O(\log(r))$, and therefore,

$$qT_R(r, f) \leq \sum_{j=1}^{q} \overline{Z}_R(r, f-a_j) + Z_R(r, f') - Z_R^W(r, f') + O(\log(r)),$$

which ends the proof of Theorem 46.14.

Consider now the situation in Theorem 46.15. By Theorem 46.9, for each $j = 1, \ldots, q$, there exists a constant $B_j > 0$ such that $Z_R(r, f') \leq Z_R(r, f-a_j) + \overline{N}_R(r, f-a_j) + B_j\log(r)$. Consequently, there exists a constant $C > 0$ such that $Z_R(r, f') \leq Z_R(r, f-a_{k(r)}) + \overline{N}_R(r, f-a_{k(r)}) + C\log(r)$ $\forall r > R$.

Therefore, by Relation (9), that remains true in Theorem 46.15, we can derive

$$(q-1)T_R(r,f) \le \sum_{j=1}^{q} \overline{Z}_R(r,f-\alpha_j) + \overline{N}_R(r,f) - Z_R^W(r,f') + O(\log(r)) \quad \forall r \in I.$$

□

Corollary 46.16. *Let $f \in \mathcal{M}(\mathbb{K})$ and let $a_1,\dots,a_q \in \mathbb{K}$ be distinct. Then, $(q-1)T_R(r,f) \le \sum_{j=1}^{q} Z_R(r,f-a_j) + O(\log(r))$ $(r \in I)$.*

Corollary 46.17. *Let $\mathbb{K}$ be of characteristic 0. Let $f \in \mathcal{M}(D)$, let $\alpha_1,\dots,\alpha_q \in \mathbb{K}$, with $q \ge 2$, and let $W = \{\alpha_1,\dots,\alpha_q\}$. Then*

$$\sum_{j=1}^{q} \left(Z_R(r,f-\alpha_j) - \overline{Z}_R(r,f-\alpha_j)\right) \le T_R(r,f) + \overline{N}_R(r,f) - Z_R^W(r,f') + O(\log(r)) \quad (r \in I).$$

Chapter 47

Immediate Applications of the Nevanlinna Theory

Throughout this chapter, the field $\mathbb{K}$ is supposed to be of characteristic 0.

Notation. In all the chapters, the field $\mathbb{K}$ is supposed to be of characteristic 0. As in Chapter 41, we denote by D the set $\mathbb{K}\backslash d(0, R^-)$ with R a positive number. The definitions of $\mathcal{A}(D)$, $\mathcal{A}^c(D)$, $\mathcal{M}(D)$, and $\mathcal{M}^c(D)$ are those given in Chapter 46.

As immediate applications of the second main theorem, we can note Theorems 47.1–47.4.

Theorem 47.1. *Let $a_1, a_2 \in \mathbb{K}(a_1 \neq a_2)$ and let f, $g \in \mathcal{A}(\mathbb{K})$ satisfy $f^{-1}(\{a_i\}) = g^{-1}(\{a_i\})(i = 1, 2)$. Then, $f = g$.*

Remark. Theorem 47.1 does not hold in complex analysis. Indeed, let $f(z) = e^z$, $g(z) = e^{-z}$, and let $a_1 = 1$, $a_2 = -1$. Then, $f^{-1}(\{a_i\}) = g^{-1}(\{a_i\})$ $(i = 1,\ 2)$, though $f \neq g$.

Theorem 47.2. *Let a_1, a_2, $a_3 \in \mathbb{K}$ $(a_i \neq a_j\ \forall i \neq j)$ and let f, $g \in \mathcal{A}_u(d(a, R^-))$ (respectively, $g \in \mathcal{A}^c(D)$) satisfy $f^{-1}(\{a_i\}) = g^{-1}(\{a_i\})(i = 1,\ 2,\ 3)$. Then, $f = g$.*

Theorem 47.3. *Let a_1, a_2, a_3, $a_4 \in \mathbb{K}(a_i \neq a_j\ \forall i \neq j)$ and let f, $g \in \mathcal{M}(\mathbb{K})$ satisfy $f^{-1}(\{a_i\}) = g^{-1}(\{a_i\})(i = 1,\ 2,\ 3,\ 4)$. Then, $f = g$.*

Theorem 47.4. *Let* $a_1,\ a_2,\ a_3,\ a_4,\ a_5 \in \mathbb{K}(a_i \neq a_j\ \forall i \neq j)$ *and let* $f,\ g \in \mathcal{M}_u(d(a,R^-))$ *(respectively* $f,\ g \in \mathcal{M}^c(D)$*) satisfy* $f^{-1}(\{a_i\}) = g^{-1}(\{a_i\})$ $(i = 1,\ 2,\ 3,\ 4,\ 5)$. *Then,* $f = g$.

Remark. Let $f(x) = \frac{x}{3x-1}$, $g(x) = \frac{x^2}{x^2+2x-1}$. Let $a_0 = 0$, $a_1 = 1$, $a_2 = \frac{1}{2}$. Then we can check that $f^{-1}(\{a_i\}) = g^{-1}(\{a_i\})$, $i = 1,\ 2,\ 3$. So, Theorem 47.3 is sharp.

Proof (Theorems 47.1–47.4). Let $I =]0, +\infty[$ in Theorems 47.1 and 47.3 and let $I =]0, R[$ in Theorems 47.2 and 47.4. For each $j = 1, \dots, n$, let S_j be the set of all zeros of $f - a_j$ (without taking multiplicities into account). Since $a_i \neq a_j\ \forall i \neq j$, we have $S_i \cap S_j = \emptyset\ \forall i \neq j$. Next, we note that $f(x) = a_j$ implies $f(x) - g(x) = 0$. Consequently, we check the following:

(1) $\sum_{j=1}^{n} \overline{Z}(r, f - a_j) \leq \overline{Z}(r, f - g)$.

Suppose first that f and g either belong to $\mathcal{A}(K)$ or belong to $\mathcal{A}(d(0, R^-))$.

By applying Theorem 45.7 to f, we obtain

$$\begin{aligned}(n-1)T(r,f) &\leq \sum_{j=1}^{n} \overline{Z}(r, f - a_j) + \overline{N}(r,f) - \log(r) + O(1)\\ &\leq n\overline{Z}(r, f-g) + \overline{N}(r,f) - \log(r) + O(1) \quad (r \in I),\end{aligned}$$

hence by (1),

$$(n-1)T(r,f) \leq T(r, f-g) + \overline{N}(r,f) - \log(r) + O(1)\ (r \in I),$$

and finally

$$(n-1)T(r,f) \leq T(r, f-g) + N(r,f) - \log(r) + O(1)\ (r \in I).$$

Similarly,

$$(n-1)T(r,g) \leq T(r, f-g) + N(r,g) - \log(r) + O(1)\ (r \in I),$$

therefore we obtain

(2) $(n-1)\max(T(r,f), T(r,g)) \leq T(r, f-g)) + \max(N(r,f), N(r,g)) - \log(r) + O(1)\ (r \in I)$.

Assume we are in the hypothesis of Theorem 47.1. We have $N(r,f) = N(r,g) = 0$, and by Theorem 43.3, $T(r,f-g) \leq \max(T(r,f),T(r,g)) + O(1)$. Consequently, by (2),

$$\begin{aligned}&(n-1)\max(T(r,f),T(r,g))\\&\quad\leq \max(T(r,f),T(r,g)) - \log(r) + O(1) \quad (r \in I).\end{aligned}$$

Since r is not bounded, we can see that the inequality does not hold with $n = 2$, when r goes to $+\infty$.

Now, assume the hypothesis of Theorem 47.2. Again, we have $N(r,f) = N(r,g) = 0$, and by Theorem 43.3, $T(r,f-g) \leq \max(T(r,f),T(r,g)) + O(1)$, hence by (2),

$$(n-1)\max(T(r,f),T(r,g)) \leq \max(T(r,f),T(r,g)) + O(1) \ (r \in I).$$

Since f, g are unbounded, by Theorem 43.10, so are $T(r,f)$, $T(r,g)$ in intervals $]r_0, R[$, hence the inequality does not hold with $n = 3$.

Suppose now that f and g belong to $\mathcal{A}^c(D)$. We then obtain

$$\begin{aligned}(n-1)T_R(r,f) &\leq T_R(r,f-g) + O(\log(r)) \quad (r > R)\\&\leq \max(T_R(r,f),T_R(r,g)) + O(\log(r),\end{aligned}$$

and similarly,

$$\begin{aligned}(n-1)T_R(r,g) &\leq T_R(r,f-g) + O(\log(r)) \quad (r > R)\\&\leq \max(T_R(r,f),T_R(r,g)) + O(\log(r)),\end{aligned}$$

therefore

$$(n-1)\max(R_R(r,f),T(r,g)) \leq \max(T_R(r,f),T_R(r,g)) + O(\log(r)),$$

and hence $n \leq 2$, which proves the conclusion whenever $n \geq 3$.

Assume now the hypothesis of Theorem 47.3. Since

$$\max(N(r,f),N(r,g)) \leq \max(T(r,f),T(r,g)),$$

by (2) and Theorem 43.3, we have

$$\begin{aligned}&(n-1)\max(T(r,f),T(r,g))\\&\quad\leq 3\max(T(r,f),T(r,g)) - \log(r) + O(1) \ (r \in I).\end{aligned}$$

Since r is not bounded, the inequality does not hold with $n = 4$, when r goes to ∞.

Finally, assume we are in the hypothesis of Theorem 47.4. Suppose first that f and g belong to $\mathcal{M}_u(d(0, R^-))$. By (2) and Theorem 43.3, we have

$$(n-1)\max(T(r,f),T(r,g)) \leq 3\max(T(r,f),T(r,g))+O(1)\ (r \in I).$$

Since $T(r, f)$, $T(r, g)$ are not bounded, the inequality does not hold with $n = 5$.

And now, suppose that f and g belong to $\mathcal{M}^c(D)$. We then obtain

$$\begin{aligned}(n-1)T_R(r,f) &\leq T_R(r, f-g) + O(\log(r)) \quad (r > R)\\ &\leq 3(T_R(r,f) + T_R(r,g)) + O(\log(r)),\end{aligned}$$

and similarly,

$$\begin{aligned}(n-1)T_R(r,g) &\leq T_R(r, f-g) + O(\log(r)) \quad (r > R)\\ &\leq 3(T_R(r,f) + T_R(r,g)) + O(\log(r)),\end{aligned}$$

therefore

$$(n-1)(T_R(r,f) + T_R(r,g)) \leq 3(T_R(r,f) + T_R(r,g)) + O(\log(r)),$$

and hence $n \leq 4$, which proves the conclusion whenever $n \geq 5$. That finishes the proof of Theorem 47.4. □

Definitions. Let $f \in \mathcal{M}(\mathbb{K})$. The function f is called *a function of uniqueness* (respectively *a function of strong uniqueness*) for a family $\mathcal{F}$ of functions defined in a suitable subset of $\mathbb{K}$ if given any two functions $f,\ g \in \mathcal{F}$ satisfying $h \circ f = h \circ g$ (respectively $h \circ f = b(h \circ g)$ with $b \in \mathbb{K}^*$), then f and g are identical.

Similarly, we consider the same question in the purely algebraic context. Let E be an algebraically closed field, let $h \in E(x)$, and let $\mathcal{F}$ be a subset of $E(x)$. Then h is called *a function of uniqueness for $\mathcal{F}$* (respectively *a function of strong uniqueness for $\mathcal{F}$*) if given any two functions $f,\ g \in \mathcal{F}$ satisfying $h \circ f = h \circ g$ (respectively $h \circ f = b(h \circ g)$, with $b \in E^*$), f and g are identical.

Particularly in each case, if h is a polynomial, it is called *a polynomial of uniqueness for the family $\mathcal{F}$* (respectively *a polynomial of strong uniqueness for the family $\mathcal{F}$*).

In Theorem 47.6, we need the following basic lemma [59]:

Lemma 47.5. *Let E be an algebraically closed field of characteristic 0 and let $P(x) = (n-1)^2(x^n-1)-n(n-2)(x^{n-1}-1)^2 \in E[x]$. Then P admits 1 as a zero of order 4 and all other zeros u_j $(1 \leq j \leq 2n-6)$ are simple.*

Theorem 47.6. *Let*

$$Q(x) = b((n+2)(n+1)x^{n+3} - 2(n+3)(n+1)x^{n+2} + (n+3)(n+2)x^{n+1})$$

with $b \in \mathbb{K}^*$. Let $R \in]0, +\infty[$. Then Q is a polynomial of uniqueness for $\mathcal{M}(\mathbb{K})$ for every $n \geq 2$ and is a polynomial of uniqueness for $\mathcal{M}_u(d(0,R^-))$ and for $\mathcal{M}^c(D)$ for every $n \geq 3$.

Proof. Suppose $f,\ g \in \mathcal{M}(\mathbb{K})$ (respectively $f,\ g \in \mathcal{M}_u(d(0,R^-))$, respectively $f,\ g \in \mathcal{M}^c(D)$) and suppose that $Q(f) = Q(g)$. Let $h = \frac{f}{g}$. We can derive

$$(n+2)(n+1)(h^{n+3}-1)g^2 - 2(n+3)(n+1)(h^{n+2}-1)g + (n+3)(n+2)(h^{n+1}-1) = 0.$$

If h is a constant, it is 1, a contradiction. So, we suppose h is not constant. If g lies in $\mathcal{M}(\mathbb{K})$, so does h. Now, if g belongs to $\mathcal{M}_u(d(0,R^-))$ or to $\mathcal{M}^c(D)$, so does h, respectively. Indeed, suppose that $h \in \mathcal{M}_b(d(0,R^-))$. Then clearly we have $T(r,(n+2)(n+1)(h^n+2)g^2) \geq 2T(r,g)+O(1)$, while $T(r,-2(n+3)(n+1)(h^{n+2}-1)g+(n+3)(n+2)(h^{n+1}-1)) \leq T(r,g)+O'1$, a contradiction. Similarly, if h has finitely many zeros and poles, we have the same contradiction.

Let $P(x) = (n+2)^2(x^{n+3})-1)-(n+3)(n+1)(x^{n+2}-1)^2 \in \mathbb{K}[x]$. By Lemma 47.5, P admits 1 as a zero of order 4 and all other zeros u_j $(1 \leq j \leq 2n)$ are simple. By change of variable, we can obviously assume that $h - u_j$ has no zero and no pole at 0. Consequently, we check that

$$\left(g - \left(\frac{n+3}{n+2}\right)\left(\frac{h^{n+2}-1}{h^{n+3}-1}\right)\right)^2 = \frac{(n+3)(h-1)^4\prod_{j=1}^{2n}(h-u_j)}{(n+2)^2(n+1)(h^{n+3}-1)^2}.$$

Since $\frac{(n+3)(h-1)^4\prod_{j=1}^{2n}(h-u_j)}{(n+2)^2(n-+1)(h^{n+3}-1)^2}$ is equal to a square, clearly each zero of $h-u_j$ $(1 \leq j \leq 2n)$ has order at least 2. Let $J =]0,+\infty[$ (respectively

$J =]0, R[$, respectively $J = [R, +\infty[$). Consequently,

$$\sum_{j=1}^{2n} \overline{Z}(r, h - u_j) \leq \frac{1}{2} \sum_{j=1}^{2n} Z(r, h - u_j) \leq \frac{1}{2}(2n)T(r, h) + O(1) \ (r \in J).$$

Suppose first that f and g belong to $\mathcal{M}(\mathbb{K})$ or to $\mathcal{M}(d(0, R^-))$. Then, applying Theorem 45.7 to h at the points u_j ($1 \leq j \leq 2n$), we obtain

$$\begin{aligned}(2n-1)T(r,h) &\leq \sum_{j=1}^{2n} \overline{Z}(r, h - u_j) + \overline{N}(r, h) - \log(r) + O(1) \\ &\leq \frac{1}{2} \sum_{j=1}^{2n} Z(r, h - u_j) + \overline{N}(r, h) + O(1) \\ &\leq \frac{1}{2}(2n)T(r, h) + \overline{N}(r, h) - \log(r) + O(1) \quad (r \in J),\end{aligned}$$

and therefore $(2n - 1)T(r, h) \leq nT(r, h) + T(r, h) - \log(r) + O(1)$. If f, g belong to $\mathcal{M}(\mathbb{K})$, we conclude that $n \leq 1$. And if f, g belong to $\mathcal{M}_u(d(0, R^-))$, we conclude that $n \leq 2$.

Suppose now that f and g belong to $\mathcal{M}^c(D)$. Then we can apply Theorem 46.13 and we have $(2n-1)T_R(r, h) \leq nT_R(r, h) + T_R(r, h) + O(\log(r))$, therefore we have again $n \leq 2$, which ends the proof. □

Corollary 47.7. *Let $P(x) \in \mathbb{K}[x]$ have a derivative of the form $c(x - a)^n(x - b)^2$. Then P is a polynomial of uniqueness for $\mathcal{M}(\mathbb{K})$ $\forall n \geq 2$ and is a polynomial of uniqueness for $\mathcal{M}_u(d(\alpha, R^-))$ and for $\mathcal{M}^c(D)$ for every $n \geq 3$.*

Theorem 47.8. *Let $Q(x) = x^{n+1} - x^n$. Let $a \in \mathbb{K}$ and $R \in]0, +\infty[$. Then Q is a polynomial of uniqueness for $\mathcal{A}(\mathbb{K})$, for $\mathcal{A}_u(d(0, R^-))$, and for $\mathcal{A}^c(D)$ for every $n \geq 2$.*

Proof. Let f, $g \in \mathcal{A}(\mathbb{K})$ (respectively f, $g \in \mathcal{A}_u(d(0, R^-))$, respectively f, $g \in \mathcal{A}^c(D)$. Suppose $f^n(f - 1) = g^n(g - 1)$. Let $h = \frac{f}{g}$ and suppose h is not the constant 1. Then we have $g = \left(\frac{h^n - 1}{h^{n+1} - 1}\right)$. Consequently, if h belongs to $\mathbb{K}$ (respectively to $\mathcal{M}_b(d(0, R^-))$, respectively to $\mathcal{M}_0(D)$), so does g, a contradiction. Thus, h belongs to $\mathcal{M}(\mathbb{K}) \setminus \mathbb{K}$ (respectively to $\mathcal{M}_u(d(0, R^-))$, respectively to $\mathcal{M}^c(D)$).

Now, since $n \geq 2$, by Corollary 36.23, h has to take at least one of the $(n+1)$th roots of 1 other than 1 and such an $(n+1)$th root of 1 cannot be an nth root of 1. Consequently, g admits a pole, a contradiction. Therefore, h is identically equal to 1 and hence $f = g$. □

Now, we must examine the situation in $\mathcal{M}(D)$ in order to obtain a result similar to Theorem 47.8

Theorem 47.9. *Let $Q(x) = x^{n+1} - x^n$. Let $a \in \mathbb{K}$ and $R \in]0, +\infty[$. Then, Q is a polynomial of uniqueness for $\mathcal{A}^c(D)$ for every $n \geq 2$.*

Proof. Let $f, g \in \mathcal{A}^c(D)$. Suppose $f^n(f-1) = g^n(g-1)$. Let $h = \frac{f}{g}$ and suppose h is not the constant 1. Then, we have $g = \left(\frac{h^n-1}{h^{n+1}-1}\right)$. Consequently, if h does not belong to $\mathcal{M}^c(D)$, neither does g, a contradiction. Thus, h belongs to $\mathcal{M}^c(D)$).

Now, since $n \geq 2$, by Theorem 39.4, h has to take at least one of the $(n+1)$th roots of 1 other than 1 and such an $(n+1)$th root of 1 cannot be an nth root of 1. Consequently, g admits a pole, a contradiction. Therefore, h is identically equal to 1, and hence, $f = g$. □

We must also note Theorems 47.10 and 47.11:

Theorem 47.10. *Let $f, g \in \mathcal{M}(\mathbb{K})$ satisfy $g^m + f^n = 1$, with $\min(m, n) \geq 2$, $\max(m, n) \geq 3$. Then f and g are constant. Moreover, if $f, g \in \mathcal{A}(\mathbb{K})$ and satisfy $g^m + f^n = 1$, with $\min(m, n) \geq 2$, then f and g are constant.*

Theorem 47.11. *Let $f, g \in \mathcal{M}(d(a, R^-))$ satisfy $g^m + f^n = 1$, with $\min(m, n) \geq 3$ and $\max(m, n) \geq 4$. Then f and g belong to $\mathcal{M}_b(d(a, R^-))$. Moreover, if $\mathbb{K}$ has characteristic different from 2, and if f and g belong to $\mathcal{A}(d(a, R^-))$ and satisfy $g^m + f^n = 1$, with $\min(m, n) \geq 2$, then f and g belong to $\mathcal{A}_b(d(a, R^-))$.*

Proof (Theorems 47.10 and 47.11). We first assume $\max(m, n) \geq 3$ and that f, g belong to $\mathcal{M}(K) \setminus \mathbb{K}$ in Theorem 46.3 and belong to $\mathcal{M}_u(d(a, R^-))$ in Theorem 47.11.

Let $a_1, \ldots, a_n$ be the nth roots of 1. Applying Theorem 45.7 to f at the points $a_1, \ldots, a_n$, we obtain the following:

$$(1)\ (n-1)T(r, f) \leq \sum_{j=1}^{n} \overline{Z}(r, f - a_i) + \overline{N}(r, f) - \log r + O(1),\ (r \in I).$$

Clearly, each zero of $f^n - 1$ has order $\geq m$. Hence, for each $j = 1, \ldots, n$, each zero of $f - a_j$ has order $\geq m$. Consequently,

$$(2)\ \overline{Z}(r, f - a_i) \leq \frac{1}{m} Z(r, f - a_i).$$

Therefore, in Theorem 47.10, by (1), we have

$$(n-1)T(r, f) \leq \left(\frac{1}{m}\right) \sum_{j=1}^{n} Z(r, f - a_i) + \overline{N}(r, f) - \log r + O(1)\ (r \in I),$$

and therefore, we can see that $nm < n + m + 1$, hence $\max(n, m) \leq 2$, which contradicts the general hypothesis of Theorem 47.10. Consequently, f and g are constants.

In the proof of Theorem 47.11, without loss of generality, we can obviously suppose that $a = 0$. Assuming now that f, g belong to $\mathcal{M}_u(d(0, R^-))$, Relation (1) holds again, whereas r now is bounded and hence we now have $nm \leq n + m + 1$, hence $\max(n, m) \leq 3$ and $\min(m, n) \leq 2$, a contradiction to the general hypothesis of Theorem 47.11, therefore f and g belong to $\mathcal{M}_b(d(a, R^-))$.

Now, in Theorem 47.10, suppose that f and g belong to $\mathcal{A}(\mathbb{K})$. Instead of (1), we obtain

$$(3)\ (n-1)T(r, f) \leq \sum_{j=1}^{n} \overline{Z}(r, f - a_i) - \log r + O(1) \quad (r \in I),$$

and hence by (2), we derive $nm < n + m$, a contradiction as soon as $\min(m, n) \geq 2$. Hence, f and g are constants.

Finally, in Theorem 47.11, suppose that f and g belong to $\mathcal{A}_u(d(0, R^-))$. Then, by (1) again, we have $nm \leq n + m$, therefore $\max(m, n) \leq 2$, hence $m = n = 2$. But then, by Theorem 14.22, we know that f and g are bounded in $d(0, R^-)$. This finishes the proof. □

Chapter 48

Applications to Curves

Throughout this chapter, the field $\mathbb{K}$ is supposed to be of characteristic 0.

We examine particular cases where curves are defined by their equations so that, for most of them, the p-adic Nevanlinna theory lets us find easy proofs. Most of results come from [24].

Definition. Let $F(x,y) \in \mathbb{K}[x,y]$. A point (a,b) of the algebraic curve of equation $F(x,y) = 0$ is called a *singular point* if $\frac{\partial F}{\partial x} = \frac{\partial F}{\partial y} = 0$. An algebraic curve is said to be *degenerate* if it admits a singular point. An algebraic curve of degree 2 (respectively 3) is called *a conic* (respectively *an elliptic curve*).

Throughout this chapter, the field $\mathbb{K}$ is supposed to have characteristic 0.

Remark. Let $P,\ Q \in \mathbb{K}[x]$. A point $(\alpha,\ \beta) \in \mathbb{K}^2$ is a singular point of the curve of equation $P(x) = Q(y)$ if and only if $P(\alpha) = Q(\beta)$ and $P'(\alpha) = Q'(\beta) = 0$.

First, Theorem 48.1 is classical:

Theorem 48.1. *Let Λ be a non-degenerate conic with a center in $\mathbb{K}$ and let $f,\ g \in \mathcal{A}(d(a,R^-))$ be such that $\Big(f(u),g(u)\Big) \in \Lambda \ \ \forall u \in D$. Then f and g belong to $\mathcal{A}_b(d(a,R^-))$. Moreover, if both f and g lie in $\mathcal{A}(\mathbb{K})$, then they are constant.*

Proof. If the equation of Λ is of the form $ax^2 + by^2 = t$, with $t \in \mathbb{K}$, the proof is provided by Theorem 14.22. Now, we consider the general case. Since Λ has a center, by a suitable change of variable of the form $u = x - \alpha$, $w = y - \beta$, we may assume that the equation is of the form $au^2 + bw^2 + 2cuw = M$. Moreover, since Λ is non-degenerate, we know that $M \neq 0$. We first note that if $a = b = 0$, then f and g are invertible in $\mathcal{A}(\mathbb{K})$ and therefore are constant. Thus, since a and b play the same role, we may assume, for instance, $b \neq 0$. In this way, g is a solution (in $\mathcal{A}(\mathbb{K})$) of the equation $bY^2 + (2cf)Y + (af^2 - M) = 0$. Thereby, the reduced discriminant $f^2(c^2 - ab) + bM$ must be equal to $(bg - cf)^2$. Consequently, putting $h = bg - cf$, we check that $h^2 = f^2(c^2 - ab) + bM$. Thus, f and h satisfy an equation of the form $\lambda f^2 + \mu h^2 = l \in \mathbb{K}$. But since $bM \neq 0$, and since h lies in $\mathcal{A}(\mathbb{K})$, we come back to the first case, proving that f and h are constant and therefore so is g. □

Remark. The p-adic functions sin and cos are bounded inside $d(0, (p^{-\frac{1}{p-1}})^-)$ when the residue characteristic is p (respectively inside $d(0, 1^-)$ when the residue characteristic is 0) and satisfy $\sin^2 x + \cos^2 x = 1$. Throughout this chapter, we denote by D an infinite bounded set included in a disk $d(a, r)$, for some $r < R$.

The second main theorem gives an easy proof of the impossibility to parametrize elliptic and hyperelliptic curves.

Theorem 48.2. *Let $a_i \in \mathbb{K}, i = 1, 2, 3$, be pairwise distinct. There do not exist $f,\ g \in \mathcal{M}(\mathbb{K}) \backslash \mathbb{K}$ and there do not exist $f,\ g \in \mathcal{A}_u(d(a, R^-))$ such that $(g(x))^2 = (f(x) - a_1)(f(x) - a_2)(f(x) - a_3)$.*

Proof. Suppose two functions $f,\ g \in \mathcal{M}(\mathbb{K}) \backslash \mathbb{K}$ satisfy $(g(x))^2 = (f(x)-a_1)(f(x)-a_2)(f(x)-a_3)$. Then each zero of $f - a_i$, $i = 1, 2, 3$, must be of order at least 2. And each pole of $(f - a_1)(f - a_2)(f - a_3)$ is a pole of g^2 and hence is of even order, and hence each pole of f is at least of order 2. Consequently, $\overline{Z}(r, f) \leq \frac{1}{2}Z(r, f)$ and $\overline{N}(r, f) \leq \frac{1}{2}N(r, f)$. Now, applying the second main theorem, we have

$$2T(r, f) \leq \sum_{i=1}^{3} \overline{Z}(r, f - a_i) + \overline{N}(r, f) - \log r + O(1),$$

and hence $2T(r, f) \leq 2T(r, f) - \log r + O(1)$, a contradiction when r goes to $+\infty$. Next, if $f,\ g$ belong to $\mathcal{A}_u(d(a, R^-))$, then assuming

$a = 0$, we have

$$2T(r, f) \leq \sum_{i=1}^{3} \overline{Z}(r, f - a_i) + O(1),$$

hence

$$2T(r, f) \leq \frac{\sum_{i=1}^{3} T(r, f - a_i)}{2} + O(1),$$

a contradiction. □

Theorem 48.3. *Let P, Q be in $\mathbb{K}[x]$ with $2 \leq \deg(P) \leq \deg(Q) = 3$. There exist non-constant meromorphic functions $f,\ g \in \mathcal{M}(\mathbb{K})$ such that $P(f) = Q(g)$ if and only if P' has a zero α and Q' has a zero β satisfying $P(\alpha) = Q(\beta)$.*

Proof. Let Λ be the curve of equation $P(x) = Q(y)$. Since Λ is a curve of degree 3, it is an elliptic curve, and hence by definition, it has a singular point if and only if there exists $(\alpha, \beta) \in \Lambda$ such that $P'(\alpha) = Q'(\beta) = 0$. By results in algebraic geometry, its genus is 0 if and only if the curve has a singular point. If the curve has genus zero, then there exist $f,\ g \in \mathbb{K}(x)$ such that $P(f(t)) = Q(g(t))$. If the curve has no singular point, it is of genus 1, and then by classical results in algebraic geometry, it is birationally equivalent to a curve of equation $y^2 = G(f(t))$ with G a polynomial of degree 3 with three distinct zeros. Consequently, by Theorem 48.2, there exist no non-constant meromorphic functions $f,\ g \in \mathcal{M}(\mathbb{K})$ such that $(g(t))^2 = G(f(t))$, and therefore, by birational equivalence, there exist no no-constant meromorphic functions $f,\ g \in \mathcal{M}(\mathbb{K})$ such that $P(f) = Q(g)$. □

Theorem 48.4. *Let $a_i \in \mathbb{K}$, $i = 1, \dots, q$, $q \geq 5$, be pairwise distinct, and let $R \in \mathbb{R}_+^*$. Let $m \in \mathbb{N}^*$ be ≥ 2 and prime to q. There do not exist $f,\ g \in \mathcal{M}(\mathbb{K}) \setminus \mathbb{K}$ and there do not exist $f,\ g \in \mathcal{M}_u(d(0, R^-))$ such that $(g(x))^m = \prod_{i=1}^{q}(f(x) - a_i)$.*

Proof. The proof is similar to that of Theorem 48.2. Suppose two functions $f,\ g \in \mathcal{M}(\mathbb{K})) \setminus \mathbb{K}$ satisfy $(g(x))^m = \prod_{i=1}^{q}(f(x) - a_i)$. Then each zero of $f - a_i$, $i = 1, \dots, q$, must be of order at least 2. And each pole of $\prod_{i=1}^{q}(f(x) - a_i))$ is a pole of g^m. But since m is prime to q, each pole of f is of order at least 2. Consequently, we have again

$\overline{Z}(r,f) \leq \frac{1}{2}Z(r,f)$ and $\overline{N}(r,f) \leq \frac{1}{2}N(r,f)$. Now, applying Theorem 45.7, we have

$$(q-1)T(r,f) \leq \sum_{i=1}^{q} \overline{Z}(r,f-a_i) + \overline{N}(r,f) - \log r + O(1),$$

and hence $(q-1)T(r,f) \leq (\frac{q+1}{2})T(r,f) - \log r + O(1)$, a contradiction whenever $f,\ g \in \mathcal{M}(\mathbb{K})$ or whenever $f,\ g \in \mathcal{M}_u(d(0,R^-))$. □

Corollary 48.5. *Let Λ be an algebraic curve on $\mathbb{K}$ of genus 1 and let $f,\ g \in \mathcal{M}(\mathbb{K})$ be such that $(f(u),g(u)) \in \Lambda\ \forall u \in D$. Then f and g are constants.*

Corollary 48.6. *Let Λ be an algebraic curve on $\mathbb{K}$ of genus 1 or 2 on $\mathbb{K}$ and let $f,\ g \in \mathcal{A}(d(a,R^-))$ be such that $(f(u),g(u)) \in \Lambda\ \forall u \in D$. Then f and g are bounded.*

Proof (Corollaries 48.5 and 48.6). Indeed, every algebraic curve of genus 1 is birationally equivalent to an non-degenerate elliptic curve. So, we can apply Theorem 48.2 to prove Corollaries 48.5 and 48.6. □

Here we take this opportunity to recall that there exists no parameterization of conics with a center, by entire functions, on the field $\mathbb{K}$. Such a result cannot be extended to bounded analytic functions as shown by the functions sin and cos defined in $d(0,(p^{-\frac{1}{p-1}})^-)$ when the residue characteristic of $\mathbb{K}$ is p (respectively $d(0,1^-)$ when the residue characteristic of $\mathbb{K}$ is 0).

We now consider more general algebraic equations of curves.

Theorem 48.7. *Let $P,\ Q \in \mathbb{K}[X]$ be two relatively prime polynomials of degrees s and t respectively, let n be the number of distinct zeros of Q, let $m \in \mathbb{N}^*$, and let $g \in \mathcal{M}(\mathbb{K})$ (respectively $g \in \mathcal{M}(d(a,R^-))$) be a non-constant function all poles of which have order $\geq m$. Suppose that there exists a function $f \in \mathcal{M}(\mathbb{K})$ (respectively $f \in \mathcal{M}(d(a,R^-))$) satisfying $g(x)Q(f(x)) = P(f(x))$ for every $x \in D$ which is not a pole of f or g:*

(i) *Assume that $f \in \mathcal{M}_u(d(a,R^-))$. Then $mn \leq t+2m$. Moreover, if $s > t$, then $mn \leq \min(t+2m, s+m)$.*
(ii) *Assume $f,\ g \in \mathcal{M}(\mathbb{K})$. Then $mn < t+2m$. Moreover, if $s > t$, then $mn < \min(t+2m, s+m)$.*

Proof. If $n < 2$, the inequality $mn < t + 2m$ is trivial. So, we suppose $n \geq 2$. Let $Q(X) = \prod_{j=1}^{n}(X - b_j)^{t_j}$. Since P and Q have no common zeros, each zero α of $Q(f(t))$ is a pole of $g(t)$, and therefore, it is a zero of order at least m of $Q(f(t))$. As a consequence, for each zero α of $f - b_j$, we have $t_j\omega_\alpha(f - b_j) \geq m$, hence

(1) $\overline{Z}(r, f - b_j) \leq \dfrac{t_j}{m} Z(r, f - b_j)$.

But since $Z(r, f - b_j) \leq T(r, f) + O(1)$ $(r \in I)$, by (1), we have

(2) $\sum_{j=1}^{n} \overline{Z}(r, f - b_j) \leq \dfrac{t}{m} T(r, f) + O(1) \quad (r \in I)$.

Then, by Theorem 45.7, we obtain

(3) $(n - 1)T(r, f) \leq \dfrac{t}{m} T(r, f) + \overline{N}(r, f) - \log r + O(1) \quad (r \in I)$.

In particular, this implies

(4) $(n - 2)T(r, f) \leq \dfrac{t}{m} T(r, f) - \log r + O(1) \quad (r \in I)$.

Suppose that f, g belong to $\mathcal{M}(d(0, R^-))$. By Theorem 42.10, if f does not lie in $\mathcal{M}_b(d(0, R^-))$, we have $m(n - 2) \leq t$. And if f, g lies in $\mathcal{M}(\mathbb{K})$, then when $-\log r$ tends to $+\infty$, it is seen that inequality (4) implies $(n - 2) < \frac{t}{m}$.

Now, suppose that $s > t$. Then each pole α of f is a pole of g and therefore satisfies $\omega_\alpha(f)(s - t) = \omega_\alpha(g)$, hence $\overline{N}(r, f) \leq N(r, f)(\frac{s-t}{m})$. Consequently, Relation (3) becomes

(5) $(n - 1)T(r, f) \leq \frac{t}{m} T(r, f) + (\frac{s-t}{m}) N(r, f) - \log r + O(1)$ $(r \in I)$ and so we have $(n - 1) \leq \frac{t}{m} + \frac{s-t}{m}$, thereby $mn \leq \min(s + m, t + 2m)$. Finally, if in addition, f and g lie in $\mathcal{M}(\mathbb{K})$, as r tends to $+\infty$, inequality (5) becomes strict, and therefore, $mn < \min(s + m, t + 2m)$, which finishes the proof. □

Examples. (1) Let Λ be the curve of equation $y^4(x - b')(x - b'') = (x - c)^3$ (with b', b'', c all distinct) and let f, $g \in \mathcal{M}(d(a, R^-))$ be such that $(f(u), g(u)) \in \Lambda$ for all $u \in D$. Then, by Theorem 48.7, f, $g \in \mathcal{M}_b(d(a, R^-))$. (2) Let Λ be the curve of equation $y^3(x - b')(x - b'') = (x - c)^3$ (with b', b'', c all distinct) and let f, $g \in \mathcal{M}(\mathbb{K})$ be such that $(f(u), g(u)) \in \Lambda$ for all $u \in D$. Then, by Theorem 48.7, f, g are constant.

Notation. Henceforth, given two integers m, n, we denote by g.c.d.(m, n) the greatest common divisor of m and n.

In the proof of Theorem 48.9, we use the following arithmetical lemma:

Lemma 48.8. *If $a, b, m, s \in \mathbb{N}^*$ satisfy $sa = mb$, then $a \geq \frac{m}{\text{g.c.d.}(s,m)}$.*

Theorem 48.9. *Let $P(X) = A\prod_{i=1}^{k}(X - a_i)^{s_i}$, $Q(X) = B\prod_{j=1}^{n}(X - b_j)^{t_j}$ be two relatively prime polynomials of $\mathbb{K}[X]$ of respective degrees s and t. Let $m \in \mathbb{N}^*$ and let $f,\ g \in \mathcal{M}(\mathbb{K})$ (respectively $f,\ g \in \mathcal{M}(d(a, R^-))$) satisfy $(g(x))^m Q(f(x)) = P(f(x))$ for all $x \in D$.*
(α) Suppose $f,\ g \in \mathcal{M}(d(a, R^-))$. If

$$k + n > 1 + \frac{1}{m}\left(g.c.d.(m, |s-t|_\infty) + \sum_{i=1}^{k} g.c.d.(m, s_i)\right.$$
$$\left. + \sum_{j=1}^{n} g.c.d.(m, t_j)\right),$$

then both f and g lie in $\mathcal{M}_b(d(a, R^-))$.

Moreover, if f lies in $\mathcal{A}(d(a, R^-))$, and if

$$k + n > 1 + \frac{1}{m}\left(\sum_{i=1}^{k} g.c.d.(m, s_i) + \sum_{j=1}^{n} g.c.d.(m, t_j)\right),$$

then $f \in \mathcal{A}_b(d(a, R^-))$ and $g \in \mathcal{M}_b(d(a, R^-))$.
(β) Suppose both $f,\ g$ lie in $\mathcal{M}(\mathbb{K})$. If

$$k + n \geq 1 + \frac{1}{m}\left(g.c.d.(m, |s-t|_\infty) + \sum_{i=1}^{k} g.c.d.(m, s_i)\right.$$
$$\left. + \sum_{j=1}^{n} g.c.d.(m, t_j)\right),$$

then both f and g are constant.

Finally, if $f,\ g \in \mathcal{M}(\mathbb{K})\setminus\mathbb{K}$ and if

$$k + n \geq 1 + \frac{1}{m}\left(1 + \sum_{i=1}^{k} g.c.d.(m, s_i) + \sum_{j=1}^{n} g.c.d.(m, t_j)\right),$$

then f admits at least one pole of order $< m$.

Proof. When f, g belong to $\mathcal{M}(d(a, R^-))$, we can obviously assume $a = 0$. It is clear that if f is constant, so is g. Suppose that f is not constant. Then we have $s = \sum_{i=1}^{k} s_i, t = \sum_{j-1}^{n} t_j$. For each $i = 1, \dots, k$, every zero α of $f - a_i$ is a zero of g and therefore is a zero of order $\omega_\alpha(f - a_i) = \frac{m\omega_\alpha(g)}{s_i}$. But by Lemma 48.8, we have $\omega_\alpha(f - a_i) \geq \frac{m}{g.c.d.(m,s_i)}$. In the same way, for each $j = 1, \dots, n$, every zero β of $f - b_j$ is a pole of g and therefore is a zero of order $\omega_\beta(f - b_j) = \frac{m\omega_\beta(g)}{t_j}$, and by Lemma 48.8, we have $\frac{m\omega_\beta(g)}{t_j} \geq \frac{m}{g.c.d.(m,t_j)}$. So, we have

$$(6)\ \overline{Z}(r, f - a_i) \leq \frac{g.c.d.(m, s_i)}{m} T(r, f) + O(1) \quad (r \in I),$$

$$(7)\ \overline{Z}(r, f - b_j) \leq \frac{g.c.d.(m, t_j)}{m} T(r, f) + O(1) \quad (r \in I).$$

Then, applying Theorem 45.7 to f at the points $a_1, \dots, a_k,\ b_1, \dots, b_n$ and using (6) and (7), we obtain

$$\begin{aligned}(8)\ &(k + n - 1)T(r, f)\\ &\leq \frac{1}{m}\Big(\sum_{i=1}^{k} g.c.d.(m, s_i) + \sum_{j=1}^{n} g.c.d.(m, t_j)\Big)T(r, f)\\ &\quad + \overline{N}(r, f) - \log r + O(1) \quad (r \in I).\end{aligned}$$

We now find an upper bound of $\overline{N}(r, f)$. Let γ be a pole of f. We have

$$(9)\ (s - t)\omega_\gamma(f) = m\omega_\gamma(g).$$

In the same way, from (9), we obtain $|\omega_\gamma(f)|_\infty \geq \frac{m}{g.c.d.(m,|s-t|_\infty)}$, and therefore,

$$\begin{aligned}(10)\quad \overline{N}(r, f) &\leq \frac{g.c.d.(m, |s - t|_\infty)}{m} N(r, f)\\ &\leq \frac{g.c.d.(m, |s - t|_\infty)}{m} T(r, f).\end{aligned}$$

We prove (11):

(11)

$$(k+n-1)T(r,f) \le \frac{1}{m}\left(g.c.d.(m,|s-t|_\infty)+\sum_{i=1}^{k} g.c.d.(m,s_i)+\sum_{j=1}^{n} g.c.d.(m,t_j)\right) \times T(r,f) - \log r + O(1)(r \in I).$$

If $s=t$, we just have the inequality $\overline{N}(r,f) \le N(r,f) \le T(r,f)$. Then From (8), we obtain

$$(k+n-1)T(r,f) \le \frac{1}{m}\left(m+\sum_{i=1}^{k} g.c.d.(m,s_i)+\sum_{j=1}^{n} g.c.d.(m,t_j)\right) \times T(r,f) - \log r + O(1) \quad (r \in I).$$

But in this case, we have $g.c.d.(m,|s-t|_\infty)=m$. So, we obtain (11). And if $s \neq t$, by (8), we obtain relation (11) again.

Thus, from relation (11), using Theorem 43.10, if f does not lie in $\mathcal{M}_b(d(0,R^-))$, we can get

(12)

$$(k+n-1) \le \frac{1}{m}\left(g.c.d.(m,|s-t|_\infty)+\sum_{i=1}^{k} g.c.d.(m,s_i)+\sum_{j=1}^{n} g.c.d.(m,t_j)\right),$$

a contradiction with the hypothesis. Hence, if f belongs to $\mathcal{M}(d(0,R^-))$, it must lie in $\mathcal{M}_b(d(0,r^-))$ and so must g.

Now, suppose $f,\ g \in \mathcal{M}(\mathbb{K})$. When r tends to $+\infty$, we can see that the inequality (12) becomes strict. Therefore, if the inequality is not satisfied, f and g are constant. Conversely, suppose that f is not constant and that all poles of f have order $\ge m$. By (11), we have

$$(k+n-1)T(r,f) \le \frac{1}{m}\left(\sum_{i=1}^{k} g.c.d.(m,s_i)+\sum_{j=1}^{n} g.c.d.(m,t_j)\right) \times T(r,f) + \frac{1}{m}T(r,f) - \log r + O(1) \quad (r \in I),$$

which implies $k+n-1 < \frac{1}{m}(1+\sum_{i=1}^{k} g.c.d.(m,s_i)+\sum_{j=1}^{n} g.c.d.(m,t_j))$. Hence, if this inequality is not true, f must admit at least one pole of order $< m$.

Finally, suppose $f \in \mathcal{A}(d(0,R^-))$. Then (8) becomes

(13)
$$(k+n-1)T(r,f) \leq \frac{1}{m}\left(\sum_{i=1}^{k} g.c.d.(m,s_i)+\sum_{j=1}^{n} g.c.d.(m,t_j)\right) \times T(r,f) - \log r + O(1) \quad (r \in I),$$

and then, if f does not lie in $\mathcal{A}_b(d(0,R^-))$, by Theorem 43.10, we obtain

$$k+n-1 \leq \frac{1}{m}\left(\sum_{i=1}^{k} g.c.d.(m,s_i)+\sum_{j=1}^{n} g.c.d.(m,t_j)\right),$$

hence $k+n-1 \leq \frac{1}{m}(s+t)$, a contradiction. That finishes the proof. □

Examples. (3) Let c', $c'' \in \mathbb{K}$ (with $c' \neq c''$) and let Λ be the curve of equation $y^3 = (x-c')^2(x-c'')$. Let f, $g \in \mathcal{M}(\mathbb{K})$ be such that $(f(u),g(u)) \in \Lambda$ for all $u \in D$. If f and g are not constant, then f admits at least one pole of order 1 or 2. Here the genus is clearly 0, therefore there exist f, $g \in \mathbb{K}(u)$ satisfying $g^3 = (f-c')^2(f-c'')$ and hence f admits at least one pole of order 1 or 2.

(4) Let Λ be the curve of equation $y^3(x-b)^2 = (x-c')^2(x-c'')$ (with b, c', c'' all distinct) and let f, $g \in \mathcal{M}(d(a,R^-))$ be such that $(f(u),g(u)) \in \Lambda$ for all $u \in D$. Then by Theorem 48.9, f, $g \in \mathcal{M}_b(d(a,R^-))$.

(5) Let Λ be the curve of equation $y^2(x-b')(x-b'') = (x-c)$ (with b', b'', c all distinct) and let $f \in \mathcal{A}(d(a,R^-))$ and let $g \in \mathcal{M}(d(a,R^-))$ be such that $(f(u),g(u)) \in \Lambda$ for all $u \in D$. Then by Theorem 48.9, $f \in \mathcal{A}_b(d(a,R^-))$ and $g \in \mathcal{M}_b(d(a,R^-))$.

Remark. In [24], it was proved that the equation $f^m + g^n = 1$ in $\mathcal{M}(\mathbb{K})$ leads to f, $g \in \mathbb{K}$ as soon as the least common multiple q of m and n satisfies $\frac{1}{m}+\frac{1}{n}+\frac{1}{q} \geq 1$. Further, if f, $g \in \mathcal{A}(\mathbb{K})$, that leads to f, $g \in \mathbb{K}$ as soon as $\min(m,n) \geq 2$. Here we are now able to generalize these conclusions.

Chapter 49

Branched Values

Throughout this chapter, the field $\mathbb{K}$ *is supposed to be of characteristic* 0.

In complex functions theory, a notion closely linked to Picard's exceptional values was introduced: the notion of "perfectly branched value" [29]. Here we consider the same notion on $\mathcal{M}(\mathbb{K})$, on $\mathcal{M}(d(a, R^-))$, and on $\mathcal{M}(D)$. Most of results come from [51].

Definition. Let $f \in \mathcal{M}(\mathbb{K})$ (respectively let $f \in \mathcal{M}(d(a, R^-))$, respectively let $f \in \mathcal{M}(D)$) and let $b \in \mathbb{K}$. The value b is said to be *a perfectly branched value for* f if all zeros of $f - b$ are multiple zeros, except finitely many. And in this book, b is said to be *a totally branched value for* f if all zeros of $f - b$ are multiple zeros, without exception. Similarly, ∞ is called *a perfectly branched value for* f if all poles of f are multiple but finitely many and it is called *a totally branched value for* f if all poles of f are multiple, without exception.

In $\mathbb{C}$, it is known that a transcendental meromorphic function admits at most four perfectly branched values and an entire function admits at most two perfectly branched values. As explained by Charak in [29], these numbers, respectively 4 and 2, are sharp. The Weierstrass function $\mathcal{P}$ has four totally branched values (considering ∞ as a value), and of course, sine and cosine functions admit two totally branched values: 1 and -1.

Here we do a similar study on p-adic functions and obtain sometimes certain better results. Particularly, an entire function admits at most one perfectly branched value.

Lemma 49.1 is immediate:

Lemma 49.1. *Let $f \in \mathcal{M}(\mathbb{K})$ (respectively let $f \in \mathcal{M}_u(d(0, R^-))$, respectively let $f \in \mathcal{M}^c(D)$), admitting a perfectly branched value $b \neq 0$. Then $\frac{1}{f}$ admits $\frac{1}{b}$ as a perfectly branched value. If f admits 0 as a perfectly branched value, $\frac{1}{f}$ admits ∞ as a perfectly branched value. If f admits ∞ as a perfectly branched value, $\frac{1}{f}$ admits 0 as a perfectly branched value.*

We have an immediate application of the definition with meromorphic functions whose denominator is a small function with respect to the numerator or vice versa.

Theorem 49.2. *Let $f,\ g \in \mathcal{A}(\mathbb{K}) \setminus \mathbb{K}[x]$ (respectively $f,\ g \in \mathcal{A}_u(d(0, R^-))$, respectively $f,\ g \in \mathcal{A}^c(D)$) be such that $\limsup_{r\to+\infty} \frac{T(r,f)}{T(r,g)} > 2$ (respectively $\limsup_{r\to R^-} \frac{T(r,f)}{T(r,g)} > 2$, respectively $\limsup_{r\to+\infty} \frac{T_R(r,f)}{T_R(r,g)} > 2$). Then both $\frac{f}{g}$ and $\frac{g}{f}$ have at most two perfectly branched values.*

Proof. Set $\phi = \frac{f}{g}$. Without loss of generality, we can suppose that f and g have no common zero. Indeed, suppose first that $f,\ g \in \mathcal{A}(\mathbb{K})\setminus\mathbb{K}[x]$ or $f,\ g \in \mathcal{A}^c(D)$. By Lemma 36.3, we can write f and g in the form $f = \widetilde{f}\,.\,h$ and $g = \widetilde{g}\,.\,h$, where $\widetilde{f}$ and $\widetilde{g}$ have no common zero and then $Z(r, f) = Z(r, \widetilde{f}) + Z(r, h)$, $Z(r, g) = Z(r, \widetilde{g}) + Z(r, h)$, and so much the more, we have $\frac{T(r,\widetilde{f})}{T(r,\widetilde{g})} > 2$.

Now, if $f,\ g \in \mathcal{A}_u(d(0, r)^-)$, we can place ourselves in an algebraically closed spherically complete extension to obtain the same conclusion because the Nevanlinna functions are the same in such an extension. Therefore, we assume that f and g have no common zero.

Suppose first that f and g belong to $\mathcal{A}(\mathbb{K})\setminus\mathbb{K}[x]$ or to $\mathcal{A}_u(d(0, R^-))$, Since $f,\ g$ have no common zero, we have $Z(r, \phi) = Z(r, f)$ and $N(r, \phi) = Z(r, g)$, hence $T(r, \phi) = \max(Z(r, f), Z(r, g)) + O(1)$.

Now, by hypothesis, there exists $\lambda < \frac{1}{2}$ and a sequence $(r_n)_{n\in\mathbb{N}}$ such that $\lim_{n\to+\infty} r_n = +\infty$ (respectively $\lim_{n\to+\infty} r_n = R$) and such that

$$T(r_n, g) \leq \lambda T(r_n, f) \quad \forall n \in \mathbb{N}. \tag{1}$$

Suppose that ϕ has three perfectly branched values b_j, $j = 1, 2, 3$. Applying Theorem 45.7, we have

$$(2) \qquad 2T(r,\phi) \le \sum_{j=1}^{3} \overline{Z}(r,\phi - b_j) + \overline{N}(r,\phi) - \log r + O(1).$$

But here, for each $j = 1, 2, 3$, we note that $\overline{Z}(r,\phi - b_j) \le \frac{Z(r,\phi-b_j)}{2} + q_j \log(r)$ with $q_j \in \mathbb{N}$ and $Z(r,\phi - b_j) = Z(r, f - b_j g) \le \max(T(r,f), T(r,g))$. But since $T(r_n, f) > T(r_n, g)$, we have $T(r_n, \phi - b_j) \le T(r_n, f) + O(1)$, hence $\overline{Z}(r,\phi - b_j) \le \frac{T(r_n,f)}{2} + q_j \log(r_n) + O(1)$. Now, putting $q = q_1 + q_2 + q_3$, by (2), we obtain

$$2T(r_n, f) \le \frac{3T(r_n, f)}{2} + T(r_n, g) + 2q \log(r_n) + O(1),$$

hence

$$T(r_n, f) \le 2T(r_n, g) + q \log(r_n) + O(1),$$

a contradiction to (1).

Similarly, if f and g belong to $\mathcal{A}^c(D)$, we can make the same reasoning by replacing T by T_R, Z by Z_R, and N by N_R. □

Theorem 49.3. *Let f, $g \in \mathcal{A}(\mathbb{K}))$ be such that $+\infty > \rho(f) > \rho(g)$. Then,*

$$\liminf_{r\to+\infty} \frac{T(r,g)}{T(r,f)} = 0.$$

Proof. Suppose first f, $g \in \mathcal{A}(\mathbb{K}))$. Let $\gamma = \frac{\rho(g)}{\rho(f)}$ and let $(r_n)_{n\in\mathbb{N}}$ be a sequence in $]0, +\infty[$ such that $\lim_{n\to+\infty} r_n = +\infty$ and $\lim_{n\to+\infty} \frac{\log(\log(|f|(r_n)))}{-\log(R)} = \rho(f)$. By hypothesis, we have

$$\lim_{n\to+\infty} \frac{\log(\log(|g|(r_n)))}{\log(\log(|f|(r_n)))} \le \gamma,$$

hence

$$\lim_{n\to+\infty} \frac{\log(T(r_n, g))}{\log(T(r_n, f))} \le \gamma.$$

Take $\beta \in]\gamma, 1[$. Then when n is big enough, we can get

$$\frac{T(r_n, g)}{T(r_n, f)} \leq (T(r_n, f))^{\beta-1}.$$

But since $\beta < 1$ and since $\lim_{n\to\infty} T(r_n, f) = +\infty$, one sees that $\lim_{n\to\infty}(T(r_n, f))^{\beta-1} = 0$, which ends the proof when f and g belong to $\mathcal{A}(\mathbb{K})$. □

Corollary 49.4. *Let $f,\ g \in \mathcal{A}(\mathbb{K})$ be such that $\rho(f) \neq \rho(g)$. Then both $\frac{f}{g}$ and $\frac{g}{f}$ have at most two perfectly branched values.*

The proof of the following theorems require several basic lemmas.

Lemma 49.5. *Let $(\alpha_i)_{1\leq i\leq t}$, $(\beta_i)_{1\leq i\leq t}$ be two finite sequences of $\mathbb{K}$ such that $|\alpha_i| < R$, $|\beta_i| < R$ $\forall i = 1, \dots, t$. Let $\Theta(x) = \prod_{i=1}^t \left(\frac{1-\frac{\beta_i}{x}}{1-\frac{\alpha_i}{x}}\right)$. Then the function $\sqrt{\Theta(x)}$ is defined and belongs to $\mathcal{A}(\mathbb{K} \setminus d(0, 4R))$. Moreover, if $p \neq 2$, it belongs to $\mathcal{A}(\mathbb{K} \setminus d(0, R))$.*

Proof. By Theorem 29.23, there exists a unique function $\ell \in \mathcal{A}(d(1, (\frac{1}{4})^-))$ with value in $d(1, 1^-)$ such that $(\ell(u))^2 = u$ $\forall u \in d(1, (\frac{1}{4})^-)$. Moreover, if $p \neq 2$, then ℓ has continuation to a function $\overline{\ell} \in \mathcal{A}(d(1, 1^-))$ with value in $d(1, 1^-)$ again and such that $(\overline{\ell}(u))^2 = u$ $\forall u \in d(1, 1)$. Here, we put $u = \prod_{i=1}^t \left(\frac{1-\frac{\beta_i}{x}}{1-\frac{\alpha_i}{x}}\right)$. □

Lemma 49.6. *Let $g,\ h \in \mathcal{A}(\mathbb{K})$ with $\frac{g}{h}$ transcendental and let $\Theta(x) = \prod_{i=1}^t \left(\frac{1-\frac{\beta_i}{x}}{1-\frac{\alpha_i}{x}}\right)$ with $|\alpha_i| < R$ and $|\beta_i| < R$ $\forall i = 1, \dots, t$. Then the function $g(x)^2 - h(x)^2\Theta(x)$ belongs to $\mathcal{A}(\mathbb{K} \setminus d(0, 4R))$ and satisfies $\lim_{r\to+\infty} \frac{|g^2-h^2\Theta|(r)}{r^m} = +\infty$ $\forall m \in \mathbb{N}$.*

Proof. We first note that $g^2 - h^2\Theta$ obviously belongs to $\mathcal{A}(\mathbb{K} \setminus d(0, R))$. Let us fix $m \in \mathbb{N}$. By Lemma 49.6, $\sqrt{\Theta}$ is defined in $\mathbb{K} \setminus d(0, R)$ and belongs to $\mathcal{A}(\mathbb{K} \setminus d(0, 4R))$. Let $\ell = \sqrt{\Theta}$. Then can write $g^2 - h^2\Theta = (g - h\ell)(g + h\ell)$.

Since $\frac{g}{h}$ is transcendental, $g^2 - h^2\Theta$ is not identically zero. So, by Theorem 22.22, there exists $a > 0$ and $q \in \mathbb{Z}$ such that $|g - h\ell|(r) \geq ar^q$ $\forall r > R$.

Suppose first h is transcendental. Since h is entire and since $|\ell|(r) = 1\ \forall r > R$, by Theorem 22.22, we have

$$(1) \qquad \forall m \in \mathbb{N},\ \lim_{r\to\infty} \frac{|h\ell|(r)}{r^m} = +\infty.$$

Consequently, $|h\ell|(r) > |g - h\ell|(r)$, and hence $|h\ell|(r) = |g + h\ell|(r)$. Thus,

$$|g - h\ell|(r)|g + h\ell|(r) \geq ar^q|h\ell|(r).$$

Then the conclusion comes from (1). Suppose now h is not transcendental, hence it is a polynomial and then g is transcendental. Consequently, when r is big enough, we have $|g - h\ell|(r) = |g|(r) = |g + h\ell|(r)$, and hence $|g^2 - h^2\Theta|(r) = (|g|(r))^2$, which yields the same conclusion. □

Theorem 49.7. *Let $f \in \mathcal{M}(\mathbb{K})$ be transcendental (respectively let $f \in \mathcal{M}_u(d(a, R^-))$, respectively let $f \in \mathcal{M}^c(D)$). Then f has at most four perfectly branched values. Moreover, any function $g \in \mathcal{M}(\mathbb{K})$ has at most three totally branched values.*

Remark. Let $f \in \mathcal{M}(\mathbb{K})$. If $a \in \mathbb{K}$ is a perfectly branched value for f, then for every $b \in \mathbb{K}$, $a + b$ is a perfectly branched value for $f + b$. Moreover, if $a \neq 0$, then $\frac{1}{a}$ is a perfectly branched value for $\frac{1}{f}$. So, we are going to construct a function $f \in \mathcal{M}(\mathbb{K})$ admitting three distinct totally branched values. Let $\ell = \prod_{j=1}^{\infty}\left(1 - \frac{x}{a_j}\right) \in \mathcal{A}(\mathbb{K})$ with $\lim_{j\to+\infty}|a_j| = +\infty$ and $a_j \neq a_k\ \forall j \neq k$. Let $u = \prod_{k=1}^{\infty}\left(1 - \frac{x}{a_{2k}}\right)$ and let $w = \prod_{k=1}^{\infty}\left(1 - \frac{x}{a_{2k-1}}\right)$. So, both u and w belong to $\mathcal{A}(\mathbb{K})$ and satisfy $uw = \ell$. Now, let $\phi = \frac{u^2+w^2}{2}$ and $\theta = \frac{w^2-u^2}{2}$. Then, $\phi^2 - \theta^2 = \ell^2$. Now, let $g = \left(\frac{\phi}{\ell}\right)^2$. Note that g admits 0 and 1 as totally branched values. Consequently, $g + 1$ admits 1 and 2 as totally branched values and hence the function $f = \frac{1}{g+1}$ admits 1 and $\frac{1}{2}$ as totally branched values. But, on the other hand, all poles of g are multiple, hence so are those of $g + 1$. Consequently, f also admits 0 as a totally branched value. Thus, Theorem 49.8 is sharp as far as totally branched values are concerned for meromorphic functions. One can only ask whether there exist meromorphic functions admitting four perfectly branched values where some of them are not totally branched values.

Theorem 49.8. *Let $f \in \mathcal{M}(\mathbb{K})$ be transcendental and have finitely many poles. Then f has at most one perfectly branched value.*

Corollary 49.9. *Let $f \in \mathcal{A}(\mathbb{K})$ be transcendental. Then f has at most one perfectly branched value.*

Remark. However, a polynomial can admit two values looking like "perfectly branched values". Yet, the definition of a perfectly branched value does not really apply to a polynomial or a rational function.

Example. Let $P(x) = x^3 - x^2 + \frac{4}{27}$. Then 0 and $-\frac{4}{27}$ are two perfectly branched values that are not totally branched. Indeed, on one hand, $-\frac{4}{27}$ is perfectly but not totally branched since $P(x) - \frac{4}{27} = x^2(x-1)$. On the other hand, we can check that $P(x) = \left(x - \frac{2}{3}\right)^2 \left(x + \frac{1}{3}\right)$.

Theorem 49.10. *Let $f \in \mathcal{M}_u(d(a, R^-))$ have finitely many poles. Then f has at most two perfectly branched values. Let $f \in \mathcal{M}^c(D)$ have finitely many poles in D. Then f has at most two perfectly branched values.*

Corollary 49.11. *Let $f \in \mathcal{A}_u(d(a, R^-))$ (respectively let $f \in \mathcal{A}^c(D)$). Then f has at most two perfectly branched values.*

Theorem 49.8 gives an easy proof of the impossibility to parametrize elliptic and hyperelliptic curves by meromorphic functions in all $\mathbb{K}$.

Corollary 49.12. *Let $a_i \in \mathbb{K}, i = 1, 2, 3$, be pairwise distinct. There do not exist $f,\ g \in \mathcal{M}(\mathbb{K})\setminus\mathbb{K}$, there do not exist $f,\ g \in \mathcal{A}_u(d(a, R^-))$, and there do not exist $f,\ g \in \mathcal{A}^c(D)$ such that $(g(x))^2 = (f(x) - a_1)(f(x) - a_2)(f(x) - a_3)$.*

Proof. Suppose two functions $f,\ g \in \mathcal{M}(\mathbb{K}) \setminus \mathbb{K}$ satisfy $(g(x))^2 = (f(x)-a_1)(f(x)-a_2)(f(x)-a_3)$. Then each zero of $f - a_i$, $i = 1,\ 2,\ 3$, must be of order at least 2. And each pole of $(f-a_1)(f-a_2)(f-a_3)$ is a pole of g^2 and hence is of even order, and hence each pole of f is at least of order 2. f admits four totally branched values: a_1, a_2, a_3, ∞, what is impossible by Theorem 49.8. □

Remark. We don't know whether there exists a function $f \in \mathcal{A}_u(d(a, R^-))$ admitting two perfectly branched values. The only case when we can improve Theorem 49.11 is the case when $\mathbb{K}$ has residue characteristic 0.

In the proofs of Theorems 49.8 and 49.9, without loss of generality, we can obviously assume that all supposed perfectly branched values of the functions we consider are finite, what we do for simplicity.

Proof (Theorems 49.8, 49.9, and 49.11). If f lies in $\mathcal{M}_u(d(a, R^-))$, we assume that $a = 0$. Suppose f has q perfectly branched values b_j with $j = 1, \dots, q$. For each j, let s_j be the number of simple zeros of $f - b_j$ and let $s = \sum_{j=1}^{q} s_j$. Applying Theorem 45.7, we have

$$(1) \qquad (q-1)T(r, f) \le \sum_{j=1}^{q} \overline{Z}(r, f - b_j) + \overline{N}(r, f) - \log r + O(1).$$

But since $f - b_j$ has s_j simple zeros, we have

$$\overline{Z}(r, f - b_j) \le \frac{Z(r, f - b_j) + s_j \log r}{2} + O(1) \le \frac{T(r, f) + s_j \log r}{2} + O(1) \quad \forall j = 1, \dots, q,$$

hence

$$(2) \qquad (q-1)T(r, f) \le \frac{qT(r, f)}{2} + T(r, f) + \left(\frac{s}{2} - 1\right) \log r + O(1).$$

By (2), clearly, we have $q \le 4$ in all cases, which shows the first statement of Theorem 49.8 whenever $f \in \mathcal{M}(\mathbb{K})$ or $f \in \mathcal{M}_u(d(0, R^-))$. Now, when $f \in \mathcal{M}^c(D)$, then (2) becomes $(q-1)T_R(r, f) \le \frac{qT_R(r,f)}{2} + T_R(r, f) + (O(\log(r)))$ and we have the same conclusion.

Further, suppose that f lies in $\mathcal{M}(\mathbb{K})$ or in $\mathbb{K}(x)$ and that $b_1, \dots, b_q$ are totally branched values. Then, $s = 0$, hence by (2), we have

$$(3) \qquad \left(\frac{q}{2} - 2\right) T(r, f) \le -\log r + O(1).$$

In Theorem 49.8, since f is transcendental, we have $\log r = o(T(r, f))$, hence $q \le 3$.

Consider now the hypotheses of Theorems 49.9 and 49.11. Let t be the number of poles of f, taking multiplicity into account. We have $N(r,f) \leq t\log r + O(1)$, hence by (1), we obtain $(q-1)T(r,f) \leq \frac{q}{2}T(r,f) + O(\log r)$, and hence $q \leq 2$. When $f \in \mathcal{M}^c(D)$, we just replace N by N_R.

Thus, Theorem 49.11 is proved.

For the proof of Theorem 49.9, without loss of generality, we may assume that these perfectly branched values are 0 and b. Suppose first that f has infinitely many zeros of order ≥ 3. Then $Z(r,f) - 2\overline{Z}(r,f)$ is a function $\theta(r)$ such that

$$\lim_{r\to+\infty} \frac{\theta(r)}{\log r} = +\infty, \tag{4}$$

therefore

$$\overline{Z}(r,f) \leq \frac{T(r,f) - \theta(r)}{2}. \tag{5}$$

On the other hand, by (1), with $q = 2$, we have

$$T(r,f) \leq T(r,f) - \frac{\theta(r)}{2} + O(\log r),$$

and then, by (4), we can see a contradiction proving that f cannot admit 0 and b as branched values.

Suppose now that all zeros of both f and $f-b$ are of order 2 except finitely many. So, there exists $S > 0$ satisfying the following properties:

(i) All poles of f lie in $d(0,S)$.
(ii) $|f|(r) > |b|\ \forall r > S$.
(iii) All zeros of f and of $f-b$ in $\mathbb{K}\setminus d(0,S)$ are of order 2 exactly.

We can then write f in the form $\frac{Pg^2}{V}$ with $P, V \in \mathbb{K}[x]$, $g \in \mathcal{A}(\mathbb{K})$ and $\deg(P) = k$. Similarly, $f-b$ is in the form $\frac{Qh^2}{V}$ with $Q \in \mathbb{K}[x]$, $h \in \mathcal{A}(\mathbb{K})$, where all zeros of P, Q, V lie in $d(0,S)$ and all zeros of g, h lie in $\mathbb{K}\setminus d(0,S)$ and are simple zeros. Set $\deg(V) = t$. We note that g is transcendental.

By (ii), we have $|f|(r) = |f-b|(r)\ \forall r > S$. Consequently, by Lemma 36.20, f and $f-b$ have the same number of zeros in $d(0,S)$, and hence, $\deg(P) = \deg(Q)$. Let $P(x) = \prod_{i=1}^{k}(x-\alpha_i)$ and

$Q(x) = \prod_{i=1}^{k}(x - \beta_i)$. Let $\Theta(x) = \prod_{i=1}^{k}\left(\frac{1-\frac{\beta_i}{x}}{1-\frac{\alpha_i}{x}}\right)$ and let $\Xi(x) = \frac{V(x)}{P(x)}$. Of course, $\lim_{r\to+\infty}\frac{|\Xi|(r)}{r^{t+1}} = 0$.

Now, we have

$$g(x)^2 = h(x)^2\Theta(x) + b\Xi(x).$$

By Lemma 49.7, we can derive

$$\lim_{r\to+\infty}\frac{\left(|g^2 - h^2\Theta|(r)\right)}{r^m} = +\infty \quad \forall m \in \mathbb{N},$$

a contradiction to $\lim_{r\to+\infty}\frac{|\Xi|(r)}{r^{t+1}} = 0$. This completes the proof of Theorem 49.9. □

Remark. Corollary 49.12 may suggest that Theorem 28.4 is wrong. Indeed, by Theorem 28.4, given sequences $(a_n)_{n\in\mathbb{N}}$, $(b_n)_{n\in\mathbb{N}}$, $(c_n)_{n\in\mathbb{N}}$ in $d(0, R^-)$ such that

$$\lim_{n\to+\infty}|a_n| = \lim_{n\to+\infty}|b_n| = \lim_{n\to+\infty}|c_n| = R,$$

together with $\prod_{n=0}^{\infty}\frac{R}{|a_n|} = +\infty$, there exists functions $f \in \mathcal{A}(d(0, R^-))$ admitting each a_n as a zero of order 2 and such that $f - 1$ admits each b_n as a zero of order 2 and such that $f - 2$ admits each c_n as a zero of order 2. Moreover, since $\prod_{n=0}^{\infty}\frac{R}{|a_n|} = +\infty$, we can check that $f \in \mathcal{A}_u(d(0, R^-))$.

The explanation is that, given such a function f, either f or $f - 1$ or $f - 2$ has infinitely many other zeros of order 1, and therefore, at least one of these three values (0, 1, 2) is not perfectly branched.

Chapter 50

Exceptional Values of Functions and Derivatives

Throughout this chapter, the field $\mathbb{K}$ is supposed to be of characteristic 0.

This chapter is aimed at studying various properties of derivatives of meromorphic functions, particularly their sets of zeros. Many important results are due to Jean-Paul Bézivin [10, 11, 12].

We first note a general property concerning quasi-exceptional values of meromorphic functions and derivatives.

Notation. Let $f \in \mathcal{M}(\mathbb{K})$ (respectively $f \in \mathcal{M}(d(0, R^-)))$ and let $\mathcal{T}$ be a property satisfied by f at certain points. Let $r \in]0, R[$. Assume that $f(0) \neq 0, \infty$. We denote by $Z(r, f \mid \mathcal{T})$ the counting function of zeros of f in $\mathbb{K}$ (respectively in $d(0, r)$) at the points where f satisfies $\mathcal{T}$ (counting multiplicity) i.e. if (a_n) is the finite or infinite sequence of zeros of f in $d(0, R^-)$ with respective multiplicity order s_n, where $\mathcal{T}$ is satisfied, we put $Z(r, f) = \sum_{|a_n| \leq r, \mathcal{T}} s_n(\log r - \log |a_n|)$.

And we denote by $\overline{Z}(r, f \mid \mathcal{T})$ the counting function of zeros of f in $\mathbb{K}$ (respectively in $d(0, r)$) at the points where f satisfies $\mathcal{T}$, (ignoring multiplicity).

Similarly, we denote by $N(r, f \mid \mathcal{T})$ the counting function of poles of f in $\mathbb{K}$ (respectively in $d(0, r)$) at the points where f satisfies $\mathcal{T}$ (counting multiplicity) and we denote by $\overline{N}(r, f \mid \mathcal{T})$ the counting function of poles of f in $\mathbb{K}$ (respectively in $d(0, r)$) at the points where f satisfies $\mathcal{T}$ (ignoring multiplicity).

Given two meromorphic functions $f,\ g \in \mathcal{M}(\mathbb{K})$ or $f,\ g \in \mathcal{M}(d(a, R^-))$ ($a \in \mathbb{K},\ R > 0$), we denote by $W(f, g)$ the Wronskian of f and g: $f'g - fg'$.

Theorem 50.1. *Let $f \in \mathcal{M}(\mathbb{K}) \setminus \mathbb{K}(x)$ (respectively let $f \in \mathcal{M}_u(d(\alpha, R^-))$). If f admits a quasi-exceptional value, then f' has no quasi-exceptional value different from* 0.

Proof. Without loss of generality, we may assume $\alpha = 0$ and that f has no zero and no pole at 0. Let $b \in \mathbb{K}$ and suppose that b is a quasi-exceptional value of f. There exist $P \in \mathbb{K}[x]$ and $l \in \mathcal{A}(\mathbb{K})\setminus\mathbb{K}[x]$ (respectively and $l \in \mathcal{A}_u(d(0, R^-))$) without common zeros, with P, such that $f = b + \frac{P}{l}$.

Let $c \in \mathbb{K}^*$. Remark that $f' - c = \frac{P'l - Pl' - cl^2}{l^2}$. Let $a \in \mathbb{K}$ (respectively let $a \in d(0, R^-)$). If a is a pole of f, it is a pole of $f' - c$ and we can check that

(1) $\omega_a(P'l - Pl' - cl^2) = \omega_a(l') = \omega_a(l) - 1$

because a is not a zero of P.

Now, suppose that a is not a pole of f. Then,

(2) $\omega_a(f' - c) = \omega_a(P'l - Pl' - cl^2)$

Consequently, $Z(r, f' - c) = Z(r, (P'l - Pl' - cl^2) \mid l(x) \neq 0)$. But, by (1), we have

(3) $Z(r, (P'l - Pl' - cl^2) \mid l(x) = 0) < Z(r, l)$,

and therefore by (2) and (3), we obtain

(4) $Z(r, f' - c) = Z(r, (P'l - Pl' - cl^2) \mid l(x) \neq 0) > Z(r, P'l - Pl' - cl^2) - Z(r, l)$.

Now, let us examine $Z(r, P'l - Pl' - cl^2)$. Let $r \in]0, +\infty[$ (respectively let $r \in]0, R[$). Since $l \in \mathcal{A}(\mathbb{K})$ is transcendental (respectively since $l \in \mathcal{A}_u(d(0, R^-))$), we can check that when r is big enough, we have $|Pl'|(r) < |c|\big(|l|(r)\big)^2$ and $|Pl|(r) < |c|\big(|l|(r)\big)^2$, hence clearly $|P'l - Pl'|(r) < |c|\big(|l|(r)\big)^2$, and hence

$|P'l - Pl' - cl^2|(r) = |c|\big(|l|(r)\big)^2$. Consequently, when r is big enough, by Theorem 43.2, we have $Z(r, P'l - Pl' - cl^2) = Z(r, l^2) + O(1)$. But $Z(r, l^2) = 2Z(r, l)$, hence $Z(r, P'l - Pl' - cl^2) = 2Z(r, l) + O(1)$, and therefore by (4), we check that when r is big enough,

(5) $Z(r, f' - c) > Z(r, l)$.

Now, if $l \in \mathcal{A}(\mathbb{K})$, since l is transcendental, by (5), for every $q \in \mathbb{N}$, we have $Z(r, f' - c) > Z(r, l) > q \log r$, when r is big enough, hence $f' - c$ has infinitely many zeros in $\mathbb{K}$. And similarly if $l \in \mathcal{A}_u(d(0, R^-))$, then by (5), $Z(r, f' - c)$ is unbounded when r tends to R, hence $f' - c$ has infinitely many zeros in $d(0, R^-)$. □

We now note a property of differential equations of the form $y^{(n)} - \theta y = 0$ that is almost classical.

The problem of a constant Wronskian is involved in several questions.

Theorem 50.2. *Let $h,\ l \in \mathcal{A}(\mathbb{K})$ (respectively $h,\ l \in \mathcal{A}(d(\alpha, R^-))$) and satisfy $h'l - hl' = c \in \mathbb{K}$, with h non-affine. If $h,\ l$ belong to $\mathcal{A}(\mathbb{K})$, then $c = 0$ and $\frac{h}{l}$ is a constant. If $c \neq 0$ and if $h,\ l \in \mathcal{A}(d(\alpha, R^-))$, there exists $\phi \in \mathcal{A}(d(\alpha, R^-))$ such that $h'' = \phi h,\ l'' = \phi l$.*

Proof. Suppose $c \neq 0$. If $h(a) = 0$, then $l(a) \neq 0$. Next, h and l satisfy

(1) $$\frac{h''}{h} = \frac{l''}{l}.$$

Remark first that since h is not affine, h'' is not identically zero. Next, every zero of h or l of order ≥ 2 is a trivial zero of $h'l - hl'$, which contradicts $c \neq 0$. So, we can assume that all zeros of h and l are of order 1.

Now suppose that a zero a of h is not a zero of h''. Since a is a zero of h of order 1, $\frac{h''}{h}$ has a pole of order 1 at a and so does $\frac{l''}{l}$, hence $l(a) = 0$, a contradiction. Consequently, each zero of h is a zero of order 1 of h and is a zero of h'', and hence, $\frac{h''}{h}$ is an element ϕ of $\mathcal{M}(\mathbb{K})$ (respectively of $\mathcal{M}(d(\alpha, R^-))$) that has no pole in $\mathbb{K}$ (respectively in $d(\alpha, R^-)$). Therefore, ϕ lies in $\mathcal{A}(\mathbb{K})$ (respectively in $\mathcal{A}(d(\alpha, R^-))$).

The same holds for l, and so, l'' is of the form θl with $\theta \in \mathcal{A}(\mathbb{K})$ (respectively in $\mathcal{A}(d(\alpha, R^-))$). But since $\frac{h''}{h} = \frac{l''}{l}$, we have $\phi = \theta$.

Now, suppose h, l belong to $\mathcal{A}(\mathbb{K})$. Since h'' is of the form ϕh with $\phi \in \mathcal{A}(\mathbb{K})$, we have $|h''|(r) = |\phi|(r)|h|(r)$. But by Theorem 37.10, we know that $|h''|(r) \leq \frac{1}{r^2}|h|(r)$, a contradiction when r tends to $+\infty$. Consequently, $c = 0$. But then $h'l - hl' = 0$ implies that the derivative of $\frac{h}{l}$ is identically zero, hence $\frac{h}{l}$ is constant. □

Corollary 50.3. *Let h, $l \in \mathcal{A}(\mathbb{K})$ with coefficients in $\mathbb{Q}$ also be entire functions in $\mathbb{C}$, with h non-affine. If $h'l - hl'$ is a constant c, then $c = 0$.*

Theorem 50.4. *Let $\theta \in \mathcal{M}(\mathbb{K})$ (respectively let $\theta \in \mathcal{M}(d(\alpha, R^-))$) and let $(\mathcal{E})$ be the differential equations $y^{(n)} - \theta y = 0$. Let E be the subvector space of $\mathcal{M}(\mathbb{K})$ (respectively of $\mathcal{M}(d(\alpha, R^-))$) of the solutions of $(\mathcal{E})$:*

(i) *If $n = 1$, then the dimension of E is at most 1.*
(ii) *If θ belongs to $\mathcal{A}(\mathbb{K})$ (respectively if θ belongs to $\mathcal{A}_u(d(\alpha, R^-))$), then $E = \{0\}$.*
(iii) *If θ belongs to $\mathcal{A}_b(d(\alpha, R^-))$ and satisfies $\|\theta\|_{d(\alpha,R^-)} > \dfrac{1}{R^n}$, then $E = \{0\}$ again.*

Proof. In each case, we assume that $(\mathcal{E})$ admits a non-identically zero solution h. Then $h^{(n)}$ may not be identically zero.

Suppose first that $n = 1$. Suppose that $g \in E$. Let $u = \frac{h}{g}$. Since $h' = \theta h$, we have $u'g + ug' = \theta ug$, therefore $u\frac{g'}{g} = u\theta = u' + u\frac{g'}{g}$, and hence $u' = 0$ i.e. u is a constant. Consequently, E is at most of dimension 1.

Suppose now that θ lies in $\mathcal{A}(\mathbb{K})$. Then $|\theta|(r) = \frac{|h^{(n)}|(r)}{|h|(r)}$ is an increasing function in r in $]0, +\infty[$, a contradiction to the inequality $\frac{|h^{(n)}|(r)}{|h|(r)} \leq \frac{1}{r^n}$ coming from Theorem 37.10.

Now, suppose that $\theta \in \mathcal{A}_u(d(\alpha, R^-))$. Without loss of generality, we may assume $\alpha = 0$. Similarly, $|\theta|(r)$ is unbounded in $]0, R^-[$, a contradiction to $\frac{|h^{(n)}|(r)}{|h|(r)} \leq \frac{1}{r^n}$ again.

In the same way, if θ belongs to $\mathcal{A}_b(d(0, R^-))$ and satisfies $\|\theta\|_{d(0,R^-)} > \frac{1}{R^n}$, the inequality $\frac{|h^{(n)}|(r)}{|h|(r)} \leq \frac{1}{r^n}$ is then violated when r tends to R. □

Remark. The hypothesis θ *unbounded in* $d(\alpha, R^-)$ is indispensable to show that the space E is of dimension 0 or 1, as shown by the example given again by the p-adic hyperbolic functions $h(x) = \cosh(x)$ and $l(x) = \sinh(x)$. The radius of convergence of both h, l is $p^{\frac{-1}{p-1}}$ when $\mathbb{K}$ has residue characteristic p and is 1 when $\mathbb{K}$ has residue characteristic 0. Of course, both functions are solutions of $y'' - y = 0$, but they are bounded.

Theorem 50.5 given in [10] is an improvement of Theorem 50.2.

Theorem 50.5 (Bézivin). Let $f,\ g \in \mathcal{A}(\mathbb{K})$ be such that $W(f, g)$ is a non-identically zero polynomial. Then both f, g are polynomials.

Proof. First, by Theorem 50.2, we check that the claim is satisfied when $W(f, g)$ is a polynomial of degree 0. Now, suppose the claim holds when $W(f, g)$ is a polynomial of certain degree n. We show it for $n + 1$. Let $f, g \in \mathcal{A}(\mathbb{K})$ be such that $W(f, g)$ is a non-identically zero polynomial P of degree $n + 1$.

Thus, by hypothesis, we have $f'g - fg' = P$, hence $f''g - fg'' = P'$. We can extract g' and get $g' = \frac{(f'g-P)}{f}$. Now consider the function $Q = f''g' - f'g''$ and replace g' by what we just found: we can get $Q = f'(\frac{(f''g-fg'')}{f}) - \frac{Pf''}{f}$.

Now, we can replace $f''g - fg''$ by P' and obtain $Q = \frac{(f'P'-Pf'')}{f}$. Thus, in that expression of Q, we can write $|Q|(R) \leq \frac{|f|(R)|P|(R)}{R^2|f|(R)}$, hence $|Q|(R) \leq \frac{|P|(R)}{R^2}$ $\forall R > 0$. But by definition, Q belongs to $\mathcal{A}(\mathbb{K})$. Consequently, Q is a polynomial of degree $t \leq n - 1$.

Now, suppose Q is not identically zero. Since $Q = W(f', g')$ and since $\deg(Q) < n$, by the induction, hypothesis f' and g' are polynomials and so are f, g. Finally, suppose $Q = 0$ meromorphic functions in $\mathbb{K}$: $(\mathcal{E})$ $y' = \theta y$ with $\theta = \frac{P'}{P}$, whereas y belongs to $\mathcal{A}(\mathbb{K})$. By Theorem 50.4, the space of solutions of $(\mathcal{E})$ is of dimension 0 or 1. Consequently, there exists $\lambda \in \mathbb{K}$ such that $f' = \lambda P$, hence f is a polynomial. The same holds for g. □

Then, $P'f' - Pf'' = 0$, and therefore, f', P are two solutions of the differential equation of order 1 for the following:

Remark. In Archimedean analysis, Theorem 50.5 does not hold. For example, take $f(x) = e^x$, $g(x) = e^{-x}$. Then, $W(f, g) = 2$. We can also consider $f(x) = xe^x$, $g(x) = e^{-x}$. Then, $W(f, g) = 2x + 1$.

Lemma 50.6. *Let $U, V \in \mathcal{A}(\mathbb{K})$ have no common zero and let $f = \frac{U}{V}$. If f' has finitely many zeros, there exists a polynomial $P \in \mathbb{K}[x]$ such that $U'V - UV' = P\widetilde{V}$.*

Proof. If V is a constant, the statement is obvious. So, we assume that V is not a constant. Now $\widetilde{V}$ divides V' and hence V' factorizes in the way $V' = \widetilde{V}Y$ with $Y \in \mathcal{A}(\mathbb{K})$. Then no zero of Y can be a zero of V. Consequently, we have

$$f'(x) = \frac{U'V - UV'}{V^2} = \frac{U'\overline{V} - UY}{\overline{V}^2\widetilde{V}}.$$

The two functions $U'\overline{V} - UY$ and $\overline{V}^2\widetilde{V}$ have no common zero since neither have U and V. So, the zeros of f' are those of $U'\overline{V} - UY$ which therefore has finitely many zeros and consequently is a polynomial P, hence $U'V - UV' = P\widetilde{V}$. □

Theorem 50.7. *Let $f \in \mathcal{M}(\mathbb{K})\backslash\mathbb{K}(x)$ have finitely many poles of order ≥ 3 and admit primitives. Then f has no quasi-exceptional value.*

Proof. Suppose that f admits a quasi-exceptional value. Without loss of generality, we can assume that this value is 0. Let F be a primitive of f and let $F = \frac{U}{V}$, with U, $V \in \mathcal{A}(\mathbb{K})$, having no common zero. By Lemma 50.7, there exists a polynomial P such that $U'V - UV' = P\widetilde{V}$. But since f has finitely many poles of order ≥ 3, F has finitely many poles of order ≥ 2, hence $\widetilde{V}$ has finitely many zeros, hence it is a polynomial. But then $P\widetilde{V}$ is a polynomial and then, by Theorem 50.5, both U, V are polynomials, therefore $F \in \mathbb{K}(x)$ a contradiction. □

Theorem 50.8. *Let $f \in \mathcal{M}(\mathbb{K})$ have finitely many multiple poles such that for certain $b \in \mathbb{K}$, $f' - b$ has finitely many zeros. Then, f belongs to $\mathbb{K}(x)$.*

Proof. Suppose first $b = 0$. Let us write $f = \frac{U}{V}$ with $U, V \in \mathcal{A}(\mathbb{K})$, having no common zeros. By Lemma 50.7, there exists a polynomial $P \in \mathbb{K}[x]$ such that $U'V - UV' = P\widetilde{V}$. Since f has finitely many multiple poles, $\widetilde{V}$ is a polynomial, hence so is $U'V - UV'$. But then by Theorem 50.5, both U, V are polynomials, which ends the proof when $b = 0$. Consider now the general case. $f' - b$ is the derivative of $f - bx$ that satisfies the same hypothesis, so the conclusion is immediate. □

Theorem 50.9. *Let $f \in \mathcal{M}(\mathbb{K})$ admit primitives. If f has two perfectly branched values, it has no quasi-exceptional value. Moreover, if f has one totally branched value, then it has no exceptional value.*

Proof. Suppose that f has two perfectly branched values a and b and a quasi-exceptional value c. Since f admits primitives, $N(r, f)$ satisfies $\overline{N}(r, f) \leq \frac{N(r,f)}{2} + o(T(r, f))$, hence by Theorem 45.7, we have

$$2T(r, f) \leq \frac{(Z(r, f - a) + Z(r, f - b) + N(r, f))}{2} + o(T(r, f)),$$

hence $2T(r, f) \leq \frac{3T(r,f)}{2} + o(T(r, f))$, a contradiction.

Suppose now that f has one totally branched and an exceptional value c. Then $N(r, f)$ satisfies

$$2T(r, f) \leq \frac{(Z(r, f - a) + N(r, f))}{2} - \log(r),$$

a contradiction. □

Notation. For each $n \in \mathbb{N}^*$, we set $\lambda_n = \max\{\frac{1}{|k|},\ 1 \leq k \leq n\}$. Given positive integers n, q, we denote by C_n^q the combination $\frac{n!}{q!(n-q)!}$.

For convenience, in this chapter, log is the Neperian logarithm and we denote by e the number such that $\log(e) = 1$ and Exp is the real exponential function.

Remark. For every $n \in \mathbb{N}^*$, we have $\lambda_n \leq n$ because $k|k| \geq 1\ \forall k \in \mathbb{N}$. The equality holds for all n of the form p^h.

Lemmas 50.10 and 50.11 are due to Jean-Paul Bézivin [10]:

Lemma 50.10. *Let $U,\ V \in \mathcal{A}(d(0, R^-))$. Then for all $r \in]0, R[$ and $n \geq 1$, we have*

$$|U^{(n)}V - UV^{(n)}|(r) \leq |n!|\lambda_n \frac{|U'V - UV'|(r)}{r^{n-1}}.$$

More generally, given $j,\ l \in \mathbb{N}$, we have

$$|U^{(j)}V^{(l)} - U^{(l)}V^{(j)}|(r) \leq |(j!)(l!)|\lambda_{j+l} \frac{|U'V - UV'|(r)}{r^{j+l-1}}.$$

Proof. Set $g = \frac{U}{V}$ and $f = g'$. Applying Theorem 37.10 to f for $k-1$, we obtain

$$|g^{(k)}|(r) = |f^{(k-1)}|(r) \leq |(k-1)!| \frac{|f|(r)}{r^{k-1}} = |(k-1)!| \frac{|U'V - UV'|(r)}{|V^2|(r)r^{k-1}}.$$

As in the proof of Theorem 37.10, we set $U = V\left(\frac{U}{V}\right)$. By Leibniz formula again, now we can obtain

$$U^{(n)} = \sum_{q=1}^{n} C_n^q V^{(n-q)} \left(\frac{U}{V}\right)^{(q)} + V^{(n)}\left(\frac{U}{V}\right),$$

hence

$$U^{(n)} - V^{(n)}\left(\frac{U}{V}\right) = \sum_{q=1}^{n} C_n^q V^{(n-q)} \left(\frac{U}{V}\right)^{(q)}. \tag{1}$$

Now, we have

$$\left|\left(\frac{U}{V}\right)^{(q)}\right|(r) = |g^{(q)}|(r) \leq |(q-1)!| \frac{|U'V - UV'|(r)}{|V^2|(r)r^{q-1}}$$

and

$$|V^{(n-q)}|(r) \leq |(n-q)!| \frac{|V|(r)}{r^{n-q}}.$$

Consequently, the general term in (1) is upper bounded as

$$\left|C_n^q V^{(n-q)}\left(\frac{U}{V}\right)^{(q)}\right|(r) \leq \frac{|(n!)((n-q)!)((q-1)!)|}{|(q!)((n-q)!)|}\frac{|U'V-UV'|(r)}{|V|(r)r^{n-1}}$$
$$\leq \lambda_n \frac{|n!||U'V-UV'|(r)}{|V|(r)r^{n-1}}.$$

Therefore, by (1), we obtain

$$\left|U^{(n)} - V^{(n)}\left(\frac{U}{V}\right)\right|(r) \leq |n!|\lambda_n \frac{|U'V-UV'|(r)}{|V|(r)r^{n-1}},$$

and finally

$$\left|U^{(n)}V - V^{(n)}U\right|(r) \leq |n!|\lambda_n \frac{|U'V-UV'|(r)}{r^{n-1}}.$$

We can now generalize the first statement. Set $P_j = U^{(j)}V - UV^{(j)}$. By induction, we can show the following equality that already holds for $l \leq j$:

$$U^{(j)}V^{(l)} - U^{(l)}V^{(j)} = \sum_{h=0}^{l} C_l^h (-1)^h P_{j+h}^{(l-h)}.$$

Then, the second statement gets just an application of the first. □

Lemma 50.11. *Let $U, V \in \mathcal{A}(\mathbb{K})$ and let $r,\ R \in]0, +\infty[$ satisfy $r < R$. For all $x, y \in \mathbb{K}$ with $|x| \leq R$ and $|y| \leq r$, we have the following inequality:*

$$|U(x+y)V(x) - U(x)V(x+y)| \leq \frac{R|U'V-UV'|(R)}{e(\log R - \log r)}$$

Proof. By Taylor's formula at the point x, we have

$$U(x+y)V(x) - U(x)V(x+y) = \sum_{n\geq 0} \frac{U^{(n)}(x)V(x) - U(x)V^{(n)}(x)}{n!}y^n.$$

Now, by Lemma 50.9, we have

$$\left|\frac{U^{(n)}(x)V(x) - U(x)V^{(n)}(x)}{n!}y^n\right| \leq \lambda_n \frac{|U'V - UV'|(R)}{R^{n-1}}r^n$$
$$= \lambda_n R|U'V - UV'|(R)\left(\frac{r}{R}\right)^n.$$

Consequently, $\lim_{n\to+\infty}\left|\frac{U^{(n)}(x)V(x)-U(x)V^{(n}(x)}{n!}y^n\right| = 0$, therefore we can define $B = \max_{n\geq 1}\{\lambda_n\left(\frac{r}{R}\right)^n\}R|U'V - UV'|(R)$ and we have $|U(x+y)V(x) - U(x)V(x+y)| \leq B$. Now, as remarked above, we have $\lambda_n \leq n$. We can check that the function h defined in $]0,+\infty[$ as $h(t) = t\left(\frac{r}{R}\right)^t$ reaches it maximum at the point $u = \frac{1}{e(\text{Log}R-\text{Log}r)}$. Consequently, $B \leq \frac{1}{e(\text{Log}R-\text{Log}r)}$, and therefore,

$$|U(x+y)V(x) - U(x)V(x+y)| \leq \frac{R|U'V - UV'|(R)}{e(\text{Log}R - \text{Log}r)}. \qquad \square$$

Notation. Let $f \in \mathcal{M}(d(0,R^-))$. For each $r \in]0,R[$, we denote by $s(r,f)$ the number of zeros of f in $d(0,r)$, each counted with its multiplicity and we set $t(r,f) = s(r,\frac{1}{f})$. Similarly, we denote by $\beta(r,f)$ the number of multiple zeros of f in $d(0,r)$, each counted with its multiplicity.

Theorem 50.12. *Let $f \in \mathcal{M}(\mathbb{K})$ be such that for some $c,q \in]0,+\infty[$, $t(r,f)$ satisfies $t(r,f) \leq cr^q$ in $[1,+\infty[$. If f' has finitely many zeros, then $f \in \mathbb{K}(x)$.*

Proof. Suppose f' has finitely many zeros and set $f = \frac{U}{V}$. If V is a constant, the statement is immediate. So, we suppose V is not a constant and hence it admits at least one zero a. By Lemma 50.7, there exists a polynomial $P \in \mathbb{K}[x]$ such that $U'V - UV' = P\widetilde{V}$. Next, we take $r,R \in [1,+\infty[$ such that $|a| < r < R$ and $x \in d(0,R)$, $y \in d(0,r)$. By Lemma 50.11, we have

$$|U(x+y)V(x) - U(x)V(x+y)| \leq \frac{R|U'V - UV'|(R)}{e(\text{Log}R - \text{Log}r)}.$$

Note that $U(a) \neq 0$ because U and V have no common zero. Now set $l = \max(1, |a|)$ and take $r \geq l$. Setting $c_1 = \frac{1}{e|U(a)|}$, we have

$$|V(a+y)| \leq c_1 \frac{R|P|(R)|\widetilde{V}|(R)}{\mathrm{Log}R - \mathrm{Log}r}.$$

Then taking the supremum of $|V(a+y)|$ inside the disk $d(0,r)$, we can derive

$$|V|(r) \leq c_1 \frac{R|P|(R)|\widetilde{V}|(R)}{\mathrm{Log}R - \mathrm{Log}r}. \tag{1}$$

Let us apply Corollary 22.30, by taking $R = r + \frac{1}{r^q}$, after noting that the number of zeros of $\widetilde{V}(R)$ is bounded by $s(r,V)$. So, we have

$$|\widetilde{V}|(R) \leq \left(1 + \frac{1}{r^{q+1}}\right)^{\beta((r+\frac{1}{r^q}),V)} |\widetilde{V}|(r). \tag{2}$$

Now, due to the hypothesis, $s(r,V) = t(r,f) \leq cr^q$ in $[1, +\infty[$, we have

$$\begin{aligned} \left(1 + \tfrac{1}{r^{q+1}}\right)^{\beta((r+\frac{1}{r^q}),V)} &\leq \left(1 + \tfrac{1}{r^{q+1}}\right)^{[c(r+\frac{1}{r^q})^m]} \\ &= \mathrm{Exp}\Big[c(r + \tfrac{1}{r^q})^q \mathrm{Log}(1 + \tfrac{1}{r^{q+1}})\Big]. \end{aligned} \tag{3}$$

The function $h(r) = c(r+\frac{1}{r^m})^m \mathrm{Log}(1+\frac{1}{r^{m+1}})$ is continuous on $]0, +\infty[$ and equivalent to $\frac{c}{r}$ when r tends to $+\infty$. Consequently, it is bounded on $[l, +\infty[$. Therefore, by (2) and (3), there exists a constant $M > 0$ such that, for all $r \in [l, +\infty[$, by (3), we obtain

$$|\widetilde{V}|\left(r + \frac{1}{r^q}\right) \leq M|\widetilde{V}|(r). \tag{4}$$

On the other hand,

$$\mathrm{Log}\Big(r + \frac{1}{r^q}\Big) - \mathrm{Log}r = \mathrm{Log}\Big(1 + \frac{1}{r^{q+1}}\Big)$$

clearly satisfies an inequality of the form

$$\mathrm{Log}\Big(1 + \frac{1}{r^{q+1}}\Big) \geq \frac{c_2}{r^{q+1}}$$

in $[l, +\infty[$ with $c_2 > 0$. Moreover, we can obviously find positive constants c_3, c_4 such that

$$\left(r + \frac{1}{r^q}\right) |P| \left(r + \frac{1}{r^q}\right) \leq c_3 r^{c_4}.$$

Consequently, by (1) and (4), we can find positive constants c_5, c_6 such that $|V|(r) \leq c_5 r^{c_6} |\widetilde{V}|(r)$ $\forall r \in [l, +\infty[$. Thus, writing again $V = \overline{V}\widetilde{V}$, we have $|\overline{V}|(r)|\widetilde{V}|(r) \leq c_5 r^{c_6} |\widetilde{V}|(r)$, and hence

$$|\overline{V}|(r) \leq c_5 r^{c_6} \quad \forall r \in [l, +\infty[.$$

Consequently, by Corollary 22.31, $\overline{V}$ is a polynomial of degree $\leq c_6$ and hence it has finitely many zeros and so does V. But then, by Theorem 50.8, f must be a rational function. □

Corollary 50.13. *Let f be a meromorphic function on $\mathbb{K}$ such that, for some $c, q \in]0, +\infty[$, $t(r, f)$ satisfies $t(r, f) \leq cr^q$ in $[1, +\infty[$. If for some $b \in \mathbb{K}$ $f' - b$ has finitely many zeros, then f is a rational function.*

Proof. Suppose $f' - b$ has finitely many zeros. Then $f - bx$ satisfies the same hypothesis as f, hence it is a rational function and so is f. □

Corollary 50.14. *Let $f \in \mathcal{M}(\mathbb{K}) \setminus \mathbb{K}(x)$ admitting primitives and be such that* $\log(t(r, f)) = O(\log r)$. *Then f has no quasi-exceptional value.*

Proof. Indeed, since f admits primitives, all poles are multiple, hence, given a primitive F of f, we have $t(r, F) \leq t(r, f)$. Consequently, by hypothesis, we have $\log(t(r, F)) \leq O(\log(r))$, hence thanks to Corollary 50.13, F' has no quasi-exceptional value. □

Thanks to Theorem 31.5, we can derive Corollaries 50.15 and 50.16:

Corollary 50.15. *Let $f \in \mathcal{M}(\mathbb{K}) \setminus \mathbb{K}(x)$, admitting primitives, be of the form $\frac{g}{h}$ and be such that $\rho(h) < +\infty$ and $\psi(h) < +\infty$. Then f has no quasi-exceptional value.*

Corollary 50.16. *Let $f \in \mathcal{M}(\mathbb{K}) \setminus \mathbb{K}(x)$, admitting primitives, be of the form $\frac{g}{h}$ with g, $h \in \mathcal{A}(\mathbb{K})$ and be such that $\rho(h) < +\infty$ and $\sigma(h) < +\infty$. Then f has no quasi-exceptional value.*

Corollary 50.17. *Let $f \in \mathcal{M}(\mathbb{K})\setminus\mathbb{K}(x)$ be of the form $\frac{g}{h}$ with $g,\ h \in \mathcal{A}(\mathbb{K})$ and be such that $\rho(h) < +\infty$ and $\sigma(h) < +\infty$. Then f' has no quasi-exceptional value.*

Proof. Let $f = \frac{g}{h}$ with $\sigma(h) < +\infty$. Then, $f' = \frac{g'h-h'g}{h^2}$ and $\sigma(h^2) = 2\sigma(h)$. Thus, one can apply Corollary 50.16 to f'. □

Corollary 50.18. *Let $f \in \mathcal{M}(\mathbb{K})\setminus\mathbb{K}(x)$ be of the form $\frac{g}{h}$ with $g,\ h \in \mathcal{A}(\mathbb{K})$ and be such that $\rho(h) < +\infty$ and $\sigma(h) < +\infty$. If f has a quasi-exceptional value, then it has a non-zero residue.*

Example 1. Let $(a_n)_{n\in\mathbb{N}}$ be a sequence in $\mathbb{K}$ such that $|a_n| \leq |a_{n+1}|$ and $\lim_{n\to+\infty} |a_n| = +\infty$ and let

$$f(x) = \sum_{n=0}^{\infty} \frac{b_n}{(x-a_n)^{s_n}} \quad \text{with } |b_n| \leq 1,\ s_n \geq 2\ \forall n \text{ and } s_n = 2\ \forall n \geq t.$$

Then the function $f(x) = \sum_{n=0}^{\infty} \frac{b_n}{(x-a_n)^{s_n}}$ admits primitives and has no quasi-exceptional value by Theorem 50.7.

Example 2. Let $(a_n)_{n\in\mathbb{N}}$ be a sequence in $\mathbb{K}$ such that $|a_n| < |a_{n+1}|$ and $\lim_{n\to+\infty} |a_n| = +\infty$ and suppose that $\log(n) = O(\log|a_n|)$. Then the function $f(x) = \sum_{n=0}^{\infty} \frac{b_n}{(x-a_n)^{s_n}}$ with $|b_n| \leq 1,\ s_n \geq 2\ \forall n$, admits primitives and has no quasi-exceptional value by Corollary 50.14.

Example 3. Let $h \in \mathcal{A}(\mathbb{K}) \setminus \mathbb{K}[x]$ be a function having only zeros of order 1 and let $P(x) \in \mathbb{K}[x]$. Let $f(x) = \frac{P(x)}{h(x)^2}$. Then f has no primitive.

Indeed, suppose that f have a primitive $F = \frac{U}{V}$, where U and V lie in $\mathcal{A}(\mathbb{K})$ have no common zeros. Since the zeros of h are of order 1, it is seen that all zeros of V are of order 1 and are all the zeros of h. Consequently, $\widetilde{V} = 1$, $\overline{V} = V$, and $F' = \frac{U'V-UV'}{V^2}$ admits no simplification. Therefore, $U'V - UV' = P$. But then, by Theorem 50.5, U and V are polynomials and $V^2 = h^2$, a contradiction to the hypothesis $h \in \mathcal{A}(\mathbb{K}) \setminus \mathbb{K}[x]$.

Remark 1. In Example 3, the function f certainly has residues different from 0 because if all residues were null, the function then would have primitives.

Remark 2. Let f be a non-rational meromorphic function in $\mathbb{C}$ admitting primitives. Then the second main theorem lets easily show that f has at most one Picard value [66].

Now, by Theorems 50.7 and 50.9 and Corollary 50.14, it seems that a function admitting primitives has no quasi-exceptional value. Actually, as we saw in Theorem 37.12, a meromorphic function in $\mathbb{K}$ admitting primitives is a meromorphic function whose residues are all null. Thus, the following conjecture appears likely:

Conjecture. *A transcendental meromorphic function in* $\mathbb{K}$ *whose residues are all null has no quasi-exceptional value.*

Chapter 51

Small Functions

Throughout this chapter, the field $\mathbb{K}$ is supposed to be of characteristic 0.

Small functions with respect to a meromorphic function are well known in the general theory of complex functions. Particularly, one knows the Nevanlinna theorem on three small functions. Here we construct a similar theory.

Definition and notation. Throughout this chapter, we set $a \in \mathbb{K}$ and $R \in]0, +\infty[$ and we still denote by D the set $\mathbb{K}\setminus d(0, R^-)$. For each $f \in \mathcal{M}(\mathbb{K})$ (respectively $f \in \mathcal{M}(d(a, R^-))$, respectively $f \in \mathcal{M}(D)$), we denote by $\mathcal{M}_f(\mathbb{K})$ (respectively $\mathcal{M}_f(d(a, R^-))$, respectively $\mathcal{M}_f(D)$) the set of functions $h \in \mathcal{M}(\mathbb{K})$ (respectively $h \in \mathcal{M}(d(a, R^-))$, respectively $\mathcal{M}(D)$) such that $T(r, h) = o(T(r, f))$ when r tends to $+\infty$ (respectively $T(r, h) = o(T(r, f))$ when r tends to R, respectively $T_R(r, h) = o(T_R(r, f))$ when r tends to $+\infty$). Similarly, if $f \in \mathcal{A}(\mathbb{K})$ (respectively $f \in \mathcal{A}(d(a, R^-))$, $f \in \mathcal{A}(D)$,) we denote by $\mathcal{A}_f(\mathbb{K})$ (respectively $\mathcal{A}_f(d(a, R^-))$, respectively $\mathcal{A}_f(D)$) the set $\mathcal{M}_f(\mathbb{K}) \cap \mathcal{A}(\mathbb{K})$ (respectively $\mathcal{M}_f(d(a, R^-)) \cap \mathcal{A}(d(a, R^-))$, respectively $\mathcal{M}_f(D) \cap \mathcal{A}(D)$).

The elements of $\mathcal{M}_f(\mathbb{K})$ (respectively $\mathcal{M}_f(d(a, R^-))$, respectively $\mathcal{M}_f(D)$) are called *small meromorphic functions with respect to f*, *small functions* in brief. Similarly, if $f \in \mathcal{A}(\mathbb{K})$ (respectively $f \in \mathcal{A}(d(a, R^-))$, respectively $f \in \mathcal{A}(D)$), the elements of $\mathcal{A}_f(\mathbb{K})$ (respectively $\mathcal{A}_f(d(a, R^-))$, respectively $\mathcal{A}_f(D)$) are called *small analytic functions with respect to f, small functions in brief.*

Theorems 51.1 and 51.2 are immediate consequences of Theorems 43.3 and 46.8:

Theorem 51.1. *Let $a \in \mathbb{K}$ and $r > 0$. $\mathcal{A}_f(\mathbb{K})$ is a $\mathbb{K}$-subalgebra of $\mathcal{A}(\mathbb{K})$, $\mathcal{A}_f(d(a,R^-))$ is a $\mathbb{K}$-subalgebra of $\mathcal{A}(d(a,R^-))$, $\mathcal{A}_f(D)$ is a $\mathbb{K}$-subalgebra of $\mathcal{A}(D)$, $\mathcal{M}_f(\mathbb{K})$ is a subfield field of $\mathcal{M}(\mathbb{K})$, $\mathcal{M}_f(d(a,R^-))$ is a subfield of field of $\mathcal{M}(a,R^-)$, and $\mathcal{M}_f(D)$ is a subfield field of $\mathcal{M}(D)$. Moreover, $\mathcal{A}_b(d(a,R^-))$ is a subalgebra of $\mathcal{A}_f(d(a,R^-))$ and $\mathcal{M}_b(d(a,R^-))$ is a subfield of $\mathcal{M}_f(d(a,R^-))$.*

Theorem 51.2. *Let $f \in \mathcal{M}(\mathbb{K})$ (respectively $f \in \mathcal{M}(d(0,R^-))$, respectively $f \in \mathcal{M}(\mathbb{K})$) and let $g \in \mathcal{M}_f(\mathbb{K})$ (respectively $g \in \mathcal{M}_f(d(0,R^-))$, respectively $g \in \mathcal{M}_f(D)$). Then, $T(r,fg) = T(r,f) + o(T(r,f))$ and $T(r,\frac{f}{g}) = T(r,f) + o(T(r,f))$ (respectively $T(r,fg) = T(r,f) + o(T(r,f))$ and $T(r,\frac{f}{g}) = T(r,f) + o(T(r,f))$, respectively $T_R(r,fg) = T_R(r,f) + o(T_R(r,f))$ and $T_R(r,\frac{f}{g}) = T_R(r,f) + o(T_R(r,f))$).*

Here we can mention some precisions to Theorem 51.1 that are useful later:

Theorem 51.3. *Let $f \in \mathcal{A}(\mathbb{K})$ (respectively let $f \in \mathcal{A}_u(d(a,r^-))$, respectively let $f \in \mathcal{A}(D)$). Let $g,\ h \in \mathcal{A}_f(\mathbb{K})$ (respectively let $g,\ h \in \mathcal{A}_f(d(a,r^-))$, respectively let $g,\ h \in \mathcal{A}_f(D)$) with g and h not identically zero. If gh belongs to $\mathcal{A}_f(\mathbb{K})$ (respectively to $\mathcal{A}_f(d(a,r^-))$, respectively to $\mathcal{A}_f(D)$), then so do g and h.*

Proof. Concerning the claim on $f \in \mathcal{A}_u(d(a,r^-))$, we can obviously assume $a = 0$. By Theorems 43.3 and 46.7, we have $T(r,g.h) = T(r,g) + T(r,h) + O(1)$. Consequently, $T(r,g.h) = o(T(r,f))$ if and only if $T(r,g) = o(T(r,f))$ and $T(r,h) = o(T(r,f))$. □

Theorem 51.4. *Let $f,\ g \in \mathcal{A}(\mathbb{K})$ (respectively let $f,\ g \in \mathcal{A}_u(d(0,r^-))$, respectively let $f,\ g \in \mathcal{A}(D)$) and let $q \in \mathbb{N}^*$. If $\frac{f}{g}$ is not a qth root of 1, then $f^q - g^q$ does not belong to $\mathcal{A}_f(\mathbb{K})$ (respectively to $\mathcal{A}_f(d(0,r^-))$, respectively to $\mathcal{A}_f(D)$).*

Proof. Concerning the claim on $f \in \mathcal{A}_u(d(a,r^-))$, we can obviously assume $a = 0$. Let $h = \frac{f}{g}$. Since h is not a qth root of 1, neither $f - g$ nor the function $F(x) = \sum_{j=0}^{q-1} f^j g^{q-1-j}$ is identically zero.

Suppose that $f^q - g^q \in \mathcal{A}_f(\mathbb{K})$ (respectively $f^q - g^q \in \mathcal{A}_f(d(0, r^-))$, respectively $\mathcal{A}_f(D)$). So, by Theorem 51.3, both $f - g$ and F belong to $\mathcal{A}_f(\mathbb{K})$ (respectively to $\mathcal{A}_f(d(0, r^-))$, respectively to $\mathcal{A}_f(D)$). Let $w = f - g$, hence $g = f + w$. Then, $F(x) = \sum_{j=0}^{q-1} f^j (f + w)^{q-1-j}$. Thus, we can check that $F(x)$ is of the form $f^{q-1} + P(f(x))$ with $P(Y)$ a polynomial in Y of degree at most $q - 2$, with coefficients in $\mathcal{A}_f(\mathbb{K})$ (respectively in $\mathcal{A}_f(d(0, r^-))$, respectively in $\mathcal{A}_f(D)$). Consequently, by Theorems 43.3 and 46.7, $T(r, F(x))$ is of the form $(q - 1)T(r, f) + o(T(r, f))$ because $T(r, P(f(x))) < (q - 2)T(r, f) + o(T(r, f))$, which proves that F does not belong to $\mathcal{A}_f(\mathbb{K})$ (respectively to $\mathcal{A}_f(d(0, r^-))$, respectively to $\mathcal{A}_f(D)$). □

In the proof of Theorem 51.5 and in the sequel, we have to use the following notation:

Notation. Let $h \in \mathcal{M}(\mathbb{K}) \setminus \mathbb{K}$ (respectively $h \in E(x) \setminus E$) and let $\mathcal{Z}(h)$ be the set of zeros c of h' such that $h(c) \neq h(d)$ for every zero d of h' other than c. If $\mathcal{Z}(h)$ is finite, we denote by $\Upsilon(h)$ its cardinal, and if $\mathcal{Z}(h)$ is not finite, we put $\Upsilon(h) = +\infty$.

Theorem 51.5 is a wide generalization of Theorem 43.8. It consists of the following claim: given a meromorphic function f and a rational function G of degree n whose coefficients are small functions with respect to f, then $T(r, G(f))$ is equivalent to $nT(r, f)$. The big difficulty consists of showing that $T(r, G(f))$ is not smaller than $nT(r, f)$. The proof, based on an elementary property of Bezout's theorem, was given in $\mathbb{C}$ by Gackstatter and Laine and was made in a field such as $\mathbb{K}$ by Yang and Peichu Hu [67].

Theorem 51.5. *Let $f \in \mathcal{M}(\mathbb{K})$ (respectively let $f \in \mathcal{M}(d(0, R^-))$, respectively let $f \in \mathcal{M}(D)$). Let $G(Y) \in \mathcal{M}_f(\mathbb{K})(Y)$ (respectively $G \in \mathcal{M}_f(d(0, R^-))(Y)$, respectively $G(Y) \in \mathcal{M}_f(D)(Y)$) and let $n = \deg(G)$. Then $T(r, G(f)) = nT(r, f) + o(T(r, f))$ (respectively $T(r, G(f)) = nT(r, f) + o(T(r, f))$, respectively $T_R(r, G(f)) = nT_R(r, f) + o(T_R(r, f))$).*

Proof. Let $G = \frac{P}{Q}$ with P, Q relatively prime in the ring $\mathcal{M}_f(\mathbb{K})$ (respectively $\mathcal{M}_f(d(a, R^-))$, $\mathcal{M}_f(D)$). Suppose first $G(Y) \in \mathcal{M}_f(\mathbb{K})[Y]$ (respectively $G \in \mathcal{M}_f(d(0, R^-))[Y]$, respectively $G(Y) \in \mathcal{M}_f(D)[Y]$), hence $G = P$. Let $P(X) = \sum_{j=0}^{n} b_j X^j$ with $c_j \in \mathcal{M}_f(\mathbb{K})$ (respectively $c_j \in \mathcal{M}_f(d(0, R))$, $c_j \in \mathcal{M}_f(D)$). By Theorems 43.8

and 46.7, we have $T(r, P(f)) = T(r, b_n^{-1}P(f)) + o(T(r,f))$ (respectively $T(r, P(f)) = T(r, b_n^{-1}P(f)) + o(T(r,f))$, respectively $T_R(r, P(f)) = T_R(r, b_n^{-1}P(f)) + o(T_R(r,f)))$. Consequently, without loss of generality, we may also assume that P is monic.

Let $\widehat{\mathbb{K}}$ be an algebraically closed spherically complete extension of $\mathbb{K}$. Given $a \in \mathbb{K}$, we denote by $\widehat{d}(a, R^-)$ the disk $\{x \in \widehat{\mathbb{K}} \mid |x-a| < R\}$.

Suppose first $f \in \mathcal{M}(\mathbb{K})$ or $f \in \mathcal{M}(D)$. We can write $f = \frac{h}{l}$ with $h, l \in \mathcal{A}(\mathbb{K})$ having no common zero. Now suppose that $f \in \mathcal{M}(d(0, R^-))$. Since f has continuation to a function $\widehat{f}$ meromorphic in the disk $\widehat{d}(0, R^-)$ of the field $\widehat{\mathbb{K}}$, by Lemma 43.1, we know that the Nevanlinna function $T(r, \widehat{f})$ in $\widehat{\mathbb{K}}$ is exactly this of f in $\mathbb{K}$. Consequently, without loss of generality, we may assume that $\mathbb{K}$ is spherically complete. Thus, we can write f in the form $f = \frac{h}{l}$ with $h, l \in \mathcal{A}(d(0, R^-))$ having no common zero. Then $P(f)$ is in the form $\frac{\sum_{j=0}^{n} b_j h^j l^{n-j}}{B l^n}$ with $B,\ b_j \in \mathcal{A}_f(\mathbb{K})$ (respectively $B,\ b_j \in \mathcal{A}_f(d(0, R^-))$, respectively $B,\ b_j \in \mathcal{A}_f(D)$). Clearly, we have $Z\Big(r, \frac{\sum_{j=0}^{n} b_j h^j l^{n-j}}{l^n}\Big) \le Z\Big(r, \sum_{j=0}^{n} b_j h^j l^{n-j}\Big)$ and by Theorem 43.8 (respectively by Theorem 43.8, respectively by Theorem 46.7)

$$Z\left(r, \sum_{j=0}^{n} b_j h^j l^{n-j}\right) \le \max_{0 \le j \le n} Z(r, b_j h^j l^{n-j}) \le nT(r,f) + o(T(r,f)$$

(respectively

$$Z\left(r, \sum_{j=0}^{n} b_j h^j l^{n-j}\right) \le \max_{0 \le j \le n} Z(r, b_j h^j l^{n-j}) \le nT(r,f) + o(T(r,f),$$

respectively

$$\begin{aligned} Z_R\left(r, \sum_{j=0}^{n} b_j h^j l^{n-j}\right) &\le \max_{0 \le j \le n} Z_R(r, b_j h^j l^{n-j}) \\ &\le nT_R(r,f) + o(T_R(r,f))). \end{aligned}$$

On the other hand, $N(r, P(f)) \leq nN(r, f) + o(T(r, f))$ (respectively $N(r, P(f)) \leq nN(r, f)+o(T(r, f))$, respectively $N_R(r, P(f)) \leq nN_R(r, f) + o(T_R(r, f)))$, hence $T(r, P(f)) \leq nT(r, f) + o(T(r, f))$ (respectively $T(r, P(f)) \leq nT(r, f) + o(T(r, f))$, respectively $T_R(r, P(f)) \leq nT_R(r, f) + o(T_R(r, f)))$.

So, now it remains us to prove the reverse inequality. Indeed, suppose that inequality does not hold. For simplicity, we first suppose that f lies either in $\mathcal{M}(\mathbb{K})$ or in $\mathcal{M}(d(0, R^-))$.

Then there exists $\rho \in]0, 1[$ and a sequence of intervals $]r'_m, r''_m[$ such that $\lim_{m\to+\infty} r'_m = +\infty$ (respectively $\lim_{m\to+\infty} r'_m = R$, respectively $\lim_{m\to+\infty} r'_m = +\infty$) and that $T(r, P(f)) \leq \rho nT(r, f)$ $\forall r \in \bigcup_{m=1}^{\infty}]r'_m, r''_m[$. Particularly, we have $N(r, P(f)) \leq \rho nT(r, f)$ $\forall r \in \bigcup_{m=1}^{\infty}]r'_m, r''_m[$.

Let us write again $P(f)$ in the form $\sum_{j=0}^{n} c_j f^j$. We shall prove the following

$$(1)\quad N(r, P(f) \geq nN(r, f) - n\sum_{j=0}^{n} N(r, c_j) = nN(r, f) + o(T(r, f)).$$

Indeed, let α be a pole of f and suppose it is not a pole of order $\geq -n\omega_\alpha(f)$ of $P(f)$. We can check that there must exist $j \in \{0, \ldots, n-1\}$ such that $-\omega_\alpha(c_j) \geq -\omega_\alpha(f)$. Consequently, at α, we have $-\omega_\alpha(P(f)) \geq -n\omega_\alpha(f) - \max_{j=0,\ldots,n-1}(-n\omega_\alpha(c_j))$, hence

$$N\big(r, f \mid \omega_\alpha(P(f)) < n\omega_\alpha(f)\big) \leq n\sum_{j=0}^{n-1} N(r, c_j),$$

hence (1) follows clearly.

Consequently, there exists $q \in \mathbb{N}$ such that $nN(r, f) \leq \rho nT(r, f)$ $\forall r \in \bigcup_{m=q}^{\infty}]r'_m, r''_m[$ and therefore there exists $s \in \mathbb{N}$ and $\lambda \in]0, 1[$ such that

$$(2)\qquad N(r, Bl^n) < n\lambda T(r, f) \quad \forall r \in \bigcup_{m=s}^{\infty}]r'_m, r''_m[.$$

In particular, there exists $t \in \mathbb{N}$ such that

$$T(r, b_j h^j l^{n-j}) < \lambda T(r, b_n h^n)\ \forall j = 0, \ldots, n-1, \forall r \in \bigcup_{m=t}^{\infty}]r'_m, r''_m[.$$

Now, since $T(r, b_n h^n) > T(r, b_j h^j l^{n-j})$ $\forall j = 0, \ldots, n-1, \forall r \in \bigcup_{m=t}^{\infty}]r'_m, r''_m[$, we note that $|\sum_{j=0}^{n} b_j h^j l^{n-j}|(r) = |b_n|(r)$ $\forall r \in \bigcup_{m=t}^{\infty}]r'_m, r''_m[$,

hence

$$T\left(r, \sum_{j=0}^{n} b_j h^j l^{n-j}\right) = T(r, b_n h^n) \quad \forall r \in \bigcup_{m=t}^{\infty}]r'_m, r''_m[.$$

Consequently, there exists $\ell \in]\rho, 1[$ and $u \in \mathbb{N}$ $(u \geq t)$ such that

$$(3) \qquad T\left(r, \sum_{j=0}^{n} b_j h^j l^{n-j}\right) > \ell n T(r, h) \quad \forall r \in \bigcup_{m=u}^{\infty}]r'_m, r''_m[.$$

Now, by (2), we have $T(r, h) = T(r, f)$ $\forall r \in \bigcup_{m=u}^{\infty}]r'_m, r''_m[$ and

$$T(r, P(f)) = T\left(r, \sum_{j=0}^{n} b_j h^j l^{n-j}\right) \quad \forall r \in \bigcup_{m=u}^{\infty}]r'_m, r''_m[,$$

hence finally, by (3), we obtain $T(r, P(f)) > \ell n T(r, f)$ $\forall r \in \bigcup_{m=u}^{\infty}]r'_m, r''_m[$, a contradiction which proves the claim when G is a polynomial.

Similarly, replacing N by N_R, T by T_R, we can make the same reasoning when f belongs to $\mathcal{M}(D)$.

We now consider the general case $G(Y) \in \mathcal{M}_f(\mathbb{K})(Y)$ (respectively $G \in \mathcal{M}_f(d(0, R^-))(Y)$). Without loss of generality, we may assume $\deg(G) = \deg(P)$.

Since P and Q are relatively prime, by Bezout's theorem in a ring of polynomials on a field, we can find A, $B \in \mathcal{M}_f(\mathbb{K})[Y]$ (respectively $A, B \in \mathcal{M}_f(d(0, R^-))[Y]$) such that $AQ + PB = 1$. Since $\deg(Q) \leq \deg(P)$, of course, $\deg(B) \leq \deg(A)$, hence $\deg\left(\frac{A}{B}\right) = \deg(A)$. Now,

$$T\left(r, \frac{B(f)}{A(f)} + \frac{Q(f)}{P(f)}\right) = T\left(r, \frac{1}{A(f)P(f)}\right) = T(r, A(f)P(f)) + O(1).$$

Consequently, by the theorem already proven when G is a polynomial, we have

$$T\left(r, \frac{B(f)}{A(f)} + \frac{Q(f)}{P(f)}\right) = (\deg(A) + \deg(P))T(r, f) + o(T(r, f)),$$

and since $\deg(P) = \deg(G)$, actually we have

$$(4) \ T\left(r, \frac{B(f)}{A(f)} + \frac{Q(f)}{P(f)}\right) = (\deg(A) + \deg(G))T(r, f) + o(T(r, f)).$$

Now, $T\big(r, \frac{B(f)}{A(f)} + \frac{Q(f)}{P(f)}\big) \leq T(r, \frac{B(f)}{A(f)}) + T\big(r, \frac{Q(f)}{P(f)}\big) + o(T(r, f)$, and by the first inequality already proven above, we obtain

$$T\left(r, \frac{B(f)}{A(f)} + \frac{Q(f)}{P(f)}\right) \\ \leq \deg\left(\frac{B}{A}\right) T(r, f) + T\left(r, \frac{Q(f)}{P(f)}\right) + o(T(r, f).$$

But since $\deg(B) \leq \deg(A)$, actually we have

$$T\left(r, \frac{B(f)}{A(f)} + \frac{Q(f)}{P(f)}\right) \leq \deg(A)T(r, f) + T\left(r, \frac{Q(f)}{P(f)}\right) + o(T(r, f))$$

i.e.

$$T\left(r, \frac{B(f)}{A(f)} + \frac{Q(f)}{P(f)}\right) \\ \leq \deg(A)T(r, f) + T(r, G(f)) + o(T(r, f)). \tag{5}$$

Now, by (4) and (5), we can see that $\deg(G)T(r, f) \leq T(r, G(f)) + o(T(r, f))$. Similarly, when f belongs to $\mathcal{M}(D)$ and G belongs to $\mathcal{M}_f(D)(Y)$, we can make the same reasoning, as above. This completes the proof. □

Theorem 51.6. *Let $a \in \mathbb{K}$ and $r > 0$. Let $f \in \mathcal{M}(\mathbb{K})\backslash\mathbb{K}(x)$ (respectively $f \in \mathcal{M}_u(d(a, R^-))$, respectively $f \in \mathcal{M}^c(D)$). Then, f is transcendental over $\mathcal{M}_f(\mathbb{K})$ (respectively over $\mathcal{M}_f(d(a, R^-))$, respectively over $\mathcal{M}_f(D)$).*

Proof. Suppose there exists a polynomial $P(Y) = \sum_{j=0}^{n} a_j Y^j \in \mathcal{M}_f(\mathbb{K})[Y] \neq 0$ such that $P(f) = 0$. If f belongs to $\mathcal{M}_u(d(a, R^-))$, we may obviously suppose that $a = 0$. By Theorem 51.5, we have $T(r, a_n f^n) = nT(r, f) + o(T(r, f))$ whenever f belongs to $\mathcal{M}(\mathbb{K})\backslash\mathbb{K}(x)$ or to $\mathcal{M}_f(d(0, R^-))$ and $T_R(r, a_n f^n) = nT_R(r, f) + o(T_R(r, f))$ whenever f belongs to $\mathcal{M}^c(\mathbb{K})$, whereas $T(r, \sum_{j=0}^{n-1} a_j f^j) = (n-1)T(r, f) + o(T(r, f))$, a contradiction. □

Corollary 51.7. *Let $a \in \mathbb{K}$ and $r > 0$. Let $f \in \mathcal{M}(\mathbb{K})\backslash\mathbb{K}(x)$ (respectively $f \in \mathcal{M}_u(d(a, R^-))$, respectively $f \in \mathcal{M}^c(D)$). Then, f is transcendental over $\mathbb{K}(x)$.*

A function $h \in \mathcal{M}_b(d(a, R^-))$ is obviously small with respect to any function $f \in \mathcal{M}_u(d(a, R^-))$. So, we have the following corollary:

Corollary 51.8. *Let $a \in \mathbb{K}$ and $r > 0$ and let $f \in \mathcal{M}_u(d(a, R^-))$. Then, f is transcendental over $\mathcal{M}_b(d(a, R^-))$. Let $f \in \mathcal{M}^c(D)$. Then f is transcendental over $\mathcal{M}_0(D)$.*

By Corollary 36.23, we know that a meromorphic function $f \in \mathcal{M}(\mathbb{K})$ or $f \in \mathcal{M}(d(0, R^-))$ admits at most one quasi-exceptional value. Here we generalize that statement.

Theorem 51.9. *Let $a \in \mathbb{K}$ and $r > 0$. Let $f \in \mathcal{M}(\mathbb{K})\setminus\mathbb{K}(x)$ (respectively $f \in \mathcal{M}_u(d(a, R^-))$, respectively $f \in \mathcal{M}^c(D)$). There exists at most one function $g \in \mathcal{M}_f(\mathbb{K})$ (respectively $g \in \mathcal{M}_f(d(a, R^-))$, respectively $g \in \mathcal{M}_f(D)$) such that $f - g$ have finitely many zeros. Moreover, if f belongs to $\mathcal{A}(\mathbb{K})\setminus\mathbb{K}[x]$ (respectively to $\mathcal{A}_u(d(a, R^-))$, respectively to $\mathcal{A}^c(D)$), then there exists no function $g \in \mathcal{M}_f(\mathbb{K})\setminus\mathbb{K}(x)$ (respectively $g \in \mathcal{M}_f(d(a, R^-))$, respectively $g \in \mathcal{M}_f(D)$) such that $f - g$ have finitely many zeros.*

Proof. Concerning claims on $\mathcal{M}_u(d(a, R^-))$, we can obviously assume $a = 0$. Suppose that there exist two distinct functions $g_1,\ g_2 \in \mathcal{M}_f(\mathbb{K})$ (respectively $g_1,\ g_2 \in \mathcal{M}_f(d(0, R^-))$, respectively $g_1,\ g_2 \in \mathcal{M}_f(D)$) such that $f - g_k$ has finitely many zeros. So, there exist $P_1,\ P_2 \in \mathbb{K}[x]$ and $h_1,\ h_2 \in \mathcal{A}(\mathbb{K})$ (respectively $h_1,\ h_2 \in \mathcal{A}(d(0, R^-))$) such that $f - g_k = \frac{P_k}{h_k}$, $k = 1, 2$, and hence we note that

$$\text{(1)} \qquad T(r, f) = T\left(r, \frac{P_k}{h_k}\right) + o(T(r, f))$$
$$= T(r, h_k) + o(T(r, f)) \ k = 1, 2.$$

Consequently, putting $g = g_2 - g_1$, we have

$$\frac{P_1}{h_1} = \frac{P_2}{h_2} + g,$$

and by Theorem 51.1, g belongs to $\mathcal{M}_f(\mathbb{K})$ (respectively to $\mathcal{M}_f(d(0, R^-))$). Therefore, $P_1 h_2 - P_2 h_1 = g h_1 h_2$, and hence,

$$\text{(2)} \qquad T(r, P_1 h_2 - P_2 h_1) = T(r, g h_1 h_2).$$

Now, by Theorem 43.3, we have

$$\begin{aligned} T(r, P_1h_2 - P_2h_1) &\le \max(T(r, P_1h_2), T(r, P_2h_1)) \\ &\le \max(T(r, h_1), T(r, h_2)) + o(T(r, f)), \end{aligned}$$

and hence by (1), we obtain

$$(3) \qquad T(r, P_1h_2 - P_2h_1) \le T(r, f) + o(T(r, f)).$$

On the other hand, by Theorem 51.2, we have

$$T(r, gh_1h_2) = T(r, h_1h_2) + o(T(r, h_1h_2)) = 2T(r, f) + o(T(r, f)),$$

a contradiction to (3).

Now, if f belongs to $\mathcal{M}^c(D)$, we can make the same reasoning with T_R instead of T.

Suppose now that f belongs to $\mathcal{A}(\mathbb{K})\setminus\mathbb{K}[x]$ and that there exists a function $w \in \mathcal{M}_f(\mathbb{K})$ such that $f - w$ has finitely many zeros. Set $w = \frac{l}{t}$, where l and t belong to $\mathcal{A}_f(\mathbb{K})$ and have no common zeros. Thus, $f - w = \frac{tf-l}{t}$ and each zero of $tf - l$ cannot be a zero of t, hence is zero of $f - w$. Consequently, since $f - w$ has finitely many zeros, $tf - l$ has finitely many zeros and hence is a polynomial. But since l belongs to $\mathcal{A}_f(\mathbb{K})$, when r is big enough, we have $|f|(r) > |l|(r)$ and hence $|tf|(r) > |l|(r)$, therefore $|tf - l|(r) = |tf|(r)$. And since f is transcendental, by Corollary 22.23, for every fixed $q \in \mathbb{N}$, $|f|(r) > r^q$ when r is big enough. Similarly, $|tf - l|(r) > r^q$ when r is big enough. Consequently, by Corollary 22.23, $tf - l$ is not a polynomial, which proves that w does not exist.

Suppose now that f belongs to $\mathcal{A}_u(d(0, R^-))$ and that there exists a function $w \in \mathcal{M}_f(d(0, R^-))$ such that $f - w$ has finitely many zeros. Without loss of generality, we can assume that the field $\mathbb{K}$ is spherically complete because both f and w have continuation to an algebraically closed spherically complete extension of $\mathbb{K}$ where their zeros are the same as in $\mathbb{K}$. Consequently, we can write $w = \frac{l}{t}$, where l and t have no common zeros. Now, the zeros of $f - w$ are those of $tf - l$, hence $tf - l$ has finitely many zeros and hence is bounded in $d(0, R^-)$. But since w belongs to $\mathcal{M}_f(d(0, R^-))$, so does l and hence $|tf|(r) > |l|(r)$ when r tends to R. Consequently, $|tf - l|(r) = |tf|(r)$ is not bounded in $d(0, R^-)$, a contradiction proving again that w does not exist.

Suppose finally that f belongs to $\mathcal{A}^c(D)$. We can make the same reasoning as in $\mathcal{A}(\mathbb{K})$ by replacing T by T_R. □

Theorem 51.10 is known as second main theorem on three small functions in p-adic analysis [67]. It holds as well as in complex analysis, where it was showed first [67]. Note that this theorem was generalized to any finite set of small functions by Yamanoi in complex analysis [94], through methods that have no equivalent on a p-adic field.

Theorem 51.10. *Let $f \in \mathcal{M}(\mathbb{K})$ (respectively $f \in \mathcal{M}_u(d(0, R^-))$, respectively $f \in \mathcal{M}^c(D)$) and let w_1, $w_2, w_3 \in \mathcal{M}_f(\mathbb{K})$ (respectively $w_1, w_2, w_3 \in \mathcal{M}_f(d(0, R^-))$, respectively $w_1, w_2, w_3 \in \mathcal{M}_f(D)$) be pairwise distinct and let $S(r) = \max(T(r, w_j),\ j = 1, 2, 3)$. Then, $T(r, f) \leq \sum_{j=1}^{3} \overline{Z}(r, f - w_j) + 13S(r)$, respectively $T(r, f) \leq \sum_{j=1}^{3} \overline{Z}(r, f - w_j) + 13S(r)$, respectively $T_R(r, f) \leq \sum_{j=1}^{3} \overline{Z}_R(r, f - w_j) + 13S(r)$.*

Proof. We make the proof when f belongs to $\mathcal{M}(\mathbb{K})$ or to $\mathcal{M}_u(d(0, R^-))$. Let $\phi(x) = \frac{(f(x)-w_1(x))(w_2(x)-w_3(x))}{(f(x)-w_3(x))(w_2(x)-w_1(x))}$.

By Theorem 45.7, we have

(1) $T(r, \phi) \leq \overline{Z}(r, \phi) + \overline{Z}(r, \phi - 1) + \overline{N}(r, \phi) + O(1)$.

On the other hand, we have $T(r, f) \leq T(r, f - w_3) + T(r, w_3)$, hence

$$\begin{aligned} T(r, f) &\leq T\left(r, \frac{1}{f - w_3}\right) + T(r, w_3) \\ &\leq T\left(r, \frac{w_3 - w_1}{f - w_3}\right) + T(r, w_1 - w_3) + T(r, w_3) + 0(1), \end{aligned}$$

thereby

$$\begin{aligned} (2)\ T(r, f) &\leq T\left(r, \frac{w_3 - w_1}{f - w_3} + 1\right) + 3S(r) + O(1) \\ &= T\left(r, \frac{f - w_1}{f - w_3}\right) + 3S(r) + O(1). \end{aligned}$$

Now, $T\big(r, \frac{w_2-w_1}{w_2-w_3}\big) \leq 4S(r)+O(1)$. Consequently, by writing $\frac{f-w_1}{f-w_3} = \phi\big(\frac{w_2-w_1}{w_2-w_3}\big)$, we have $T\big(r, \frac{f-w_1}{f-w_3}\big) \leq T(r,\phi) + T\big(r, \frac{w_2-w_1}{w_2-w_3}\big) \leq T(r,\phi) + 4S(r) + O(1)$, and finally $T(r,f) \leq T(r,\phi) + 7S(r) + O(1)$. Thus, by (1), we obtain

(3) $T(r,f) \leq \overline{Z}(r,\phi) + \overline{Z}(r,\phi-1) + \overline{N}(r,\phi) + 7S(r) + O(1)$.

Now, we can check that

$$\begin{aligned}
&\overline{Z}(r,\phi) + \overline{Z}(r,\phi-1) + \overline{N}(r,\phi)\\
&\quad\leq \sum_{j=1}^{3} \overline{Z}(r,f-w_j) + \sum_{1\leq j<k\leq 3} \overline{Z}(r,w_k-w_j)\\
&\quad\leq \sum_{j=1}^{3} \overline{Z}(r,f-w_j) + 6S(r),
\end{aligned}$$

which, by (3), completes the proof when f belongs to $\mathcal{M}(\mathbb{K})$ or to $\mathcal{M}_u(d(0,R^-))$. When f belongs to $\mathcal{M}(D)$, we can make a similar proof just by replacing T by T_R and Z by Z_R. □

Theorem 51.11. *Let $f \in \mathcal{M}(\mathbb{K})$ (respectively $f \in \mathcal{M}_u(d(0,R^-))$, respectively $f \in \mathcal{M}^c(D)$) and let w_1, $w_2 \in \mathcal{M}_f(\mathbb{K})$ (respectively $w_1, w_2 \in \mathcal{M}_f(d(0,R^-))$, respectively $w_1, w_2 \in \mathcal{M}_f(D)$) be distinct. Let $S(r) = \max(T(r,w_1), T(r,w_2))$ (respectively $S(r) = \max(T(r, w_1), T(r,w_2))$, respectively $S(r) = \max(T_R(r,w_1), T_R(r,w_2))$). Then, $T(r,f) \leq \overline{Z}(r,f-w_1) + \overline{Z}(r,f-w_2) + \overline{N}(r,f) + 7S(r)$, respectively $T(r,f) \leq \overline{Z}(r,f-w_1) + \overline{Z}(r,f-w_2) + \overline{N}(r,f) + 7S(r)$, respectively $T_R(r,f) \leq \overline{Z}_R(r,f-w_1) + \overline{Z}_R(r,f-w_2) + \overline{N}_R(r,f) + 7S(r)$).*

Proof. Suppose first $f \in \mathcal{M}(\mathbb{K})$ or $f \in \mathcal{M}_u(d(0,R^-))$. Let $g = \frac{1}{f}$, $h_j = \frac{1}{w_j}$, $j = 1,2$, $h_3 = 0$. Clearly,

$$T(r,g) = T(r,f) + O(1),\ T(r,h_j) = T(r,w_j), \quad j = 1,2,$$

so we can apply Theorem 51.10 to g, h_1, h_2, h_3. So, we have $T(r,g) \leq \overline{Z}(r,g-h_1) + \overline{Z}(r,g-h_1) + \overline{Z}(r,g) + 7S(r)$. But we note that $\overline{Z}(r,g-h_j) = \overline{Z}(r,f-w_j)$ for $j = 1,2$ and that $\overline{Z}(r,g) = \overline{N}(r,f)$. Moreover, we know that $T(r,g) = T(r,f) + O(1)$. Consequently, when $w_1 w_2$ is not identically zero, the claim is proved.

Now, suppose that $w_2 = 0$. Take $\lambda \in \mathbb{K}^*$, let $l = f + \lambda$ and $\theta_j = w_j + \lambda$ $j = 1, 2$. Thus, we have $T(r, l) = T(r, f) + O(1)$, $T(r, \theta_j) = T(r, w_j) + O(1)$, $j = 1, 2$, and $\overline{N}(r, l) = \overline{N}(r, f)$. We note that $\max_{j=1,2} T(r, \theta_j) = S(r) + O(1)$.

By the claim already proved when $w_1 w_2 \neq 0$, we can write $T(r, l) \leq \overline{Z}(r, l - \theta_1) + \overline{Z}(r, l - \theta_2) + \overline{N}(l, r) + 7S(r)$, hence $T(r, f) \leq \overline{Z}(r, f - w_1) + \overline{Z}(r, f) + \overline{N}(f, r) + 7S(r) + O(1)$.

Suppose now $f \in \mathcal{M}(D)$. By replacing T by T_R, Z by Z_R, and N by N_R, we can check that the same reasoning applies. □

Corollary 51.12. *Let $f \in \mathcal{A}(\mathbb{K})$ (respectively $f \in \mathcal{A}_u(d(0, R^-))$, respectively $f \in \mathcal{A}^c(D)$) and let w_1, $w_2 \in \mathcal{A}_f(\mathbb{K})$ (respectively $w_1, w_2 \in \mathcal{A}_f(d(0, R^-))$, respectively $w_1, w_2 \in \mathcal{A}_f(D)$) be distinct. Let $S(r) = \max(T(r, w_1), T(r, w_2))$ (respectively $S(r) = \max(T(r, w_1), T(r, w_2))$, respectively $S(r) = \max(T_R(r, w_1), T_R(r, w_2))$). Then, $T(r, f) \leq \overline{Z}(r, f - w_1) + \overline{Z}(r, f - w_2) + 7S(r) + O(1)$, respectively $T(r, f) \leq \overline{Z}(r, f - w_1) + \overline{Z}(r, f - w_2) + 7S(r) + O(1)$, respectively $T_R(r, f) \leq \overline{Z}_R(r, f - w_1) + \overline{Z}_R(r, f - w_2) + 7S(r) + O(1)$).*

Here is now an application of that theory:

Theorem 51.13. *Let h, $w \in \mathcal{A}_b(d(a, R^-))$ and let m, $n \in \mathbb{N}^*$ be such that $\min(m, n) \geq 2$, $\max(m, n) \geq 3$. Then the functional equation*

$$(\mathcal{E}) \quad (g(x))^n = h(x)(f(x))^m + w(x)$$

has no solution in $\mathcal{A}_u(d(a, R^-))$.

Proof. Without loss of generality, we can obviously assume $a = 0$. Let $F(x) = g(x)^n$. Thanks to Corollary 51.12, we can write

$$(1) \qquad T(r, F) \leq \overline{Z}(r, F) + \overline{Z}(r, F - w) + o(T(r, F)).$$

Now, it appears that $\overline{Z}(r, F) \leq \frac{1}{n} Z(r, F)$. Moreover, since h is bounded, $Z(r, h)$ is bounded, hence $\overline{Z}(r, hf^m) \leq Z(r, f) + Z(r, h) = Z(r, f) + O(1)$, hence

$$(2) \qquad \overline{Z}(r, hf^m) \leq \frac{1}{m} Z(r, hf^m) + O(1) = \frac{1}{m} Z(r, F) + O(1).$$

On the other hand, $Z(r, F) = Z(r, F - w) + O(1) = T(r, F) + O(1)$. Consequently, by (1) and (2), we can derive

$$T(r, F) \leq \left(\frac{1}{m} + \frac{1}{n}\right) T(r, F) + o(T(r, F)).$$

Therefore, we have $\frac{1}{m} + \frac{1}{n} \geq 1$, a contradiction to the hypothesis which implies $\frac{1}{m} + \frac{1}{n} \leq \frac{5}{6}$. □

Theorem 51.14. *Let $f \in \mathcal{M}(\mathbb{K})$ be transcendental (respectively $f \in \mathcal{M}_u(d(0, R^-))$, respectively $f \in \mathcal{M}^c(D)$) and let $w_j \in \mathcal{M}_f(\mathbb{K})$ ($j = 1, \ldots, q$) (respectively $w_j \in \mathcal{M}_f(d(a, R^-))$, respectively $w_j \in \mathcal{M}_f(D)$) be q distinct small functions. Then,*

$$qT(r, f) \leq 3 \sum_{j=1}^{q} \overline{Z}(r, f - w_j) + o(T(r, f)),$$

(respectively

$$qT(r, f) \leq 3 \sum_{j=1}^{q} \overline{Z}(r, f - w_j) + o(T(r, f)),$$

respectively

$$qT_R(r, f) \leq 3 \sum_{j=1}^{q} \overline{Z}_R(r, f - w_j) + o(T_R(r, f))).$$

Moreover, if f has finitely many poles in $\mathbb{K}$ (respectively in $d(0, R^-)$, respectively in D), then

$$qT(r, f) \leq 2 \sum_{j=1}^{q} \overline{Z}(r, f - w_j) + o(T(r, f)),$$

(respectively

$$qT(r, f) \leq 2 \sum_{j=1}^{q} \overline{Z}(r, f - w_j) + o(T(r, f)),$$

respectively

$$qT_R(r, f) \leq 2 \sum_{j=1}^{q} \overline{Z}_R(r, f - w_j) + o(T_R(r, f))).$$

Proof. Suppose first that f and w_j ($j = 1, \dots, q$) belong to $\mathcal{M}(\mathbb{K})$ or to $\mathcal{M}(d(0, R^-))$. By Theorem 51.10, for every triplet (i, j, k) such that $1 \leq i \leq j \leq k \leq q$, we can write

$$T(r, f) \leq \overline{Z}(r, f - w_i) + \overline{Z}(r, f - w_j) + \overline{Z}(r, f - w_k) + o(T(r, f)).$$

The number of such inequalities is C_q^3. Summing up, we obtain

$$(1) \quad C_q^3 T(r, f) \leq \sum_{(i,j,k),\ 1 \leq i \leq j \leq k \leq q} \overline{Z}(r, f - w_i) + \overline{Z}(r, f - w_j) + \overline{Z}(r, f - w_k) + o(T(r, f)).$$

In this sum, for each index i, the number of terms $\overline{Z}(r, f - w_i)$ is clearly C_{q-1}^2. Consequently, by (1), we obtain

$$C_q^3 T(r, f) \leq C_{q-1}^2 \sum_{i=1}^{q} \overline{Z}(r, f - w_i) + o(T(r, f)),$$

and hence

$$\frac{q}{3} T(r, f) \leq \sum_{i=1}^{q} \overline{Z}(r, f - w_i) + o(T(r, f)).$$

Suppose now that f has finitely many poles. By Theorem 51.11, for every pair (i, j) such that $1 \leq i \leq j \leq q$, we have

$$T(r, f) \leq \overline{Z}(r, f - w_i) + \overline{Z}(r, f - w_j) + o(T(r, f)).$$

The number of such inequalities is then C_q^2. Summing up we now obtain

$$(2) \ C_q^2 T(r, f) \leq \sum_{(i,j,\ 1 \leq i \leq j \leq q)} \overline{Z}(r, f - w_i) + \overline{Z}(r, f - w_j) + o(T(r, f)).$$

In this sum, for each index i, the number of terms $\overline{Z}(r, f - w_i)$ is clearly $C_{q-1}^1 = q - 1$. Consequently, by (1), we obtain

$$C_q^2 T(r, f) \leq (q - 1) \sum_{i=1}^{q} \overline{Z}(r, f - w_i) + o(T(r, f)),$$

and hence

$$\frac{q}{2} T(r, f) \leq \sum_{i=1}^{q} \overline{Z}(r, f - w_i) + o(T(r, f)).$$

Now, if f and w_j $(j = 1, \dots, q)$ belong to $\mathcal{M}(D)$, we can make the same reasoning with T_R instead of T and Z_R instead of Z. $\square$

Definition. A small function θ with respect to a function $f \in \mathcal{M}(\mathbb{K})$ is said to be *perfectly branched for* f if all zeros of $f - \theta$ are multiple except finitely many.

Theorem 51.15. *Let $f \in \mathcal{M}(\mathbb{K})$ be transcendental and have finitely many poles. Then f has at most one perfectly branched rational function.*

Proof. Suppose that f has two distinct perfectly branched rational functions P and Q. Then, $f - Q$ has two distinct perfectly branched rational functions $P - Q$ and 0. So, without loss of generality, we can assume $Q = 0$. Now, let $g = \frac{f}{P}$. We can see that g satisfies all hypotheses of Theorem 49.9 and has two perfectly branched values: 0 and 1. Consequently, g is not transcendental and therefore neither is f, a contradiction. $\square$

Concerning entire functions, it is useful to examine the link between small functions and order of growth. Let us recall that when $f \in \mathcal{A}(\mathbb{K})$, then $T(r, f) = \text{Log}(|f|(r))$.

Theorem 51.16. *Let $f,\ h \in \mathcal{A}(\mathbb{K})$ be such that $\rho(h) < \widetilde{\rho}(f)$. Then h is a small function with respect to f.*

Proof. By hypothesis, there exists $\lambda > 0$ and $R > 0$ such that

$$\frac{\text{Log}(\text{Log}(|h|(r)))}{\text{Log}(r)} + 2\lambda < \widetilde{\rho}(f) \quad \forall r > R,$$

and hence there exists $R' > R$ such that

$$\text{Log}(\text{Log}(|h|(r))) + \lambda Log(r) < \text{Log}(\text{Log}(|f|(r))) \quad \forall r > R',$$

therefore $r^{\lambda}\text{Log}(|h|(r)) < \text{Log}(|f|(r))$, which proves that

$$\lim_{r \to +\infty} \frac{\text{Log}(|h|(r))}{\text{Log}(|f|(r))} = 0. \qquad \square$$

Corollary 51.17. *Let $f,\ h \in \mathcal{A}(\mathbb{K})$, with f regular and such that $\rho(h) < \rho(f)$. Then h is a small function with respect to f.*

Theorem 51.18. *Let $f,\ h \in \mathcal{A}(\mathbb{K})$ be such that $\rho(h) = \rho(f)$ and $\sigma(h) = 0 < \widetilde{\sigma}(f)$. Then h is a small function with respect to f.*

Proof. By hypothesis, we have $\lim\limits_{r\to+\infty} \dfrac{\sigma(h,r)}{\widetilde{\sigma}(f,r)} = 0$, hence

$$\lim_{r\to+\infty} \frac{\mathrm{Log}(|h|(r))}{\mathrm{Log}(|f|(r))} = 0,$$

therefore h is a small function with respect to f. □

Theorem 51.19. *Let $f \in \mathcal{A}(\mathbb{K})$ and let $h \in \mathcal{A}(\mathbb{K})$ satisfy $|f|(r) > |h|(r)\ \forall r \geq R$. Then, $\rho(f+h) = \rho(f)$ and $\sigma(f+h) = \sigma(f)$. Moreover, $\psi(f+h) = \psi(f)$.*

Proof. By construction, we have $\rho(f+h) = \rho(f)$. Then,

$$\limsup_{r\to\infty} \frac{\mathrm{Log}(|f+h|(r))}{r^{\rho(f+h)}} = \limsup_{r\to\infty} \frac{\mathrm{Log}(|f|(r))}{r^{\rho(f)}} = \sigma(f),$$

hence $\sigma(f+h) = \sigma(f)$.

Now, by Corollary 22.8, the number of zeros of $f+h$ in each disk $d(0,r)$ equals the number of zeros of f in $d(0,r)$, for every $r > R$. Consequently, since $\rho(f+h) = \rho(f)$, we have $\psi(f+h) = \psi(f)$. □

Corollary 51.20. *Let $f \in \mathcal{A}(\mathbb{K})$ and let $h \in \mathcal{A}(\mathbb{K})$ be a small function with respect to f. Then, $\rho(f+h) = \rho(f)$ and $\sigma(f+h) = \sigma(f)$. Moreover, $\psi(f+h) = \psi(f)$.*

Chapter 52

Meromorphic Functions Sharing Some Small Ones

Throughout this chapter, the field $\mathbb{K}$ is supposed to be of characteristic 0.

Definition. Two functions f, g defined in a set E are said *to share a function h, ignoring multiplicity* if $f(x) - h(x) = 0$ is equivalent to $g(x) - h(x) = 0$.

This kind of problem was considered in [95].

Theorem 52.1. *Let $f,\ g \in \mathcal{M}(\mathbb{K})$ be transcendental (respectively $f,\ g \in \mathcal{M}_u(d(a, R^-))$, respectively $f,\ g \in \mathcal{M}^c(D)$), be distinct and share q distinct small functions ignoring multiplicity. $w_j \in \mathcal{M}_f(\mathbb{K}) \cap \mathcal{M}_g(\mathbb{K})$ $(j = 1, \ldots, q)$ (respectively $w_j \in \mathcal{M}_f(d(a, R^-)) \cap \mathcal{M}_g(d(a, R^-))$ $(j = 1, \ldots, q)$ (respectively $w_j \in \mathcal{M}_f(D) \cap \mathcal{M}_g(D)$ $(j = 1, \ldots, q)$), other than the constant ∞. Then,*

$$\sum_{j=1}^{q} \overline{Z}(r, f - w_j) \leq \overline{Z}(r, f - g) + o(T(r, f)) + o(T(r, g)).$$

Proof. Suppose that f and g belong to $\mathcal{M}(\mathbb{K})$, are distinct, and share q distinct small functions I.M. $w_j \in \mathcal{M}_f(\mathbb{K}) \cap \mathcal{M}_g(\mathbb{K})$ $(j = 1, \ldots, q)$.

Let b be a zero of $f - w_i$ for a certain index i. Then it is also a zero of $g - w_i$. Suppose that b is counted several times in the sum $\sum_{j=1}^{q} \overline{Z}(r, f - w_j)$, which means that it is a zero of another function $f - w_h$ for a certain index $h \neq i$. Then we have $w_i(b) = w_h(b)$ and hence b is a zero of the function $w_i - w_h$ which belongs to $\mathcal{M}_f(\mathbb{K})$. Now, put $\widetilde{Z}(r, f - w_1) = \overline{Z}(r, f - w_1)$, and for each $j > 1$, let $\widetilde{Z}(r, f - w_j)$ be the counting function of zeros of $f - w_j$ in the disk $d(0, r^-)$ ignoring multiplicity and avoiding the zeros already counted as zeros of $f - w_h$ for some $h < j$. Consider now the sum $\sum_{j=1}^{q} \widetilde{Z}(r, f - w_j)$. Since the functions $w_i - w_j$ belong to $\mathcal{M}_f(\mathbb{K})$, clearly, we have

$$\sum_{j=1}^{q} \overline{Z}(r, f - w_j) = \sum_{j=1}^{q} \widetilde{Z}(r, f - w_j) = o(T(r, f)).$$

It is clear, from the assumption, that $f(x) - w_j(x) = 0$ implies $g(x) - w_j(x) = 0$ and hence $f(x) - g(x) = 0$. Since $f - g$ is not the identically zero function, it follows that

$$\sum_{j=1}^{q} \overline{Z}(r, f - w_j) \leq \overline{Z}(r, f - g).$$

Consequently,

$$\sum_{j=1}^{q} \overline{Z}(r, f - w_j) \leq \overline{Z}(r, f - g) + o(T(r, f)) + o(T(r, g)).$$

Now, if f and g belong to $\mathcal{M}(d(0, R^-))$ or to $\mathcal{M}(D)$, the proof is exactly the same. □

Theorem 52.2. *Let $f,\ g \in \mathcal{M}(\mathbb{K})$ be transcendental (respectively $f,\ g \in \mathcal{M}_u(d(a, R^-))$, respectively $f,\ g \in \mathcal{M}^c(D)$), be distinct, have finitely many poles, and share three distinct small functions, ignoring multiplicity. $w_j \in \mathcal{M}_f(\mathbb{K}) \cap \mathcal{M}_g(\mathbb{K})$ $(j = 1, 2, 3)$ (respectively $w_j \in \mathcal{M}_f(d(a, R^-)) \cap \mathcal{M}_g(d(a, R^-))$ $(j = 1, 2, 3)$, respectively $w_j \in \mathcal{M}_f(D) \cap \mathcal{M}_g(D)$ $(j = 1, 2, 3)$). Then, $f = g$.*

Proof. We put $V(r) = \max(T(r, f), T(r, g))$. Suppose that f and g are distinct and share q small function I.M. w_j $(1 \leq j \leq q)$. By Theorem 51.14, we have

$$qT(r, f) \leq 3\sum_{j=1}^{q} \overline{Z}(r, f - w_j) + o(T(r, f)).$$

But thanks to Theorem 52.1, we can derive

$$qT(r,f) \leq 3T(r,f-g) + o(T(r,f)),$$

and similarly

$$qT(r,g) \leq 3T(r,f-g) + o(T(r,g)),$$

hence

$$qV(r) \leq 3T(r,f-g) + o(V(r)). \tag{1}$$

Suppose now that f and g have finitely many poles. By Theorems 51.14, Relation (1) gives us

$$qV(r) \leq 2V(r) + o(V(r)),$$

which is obviously absurd whenever $q \geq 3$ and proves that $f = g$ when f and g belong to $\mathcal{M}(\mathbb{K})$ as well as when f and g belong to $\mathcal{M}_u(d(0,R^-))$ or to $\mathcal{M}^c(D)$, after replacing T by T_R and Z by Z_R. □

By Theorem 51.16, Corollary 52.3 is easy:

Corollary 52.3. *Let $f,\ g \in \mathcal{A}(\mathbb{K})$ be transcendental, distinct, and share three distinct functions, ignoring multiplicity, $\omega_j \in \mathcal{A}(\mathbb{K})$, $j = 1,2,3$, such that $\max_{1\leq j\leq 3} \rho(\omega_j) < \min(\widetilde{\rho}(f), \widetilde{\rho}(g))$. Then, $f = g$.*

Corollary 52.4. *Let $f,\ g \in \mathcal{A}(\mathbb{K})$ be transcendental, regular, distinct, and share three distinct functions, ignoring multiplicity, $\omega_j \in \mathcal{A}(\mathbb{K})$, $j = 1,2,3$, such that $\max_{1\leq j\leq 3} \rho(\omega_j) < \min(\rho(f), \rho(g))$. Then, $f = g$.*

In the same way, by Theorem 51.18, Corollary 52.5 is easy:

Corollary 52.5. *Let $f,\ g \in \mathcal{A}(\mathbb{K})$ be transcendental, clean, and share three distinct functions, ignoring multiplicity, $\omega_j \in \mathcal{A}(\mathbb{K})$, $j = 1,2,3$, such that $\rho(\omega_j) = \rho(f) = \rho(g)$, $j = 1,2,3$, and $\sigma(\omega_j) = 0 < \min(\sigma(f), \sigma(g))$. Then, $f = g$.*

After this first result in [56] letting find that two meromorphic functions sharing seven small functions are equal, Ta Thi Hoai An and Nguyen Phuong obtained a much better estimation. We denote by E the disk $d(0,R^-)$.

Lemma 52.6 (Ta Thi Hoai An-Nguyen Phuong [5]). *Let f, g be distinct and belong to $\mathcal{M}(\mathbb{K})$ (respectively to $\mathcal{M}_u(E)$, respectively to $\mathcal{M}^c(D)$) sharing q distinct small functions ignoring multiplicity: $h_1, \dots, h_q$ with $q \geq 5$. Then, for every subset $\{i_1, i_2, i_3, i_4\}$ of $\{1, \dots, q\}$, we have*

$$\sum_{j \in \{1,\dots,q\} \setminus \{i_1,\dots,i_4\}} \overline{Z}(r, f - h_j) + \overline{Z}(r, g - h_j)$$
$$\leq o(T((r, f)) + = o(T((r, g))$$

(respectively

$$\sum_{j \in \{1,\dots,q\} \setminus \{i_1,\dots,i_4\}} \overline{Z}(r, f - h_j) + \overline{Z}(r, g - h_j) \leq o(T(r, f)) + o(T(r, g)),$$

respectively

$$\sum_{j \in \{1,\dots,q\} \setminus \{i_1,\dots,i_4\}} \overline{Z}_R(r, f - h_j) + \overline{Z}_R(r, g - h_j)$$
$$\leq o(T(r, f)) + o(T(r, g))).$$

Proof. Let us first suppose f, $g \in \mathcal{M}(\mathbb{K})$, or f, $g \in \mathcal{M}_u(E)$. Clearly, we only have to prove that

$$\sum_{j=5}^{q} \overline{Z}(r, f - h_j) \leq o(T(r, f)) + o(T(r, g)).$$

Thus, we assume that

$$\sum_{j=5}^{q} \overline{Z}(r, f - h_j) \neq o(T(r, f) + o(T(r, g)).$$

Given any meromorphic function $w \in \mathcal{M}(\mathbb{K})$ (respectively in $\mathcal{M}_u(E)$), we put

$$L(w) = \frac{(w - h_1)(h_3 - h_2)}{(w - h_2)(h_3 - h_4)}$$

and $F = L(f)$, $G = L(g)$.

Without loss of generality, we may assume that $h_1 = \infty,\ h_2 = 0$, $h_3 = 1$, and we put $h = h_4$. Now, we set

$$B = \frac{f'(h'g - hg')(f-g)}{f(f-1)(g-h)} - \frac{g'(h'f - hf')(f-g)}{g(g-1)(f-h)}$$

and

$$\begin{aligned} Q &= f'(h'g - hg')(f-h)(g-1) - g'(h'f - hf')(g-h)(f-1) \\ &= h'ff'g^2 - h'f'g - h(h-1)ff'g' - hh'f'g^2 + hh'f'g \\ &\quad - h'f^2gg' + h'fgg' + h(h-1)f'gg' + hh'f^2g' - hh'fg'. \end{aligned}$$

Then,

$$B = \frac{(f-g)Q}{f(f-1)g(g-1)}.$$

Suppose first that B is identically zero. Then we have

$$\frac{f'(h'g - hg')(f-g)}{f(f-1)(g-h)} = \frac{g'(h'f - hf')(f-g)}{g(g-1)(f-h)}. \tag{1}$$

If h is a constant, then $f = g$ which contradicts our hypothesis. Hence, h is not a constant, and then, by (1), we have

$$\frac{(f-1)(g-h)}{g-1)(f-h)} - 1 = \frac{f'(h'g - hg')}{g'(h'f - hf')} - 1,$$

therefore

$$\frac{(f-g)(g-h)}{(g-1)(f-h)} = \frac{h'(f'-g')g - (f-g)g'}{g'(h'f - hf')},$$

and hence

$$\frac{f'-g'}{f-g} = \frac{(1-h)g'(h'f - hf')}{h'g(g-1)(f-h)} + \frac{g'}{g}. \tag{2}$$

Let us fix $j \geq 5,\ j \leq q$. By (1), there exists a common zero $\alpha \in \mathbb{K}$ (respectively $\alpha \in D$) of $f - h_j$ and $g - h_j$ which is not a zero or a pole of $h, h', h_j - 1, h_j - h$. Then, α must be a pole of the left-hand side of (2) but cannot be a pole of the right-hand side of (2). This is a contradiction showing that B cannot be identically zero.

Now, let us fix $j \geq 5$ such that $j \leq q$ and suppose that β is a common zero of $f - h_j$ and $g - h_j$ but is not a zero or a pole of $h, h_j, h_j - 1, h_j - h$. Then it is a zero of $f - g$ and it is not a pole of

$$\frac{Q}{f(f-1)(f-h)g(g-1)(g-h)},$$

and hence it is a zero of B.

Now, since f and g share each h_j ignoring multiplicity, we have

$$\begin{aligned} (3) \quad & \sum_{j=1}^{5} \overline{Z}(r, g - h_j) \\ & = \sum_{j=1}^{5} \overline{Z}(r, f - h_j) \leq Z(r, B) + o(T(r, f)) + o(T(r, g)) \\ & \leq T(r, B) + o(T(r, f)) + o(T(r, g)). \end{aligned}$$

We estimate $T(r, B) - N(r, B)$. By computation, B can be written as

$$\begin{aligned} B = & \frac{f'(h'g - hg')}{(f-1)g(g-h)} - \left(\frac{f'}{f-1} - \frac{f'}{f}\right)\left(\frac{h'g - hg'}{g-h}\right) \\ & + \frac{g'(h'f - hf')}{(g-1)f(f-h)} - \left(\frac{g'}{g-1} - \frac{g'}{g}\right)\left(\frac{h'f - hf'}{f-h}\right) \\ = & \left(\frac{f'}{f-1}\right)\left(\frac{g'}{g} - \frac{g' - h'}{g-h}\right) - \left(\frac{f'}{f-1} - \frac{f'}{f}\right)\left(h' - \frac{h(g' - h')}{g-h}\right) \\ & + \left(\frac{g'}{g-1}\right)\left(\frac{f'}{f} - \frac{f' - h'}{f-h}\right) - \left(\frac{g'}{g-1} - \frac{g'}{g}\right)\left(h' - \frac{h(f' - h')}{f-h}\right). \end{aligned}$$

In this last relation, B appears as a sum of logarithmic derivative, and hence, by Corollary 43.18, we can see that

$$(4) \qquad T(r, B) - N(r, B) \leq 0.$$

Next, we estimate $N(r, B)$. Clearly, the poles of B can only occur at the zeros of $f, g, f-1, g-1, f-h, g-h$ and at the poles of f, g, h.

Let A be the set of all zeros, 1-points, and poles of h. Then we first estimate the counting function of poles of B when we are in A. Recall that f and g share the constants 0, 1, and ∞.

Let γ be a zero of h. Then we can check that B has no pole at γ.

Let γ be a 1-value of h. If B had a pole, then $f-1$ would have a zero of order s, $g-1$ would have a zero of order t, $f-g$ would have a zero, and Q would have a zero of order $s+t$, hence B would have a zero, not a pole.

Now, let γ be a pole of h of order n.

If $f(\gamma)=1$, then $g(\gamma)=1$ because f and g share h_3, hence B has no pole at γ.

If f has a pole of order s at γ, then g also has a pole of order t because f and g share h_1. We can assume $s \geq t$, then B has a pole of order at most $n+2$, hence the counting function of poles of B when γ is a pole of f, g, and h and hence is a pole of B of order $n+2$, is bounded by thrice the counting function of the poles of h, and hence is of the form $o(T(r,f))+o(T(r,g))$.

Next, suppose that $f(\gamma) \neq 1, \infty$, hence $g(\gamma) \neq 1, \infty$. If f and g have no zero at γ, it is clear that B has at most a pole of order n, hence the counting function of poles of B when $f(\gamma) \neq 0, 1, \infty$ (hence $g(\gamma) \neq 0, 1, \infty$) is of the form $o(T(r,f))+o(T(r,g))$. Finally, if $f(\gamma)=0$, then $g(\gamma)=0$ because f and g share h_2 and then we can check that B may admit a pole at γ of order at most $2n+2$. Consequently, the counting function of poles of B when γ lies in A and is a pole of h is bounded by thrice the counting function of the poles of h and hence is of the form $o(T(r,f))+o(T(r,g))$.

Thus, the counting function of poles of B when γ lies in A is of the form $o(T(r,f))+o(T(r,g))$.

Consider now the poles of B when $\gamma \notin A$.

Let $\gamma \notin A$ be a common pole of f and g of order s_1 and t_1, respectively. Then, γ is a pole of Q of order at most $2s_1+2t_1+1$. Next, γ is a pole of $f-g$ of order $\max(s_1,t_1)$. Hence, from the definition of B, γ is a pole of the numerator of B of order at most $2s_1+2t_1+1+\max(s_1,t_1)$, and on the other hand, it is a pole of the denominator of B of order $3s_1+3t_1$. But since

$$2s_1+2t_1+1+\max(s_1,t_1)-3(s_1+t_1)=1+\max(s_1,t_1)-s_1-t_1 \leq 0,$$

it follows that γ is not a pole of B.

Let $\gamma \notin A$ be a common zero of f and g of order s_2 and t_2, respectively. We can check that γ is a zero of Q of order at least s_2+t_2+1 and that γ is a zero of $f-g$ of order at least $\min(s_2,t_2)$.

On the other hand, we can check that the denominator of B has a zero at γ of order at most $s_2 + t_2$. Consequently, B has no pole at γ.

Suppose now that $\gamma \notin A$ is a common zero of $f - h$ and $g - h$. Then, γ is a zero of $f - g$. On the other hand, γ is a pole of order 1 of $\frac{f'}{f-h}$ and $\frac{g'}{g-h}$. Consequently, by the definition of B, we can check that γ is not a pole of B. Similarly, when γ is a common zero of $f-1$ and $g-1$, γ is not a pole of B.

Suppose now that $\gamma \notin A$ is a pole of f but is not a pole of g and is not a zero of g, $g-1$, and $g-h$. Since f and g share h_0, that situation does not occur.

Suppose now that $\gamma \notin A$ is a zero of f but is not a pole of g and is not a zero of g. Since f and g share h_0, that situation does not occur.

Suppose now that $\gamma \notin A$ is a zero of $f - h_i$ with $i = 3,\ 4$ but is not a pole of g and is not a zero of g, $g-1$ and $g-h$. Then γ is a pole of B of order at most 1. Similarly, if γ is a zero of $g - h_i$ with $i = 3,\ 4$ but is not a pole of f and is not a zero of f, $f-1$ and $f-h$. Then γ is a pole of B of order at most 1. However, since f and g share h_3 and h_4, such a situation does not occur.

Suppose now that $\gamma \notin A$ is a common zero of $f - h_i$ and $g - h_j$ for some $1 \leq i \leq 4$ and $1 \leq j \leq 4$ with $i \neq j$ and $i \neq 2,\ j \neq 2$. Then γ is a pole of B of order at most 2. However, in such a situation, since f and g share h_i, then γ is a zero of $h_i - h_j$. But since h_i and h_j are small with respect to f and g, the counting function of points γ that are common zeros of $f - h_i$ and $g - h_j$ for some $1 \leq i \leq 4$ and $1 \leq j \leq 4$ with $i \neq j$ and $i \neq 2,\ j \neq 2$ is bounded by $o(T(r, f)) + o(T(r, g))$.

Finally, suppose that $\gamma \notin A$ is a zero of $f - h_i$ and a pole of g for some $i = 2, 3, 4$ or a zero of $g - h_i$ and a pole of f for some $i = 2, 3, 4$. But since f and g share $h_2,\ h_3,\ h_4$, such a situation is impossible.

In conclusion, we have

$$N(r, B) \leq o(T(r, f)) + o(T(r, g)),$$

and therefore, by (3) and (4), that completes the proof of Lemma 52.6 when $f,\ g \in \mathcal{M}(\mathbb{K})$ or $f,\ g \in \mathcal{M}_u(E)$. Now, when $f,\ g \in \mathcal{M}^c(D)$, the proof is similar. □

Corollary 52.7. *Let $f,\ g$ be distinct and belong to $\mathcal{M}(\mathbb{K})$ (respectively to $\mathcal{M}_u(E)$, respectively to $\mathcal{M}^c(D)$) sharing q distinct small functions ignoring multiplicity: $h_1, \ldots, h_q$ (with $q \geq 5$). Then, for* *every subset* $\{i_1, i_2, i_3, i_4\}$ *of* $\{1, \ldots, q\}$*, we have*

$$\overline{Z}(r, f - h_j) + \overline{Z}(r, g - h_j) \le o(T((r, f)) + o(T(r, g)) \ \forall j = 1, \ldots, q$$

(respectively

$$\overline{Z}(r, f - h_j) + \overline{Z}(r, g - h_j) \le o(T(r, f)) + o(T(r, g)) \ \forall j = 1, \ldots, q$$

respectively

$$\overline{Z}_R(r, f - h_j) + \overline{Z}_R(r, g - h_j) \le o(T(r, f)) + o(T(r, g)) \ \forall j = 1, \ldots, q).$$

Theorem 52.8 (Ta Thi Hoai An-Nguyen Phuong [5]). *Let* f, $g \in \mathcal{M}(\mathbb{K})$ *be transcendental (respectively* f, $g \in \mathcal{M}_u(E)$*, respectively* f, $g \in \mathcal{M}^c(D)$*), be distinct, and share five distinct small meromorphic functions ignoring multiplicity,* $w_j \in \mathcal{M}_f(\mathbb{K}) \cap \mathcal{M}_g(\mathbb{K})$ $(j = 1, \ldots, 5)$ *(respectively* $w_j \in \mathcal{M}_f(D) \cap \mathcal{M}_g(D)$ $(j = 1, \ldots, 5)$*, respectively* $w_j \in \mathcal{M}_f(E) \cap \mathcal{M}_g(E)$ $(j = 1, \ldots, 5)$*). Then,* $f = g$.

Proof. Suppose first that f, $g \in \mathcal{M}(\mathbb{K})$ or $f, g \in \mathcal{M}_u(E)$. Suppose that f and g are not identic. Applying Corollary 52.7 with $q = 5$, we have $\overline{Z}(r, f - h_j) = o(T(r, f)) + o(T(r, g))$ and $\overline{Z}(r, g - h_j) = o(T(r, f)) + o(T(r, g))$ for every $j \in \{1, \ldots, 5\}$. Therefore,

$$(1) \qquad \sum_{j=1}^{5} \overline{Z}(r, f - h_j) + \overline{Z}(r, g - h_j) = o(Tf, r)) + o(T(r, g)).$$

Now, by Theorem 51.14, we have

$$\begin{aligned} &5(T(r, f) + T(r, g)) \\ &\quad \le 3 \sum_{j=1}^{5} \overline{Z}(r, f - h_j) + \overline{Z}(r, g - h_j) + o(T(r, f)) + o(T(r, g)), \end{aligned}$$

hence, by (1), $5(T(r, f) + T(r, g)) \le 2(o(T(r, f)) + o(T(r, g)))$, which is absurd and proves that $f = g$.

Suppose now that $f, g \in \mathcal{M}^c(D)$. Replacing each symbol $T(r, .)$ by $T_R(r, .)$ and $\overline{Z}(r, .)$ by $\overline{Z}_R(r, .)$, we can make the same reasoning and conclude in the same way $f = g$. □

We now show an improvement of Theorem 51.14 in the case when $q \geq 5$.

Notation. Given $f \in \mathcal{M}(\mathbb{K})$ (respectively let $f \in \mathcal{M}_u(E)$, respectively $f \in \mathcal{M}^c(D)$), we put $M(r, f) = T(r, f) - N(r, f)$.

Lemma 52.9 (Ta Thi Hoai An-Nguyen Phuong). *Let $f \in \mathcal{M}(\mathbb{K})$ (respectively let $f \in \mathcal{M}_u(E)$, respectively let $f \in \mathcal{M}^c(D)$) be non-constant and let $h_1, \ldots, h_5$ be distinct small functions with respect to f. We have*

$$2T(r, f) \leq \sum_{i=1}^{5} \overline{Z}(r, f - h_i) + S(r, f).$$

Proof. We first suppose $f \in \mathcal{M}(\mathbb{K})$, or $f \in \mathcal{M}_u(E)$. Let $g = \frac{(f-h_2)(h_3-h_1)}{(f-h_1)(h_3-h_2)}$. Then it is easily seen that

$$\begin{aligned}
&\overline{Z}(r, g) = \overline{Z}(r, f - h_2) + o(T(r, f)),\\
&\overline{N}(r, g) = \overline{Z}(r, f - h_1) + o(T(r, f)),\\
&\overline{Z}(r, g - 1) = \overline{Z}(r, f - h_3) + o(T(r, f)).
\end{aligned}$$

Consequently, in order to prove Lemma 52.9, it is sufficient to prove the following inequality:

$$\begin{aligned}
(1)\quad 2T(r, f) \leq\ & \overline{N}(r, f) + \overline{Z}(r, f) + \overline{Z}(r, f - 1) + \overline{Z}(r, f - h_4)\\
&+ \overline{Z}(r, f - h_5) + o(T(r, f)),
\end{aligned}$$

where $h_1 = \infty$, $h_2 = 0$, $h_3 = 1$, and h_4 and h_5 are two small functions with respect to f different from $0, 1, \infty$, and $h_4 \neq h_5$.

If one of the functions h_4, h_5 is a constant, then (1) is immediate by Theorem 45.7. Consequently, now we can assume that both h_4, h_5 are non-constant small functions.

Let

$$H = \det \begin{pmatrix} ff' & f' & f(f-1) \\ h_4 h_4' & h_4' & h_4(h_4 - 1) \\ h_5 h_5' & h_5' & h_5(h_5 - 1) \end{pmatrix}.$$

By a simple computation, we get

$$
\begin{aligned}
(2) \qquad H = {} & f(f-1)h_4(h_4-1)h_5(h_5-1) \\
& \times \left(\left(\frac{(h_4'}{h_4} - \frac{h_5'}{h_5}\right)\left(\frac{f'}{f-1} - \frac{h_5'}{h_5-1}\right)\right. \\
& \left. - \left(\frac{h_4'}{h_4-1} - \frac{h_5'}{h_5-1}\right)\left(\frac{f'}{f} - \frac{h_5'}{h_5}\right)\right).
\end{aligned}
$$

Suppose first that H is identically zero. Since f is not a constant and since h_4, h_5 are not identically 0 or 1, it follows from the definition of H that

$$
\begin{aligned}
(3) \qquad & \left(\frac{h_4'}{h_4} - \frac{h_5'}{h_5}\right)\frac{f'}{f-1} - \left(\frac{h_4'}{h_4-1} - \frac{h_5'}{h_5-1}\right)\frac{f'}{f} \\
& = \left(\frac{h_4'}{h_4} - \frac{h_5'}{h_5}\right)\frac{fh_5'}{h_5-1} - \left(\frac{h_4'}{h_4-1} - \frac{h_5'}{h_5-1}\right)\frac{h_5'}{h_5}.
\end{aligned}
$$

We must now distinguish four cases.

Case 1. $\frac{h_4'}{h_4} = \frac{h_5'}{h_5}$. It follows from (3) that either $\frac{h_4'}{h_4-1} = \frac{h_5'}{h_5-1}$ or $\frac{f'}{f} = \frac{h_5'}{h_5}$. If $\frac{h_4'}{h_4-1} = \frac{h_5'}{h_5-1}$, then h_4 and h_5 are constants, which is excluded by hypothesis. Hence, $\frac{f'}{f} = \frac{h_5'}{h_5}$ and then f is of the form $c \cdot h_5$, where c is a constant, which contradicts our hypothesis: h_5 is a small function with respect to f.

Case 2. $\frac{h_4'}{h_4-1} = \frac{h_5'}{h_5-1}$. Similar to Case 1, we have a contradiction.

Case 3. $\frac{h_4'}{h_4} - \frac{h_5'}{h_5} = \frac{h_4}{h_4-1} - \frac{h_5'}{h_5-1} \neq 0$. It follows from (3) that

$$\frac{f'}{f-1} - \frac{f'}{f} = \frac{h_5'}{h_5-1} - \frac{h_5'}{h_5},$$

which implies

$$\frac{f-1}{f} = \frac{C(h_5-1)}{h_5},$$

where C is a constant. Thus, we obtain

$$\frac{1}{f} = 1 - \frac{C(h_5-1)}{h_5},$$

and hence $T(r, \frac{1}{f}) = o(T(r, f))$, which is absurd because $T(r, \frac{1}{f}) = T(r, f) + o(T(r, f))$.

Case 4.

$$\frac{h_4'}{h_4} \neq \frac{h_5'}{h_5}, \ \frac{h_4'}{h_4 - 1} \neq \frac{h_5'}{h_5 - 1}, \quad \text{and} \quad \frac{h_4'}{h_4} - \frac{h_5'}{h_5} \neq \frac{h_4'}{h_4 - 1} - \frac{h_5'}{h_5 - 1}.$$

Then it follows from (3) that the zeros of $f - 1$ can only occur when h_j takes value 0 or 1, for $j = 4,\ 5$ or when $\frac{h_4'}{h_4} - \frac{h_5'}{h_5}$ has a zero. In the same way, the zeros of f can only occur when h_j takes value 0 or 1, for $j = 4,\ 5$ and when $\frac{h_4'}{h_4-1} - \frac{h_5'}{h_5-1}$ has a zero. Moreover, by (3), we can see that the poles of f can only occur when h_j takes value 1 or when h_j has a pole, for $j = 4,\ 5$, or when $\frac{h_4'}{h_4} - \frac{h_5'}{h_5} - \frac{h_4'}{h_4-1} + \frac{h_5'}{h_5-1}$ has a zero. Therefore, we have

$$\overline{N}(r, f) + \overline{Z}(r, f) + \overline{Z}(r, f - 1) = o(T(r, f)). \tag{4}$$

Applying the second main Theorem 45.7 to f, we can derive

$$T(r, f) \leq \overline{Z}(r, f) + \overline{Z}(r, f - 1) + \overline{N}(r, f) - \log(r) = o(T(r, f)),$$

which is absurd and finishes proving that $H \not\equiv 0$.

Now, given $r > 0$, we put

$$\delta(r) = \min(1,\ |h_4|(r),\ |h_5|(r),\ |h_4 - 1|(r),\ |h_5 - 1|(r), |h_4 - h_5|(r)).$$

Then we have

$$\begin{aligned}
&\log^+\left(\frac{1}{\delta(r)}\right) \\
&\quad \leq \log^+\left(\max\left(1,\ \frac{1}{|h_4|(r)},\ \frac{1}{|h_5|(r)},\ \frac{1}{|h_4 - 1|(r)},\right.\right. \\
&\qquad \left.\left.\frac{1}{|h_5 - 1|(r)},\ \frac{1}{|h_4 - h_5|(r)}\right)\right) \\
&\quad \leq \log^+\left(\frac{1}{|h_4|(r)}\right) + \log^+\left(\frac{1}{|h_5|(r)}\right) + \log^+\left(\frac{1}{|h_4 - 1|(r)}\right), \\
&\qquad + \log^+\left(\frac{1}{|h_5 - 1|(r)}\right) + \log^+\left(\frac{1}{|h_4 - h_5|(r)}\right) + \log(6).
\end{aligned}$$

Then, by Theorem 43.2, we have

$$\log^+\left(\frac{1}{\delta(r)}\right) \leq M\left(r, \frac{1}{h_4}\right) + M\left(r, \frac{1}{h_5}\right) + M\left(r, \frac{1}{h_4 - 1}\right)$$
$$+ M\left(r, \frac{1}{h_5 - 1}\right) + M\left(r, \frac{1}{h_4 - h_5}\right) + \log(6).$$

Consequently, we can see that

$$\log^+\left(\frac{1}{\delta(r)}\right) = o(T(r, f)). \tag{5}$$

First, we consider the case when $|f - h_j|(r) > \frac{\delta(r)}{2}$ for $2 \leq j \leq 5$. Then, we have

$$M\left(r, \frac{1}{f}\right) + M\left(r, \frac{1}{f-1}\right) + M\left(r, \frac{1}{f - h_4}\right) + M\left(r, \frac{1}{f - h_5}\right) \tag{6}$$
$$\leq 5\log^+\left(\tfrac{1}{\delta(r)}\right) + O(1) = s(r, f).$$

Now, let i be an index such that $2 \leq i \leq 5$ such that $|f - h_i|(r) \leq \frac{\delta(r)}{2}$. Then for every $j \neq i$, with $2 \leq j \leq 5$, we have

$$\delta(r) \leq |h_i - h_j|(r) \leq |f - h_i|(r) + |f - h_j|(r) \leq \frac{\delta(r)}{2} + |f - h_j|(r),$$

hence

$$\frac{\delta(r)}{2} \leq |f - h_j|(r).$$

Therefore, for $i \neq j$, we have

$$\sum_{\substack{j=2\\ j\neq i}}^{5} M\left(r, \frac{1}{f - h_j}\right) = \sum_{\substack{j=2\\ j\neq i}}^{5} \log^+\left(\frac{1}{|f - h_j|(r)}\right) \leq 3\log^+\left(\frac{1}{\delta(r)}\right).$$

Combining (5) and the last inequality, we have

$$\sum_{\substack{j=2,\\ j\neq i}}^{5} M\left(r, \frac{1}{f - h_j}\right) = o(T(r, f)). \tag{7}$$

On the other hand, for $2 \le i \le 5$, we can write

$$ff' = (f - h_i)(f' - h_i') + h_i'(f - h_i) + h_i(f' - h_i') + h_i h_i'$$
$$f' = (f' - h_i') + h_i',$$
$$f(f-1) = f^2 - f = (f - h_i)^2 + (2h_i - 1)(f - h_i) + h_i^2 - h_i.$$

And now we put $g_i = (f - h_i)(f' - h_i') + h_i'(f - h_i) + h_i(f' - h_i')$ and $l_i = (f - h_i)^2 + (2h_i - 1)(f - h_i)$ and then, thanks to properties of determinants, we obtain

$$H = \det \begin{pmatrix} g_i' & f' - h_i' & l_i \\ h_4 h_4' & h_4' & h_4(h_4 - 1) \\ h_5 h_5' & h_5' & h_5(h_5 - 1) \end{pmatrix}. \tag{8}$$

Now, we have $\log^+(\delta(r)) \le \log^+(1 + |\, h_i \,|(r))$ $\forall i = 2, \ldots, 5$ because it is obvious from the definition for $i = 4,\ 5$ and also for $i = 2$ and $i = 3$ because $h_2 = 0$ and $h_3 = 1$. Consequently, for every $i = 2, \ldots, 5$, we have $\log^+(\delta(r)) \le \log^+(|h_i|(r)) + \log(2) = M(r, h_i) + \log(2) = o(T(r, f))$. Then, by (8) and using Lemma 3.5, we obtain for every $i = 2, \ldots, 5$:

$$\begin{aligned}
\log^+ \left| \frac{H}{f - h_i} \right| &\le \log^+ \left(\left| \frac{f' - h_i'}{f - h_i} \right| (r) \right) + \log^+(|f - h_i|(r)) \\
&\quad + O\Big(\log^+(|h_i|(r) + \log^+(|h_i'|(r)) + \log^+(|h_4(r)) \\
&\quad + \log^+(|h_4'|(r)) + \log^+(|h_5|(r)) + \log^+(|h_5'|(r)) \Big) \\
&\le m\left(r, \frac{f' - h_i'}{f - h_i}\right) + \log^+(\delta(r)) + o(T(r, f)) \\
&= o(T(r, f)).
\end{aligned}$$

Hence, for every $i = 2, \ldots, 5$, we obtain

$$\begin{aligned}
m\left(\frac{1}{f - h_i}\right) &= \log^+\left(\frac{1}{|f - h_i|(r)}\right) \\
&\le \log^+ \left| \frac{H}{f - h_i} \right| (r) + \log^+ \left| \frac{1}{H} \right| (r) \\
&\le m\left(r, \frac{1}{H}\right) + o(T(r, f)).
\end{aligned} \tag{9}$$

Then, by (6), (7), (8), and (9), we can check that in all cases we have

(10)
$$M\left(r,\frac{1}{f}\right)+M\left(r,\frac{1}{f-1}\right)+M\left(r,\frac{1}{f-h_4}\right)+M\left(r,\frac{1}{f-h_5}\right)$$
$$\leq M\left(r,\frac{1}{H}\right)+o(T(r,f)).$$

Now, by Theorem 43.3, we can write

$$4T(r,f)=M\left(r,\frac{1}{f}\right)+M\left(r,\frac{1}{f-1}\right)+M\left(r,\frac{1}{f-h_4}\right)$$
$$+M\left(r,\frac{1}{f-h_5}\right)+N\left(r,\frac{1}{f}\right)+N\left(r,\frac{1}{f-1}\right)$$
$$+N\left(r,\frac{1}{f-h_4}\right)+N\left(r,\frac{1}{f-h_5}\right)+o(T(r,f)),$$

hence by (10),

(11)
$$4T(r,f)\leq M\left(r,\frac{1}{H}\right)+N\left(r,\frac{1}{f}\right)+N\left(r,\frac{1}{f-1}\right)$$
$$+N\left(r,\frac{1}{f-h_4}\right)+N\left(r,\frac{1}{f-h_5}\right)+o(T(r,f))$$
$$\leq T(r,H)-Z(r,H)+Z(r,f)+Z(r,f-1)$$
$$+Z(r,f-h_4)+Z(r,f-h_5)+o(T(r,f)).$$

Now, given a zero α of order s of any function $f-h_i$ $(2\leq i\leq 5)$, then α is also a zero of H of order at least $s-1$. Then, from (11), we can see that we have

(12)
$$4T(r,f)\leq \overline{Z}(r,f)+\overline{Z}(r,f-1)+\overline{Z}(r,f-h_4)+\overline{Z}(r,f-h_5)$$
$$+T(r,H)+o(T(r,f)).$$

Next, by (2), we can check that

$$M(r,H)\leq 2m(r,f)+o(T(r,f))$$
$$N(r,H)\leq 2N(r,f)+\overline{N}(r,f)+o(T(r,f)).$$

Consequently,

$$T(r,H) \leq 2(T(r,f) + \overline{N}(r,f) + o(T(r,f)). \tag{13}$$

Then, by (12) and (13), we obtain

$$\begin{aligned}2T(r,f) \leq \overline{N}(r,f) + \overline{Z}(r,f) + \overline{Z}(r,f-1) + \overline{Z}(r,f-h_4) \\ + \overline{Z}(r,f-h_5) + o(T(r,f)),\end{aligned}$$

which finishes proving (1) and hence ends the proof of Lemma 4.1 when f belongs to $\mathcal{M}(\mathbb{K})$ or $\mathcal{M}_u(E)$. Next, when f belongs to $\mathcal{M}^c(D)$, we have a similar proof, writing corresponding counting functions $T_R(r,.)$ instead of $T(r,.)$, $Z_R(r,.)$ instead of $Z(r,.)$, $N_R(r,.)$ instead of $N(r,.)$, etc. □

Theorem 52.10. *Let $f \in \mathcal{M}(\mathbb{K})$ (respectively let $f \in \mathcal{M}_u(E)$, respectively let $f \in \mathcal{M}^c(D)$) be non-constant and let $h_1, \ldots, h_q (q \geq 5)$ be q distinct small functions with respect to f. We have*

$$\frac{2q}{5}T(r,f) \leq \sum_{i=1}^{q} \overline{Z}(r,f-h_i) + o(T(r,f))$$

(respectively

$$\frac{2q}{5}T(r,f) \leq \sum_{s=1}^{q} \overline{Z}(r,f-h_{i_s}) + o(T(r,f)),$$

respectively

$$\frac{2q}{5}T_R(r,f) \leq \sum_{s=1}^{q} \overline{Z}_R(r,f-h_{i_s}) + o(T(r,f)).$$

Proof. By Lemma 52.9, for every subset $\{i_1, \ldots, i_5\}$ of $\{1, \ldots, q\}$ such that $1 \leq i_1 < \cdots < i_5 \leq q$, we have

$$2T(r,f) \leq \sum_{s=1}^{q} \overline{Z}(r,f-h_{i_s}) + o(T(r,f))$$

(respectively

$$2T(r,f) \leq \sum_{s=1}^{q} \overline{Z}(r,f-h_{i_s}) + o(T(r,f)),$$

respectively

$$2T_R(r,f) \leq \sum_{s=1}^{q} \overline{Z}_R(r, f-h_{i_s}) + o(T(r,f))).$$

The number of such inequalities is C_q^5. Summing up over all subsets $\{i_1,\ldots,i_5\}$ of $\{1,\ldots,q\}$, we can get

$$\begin{aligned}
&2C_q^5 T(r,f) \\
&\leq \sum_{\substack{\{i_1,\ldots,i_5\}\subset\{1,\ldots,q\} \\ 1\leq i_1<\cdots<i_5\leq q}} \Big(\overline{Z}(r,f-h_{i_1}) + \overline{Z}(r,h_{i_2}) + \overline{Z}(r,h_{i_3}) \\
&\quad + \overline{Z}(r,h_{i_4}) + \overline{Z}(r,h_{i_5})\Big) + o(T(r,f))
\end{aligned}$$

(respectively

$$\begin{aligned}
&2C_q^5 T(r,f) \\
&\leq \sum_{\substack{\{i_1,\ldots,i_5\}\subset\{1,\ldots,q\} \\ 1\leq i_1<\cdots<i_5\leq q}} \Big(\overline{Z}(r,f-h_{i_1}) + \overline{Z}(r,h_{i_2}) + \overline{Z}(r,h_{i_3}) \\
&\quad + \overline{Z}(r,h_{i_4}) + \overline{Z}(r,h_{i_5})\Big) + o(T(r,f)),
\end{aligned}$$

respectively

$$\begin{aligned}
&2C_q^5 T_R(r,f) \\
&\leq \sum_{\substack{\{i_1,\ldots,i_5\}\subset\{1,\ldots,q\} \\ 1\leq i_1<\cdots<i_5\leq q}} \Big(\overline{Z}_R(r,f-h_{i_1}) + \overline{Z}_R(r,h_{i_2}) + \overline{Z}(_R(r,h_{i_3}) \\
&\quad + \overline{Z}_R(r,h_{i_4}) + \overline{Z}_R(r,h_{i_5})\Big) + o(T(r,f))).
\end{aligned}$$

In each one of the last inequalities, we can check that for each index i_k, the term $\overline{Z}(r,f-h_{i_k})$ (respectively $\overline{Z}(r,f-h_{i_k})$, respectively $\overline{Z}_R(r,f-h_{i_k})$) intervenes C_{q-1}^4 times. Consequently, we can derive that

$$2C_q^5 T(r,f) \leq C_{q-1}^4 \sum_{i=1}^{q} \overline{Z}(r,f-h_i) + o(T(r,f))$$

(respectively

$$2C_q^5 T(r,f) \le C_{q-1}^4 \sum_{i=1}^{q} \overline{Z}(r, f-h_i) + o(T(r,f)),$$

respectively

$$2C_q^5 T_R(r,f) \le C_{q-1}^4 \sum_{i=1}^{q} \overline{Z}_R(r, f-h_i) + o(T(r,f))).$$

Consequently, it follows that

$$\frac{2q}{5} T(r,f) \le \sum_{i=1}^{q} \overline{Z}(r, f-h_i) + o(T(r,f))$$

(respectively

$$\frac{2q}{5} T(r,f) \le \sum_{i=1}^{q} \overline{Z}(r, f-h_i) + o(T(r,f)),$$

respectively

$$\frac{2q}{5} T_R(r,f) \le \sum_{i=1}^{q} \overline{Z}_R(r, f-h_i) + o(T(r,f))).$$

This completes the proof of Theorem 52.10. □

Chapter 53

The p-adic Hayman Conjecture

Throughout this chapter, the field $\mathbb{K}$ is supposed to be of characteristic 0.

In the 1950s, Walter Hayman asked the question of whether, given a meromorphic function in $\mathbb{C}$, the function $g'g^n$ might admit a quasi-exceptional value $b \neq 0$ [66]. Hayman showed that $g'g^n$ has no quasi-exceptional value, whenever $n \geq 3$. Henceforth, the problem was solved for $n = 2$ by Mues in 1979 [77] and next for $n \geq 1$, in 1995 by Bergweiler and Eremenko [8] and separately by Chen and Fang [30]. See also [95]. The same problem is posed on the field $\mathbb{K}$, both in $\mathcal{M}(\mathbb{K})$ and in a field $\mathcal{M}(d(a, R^-))$ ($a \in \mathbb{K}$, $R > 0$). These are the studies in [52, 80].

We still denote by D the set $\mathbb{K} \setminus d(0, R^-)$ and use the sets $\mathcal{A}(D), \mathcal{A}^c(D), \mathcal{M}(D), \mathcal{M}^c(D)$.

The following lemma is immediate.

Lemma 53.1. *Let $g \in \mathcal{M}(\mathbb{K})$ (respectively let $g \in \mathcal{M}(d(a, R^-))$, $a \in \mathbb{K}$, $R > 0$, respectively let $g \in \mathcal{M}(D)$), set $f = \frac{1}{g}$, and let $n \in \mathbb{N}^*$. Then $g'g^n$ admits a quasi-exceptional value $b \in \mathbb{K}^*$ if and only if $f' + bf^{n+2}$ has finitely many zeros that are not zeros of f.*

Remark. We can also consider the same problem when $n = -1$ i.e. the question of whether $f' + bf$ has infinitely many zeros. We examine this in p-adic analysis. When $n = 0$, in $\mathbb{C}$, the well-known counter-example furnished by the function $\tan x$ shows that $f' - f^2$ may have no zero. On the field $\mathbb{K}$, we examine the cases $n = -1$ and $n = 0$.

Henceforth, we examine that problem by considering the set of zeros of $f' + bf^m$, with $b \neq 0$. In the field $\mathbb{K}$, two theorems are specific to p-adic analysis. Both are based on the following lemma.

Lemma 53.2. *Let $f \in \mathcal{M}(D)$ (respectively $f \in \mathcal{M}(d(0, R^-))$, $R > 0$), suppose that f admits infinitely many zeros and suppose that there exists a sequence of intervals $[r'_n, r''_n]$ such that $\lim_{n\to+\infty} r'_n = +\infty$ (respectively $\lim_{n\to+\infty} r'_n = \lim_{n\to+\infty} r''_n = R$) and such that $|(f' + f^m)|(r) = |f^m|(r)$ $\forall r \in \bigcup_{n\in\mathbb{N}}[r'_n, r''_n]$. Let $m \in \mathbb{N}^*$ be $\neq 2$. Then $f' + f^m$ has infinitely many zeros that are not zeros of f.*

Proof. Let $J = \bigcup_{n\in\mathbb{N}}[r'_n, r''_n]$. By Corollary 22.6, we have

$$\nu^+(f' + f^m, \log r) = \nu^+(f^m, \log r), \ \nu^-(f' + f^m, \log r)$$
$$= \nu^-(f^m, \log r) \ \forall r \in J.$$

Consequently, when $f \in \mathcal{M}(D)$, in each annulus $\Gamma(0, R, r)$ with $r \in J$, f^m and $f' + f^m$ have the same difference between the number of zeros and poles. And when $f \in \mathcal{M}(d(0, R^-))$, in each disk $d(0, r)$ with $r \in J$, f^m and $f' + f^m$ have the same difference between the number of zeros and poles.

Now, if $m \geq 3$, the poles of $f' + f^m$ and f^m are the same taking multiplicity into account. And when $m = 1$, each pole of f is a pole of $f' + f$ with a strictly greater order. Consequently, when $f \in \mathcal{M}(D)$ for each $r \in J$, the number of zeros of $f' + f^m$ in when $\Gamma(0, R, r)$ is superior or equal to this of f^m. And similarly, when $f \in \mathcal{M}(d(0, R^-))$, the number of zeros of $f' + f^m$ in when $d(0, r)$ is superior or equal to this of f^m.

Now, for each $n \in \mathbb{N}$, let s_n be the number of distinct zeros of f in $\Gamma(0, R, r)$ (respectively in $d(0, r''_n)$). Since f has infinitely many zeros, the sequence s_n is increasing and tends to $+\infty$. On the other hand, for each zero α of order u of f, either α is not a zero of $f' + f^m$ (when $u = 1$) or it is a zero of order $u - 1$. Consequently, the number of zeros of $f' + f^m$ in $\Gamma(0, R, r''_n)$ (respectively in $d(0, r''_n)$) which are not zeros of f is at least s_n. Thus, we have proved that $f' + f^m$ has infinitely many zeros that are not zeros of f. □

Remark. Lemma 53.2 particularly applies to functions $f \in \mathcal{M}(\mathbb{K})$.

We prove together Theorems 53.3 and 53.4.

Theorem 53.3. *Let $f \in \mathcal{M}^c(D)$ satisfy* $\limsup_{r\to\infty} |f|(r) > 0$ *and let $b \in \mathbb{K}^*$. Let $m \in \mathbb{N}^*$ $bc \geq 3$. Then $f' + bf^m$ has infinitely many zeros that are not zeros of f.*

Proof. Without loss of generality, we can assume $b = 1$. By hypotheses, there exists a sequence of intervals $[r'_n, r''_n]$ such that $\lim_{n\to+\infty} r'_n = +\infty$ and such that, putting $J = \bigcup_{n\in\mathbb{N}} [r'_n, r''_n]$, we have $\liminf_{\substack{r\to\infty,\\ r\in J}} |f|(r) > 0$.

We prove that there exists $t > 0$ such that $|f' + f^m|(r) = |f^m|(r)\ \forall r \in J \cap [t, +\infty[$. By Theorem 37.10, we have $|f'|(r) \leq \frac{|f|(r)}{r}$. Thus, when $r \in J$ is big enough, we have $r(|f|(r))^{m-1} > 1$, hence $(|f|(r))^m > |f'|(r)$. Thus, there exists $t \geq s$ such that $(|f|(r))^m > |f'|(r)\ \forall r \in J \cap [t, +\infty[$. Let $J' = J \cap [t, +\infty[$. So we have $|f' + f^m|(r) = |f^m|(r)\ \forall r \in J'$.

We can now conclude. For each $n \in \mathbb{N}$, let q_n be the number of zeros of f in $\Gamma(0, R, r''_n)$. Suppose the sequence $(q_n)_{n\in\mathbb{N}}$ is bounded. Then, f has finitely many zeros, hence it is of the form $\frac{P}{h}$ with $P \in \mathbb{K}[x]$ and $h \in \mathcal{A}(E)$. Consequently, we have $\lim_{r\to+\infty} |f|(r) = 0$, a contradiction to the hypothesis. Therefore, the sequence $(q_n)_{n\in\mathbb{N}}$ which is increasing by definition, tends to $+\infty$. Now, we may apply Lemma 53.2 showing that $f' + f^m$ has infinitely many zeros that are not zeros of f. □

We have a similar version for functions inside a disk $d(a, R^-)$.

Theorem 53.4. *Let $f \in \mathcal{M}_u(d(a, R^-))$ satisfy* $\limsup_{r\to R} |f|(r) = +\infty$ *and let $b \in \mathbb{K}^*$. Let $m \in \mathbb{N}^*$ be ≥ 3. Then $f' + bf^m$ has infinitely many zeros that are not zeros of f.*

Proof. Without loss of generality, we can assume $b = 1$. By hypotheses, there exists a sequence of intervals $[r'_n, r''_n]$ such that $\lim_{n\to+\infty} r'_n = \lim_{n\to+\infty} r''_n = R$ and such that, putting $J = \bigcup_{n\in\mathbb{N}} [r'_n, r''_n]$, we have $\limsup_{\substack{r\to\infty,\\ r\in J}} |f|(r) + \infty$. Thus, we have $|f'|(r) \leq \frac{|f|(r)}{r} \leq \frac{|f|(r)}{R}$. Now, when r is close enough to R with $r \in J$, we have $r(|f|(r))m - 1 > 1$, hence $(|f|(r))^m > |f'|(r)$. Thus, there exists $t > 0$ such that $(|f|(r))^m > |f'|(r)\ \forall r \in [t, +\infty[\cap J$. We can set $J' = J \cap [t, R[$ and then we have $|f' + f^m|(r) = |f^m|(r)\ \forall r \in J'$.

We can now conclude in Theorem 53.4 as in Theorem 53.3. For each $n \in \mathbb{N}$, let q_n be the number of zeros of f in $d(0, r''_n)$. Suppose the sequence $(q_n)_{n\in\mathbb{N}}$ is bounded. Then, f has finitely many zeros, hence it is of the form $\frac{P}{h}$ with $P \in \mathbb{K}[x]$ and $h \in \mathcal{A}(d(0, R^-))$. Consequently, we have $\limsup_{r\to+\infty} |f|(r) < \infty$, a contradiction to the hypothesis. Therefore, the sequence $(q_n)_{n\in\mathbb{N}}$ which is increasing by definition, tends to $+\infty$. Now, we may apply Lemma 53.2 showing that $f' + f^m$ has infinitely many zeros that are not zeros of f. $\square$

In the case $m = 1$, we can have a better conclusion in $\mathcal{M}(\mathbb{K})$.

Theorem 53.5. *Let $f \in \mathcal{M}(\mathbb{K})\setminus\mathbb{K}(x)$. For each $b \in \mathbb{K}^*$, $f' + bf$ has infinitely many zeros that are not zeros of f.*

Proof. Without loss of generality, we can assume again $b = 1$. By Theorem 37.10, we have $|f'|(r) < |f|(r)$ when r is big enough, and hence $|f' + f|(r) = |f|(r)$ in an interval $I = [s, +\infty[$. Suppose first that f has infinitely many zeros. We can then apply Lemma 53.2 and get the conclusion.

Suppose now that f has finitely many zeros. Then f has infinitely many poles c_n of respective order t_n. Since $\mathbb{K}$ has characteristic zero, f' admits each c_n as a pole of order $t_n + 1$ and similarly, $f' + f$ also admits each c_n as a pole of order $t_n + 1$. Thus, we have $N(r, f' + f) = N(r, f) + \overline{N}(r, f)$. But since $|f' + f|(r) = |f|(r)$ holds in I, we have $\nu(f' + f, \log r) = \nu(f, \log r)\ \forall r \in I$, and hence $Z(r, f' + f) - N(r, f' + f) = Z(r, f) - N(r, f)$, therefore $Z(r, f' + f) - (N(r, f) + \overline{N}(r, f)) = Z(r, f) - N(r, f)$, and hence $Z(r, f' + f) = Z(r, f) + \overline{N}(r, f)$. Since we have supposed that f has finitely many zeros and since f has infinitely many poles, $f' + f$ has infinitely many zeros and all but finitely many are not zeros of f. $\square$

Concerning functions $f' + bf^2$, we can obtain a first conclusion when f is analytic:

Theorem 53.6. *Let $f \in \mathcal{A}^c(D)$ have infinitely many zeros in D (respectively let $a \in \mathbb{K}$, let $R \in]0, +\infty[$, and let $f \in \mathcal{A}_u(d(a, R^-))$). For each $b \in \mathbb{K}^*$, $f' + bf^2$ has infinitely many zeros that are not zeros of f.*

Proof. Without loss of generality, we can assume $b = 1$ and $a = 0$. Clearly, $\lim_{r\to+\infty} |f|(r) = +\infty$ (respectively $\lim_{r\to R} |f|(r) = +\infty$), hence when r is big enough, in $]0, +\infty[$ (respectively in $]0, R[$), we have $|f' + f^2|(r) = |f^2|(r)$, therefore, by Corollary 22.6, f^2 and $f' + f^2$ have the same number of zeros in $C(0, r)$ (taking multiplicity into account). Let $\alpha \in C(0, r)$ be a zero of f of order q. When r is big enough, it is a zero of order $2q$ for f^2 and it is a zero of order $q-1$ for $f'+f^2$. Consequently, by Corollary 22.6, $f'+f^2$ has at least $q+1$ zero in $C(0, r)$ that are not zeros of f (taking multiplicity into account). This is true for every such zeros of f and hence $f'+f^2$ has infinitely many zeros that are not zeros of f. □

Corollary 53.7. *Let $m \in \mathbb{N}^*$ be ≥ 1, and let $f \in \mathcal{A}(\mathbb{K}) \setminus \mathbb{K}(x)$. For each $b \in \mathbb{K}^*$, $f' + bf^m$ has infinitely many zeros that are not zeros of f.*

Corollary 53.8. *Let $m \in \mathbb{N}$ be ≥ 2, let $a \in \mathbb{K}$, let $R \in]0, +\infty[$, and let $f \in \mathcal{A}_u(d(a, R^-))$ (respectively let $f \in \mathcal{A}(D)$ have infinitely many zeros in D). For each $b \in \mathbb{K}^*$, $f' + bf^m$ has infinitely many zeros in $d(a, R^-)$ (respectively in D) that are not zeros of f.*

Proposition 53.9. *Let $f \in \mathcal{M}^c(D)$ (respectively let $a \in \mathbb{K}$ and $R \in \mathbb{R}_+^*$ and let $f \in \mathcal{M}_u(d(a, R^-)))$ and let $m \in \mathbb{N}$. If $m \geq 5$, then for each $b \in \mathbb{K}^*$, $f' + bf^m$ has infinitely many zeros that are not zeros of f. If $m = 4$, if $f \in \mathcal{M}(\mathbb{K}) \setminus \mathbb{K}(x)$, and if f admits at least s multiple zeros and at least t multiple poles, then $f' + bf^4$ admits a number of zeros that are not zeros of f (taken account of multiplicity) which is strictly superior to $\frac{s+t}{14}$.*

Proof. For convenience, we denote $Z(r, \phi)$ instead of $Z_R(r, \phi)$ and $N(r, \phi)$ instead of $N_R(r, \phi)$, when $\phi \in \mathcal{M}(D)$.

By Corollary 22.2, the zeros of $f' + bf^m$ in $\mathbb{K}$ are the same as in a spherically complete algebraically closed extension $\widehat{\mathbb{K}}$ of $\mathbb{K}$. So, for simplicity, we can suppose that the field $\mathbb{K}$ is spherically complete without loss of generality. We can also suppose that $b = 1$. Then if $f \in \mathcal{M}^c(D)$, we can obviously write $f = \frac{h}{l}$ with $h,\ l \in \mathcal{A}(D)$, having no common zeros, and if $f \in \mathcal{M}(d(a, R^-))$, since $\mathbb{K}$ is spherically complete, we can also write $f = \frac{h}{l}$ with $h,\ l \in \mathcal{A}(d(a, R^-))$, having no common zeros again.

Let $g = \frac{1}{f}$ and let $n = m - 2$. So, by Lemma 53.1, the problem is reduced to show that $g'g^n - 1$ has infinitely many zeros. Suppose that $g'g^n - 1$ has finitely many zeros in D (respectively in $d(0, R^-)$). Then, $g'g^n - 1 = \frac{(l'h - h'l)l^n - h^{n+2}}{h^{n+2}}$, and since $h,\ l$ have no common zeros, this is of the form $\frac{P}{h^{n+2}}$, where P is a polynomial of degree q. Now, set $F = (l'h - h'l)l^n$. Applying Corollary 51.12 to F, we have

(1) $T(r,F) = Z(r,F) + O(1) \leq \overline{Z}(r,F) + \overline{Z}(r,F - P) + 7T(r,P) + O(1)$.

By (1), we can derive
$Z(r,l'h - h'l) + nZ(r,l) \leq \overline{Z}(r,l'h - h'l) + \overline{Z}(r,l) + \overline{Z}(r,F - P) + 7T(r,P) + O(1)$. Actually, $\overline{Z}(r,F - P) = \overline{Z}(r,h)$, hence we have $nZ(r,l) \leq \overline{Z}(r,l) + \overline{Z}(r,h) + 7T(r,P) + O(1)$, and hence $(n-1)Z(r,l) \leq Z(r,h) + 7T(r,P) + O(1)$. But since $T(r,P) = q\log r + O(1)$, we have

(2) $(n-1)Z(r,l) \leq Z(r,h) + 7q\log r + O(1)$.

Consider the hypothesis $f \in \mathcal{M}^c(D)$. By Theorem 53.3, if $\liminf_{r\to+\infty} |f|(r) > 0$ i.e. if $\liminf_{r\to+\infty} Z(r,f) - N(r,f) > -\infty$, the claim is proved. Consequently, if the claim is not true, we can assume $\liminf_{r\to+\infty} Z(r,f) - N(r,f) = -\infty$ i.e.

(3) $\liminf\limits_{r\to+\infty} Z(r,l) - Z(r,h) = +\infty$.

Since $f \in \mathcal{M}^c(D)$, by (3) we notice that $l \in \mathcal{A}^c(D)$, Consequently, (2) is impossible whenever $n \geq 3$, i.e. $m \geq 5$.

Consider now the hypothesis $f \in \mathcal{M}_u d(0,R^-))$. We have the same reasonings with limits to R^- instead of $+\infty$ and $h,\ l \in \mathcal{A}(d(0,R^-))$ and we find again relations (2) and (3) with R^- instead of $+\infty$. Then, by Theorem 53.4, if $\limsup_{r\to R^-} |f|(r) = +\infty$, i.e. if $\limsup_{r\to R^-} Z(r,f) - N(r,f) = +\infty$, the claim is proved. So, we assume that $\limsup_{r\to R^-} Z(r,f) - N(r,f) < +\infty$. But then (2) is impossible, whenever $n \geq 3$ i.e. $m \geq 5$, which ends the proof when $f \in \mathcal{M}_u(d(0,R^-))$.

Now, suppose $m = 4$ i.e $n = 2$ and more precisely, suppose $\overline{Z}(r,l) \leq Z(r,l)) - \frac{s\log(r)}{2}$ and $\overline{Z}(r,h) \leq Z(r,h) - \frac{t\log(r)}{2}$, hence

(4) $(n-1)Z(r,l) \leq Z(r,h) + \left(7q - \dfrac{s+t}{2}\right)\log r + O(1)$.

Then Relation (3) implies $7q - \frac{s+t}{2} > 0$, and hence $f'f^n$ admits a number of zeros strictly superior to $\frac{s+t}{14}$ which ends the proof. □

Corollary 53.10. *Let $a \in \mathbb{K}$, $R > 0$, and let $f \in \mathcal{M}_u(d(a, R^-))$. For every $n \in \mathbb{N}$, $n \geq 3$, and for every $b \in \mathbb{K}^*$, $f'f^n - b$ has infinitely many zeros.*

Corollary 53.11. *Let $f \in \mathcal{M}(\mathbb{K})\backslash\mathbb{K}(x)$. Let $b \in \mathbb{K}^*$. Then if f has infinitely many multiple zeros or poles, then $f' + bf^4$ has infinitely many zeros that are not zeros of f.*

We now thoroughly examine the situation when $m = 4$, i.e. $n = 2$, as made in [52]. This requires several basic lemmas.

Lemma 53.12. *Let $f \in \mathcal{M}(\mathbb{K})$ be transcendental and such that f' has finitely many multiple zeros. Then $\frac{f''f}{(f')^2}$ has no quasi-exceptional value.*

Proof. Let $g = \frac{f}{f'}$. A pole of g is a zero of f', hence by hypothesis, g has finitely many multiple poles. Consequently, by Theorem 50.8, g' has no quasi-exceptional value. And hence neither has $1 - g'$. But $g' = \frac{(f')^2 - f''f}{(f')^2} = 1 - \frac{f''f}{(f')^2}$. Therefore, $\frac{f''f}{(f')^2}$ has no quasi-exceptional value. □

Lemma 53.13. *Let $f \in \mathcal{M}(\mathbb{K})$ be transcendental and have finitely many multiple zeros. Then $f''f + 2(f')^2$ has infinitely many zeros that are not zeros of f.*

Proof. Suppose first that f' has infinitely many multiple zeros. Since f has finitely many multiple zeros, the zeros of f' are not zeros of f except at most finitely many. Hence, f' has infinitely many multiple zeros that are not zeros of f. And then, they are zeros of f'', hence of $f''f + 2(f')^2$, which proves the statement.

So, we are now led to assume that f' has finitely many multiple zeros. By Lemma 53.12, $\frac{f''f+2(f')^2}{(f')^2}$ has infinitely many zeros. Let $c \in \mathbb{K}$ be a pole of order q of f. Without loss of generality, we can suppose $c = 0$. The beginning of the Laurent development of f at 0 is of the form $\frac{a_{-q}}{x^q} + \frac{\varphi(x)}{x^{q-1}}$ whereas $\varphi \in \mathcal{M}(\mathbb{K})$ has no pole at 0. Consequently, $\frac{f''f+2(f')^2}{(f')^2}$ is of the form

$$\frac{(a_{-q})^2(3q^2 + q) + x\phi(x)}{(a_{-q})^2(q^2) + x\theta(x)}$$

whereas ϕ, $\theta \in \mathcal{M}(\mathbb{K})$ have no pole at 0. So, the function $\frac{f''f+2(f')^2}{(f')^2}$ has no zero at 0. Therefore, each zero of $\frac{f''f+2(f')^2}{(f')^2}$ is a zero of $f''f + 2(f')^2$, and hence $f''f + 2(f')^2$ has infinitely many zeros.

Now, let us show that the zeros of $f''f + 2(f')^2$ are not zeros of f, except maybe finitely many. Let c be a zero of $f''f + 2(f')^2$ and suppose that c is a zero of f. Then, it is a zero of f', and hence it is a multiple zero of f. But by hypotheses, f has finitely many multiple zeros, hence the zeros of $f''f + 2(f')^2$ are not zeros of f, except at most finitely many. That finishes proving the claim. □

Lemma 53.14. *Let $f \in \mathcal{M}(\mathbb{K})$ be transcendental and let $b \in \mathbb{K}^*$ be such that $f^2f' - b$ has finitely many zeros. Then, $N(r, f) \leq Z(r, f) + O(1)$.*

Proof. Let $F = f^2f'$. Since $F - b$ is transcendental and has finitely many zeros, it is of the form $\frac{P(x)}{h(x)}$ with $h \in \mathcal{A}(\mathbb{K})\setminus\mathbb{K}[x]$. Consequently, $|F|(r)$ is a constant when r is big enough, and therefore, by Theorem 43.2, we have $Z(r, F) = N(r, F) + O(1)$ when r is big enough. Now, $Z(r, F) = 2Z(r, f) + Z(r, f')$ and, by Theorem 43.13, $Z(r, f') \leq Z(r, f) + \overline{N}(r, f) - \log r + O(1)$. On the other hand, by Theorem 43.13 again, we have $N(r, F) = 3N(r, f) + \overline{N}(r, f)$. Consequently, $3N(r, f) + \overline{N}(r, f) \leq 3Z(r, f) + \overline{N}(r, f) - \log r + O(1)$, which proves the claim. □

We can now state the solution of the Hayman conjecture in a p-adic field. Theorem 53.15 was published in [52].

Theorem 53.15. *Let $f \in \mathcal{M}(\mathbb{K})\setminus\mathbb{K}(x)$. Then for each $b \in \mathbb{K}^*$, $f'f^2 - b$ has infinitely many zeros.*

Proof. Let $b \in \mathbb{K}^*$ and suppose that the claim is wrong, i.e. $f^2f' - b$ has k zeros, taking multiplicity into account. By Proposition 53.9, we may assume that f has finitely many multiple zeros and finitely multiple poles. Set $F = f^2f'$. Then, $F' = f(f''f + 2(f')^2)$. By Lemma 53.13, $f''f + 2(f')^2$ has infinitely many zeros that are not zeros of f. Consequently, F' admits for zeros: the zeros of f and the zeros of $f''f + 2(f')^2$. And by Lemma 53.13, there exists a sequence of zeros of $f''f + 2(f')^2$ that are not zeros of f.

Let $S = \{0, b\}$ and let $Z_0^S(r, F')$ be the counting function of zeros of F' when $F(x)$ is different from 0 and b. Since $F - b$ has finitely many zeros, the zeros c of F' which are not zeros of f cannot satisfy $F(c) = b$ except at most finitely many. Consequently, there are infinitely many zeros of F' counted by the counting function $Z_0^S(r, F')$, and hence for every fixed integer $q \in \mathbb{N}$, we have

$$Z_0^S(r, F') \geq q \log r + O(1). \tag{1}$$

Let us apply Theorem 45.7 to F. We have

$$\begin{aligned} T(r, F) \leq \overline{Z}(r, F) + \overline{Z}(r, F - b) + \overline{N}(r, F) - Z_0^S(r, F') \\ - \log(r) + O(1). \end{aligned} \tag{2}$$

Now, we have

$$\overline{Z}(r, F) \leq Z(r, f) + Z(r, f'), \tag{3}$$

$$\overline{N}(r, F) = \overline{N}(r, f), \tag{4}$$

and since the number of zeros of $F - b$ is k, taking multiplicity into account, we have

$$\overline{Z}(r, F - b) \leq k \log r + O(1). \tag{5}$$

Consequently, by (2), (3), (4), and (5), we obtain

$$\begin{aligned} T(r, F) \leq Z(r, f) + Z(r, f') + \overline{N}(r, f) - Z_0^S(r, F') \\ + (k - 1) \log r + O(1). \end{aligned} \tag{6}$$

On the other hand, by construction, $T(r, F) \geq Z(r, F) = 2Z(r, f) + Z(r, f')$, hence by (6), we obtain

$$Z(r, f) \leq \overline{N}(r, f) - Z_0^S(r, F') + (k - 1) \log r + O(1). \tag{7}$$

Now, by Lemma 53.14, we have $N(r, f) \leq Z(r, f) + O(1)$, hence by (7), we obtain $0 \leq (k-1) \log r - Z_0^S(r, F') + O(1)$, and hence, by (1), fixing $q > k - 1$ we can derive $0 \leq (k - 1) \log r - q \log r + O(1)$, a contradiction. That finishes the proof of Theorem 53.15. □

By Lemma 53.1 and Theorems 53.9 and 53.15, we can now state the general result on the p-adic Hayman conjecture:

Corollary 53.16 (52). *Let $f \in \mathcal{M}(\mathbb{K})$ be transcendental and let $b \in \mathbb{K}^*$. Then for every $n \geq 2$, $f'f^n - b$ has infinitely many zeros. For every $m \geq 4$, $f' + bf^4$ has infinitely many zeros that are not zeros of f.*

Concerning the case $m = 3$ i.e. $n = 1$ which remains unsolved, thanks to Theorem 50.11, Corollary 53.16 has an immediate application to the conjecture with additional hypotheses.

Corollary 53.17. *Let $f \in \mathcal{M}(\mathbb{K})$. Suppose that there exists $c, q \in]0, +\infty[$ such that $N(r, f) \leq cr^q$ $\forall r \in [1, +\infty[$. If $f'f^n - b$ has finitely many zeros for some $b \in \mathbb{K}^*$, with $n \in \mathbb{N}$, then $f \in \mathbb{K}(x)$.*

Proof. Suppose f is transcendental. By hypothesis, f^{n+1} satisfies $Z(r, \frac{1}{f^{n+1}}) = \xi(r, f^{n+1}) \leq c(n+1)r^q$ $\forall r \in [1, +\infty[$, hence by Corollary 53.16 and Theorem 50.11, $f'f^n$ has no quasi-exceptional value different from 0. □

Corollary 53.17 may be written in another way:

Corollary 53.18. *Let $f \in \mathcal{M}(\mathbb{K}) \setminus \mathbb{K}(x)$. Suppose that there exists $c, q \in]0, +\infty[$ such that $Z(r, f) \leq cr^q$ $\forall r \in [1, +\infty[$. Then for all $m \in \mathbb{N}$, $m \geq 3$, and for all $b \in \mathbb{K}$, $f' - bf^m$ admits infinitely many zeros that are not zeros of f.*

Proof. We set $g = \frac{1}{f}$. Then by Corollary 53.17, $g'g^{m-2}$ has no quasi-exceptional value. Consequently, given $b \in \mathbb{K}^*$, $g'g^{m-2} + b$ has infinitely many zeros, and hence, $f' - bf^m$ has infinitely many zeros that are not zeros of f. Next, if $b = 0$, by Theorem 50.11, f' has infinitely many zeros. □

Consider now the case $m = 3$, i.e. $n = 1$.

Theorem 53.19. *Let $f \in \mathcal{M}(\mathbb{K})$. Suppose that there exists $c, q \in]0, +\infty[$ such that $s(r, f) \leq cr^q$ $\forall r \in [1, +\infty[$. Then, for all $b \in \mathbb{K}$, $\frac{f'}{f^2} - b$ has infinitely many zeros.*

Proof. Set $g = \frac{1}{f}$ again. Since the poles of g are the zeros of f, we have $t(r, g) \leq cr^q$. Consequently, by Corollary 50.12, g' has no quasi-exceptional value. □

Remark. Using Theorem 53.19 to study the zeros of $f' - bf^2$ that are not zeros of f is not so immediate, as we see in the following because of residues of f at poles of order 1. Of course, if $\frac{1}{f}$ is an affine function, $f' + f^2$ has no zeros, except if it is identically zero. And if it is not identically zero, the residue at the pole is not 1 in the general case.

Lemma 53.20. *Let $f = \frac{h}{l} \in \mathcal{M}(\mathbb{K})$ with $h,\ l \in \mathcal{A}(\mathbb{K})$ having no common zero, let $b \in \mathbb{K}^*$, and let $a \in \mathbb{K}$ be a zero of $h'l - hl' + bh^2$ that is not a zero of $f' + bf^2$. Then a is a pole of order 1 of f and* $\mathrm{res}(f, a) = \frac{1}{b}$.

Proof. Clearly, if $l(a) \neq 0$, a is a zero of $f' + bf^2$. Hence, a zero a of $h'l - hl' + bh^2$ that is not a zero of $f' + bf^2$ is a pole of f. Now, when $l(a) = 0$, we have $h(a) \neq 0$, hence $l'(a) = bh(a) \neq 0$ and therefore a is a pole of order 1 of f such that $\frac{h(a)}{l'(a)} = \frac{1}{b}$. But since a is a pole of order 1, we have $\mathrm{res}(f, a) = \frac{h(a)}{l'(a)}$, which ends the proof. □

Theorem 53.21 is not a result specific to p-adic analysis but it is useful in Theorem 53.23.

Theorem 53.21. *Let $f \in \mathcal{M}(\mathbb{K})$ (respectively let $a \in \mathbb{K}$, and let $f \in \mathcal{M}(d(a, R^-))$), let $b \in \mathbb{K}^*$ and let $\alpha \in \mathbb{K}$ (respectively let $\alpha \in d(a, R^-)$) be a point that is not a zero of f and such that the residue of f at α is different from $\frac{1}{b}$. Then α is a zero of $f' + bf^2$ if and only if it is a zero of $\frac{f'}{f^2} + b$. Moreover, if it is a zero of both functions, it has the same multiplicity with both.*

Proof. Suppose first α is a zero of $f' + bf^2$. If α is not a pole of f, of course it is a zero of $\frac{f'}{f^2} + b$ with same multiplicity. Suppose now that α is a pole of f: since it is not a pole of $f' + bf^2$, it must be a pole of order 1 of f. Without loss of generality, we may assume that $\alpha = 0$ (respectively $a = \alpha = 0$). Consider the Laurent series of f at 0: $f(x) = \frac{a_{-1}}{x} + a_0 + a_1 x + x^2\phi(x)$ with $\phi \in \mathcal{M}(\mathbb{K})$ (respectively $\phi \in \mathcal{M}(d(0, R^-))$ and $\phi(0) \neq \infty$. Then $f' + bf^2$ is of the form

$$f'(x) + bf(x)^2 = \frac{a_{-1}(-1 + ba_{-1})}{x^2} + \frac{2ba_0 a_1}{x} + a_1 + b(a_0^2 + 2a_1 a_{-1}) + x\eta(x)$$

with $\eta \in \mathcal{M}(\mathbb{K})$ (respectively $\eta \in \mathcal{M}(d(0, R^-))$ and $\eta(0) \neq \infty$, and hence, we have $a_{-1}(-1 + ba_{-1}) = 0$, $a_0 a_{-1} = 0$, $a_0^2 + 2a_1 a_{-1} = 0$. Since by hypothesis $\text{res}(f, \alpha) \neq -\frac{1}{b}$, we have $(1 + ba_{-1}) \neq 0$, hence $a_{-1} = 0$, a contradiction. Consequently, every zero of $f' + bf^2$ that is not a zero of f is a zero of $\frac{f'}{f^2} + b$ with the same multiplicity.

Conversely, suppose now that α is a zero of $\frac{f'}{f^2} + b$. If α is not a pole of f, it is a zero of $f' + bf^2$, with the same multiplicity, because by hypothesis it is not a zero of f. Now suppose that α is a zero of $\frac{f'}{f^2} + b$ and is a pole of f. Clearly, it is a pole of order 1, and again, we may assume that $\alpha = 0$.

Consider again the Laurent series of f at 0: $f(x) = \frac{a_{-1}}{x} + a_0 + a_1 x + x^2 \phi(x)$ with $\phi \in \mathcal{M}(\mathbb{K})$ and $\phi(0) \neq \infty$. Then,

$$\frac{f'}{f^2} = \frac{\frac{-a_{-1}}{x^2} + a_1 + x\theta(x)}{\frac{(a_{-1})^2}{x^2} + \frac{2a_0 a_1}{x} + a_0^2 + 2a_1 a_{-1} + x\eta(x)}$$

where both θ, $\eta \in \mathcal{M}(\mathbb{K})$ have no pole at 0. Clearly, $\frac{f'}{f^2}$ is analytic at 0 and its value is $\frac{-1}{a_{-1}}$. But since 0 is a zero of $\frac{f'}{f^2} + b$, we have $a_{-1} = \frac{1}{b}$, what is excluded by hypothesis. Thus, we have proved that every zero of $\frac{f'}{f^2} + b$ is a zero of $f' + bf^2$ (that is not a zero of f) with the same multiplicity, and this ends the proof of Theorem 53.21.

□

Theorem 53.22. *Let $b \in \mathbb{K}^*$ and let $f \in \mathcal{M}(\mathbb{K})$ have finitely many zeros and finitely many residues at its simple poles equal to $\frac{1}{b}$ and be such that $f' + bf^2$ has finitely many zeros. Then f belongs to $\mathbb{K}(x)$.*

Proof. Let $f = \frac{P}{l}$ with $P \in \mathbb{K}[x]$, $l \in \mathcal{A}(\mathbb{K})$ having no common zero with P. Then, $f' + bf^2 = \frac{P'l - l'P + bP^2}{l^2}$. By hypothesis, this function has finitely many zeros. Moreover, if a is a zero of $P'l - l'P + bP^2$ but is not a zero of $f' + bf^2$, then by Lemma 53.20 it is a pole of order 1 of f such that $\text{res}(f, a) = \frac{1}{b}$. Consequently, $P'l - l'P + bP^2$ has finitely many zeros, and hence, we can write $\frac{P'l - l'P + bP^2}{l^2} = \frac{Q}{l^2}$ with $Q \in \mathbb{K}[x]$, hence $P'l - l'P = -bP^2 + Q$. But then, by Theorem 50.5, l is a polynomial, which ends the proof. □

Remark. If $f(x) = \frac{1}{x}$, the function $f' + bf^2$ has no zero whenever $b \neq 1$.

Theorem 53.23. *Let $f \in \mathcal{M}(\mathbb{K})$ be transcendental and have finitely many zeros of order ≥ 2 and let $b \in \mathbb{K}$. Then $\frac{f'}{f^2} + b$ has infinitely many zeros. Moreover, if $b \neq 0$, every zero α of $\frac{f'}{f^2} + b$ that is not a zero of $f' + bf^2$ is a pole of f of order 1 such that the residue of f at α is equal to $\frac{1}{b}$.*

Proof. Let $g = \frac{f'}{f^2} + b$. Since all zeros of f are of order 1 except maybe finitely many, g has finitely many poles of order ≥ 3, hence a primitive G of g has finitely many poles of order ≥ 2. Consequently, by Theorem 50.8, g has infinitely many zeros.

Now, suppose $b \neq 0$. Let α be a zero of g. If α is not a pole of f, it is a zero of $f' + bf^2$ and we can see that it is not a zero of f.

Finally, suppose that α is a pole of f. Then it must be a pole of order 1, and then, by Lemma 53.20, the residue of f at α is $\frac{1}{b}$. $\square$

Corollary 53.24. *Let $f \in \mathcal{M}(\mathbb{K}) \setminus \mathbb{K}(x)$ have finitely many zeros of order ≥ 2 and finitely many poles of order 1 and let $b \in \mathbb{K}^*$. Then $f' + bf^2$ has infinitely many zeros that are not zeros of f.*

Chapter 54

Composition of Meromorphic Functions

Throughout this chapter, the field $\mathbb{K}$ *is supposed to be of characteristic* 0.

In order to look for meromorphic functions of uniqueness in the following chapter, here we first examine the composition of meromorphic functions which, in the general case, is not a meromorphic function.

Throughout this chapter, $\mathbb{E}$ *is an algebraically closed field of characteristic* 0 *without any assumption on the existence of an absolute value. The field* $\mathbb{K}$ *is supposed to have characteristic zero.*

In each main claim, instead of assuming that the equality $h(f(x)) = h(g(x))$ holds in the whole set of definition, thanks to properties of analytic sets, we check that it is sufficient to have the equality on a bounded sequence having no cluster point at the poles of f and g. For that, we use properties of analytic elements on infraconnected subsets of $\mathbb{K}$.

In the field $\mathbb{K}$ (as in $\mathbb{C}$), the composition of two meromorphic functions $h \circ f$ is not a meromorphic function, in the general case: a pole of f is currently narrowed by poles of $h \circ f$ coming from the poles of h. This is why we first have to study general and basic properties of such functions in the ultrametric context.

Remark and definition. Consider a Laurent series $\sum_{n=-\infty}^{+\infty} a_n(x-\alpha)^n$ converging for $0 < |x - \alpha| < r$.

If the Laurent series is a power series, it then defines an analytic function $f(x)$ in the disk $d(\alpha, r^-)$. The point α is called *a regular point for f*. Else, α is called *a singular point for f*.

If the Laurent series is of the form $\sum_{n=-q}^{\infty} a_n(x-\alpha)^n$ with $a_{-q} \neq 0$, then it defines a meromorphic function in $d(\alpha, r^-)$ and α is a pole of order q of that function.

If α is a singular point but is not a pole, it is called *a point of high singularity for f*. This is the case if and only if infinitely many terms a_n with $n < 0$ are not zero.

Notation. Let $h \in \mathcal{M}(\mathbb{K})$ (respectively $h \in \mathcal{M}(d(a, R^-))$). We denote by $\mathcal{P}(h)$ the set of poles of h and by $\mathcal{C}(h)$ the set of zeros of h. Let $f \in \mathcal{M}(\mathbb{K})$ (respectively let $f \in \mathcal{M}(d(a, r^-))$). We set $\mathcal{T}(f, h) = \{x \in \mathbb{K} \mid f(x) \in \mathcal{P}(h)\}$ (respectively $\mathcal{T}(f, h) = \{x \in d(a, R^-) \mid f(x) \in \mathcal{P}(h)\}$) and we denote by $\mathcal{S}(f, h)$ the set $\mathcal{P}(f) \cup \mathcal{T}(f, h)$.

Proposition 54.1 is classical in $\mathbb{C}$ as in the field $\mathbb{K}$:

Proposition 54.1. *Let $h \in \mathcal{M}(\mathbb{K})$ and let $f \in \mathcal{M}(\mathbb{K})$ (respectively let $f \in \mathcal{M}(d(a, R^-))$). Then $\mathbb{K} \setminus (\mathcal{S}(f, h))$ is an open subset of $\mathbb{K}$ dense in $\mathbb{K}$ (respectively $d(a, R^-) \setminus (\mathcal{S}(f, h))$ is an open subset of $d(a, R^-)$ dense in $d(a, R^-)$). For each $\alpha \in \mathbb{K} \setminus \mathcal{P}(f)$ (respectively $\alpha \in d(a, R^-) \setminus \mathcal{P}(f)$), $h \circ f(x)$ is equal to a Laurent series in $x - \alpha$ in any set of the form $d(\alpha, r) \setminus \{\alpha\}$ included in $\mathbb{K} \setminus \mathcal{P}(\alpha)$ (respectively included in $d(a, R^-) \setminus (\mathcal{P}(f))$. If $\alpha \notin \mathcal{T}(f, h)$, the Laurent series in $x - \alpha$ of $h \circ f$ has no terms of negative index. If $\alpha \in \mathcal{T}(f, h)$, then the Laurent series of $h \circ f$ in $x - \alpha$ has finitely many terms of negative indices (i.e. $h \circ f$ is meromorphic in a disk of $\mathbb{K}$ of center α and has a pole at α).*

Definition. Let $h \in \mathcal{M}(\mathbb{K})$ and let $f \in \mathcal{M}(\mathbb{K})$ (respectively let $f \in \mathcal{M}(d(a, R^-))$). Let $\alpha \in \mathbb{K} \setminus \mathcal{P}(f)$ (respectively $\alpha \in d(a, R^-) \setminus \mathcal{P}(f)$). We call *Laurent series of $h \circ f$ at α* the Laurent series in $x - \alpha$ equal to $h \circ f(x)$ in a neighborhood of α, excepting α when the function $h \circ f$ is not defined at α.

Remark. Under the notation of the last definition, if the Laurent series of $h \circ f$ at α is a power series, the series then defines an analytic function in a disk $d(\alpha, r^-)$. The point α is then a regular point for $h \circ f$. Else, α is a singular point for $h \circ f$.

If the Laurent series at $\alpha \in \mathcal{T}(f, h)$ is of the form $\sum_{n=-q}^{\infty} a_n(x - \alpha)^n$ with $a_{-q} \neq 0$, then the Laurent series defines a meromorphic function in a disk $d(\alpha, r^-)$ and α is a pole of order q of that function.

Next, by Proposition 54.1, a singular point for $h \circ f$ which is not a pole of $h \circ f$ belongs to $\mathcal{P}(f)$ and is a point of high singularity for $h \circ f$.

Proposition 54.2 is immediate:

Proposition 54.2. *A point of high singularity for $h \circ f$ is a pole of f. Conversely, a pole of f is just a pole for $h \circ f$ when h is a (non-constant) polynomial. And when h is a rational function tending to a finite limit at infinity, then a pole of f is a regular point for $h \circ f$.*

Notation. Given a subset A of $\mathbb{K}$ and positive numbers $t, r \in]0, t[$, we set $\mathcal{D}(t, r, A) = d(0, t) \setminus \bigcup_{\alpha \in A} d(\alpha, r^-)$.

Lemma 54.3. *Let $h \in \mathcal{M}(\mathbb{K})$, let $f \in \mathcal{M}(\mathbb{K})$ (respectively let $f \in \mathcal{M}(d(a, R^-))$). Then for every $t > 0$ (respectively $t \in]0, R[$) and $r \in]0, t[$, both $\mathcal{P}(f) \cap d(0, t)$ and $\mathcal{T}(f, h) \cap \mathcal{D}(t, r, \mathcal{P}(f))$ are finite, any cluster point of $\mathcal{S}(f, h)$ is a pole of f, the number of holes of $\mathcal{D}(t, r, \mathcal{S}(f, h))$ is finite, and $h \circ f$ belongs to $H(\mathcal{D}(t, r, \mathcal{S}(f, h)))$.*

Proof. We assume $a = 0$. Let $t > 0$ (respectively $t \in]0, R[$) be fixed and let $r \in]0, t[$. Let $\{\alpha_1, \ldots, \alpha_q\}$ be the finite set of all poles of f and g in $d(0, t)$.

Now consider $\mathcal{T}(f, h) \cap \mathcal{D}(t, r, \mathcal{P}(f))$. The set $E = \mathcal{D}(t, r, \mathcal{P}(f))$ is a closed bounded infraconnected set with finitely many holes, hence by Corollary 21.3, f is bounded and takes every value finitely many times. Consequently, h has finitely many poles in $f(E)$ and $\mathcal{T}(f, h) \cap \mathcal{D}(t, r, \mathcal{P}(f))$ is finite. Therefore, the number of holes of $\mathcal{D}(t, r, \mathcal{S}(f, h))$ is finite.

We note that any cluster point of $\mathcal{S}(f, h)$ in $d(0, t)$ is necessarily one of the α_j. Indeed, let $\alpha \in \mathcal{S}(f, h) \cap d(0, t)$ with $\alpha \neq \alpha_j \ \forall j = 1, \ldots, q$. Then $f(\alpha)$ is a pole of h, hence there exists a disk $d(f(\alpha), s)$ such that h has no singularity in $d(f(\alpha), s)$ except $f(\alpha)$ i.e. h is of the form $\frac{g}{(x - f(\alpha))^u}$ with $g \in H(d(f(\alpha), s))$. And of course there exists a disk $d(\alpha, \ell)$ such that $f(d(\alpha, \ell)) \subset d(f(\alpha), s)$, which shows that $h \circ f$ has no singularity other than α in $d(\alpha, \ell)$, hence α is not a cluster point of $\mathcal{S}(f, h)$. □

Lemma 54.4. *Let $h \in \mathcal{A}(\mathbb{K})$ and let $f \in \mathcal{A}(\mathbb{K})\setminus\mathbb{K}$ (respectively $f \in \mathcal{A}_u(d(0,R^-))$). There exists $s > 0$ (respectively $s \in]0,R[$) such that $|h \circ f|(r) = |h|(|f|(r))\ \forall r \geq s$ (respectively $\forall r \in [s,R[$).*

Proof. Let $b = f(0)$. By Theorem 22.9, $f(d(0,r))$ is a disk $d(b,t)$. Suppose f belongs to $\mathcal{A}(\mathbb{K})\setminus\mathbb{K}$, hence it admits a zero $\alpha \in \mathbb{K}$. Let $s = |\alpha|$ and let us take $r \geq s$. Then $f(d(0,r))$ is a disk $d(b,t)$ equal to $d(0,t)$, hence $t = |f|(r)$. Consequently, $|h \circ f|(r) = \sup\{|h(f(x))| \mid x \in d(0,r)\} = \sup\{|h(u)| \mid u \in d(0,|f|(r))\}$.

Similarly, suppose f belongs to $\mathcal{A}_u(d(a,R^-))\setminus\mathbb{K}$. Since f is unbounded, it admits a zero $\alpha \in d(a,R^-)$. By putting again $s = |\alpha|$ we can go on as in the previous case. □

Corollary 54.5. *Let $h \in \mathcal{A}(\mathbb{K})$ and let $f \in \mathcal{A}(\mathbb{K})$ (respectively $f \in \mathcal{A}(d(a,R^-))$). Then there exists $s > 0$ (respectively $s \in]0,R[$) such that $Z(r, h \circ f) - Z(|f|(r), h)$ is a constant for all $r \geq s$.*

Lemma 54.6. *Let $h \in \mathcal{M}(\mathbb{K})$ and let $f \in \mathcal{A}(\mathbb{K})$ (respectively $f \in \mathcal{A}(d(a,R^-))$). Then, $T(r, h \circ f) = T(|f|(r), h) + O(1)$.*

Proof. By Lemma 36.3, there exist $\phi, \theta \in \mathcal{A}(\mathbb{K})$ with no common zeros such that $h = \frac{\phi}{\theta}$. Then $Z(r, \phi \circ f) = Z(r, h \circ f)$ and $Z(r, \theta \circ f) = N(r, h \circ f)$. On the other hand, by Corollary 54.5, we have $Z(r, \phi \circ f) = Z(|f|(r), \phi) + O(1)$, $Z(r, \theta \circ f) = Z(|f|(r), \theta) + O(1)$. Consequently, $T(r, h \circ f) = \max(Z(|f|(r), \phi), Z(|f|(r), \theta)) + O(1)$. But now, we have $\max(Z(|f|(r), \phi), Z(|f|(r), \theta)) = T(|f|(r), h) + O(1)$, which ends the proof. □

Lemma 54.7. *Let $\phi \in \mathcal{A}(\mathbb{K})$ satisfy $h(0) \neq 0$, $s > 0$ (respectively $s \in]0,R[$) and let η, υ be increasing continuous functions from $]0,+\infty[$ (respectively from $[0,R[$) to $]0,+\infty[$ satisfying $\lim_{r\to+\infty} \eta(r) = \lim_{r\to+\infty} \upsilon(r) = +\infty$ (respectively $\lim_{r\to R} \eta(r) = \lim_{r\to R} \upsilon(r) = +\infty$) and such that $Z(\eta(r), \phi) - Z(\upsilon(r), \phi)$ is a constant in $[s,+\infty[$ (respectively in $[s,R[$). Then $\log(\eta(r)) - \log(\upsilon(r))$ is bounded in $[s,+\infty[$ (respectively in $[s,R[$). Moreover, if h is not a polynomial (respectively if ϕ belongs to $\mathcal{A}_u(d(a,R^-))$), then $\log(\eta(r)) - \log(\upsilon(r))$ tends to 0 when r tends to $+\infty$ (respectively to R).*

Proof. Since $Z(\eta(r), \phi) = \log(|\phi|(\eta(r))) - \log(|\phi(0)|)$ and $Z(\upsilon(r), \phi) = \log(|\phi|(\upsilon(r))) - \log(|\phi(0)|)$, we can see that $\log(|\phi|(\eta(r))) - \log(|\phi|(\upsilon(r)))$ is a constant C in $[s,+\infty[$ (respectively in $[s, R[$). Then by Corollary 22.30, we know that $|\log(|\phi|(\eta(r))) - \log(|\phi|(\upsilon(r))|_\infty$

$$\geq \min(\nu^+(\phi, \log(\upsilon(r))), \nu^+(\phi, \log(\eta(r))))|\log(\eta(r)) - \log(\upsilon(r))|_\infty.$$

Suppose first $\phi \in \mathcal{A}(\mathbb{K})$. If ϕ is not a polynomial, we have

$$\lim_{r\to+\infty} \nu^+(\phi, \log(\eta(r))) = \lim_{r\to+\infty} \nu^+(\phi, \log(\upsilon(r))) = +\infty,$$

hence $\lim_{r\to\infty} \log(\eta(r)) - \log(\upsilon(r)) = 0$. And if ϕ is a polynomial of degree q, then when r is big enough, we have $\nu^+(\phi, \log(\eta(r))) = \nu^+(\phi, \log(\upsilon(r))) = q$, hence $\log(\eta(r)) - \log(\upsilon(r))$ is constant.

Now, if ϕ belongs to $\mathcal{A}_u(d(a, R^-))$, then

$$\lim_{r\to R} \nu^+(\phi, \log(\eta(r))) = \lim_{r\to R} \nu^+(\phi, \log(\upsilon(r))) = +\infty,$$

hence $\lim_{r\to R}(\log(\eta(r)) - \log(\upsilon(r))) = 0$ again. In conclusion, $\log(\eta(r)) - \log(\upsilon(r)))$ is bounded in all cases. □

Lemma 54.8. *Let $h \in \mathcal{M}(\mathbb{K})$ and let $f,\ g \in \mathcal{A}(\mathbb{K})$ (respectively $f,\ g \in \mathcal{A}(d(a, R^-)))$ satisfy $h \circ f = h \circ g$. Then $T(r, f) - T(r, g)$ is bounded in any interval $[b, +\infty[$ (respectively $[b, R[$).*

Proof. Let $\phi, \theta \in \mathcal{A}(\mathbb{K})$ with no common zeros such that $h = \frac{\phi}{\theta}$. Then $Z(r, \phi \circ f) = Z(r, h \circ f)$ and $Z(r, \phi \circ g) = Z(r, h \circ g)$. On the other hand, by Lemma 54.5, there exists $s', s'' > 0$ (respectively $s', s'' \in]0, R[$) such that $Z(r, \phi \circ f)$ is of the form $Z(|f|(r), \phi) + C'$ with $C' \in \mathbb{R}$ whenever $r \geq s'$ (respectively whenever $r \in [s', R[$) and similarly $Z(r, \phi \circ g)$ is of the form $Z(|g|(r), \phi) + C''$ with $C'' \in \mathbb{R}$, whenever $r \geq s''$ (respectively whenever $r \in [s'', R[$). Consequently, putting $s = \max(s', s'')$, we have $Z(|f|(r), \phi) + C' = Z(|g|(r), \phi) + C''\ \forall r \geq s$, hence $Z(|f|(r), \phi) - Z(|g|(r), \phi)$ is a constant C whenever $r \geq s$. Now, since the functions $|f|(r)$, $|g|(r)$ are continuous strictly increasing functions of r, tending to $+\infty$ when r tends to $+\infty$ (respectively when r tends to R), then by Lemma 54.7, $\log(|f|(r)) - \log(|g|(r))$ is bounded in $[b, +\infty[$ (respectively in $[b, R[$) hence so is $T(r, f) - T(r, g)$. □

Lemma 54.9. *Let $h \in \mathcal{A}(\mathbb{K})\setminus\mathbb{K}$ and let $f \in \mathcal{M}(\mathbb{K})$ (respectively $f \in \mathcal{M}(d(a, R^-)))$. Let α be a pole of f in $\mathbb{K}$ (respectively in $d(a, r^-)$). Then α is a singular point for $h \circ f$.*

Let $h \in \mathcal{M}(\mathbb{K})\setminus\mathbb{K}(x)$ and let $f \in \mathcal{A}(\mathbb{K})$ (respectively $f \in \mathcal{A}(d(a, R^-))$). Let α be a pole of f in $\mathbb{K}$ (respectively in $d(a, r^-)$). Then α is a point of high singularity for $h \circ f$.

Proof. The first statement is immediate. Now suppose that h is a meromorphic function and is not a rational function. Let us show the second statement. Let q be the order of the pole α. Without loss of generality, we can assume that $\alpha = 0$. Putting $u = \frac{1}{x}$ and $g(u) = f(x)$, we can see that $g(u)$ is equal to a Laurent series converging in a set $\mathbb{K}\setminus d(0, S)$. Moreover, by taking S big enough, we have $\nu(g, \mu) = q \ \forall \mu \geq \log(S)$, while g has no zero and no pole inside $\mathbb{K}\setminus d(0, S)$. Consequently, $g(u)$ is of the form $\sum_{-\infty}^{q} b_n u^n$ and $|g(u)| = |b_q||u_q|^q$, $g(C(0, r)) = C(0, |b_q|r^q)$.

Suppose that h has a sequence of zeros (respectively poles) $(a_n)_{n\in\mathbb{N}}$ such that $|a_n| > |b_q|S^q$, $\lim_{n\to+\infty} |a_n| = +\infty$. There exists a sequence $(c_n)_{n\in\mathbb{N}}$ such that $|b_q||c_n|^q = |a_n|$ and $g(c_n) = a_n$. Now, setting $u_n = \frac{1}{c_n}$, the sequence (u_n) is a sequence of zeros (respectively poles) of $h \circ f$ that converges to 0, which shows that 0 is neither a regular point nor a pole for $h \circ f$. Consequently, if h is not a rational function, $h \circ f$ admits 0 as a point of high singularity. □

We can now state Proposition 54.10.

Proposition 54.10. *Let $h \in \mathcal{M}(\mathbb{K})$ and let $f, g \in \mathcal{M}(\mathbb{K})$ (respectively let $f, g \in \mathcal{M}(d(a, R^-))$). Let $(a_m)_{m\in\mathbb{N}}$ be a bounded sequence of $\mathbb{K}\setminus(\mathcal{S}(f, h) \cup \mathcal{S}(g, h))$ (respectively a sequence in $d(a, R^-)\setminus(\mathcal{S}(f, h) \cup \mathcal{S}(g, h))$ such that $\sup_{m\in\mathbb{N}} |a_m - a| < R$) admitting no cluster point in $\mathcal{P}(f) \cup \mathcal{P}(g)$, satisfying further $h \circ f(a_m) = h \circ g(a_m)\ \forall m \in \mathbb{N}$. Then, $h \circ f(x) = h \circ g(x)\ \forall x \in \mathbb{K}\setminus(\mathcal{S}(f, h) \cup \mathcal{S}(g, h))$ (respectively $\forall x \in d(a, R^-)\setminus(\mathcal{S}(f, h) \cup \mathcal{S}(g, h))$).*

Proof. We assume $a = 0$. Suppose that $h \circ f$ and $h \circ g$ are two distinct functions. Let $s = \sup |a_m|$. Let $t > s$ (respectively $t \in]s, R[$) be fixed.

We now prove that infinitely many terms of the sequence (a_m) belong to $d(0, t)$. Suppose it is not true.

Let us first show that the sequence $(a_m)_{m\in\mathbb{N}}$ does not admit any point of $\mathcal{S}(f, h) \cup \mathcal{S}(g, h)$ as a cluster point. By hypothesis, it may not admit a cluster point in $\mathcal{P}(f) \cup \mathcal{P}(g)$. We show that it does not admit any $b \in \mathcal{T}(f, h)$ as a cluster point, either. Indeed, suppose that a subsequence of the sequence $(a_m)_{m\in\mathbb{N}}$ converges to $b \in \mathcal{T}(f,h)$. So, $f(b)$ is a pole of order q' of h, and hence, $h(f(x))$ is equal to a Laurent series of the form $\sum_{n=q'}^{\infty} \lambda'_n(x-b)^n$ in a disk of center b. In the same way, $h(g(x))$ is equal to a Laurent series of the form $\sum_{n=q''}^{\infty} \lambda''_n(x-b)^n$ in a disk of center b, therefore $h(g(x))-h(f(x))$ is a Laurent series of the form $\sum_{n=q}^{\infty} \lambda_n(x-b)^n$ in a disk $d(b,u^-)$ and therefore belongs to $\mathcal{M}(d(b,u^-))$. But since b is a cluster point of the sequence (a_m), the meromorphic function $h(g(x)) - h(f(x))$ admits b as a cluster point of zeros, a contradiction. Consequently, the sequence (a_m) does not admit b as a cluster point of $\mathcal{S}(f,h)$. Similarly, it does not admit b as a cluster point of $\mathcal{S}(g,h)$. Thus, the sequence (a_m) has no cluster point in $\mathcal{S}(f,h)\cup\mathcal{S}(g,h)$.

Now, since the sequence (a_m) is bounded, by Corollary 2.19, we can extract from that sequence a subsequence thinner than a circular filter $\mathcal{F}$. Since $\mathcal{F}$ does not converge to a point of $\mathcal{S}(f,h)\cup\mathcal{S}(g,h)$, either the filter converges to a point of $d(0,t)\setminus\mathcal{S}(f,h)\cup\mathcal{S}(g,h)$ or it is a large circular filter secant with $d(0,t)\setminus\mathcal{S}(f,h)\cup\mathcal{S}(g,h)$.

If $\mathcal{F}$ converges to a point $\alpha \in d(0,t)\setminus\mathcal{S}(f,h)\cup\mathcal{S}(g,h)$, the function $h\circ f - h\circ g$ has a sequence of zeros converging at α, hence it is identically zero in $d(0,\ell)$. Consider the set $E(r,t) := \mathcal{D}(t,r,\mathcal{S}(f,h))\cup \mathcal{S}(g,h)$ for some $r \in]0,\ell[$. By Lemma 54.3, $E(r,t)$ has finitely many holes, hence, by Corollary 21.4, it is an analytic subset of $\mathbb{K}$. But by Lemma 36.1, $h\circ f - h\circ g$ belongs to $H(E(r,t))$. Consequently, $h\circ f - h\circ g$ is identically zero in $E(r,t)$. Since we can choose r arbitrary small and t arbitrary big (with $t < R$), we can conclude that $h\circ f - h\circ g$ is identically zero.

Suppose now that $\mathcal{F}$ is a large circular filter secant with $d(0,t)\setminus \mathcal{S}(f,h)\cup\mathcal{S}(g,h)$. Then, when r is small enough, $\mathcal{F}$ is secant with $\mathcal{D}(t,r,\mathcal{S}(f,h))\cup\mathcal{S}(g,h)$ which has finitely many holes and hence is an analytic set, hence $h\circ f - h\circ g$ is identically zero in $\mathcal{D}(t,r,\mathcal{S}(f,h))\cup \mathcal{S}(g,h)$ again so we are led to the same conclusion. □

Proposition 54.11. *Let $h \in \mathcal{M}(\mathbb{K})\setminus\mathbb{K}$ and let $f,\ g \in \mathcal{M}(\mathbb{K})$ (respectively $f,\ g \in \mathcal{M}(d(a,R^-)))$ satisfy $h\circ f(x) = h\circ g(x)\ \forall x \in \mathbb{K}\setminus(\mathcal{S}(f,h)\cup\mathcal{S}(g,h))$ (respectively $h\circ f(x) = h\circ g(x)\ \forall x \in d(a,R^-)\setminus(\mathcal{S}(f,h)\cup\mathcal{S}(g,h)))$. Moreover, if f or g does not belong to $\mathcal{A}(\mathbb{K})$ (respectively if f or g does not belong to $\mathcal{A}(d(a,R^-)))$, we assume that $h \notin \mathbb{K}(x)\setminus\mathbb{K}[x]$. Then $f,\ g$ satisfy $\mathcal{P}(f) = \mathcal{P}(g)$, $\mathcal{S}(f,h) = \mathcal{S}(g,h)$.*

Proof. First we show that $\mathcal{P}(f) = \mathcal{P}(g)$. Suppose α is a pole of f. For this, we obviously suppose that $f \notin \mathcal{A}(\mathbb{K})$ (respectively $f \notin \mathcal{A}(d(a, R^-))$). Consequently, by hypothesis, $h \notin \mathbb{K}(x) \setminus \mathbb{K}[x]$. If $h \in \mathcal{A}(\mathbb{K})$ (respectively if $h \in \mathcal{A}(d(a, R^-))$), then by Lemma 54.9, α is a singular point for $h \circ f$, hence for $h \circ g$, therefore α is a singular point for g and hence it is a pole for g. Now suppose that $h \notin \mathcal{A}(\mathbb{K})$, hence $h \in \mathcal{M}(\mathbb{K}) \setminus \mathbb{K}(x)$ (respectively $h \notin \mathcal{A}(d(a, R^-))$), hence $h \in \mathcal{M}(d(a, R^-)) \setminus \mathbb{K}(x)$. Since h does not lie in $\mathbb{K}(x)$, by Lemma 54.9, α is a point of high singularity for $h \circ f$, hence for $h \circ g$. But if α is not a pole for g, it is a regular point for g, hence it is either a regular point or a pole for $h \circ g$, a contradiction. Consequently, α is a pole for g, and therefore, since f and g play the same role, we have $\mathcal{P}(f) = \mathcal{P}(g)$.

Now, suppose $\alpha \in \mathcal{T}(f, h)$, hence $h \circ f$ has a pole at α and so does $h \circ g$, thereby $\alpha \in \mathcal{T}(g, h)$. Consequently, $\mathcal{T}(f, h) \subset \mathcal{T}(g, h)$ hence $\mathcal{T}(f, h) = \mathcal{T}(g, h)$ and therefore $\mathcal{S}(f, h) = \mathcal{S}(g, h)$, which completes the proof. $\square$

Remark. In order to avoid the restriction: *if f, g do not belong to $\mathcal{A}(\mathbb{K})$ (respectively if f, g do not belong to $\mathcal{A}(d(a, r^-))$), we assume that $h \notin \mathbb{K}(x) \setminus \mathbb{K}[x]$*, we would like to show $\mathcal{P}(f) = \mathcal{P}(g)$ when h is a rational function. But it is hopeless as shows the following situation. Suppose $h \in \mathbb{K}(x)$ is not a function of uniqueness for meromorphic functions and let f, $g \in \mathcal{M}(\mathbb{K})$ satisfy $h \circ f = h \circ g$ and $f(c) \neq g(c)$ for some $c \in \mathbb{K}$ (respectively $c \in d(a, r^-)$). If c is a pole of f, this just shows $\mathcal{P}(f) \neq \mathcal{P}(g)$. Suppose c is not a pole for f and g. Let $b = f(c)$, let $\phi(x) = \frac{1}{f(x)-b}$, let $\theta(x) = \frac{1}{g(x)-b}$, and let $G(u) = h(b + \frac{1}{u})$. Then G belongs to $\mathbb{K}(x)$ and we can check that $G \circ \theta = G \circ \phi$ and that c lies in $\mathcal{P}(\phi)$ but not in $\mathcal{P}(\theta)$.

Chapter 55

Functions of Uniqueness

Throughout this chapter, $\mathbb{K}$ *is of characteristic* 0.

We want to study sufficient conditions on a meromorphic function h assuring that if the composition of meromorphic functions of the form $h \circ f$ and $h \circ g$ are equal, then f and g are equal. This kind of problem follows many other problems of uniqueness studied, particularly on unique range sets with (or without) multiplicities and polynomials of uniqueness for analytic or meromorphic functions in the complex field and in an ultrametric field [3, 47, 93, 98]. Polynomials of uniqueness were introduced and studied in $\mathbb{C}$ and in a p-adic field by Yi and Yang [98], Fujimoto [60], and Li and Yang [97]. Here most results come from [47].

Throughout this chapter, $\mathbb{E}$ *is an algebraically closed field of characteristic* 0 *without any assumption on the existence of an absolute value. The field* $\mathbb{K}$ *is supposed to have characteristic zero.*

In Chapter 41, we have showed that certain polynomials whose derivative has two distinct zeros are polynomials of uniqueness for meromorphic functions or for analytic functions in $\mathbb{K}$ or inside a disk. Here we look for a more systematic process in order to obtain functions of uniqueness.

Previous results obtained in $\mathbb{C}$ by Fujimoto in $\mathbb{C}$ and in $\mathbb{K}$ by An and Khoai and next by An, Wang and Wong concerned polynomials. Here we consider a meromorphic function h instead of a polynomial P and we only assume that a few zeros $c_1, \dots, c_k$ of h' satisfy $h(c_j) \neq h(d)$ for every other zero d of h'. We then examine the situation in four cases in $\mathbb{K}$: f, g entire or meromorphic functions in the whole p-adic field $\mathbb{K}$ or "unbounded" meromorphic function inside an "open" disk of $\mathbb{K}$. We also consider the problem for rational functions on the field $\mathbb{E}$.

On the other hand, in each main claim, instead of assuming that the equality $h(f(x)) = h(g(x))$ holds in the whole set of definition, thanks to properties of analytic sets, we check that it is sufficient to have the equality on a bounded sequence having no cluster point at the poles of f and g. For that, we use properties of analytic elements on infraconnected subsets of $\mathbb{K}$.

Notation. In the following, we denote by h a meromorphic function either in $\mathbb{K}$ or inside a disk $d(a, R^-)$ and we denote by $c_1, \dots, c_n, \dots$ the finite or infinite sequence of zeros of h'.

Lemma 55.1. *Let $h \in \mathcal{M}(\mathbb{K})\backslash\mathbb{K}$ and let $f,\ g \in \mathcal{M}(\mathbb{K})$ (respectively $f,\ g \in \mathcal{M}(d(0, r^-))$) satisfy $h \circ f(x) = h \circ g(x)\ \forall x \in \mathbb{K}\backslash\mathcal{S}(f, h) \cup \mathcal{S}(g, h)$ (respectively $h \circ f(x) = h \circ g(x)\ \forall x \in d(0, R^-)\backslash\mathcal{S}(f, h) \cup \mathcal{S}(g, h)$). Moreover, if $f,\ g$ do not belong to $\mathcal{A}(\mathbb{K})$ (respectively if $f,\ g$ do not belong to $\mathcal{A}(d(0, r^-))$, we assume that $h \notin \mathbb{K}(x)\backslash\mathbb{K}[x]$. Then $f,\ g$ satisfy $\mathcal{P}(f) = \mathcal{P}(g)$, $\mathcal{S}(f, h) = \mathcal{S}(g, h)$.*

For each $j = 1, \dots, k$, let $q_j = \omega_{c_j}(h')$. We assume that $h(c_j) \neq h(c_n)\ \forall j = 1, \dots, k,\ \forall n \neq j$. Then $f,\ g$ satisfy

$$\overline{N}(r, f) + \sum_{j=1}^{k} \overline{Z}(r, f - c_j) \leq \overline{Z}\left(r, \frac{1}{f} - \frac{1}{g}\right) + \sum_{j=1}^{k} \frac{1}{q_j} Z(r, g' \mid f(x) = c_j,\ g(x) \notin \mathcal{C}(h')).$$

Furthermore, if $f,\ g \in \mathcal{A}(\mathbb{K})$ (respectively if $f,\ g \in \mathcal{A}(d(0, r^-))$), then

$$\sum_{j=1}^{k} \overline{Z}(r, f - c_j) \leq \overline{Z}(r, f - g) + \sum_{j=1}^{k} \frac{1}{q_j} Z(r, g' \mid f(x) = c_j,\ g(x) \notin \mathcal{C}(h')).$$

Proof. First, we show that $\mathcal{P}(f) = \mathcal{P}(g)$. Suppose α is a pole of f. If $h \in \mathcal{A}(\mathbb{K})$ (respectively if $h \in \mathcal{A}(d(0, R^-))$), then by Lemma 54.9, α is a singular point for $h \circ f$, hence for $h \circ g$, hence α is a singular point for g, hence it is a pole for g. Now suppose that $h \notin \mathcal{A}(\mathbb{K})$, hence $h \in \mathcal{M}(\mathbb{K})\backslash\mathbb{K}(x)$ (respectively suppose $h \notin \mathcal{A}(d(0, R^-))$, hence

$h \in \mathcal{M}(d(0, R^-))\backslash\mathbb{K}(x))$. Since h does not lie in $\mathbb{K}(x)$, by Lemma 54.9, α is a point of high singularity for $h \circ f$, hence for $h \circ g$. But if α is not a pole for g, it is a regular point for g, hence it is either a regular point or a pole for $h \circ g$, a contradiction. Consequently, α is a pole for g and therefore, since f and g play the same role, we have $\mathcal{P}(f) = \mathcal{P}(g)$.

Now, suppose $\alpha \in \mathcal{T}(f, h)$, hence $h \circ f$ has a pole at α and so does $h \circ g$, thereby $\alpha \in \mathcal{T}(g, h)$. Consequently, $\mathcal{T}(f, h) \subset \mathcal{T}(g, h)$, hence $\mathcal{T}(f, h) = \mathcal{T}(g, h)$, and therefore, $\mathcal{S}(f, h) = \mathcal{S}(g, h)$.

We now assume that $h(c_j) \neq h(c_n)$ $\forall j = 1, \dots, k$, $\forall n \neq j$. Without loss of generality, we may assume that $0 \notin \mathcal{C}(h')$. Indeed, if $0 \in \mathcal{C}(h')$, we can find $b \in \mathbb{K}$ such that $c_n + b \neq 0$ $\forall n \in \mathbb{N}^*$. Setting $\overline{f}(x) = f(x) - b$, $\overline{g}(x) = g(x) - b$ and $\overline{h}(z) = h(z + b)$, we have $\overline{h}(\overline{f}(x) = \overline{h}(\overline{g}(x)$ $\forall x \in \mathbb{K}\backslash\mathcal{S}(\overline{f}, \overline{h})$ (respectively $\overline{h}(\overline{f}(x)) = \overline{h}(\overline{g}(x))$ $\forall x \in d(0, R^-)\backslash\mathcal{S}(\overline{f}, \overline{h})$), thereby we may process in the same way with $\overline{f}$, $\overline{g}, \overline{h}$. Moreover, we note that if f, g lie in $\mathcal{A}(\mathbb{K})$ (respectively in $\mathcal{A}(d(0, r^-))$), then so do $\overline{f}$, $\overline{g}$. Consequently, in order to simplify a deduction, we assume that $c_n \neq 0$ $\forall n \in \mathbb{N}^*$.

Let $\phi = \frac{1}{f} - \frac{1}{g}$. Since $\mathcal{P}(f) = \mathcal{P}(g)$, for each pole α of f, we have $\phi(\alpha) = 0$, therefore

$$(1) \quad \overline{N}(r, f) \leq \overline{Z}(r, \phi \mid x \in \mathcal{P}(f)).$$

Let us fix $j \in \{1, \dots, k\}$ and let $\alpha \in \mathbb{K}$ (respectively $\alpha \in d(0, R^-)$) satisfy $f(\alpha) = c_j$. Suppose first that $g(\alpha)$ lies in $\mathcal{C}(h')$. Thanks to the hypothesis $h(c_n) \neq h(c_j)$ $\forall n \neq j$, if $g(\alpha) \neq c_j$, then $h(g(\alpha)) \neq h(c_j)$, a contradiction to $h(g(\alpha)) = h(f(\alpha))$. So we have $g(\alpha) = f(\alpha) = c_j$ and since $c_j \neq 0$, then $\phi(\alpha) = 0$. Consequently,

$$(2) \quad \overline{Z}(r, f - c_j \mid g(x) \in \mathcal{C}(h')) \leq \overline{Z}(r, \phi \mid f(x) = c_j),$$

and similarly if f, $g \in \mathcal{A}(\mathbb{K})$ or if f, $g \in \mathcal{A}_u(d(0, R^-))$,

$$(3) \quad \overline{Z}(r, f - c_j \mid g(x) \in \mathcal{C}(h')) \leq \overline{Z}(r, f - g \mid f(x) = c_j).$$

Consequently, since $\mathcal{P}(f) \cap \mathcal{C}(h') = \emptyset$, in the general case, by (1) and (2), we can derive

$$(4) \quad \overline{N}(r, f) + \sum_{j=1}^{k} \overline{Z}(r, f - c_j \mid g(x) \in \mathcal{C}(h')) \leq \overline{Z}(r, \phi).$$

Similarly, if $f,\ g \in \mathcal{A}(\mathbb{K})$ (respectively if $f,\ g \in \mathcal{A}(d(0,r^-))$), then by (3), we have

$$(5)\quad \sum_{j=1}^{k} \overline{Z}(r, f - c_j \mid g(x) \in \mathcal{C}(h')) \leq \overline{Z}(r, f - g).$$

In order to complete the proof, we will show

$$(6)\quad \overline{Z}(r, f - c_j \mid g(x) \notin \mathcal{C}(h')) \leq \frac{1}{q_j} Z(r, g' \mid f(x) = c_j,\ g(x) \notin \mathcal{C}(h')).$$

Indeed, consider α such that $g(\alpha) \notin \mathcal{C}(h')$. Since $h'(f(\alpha)) = h'(c_j) = 0$, we note that $f'(\alpha)h'(f(\alpha)) = g'(\alpha)h'(g(\alpha)) = 0$. But since $g(\alpha) \notin \mathcal{C}(h')$, we have $h'(g(\alpha)) \neq 0$, hence $g'(\alpha) = 0$. Consequently, we obtain

$$(7)\quad \overline{Z}(r, f - c_j \mid g(x) \notin \mathcal{C}(h')) \leq \overline{Z}(r, g' \mid f(x) = c_j,\ g(x) \notin \mathcal{C}(h')).$$

On the other hand, since $f(\alpha) = c_j$, we see that $\omega_\alpha(f'(x)h'(f(x))) \geq q_j$, hence

$$(8)\quad \omega_\alpha(g'(x)h'(g(x))) \geq q_j.$$

But since $g(\alpha) \notin \mathcal{C}(h')$, we have $h'(g(\alpha)) \neq 0$, hence by (8), $\omega_\alpha(g') \geq q_j$, and consequently,

$$(9)\quad \overline{Z}(r, g' \mid f(x) = c_j,\ g(x) \notin \mathcal{C}(h')) \leq \frac{1}{q_j} Z(r, g' \mid f(x) = c_j,\ g(x) \notin \mathcal{C}(h'))$$

Thus, by (7) and (9), we obtain (6) which by (4) proves

$$\overline{N}(r, f) + \sum_{j=1}^{k} \overline{Z}(r, f - c_j) \leq \overline{Z}(r, \frac{1}{f} - \frac{1}{g}) + \sum_{j=1}^{k} \frac{1}{q_j} Z(r, g' \mid f(x) = c_j,\ g(x) \notin \mathcal{C}(h')).$$

Similarly, by (5) and (6), we have

$$\sum_{j=1}^{k} \overline{Z}(r, f - c_j) \leq \overline{Z}(r, f - g) + \sum_{j=1}^{k} \frac{1}{q_j} Z(r, g' \mid f(x) = c_j,\ g(x) \notin \mathcal{C}(h')).$$

□

Theorem 55.2. *Let $h \in \mathcal{M}(\mathbb{K})\backslash\mathbb{K}$, let $f,\ g \in \mathcal{A}(\mathbb{K})\backslash\mathbb{K}$, and let $(a_m)_{m\in\mathbb{N}}$ be a bounded sequence of $\mathbb{K}\backslash(\mathcal{T}(f,h) \cup \mathcal{T}(g,h))$ satisfying $h \circ f(a_m) = h \circ g(a_m)\ \forall m \in \mathbb{N}$. Then, $\mathcal{T}(f,h) = \mathcal{T}(g,h)$ and $h \circ f(x) = h \circ g(x)\ \forall x \in \mathbb{K}\backslash\mathcal{T}(f,h)$. Moreover, if $\Upsilon(h) \geq 2$, then $f = g$.*

Remark. Throughout Corollaries 55.3, 55.4, and 55.11, we apply to polynomials (respectively rational functions) with coefficients in $\mathbb{E}$ results proven for analytic (respectively meromorphic) functions in $\mathbb{K}$. Indeed, as it was often previously done, since $\mathbb{E}$ has characteristic 0, there exists a finite extension $\mathbb{M}$ of $\mathbb{Q}$ containing all coefficients, zeros, and poles of all functions involved and consequently we can consider $\mathbb{M}$ as a subfield of $\mathbb{C}_p$.

Corollary 55.3. *Let $h \in \mathcal{M}(\mathbb{K})$ (respectively $h \in \mathbb{E}(x)$) satisfy $\Upsilon(h) \geq 2$. Then h is a function of uniqueness for $\mathcal{A}(\mathbb{K})$ (respectively for $\mathbb{E}[x]$).*

Notation. Given a meromorphic function $f \in \mathcal{M}(\mathbb{K})$ or $f \in \mathcal{M}(d(0,R^-))$, we denote by (F) the hypothesis: *the restriction of f to the set of zeros of f' is injective* [60, 3, 49, 50].

Thus, particulary, if a function f satisfies Hypothesis (F) and if f' has q zeros, then $\Upsilon(f) = q$.

Examples. (1) Let $h(x) = \frac{x(x-1)}{x-2}$. Hence, $h'(x) = \frac{x^2-4x+2}{(x-2)^2}$. Let $\sqrt{2}$ denote a square root of 2 in the field $\mathbb{K}$. The zeros of h' are $c_1 = 2+\sqrt{2}$, $c_2 = 2-\sqrt{2}$. Thus, $h(c_1) = 3-2\sqrt{2}$, $h(c_2) = 3+\sqrt{2}$, hence h satisfies the hypothesis of Corollary 55.3.

(2) Let $b \in \mathbb{E}^*$ be a zero of the polynomial $Q(x) = \frac{x^5}{20} - \frac{x^3}{6} + \frac{x}{4} + \frac{2}{15}$ and let $h(x) = \frac{x^5}{5} - \frac{bx^4}{4} - \frac{x^3}{3} + \frac{bx^2}{2}$. Then, $h'(x) = x(x-1)(x+1)(x-b)$. Now, we note that $h(0) = 0$, $h(1) = \frac{2}{15} + \frac{b}{4}$, $h(-1) = -\frac{2}{15} + \frac{b}{4}$, $h(b) = -\frac{b^5}{20} + \frac{b^3}{6}$. Since $Q(b) = 0$, we have $h(1) = h(b)$ and clearly $h(1) \neq h(0)$, $h(-1) \neq h(1)$, $h(0) \neq h(b)$, $h(-1) \neq h(b)$. Consequently, h' has four zeros $c_1 = 0$, $c_2 = -1$, $c_3 = 1$, $c_4 = b$ satisfying $h(c_j) \neq h(c_l)\ \forall j = 1,2,\ l \neq j,\ 1 \leq l \leq 4$. Therefore, h satisfies the hypothesis of Corollary 55.3. However, h does not satisfy Hypothesis (F) because $h(c_3) = h(c_4)$.

Corollary 55.4. *Let $P(x) \in \mathbb{K}[x]$ be of degree $n \geq 3$ and assume that P' has exactly two distinct zeros. Then P is a polynomial of uniqueness for $\mathcal{A}(\mathbb{K})$.*

Proof. Let $a_1,\ a_2$ be the two zeros of P', of order $s_1,\ s_2$ respectively. Suppose $\Upsilon(P) < 2$. Then, $P(a_1) = P(a_2)$. Let $b = P(a_1)$. Then, $P - b$ admits each a_j as a zero of order $s_j + 1,\ j = 1,\ 2$. Consequently, $\deg(P) = \deg(P - b) \geq s_1 + s_2 + 2$, and hence, $\deg(P') = s_1 + s_2 + 1$, a contradiction since $a_1,\ a_2$ are the only zeros of P'. Consequently, $\Upsilon(P) = 2$. □

Corollary 55.5. *Let $P(x) \in \mathbb{E}[x]$ be a of degree 3. Then P is a polynomial of uniqueness for $\mathcal{A}(\mathbb{K})$ if and only if it is not of the form $c(x-a)^3 + b,\ a,\ b,\ c \in \mathbb{K}$.*

Proof. Indeed, P' has one or two distinct zeros. On one hand, if P is of the form $c(x-a)^3 + b,\ a,\ b,\ c \in \mathbb{K}$, obviously it is not a polynomial of uniqueness for $\mathcal{A}(\mathbb{K})$. On the other hand, if P is not of the form $c(x-a)^3 + b,\ a,\ b,\ c \in \mathbb{K}$, then P' has two distinct zeros, hence by Corollary 55.4, P is a polynomial of uniqueness for $\mathcal{A}(\mathbb{K})$. □

Corollary 55.6. *Let $h,\ f,\ g \in \mathcal{A}(\mathbb{C})$ have all coefficients in $\mathbb{Q}$ and also lie in $\mathcal{A}(\mathbb{C}_p)$ for some prime p. Let $(a_m)_{m\in\mathbb{N}}$ be a bounded sequence of $\mathbb{C}$ satisfying $h \circ f(a_m) = h \circ g(a_m)\ \forall m \in \mathbb{N}$. Then, $h \circ f(x) = h \circ g(x)\ \forall x \in \mathbb{C}$. Moreover, if $\Upsilon(h) \geq 2$, then $f = g$.*

Proof. We know that the identity $h \circ f = h \circ g$ is obvious in $\mathcal{A}(\mathbb{C})$, which means the coefficients of the two functions are the same, hence the identity also holds in $\mathbb{C}_p$. Therefore, if $\Upsilon(h) \geq 2$, by Theorem 55.2, we have $f = g$ in $\mathcal{A}(\mathbb{C}_p)$ i.e. the coefficients of $f,\ g$ are the same, hence this identity obviously holds in $\mathcal{A}(\mathbb{C})$. □

Remarks. Conversely, a polynomial of degree 2 is never a function of uniqueness for any family of functions because through a suitable translation of the variable, it is possible to put it in the form of an even polynomial.

The condition $\Upsilon(h) \geq 2$ is not a necessary condition to assure that h is a function of uniqueness for entire or meromorphic functions: for instance, a linear fractional function has a derivative which has no zero but obviously is a function of uniqueness for meromorphic functions in $\mathbb{K}$ or in $d(a, R^-)$.

The proofs of Theorems 55.7, 55.10, and 55.12 are gathered in the following:

Theorem 55.7. *Let $h \in \mathcal{M}(\mathbb{K})\backslash\mathbb{K}$, let $f,\ g \in \mathcal{A}_u(d(a, R^-))$, and let $(a_m)_{m\in\mathbb{N}}$ be a sequence of $d(a, R^-)$ such that $\sup_{n\in\mathbb{N}} |a_n - a| < R$,* *satisfying* $h \circ f(a_m) = h \circ g(a_m)$ $\forall m \in \mathbb{N}$. *Then,* $\mathcal{S}(f,h) = \mathcal{S}(g,h)$ *and* $h \circ f(x) = h \circ g(x)$ $\forall x \in d(a, R^-)\backslash\mathcal{S}(f,h)$. *Moreover, if* $\Upsilon(h) \geq 3$, *then* $f = g$.

Example. Let $h(x) = \frac{x^3-x^2+x-2}{x-2}$. Hence, $h'(x) = \frac{2x^3-7x^2+4x}{(x-2)^2}$. Let $\sqrt{17}$ denote a square root of 17 in the field $\mathbb{K}$. The zeros of h' are $c_1 = 0$, $c_2 = \frac{7-\sqrt{17}}{4}$, $c_3 = \frac{7+\sqrt{17}}{4}$. Thus, $h(c_1) = 1$, $h(c_2) = \frac{73-17\sqrt{17}}{2}$, $h(c_3) = \frac{73+17\sqrt{17}}{2}$, hence h satisfies the hypothesis of Theorem 55.7.

Corollary 55.8. *Let* $h \in \mathcal{A}(\mathbb{K})$ *satisfy* $\Upsilon(h) \geq 3$. *Then* h *is a function of uniqueness for* $\mathcal{A}_u(d(a, R^-))$.

Similar to Corollary 55.6, we can state Corollary 55.9:

Corollary 55.9. *Let* h, f, $g \in \mathcal{A}(\mathbb{C})$ *have all coefficients in* $\mathbb{Q}$ *and assume that* h *also lies in* $\mathcal{A}(\mathbb{C}_p)$ *for some prime* p *and* f, g *lie in* $\mathcal{A}_u(d(a, R^-))$ *(with respect to the field* $\mathbb{C}_p$*). Let* $(a_m)_{m\in\mathbb{N}}$ *be a bounded sequence of* $\mathbb{C}$ *satisfying* $h \circ f(a_m) = h \circ g(a_m)$ $\forall m \in \mathbb{N}$. *Then,* $h \circ f(x) = h \circ g(x)$ $\forall x \in \mathbb{C}$. *Moreover, if* $\Upsilon(h) \geq 3$, *then* $f = g$.

Remark. In this chapter, we see that if P is of the form $x^n - bx^{n-1} + t$ ($t \in \mathbb{K}$), then it is not a function of uniqueness for $\mathcal{M}(\mathbb{K})$ and an immediate generalization shows that the same holds when P' has exactly two distinct zeros, one of them being of order 1.

Assuming again that the set of zeros S of a polynomial P is affinely rigid and P satisfies Hypothesis (F), it is shown in [3] (Theorem 1) that if P' has exactly two distinct zeros c_j of order m_j ($j = 1, 2$), then it is a function of uniqueness for $\mathcal{M}(\mathbb{K})$ if and only if $\min(m_1, m_2) \geq 2$.

On the other hand, Theorem 47.6 shows that a polynomial such that $\Upsilon(P) = 2$ may be a polynomial of uniqueness for both $\mathcal{M}(\mathbb{K})$ and $\mathcal{M}(d(0, R^-))$.

Theorem 55.10. *Let* $h \in \mathcal{M}(\mathbb{K})\backslash(\mathbb{K} \cup (\mathbb{K}(x)\backslash\mathbb{K}[x]))$, *let* f, $g \in \mathcal{M}(\mathbb{K})\backslash\mathbb{K}$, *and let* $(a_m)_{m\in\mathbb{N}}$ *be a bounded sequence of* $\mathbb{K}\backslash(\mathcal{S}(f,h) \cup \mathcal{S}(g,h))$ *of diameter* $< R$ *admitting no cluster point in* $\mathcal{P}(f) \cup \mathcal{P}(g)$, *satisfying further* $h \circ f(a_m) = h \circ g(a_m)$ $\forall m \in \mathbb{N}$. *Then,* $\mathcal{S}(f,h) = \mathcal{S}(g,h)$ *and* $h \circ f(x) = h \circ g(x)$ $\forall x \in \mathbb{K}\backslash\mathcal{S}(f,h)$. *Moreover, if* $\Upsilon(h) \geq 3$, *then* $f = g$.

Corollary 55.11. *Let $h \in \mathcal{A}(\mathbb{K})$ (respectively $h \in \mathbb{E}[x]$) satisfy $\Upsilon(h) \geq 3$. Then h is a function of uniqueness for $\mathcal{M}(\mathbb{K})$ (respectively for $\mathbb{E}(x)$).*

Remark. In [3], it is shown that a polynomial $P \in \mathbb{E}[x]$ satisfying (F), whose set of zeros is affinely rigid, is a function of uniqueness for $\mathcal{M}(\mathbb{K})$ if and only if either P' has at least three distinct zeros, or P' has just two distinct zeros, both of order ≥ 2. By Theorems 55.10, we can find other polynomials of uniqueness for $\mathcal{M}(\mathbb{K})$ having a set of zeros which is not affinely rigid.

Example. Suppose $p \neq 3$. Let $P(x) = x^4 - 4x$, let j be a cubic root of 1 different from 1, and let $a \in \mathbb{K}$ be a cubic root of 4. Then P has four distinct zeros $\{0,\ a,\ ja,\ j^2a\}$. Thus, the set of zeros of P is not affinely rigid (but is centered). Next, the set of zeros of P' is $\{1,\ j,\ j^2\}$ and we can check that P satisfies (F), hence $\Upsilon(P) = 3$, therefore P is a function of uniqueness for $\mathcal{M}(\mathbb{K})$.

Remark. By Theorem 1.10, given a polynomial $P(x) \in \mathbb{E}[x]$ of degree 4, we can check that P satisfies the hypotheses of Theorem 55.10 and Corollaries 55.3 and 55.11, if and only if it is not of the form $[(x-a+l)(x-a-l)]^2 + A$ with $A \in E,\ l \in E^*$ (which means P is not an even function, up to some affine change of variable).

Example. $P(x) = (x^2 - a^2)^2 + x$ satisfies the hypotheses of Corollaries 55.8 and 55.11.

Now, let P be a polynomial of degree 5 such that P' admits four distinct zeros $c_1,\ c_2,\ c_3,\ c_4$. If $\Upsilon(P) > 0$, then $\Upsilon(P) \geq 2$. Indeed, suppose $\Upsilon(P) = 1$. We may assume that $P(c_1) = P(c_2) = P(c_3)$ and $P(c_1) \neq P(c_4)$. But then, $P - P(c_1)$ admits three zeros of order 2, a contradiction with $\deg(P) = 5$.

Similarly, if $\Upsilon(P) = 0$, then up to a good indexation we have $P(c_1) = P(c_2)$ and $P(c_3) = P(c_4)$.

Theorem 55.12. *Let $h \in \mathcal{M}(\mathbb{K}) \backslash (\mathbb{K} \cup (\mathbb{K}(x) \backslash \mathbb{K}[x]))$, let $f,\ g \in \mathcal{M}_u(d(a, R^-))$, and let $(a_m)_{m \in \mathbb{N}}$ be a bounded sequence of $d(a, R^-) \backslash (\mathcal{S}(f,h) \cup \mathcal{S}(g,h))$ admitting no cluster point in $\mathcal{P}(f) \cup \mathcal{P}(g)$, satisfying further $h \circ f(a_m) = h \circ g(a_m)\ \forall m \in \mathbb{N}$. Then, $\mathcal{S}(f,h) = \mathcal{S}(g,h)$ and $h \circ f(x) = h \circ g(x)\ \forall x \in d(a, R^-) \backslash \mathcal{S}(f,h)$. Moreover, if $\Upsilon(h) \geq 4$, then $f = g$.*

Corollary 55.13. *Let $h \in \mathcal{A}(\mathbb{K})$ satisfy $\Upsilon(h) \geq 4$. Then h is a function of uniqueness for $\mathcal{M}_u(d(a, R^-))$.*

Proof. *of Theorems* 55.2, 55.7, 55.10, *and* 55.12. By Proposition 54.10, we have $h \circ f(x) = h \circ g(x)$ $\forall x \in \mathbb{K} \setminus (\mathcal{S}(f,h) \cup \mathcal{S}(g,h))$ in Theorems 55.2 and 55.7 and $h \circ f(x) = h \circ g(x)$ $\forall x \in d(a, R^-) \setminus (\mathcal{S}(f,h) \cup \mathcal{S}(g,h))$ in Theorems 55.10 and 55.12. Suppose that f and g are not identical.

Then by Lemma 55.1, we have $\mathcal{P}(f) = \mathcal{P}(g)$ and $\mathcal{S}(f,h) = \mathcal{S}(g,h)$. Without loss of generality, we can obviously assume that $a = 0$ in Theorems 55.10 and 55.12. Since $c_1, \ldots, c_k$ lie in $\mathcal{C}(h')$, clearly by applying Theorem 45.7, we obtain in Theorems 55.2 and 55.10

(1)

$$(k-1)T(r,f) \leq \sum_{j=1}^{k} \overline{Z}(r, f - c_j) + \overline{N}(r,f) - Z(r, f' \mid f(x) \notin \mathcal{C}(h'))$$
$$- \log r + O(1) \ (r > 0),$$

(2)

$$(k-1)T(r,g) \leq \sum_{j=1}^{k} \overline{Z}(r, g - c_j) + \overline{N}(r,g) - Z(r, g' \mid g(x) \notin \mathcal{C}(h'))$$
$$- \log r + O(1) \ (r > 0).$$

In Theorems 55.7 and 55.12, we have

(1bis)

$$(k-1)T(r,f) \leq \sum_{j=1}^{k} \overline{Z}(r, f - c_j) + \overline{N}(r,f) - Z(r, f' \mid f(x) \notin \mathcal{C}(h'))$$
$$+ O(1) \ (r \in]0, R[),$$

(2bis)

$$(k-1)T(r,g) \leq \sum_{j=1}^{k} \overline{Z}(r, g - c_j) + \overline{N}(r,g) - Z(r, g' \mid g(x) \notin \mathcal{C}(h'))$$
$$+ O(1) \ (r \in]0, R[).$$

Now, let $\phi = \frac{1}{f} - \frac{1}{g}$, and for each $j = 1, \ldots, k$, let $q_j = \omega_{c_j}(h')$. By Lemma 55.1, in Theorems 55.2 and 55.10, we obtain

$$(3) \qquad (k-1)T(r,f) \leq \overline{Z}(r,\phi) + \sum_{j=1}^{k} \frac{1}{q_j} Z(r, g' \mid f(x) = c_j,$$
$$g(x) \notin \mathcal{C}(h')) - Z(r, f' \mid f(x) \notin \mathcal{C}(h')) - \log r + O(1),$$

and similarly,

$$(k-1)T(r,g) \leq \overline{Z}(r,\phi) + \sum_{j=1}^{k} \frac{1}{q_j} Z(r, f' \mid g(x) = c_j,$$
$$f(x) \notin \mathcal{C}(h')) - Z(r, g' \mid g(x) \notin \mathcal{C}(h')) - \log r + O(1).$$

In Theorems 55.7 and 55.12, we obtain

$$(3\text{bis}) \qquad (k-1)T(r,f) \leq \overline{Z}(r,\phi) + \sum_{j=1}^{k} \frac{1}{q_j} Z(r, g' \mid f(x) = c_j,$$
$$g(x) \notin \mathcal{C}(h')) - Z(r, f' \mid f(x) \notin \mathcal{C}(h')) + O(1),$$
$$(4\text{bis}) \qquad (k-1)T(r,g) \leq \overline{Z}(r,\phi) + \sum_{j=1}^{k} \frac{1}{q_j} Z(r, f' \mid g(x) = c_j,$$
$$f(x) \notin \mathcal{C}(h')) - Z(r, g' \mid g(x) \notin \mathcal{C}(h')) + O(1).$$

By adding in each case the two inequalities we have respectively obtained, in Theorems 55.2 and 55.10, by (3) and (4), we obtain

(5)
$$(k-1)(T(r,f) + T(r,g)) \leq 2\overline{Z}(r,\phi) + \sum_{j=1}^{k} \frac{1}{q_j} [Z(r, f' \mid g(x) = c_j,$$
$$f(x) \notin \mathcal{C}(h')) + Z(r, g' \mid f(x) = c_j,\ g(x) \notin \mathcal{C}(h'))]$$
$$- Z(r, f' \mid f(x) \notin \mathcal{C}(h'))$$
$$- Z(r, g' \mid g(x) \notin \mathcal{C}(h'))$$
$$- 2\log r + O(1),$$

and in Theorems 55.7 and 55.12, by (3bis) and (4bis), we have

(5bis)

$$(k-1)(T(r,f)+T(r,g)) \leq 2\overline{Z}(r,\phi) + \sum_{j=1}^{k} \frac{1}{q_j}[Z(r,f' \mid g(x)=c_j,$$

$$f(x) \notin \mathcal{C}(h')) + Z(r,g' \mid f(x)=c_j,\ g(x) \notin \mathcal{C}(h'))]$$
$$- Z(r,f' \mid f(x) \notin \mathcal{C}(h'))$$
$$-Z(r,g' \mid g(x) \notin \mathcal{C}(h')) + O(1).$$

Now, in each inequality, we note that in the left-side member we have the term $\sum_{j=1}^{k} \frac{1}{q_j}[Z(r,f' \mid g(x) = c_j,\ f(x) \notin \mathcal{C}(h'))] - Z(r,f' \mid f(x) \notin \mathcal{C}(h'))$ which is clearly inferior or equal to zero and similarly $\sum_{j=1}^{k} \frac{1}{q_j}[Z(r,g' \mid f(x) = c_j,\ g(x) \notin \mathcal{C}(h')) - Z(r,g' \mid g(x) \notin \mathcal{C}(h')) \leq 0$. Consequently, in Theorems 55.2 and 55.10, we obtain

(6) $(k-1)(T(r,f)+T(r,g)) \leq 2\overline{Z}(r,\phi) - 2\log r + O(1)$, and

in Theorems 55.7 and 55.12, we have

(6bis) $(k-1)(T(r,f)+T(r,g)) \leq 2\overline{Z}(r,\phi) + O(1)$.

Now, by Theorem 43.3, in Theorems 55.10 and 55.12, we have $Z(r,\phi) \leq T(r,f) + T(r,g) + O(1)$. Consequently, in Theorem 55.10, we have $k \leq 2$, and in Theorem 55.12, we have $k \leq 3$.

Now, assume the hypotheses of Theorems 55.2 and 55.7. By Lemma 55.1, we can replace $Z(r,\phi)$ by $Z(r,f-g)$. Next, by Lemma 54.8, $T(r,f) - T(r,g)$ is bounded, and hence, by Theorem 43.8, we derive $T(r,f-g) \leq T(r,f) + O(1) = T(r,g) + O(1)$. Consequently, in place of (6), in Theorem 55.2, we can obtain

(7) $(k-1)(T(r,f)+T(r,g)) \leq 2\overline{Z}(r,f-g)$

$$-2\log r + O(1) \leq T(r,f) + T(r,g) - 2\log r + O(1),$$

and in place of (6bis), in Theorem 55.7, we have

(7bis) $(k-1)(T(r,f)+T(r,g)) \leq 2\overline{Z}(r,f-g)$

$$-2\log r + O(1) \leq T(r,f) + T(r,g) + O(1).$$

Thus, we can conclude that $k \leq 1$ in Theorem 55.2 and $k \leq 2$ in Theorem 55.7. □

Example. Let Ω be an algebraic closure of $\mathbb{Q}$. Let $R \in \mathbb{Q}(x)$, let $\mathcal{C}(R') = \{c_1, \dots, c_q\}$, and assume that $R(c_j) \neq R(c_n)$ $\forall j = 1, \dots, k$, $\forall n \leq q$. Let $h(x) = e^{R(x)}$. Then, $\mathcal{C}(h') = \{c_1, \dots, c_q\}$ and we check that $h(c_j) \neq h(c_n)$ $\forall j = 1, \dots, k$, $\forall n \leq q$.

Indeed, suppose that $h(c_j) = h(c_l)$ with $j \neq l$ and $j \leq k$. Then, $R(c_j) - R(c_l)$ is of the form $2im\pi$ with $m \in \mathbb{Z}$, which is impossible because $R(c_j) - R(c_l)$ lies in Ω.

For instance, let α be a zero of the polynomial $D(x) = \frac{x^4}{42} - \frac{x^2}{4} + \frac{2}{3}$ and let $P(x) = x^7 - 7x^5 + \dfrac{28}{3}x^3 - \dfrac{7\alpha}{6}x^6 + \dfrac{35\alpha}{4}x^4 - 14\alpha x^2 + A$ (with $A \in \Omega$).

We check that $\alpha \notin \mathbb{Q}$ (because it is a square root of a zero of $\frac{u^2}{42} - \frac{u}{4} + \frac{2}{3}$). Then, $P'(x) = 7x^6 - 35x^4 + 28x^2 + 7\alpha x^5 + 35\alpha x^3 - 28\alpha x$ admits six distinct zeros: $c_1 = 1$, $c_2 = -1$, $c_3 = 2$, $c_4 = -2\, c_5 = 0$, $c_6 = \alpha$. We note that $P(c_5) = P(c_6) = A$. Next, for all $j, l, (1 \leq j < l \leq 6)$, $P(c_j) - P(c_l)$ is of the form $s + t\alpha$ with $s,\ t \in \mathbb{Q}$ and $t \neq 0$, except if $j = 5$ and $l = 6$. Consequently, $P(c_j) \neq P(c_l)$ for all $j, l, 1 \leq j < l \leq 6$ such that $j < 5$. Therefore, P (playing the role of h) satisfies the hypothesis of Theorem 55.12. However, P does not satisfy Hypothesis (F).

Theorem 55.14. *Let $P \in \mathbb{K}[x]$ (respectively $P \in \mathbb{E}[x]$) be of degree 4. Then P is not a polynomial of uniqueness for $\mathcal{M}(\mathbb{K})$ (respectively for $\mathcal{A}(\mathbb{K})$, respectively $\mathbb{E}(x)$, respectively for $\mathbb{E}[x]$) if and only if the set of zeros of P' is of the form $\{a - l, a, a + l\}$.*

Proof. Let S be the set of zeros of P'. Suppose first that P is not a polynomial of uniqueness for $\mathcal{M}(\mathbb{K})$ (respectively for $\mathbb{E}(x)$). Then, by Corollary 55.11, P we have $\Upsilon(P) < 3$, hence P' admits two zeros a, b such that $P(a) = P(b)$. Through an affine change of variable, we can assume that $a = 1$, $b = -1$ and we can also assume that P is monic and satisfies $P(-1) = P(1) = 0$. Consequently, P admits 1 and -1 as zeros of order 2 and hence $P(x) = (x^2 - 1)^2$. Then the zeros of P' are -1, 0, 1. Thus, due to the affine change of variable we made, in general, S is of the form $\{a - l, a, a + l\}$. Conversely, if S is of the form $\{a - l, a, a + l\}$, then through a similar change of variable we can assume $S = \{-1,\ 0,\ 1\}$, so $P(x) = (x^2 - 1)^2 + C$ with $C \in \mathbb{K}$ and hence $P(f) = P(-f))\ \forall f \in \mathcal{M}(\mathbb{K})$ (respectively $\forall f \in E(x))$. □

Lemma 55.15 is useful for the following theorem. It is an immediate consequence of properties of zeros of power series (see Corollaries 22.3 and 22.34).

Lemma 55.15. *Let* $h(x) = \sum_{n=0}^{\infty} a_n x^n \in \mathcal{A}(\mathbb{K})$ *satisfy* $\left|\frac{a_n}{a_{n+1}}\right| < \left|\frac{a_{n+1}}{a_{n+2}}\right|\ \forall n \geq t$ *and* $|a_t|\left|\frac{a_t}{a_{t+1}}\right|^t \geq |a_n|\left|\frac{a_t}{a_{t+1}}\right|^n\ \forall n < t$. *Then* h *admits* t *zeros in* $d(0, \left|\frac{a_t}{a_{t+1}}\right|)$ *(taking multiplicities into account), admits a unique zero of order 1 in each circle* $C(0, \left|\frac{a_m}{a_{m+1}}\right|)$ *for each* $m > t$*, and admits no other zero in* $\mathbb{K}$.

Theorem 55.16. (1) *Let* $(b_n)_{n\in\mathbb{N}}$ *be a sequence of* $\mathbb{K}$ *satisfying* $b_0 = b_1 = 0,\ b_2 = 1,\ |2b_4| < |9(b_3)^2|,\ |3b_3b_5| < |4(b_4)^2|,\ |4b_4| \geq |5b_5|\left|\frac{b_4}{b_5}\right|,\ |4b_4| > |nb_n|\left|\frac{b_4}{b_5}\right|^{n-4}\ \forall n > 5$ *and such that the sequence* $\left|\frac{b_n}{b_{n+1}}\right|_{n\geq 2}$ *is strictly increasing of limit* $+\infty$. *Let* $h(x) = \sum_{n=0}^{\infty} b_n x^n$ *and let* $(c_n)_{n\in\mathbb{N}^*}$ *be the sequence of zeros of* h' *ordered in such a way that* $|c_n| \leq |c_{n+1}|$. *Then* h *belongs to* $\mathcal{A}(\mathbb{K})$ *and satisfies* $h(c_i) \neq h(c_n)\ \forall i = 1,\ 2,\ 3\ \forall n \neq i$.

(2) *Let* $(b_n)_{n\in\mathbb{N}}$ *be a sequence of* $\mathbb{K}$ *satisfying* $b_0 = b_1 = 0,\ b_2 = 1,\ |2b_4| < |9(b_3)^2|,\ |3b_3b_5| < |4(b_4)^2|,\ |4b_4| \geq |5b_5|\left|\frac{b_4}{b_5}\right|,\ |4b_4b_6| < |5(b_5)^2|,\ |5b_5| \geq |6b_6|\left|\frac{b_5}{b_6}\right|,\ |5b_5| > |nb_n|\left|\frac{b_5}{b_6}\right|^{n-5}\ \forall n > 6$ *and be such that the sequence* $\left|\frac{b_n}{b_{n+1}}\right|_{n\geq 2}$ *is strictly increasing of limit* $+\infty$. *Let* $h(x) = \sum_{n=0}^{\infty} b_n x^n$ *and let* $(c_n)_{n\in\mathbb{N}^*}$ *be the sequence of zeros of* h' *ordered in such a way that* $|c_n| \leq |c_{n+1}|$. *Then* h *belongs to* $\mathcal{A}(d(0, R^-))$ *and satisfies* $h(c_i) \neq h(c_n)\ \forall i = 1,\ 2,\ 3, \forall n \neq i$.

Proof. By construction, we can see that h has a zero of order 2 at 0. For each $n \geq 2$, we set $r_n = \left|\frac{b_n}{b_{n+1}}\right|$. By hypothesis, the sequence $\left(\left|\frac{b_n}{b_{n+1}}\right|\right)_{n\geq 2}$ is strictly increasing. Hence, by Lemma 55.15, h has a unique zero b_n in the circle $C(0, r_n)$ and this zero is of order 1. This is true for each $n \geq 1$ and h does not admit any other zero in $\mathbb{K}$.

Now, let $\lambda \in \mathbb{K}$ and $r > 0$ be such that $|\lambda| < |h|(r)$. We know that $\Psi(h-\lambda, \mu) = \Psi(h, \mu)\ \forall \mu \geq \log r$ and $\nu^+(h-\lambda, \mu) = \nu^+(h, \mu),\ \nu^-(h-\lambda, \mu) = \nu^-(h, \mu)\ \forall \mu \geq \log r$, hence $h - \lambda$ admits a unique zero in

$C(0, r_n)$ for each n such that $r_n \geq r$, this zero being of order 1. And then $h - \lambda$ does not admit any other zero in $\mathbb{K} \setminus d(0, r^-)$.

Next, since $|2b_4| < |9(b_3)^2|$, we see that $\left|\frac{2}{3b_3}\right| < \left|\frac{b_3}{4b_4}\right| r_2$, hence $|c_2| < r_3$. Similarly, since $|3b_3b_5| < |4(b_4)^2|$, we see that $\left|\frac{3b_3}{4b_4}\right| < \left|\frac{4b_4}{5b_5}\right| r_4$, hence $|c_3| < r_4$.

Now, suppose that there exist $m \in \mathbb{N}$ and $i \in \{1, 2, 3\}$ such that $h(c_m) = h(c_i)$ with $m \neq i$. We first note that $h(c_m) \neq 0$ $\forall m \neq 1$ because if $h(c_m) = 0$, then c_m is a zero of order 2 of h, hence $c_m = 0$. Thus, we have $m \geq 3$. Suppose $c_m \geq r_3$. Since $|c_2| < r_3$, we have $|h(c_2)| < |h|_{r_3}$, hence as it was seen, $h - h(c_2)$ admits a unique zero of order 1 in each circle $C(0, r_n)$ for each $n \geq 3$ and has no other zero in $\mathbb{K} \setminus d(0, r_3^-)$, as does h. But if $|c_m| \geq r_3$, then it is a zero of order 2 for $h - h(c_m)$, a contradiction showing that $|c_m| < r_3$. Let $\rho = \max(|c_2|, |c_m|)$. So, $\rho < r_3$ and $h - h(c_2)$ must admit at least two multiple zeros in $d(0, \rho)$. But since $|b_2| = r_3$, we know that $\nu^+(h, \mu) = \nu^-(h, \mu) = 3$ $\forall \mu \in]-\log(r_3), -\log(r_2)[$. On the other hand, when $\mu < -\log(\rho)$, we have seen that $\nu^+(h, \mu) = \nu^+(h - h(c_3), \mu)$, $\nu^-(h, \mu) = \nu^-(h - h(c_3), \mu)$, hence $\nu^+(h - h(c_3), \mu) = 3$ $\forall \mu \in]-\log(r_3), -\log(\rho)[$. Consequently, $h - h(c_3)$ admits at most three zeros in $d(0, \rho)$, taking multiplicities into account. This is a contradiction to the assumption $c_m \in d(0, r_3^-)$ and finishes showing that $h(c_m) \neq h(c_2)$ $\forall \neq 2$. Thus, we have shown that $h(c_m) \neq h(c_j)$ for $j = 1,\ 2$ and $m \neq j$. We now suppose that there exists $m \neq 3$ such that $h(c_m) = h(c_3)$. Clearly, as for $h(c_2)$, we have $h(c_3) \neq 0$, and by the above, $h(c_3) \neq h(c_2)$, hence $m > 3$. Since $|c_3| < r_4$, we have $|h(c_3)| < |h|_{r_4}$, hence $\nu^+(h - h(c_2), \mu) - \nu^-(h - h(c_2), \mu) \leq 1$ $\forall \mu \leq -\log r_4$, which shows that $c_m \in d(0, r_4^-)$. Thus, in $d(0, r_4^-)$, h' admits at least 4 zeros: c_1, c_2, c_3, c_m. But by the hypothesis $|4b_4| \geq |5b_5| \left|\frac{b_4}{b_5}\right|$ and $|4b_4| > |nb_n| \left|\frac{b_4}{b_5}\right|^{n-4}$ $\forall n > 5$, we can see that $|4b_4| r_4^3 \geq |5b_5| r_4$ and $|4b_4| r_4^3 > |nb_n| r_4^{n-1}$ $\forall n > 5$. Therefore, we have $\nu^+(h', -\log r_4) \leq 4$ and $\nu^+(h', -\log r) \leq 3$ $\forall r < r_4$. Consequently, h' admits at most three zeros in $d(0, r_4^-)$, a contradiction to the existence of a $c_m \in d(0, r_4^-)$ such that $h(c_m) = h(c_3)$ with $m \geq 4$. Thus, the first conclusion is now established.

We now assume further that $|5b_5| \geq |6b_6| \left|\frac{b_4}{b_5}\right|$, $|4b_4b_6| < |5(b_5)^2|$ and that $|5b_5| > |nb_n| \left|\frac{b_4}{b_5}\right|^{n-5}$ $\forall n > 6$. Suppose that there exists $m \neq 4$ such that $h(c_m) = h(c_4)$. By what precedes, we have $m > 4$.

Thanks to the hypothesis $|4b_4b_6| < |5(b_5)^2|$, we see that $|c_4| < r_5$. Consequently, $|h(c_m)| < |h|(r_5)$, hence c_m lies in $d(0, r_5^-)$. Therefore, $d(0, r_5^-)$ contains 5 zeros of h': c_1, c_2, c_3, c_4, c_m. But thanks to the hypotheses $|5b_5| \geq |6b_6||\frac{b_5}{b_6}|$, $|5b_5| > |nb_n||\frac{b_5}{b_6}|^{n-5}$ $\forall n > 6$, we can see that $\nu^+(h - h(c_4), -\log(r)) - \nu^-(h, -\log(r)) \leq 4$ $\forall r < r_5$, hence $h - h(c_4)$ admits at most 4 zeros in $d(0, r_5^-)$, a contradiction to the assumption $h(c_4) = h(c_m)$ for some $m \neq 4$. This ends the proof of Theorem 55.16. $\square$

Chapter 56

Urscm and Ursim

Throughout this chapter, $\mathbb{K}$ *is of characteristic* 0.

We introduce urscm and ursim for p-adic meromorphic functions. Many studies were made in the eighties and the nineties concerning functions in $\mathbb{C}$, [7, 59, 77, 78]. Studies were also made in the non-Archimedean context since the late nineties [3, 4, 21, 22, 27].

Here, we only consider the situation in an ultrametric field.

Definition and notation. Throughout this chapter, $\mathbb{E}$ is an algebraically closed field of characteristic 0 without any assumption on the existence of an absolute value and we denote by $\mathbb{P}^1(\mathbb{E})$ the projective space of dimension 1 over $\mathbb{E}$ and $\mathbb{K}$ is also of characteristic 0.

Given a family of functions $\mathcal{F}$ defined in $\mathbb{K}$ or in a subset of $\mathbb{K}$ (respectively in $\mathbb{E}$ or in a subset of $\mathbb{E}$), with values in $\mathbb{P}^1(\mathbb{K})$ (respectively in $\mathbb{P}^1(\mathbb{E})$), a subset S of $\mathbb{K}$ (respectively of $\mathbb{E}$) is called an *ursim for* $\mathcal{F}$ if for any two non-constant functions $f, g \in \mathcal{F}$ satisfying $f^{-1}(S) = g^{-1}(S)$, these functions are equal. This definition particularly applies to $\mathcal{A}(\mathbb{K})$, $\mathcal{M}(\mathbb{K})$, $\mathcal{A}_u(d(a, R^-))$, $\mathcal{M}_u(d(a, R^-))$, $\mathbb{K}[x]$, $\mathbb{K}(x)$, $\mathbb{E}[x]$, $\mathbb{E}(x)$.

We now define URSCM. Given a subset S of $\mathbb{E}$ and $f \in \mathbb{E}(x)$, we denote by $\mathcal{E}(f, S)$ the set in $\mathbb{E} \times \mathbb{N}^*$: $\bigcup_{a \in S}\{(z, q) \in E \times \mathbb{N}^* \mid z$ a zero of order q of $f(x) - a\}$.

And given a subset S of $\mathbb{P}^1(\mathbb{E})$ containing $\{\infty\}$ and $f \in \mathbb{E}(x)$, we denote by $\mathcal{E}(f, S)$ the subset of $E \times \mathbb{N}^*$: $\mathcal{E}(f, S \cap \mathbb{E}) \cup \{(z, q) \mid z$ a pole of order q of $f\}$.

Consider now meromorphic functions in the field $\mathbb{K}$. For a subset S of $\mathbb{K}$ and $f \in \mathcal{M}(\mathbb{K})$ (respectively $f \in \mathcal{M}(d(a, R^-)))$, we denote by $\mathcal{E}(f, S)$ the set in $\mathbb{K} \times \mathbb{N}^*$: $\bigcup_{a \in S}\{(z, q) \in \mathbb{K}\times\mathbb{N}^* \mid z\text{a zero of order } q \text{ of } f(x) - a\}$.

And given a subset of $\mathbb{P}^1(\mathbb{K})$ containing $\{\infty\}$ and $f \in \mathcal{M}(\mathbb{K})$ (respectively $f \in \mathcal{M}(d(a, R^-)))$, we denote by $\mathcal{E}(f, S)$ the subset of $\mathbb{K}\times\mathbb{N}^*$: $\mathcal{E}(f, S \cap \mathbb{K}) \cup \{(z, q) \mid z \text{ a pole of order } q \text{ of } f\}$.

Similarly, given a subset of $\mathbb{P}^1(\mathbb{E})$ containing $\{\infty\}$ and $f \in \mathbb{E}(x)$, we denote by $\mathcal{E}(f, S)$ the subset of $\mathbb{E}\times\mathbb{N}^*$: $\mathcal{E}(f, S \cap \mathbb{E}) \cup \{(z, q) \mid z \text{ a pole of order } q \text{ of } f\}$.

Let $\mathcal{F}$ be a non-empty subset of $\mathcal{M}(\mathbb{K})$ (respectively of $\mathcal{M}(d(a, R^-)))$. We say that two non-constant functions $f, g \in \mathcal{F}$ *share* S*, counting multiplicity* if $\mathcal{E}(f, S) = \mathcal{E}(g, S)$. And the set S is called *a unique range set counting multiplicity* (an *urscm* in brief) *for* $\mathcal{F}$ if for any two non-constant $f, g \in \mathcal{F}$ sharing S counting multiplicity, one has $f = g$.

Similarly, let $\mathcal{F}$ be a non-empty subset of $\mathbb{E}(x)$. We say that two non-constant functions $f, g \in \mathcal{F}$ *share* S*, counting multiplicity* if $\mathcal{E}(f, S) = \mathcal{E}(g, S)$. And the set S is called *a unique range set counting multiplicity* (an *urscm* in brief) *for* $\mathcal{F}$ if for any two non-constants $f, g \in \mathcal{F}$ sharing S counting multiplicity, one has $f = g$.

In other words, if we consider a set $S = \{a_1, \dots, a_t\} \subset \mathbb{K}$ with $a_i \neq a_j\ \forall i \neq j$, we can set $P(X) = \prod_{j=1}^{t}(X - a_j)$ and then the set $S = \{a_1, \dots, a_t\}$ is an urscm for $\mathcal{F}$ if for any two functions $f,\ g \in \mathcal{F}$ such that $P \circ f$ and $P \circ g$ have the same zeros with the same multiplicity, then $f = g$.

An immediate lemma appears:

Lemma 56.1. *If a subset of* $\mathbb{K}$ *(respectively of* $\mathbb{E}$*) is an ursim for a family* $\mathcal{F}$ *of functions, it is an urscm for* $\mathcal{F}$.

Proof. Indeed, suppose that a subset S of $\mathbb{K}$ (respectively of $\mathbb{E}$) is an ursim for a family $\mathcal{F}$ of functions and consider two functions $f,\ g \in \mathcal{F}$ such that $\mathcal{E}(f, S) = \mathcal{E}(g, S)$. So much the more, we have $f^{-1}(S) = g^{-1}(S)$, hence $f = g$. □

Similar definitions were given for meromorphic functions on $\mathbb{C}$ before these questions were examined on the field $\mathbb{K}$. Urscm of only 11 points for complex meromorphic functions in the whole field $\mathbb{C}$ where found in [59] and the same method showed the existence of urscm of only 7 points for complex entire functions. So far, they are the smallest known in $\mathbb{C}$.

In the field $\mathbb{K}$, the same method lets us find urscm of 11 points for $\mathcal{M}_u(d(a, R^-))$ and urscm of 10 points for $\mathcal{M}(\mathbb{K})$.

Concerning analytic functions, urscm of 3 points were found for entire functions and polynomials [22] and urscm of 7 points were found for unbounded analytic functions in a disk $d(a, R^-)$.

Actually, urscm for polynomials in a the field $\mathbb{E}$ where characterized as the affinely rigid sets of $\mathbb{E}$. And urscm of 3 points for entire functions in $\mathbb{K}$ where also characterized as the affinely rigid sets of 3 points in $\mathbb{K}$. That led to the conjecture: *URSCM for entire functions are the affinely rigid sets in* $\mathbb{K}$ [22]. The proof is given at the end of this chapter.

A first basic proposition is useful to understand the role of URSCM:

Proposition 56.2. *Let* $S = \{a_1, \dots, a_n\} \subset \mathbb{K}$ *(respectively* $S = \{a_1, \dots, a_n\} \subset \mathbb{E}$*), let* $a \in \mathbb{K}$*, let* $R \in \mathbb{R}_+^*$*, and let* $P(x) = \prod_{i=1}^n (x - a_i)$*. Then given any two functions* $f,\ g \in \mathcal{A}(\mathbb{K})$ *(respectively* $f,\ g \in \mathcal{A}(d(a, R^-))$*, respectively in* $\mathbb{E}[x]$*), then* $\mathcal{E}(f, S) = \mathcal{E}(g, S)$ *if and only if* $\frac{P(f)}{P(g)}$ *is a constant in* $\mathbb{K}^*$ *(respectively is an invertible function in* $\mathcal{A}(d(a, R^-))$*, respectively is a constant in* $\mathbb{E}^*$*).*

Proof. Indeed, given $f,\ g \in \mathcal{A}(\mathbb{K})$ (respectively $f, g \in \mathcal{A}(d(a, R^-))$, respectively $f, g \in \mathbb{E}[x]$), for each $\alpha \in \mathbb{K}$ (respectively $\alpha \in d(a, R^-)$, respectively $\alpha \in \mathbb{E}$), f and g share S if and only if we have $\omega_\alpha(f - f(\alpha)) = \omega_\alpha(g - g(\alpha))$, which means that $\frac{P(f)}{P(g)}$ has no zero and no pole. Consequently, $\frac{P(f)}{P(g)}$ is a constant in $\mathcal{A}(\mathbb{K})$ (respectively is an invertible element in $\mathcal{A}(d(a, R^-))$, respectively is a constant in $\mathbb{E}$). □

Corollary 56.3. *Let* $S = \{a_1, \dots, a_n\} \subset \mathbb{K}$ *(respectively let* $S = \{a_1, \dots, a_n\} \subset \mathbb{E}$*) with* $a_i \neq a_j\ \forall i \neq j$ *and let* $P(x) = \prod_{i=1}^n (x - a_i)$*. Then* P *is a polynomial of strong uniqueness for* $\mathcal{A}(\mathbb{K})$ *(respectively for* $\mathbb{E}[x]$*) if and only if* $S = \{a_1, \dots, a_n\}$ *is an URSCM for* $\mathcal{A}(\mathbb{K})$ *(respectively for* $\mathbb{E}[x]$*).*

Remark. Let $P(x) = x^4 - 4x^3$ and let j be a primitive third root of 1. Clearly, $\mathcal{C}(P') = \{1,\ j,\ j^2\}$ and $\Upsilon(P) = 3$, hence by Corollary 55.3,

P is a polynomial of uniqueness for $\mathcal{A}(\mathbb{K})$. But $P(jf) = jP(f)\ \forall f \in \mathcal{A}(\mathbb{K})$, hence P is not a polynomial of strong uniqueness for $\mathcal{A}(\mathbb{K})$ (or for $\mathbb{E}[x]$).

The equivalence for a subset $S = \{a_1, \dots, a_t\}$ to be an urscm and to define a polynomial of strong uniqueness P for $\mathcal{A}(\mathbb{K})$ does not expand to $\mathcal{M}(\mathbb{K})$ because given $f,\ g \in \mathcal{M}(\mathbb{K})$ such that $P(f)$ and $P(g)$ have the same zeros, that does not imply that they have the same poles. However, we can note the following lemma:

Lemma 56.4. *Let $P(x) = \prod_{j=1}^{t}(x - a_j)$ with $a_i \neq a_j\ \forall i \neq j$. If $\{a_1, \dots, a_t\}$ is an urscm for $\mathcal{M}(\mathbb{K})$, then P is a polynomial of strong uniqueness for $\mathcal{M}(\mathbb{K})$.*

Proof. Indeed, let $f,\ g \in \mathcal{M}(\mathbb{K})$ be such that $P \circ f = \lambda P \circ g$ with $\lambda \neq 0$. Then $P \circ f$ and $P \circ g$ obviously have the same zeros. But since $\{a_1, \dots, a_t\}$ is an urscm for $\mathcal{M}(\mathbb{K})$, we have $f = g$. □

In the proof of the following theorems, we use Lemma 56.5 that comes from [59]. It holds in $\mathcal{M}(\mathbb{K})$ as well as in $\mathcal{M}(d(a, R^-))$.

Lemma 56.5. *Let $F,\ G \in \mathcal{M}(d(a, R^-))$ (respectively $F,\ G \in \mathcal{M}(\mathbb{K})$) have the same poles, ignoring multiplicity, and let $H = \frac{F''}{F'} - \frac{G''}{G'}$. Every pole of H has multiplicity order 1. Let α be a pole of F and G. If α has same multiplicity for F and G, then H has no pole at α. Moreover, if α has a multiplicity order 1 for both F and G, then α is a zero of H.*

Proof. Without loss of generality, we can assume $\alpha = 0$. Since both $\frac{F''}{F'}$, $\frac{G''}{G'}$ are logarithmic derivative, all their poles are of order one. Therefore, all poles of H are of order one.

Now, suppose α is a pole of order q for both $\frac{F''}{F'}$ and $\frac{G''}{G'}$. Consider the Laurent series of F and G at 0: $F(x) = \frac{a_{-q}}{x^q} + \cdots + \frac{a_{-1}}{x} + a_0 + \sum_{n=1}^{\infty} a_n x^n$ and $G(x) = \frac{b_{-q}}{x^q} + \cdots + \frac{b_{-1}}{x} + b_0 + \sum_{n=1}^{\infty} b_n x^n$.

We can check that both $\frac{F''}{F'}$, $\frac{G''}{G'}$ have a residue at 0 equal to $\frac{-1}{q+1}$. Consequently, the residue of $\frac{F''}{F'} - \frac{G''}{G'}$ is null and hence $\frac{F''}{F'} - \frac{G''}{G'}$ has no pole at 0.

Suppose now that $q = 1$. Then $\frac{F''}{F'}$ is of the form

$$\left(\frac{1}{x}\right)\left(\frac{2a_{-1} + 2a_2x^2 + x^3\phi(x)}{-a_{-1} + x\theta(x)}\right)$$

with ϕ, θ meromorphic functions having no pole at 0. Similarly, $\frac{G''}{G'}$ is of the form

$$\left(\frac{1}{x}\right)\left(\frac{2b_{-1} + 2b_2x^2 + x^3\upsilon(x)}{-b_{-1} + x\eta(x)}\right)$$

with υ, η meromorphic functions having no pole at 0. Consequently, we can check that

$$H(x) = \left(\frac{1}{x}\right)\left(\frac{2a_2x^2 + x^3\phi(x)}{-a_{-1} + x\theta(x)} - \frac{2b_2x^2 + x^3\upsilon(x)}{-b_{-1} + x\eta(x)}\right),$$

and hence $H(0) = 0$. □

In the proof of the following theorems, we also need the following lemma:

Lemma 56.6. *Let $f,\ g \in \mathcal{M}(\mathbb{K}))$ (respectively $f,\ g \in \mathcal{M}(d(a, R^-)))$ be two different non-constant functions satisfying $\frac{f''}{f'} = \frac{g''}{g'}$. Then f and g are linked by a relation of the form $f = ag + b$.*

Proof. By Theorem 50.4, the space of solutions of the equation $y'g'' - yg'$ is of dimension at most 1. Consequently, $f,\ g$ satisfy a relation of the form $f' = ag'$ with $a \in \mathbb{K}$, and therefore, $f = ag + b$ with $b \in \mathbb{K}$. □

Notation. Let $n \in \mathbb{N}$ and let $c \in \mathbb{E}$ (respectively $c \in \mathbb{K}$). We denote by $S(n, c)$ the set of zeros of the polynomial $P_{n,c}(x) = (n-1)(n-2)$ $x^n - 2n(n-2)x^{n-1} + n(n-1)x^{n-3} + c$ and let $S(n, c)$ be the set of zeros of $P_{n,c}$ in $\mathbb{E}$ (respectively in $\mathbb{K}$).

Theorem 56.7. *Let $f,\ g \in \mathcal{M}(\mathbb{K})$ be two different non-constant functions satisfying $f^{-1}(S(n, c)) = g^{-1}(S(n, c))$. Then, $n \le 15$. Moreover, if $f,\ g \in \mathcal{A}(\mathbb{K})$, then $n \le 8$.*

Corollary 56.8. *Let $n \ge 16$. Then $S(n, c)$ is an ursim for $\mathcal{M}(\mathbb{K})$.*

Corollary 56.9. *Let $n \ge 9$. Then $S(n, c)$ is an ursim for $\mathcal{A}(\mathbb{K})$.*

Theorem 56.10. *Let* $f \in \mathcal{M}_u(d(a, R^-))$ *and let* $g \in \mathcal{M}(d(a, R^-))$ *be two different non-constant functions satisfying* $f^{-1}(S(n, c)) = g^{-1}(S(n, c))$*. Then,* $n \leq 16$*. Moreover, if* $f \in \mathcal{A}_u(d(a, R^-))$ *and if* $g \in \mathcal{A}(d(a, R^-))$*, then* $n \leq 9$*.*

Corollary 56.11. *Let* $n \geq 17$*. Then* $S(n, c)$ *is an ursim for* $\mathcal{M}_u(d(a, R^-))$*.*

Corollary 56.12. *Let* $n \geq 10$*. Then* $S(n, c)$ *is an ursim for* $\mathcal{A}_u(d(a, R^-))$*.*

Theorem 56.13. *Let* $f,\ g \in \mathcal{M}(\mathbb{K})$ *be two distinct non-constant functions satisfying* $\mathcal{E}(f, S(n, c)) = \mathcal{E}(g, S(n, c))$*. Then* $n \leq 9$*. Moreover, if* $f,\ g \in \mathcal{A}(\mathbb{K})$*, then* $n \leq 5$*.*

Corollary 56.14. *For every* $n \geq 10$*,* $S(n, c)$ *is an urscm for* $\mathcal{M}(\mathbb{K})$*, and for every* $n \geq 6$*,* $S(n, c)$ *is an urscm for* $\mathcal{A}(\mathbb{K})$*.*

Remark. In Theorem 47.6, we saw that $P_{n,0}$ is a polynomial of uniqueness for $\mathcal{M}(\mathbb{K})$ for every $n \geq 2$ and for $\mathcal{M}(d(a, R^-))$ for every $n \geq 3$. Here we can add some precisions:

Corollary 56.15. *For every* $n \geq 10$*,* $P_{n,c}$ *is a polynomial of strong uniqueness for* $\mathcal{M}(\mathbb{K})$*. And for every* $n \geq 6$*,* $P_{n,c}$ *is a polynomial of strong uniqueness for* $\mathcal{A}(\mathbb{K})$*.*

Theorem 56.16. *Let* $f,\ g \in \mathcal{M}_u(d(a, R^-))$ *be two different non-constant functions satisfying* $\mathcal{E}(f, S(n, c)) = \mathcal{E}(g, S(n, c))$*. Then,* $n \leq 10$*. Moreover, if* $f,\ g \in \mathcal{A}_u(d(a, R^-))$*, then* $n \leq 6$*.*

Corollary 56.17. *For every* $n \geq 11$*,* $S(n, c)$ *is an urscm for* $\mathcal{M}_u(d(a, R^-))$*. And for every* $n \geq 7$*,* $S(n, c)$ *is an urscm for* $\mathcal{A}_u(d(a, R^-))$*.*

Corollary 56.18. *For every* $n \geq 10$*,* $P_{n,c}$ *is a polynomial of strong uniqueness for* $\mathcal{M}_u(d(a, R^-))$*. And for every* $n \geq 7$*,* $P_{n,c}$ *is a polynomial of strong uniqueness for* $\mathcal{A}_u(d(a, R^-))$*.*

Proof (of Theorems 56.7, 56.10, 56.13, and 56.16). In order to simplify the notation, we just denote by P the polynomial $P_{n,c}$ and by S the set $S(n, c)$. Without loss of generality, we may assume $a = 0$ in Theorems 56.10 and 56.16.

We then consider two distinct non-constant meromorphic functions $f,\ g \in \mathcal{M}(\mathbb{K})$ in Theorems 56.7 and 56.16 and $f,\ g \in \mathcal{M}(d(0,R^-))$ in Theorems 56.10 and 56.16. And we assume $f^{-1}(S(n,c)) = g^{-1}(S(n,c))$ in Theorems 56.7 and 56.10 and $\mathcal{E}(f,S) = \mathcal{E}(g,S)$ in Theorems 56.13 and 56.16. Without loss of generality, we may assume that all functions we consider have no zero and no pole at 0.

Let $F = \frac{1}{P(f)}$, $G = \frac{1}{P(g)}$, and $H = \frac{F''}{F'} - \frac{G''}{G'}$.

We first suppose that H is identically 0. By Lemma 56.6, it is immediate to derive that there exist $A \in \mathbb{K}^*$ and $B \in \mathbb{K}$ such that $P(f) = \frac{P(g)}{BP(g)+A}$. Therefore, by Theorems 43.8, we have

(1) $T(r,f) = T(r,g) + O(1)\ (r \in]0,R[)$.

We can check that $P(X) + c$ is of the form $bX^{n-2}(X - e_1)(X - e_2) + c$, with $b, e_1, e_2 \in \mathbb{E}$ (respectively $b, e_1, e_2 \in \mathbb{K}$) and $e_1e_2 \neq 0,\ e_1 \neq e_2$.

We have to distinct three cases: (i) $B \neq 0$, (ii) $B = 0,\ A = 1$, (iii) $B = 0,\ A \neq 1$.

(i) Since $AB \neq 0$, every zero of $P(g) + \frac{A}{B}$ is a pole of $P(f)$ and therefore is a zero of order at least n of $P(g) + \frac{A}{B}$. On the other hand, we check that whenever $A,\ B \in \mathbb{K}^*$, the polynomial $P(X) + \frac{A}{B}$ admits at least two distinct zeros b_1 and b_2 of order 1 and therefore it admits another zero l of order at most $n-2$. Consequently, every zero of $g - b_1$ or $g - b_2$ has order at least n and every zero of $g - l$ has order at least 2. By a change of variable, we can obviously assume that $g,\ g - l,\ g - b_1, g - b_2$ have neither any zeros nor any poles at 0. Then, by applying Theorem 45.7, we obtain

$$\begin{aligned}2T(r,g) - \overline{N}(r,g) &\leq \overline{Z}(r,g-l) + \overline{Z}(r,g-b_1) + \overline{Z}(r,g-b_2) + O(1)\\ &\leq \frac{1}{2}Z(r,g-l) + \frac{1}{n}(Z(r,g-b_1) + Z(r,g-b_2))\\ &\quad + O(1)\ (r \in J),\end{aligned}$$

which leads to $n \leq 4$.

(ii) In this case, we have $P(f) = P(g)$. But by Theorem 47.6, P is a polynomial of uniqueness, for $\mathcal{M}_u(d(0,R^-))$, hence $f = g$.

(iii) Let $\lambda = \frac{1}{A}$. We check that at least one of the two polynomials $Q_1(X) = (n-1)(n-2)X^n - 2n(n-2)X^{n-1} + n(n-1)X^{n-2} + c(\lambda - 1)$

and $Q_2(X) = (n-1)(n-2)X^n - 2n(n-2)X^{n-1} + n(n-1)X^{n-2} + c(\frac{1}{\lambda} - 1)$ admits n distinct zeros. Indeed, suppose this not true. We note that $Q_1'(X) = Q_2'(X) = P'(X) = n(n-1)(n-2)X^{n-3}(X-1)^2$. Consequently, if both Q_1 and Q_2 admit at least one zero of order >1, then this must be 1 because it cannot be 0. But then, we have $c(\lambda - 1) = c(\frac{1}{\lambda} - 1) = -2$. Since $c \neq 0$, and since $\lambda \neq 1$, this implies $\lambda = -1$ and $c = 1$, which is excluded by hypothesis.

Now, since λ and $\frac{1}{\lambda}$ play symmetric roles, without loss of generality, we can assume that Q_1 admits n distinct zeros $b_1, \ldots, b_n$. Thus, putting $\gamma(X) = \lambda X^{n-2}(n-1)((n-2)X^2 - 2n(n-2)X + n(n-1))$, we have

(2) $Q_1(f) = \gamma(g)$.

So, applying Theorem 45.7, we obtain

$$(n-1)T(r,f) \leq \sum_{j=1}^{n} \overline{Z}(r, f - b_j) + \overline{N}(r,f) + O(1) \ (r \in J).$$

On the other hand, we have

$$\sum_{j=1}^{n} \overline{Z}(r, f - b_j) \leq \overline{Z}(r, Q_1(f)) = \overline{Z}(r,g) + \overline{Z}(r, g - e_1) + \overline{Z}(r, g - e_2)$$
$$\leq 3T(r,g).$$

But by (2), it is seen that $T(r,g) \leq T(r,f) + O(1)$ $(r \in J)$, hence finally

$$(n-1)T(r,f) \leq 3T(r,f) + \overline{N}(r,f) + O(1) \ (r \in J). \tag{3}$$

Consequently, we have $n \leq 5$. Moreover, if $f,\ g \in \mathcal{A}(d(0, R^-))$, then $N(r,f) = 0$, and then (3) leads to $n \leq 4$. This finishes proving the claims of the theorems when $H = 0$.

Henceforth, we suppose that H is not identically 0 and apply Fujimoto's method [60]. Let α be a zero of $f - a_j$, for some j. Then both $P(f)$ and $P(g)$ vanish at α. So, by Lemma 43.4, we have $\sum_{j=1}^{n} \overline{Z}(r, f - a_j) = \overline{Z}(r, P(f)) = \overline{Z}(r, P(g))$. According to Lemma 56.5, if both $P(f), P(g)$ have a zero of order one, then H also has a zero. Else, at least one of the two functions $P(f)$ and $P(g)$ admits α

as a zero of order strictly greater than 1, and then (since all zeros of P are simple), at least one of the two functions f', g' has a zero at α, while α is a zero $f - a_i$ for some i and a zero of $g - a_j$ for some j. Consequently, we obtain

(3A)
$$\sum_{j=1}^{n} \overline{Z}(r, f - a_j) \leq \overline{Z}(r, H) + \overline{Z}(r, f') + \overline{Z}(r, g') - \overline{Z}_0^S(r, f') - \overline{Z}_0^S(r, g').$$

Moreover, in the hypothesis of Theorems 56.13 and 56.16, since any zero α of $f - a_i$ is a zero of certain $g - a_j$ with the same multiplicity order, we have $\omega_\alpha(f') = \omega_\alpha(g')$. Consequently,we obtain this improvement of (3A):

$$\text{(3B)} \quad \sum_{j=1}^{n} \overline{Z}(r, f - a_j) \leq \overline{Z}(r, H) + \frac{1}{2}\Big[\overline{Z}(r, f') + \overline{Z}(r, g') - \overline{Z}_0^S(r, f') - \overline{Z}_0^S(r, g')\Big].$$

By Corollary 43.6, we know that $\overline{Z}(r, H) \leq N(r, H) + O(1)$ $(r \in J)$, hence

$$\sum_{j=1}^{n} \overline{Z}(r, f - a_j) \leq N(r, H) + \overline{Z}(r, f') + \overline{Z}(r, g') - \overline{Z}_0^S(r, f') - \overline{Z}_0^S(r, g') + O(1) \ (r \in I)$$

(respectively $r \in J$) and similarly

$$\sum_{j=1}^{n} \overline{Z}(r, g - a_j) \leq N(r, H) + \overline{Z}(r, f') + \overline{Z}(r, g') - \overline{Z}_0^S(r, f') - \overline{Z}_0^S(r, g') + O(1) \ (r \in I)$$

(respectively $r \in J$), hence

$$\text{(4A)} \quad \sum_{j=1}^{n} \overline{Z}(r, f - a_j) + \overline{Z}(r, g - a_j) \leq 2N(r, H) + 2[\overline{Z}(r, f') + \overline{Z}(r, g') - \overline{Z}_0^S(r, f') - \overline{Z}_0^S(r, g')].$$

And in the hypothesis of Theorems 56.13 and 56.16, by (3B), we obtain

$$\text{(4B)}\quad \sum_{j=1}^{n} \overline{Z}(r, f - a_j) + \overline{Z}(r, g - a_j) \le 2N(r, H) + [\overline{Z}(r, f') + \overline{Z}(r, g') - \overline{Z}_0^S(r, f') - \overline{Z}_0^S(r, g')].$$

Consider now $N(r, H)$, and let η be a pole of H. Either η is a zero of $F'G'$ or it is a pole of FG. Let φ be the counting function of the poles of H occurring when FG has a pole. So, we have

$$\text{(5)}\qquad N(r, H) = \overline{Z}(r, F'G') + \varphi(r).$$

Suppose first η is a zero of F' but is not a pole of FG. Either η is a zero of f', or it is a zero of $P'(f)$, or it is a pole of f. Consequently, η is not a zero of $P(f)$ and we have $\overline{Z}(r, F') \le \overline{Z}_0^S(r, f') + \overline{Z}(r, P'(f)) + \overline{N}(r, f)$. Since

$$P'(X) = n(n-1)(n-2)(X-1)^2 X^{n-3},$$

we have $\overline{Z}(r, P'(f)) \le 2T(r, f)$, $\overline{Z}(r, P'(g)) \le 2T(r, g)$, and therefore,

$$\text{(6)}\qquad \overline{Z}(r, F') \le \overline{Z}_0^S(r, f') + 2T(r, f) + \overline{N}(r, f).$$

Similarly, if η is a zero of G' but is not a pole of FG, we have

$$\text{(7)}\qquad \overline{Z}(r, G') \le \overline{Z}_0^S(r, g') + 2T(r, g) + \overline{N}(r, g).$$

Suppose now η is a pole of FG. Then η is a zero of $P(f)$ and $P(g)$ (we note that when η is a zero of $P(f)$, it is a zero of $P(g)$, and vice-versa), and then by Lemma 56.5, it may not be a pole of H when it is a zero of the same order of $P(f)$ and $P(g)$. Consequently, in the hypothesis of Theorems 56.7 and 56.10, $\varphi(r)$ satisfies

$$\varphi(r) \le \overline{Z}(r, f') - \overline{Z}_0^S(r, f') + \overline{Z}(r, g') - \overline{Z}_0^S(r, g'),$$

and therefore, by (5), (6), and (7), we obtain

$$N(r, H) \le \overline{Z}_0^S(r, f') + \overline{Z}_0^S(r, g') + 2T(r, f) + 2T(r, g) + \overline{N}(r, f) + \overline{N}(r, g) + \overline{Z}(r, f') - \overline{Z}_0^S(r, f') + \overline{Z}(r, g') - \overline{Z}_0^S(r, g'),$$

hence

$$N(r,H) \leq 2T(r,f) + 2T(r,g) + \overline{N}(r,f) + \overline{N}(r,g)$$
$$+ \overline{Z}(r,f') + \overline{Z}(r,g').$$

Thus, by (4A), in the hypothesis of Theorems 56.7 and 56.10, we obtain

$$\text{(8A)} \qquad \sum_{j=1}^{n} \overline{Z}(r,f-a_j) + \overline{Z}(r,g-a_j)$$
$$\leq 4(T(r,f) + T(r,g)) + 2(\overline{N}(r,f) + 2\overline{N}(r,g))$$
$$+ 4(\overline{Z}(r,f') + \overline{Z}(r,g')) + O(1).$$

Now, in the hypotheses of Theorems 56.13 and 56.16, since the order of a zero is the same for $P(f)$ and $P(g)$, the counting function φ is identically 0, so by, (5) (6), and (7), we have

$$N(r,H) \leq 2T(r,f) + 2T(r,g) + \overline{N}(r,f) + \overline{N}(r,g)$$
$$+ \overline{Z}_0^S(r,f') + \overline{Z}_0^S(r,g'),$$

and therefore, by (4B), we obtain

$$\text{(8B)} \quad \sum_{j=1}^{n} \overline{Z}(r,f-a_j) + \overline{Z}(r,g-a_j)$$
$$\leq 4(T(r,f) + T(r,g)) + 2(\overline{N}(r,f) + 2\overline{N}(r,g))$$
$$+ (\overline{Z}(r,f') + \overline{Z}(r,g')) + (\overline{Z}_0^S(r,f') + \overline{Z}_0^S(r,g')) + O(1).$$

Now, applying Theorem 45.7 to f and g at the points a_j $(1 \leq j \leq n)$, we have

$$(n-1)T(r,f) \leq \overline{N}(r,f) + \sum_{j=1}^{n} \overline{Z}(r,f-a_j) - Z_0^S(r,f')$$
$$+ O(1) - \log r \ (r \in I)$$

(respectively $r \in J$)

$$(n-1)T(r,g) \le \overline{N}(r,g) + \sum_{j=1}^{n} \overline{Z}(r,g-a_j) - Z_0^S(r,g')$$
$$- \log r + O(1), \ (r \in I)$$

(respectively $r \in J$), hence

$$(n-1)(T(r,f) + T(r,g))$$
$$\le \overline{N}(r,f) + \overline{N}(r,g) + \sum_{j=1}^{n} \overline{Z}(r,f-a_j) + \overline{Z}(r,g-a_j) - Z_0^S(r,f')$$
$$- Z_0^S(r,g') - 2\log r + O(1) \ (r \in I)$$

(respectively $r \in J$), hence by (8A), in the hypotheses of Theorems 56.7 and 56.10, we obtain

$$\text{(9A)} \quad (n-1)(T(r,f) + T(r,g))$$
$$\le 3(\overline{N}(r,f) + \overline{N}(r,g)) + 4(T(r,f) + T(r,g)) + 4(\overline{Z}(r,f')$$
$$+ \overline{Z}(r,g')) - 2\log r + O(1) \ (r \in I)$$

(respectively $r \in J$).

By Corollary 15, we can derive

$$(n-1)(T(r,f) + T(r,g))$$
$$\le 3(\overline{N}(r,f) + \overline{N}(r,g)) + 4(T(r,f) + T(r,g)) + 4(Z(r,f) + Z(r,g))$$
$$+ 4(\overline{N}(r,f) + \overline{N}(r,g)) - 10\log r + O(1) \ (r \in I)$$

(respectively $r \in J$).

Consequently, in Theorem 56.7, we have $n \le 15$, and in Theorem 56.10, we have $n \le 16$. Moreover, if f, g belong to $\mathcal{A}(\mathbb{K})$, then in Theorem 56.7, we have $n \le 8$, and if f, g belong to $\mathcal{A}_u(d(0,R^-))$, in Theorem 56.10, we obtain $n \le 9$.

By (8B), in the hypotheses of Theorems 56.13 and 56.16, we obtain

$$\begin{aligned}&(n-1)(T(r,f)+T(r,g))\\&\quad\le 3(\overline{N}(r,f)+\overline{N}(r,g))+4(T(r,f)+T(r,g))\\&\qquad+\overline{Z}(r,f')+\overline{Z}(r,g')-2\log r+O(1)\ (r\in I)\end{aligned}$$

(respectively $r \in J$).

Hence, by Corollary 43.15, we obtain $(n-1)(T(r,f)+T(r,g)) \le 3(\overline{N}(r,f)+\overline{N}(r,g))+4(T(r,f)+T(r,g))+(Z(r,f)+Z(r,g))+\overline{N}(r,f)+\overline{N}(r,g)-4\log r+O(1)$ $(r\in I)$ (resp. $r \in J$). Consequently, $n \le 9$ in Theorem 56.13 and $n \le 10$ in Theorem 56.16. Moreover, if f, g belong to $\mathcal{A}(\mathbb{K})$ or to $\mathcal{A}_u(d(0,R^-))$, then we obtain $(n-1)(T(r,f)+T(r,g)) \le 4(T(r,f)+T(r,g))+(Z(r,f)+Z(r,g))-4\log r+O(1)$ $(r\in J)$. So, if f, g belong to $\mathcal{A}(\mathbb{K})$, we have $n \le 5$, and if $f,\ g \in \mathcal{A}_u(d(0,R^-))$, we have $n \le 6$. □

We now characterize all URSCM for $\mathcal{A}(\mathbb{K})$. First, we have to recall Picard–Berkovich Theorem [9].

Theorem 56.19 ((Berkovich) [9]). *Let $F(x,y)=0$ be the equation of an algebraic curve of genus ≥ 1. There exist no functions $f,\ g \in \mathcal{M}(\mathbb{K})$ such that $F(f(t),g(t))=0$.*

We also recall a classical theorem in algebraic geometry:

Theorem 56.20 ((Picard) [81]). *Let $F(x,y) \in \mathbb{E}[x,y]$ be irreducible. If the algebraic curve of equation $F(x,y)=0$ has genus 0, there exist rational functions $\phi,\ \theta \in \mathbb{K}(t)$ and $R(x,y)\in\mathbb{K}(x,y)$ such that, setting $t = R(x,y)$, we have $F(\phi(t),\theta(t))=0\ \forall t\in\mathbb{E}$.*

We can now explain the method to characterize URSCM for $\mathcal{A}(\mathbb{K})$ in the general case.

Proposition 56.21. *Let $F(x,y)\in\mathbb{K}[x,y]$ and let $f,\ g\in\mathcal{A}(\mathbb{K})$ be non-constant and satisfy $F(f(x),g(x))=0$. There exists $h\in\mathcal{A}(\mathbb{K})$ and $A,\ B\in\mathbb{K}[x]$ such that $f=A(h),\ g=B(h)$.*

Proof. There obviously exists an irreducible factor $F_0(x,y)$ of F such that $F_0(f,g)=0$. Since f, g are not constant, by Theorem 56.19, the curve Λ defined as $F_0(x,y)=0$ has genus zero. Therefore, by Theorem 56.20, there exist rational functions ϕ, $\theta \in \mathbb{K}(t)$ and $R(x,y) \in \mathbb{K}(x,y)$ such that, setting $t = R(x,y)$, we have $F_0(\phi(t),\theta(t)) = 0 \ \forall t$. Now, let $h = R(f,g)$. Hence by definition, h belongs to $\mathcal{M}(\mathbb{K})$. Then, $f=\phi(h)$, $g=\theta(h)$.

Now, since f, g belong to $\mathcal{A}(\mathbb{K})$, h must omit $\phi^{-1}(\infty)$ and $\theta^{-1}(\infty)$. But by Theorem 29.23, h can't have two exceptional values in $\mathbb{K} \cup \{\infty\}$. Consequently, $\phi^{-1}(\infty)=\theta^{-1}(\infty)$ and this is reduced to a single point. Therefore, up to a projective linear change in coordinates, we can assume that $\phi^{-1}(\infty)=\theta^{-1}(\infty)=\{\infty\}$ and that h, ϕ, θ omit ∞. But then h belongs to $\mathcal{A}(\mathbb{K})$ and both ϕ, θ are polynomials. □

Lemma 56.22. *Let $h \in \mathcal{A}(\mathbb{K})$ be non-constant and let ϕ, θ, $P \in \mathbb{K}[x]$, with $\deg(P)=n\geq 1$, $\deg(\phi)=t\geq \deg(\theta)$ and $t\geq 1$. Suppose there exists $C\in\mathbb{K}^*$ such that $P(\phi(h))=CP(\theta(h))$. Then $\phi(h)$ is of the form $A\theta(h)+B$, with $A^n=C$.*

Proof. We can write $\phi(x)=\sum_{j=0}^t a_j x^j$, $\theta(x)=\sum_{j=0}^t b_j x^j$. Let $P(x)=\sum_{k=0}^n l_k x^k$ and let $M=\mathbb{Q}[a_0,\dots,a_t,b_0,\dots,b_t,l_0,\dots,l_n]$. Then, $\mathbb{K}$ contains elements υ transcendental over M. Since h is not constant, we can take $z\in\mathbb{K}$ such that $h(z)=\upsilon$. Then, $\phi(h(z))=\sum_{j=0}^t a_j\upsilon^j$, $\theta(x)=\sum_{j=0}^t b_j\upsilon^j$. Now, we note that all powers of υ strictly greater than $t(n-1)$ in $\sum_{j=0}^t a_j\upsilon^j$ involve l_n. First, we see that $\deg(\theta)=t$ and that $\left(\frac{a_t}{b_t}\right)^n=C$. Let $A=\frac{a_t}{b_t}$. By comparing all powers strictly greater than $t(n-1)$ in $\sum_{j=0}^t a_j\upsilon^j$ and in $\sum_{j=0}^t b_j\upsilon^j$, we can see that $a_j=Ab_j \ \forall j=1,\dots,t$, which ends the proof. □

Lemma 56.23. *Let S be an affinely rigid subset of $\mathbb{K}$. Let $g\in\mathcal{A}(\mathbb{K})$ be non-constant and let $f=ag+b$, with $a\in\mathbb{K}^*$ and $b\in\mathbb{K}$. If $f^{-1}(S)=g^{-1}(S)$, then $f=g$.*

Proof. Since f has no exceptional value, we have $f(f^{-1}(S))=S$. Suppose $f^{-1}(S)=g^{-1}(S)$. Then,

(1) $f(g^{-1}(S))=S$.

Now, $aS+b=ag(g^{-1}(S)+b=f(g^{-1}(S))$, and therefore by (1), $aS+b=S$. Thus, the similarity ϕ defined by $\phi(x)=ax+b$ preserves S. But since S is affinely rigid, ϕ is the identity, hence $f=g$. □

We can now present the conclusion of this method due to Cherry and Yang:

Theorem 56.24 ((Cherry and Yang) [31]). *A finite subset S of $\mathbb{K}$ is an URSCM for $\mathcal{A}(\mathbb{K})$ if and only if it is affinely rigid.*

Proof. On one hand, it is obvious that a finite subset S of $\mathbb{K}$ which is not affinely rigid is not an URSCM for $\mathbb{K}[x]$ and therefore for $\mathcal{A}(\mathbb{K})$. Conversely, let $S = \{s_1, \dots, s_n\}$ be a finite affinely rigid subset S of $\mathbb{K}$ and let $f,\ g \in \mathcal{A}(\mathbb{K})$ satisfy $\mathcal{E}(f, S) = \mathcal{E}(g, S)$. Let $P(x) = \prod_{j=1}^{n}(x - s_j)$. Since $\mathcal{E}(f, S) = \mathcal{E}(g, S)$, $P(f)$ and $P(g)$ have the same zeros, counting multiplicity. Consequently, $\frac{P(f)}{P(g)}$ belongs to $\mathcal{M}(\mathbb{K})$ but has no zero and no pole. Hence, it is a constant C. Now, setting $F(x, y) = P(x) - CP(y)$, we have $F(f, g) = 0$. Consequently, by Proposition 56.21, there exists $h \in \mathcal{A}(\mathbb{K})$ and $\phi,\ \theta \in \mathbb{K}[x]$ such that $f = \phi(h),\ g = \theta(h)$. Now, by Lemma 56.22, f is of the form $ag + b$ with $a \in \mathbb{K}^*$ and $b \in \mathbb{K}$, satisfying further $a^n = C$. Then, since S is affinely rigid, by Lemma 56.23, we have $f = g$, which completes the proof. □

Similarly, we can derive Theorem 56.25:

Theorem 56.25. *A finite subset $S = \{a_1, \dots, a_n\}$ of $\mathbb{E}$ is affinely rigid if and only if it is an urscm for $\mathbb{E}[x]$.*

Proof. If S is an urscm, of course, it is affinely rigid. Now suppose S is affinely rigid, let $P,\ Q \in \mathbb{E}[x]$ be such that $\mathcal{E}(P, S) = \mathcal{E}(Q, S)$ and let F be an algebraic extension of $\mathbb{Q}$ containing S and all coefficients of P and Q. There exists a $\mathbb{Q}$-isomorphism ϕ from F into $\mathbb{C}_p$ and then we can check that $\phi(S)$ is affinely rigid and hence is an urscm in $\mathbb{C}_p$ for $\mathcal{A}(\mathbb{C}_p)$ and hence for $\mathbb{C}_p[x]$. Consequently, $\phi(P) = \phi(Q)$ and hence $P = Q$. □

Corollary 56.26. *A finite subset $S = \{a_1, \dots, a_n\}$ of $\mathbb{K}$ (respectively of $\mathbb{E}$) is affinely rigid if and only if the polynomial $P(x) = \prod_{i=1}^{n}(x - a_i)$ is a polynomial of strong uniqueness for $\mathcal{A}(\mathbb{K})$ (respectively for $\mathbb{E}[x]$).*

Remark. The characterization made in characteristic 0 doesn't hold in characteristic $p \neq 0$. The following example is given in [23]. Let F be an algebraically closed field of characteristic $p \geq 3$ and let $P(X) = X^p + (X-1)^{p-1}$. Let $f(x) = x^{p-1}$ and $g(x) = (x-1)^{p-1}$. Then we can check that $P(f(x)) = P(g(x))$, hence the set S is not an URSCM for $\mathcal{A}(\mathbb{K})$.

However, S is a set of p distinct points which are the p roots of 1 of order $p+1$, other than 1, because $P(X) = \frac{X^{p+1}-1}{X-1}$. Consequently, $\sum_{s\in S} s = -1$. We check that S is affinely rigid. Suppose there exists a similarity ϕ, say $\phi(x) = ax+b$, satisfying $\phi(S) = S$. Then, $\sum_{s\in S} \phi(s) = \sum_{s\in S} s = -1$, hence $\sum_{s\in S} as + b = -1$. But $\sum_{s\in S} as + b = a\sum_{s\in S} s + pb = a\sum_{s\in S} s = -a$, hence $a = 1$. Thus, $\phi(s) = s + b$, hence $(s+b)^{p+1} = 1$, therefore $s^{p+1} + s^p b + sb^p + b^{p+1} = 1$, and hence

$$b^p + sb^{p-1} + s^p = 0. \tag{1}$$

This is true for all $s \in S$, so we can add these p relations and obtain

$$0 = pb^p + b^{p-1}\sum_{s\in S} s + \sum_{s\in S} s^p = -b^{p-1} + \left(\sum_{s\in S} s\right)^p = -b^{p-1} + (-1)^p.$$

Thus, $b^{p-1} = -1$. Consequently, by (1), we have $b = s^p - s$.

Here we note that -1 belongs to S because $p \geq 3$. Let $S^* = S\setminus\{-1\}$. Then,

$$(q-1)b = \sum_{s\in S^*} s^p - \sum_{s\in S^*} s = 0,$$

hence $b = 0$, which ends the proof (another argument lets us conclude when $p = 2$).

Chapter 57

Other Urscm, Usim, and Non-Urscm

Throughout this chapter, $\mathbb{K}$ *is of characteristic* 0.

We denote by $\mathbb{E}$ *an algebraically closed field of characteristic* 0. *The field* $\mathbb{K}$ *is supposed to have characteristic* 0.

In Chapter 56, we have constructed sets of n points which are ursims for p-adic entire functions whenever $n \geq 9$ and are ursims for p-adic meromorphic functions whenever $n \geq 16$. Here, we construct new sets which are ursims of n points for $\mathbb{E}[x]$ and for $\mathcal{A}(\mathbb{K})$ for every $n \geq 9$ but then, we show that such sets are never urscm and of course are not ursim for $\mathcal{M}(\mathbb{K})$ and for $\mathbb{E}(x)$, although they are not preserved by any Moebius functions, a contradiction to a natural expectation.

Next, we also construct urscm for unbounded analytic functions inside a disk $d(a, R^-)$, with the use of the the Nevanlinna theory on three small functions.

Lemma 57.1. *Let* $Q(x) = x^n - x^m + k \in \mathbb{E}[x]$ *with* $m < n$, g.c.d.$(m,n) = 1$, $k \in \mathbb{K}^*$. *If* Q *admits a multiple zero* c*, then* $m^m(n-m)^{(n-m)} = n^n k^{n-m}$.

Proof. Suppose Q admits a multiple zero c. Then $c \neq 0$ and we have

(1) $m = nc^{n-m}$.

Since g.c.d.$(m, n) = 1$, p cannot divide both m and n. Hence, by (1), p cannot divide m or n. Now, we can derive $P(c) = c^n - \frac{n}{m}c^n + k = 0$, and hence

(2) $(n-m)c^m = kn$.

Therefore, by (1) and (2), we have

$$c^{n(n-m)} = \frac{k^{n-m}m^{n-m}}{(n-m)^{n-m}} = \frac{m^n}{n^n},$$

hence

$$k^{n-m} = \frac{m^m(n-m)^{n-m}}{n^n}.$$

□

Corollary 57.2. *Let $P(x) = x^n - x^m + k \in \mathbb{E}[x]$ with $m < n$, g.c.d.$(m, n) = 1$, $k \neq 0$, and $m^m(n-m)^{(n-m)} \neq k^{n-m}n^n$. Then P admits n distinct zeros.*

Lemma 57.3. *Let $P(x) = x^n - x^m + k \in \mathbb{E}[x]$ with $m < n$, g.c.d.$(m, n) = 1$, $k \neq 0$ and let $c \in \mathbb{E}\setminus\{0, 1\}$. Moreover, we assume that*

$$m^m(n-m)^{(n-m)} \neq n^n(1-u)^{n-m}k^{n-m}$$

for every $u \in \mathbb{E}$ such that $u^{n-m} = (-1)^{n-m}$. Then one of the polynomials $P - ck$ and $P - \frac{k}{c}$ admits n distinct zeros.

Proof. Suppose that both $P - kc$ and $P - kc^{-1}$ have a zero of multiple order. By Corollary 57.2, we have

$$n^nk^{n-m}(1-c)^{n-m} = n^nk^{n-m}(1-c^{-1})^{n-m} = m^m(n-m)^{(n-m)},$$

hence $\left(\frac{1-c}{1-c^{-1}}\right)$ is an $(n-m)$th root t of 1. Resolving the equation shows that either $c = 1$ or $c = -t$. But $c = 1$ contradicts the relation $n^nk^{n-m}(1-c)^{n-m} = m^m(n-m)^{(n-m)}$, hence $c = -t$. But then, $1 - c = 1 + t$. Putting $u = -t$, we have $(u)^{n-m} = (-1)^{n-m}$ whereas $n^nk^{n-m}(1-u)^{n-m} = m^m(n-m)^{(n-m)}$, which contradicts the hypothesis. So, one of the two polynomials $P - kc$ and $P - \frac{k}{c}$ has two distinct zeros. □

Proposition 57.4. *Let $P(x) = x^n - x^m + k \in \mathbb{K}[x]$ with $m < n$ be such that $m^m(n-m)^{(n-m)} \neq k^{n-m}n^n(1-u)^{(n-m)}$ for every $u \in \mathbb{K}$ such that $u^{n-m} = (-1)^{n-m}$. Let $f,\ g \in \mathcal{A}(\mathbb{K})\backslash\mathbb{K}$ (respectively let $\mathcal{A}_u(d(a, R^-)))$ satisfy $P(f) = \lambda P(g)$ with $\lambda \in \mathbb{K}^*$. In the following two cases, we have $\lambda = 1$:*

(i) $f,\ g \in \mathcal{A}(\mathbb{K})\backslash\mathbb{K},\ m \geq 2$,
(ii) $f,\ g \in \mathcal{A}_u(d(\alpha, R^-)),\ m \geq 3$.

Proof. Suppose $\lambda \neq 1$. By Theorem 43.3, we first note that $T(r,f) = T(r,g) + O(1)$. If $f,\ g \in \mathcal{A}_u(d(\alpha, R^-))$, we can obviously assume $\alpha = 0$.

By Lemma 57.3, one of the two polynomials $x^n - x^m + k(1 - \frac{1}{\lambda})$ and $x^n - x^m + k(1-\lambda)$ admits n distinct zeros. Without loss of generality, we can suppose that the polynomial $Q(x) = x^n - x^m + k(1 - \frac{1}{\lambda})$ has n distinct zeros. And by hypothesis, we have $\lambda Q(g) = f^n - f^m$.

Now, $Q(x)$ is of the form $\prod_{i=1}^n (x - e_i)$ with $e_i \neq e_j\ \forall i \neq j$. Then we have $f^n - f^m = \lambda \prod_{i=1}^n (g - e_i)$. Consequently,

$$\begin{aligned} Z(r,Q(g)) - \overline{Z}(r,Q(g)) &= \sum_{j=1}^n Z(r, g - e_j) - \overline{Z}(r, g - e_j) \\ &= Z(r, f^n - f^m)) - \overline{Z}(r, f^n - f^m), \end{aligned}$$

and of course $Z(r, f^n - f^m)) - \overline{Z}(r, f^n - f^m) \geq (m-1)Z(r,f)$, hence

(1) $\quad Z(r,Q(g)) - \overline{Z}(r,Q(g)) \geq (m-1)Z(r,f)$.

But since $f \in \mathcal{A}(\mathbb{K})$ or $f \in \mathcal{A}_u(d(0, R^-))$, we have $Z(r,f) = T(r,f) + O(1)$, therefore, by (1), we obtain

(2) $\quad Z(r,Q(g)) - \overline{Z}(r,Q(g)) \geq (m-1)T(r,f) + O(1)$.

On the other hand, by Lemma 43.14, we have $Z(r,Q(g)) - \overline{Z}(r,Q(g)) \leq T(r,g) - \log r + O(1)$. Therefore, by (2), $(m-1)$ $T(r,f) \leq T(r,g) - \log r + O(1)$. And since $T(r,f) = T(r,g) + O(1)$, we can derive $(m-1)T(r,f) \leq T(r,f) - \log r + O(1)$. Hence, if $f,\ g$ belong to $\mathcal{A}(\mathbb{K})$, we have $m \leq 1$, and if $f,\ g$ belong to $\mathcal{A}_u(d(0, R^-))$, we have $m \leq 2$, a contradiction in both cases. Consequently, if $f,\ g$ belong to $\mathcal{A}(\mathbb{K})\backslash\mathbb{K}$ and if $m \geq 2$, we can conclude $\lambda = 1$. And similarly, if $f,\ g$ belong to $\mathcal{A}_u(d(0, R^-))$ and if $m \geq 3$, we have $\lambda = 1$. $\square$

Now, we use a slightly different method to consider the problem $P(f) = \lambda P(g)$ when $f,\ g$ are meromorphic.

Theorem 57.5. *Let $P(x) = x^n - x^m + k$ with $m < n$ and let $f,\ g \in \mathcal{M}(\mathbb{K})\backslash\mathbb{K}$ (respectively let $f,\ g \in \mathcal{M}_u(d(0, R^-))$) be such that $P(f) = \lambda P(g)$, with $\lambda \in \mathbb{K}^*$. If $f,\ g \in \mathcal{M}(\mathbb{K})\backslash\mathbb{K}$ and $2m \geq n + 3$, or if $f,\ g \in \mathcal{M}_u(d(0, R^-))$ and $2m \geq n + 4$, then $\lambda = 1$. Moreover, if $f,\ g \in \mathcal{A}(\mathbb{K})\backslash\mathbb{K}$ and $2m \geq n + 2$ or if $f,\ g \in \mathcal{A}_u(d(0, R^-))$ and $2m \geq n + 3$, then $\lambda = 1$.*

Proof. Suppose $\lambda \neq 1$. By Theorem 43.3, we first note that $T(r, f) = T(r, g) + O(1)$. If $f,\ g \in \mathcal{M}_u(d(\alpha, R^-))$, we can obviously assume $\alpha = 0$. Let $F(x) = f(x)^n - f(x)^m$. Applying Theorem 45.7 to F, we have $T(r, F) \leq \overline{Z}(r, F) + \overline{Z}(r, F - (\lambda - 1)) + \overline{N}(r, F) - \log r + O(1)$, hence $nT(r, f) \leq \overline{Z}(r, f) + \overline{Z}(r, f^{n-m} - 1) + \overline{Z}(r, g) + \overline{Z}(r, g^{n-m} - 1) + N(r, f) - \log r + O(1)$. Consequently, $nT(r, f) \leq (2(n - m + 1) + 1)T(r, f) - \log r + O(1)$. Finally, if $f,\ g \in \mathcal{M}(\mathbb{K})\backslash\mathbb{K}$, we have $2m \leq n + 2$, and if $f,\ g \in \mathcal{M}_u(d(0, R^-))$, we have $2m \leq n + 3$. Consequently, we have proven that $\lambda = 1$ in the following two cases:

(i) $f,\ g \in \mathcal{M}(\mathbb{K})\backslash\mathbb{K}$ with $2m \geq n + 3$,
(ii) $f,\ g \in \mathcal{M}_u(d(0, R^-))$ with $2m \geq n + 4$.

Particularly, if $f,\ g \in \mathcal{A}(\mathbb{K})\backslash\mathbb{K}$, we have $2m \leq n + 1$, and if $f,\ g \in \mathcal{A}(d(0, R^-))$, we have $2m \leq n + 2$. Consequently, we have proven that $\lambda = 1$ in the following two cases:

(i) $f,\ g \in \mathcal{A}(\mathbb{K})\backslash\mathbb{K}$ with $2m \geq n + 2$,
(ii) $f,\ g \in \mathcal{A}_u(d(0, R^-))$ with $2m \geq n + 3$. □

Remark. If $f,\ g \in \mathcal{A}(\mathbb{K})\backslash\mathbb{K}$ and if $2m \leq n + 2$, by Proposition 57.5, we have $\lambda = 1$, even if the hypothesis $m^m(n - m)^{(n-m)} \neq k^{n-m}n^n (1 - u)^{(n-m)}$ for every $u \in \mathbb{K}$ such that $u^{n-m} = (-1)^{n-m}$ is not satisfied. Similarly, when $f,\ g \in \mathcal{A}_u(d(0, R^-))$ with $2m \geq n + 3$, when $m^m(n - m)^{(n-m)} \neq k^{n-m}n^n(1 - u)^{(n-m)}$ for every $u \in \mathbb{K}$ such that $u^{n-m} = (-1)^{n-m}$ is not satisfied.

Lemma 57.6. *Let $P(x) = x^n - x^m + k$ with $k \neq 0$, $m < n$, $n \geq 3$,* g.c.d.$(m, n) = 1$. *Let $f,\ g \in \mathcal{A}(\mathbb{K})\backslash\mathbb{K}$ (respectively $f,\ g \in \mathcal{A}_u(d(\alpha, R^-))$) satisfy $P(f) = P(g)$. Then, $f = g$.*

Proof. If f, g belong to $\mathcal{A}_u(d(\alpha, R^-))$, we can obviously suppose $\alpha = 0$. Let $h = \frac{f}{g}$ and suppose $h \neq 1$. Then, $g^{n-m} = \frac{h^m-1}{h^n-1}$. We note that if f, g belong to $\mathcal{A}_u(d(0, R^-))$, then h belongs to $\mathcal{M}_u(d(0, R^-))$ because $\mathcal{M}_b(d(0, R^-))$ is a field.

Let G be the group of nth roots of 1 and let G' be the group of mth roots of 1. Since g.c.d.$(n, m) = 1$, $G \backslash G'$ contains no element different from 1. So, G has n distinct elements and $G \backslash (G \cap G')$ has $n - 1$ distinct elements. Now, $h(x)$ cannot take a value in $G \backslash G'$ because this would be a zero of $h^n - 1$ but not a zero of $h^m - 1$ and therefore it would be a pole of g^{n-m} that lies in $\mathcal{A}(\mathbb{K})$ (resp. in $\mathcal{A}_u(d(0, R^-))$). Since $G \backslash G \cap G'$ contains at least two different elements u_1 and u_2, h must take at least one of these two values, a contradiction as we just saw. Consequently, h is identically 1. □

Notation. Given $k \in \mathbb{K}^*$ and m, $n \in \mathbb{N}^*$, we set $Q_{n,m,k}(x) = x^n - x^m + k \in \mathbb{K}[x]$ and $Q_{n,k}(x) = x^n - x^{n-1} + k \in \mathbb{K}[x]$, and we denote by $Y(n, k)$ the set of zeros of $Q_{n,k}$. Moreover, we put $u_n = \frac{(n-1)^{n-1}}{n^n}$.

Theorem 57.7. *If $2 \leq m$ (respectively $3 \leq m$), with g.c.d.$(m, n) = 1$ and $m^m(n - m)^{(n-m)} \neq k^{n-m} n^n (1 - u)^{(n-m)}$ for every $u \in \mathbb{K}$ such that $u^{n-m} = (-1)^{n-m}$, then $Q_{n,m,k}$ is a polynomial of strong uniqueness for $\mathcal{A}(\mathbb{K})$ (respectively for $\mathcal{A}_u(d(\alpha, R^-))$).*

Proof. By Proposition 57.4, we have $\lambda = 1$, therefore, $f^n - f^m = g^n - g^m$, and then, by Lemma 57.6, $f = g$. That ends the proof. □

Theorem 57.8. *Let $k \in \mathbb{K} \backslash \{0, \frac{u_n}{2}\}$. Then $Y_{n,k}$ is an ursim for $\mathcal{A}(\mathbb{K})$ for every $n \geq 9$. Moreover, $Y_{n,k}$ is an ursim for $\mathcal{A}(\mathbb{K})$ for every $n \geq 3$, provided that $k \neq u_n$ whenever $3 \leq n \leq 8$.*

Proof. Suppose first that $n \geq 9$, and suppose that f, $g \in \mathcal{A}(\mathbb{K})$ are not constant and satisfy $f^{-1}(Y(n, k)) = g^{-1}(Y(n, k))$. We put $F = \frac{1}{Q_{n,k}(f)}$, $G = \frac{1}{Q_{n,k}(g)}$, $H = \frac{F''}{F'} - \frac{G''}{G'}$. Then F, G have the same poles, ignoring multiplicity. Without loss of generality, we may obviously assume that $\omega_0(F) = \omega_0(G) = \omega_0(F') = \omega_0(G') = \omega_0(f) = \omega_0(g) = 0$, and $F(0) \neq c_j$, $G(0) \neq c_j$ $\forall j = 1, 2, 3$. Following the same method as in Theorem 56.7, we can prove that H is identically 0.

So, we may assume $H = 0$. Then there exist $a, b \in \mathbb{K}$ such that $Q_{n,k}(f) = \frac{Q_{n,k}(g)}{aQ_{n,k}(g)+b}$, with $b \neq 0$. But, since f, g lie in $\mathcal{A}(\mathbb{K})$, a must be equal to zero. Thus, we have $Q_{n,k}(f) = \frac{1}{b} Q_{n,k}(g)$. Now, let $c \in \mathbb{K}$

satisfy $c^n = \frac{1}{k}$. Then we have $Q_{n,k}(x) = k((cx)^n - c(cx)^{n-1} + 1)$. Setting $P(y) = y^n - cy^{n-1} + 1$, we have $P(cf) = \frac{1}{b}P(cg)$. But as $n \geq 9$, by Proposition 57.4, we check that $b = 1$, hence $cf^n - f^{n-1} = cg^n - g^{n-1}$.

Let $h = \frac{f}{g}$. From this equality, we can easily obtain $cg = -\frac{h^{n-1}-1}{h^n-1}$. Clearly, if h is not constant, $h^n - 1$ admits zeros which are not zeros of $h^{n-1} - 1$, a contradiction with the hypothesis $g \in \mathcal{A}(\mathbb{K})$. Hence, h is constant and so are f, g, a contradiction with the hypothesis.

Suppose now that f, $g \in \mathcal{A}(\mathbb{K})$ satisfy $\mathcal{E}(f, Y(n,k)) = \mathcal{E}(g, Y(n,k))$ with $n \geq 3$ and that $k \neq u_n$ whenever $3 \leq n \leq 8$. We first note that if $n \geq 9$, $Y_{n,k}$ is an urscm for $\mathcal{A}(\mathbb{K})$, hence by Lemma 56.1, it is an ursim for $\mathcal{A}(\mathbb{K})$. Suppose now $3 \leq n \leq 8$. Then, by Proposition 56.2, there exists $\lambda \in \mathbb{K}^*$ such that $Q_{n,k}(f) = \lambda Q_{n,k}(g)$. But by Proposition 57.4, $\lambda = 1$. Consequently, $f^n - f^{n-1} = g^n - g^{n-1}$, and hence, by Lemma 57.6, we have $f = g$. Therefore, $Q_{n,k}$ is a polynomial of strong uniqueness for $\mathcal{A}(\mathbb{K})$. Moreover, by Corollary 57.2, $Q_{n,k}$ has no multiple zero since $k \neq u_n$. Consequently, by Corollary 56.3, $Y(n,k)$ is an urscm for every $n \geq 3$, provided $k \neq u_n$ whenever $3 \leq n \leq 8$. That finishes the proof of Theorem 57.8. □

Corollary 57.9. *Let $k \in \mathbb{E}\setminus\{0, \frac{u_n}{2}\}$. $Y(n,k)$ is an ursim for $\mathbb{E}[x]$ for every $n \geq 9$ and $Y(n,k)$ is an urscm for $\mathbb{E}[x]$ for every $n \geq 3$ provided that $k \neq u_n$ whenever $3 \leq n \leq 8$.*

Proof. Indeed, there exists a finite extension of $\mathbb{Q}$ containing $Y(n,k)$ and all coefficients of f and g. As we can embed this extension into $\mathbb{C}_p$ so we can transfer results from $\mathbb{C}_p[x]$ to $\mathbb{E}[x]$. □

Remarks. Concerning urscm of 3 points for $\mathcal{A}(\mathbb{K})$ or for polynomials, the last statement of Theorem 57.8 provides us with another proof that every affinely rigid subset of 3 elements in $\mathbb{K}$ is urscm for $\mathcal{A}(\mathbb{K})$ (and therefore every affinely rigid subset of 3 elements in $\mathbb{E}$ is urscm for $\mathbb{E}[x]$). Indeed, given a subset of 3 points that is not affinely rigid, by Corollary 1.7, through an affine transformation, either they are the zeros of a polynomial of the form $x^3 - k$ or they are the zeros of a polynomial of the form $x^3 - x$ which can also be transformed to $3\sqrt{3}(u^3 - u^2 + \frac{2}{27})$ and then the case $k = \frac{2}{27}$ is excluded by hypothesis on k. Consequently, if a subset of $\mathbb{K}$ of 3 elements is affinely rigid, it is an urscm for $\mathcal{A}(\mathbb{K})$, and of course, if a subset of $\mathbb{E}$ of 3 elements is affinely rigid, it is an urscm for $\mathbb{E}[x]$. Recall that more generally, in [22], it was proven just by classical techniques that all affinely rigid subsets of $\mathbb{E}$ are urscm for polynomials.

As we saw in Theorem 56.24, urscm for $\mathbb{E}[x]$ and for $\mathcal{A}(\mathbb{K})$ are characterized as the affinely rigid sets, similarly, it would be natural to think that the urscm for $\mathbb{E}(x)$ and for $\mathcal{M}(\mathbb{K})$ are the sets which are not preserved by any Moebius function other than the identity. This is definitely false, as we are going to see [43]. See also [96].

Lemma 57.10. *Let $f,\ g \in \mathcal{M}(\mathbb{K})$ such that $f^n(f-a) = g^n(g-a)$ and let $h = \frac{f}{g}$. Then, $f = \frac{a(n+2)}{n+1}\left(\frac{h^{n+1}-1}{h^{n+2}-1}\right)h$ and $g = \frac{a(n+2)}{n+1}\left(\frac{h^{n+1}-1}{h^{n+2}-1}\right)$.*

Proof. We have $g^nh^n - g^{n-1} = g^n - g^{n-1}$, hence $g^n(h^n-1) = g^{n-1}(h^{n-1}-1)$, hence $g = \frac{h^{n-1}-1}{h^n-1}$ and then $f = hg$. □

Corollary 57.11. *Let $h \in \mathbb{E}(x)$ be non-constant, and let $n \in \mathbb{N}$ be such that $n \geq 3$. Let $g = \frac{h^{n-1}-1}{h^n-1}$, and let $f = gh$. Then, $f^n(f-1) = g^n(g-1)$, $\mathcal{E}(f, Y(n,k)) = \mathcal{E}(g, Y(n,k))$. Moreover, f and g have the same poles, counting multiplicity.*

Corollary 57.12. *Let $k \in \mathbb{E}\backslash\{0, u_n\}$. Then $Y(n,k)$ is not an urscm for $\mathbb{E}(x)$ whenever $n \in \mathbb{N}^*$.*

Remark. Obviously, any set which is not an urscm for $\mathbb{E}(x)$ cannot be an urscm for $\mathcal{M}(\mathbb{K})$, cannot be an ursim for $\mathbb{E}(x)$, and cannot be an ursim for $\mathcal{M}(\mathbb{K})$. In fact, we do not know any example of an urscm for a family of functions $\mathcal{F}$ which is not also an ursim for $\mathcal{F}$. For example, as noted in [22], sets of the form $\{a, a+h, a+3h\}$, with $a \in W$, $h \in E^*$, are urscm for $\mathbb{E}[x]$ but are they also ursims for $\mathbb{E}[x]$?

Theorem 57.13. *Let $k \in \mathbb{E}\backslash\{0, u_n\}$. For every $n \geq 5$, there exists no Moebius function h preserving $Y(n,k)$, except the identity.*

Proof. We suppose that a Moebius function h defined as $h(x) = \frac{ax+b}{cx+d}$ (with $ad - bc \neq 0$) satisfies $h(Y(n,k)) = Y(n,k)$ for some $n \geq 5$. Let $R(x) = Q_{n,k}(h(x))$. Then R is a rational function of the form $\frac{\sum_{j=0}^{n} q_j x^j}{(cx+d)^n}$. Since $h(Y(n,k)) = Y(n,k)$, we note that the set of zeros of R is $Y(n,k)$ and therefore $Y(n,k)$ is the set of zeros of $\sum_{j=0}^{n} q_j x^j$. Consequently, this polynomial is of the form $\lambda Q_{n,k}$, with $\lambda \in W^*$. In particular, we have

(1) $q_1 = q_2 = q_3 = 0.$

On the other hand, we have $\sum_{j=0}^{n} q_j x^j = (ax+b)^n - (cx+d)(ax+b)^{n-1} + k(cx+d)^n$, so we obtain

$$q_1 = nab^{n-1} - (n-1)ab^{n-2}d - b^{n-1}c + kncd^{n-1},$$

$$q_2 = \binom{n}{2} a^2 b^{n-2} - \binom{(n-1)}{2} a^2 b^{n-3} d - (n-1)ab^{n-2}c + k\binom{n}{2} c^2 d^{n-2},$$

$$q_3 = \binom{n}{3} a^3 b^{n-3} - \binom{(n-1)}{3} a^3 b^{n-4} d - \binom{(n-1)}{2} a^2 b^{n-3} c + k\binom{n}{3} c^3 d^{n-3}.$$

We now compute $A = \frac{a^2 q_1}{n} - \frac{4abq_2}{n(n-1)} + \frac{6b^2 q_3}{n(n-1)(n-2)}$ and obtain $A = cd^{n-3}k(ad-bc)^2$. But of course, by (1), we have $A = 0$. Since $k(ad-bc) \neq 0$, we must have $cd = 0$.

First, suppose $d = 0$. Then, we have

(2) $bc \neq 0$.

But, as $q_1 = 0$, this implies $nab^{n-1} = b^{n-1}c$, i.e.

(3) $na = c$.

And as $q_2 = 0$, we have $\frac{n(n-1)}{2} a^2 b^{n-2} = (n-1)ab^{n-2}c$, i.e.

(4) $na^2 = 2ac$.

By (3) and (4), we check that $ac = 0$, hence by (2), $a = 0$, which contradicts (2) and (3).

Thus, we are led to the hypothesis $d \neq 0,\ c = 0$. Then, $ad \neq 0$. And then by (1) we obtain $nab^{n-1} = (n-1)ab^{n-2}d$, i.e.

(5) $nb^{n-1} = (n-1)b^{n-2}d$.

And $na^2 b^{n-2} = (n-2)a^2 b^{n-3} d$, i.e.

(6) $nb^{n-2} = (n-2)b^{n-3}d$,

hence by (5) and (6), $(n-2)b^{n-2}d = (n-1)b^{n-2}d$, and finally, $b=0$. So, $h(x) = \frac{ax}{d}$. Taking $d = 1$, we obtain $\sum_{j=0}^{n} q_j x^j = a^n x^n - a^{n-1}x^{n-1} + k$, which must be proportional to $Q_{n,k}(x)$. In particular, $x^n - \frac{x^{n-1}}{a} + \frac{k}{a^n} = x^n - x^{n-1} + k$ and therefore $a = 1$. This finishes showing that the identity is the only Moebius function preserving $Y(n,k)$. □

Remark. In Theorem 57.13, the hypothesis $n \geq 5$ is necessary as shown in Theorem 57.14:

Theorem 57.14. *Every urscm for $\mathbb{E}(x)$ or for $\mathcal{M}(\mathbb{K})$ has at least five points.*

Proof. Indeed, given any four distinct points $\alpha,\ \beta,\ \gamma,\ \delta \in E$, there does exist a Moebius function h, different from the identity, satisfying $h(\alpha) = \beta,\ h(\beta) = \alpha,\ h(\gamma) = \delta,\ h(\delta) = \gamma$. Indeed, h is of the form

$$h(x) = \frac{x(\alpha\beta - \gamma\delta) - \alpha\beta(\gamma + \delta) + \gamma\delta(\alpha + \beta)}{x(\alpha + \beta - \gamma - \delta) - \alpha\beta + \gamma\delta}.$$

And of course, every urscm for $\mathcal{M}(\mathbb{K})$ has at least five points. □

Conclusion. By Corollary 57.12, the sets $Y(n,k)$ in $\mathbb{E}$ are not urscm for $\mathbb{E}(x)$ though, by Theorem 57.13, there exist no Moebius functions h (different from the identity) preserving $Y(n,k)$ as soon as $n \geq 5$. Thus, the set of finite urscm for $\mathbb{E}(x)$ is strictly smaller than the set of finite sets S such that there exist no Moebius functions h preserving S. So, it is not easy to imagine a characterization of urscm for $\mathbb{E}(x)$. And a characterization of urscm for $\mathcal{M}(\mathbb{K})$ appears even more difficult.

Next, by Theorem 2 of [83], we know that given a compact A in $\mathbb{C}$ which is not preserved by any non-trivial similarity of the form $\varphi(x) = a + (x-a)e^{i\alpha}$ and given two monic polynomials F, G of same degree such that $F^{-1}(A) = G^{-1}(A)$, we have $F = G$. Now, we note that, by choosing for h a rational function such as $\frac{x}{x+1}$, we can obtain two rational functions f and g both equal to a quotient of monic polynomials, satisfying $f^{-1}(Y(n,k)) = g^{-1}(Y(n,k))$, though the sets $Y(n,k)$ are not preserved by similarities, at least when $n \geq 4$. Thus, one doesn't see any clear generalization of Theorem 2 of [83] to rational functions.

Example of sets $\boldsymbol{Y(5,k)}$. Let $Q(x) = x^5 - x^4 + \frac{3}{32}$. Then Q factorizes in $\mathbb{Q}[x]$ in the form $(x+\frac{1}{2})(x^4 - \frac{3}{2}x^3 + \frac{3}{2^2}x^2 - \frac{3}{2^3}x + \frac{3}{2^4})$.

In Ω_p, we can define

$$a = \frac{\sqrt{2}}{4}(\sqrt{3}\sqrt{7}-1)\sqrt{\sqrt{3}\sqrt{7}+3}, \quad b = \sqrt{2}\sqrt{3}\sqrt{\sqrt{3}\sqrt{7}-3},$$

$$\alpha = \frac{1}{8}\left(3 + a + i\sqrt{2}\sqrt{\sqrt{3}\sqrt{7} - 3a + 2b}\right),$$

$$\beta = \frac{1}{8}\left(3 - a + i\sqrt{2}\sqrt{\sqrt{3}\sqrt{7} + 3a - 2b}\right).$$

Then we have $T(5, (\frac{3}{32}) = \{\frac{-1}{2},\ \alpha,\ \overline{\alpha},\ \beta,\ \overline{\beta}\}$.

We can check that the derivative of $\frac{(h^m-1)}{h^n-1} = g^{n-m}$ is not identically zero and therefore h' is not identically zero.

Now, since g.c.d.$(m,n) = 1$, G' has m elements and $G'\backslash G$ has $m-1$ elements $u_1, \dots, u_{m-1}$. Then, for each zero β of $h - u_i$ $(1 \le i \le m-1)$, β is a zero of $h^m - 1$ and is not a zero of $h^n - 1$. Hence, it is a zero of g^{n-m} and consequently it is a zero of order at least $n-m$ of $h - u_i$. Thus, for each $i = 1, \dots, m-1$, we have

$$(1) \quad \overline{Z}(r, h - u_i) \le \frac{1}{n-m} Z(r, h - u_i).$$

Now, let us apply Theorem 45.7 to h. Since h' is not identically zero, we have

$$(m-2)T(r,h) \le \sum_{i=1}^{m-1} \overline{Z}(r, h-u_i) + N(r,h) - \log r + O(1),$$

hence by (1), we have

$$(2) \quad (m-2)T(r,h) \le \left(\frac{m-1}{n-m} + 1\right) T(r,h) - \log r + O(1).$$

Concerning urscm for $\mathcal{A}(d(0,R^-))$, we can improve the results obtained in Chapter 54 thanks to the Nevanlinna theorem on small functions.

Notation. Given $k \in \mathbb{K}^*$ and $n, m \in \mathbb{N}^*$ with $m < n$, we set by $Q_{n,m,k}(x) = x^n - x^m + k$ and we denote by $Y(n, m, k)$ the set of zeros of $Q_{n,m,k}$.

Theorem 57.15. *If* $2m \geq n + 3$*, then* $Y(n, m, k)$ *is an urscm for* $\mathcal{A}_u(d(a, R^-))$ $(a \in \mathbb{K},\ k \in \mathbb{K}^*,\ R > 0)$.

Proof. We can obviously assume that $a = 0$. Suppose $Y(n, m, k)$ is not an urscm for $\mathcal{A}(d(0, R^-))$ and let $f,\ g \in \mathcal{A}_u(d(0, R^-))$ be such that

$$\mathcal{E}(f, Y(n, m, k)) = \mathcal{E}(f, Y(n, m, k)).$$

By Proposition 56.2, $\frac{Q_{n,k}(f)}{Q_{n,k}(g)}$ is an invertible function $\phi \in \mathcal{A}_b(d(0, R^-))$. Consequently, we have $f(x)^n - f(x)^m + k = \phi(x)(g(x)^n - g(x)^m + k)$. Set $F(x) = f(x)^n - f(x)^m$. By Corollary 50.13, we can obtain

$$T(r, F) \leq \overline{Z}(r, F) + \overline{Z}(r, F - k(\phi - 1)) + o(T(r, f).$$

Now, clearly, $\overline{Z}(r, F) \leq \overline{Z}(r, f) + \overline{Z}(r, f^{n-m} - 1) + O(1)$ and $\overline{Z}(r, f^{n-m} - 1) \leq (n - m)T(r, f)$, hence

(1) $\overline{Z}(r, F) \leq (n - m + 1)T(r, f) + o(T(r, f).$

And similarly,

(2) $\overline{Z}(r, F - k(\phi - 1)) = \overline{Z}(r, g^n - g^m) \leq \overline{Z}(r, f) + \overline{Z}(r, f^{nm} - 1).$

Of course, since ϕ is bounded, we have $T(r, f) = T(r, g) + O(1)$, hence, by (1) and (2), we obtain $T(r, F) \leq 2(n - m + 1)$ $T(r, f^{n-m} - 1) + o(T(r, f).$

On the other hand, we have $T(r, F) = nT(r, f) + O(1)$, hence $nT(r, f) \leq 2(n - m + 1)T(r, f) + o(T(r, f)$. That yields $2m \leq n + 2$, a contradiction to the hypothesis. □

Corollary 57.16. $Y(5, k)$ *is an urscm for* $\mathcal{A}_u(d(a, R^-))$ $(a \in \mathbb{K},$ $k \in \mathbb{K}^*,\ R > 0)$.

Open question. *Do URSCM of* 3 *or* 4 *points for* $\mathcal{A}_u(d(a, R^-))$ *exist?*

For instance, consider an affinely rigid set of 3 or 4 points, such as $\{0,\ 1,\ 3\}$, and let $P(X) = X(X-1)(X-3)$. Do functions $f,\ g \in \mathcal{A}_u(d(0, R^-))$ exist such that $P(f) = P(g)$?

Chapter 58

Nevanlinna Theory in Characteristic p

Results in characteristic p were published in [25, 26, 27].

Notation. *In this chapter, we denote by p the characteristic of $\mathbb{K}$ and by q its characteristic exponent, i.e. $q = p$ if $p \neq 0$, and $q = 1$ if $p = 0$.*

As usual, given $a \in \mathbb{K}$ and $n \in \mathbb{N}$, we denote by $\sqrt[p^n]{a}$ the unique $b \in \mathbb{K}$ such that $b^{(p^n)} = a$.

Given m, $n \in \mathbb{N}$, we set $m \prec n$ if m divides n and $m \not\prec n$ if m does not divide n.

When $p \neq 0$, we denote by $\mathcal{S}$ the $\mathbb{F}_p$-automorphism of $\mathbb{K}$ defined by $\mathcal{S}(x) = \sqrt[p]{x}$. More generally, this mapping has continuation to a $\mathbb{K}$-algebra automorphism of $\mathbb{K}[X]$ as $\mathcal{S}(c\prod_{j=1}^{n}(X - a_j)) = \mathcal{S}(c)\prod_{j=1}^{n}(X - \mathcal{S}(a_j))$, $c \in \mathbb{K}$.

Lemma 58.1 is easily verified.

Lemma 58.1. *Let $f \in \mathcal{M}(d(0, R^-))$ be such that f' is not identically zero and let $\alpha \in d(0, R^-)$. We have $\omega_\alpha(f') = \omega_\alpha(f) - 1$ if $p \not\prec \omega_\alpha(f)$ and $\omega_\alpha(f') \geq \omega_\alpha(f)$ if $p \prec \omega_\alpha(f)$.*

Proposition 58.2. *Suppose $p \neq 0$. Let $r > 0$ and let $f \in \mathcal{M}(d(a, r^-))$. Then $\sqrt[p]{f}$ belongs to $\mathcal{M}(d(a, r^-))$ if and only if $f' = 0$. Moreover, there exists a unique $t \in \mathbb{N}$ such that $\sqrt[p^t]{f} \in \mathcal{M}(d(a, r^-))$ and $(\sqrt[p^t]{f})' \neq 0$.*

Proof. If f is of the form l^p with $l \in \mathcal{M}(d(a, r^-))$, then of course we have $f' = 0$. Now, suppose that $f' = 0$. Suppose first $f \in \mathcal{A}(d(a, r^-))$. Since $f' = 0$, $f(x)$ is of the form $\sum_{n=0}^{\infty}(a_n)^p x^n p$. Put $l(x) = \sum_{n=0}^{\infty}(a_n)x^n$. The radius of convergence of l is equal to that of f. Hence, $l \in \mathcal{A}(d(a, r^-))$, and of course, we have $f = l^p$. We now consider the general case when $f \in \mathcal{M}(d(a, r^-))$. Let $(b_n, t_n)_{n\in\mathbb{N}}$ be the sequence of poles of f inside $d(a, r^-)$ where t_n is the multiplicity order of b_n. By Theorem 27.14, we can find $h \in \mathcal{A}(d(a, r^-))$ such that $\omega_{b_n}(h) \geq t_n$ $\forall n \in \mathbb{N}$. Clearly, fh^p belongs to $\mathcal{A}(d(a, r^-))$ and satisfies $(fh^p)' = 0$. Consequently, fh^p is of the form g^p, with $g \in \mathcal{A}(d(a, r^-))$, therefore $f = \left(\frac{g}{h}\right)^p$. On the other hand, by Theorem 29.11, the set of integers s such that $\sqrt[p^s]{f}$ belongs to $\mathcal{M}(d(a, r^-))$ is obviously bounded and therefore admits a biggest element, which ends the proof. □

Definition and notation. Suppose $p \neq 0$. Given, $f \in \mathcal{M}(d(a, r^-))$, we call *ramification index of* f the integer t such that $\sqrt[p^t]{f} \in \mathcal{M}(d(a, r^-))$ and $(\sqrt[p^t]{f})' \neq 0$.

In the same way, given an algebraically closed field B of characteristic $p \neq 0$ and $P(x) \in B[x]$, we call *ramification index of* P the unique integer t such that $\sqrt[p^t]{P} \in B[x]$ and $(\sqrt[p^t]{P})' \neq 0$. This ramification index is denoted by $\mathrm{ram}(f)$ for any $f \in \mathcal{M}(d(a, r^-))$ or $f \in \mathcal{M}(\mathbb{K})$ and similarly it is denoted by $\mathrm{ram}(P)$ for any $P \in B[x]$.

Remark. Suppose $p \neq 0$ and let $f \in \mathcal{M}(d(a, r^-))$ have ramification index t. For every $r' \in]0, r[$, f has the same ramification index as an element of $\mathcal{M}(d(a, r'^-))$ because of course, on one hand, $\sqrt[p^t]{f} \in \mathcal{M}(d(a, r'^-))$, and on the other hand, by properties of analytic functions, $(\sqrt[p^t]{f})'$ is not identically zero inside $d(a, r')$.

Corollary 58.3. *Suppose $p \neq 0$ and let $f, g \in \mathcal{M}(d(a, R^-))\setminus\mathbb{K}$ satisfy an equation of the form $g(x) = G(f(x))$ and $G \in \mathbb{K}(Y)$. Then,* $\mathrm{ram}(g) \geq \mathrm{ram}(f)$ *and, putting* $t = \mathrm{ram}(f)$, $g_t = \sqrt[p^t]{g}$, $f_t = \sqrt[p^t]{f}$, $G_t = \mathcal{S}^t(G)$, *they satisfy* $g_t = G_t \circ f_t$.

Remark. This property does not hold for meromorphic functions inside a disk $d(a, r^-)$.

Corollary 58.4. *Let $p \neq 0$. Let $f(x) \in \mathcal{M}(\mathbb{K})$, let S be the set of zeros and poles of f in $\mathbb{K}$, and let d be the* g.c.d. *of* $\{\omega_\alpha(f) \mid \alpha \in S\}$. *Then,* $\mathrm{ram}(f) = -\log_p(|d|_p)$.

Notation. In the sequel, I denotes an interval of the form $[\rho, +\infty[$, with $\rho > 0$, and J denotes an interval of the form $[\rho, R[$. Let $f \in \mathcal{M}(d(0, R^-))$. The counting functions of zeros and poles of f: $Z(r, f)$ and $N(r, f)$ are defined as usual and we put again $T(r, f) = \max(Z(r, f), N(r, f))$.

Next, denoting by $E(r, f)$ the set $\{a \in d(0, r) \mid \omega_a(f) > 0, p^{\text{ram}(f)+1} \not\prec \omega_a(f)\}$, if $0 \notin E(r, f)$, we set $\widetilde{Z}(r, f) = \sum_{\alpha \in E(r,f)} \log \frac{r}{|\alpha|}$, and if $0 \in E(r, f)$, we set $\widetilde{Z}(r, f) = \log r + \sum_{\alpha \in E(r,f), \alpha \neq 0} \log \frac{r}{|\alpha|}$.

Similarly, we define $\widetilde{N}(r, f) = \widetilde{Z}(r, \frac{1}{f})$.

Now, assume that f' is not identically 0.

Let $V(r, f) = \{a \in d(0, r) \mid \omega_a(f) < 0, \ p^{\text{ram}(f)+1} \prec \omega_a(f)\}$. We put

$$N_0(r, f') = \sum_{\alpha \in V(r,f)} [\omega_\alpha(f') - \omega_\alpha(f)] \log \frac{r}{|\alpha|}.$$

Given a finite subset S of $\mathbb{K}$, we put $\Lambda'(r, f, S) = \{a \in d(0, r) \mid f'(a) = 0, \ f(a) \notin S\}$ and $\Lambda''(r, f, S) = \{a \in d(0, r) \mid p^{\text{ram}(f)+1} \prec \omega_a(f - f(a)), \ f(a) \in S\}$. Then we can define

$$Z_0^S(r, f') = \sum_{\alpha \in \Lambda'(r,f,S)} \omega_\alpha(f') \log \frac{r}{|\alpha|} + \sum_{\alpha \in \Lambda''(r,f,S)} [\omega_\alpha(f') - \omega_\alpha(f - f(\alpha))] \log \frac{r}{|\alpha|}.$$

Remarks. (1) It is easily verified that all the above functions are positive.
(2) If $p = 0$, we have $\overline{Z}(r, f) = \widetilde{Z}(r, f)$ and $\overline{N}(r, f) = \widetilde{N}(r, f)$.

Lemma 58.5. *Let $f \in \mathcal{M}(d(0, R^-))$, let $t = \text{ram}(f)$, and let $g = \sqrt[q^t]{f}$. Then, $\widetilde{Z}(r, f) = \widetilde{Z}(r, g)$ and $\widetilde{N}(r, f) = \widetilde{N}(r, g)$.*

Proof. Let a be a zero of f and let $s = \omega_a(f)$. Then s is of the form nt with $n \in \mathbb{N}^*$. If $n = 1$, then a belongs to both $E(r, f)$ and $E(r, g)$. And if $n > 1$, then $a \notin E(r, f)$. But then a is a zero of order n of g and hence a does not belong to $E(r, g)$. $\square$

Similar to Lemma 43.4, we have Lemma 58.6:

Lemma 58.6. *Let $\alpha_1, \ldots, \alpha_n \in \mathbb{K}$ be pairwise distinct, let $P(u) = \prod_{i=1}^{n}(u - \alpha_i)$, and let $f \in \mathcal{M}(d(0, R^-))$. Then, $Z(r, P(f)) = \sum_{i=1}^{n} Z(r, f - \alpha_i)$ and $\widetilde{Z}(r, P(f)) = \sum_{i=1}^{n} \widetilde{Z}(r, f - \alpha_i)$.*

Lemmas 58.7 and 58.8 are consequences of Lemma 58.1.

Lemma 58.7. *Let $f \in \mathcal{M}(d(0, R^-))$ be such that $f' \neq 0$ and let S be a finite subset of $\mathbb{K}$. Then,*

$$\sum_{b \in S}(Z(r, f - b) - \widetilde{Z}(r, f - b)) = Z(r, f') - Z_0^S(r, f').$$

Lemma 58.8. *Let $f \in \mathcal{M}(d(0, R^-))$ be such that $f' \neq 0$. Let $0 < r < R$. Then,*

$$N(r, f') = N(r, f) + \widetilde{N}(r, f) - N_0(r, f').$$

Lemma 58.9. *Let $f \in \mathcal{M}(d(0, R^-))$ be such that $f' \neq 0$. Let $0 < r < R$. Then,*

$$Z(r, f') \leq Z(r, f) + \widetilde{N}(r, f) - N_0(r, f') - \log r + O(1), (r \in J).$$

Proof. Without loss of generality, up to a change of variable, we can assume that both f, f' have no zero and no pole at 0. By Theorem 43.2, we have $Z(r, f) - N(r, f) = \Psi(f, \log r) - \Psi(f(0))$, and $Z(r, f') - N(r, f') = \Psi(f', \log r) - \Psi(f'(0))$. But, by Corollary 36.11, $\Psi(f', \log r) \leq \Psi(f, \log r) - \log r$, hence we obtain $Z(r, f') \leq N(r, f') - N(r, f) + Z(r, f) - \log r + O(1)$. And by Lemma 58.7, we have $N(r, f') - N(r, f) = \widetilde{N}(r, f) - N_0(r, f')$, which completes the proof. □

The previous results enable us to prove the ultrametric Nevanlinna main theorem in a basic form:

Theorem 58.10. *Let $\alpha_1, \ldots, \alpha_n \in \mathbb{K}$, with $n \geq 2$, and let $f \in \mathcal{M}(d(0, R^-))$ (respectively $f \in \mathcal{M}(\mathbb{K})$) of ramification index t. Let $S = \{\sqrt[q^t]{\alpha_1}, \ldots, \sqrt[q^t]{\alpha_n}\}$. Then we have*

$$\frac{(n-1)T(r, f)}{q^t} \leq \sum_{i=1}^{n} \widetilde{Z}(r, f - \alpha_i) + Z(r, (\sqrt[q^t]{f})')$$
$$- Z_0^S(r, (\sqrt[q^t]{f})') + O(1) \quad \forall r \in J$$

(respectively $\forall r \in I$).

Moreover, if f belongs to $\mathcal{A}(d(0,R^-))$ (respectively $f \in \mathcal{A}(\mathbb{K})$), then

$$\frac{nT(r,f)}{q^t} \le \sum_{i=1}^{n} \widetilde{Z}(r, f-\alpha_i) + Z(r, (\sqrt[q^t]{f})')$$
$$- Z_0^S(r, (\sqrt[q^t]{f})') + O(1) \quad \forall r \in J$$

(respectively $\forall r \in I$).

Now, following the same method as in Theorem 45.7, we can obtain that classical form of the Nevanlinna inequality where $\overline{Z}$ and $\overline{N}$ are replaced by $\widetilde{Z}$ and $\widetilde{N}$.

Theorem 58.11. *Let $\alpha_1, \dots, \alpha_n \in \mathbb{K}$, with $n \ge 2$, and let $f \in \mathcal{M}(d(0,R^-))$ (respectively $f \in \mathcal{M}(\mathbb{K})$) of ramification index t. Let $S = \{\sqrt[q^t]{\alpha_1}, \dots, \sqrt[q^t]{\alpha_n}\}$. Then we have*

$$\frac{(n-1)T(r,f)}{q^t} \le \sum_{i=1}^{n} \widetilde{Z}(r, f-\alpha_i) + \widetilde{N}(r,f) - Z_0^S(r, (\sqrt[q^t]{f})')$$
$$- N_0(r, (\sqrt[q^t]{f})') - \log r + O(1) \quad \forall r \in J$$

(respectively $\forall r \in I$).

Corollary 58.12. *Let $f \in \mathcal{M}(d(0,R^-))$ (respectively $f \in \mathcal{M}(\mathbb{K})$) be such that $f' \ne 0$ and let $\alpha_1, \dots, \alpha_n \in \mathbb{K}$, with $n \ge 2$. Then we have*

$$\sum_{i=1}^{n} Z(r, f-\alpha_i) - \widetilde{Z}(r, f-\alpha_i) \le T(r,f) + \widetilde{N}(r,f) - Z_0^S(r, f')$$
$$- N_0(r, f') - \log r + O(1) \quad \forall r \in J$$

(respectively $\forall r \in I$).

Remark. In Theorems 45.2 and 45.4 as in Corollaries 45.3 and 45.5, the field $\mathbb{K}$ is not supposed to be of characteristic 0. Consequently, all those statements apply to fields of characteristic p.

Proof of Theorems 58.10 and 58.11. For convenience, we put $g = \sqrt[q^t]{f}$, and $\beta_i = \sqrt[q^t]{\alpha_i}$ for every $i = 1, \dots, n$. So, $S = \{\beta_1, \dots, \beta_n\}$.

By Theorem 29.14, there exist ϕ, $\theta \in \mathcal{A}(d(0,R^-))$ such that $g = \frac{\phi}{\theta}$, and

(1) $Z(r, \phi) \leq Z(r, g) + 1$,

(2) $Z(r, \theta) \leq N(r, g) + 1$.

By Theorems 45.2 and 45.4, there exists $A \in \mathbb{R}$, and for any $r \in J$ (respectively $r \in I$), there exists $l(r) \in \{1, \ldots, n\}$ such that $Z(r, \phi - \beta_j\theta) \geq \max(Z(r, \phi), Z(r, \theta)) + A\ \forall j \neq l(r)$, therefore there exists $B \in \mathbb{R}$ such that

(3) $Z(r, \phi - \beta_i\theta) \geq T(r, g) + B \quad \forall i \neq l(r), \quad \forall r \in J$

(respectively $\forall r \in I$).

We check that $\mathcal{D}(\phi) - \mathcal{D}(\frac{\phi}{\theta}) = \mathcal{D}(\theta) - \mathcal{D}(\frac{\theta}{\phi})$, therefore

$$\begin{aligned}\mathcal{D}(\phi - \beta_i\theta) &= \mathcal{D}(g - \beta_i) + \mathcal{D}(\theta) - \mathcal{D}\left(\frac{1}{g - \beta_i}\right)\\ &= \mathcal{D}(g - \beta_i) + \mathcal{D}(\theta) - \mathcal{D}\left(\frac{1}{g}\right).\end{aligned}$$

Then, applying counting functions, we have $Z(r, \phi - \beta_i\theta) = Z(r, g - \beta_i) + Z(r, \theta) - N(r, g)$, and therefore, by (2), we obtain

(4) $Z(r, \phi - \beta_i\theta) \leq Z(r, g - \beta_i) + 1$.

Then, by (3) and (4), we obtain $(n - 1)(T(r, g) + B) \leq \sum_{\substack{1 \leq i \leq n,\\ i \neq l(r)}} Z(r, \phi - \beta_i\theta) \leq \sum_{\substack{1 \leq i \leq n,\\ i \neq l(r)}} Z(r, g - \beta_i) + n - 1\ \forall r \in J$ (respectively $\forall r \in I$).

Putting $M = (n - 1)(1 - B)$, we obtain

(5) $$(n - 1)T(r, g) \leq \sum_{i=1}^{n} Z(r, g - \beta_i) + M - Z(r, g - \beta_{l(r)}) \quad \forall r \in J$$

(respectively $\forall r \in I$).

By Lemma 58.7, we have

$$\sum_{i=1}^{n} Z(r, g - \beta_i) = \sum_{i=1}^{n} \widetilde{Z}(r, g - \beta_i) + Z(r, g') - Z_0^S(r, g'),$$

hence by (5), we obtain

$$(6)\quad (n-1)T(r,g) \le \sum_{i=1}^{n} \widetilde{Z}(r, g-\beta_i) + Z(r, g') - Z_0^S(r, g') - Z(r, g-\beta_{l(r)}) + O(1) \quad \forall r \in J$$

(respectively $\forall r \in I$).

Now, since $T(r,g) = \frac{T(r,f)}{q^t}$ and since $\widetilde{Z}(r, g-\beta_i) = \widetilde{Z}(r, f-\alpha_j)$ $\forall j = i, \dots, n$, we obtain

$$\frac{(n-1)T(r,f)}{q^t} \le \sum_{i=1}^{n} \widetilde{Z}(r, f-\alpha_i) + Z(r, (\sqrt[q^t]{f})') - Z_0^S(r, (\sqrt[q^t]{f})') + O(1) \quad \forall r \in J$$

(respectively $\forall r \in I$).

Suppose now that f belongs to $\mathcal{A}(d(a, R^-))$ or to $\mathcal{A}(\mathbb{K})$. Then so does g. By Theorem 43.7, we have $Z(r, g-\beta_{l(r)}) = T(r,g) + O(1)$ $\forall r \in J$ (*respectively* $\forall r \in I$) so, by (6), we obtain

$$nT(r,g) \le \sum_{i=1}^{n} \widetilde{Z}(r, g-\beta_i) + Z(r, g') - Z_0^S(r, g') + O(1) \quad \forall r \in J$$

(respectively $\forall r \in I$), and consequently,

$$\frac{nT(r,f)}{q^t} \le \sum_{i=1}^{n} \widetilde{Z}(r, f-\alpha_i) + Z(r, (\sqrt[q^t]{f})') - Z_0^S(r, (\sqrt[q^t]{f})') + O(1) \quad \forall r \in J$$

(respectively $\forall r \in I$).

Now, returning to the general case, we have $g' = (g-\beta_{l(r)})'$ and $\widetilde{N}(r,g) = \widetilde{N}(r, g-\beta_{l(r)})$. So, by Lemma 58.9, we have

$$(7)\quad Z(r,g') - Z(r, g-\beta_{l(r)}) \le \widetilde{N}(r,g) - N_0(r,g') - \log r + O(1).$$

Finally, by (6) and (7), we obtain

$$\frac{(n-1)T(r,f)}{q^t} \leq \sum_{i=1}^{n} \widetilde{Z}(r, f-\alpha_i) + \widetilde{N}(r,f) - Z_0^S(r, (\sqrt[q^t]{f})') - N_0(r, (\sqrt[q^t]{f})') - \log r \quad \forall r \in J$$

(respectively $\forall r \in I$).

That completes the proof. □

Theorem 58.13. *Let $f \in \mathcal{M}(\mathbb{K})$ (respectively $f \in \mathcal{M}_u(d(0,R^-))$) and let $u_1, u_2, u_3 \in \mathcal{M}_f(\mathbb{K})$ (respectively $u_1, u_2, u_3 \in \mathcal{M}_f(d(0,R^-))$) be pairwise distinct. Let $\phi(x) = \frac{(f(x)-u_1(x))(u_2(x)-u_3(x))}{(f(x)-u_3(x))(u_2(x)-u_1(x))}$ and let t be the ramification index of ϕ.*

Then, $\frac{T(r,f)}{q^t} \leq \sum_{j=1}^{3} \widetilde{Z}(r, f-u_j) + o(T(r,f))$.

Proof. By Theorem 58.11, we have

$$\text{(1)} \quad \frac{T(r,\phi)}{q^t} \leq \widetilde{Z}(r,\phi) + \widetilde{Z}(r,\phi-1) + \widetilde{N}(r,\phi) + O(1).$$

Next, we follow the same way as in Theorem 47.9. We have $T(r,f) \leq T(r, \frac{u_3-u_1}{f-u_3}) + o(T(r,f))$, hence

$$T(r,f) \leq T\left(r, \frac{u_3-u_1}{f-u_3} + 1\right) + o(T(r,f)) = T\left(r, \frac{f-u_1}{f-u_3}\right) + o(T(r,f)).$$

Now, $T(r, \frac{u_2-u_1}{u_2-u_3}) = o(T(r,f)$. Consequently, by writing $\frac{f-u_1}{f-u_3} = \phi(\frac{u_2-u_1}{u_2-u_3})$, we have $T(r, \frac{f-u_1}{f-u_3}) \leq T(r,\phi) + T(r, \frac{u_2-u_1}{u_2-u_3}) \leq T(r,\phi) + o(T(r,f))$ and finally $T(r,f) \leq T(r,\phi) + o(T(r,f))$. Thus, by (1), we obtain

$$\text{(2)} \quad \frac{T(r,f)}{q^t} \leq \widetilde{Z}(r,\phi) + \widetilde{Z}(r,\phi-1) + \widetilde{N}(r,\phi) + o(T(r,f)).$$

Now, we can check that

$$\widetilde{Z}(r,\phi)+\widetilde{Z}(r,\phi-1)+\widetilde{N}(r,\phi)$$
$$\leq \sum_{j=1}^{3}\widetilde{Z}(r,f-u_j)+\sum_{1\leq j<k\leq 3}\widetilde{Z}(r,u_k-u_j)$$
$$\leq \sum_{j-1}^{3}\widetilde{Z}(r,f-u_j)+o(T(r,f))$$

which, by (2), completes the proof. □

Similar to Theorem 47.10, we can derive Theorem 58.14.

Theorem 58.14. *Let $f\in\mathcal{M}(\mathbb{K})$ (respectively $f\in\mathcal{M}_u(d(0,R^-))$), let $u_1,\ u_2\in\mathcal{M}_f(\mathbb{K})$ (respectively $u_1,u_2\in\mathcal{M}_f(d(0,R^-))$) be distinct, and let t be the ramification index of $\frac{f(x)-u_1(x)}{f(x)-u_2(x)}$. Then,*

$$\frac{T(r,f)}{q^t}\leq \widetilde{Z}(r,f-u_1)+\widetilde{Z}(r,f-u_2)+\widetilde{N}(r,f)+o(T(r,f)).$$

Proof. Let $g=\frac{1}{f}$, $w_j=\frac{1}{u_j}$, $j=1,2$, $w_3=0$. Clearly,

$$T(r,g)=T(r,f)+O(1),\ T(r,w_j)=T(r,u_j),\ j=1,2,$$

so we can apply Theorem 58.13 to $g,\ w_1,\ w_2,\ w_3$. On the other hand, we note that

$$\frac{(g(x)-w_1(x))w_2(x)}{(g(x)-w_2(x))w_1(x)}=\frac{f-u_1}{f-u_2}.$$

Thus, we have $\frac{T(r,g)}{q^t}\leq \widetilde{Z}(r,g-w_1)+\widetilde{Z}(r,g-w_2)+\widetilde{Z}(r,g)+o(T(r,g))$.

But $\widetilde{Z}(r,g-w_j)=\widetilde{Z}(r,f-u_j)$ for $j=1,2$ and $\widetilde{Z}(r,g)=\widetilde{N}(r,f)$. Moreover, we know that $o(T(r,g))=o(T(r,f))$. Consequently, the claim is proved when u_1u_2 is not identically zero. Now, we can generalize exactly like in Theorem 47.10. □

Next, by setting $g = f - u_1$ and $u = u_2 - u_1$, we can write Corollary 58.15:

Corollary 58.15. *Let $g \in \mathcal{M}(\mathbb{K})$ (respectively $g \in \mathcal{M}_u(d(0, R^-))$), let $u \in \mathcal{M}_g(\mathbb{K})$ (respectively $u \in \mathcal{M}_g(d(0, R^-))$), and let t be the ramification index of $\frac{g-u}{g}$.*

Then, $\frac{T(r,g)}{q^t} \leq \widetilde{Z}(r, g) + \widetilde{Z}(r, g - u) + \widetilde{N}(r, g) + o(T(r, g)))$.

Corollary 58.16. *Let $f \in \mathcal{A}(\mathbb{K})$ (respectively $f \in \mathcal{A}_u(d(0, R^-))$) and let u_1, $u_2 \in \mathcal{A}_f(\mathbb{K})$ (respectively $u_1, u_2 \in \mathcal{A}_f(d(0, R^-))$) be distinct and let t be the ramification index of $\frac{f-u_1}{f-u_2}$. Then, $\frac{T(r,f)}{q^t} \leq \overline{Z}(r, f - u_1) + \overline{Z}(r, f - u_2) + o(T(r, f)))$.*

Corollary 58.17. *Let $f \in \mathcal{A}(\mathbb{K})$ (respectively $f \in \mathcal{A}_u(d(0, R^-))$) and let $u \in \mathcal{A}_f(\mathbb{K})$ (respectively $u \in \mathcal{A}_f(d(0, R^-))$) be non-identically zero and let t be the ramification index of $\frac{f-u}{f}$. Then, $\frac{T(r,f)}{q^t} \leq \overline{Z}(r, f) + \overline{Z}(r, f - u) + o(T(r, f)))$.*

Chapter 59

Strong Uniqueness and URSCM in Characteristic p

Notation. Throughout Chapter 54, $\mathbb{K}$ is a field of characteristic $p \neq 0$, α belongs to $\mathbb{K}$, and R belongs to $]0, +\infty[$. We try to generalize the results of Chapters 52 and 53 when the characteristic is p. Many results come from [22, 23].

Lemma 59.1. *Let $P(x) = x^n - x^m + k$ with $m < n$ and $k \neq 0$. Let $f,\ g \in \mathcal{A}(\mathbb{K})\backslash\mathbb{K}$ (respectively $f,\ g \in \mathcal{A}_u(d(\alpha, R^-))$) satisfy $P(f) = \lambda P(g)$ with $\lambda \in \mathbb{K}^*$. Then f and g have the same ramification index.*

Proof. Let s be the ramification index of f and let t be this of g. We can obviously suppose $s \leq t$. Let $f_s = \sqrt[p^s]{f}$, $g_s = \sqrt[p^s]{g}$, let $k_s = \sqrt[p^s]{k}$, let $P_s(x) = x^n - x^m + k_s$, and let $\lambda_s = \sqrt[p^s]{\lambda}$. Then f_s, g_s satisfy $P_s(f_s) = \lambda_s P_s(g_s)$, hence $f_s'(n(f_s)^{n-1} - m(f_s)^{m-1}) = g_s'(n(g_s)^{n-1} - m(g_s)^{m-1})$. Since both f_s, g_s belong to $\mathcal{A}(\mathbb{K})\backslash\mathbb{K}$ (respectively to $\mathcal{A}_u(d(\alpha, R^-))$), $n(f_s)^{n-1} - m(f_s)^{m-1}$, $n(g_s)^{n-1} - m(g_s)^{m-1}$ are not identically zero. Now, by definition of the ramification index, f_s' is not identically zero, hence neither is g_s' and therefore $t = s$. $\square$

Lemma 59.12 is a generalization of Lemma 57.6 when $p \neq 0$:

Lemma 59.2. *Let $P(x) = x^n - x^m + k$ with $k \neq 0$, $m < n$, $n \geq 3$,* g.c.d.$(m, n) = 1$ *and further, m, n satisfy one of the following two conditions:*

(A) *n is not a power of p.*
(B) *$(m - 2)(n - m) \geq n - 1$ (respectively $(m - 2)(n - m) \geq n$).*

Let $f,\ g \in \mathcal{A}(\mathbb{K})\backslash\mathbb{K}$ *(respectively* $f,\ g \in \mathcal{A}_u(d(\alpha, R^-)))$ *satisfy* $P(f) = P(g)$*. Then,* $f = g$*.*

Proof. If $f,\ g$ belong to $\mathcal{A}_u(d(\alpha, R^-))$, we can obviously suppose $\alpha = 0$. Now, set $n = p^t q$ with g.c.d.$(q,p) = 1$. Let $h = \frac{f}{g}$ and suppose $h \neq 1$. Then, $g^{n-m} = \frac{h^m-1}{h^n-1}$. As in Lemma 57.6, we note that if $f,\ g$ belong to $\mathcal{A}_u(d(0, R^-))$, then h belongs to $\mathcal{M}_b(d(0, R^-))$ because $\mathcal{M}_b(d(0, R^-))$ is a field.

Let G be the group of nth roots of 1 and let G' be the group of mth roots of 1. Since g.c.d.$(n,m) = 1$, $G\backslash G'$ contains no element different from 1.

Suppose that Condition (A) is satisfied. Let $n = qp^t$ with g.c.d.$(q,p) = 1$ and $q > 1$. Suppose first $q > 2$. Then $G\backslash G'$ contains at least two different elements u_1 and u_2, hence h must take at least one of these two values, a contradiction as we just saw. Consequently, we must have $q = 2$ and hence $p \neq 2$. Since $n \geq 3$, n is of the form $2p^t$ and G has two elements: 1 and -1. Therefore, each zero of $h^n - 1 = (h^2-1)^{p^t}$ is of order p^t. Since $-1 \notin G'$, $h(x)$ cannot take the value -1. Hence, $h(x)$ must take the value 1 at a point $\beta \in \mathbb{K}$ (respectively $\beta \in d(0, R^-)$) and then β is a zero of order p^t for $h^n - 1$. But since g.c.d.$(n,m) = 1$, it is a zero of order 1 for $h^m - 1$, and hence g^{n-m} admits β as a pole of order $p^t - 1$, a contradiction again. Thus, Condition (A) implies that h is identically 1.

Now, suppose that Condition (B) is satisfied. Since we don't need Condition (A), we can assume that $n = p^t$. If p divides $n - m$, it divides neither m nor n because g.c.d.$(n,m) = 1$. Consequently, we can assume that p divides neither m nor $n - m$. Let us first assume that $f'g'$ is not identically zero. Since $n - m$ is prime to p, we can check that the derivative of $\frac{h^m-1}{h^n-1} = g^{n-m}$ is not identically zero and therefore h' is not identically zero.

Now, since g.c.d.$(m,p) =$ g.c.d.$(m,n) = 1$, G' has m elements and $G'\backslash G$ has $m-1$ elements $u_1, \ldots, u_{m-1}$. Then, for each zero β of $h - u_i$ $(1 \leq i \leq m-1)$, β is a zero of $h^m - 1$ and is not a zero of $h^n - 1$. Hence, it is a zero of g^{n-m}, and consequently, it is a zero of order at least $n - m$ of $h - u_i$. Thus, for each $i = 1, \ldots, m-1$, we have

$$(1)\quad \widetilde{Z}(r, h - u_i) \leq \frac{1}{n-m} Z(r, h - u_i).$$

Now, let us apply Theorem 58.11 to h. Since h' is not identically zero, we have

$$(m-2)T(r,h) \le \sum_{i=1}^{m-1} \widetilde{Z}(r,h-u_i) + N(r,h) - \log r + O(1),$$

hence by (1), we have

$$(2)\quad (m-2)T(r,h) \le \left(\frac{m-1}{n-m}+1\right)T(r,h) - \log r + O(1).$$

Therefore, if $f,\ g$ belong to $\mathcal{A}(\mathbb{K})$, by (2), that yields $m-2 < \frac{n-1}{n-m}$, and if $f,\ g$ belong to $\mathcal{A}_u(d(0,R^-))$, we have $m-2 \le \frac{n-1}{n-m}$, both conditions excluded by hypotheses. Consequently, $h=1$ and therefore $f=g$.

Suppose now that the ramification index of f is s. By Lemma 59.11, the ramification index of g is s too. Let $k_s = \sqrt[p^s]{k}$, and let $P_s(x) = x^n - x^m + k_s$, $f_s = \sqrt[p^s]{f}$, $g_s = \sqrt[p^s]{g}$. Then, $P_s(f_s) = P_s(g_s)$, and we are taken back to the same problem with f_s and g_s that satisfy $(f_s)'(g_s)'$ not identically zero. So, we have $f_s = g_s$ and therefore $f=g$. □

Proposition 59.3. *Let $P(x) = x^n - x^m + k$ with $2 \le m < n$ (respectively $3 \le m < n$), g.c.d.$(m,n) = 1$ and $a \ne 0$. Assume that $m^m(n-m)^{(n-m)} \ne n^n(1-u)^{(n-m)}(\sqrt[p^s]{k})^{n-m}$ for every $s \in \mathbb{N}$ and for every $u \in \mathbb{K}$ such that $u^{n-m} = (-1)^{n-m}$. Next, we assume that one of the following two conditions (a) and (b) is satisfied:*

(a) *n is not a power of p,*
(b) *$(m-2)(n-m) \ge n-1$ (respectively $(m-2)(n-m) \ge n$).*

Let $f,\ g \in \mathcal{A}(\mathbb{K})$ (respectively for $f,\ g \in \mathcal{A}_u(d(\alpha,R^-))$) be such that $P(f) = \lambda P(g)$. Then, $\lambda = 1$.

Proof. If $f,\ g \in \mathcal{A}_u(d(\alpha,R^-))$, we can assume $\alpha = 0$. By Lemma 57.3, P has n distinct zeros, hence its set of zeros S has n elements. Suppose there exists $\lambda \in \mathbb{K}^*$ such that $P(f) = \lambda P(g)$.

If $f'g'$ is not identically zero, we can make a proof that is very similar to that of in Proposition 57.4.

Suppose first that $f'g'$ is not identically zero. We prove that $\lambda = 1$. Indeed, suppose $\lambda \neq 1$. Let $Q(x) = x^n - x^m + k - \frac{1}{\lambda}$. Then,

(1)
$$Q(g) = P(g) - \frac{1}{\lambda} = \frac{1}{\lambda}(\lambda P(g) - 1) = \frac{1}{\lambda}(P(f) - 1) = \frac{1}{\lambda}(f^n - f^m).$$

Similarly, let $R(x) = P(x) - \lambda$. Then, $R(f) = P(f) - \lambda = \lambda(\lambda^{-1}P(f) - 1) = \lambda(P(g) - 1) = \lambda(g^n - g^m)$. Now, by Lemma 57.3, at least one of the two polynomials Q, R has all its zeros distinct. Without loss of generality, we may assume that Q has all its zeros distinct. So, $Q(x)$ is of the form $\prod_{i=1}^{n}(x - e_i)$ with $e_i \neq e_j \ \forall i \neq j$. Then by (1), we have $f^n - f^m = \lambda\prod_{i=1}^{n}(g - e_i)$. Consequently, by Lemma 58.6, we have $Z(r, Q(g)) - \widetilde{Z}(r, Q(g)) = \sum_{j=1}^{n} Z(r, g - e_j) - \widetilde{Z}(r, g - e_j) = Z(r, f^n - f^m) - \widetilde{Z}(r, f^n - f^m)$ and of course $Z(r, f^n - f^m) - \widetilde{Z}(r, f^n - f^m) \geq (m-1)Z(r, f)$, hence

(2) $\quad Z(r, Q(g)) - \widetilde{Z}(r, Q(g)) \geq (m-1)Z(r, f)$.

On the other hand, since $f'g'$ is not identically zero, by Corollary 58.12, we have $Z(r, f^n - f^m) - \widetilde{Z}(r, f^n - f^m) \leq T(r, f) - \log(r) + O(1)$, hence $Z(r, Q(g)) - \widetilde{Z}(r, Q(g)) \leq T(r, g) - \log r + O(1)$, and hence by (2), we obtain

(3) $\quad (m-1)Z(r, f) \leq T(r, g) - \log r + O(1)$.

But by Theorem 43.8, we have $T(r, P(f)) = nT(r, f) + O(1)$, and by (1), we have $T(r, P(g)) = +O(1) = nT(r, g) + O(1)$, hence $T(r, f) = T(r, g) + O(1)$, and hence by (3), we have $(m-1)Z(r, f) \leq T(r, f) - \log r + O(1)$. By Theorem 43.3, this yields $(m-1)T(r, f) \leq T(r, f) - \log r + O(1)$ and therefore $m < 2$ when f, g belong to $\mathcal{A}(\mathbb{K})$. And if f, g belong to $\mathcal{A}(d(0, R^-))$, then $m \leq 2$. Consequently, since $m \geq 2$ (respectively $m \geq 3$), we have proven in both cases that $\lambda = 1$.

Suppose now that $f' = 0$ and let s be the ramification index of f. Then by Lemma 59.1, this of g is also s. Putting $k_s = \sqrt[p^s]{k}$, $P_s(x) = x^n - x^m + k_s$, $f_s = \sqrt[p^s]{f}$, $g_s = \sqrt[p^s]{g}$, and $\lambda_s = \sqrt[p^s]{\lambda}$, now we have $P(f_s) = \lambda_s P(g_s)$ and we are taken back to the same situation with P_s, f_s, g_s. On one hand, we note that the zeros of P_s are the p^sth roots of those of P. Hence, P_s also admits n distinct zeros. Next, the condition $m^m(n-m)^{(n-m)}(\sqrt[p^s]{k})^{n-m} \neq n^n(1-u)^{(n-m)}$ for every $s \in \mathbb{N}$ and for every $u \in \mathbb{K}$ such that $u^{n-m} = (-1)^{n-m}$ also applies in the same way to P_s. Consequently, we can generalize the reasoning to the case when the ramification index of f, g is s. That ends the proof. □

By Lemma 59.2 and Proposition 59.3, we have Corollary 59.4:

Corollary 59.4. *Let $P(x) = x^n - x^m + k$ with $m < n$ and $k \neq 0$, $2 \leq m < n$ (respectively $3 \leq m < n$), g.c.d.$(m, n) = 1$, $m^m(n - m)^{(n-m)} \neq n^n(1 - u)^{(n-m)}(\sqrt[p^s]{k})^{n-m}$ for every $s \in \mathbb{N}$ and for every $u \in \mathbb{K}$ such that $u^{n-m} = (-1)^{n-m}$. Assume that one of the following two conditions (a) and (b) is satisfied:*

(a) *n is not a power of p.*
(b) *$(m - 2)(n - m) \geq n - 1$ (respectively $(m - 2)(n - m) \geq n$).*

Then P is a polynomial of strong uniqueness for $\mathcal{A}(\mathbb{K})$ (respectively for $\mathcal{A}_u(d(\alpha, R^-))$). The set of zeros of P is an urscm for $\mathcal{A}(\mathbb{K})$.

Proposition 59.5. *Let $P(x) = x^n - x^m + k$ with $m < n$, $k \neq 0$ and let $f, g \in \mathcal{M}(K)\backslash\mathbb{K}$ (respectively let $f, g \in \mathcal{M}_u(d(0, R^-))$) be such that $P(f) = \lambda P(g)$, with $\lambda \in \mathbb{K}^*$. If $f, g \in \mathcal{M}(K)\backslash\mathbb{K}$ and $2m \geq n + 3$, or if $f, g \in \mathcal{M}_u(d(0, R^-))$ and $2m \geq n + 4$, then $\lambda = 1$. Moreover, if $f, g \in \mathcal{A}(K)\backslash\mathbb{K}$ and $2m \geq n + 2$ or if $f, g \in \mathcal{A}_u(d(0, R^-))$ and $2m \geq n + 3$, then $\lambda = 1$.*

Proof. The proof is the same as that of Theorem 57.5, just by replacing $\overline{Z}$ and $\overline{N}$ by $\widetilde{Z}$ and $\widetilde{N}$, respectively. □

Theorem 59.6. *Let $\mathbb{K}$ have characteristic $p \neq 2$ and let $P(x) = x^n - x^{n-2} + k$ with $k \neq 0$, $n = p^t \geq 7$ (respectively $n = p^t \geq 8$) with $t \in \mathbb{N}^*$ such that $(n - 2)^{n-2} \neq n^n k^2$.*

Then P is a polynomial of strong uniqueness for $\mathcal{A}(\mathbb{K})$ (respectively for $\mathcal{A}_u(d(\alpha, R^-))$). Moreover, the set of zeros of P has n elements and is an urscm for $\mathcal{A}(\mathbb{K})$.

Proof. In order to show that we can apply Corollary 58.4, we show first that $m^m(n - m)^{(n-m)} \neq n^n(1 - u)^{(n-m)}(\sqrt[p^s]{k})^{n-m}$, which yields

(1) $(n - 2)^{n-2} \neq n^n(1 - u)^2(\sqrt[p^s]{k})^{n-m}$ for every $s \in \mathbb{N}^*$ and for every $u \in \mathbb{K}$ such that $u^2 = (-1)^2$ i.e. for $u = 1$ or $u = -1$.

Next, we need to have $(m-2)(n-m) \geq n-1$ (respectively $(m-2)(n-m) \geq n$) i.e.

(2) $2(n-4) \geq n-1$, which is trivial when $u=1$ and becomes
(3) $4(n-2)^{n-2} \neq 4n^n(\sqrt[p^s]{k})^2$ when $u=-1$.

On the other hand, since $b^p = b\ \forall b \in \mathbb{N}$, we can see that (3) is equivalent to

(4) $(n-2)^{n-2} \neq n^n k^2$.

Now, we note that by hypothesis, either $n = p = 7$ with $t = 1$ or $n \geq 8$. Finally, (2) is equivalent to $2(n-4) \geq n-1$ (respectively $2(n-4) \geq n$). Thus, (1) and (2) are satisfied and therefore Corollary 59.4 shows the conclusion. □

Theorem 59.7. *Let $\mathbb{K}$ have characteristic 2, let $n = 2^s$, $s \geq 3$, and let $P(x) = x^n - x^{n-3} + k$ with $k \neq 0$.*

Then P is a polynomial of strong uniqueness for $\mathcal{A}(\mathbb{K})$ and for $\mathcal{A}_u(d(\alpha, R^-))$. Moreover, the set of zeros of P has n elements and is an urscm for $\mathcal{A}(\mathbb{K})$.

Proof. Indeed, in characteristic 2, all hypotheses of Theorem 59.7 involving n^n are trivially satisfied since in $\mathbb{K}$, $n = 2^s = 0$. Next, the hypotheses $n = 2^S$, $m = n-3$, and $s \geq 3$ imply $(m-2)(n-m) \geq n$. □

Theorem 59.8 shows that in characteristic $p \neq 0$, an affinely rigid set is not necessarily an urscm for $\mathcal{A}(\mathbb{K})$ and even for polynomials.

Theorem 59.8. *Let $\mathbb{E}$ be an algebraically closed field of characteristic 3. There exists no subset of three elements that is an urscm for $\mathbb{E}[x]$.*

Proof. Let S be a subset of $\mathbb{E}$ of 3 elements a_1, a_2, a_3. We can find an affine change of variable h such that the images $e_i = h(a_i)$ are the zeros of a monic polynomial $P(Y)$ of the form $Y^3 + bY + c$. Then $h(x)$ is of the form $x + m$ and therefore $(\overset{-1}{h})(x) = x - m$. Let $T = \{e_1, e_2, e_3\}$. Now, let $\alpha \in \mathbb{E}$ satisfy $\alpha^2 = -b$ and consider $f(x) = x$ and $g(x) = x + \alpha$. Then we can check that $P(f) = P(g)$.

Indeed, $P(g(x)) = (x+\alpha)^3 + b(x+\alpha) + c = x^3 + \alpha^3 + bx + c\alpha + c = P(x) + \alpha^3 + b\alpha = P(f(x)$ since $\alpha^3 + b\alpha = 0$. Now, let $\phi = (\overset{-1}{h}) \circ f$ and $\theta = (\overset{-1}{h}) \circ g$. Then,

$$\begin{aligned} P(f(x) &= \prod_{j=1}^{3} (f(x) - e_j) = \prod_{j=1}^{3} (h(\phi(x) - h(a_j)) \\ &= \prod_{j=1}^{3} (\phi(x) + m - (a_j + m)) = \prod_{j=1}^{3} (\phi(x) - a_j). \end{aligned}$$

Let $Q(Y) = \prod_{j=1}^{3}(Y - a_j)$. Then, $P(f(x)) = Q(\phi(x))$ and similarly $P(g(x)) = Q(\theta(x))$. Consequently, $Q(\phi) = Q(\theta)$. Thus, Q is not a polynomial of uniqueness for $\mathbb{E}[x]$, and hence, by Corollary 56.3, S is not an urscm for $\mathbb{E}[x]$. $\square$

Chapter 60

The Functional Equation $P(f) = Q(g)$

Throughout this chapter, the field $\mathbb{K}$ *is supposed to have characteristic* $p \geq 0$ *and characteristic exponent* q. R *is a strictly positive number.*

Most of results come from [46, 64].

In the sequel, we use the following basic lemmas:

Lemma 60.1. *Let* $\mathbb{E}$ *be a field, and let* $P \in \mathbb{E}[x]$ *satisfy* $\deg(P) = 3$ *and be such that* P' *has two distinct zeros* c_1, c_2. *Then,* $P(c_1) \neq P(c_2)$.

Proof. Suppose $P(c_1) = P(c_2)$. Without loss of generality, we can suppose $P(c_1) = P(c_2) = 0$. Then P admits two distinct zeros of order ≥ 2, a contradiction since $\deg(P) = 3$. □

Henceforth, we look for general results linked to the property $P(a) \neq Q(b)$, where a and b are zeros of P' and Q', respectively. In order to state and prove our main results (Theorems 60.4 and 60.12), we need to recall the notion of the so-called Nevanlinna characteristic function and its associated functions. However, here, for the sake of convenience, we define counting functions of zeros and poles without multiplicities which, in characteristic $p \neq 0$, are the same as in characteristic 0, contrary to some other results that require a more accurate form. Moreover, we use the ramification index for analytic functions, defined in Chapter 58.

Lemma 60.2. *Let f, g be in $\mathcal{M}(d(0,R^-))$, let P, Q be in $\mathbb{K}[x]$, and assume that f and g satisfy $P(f) = Q(g)$. Then f belongs to $\mathcal{M}_b(d(0,R^-))$ if and only if so does g.*

Proof. By Theorem 43.10, a function $h \in \mathcal{M}(d(0,R^-))$ belongs to $\mathcal{M}_b(d(0,R^-))$ if and only if $T(r,h)$ is bounded in $]0,R[$. And by Corollary 43.12, $P(h)$ belongs to $\mathcal{M}_b(d(0,r^-))$ if and only if so does h. Consequently, the claim is immediate. □

Lemma 60.3. *Let P, $Q \in \mathbb{K}[x]$ satisfy $P'Q' \neq 0$. Let f, $g \in \mathcal{M}(\mathbb{K})$ satisfy $P(f) = Q(g)$. Then, $\mathrm{ram}(f) = \mathrm{ram}(g)$ and $P_t(\sqrt[q^t]{f}) = Q_t(\sqrt[q^t]{g})$.*

Proof. Clearly, $f'P'(f) = g'Q'(g)$. Since $P'Q' \neq 0$, either $f' = g' = 0$ or $f'g' \neq 0$. Suppose $f' = g' = 0$. By Proposition 58.2, there exist f_1, $g_1 \in \mathcal{M}(\mathbb{K})$ such that $(f_1)^q = f$, $(g_1)^q = g$. Then, we have $P_1(f_1) = Q_1(g_1)$ and therefore we are led to the same problem with f_1 and g_1. Thus, by induction, after t similar operations, we finally obtain $P_t(\sqrt[q^t]{f}) = Q_t(\sqrt[q^t]{g})$ with $\sqrt[q^t]{f}$, $\sqrt[q^t]{g} \in \mathcal{M}(\mathbb{K})$ and $(\sqrt[q^t]{f})'(\sqrt[q^t]{g})' \neq 0$ and of course $t = \mathrm{ram}(f) = \mathrm{ram}(g)$. □

Theorem 60.4. *Let P, Q be in $\mathbb{K}[x]$ with $P'Q'$ not identically zero and let $m = \deg(P)$, $n = \deg(Q)$ satisfy $2 \leq \min(m,n)$. Assume that there exist k distinct zeros $c_1, \dots, c_k$ of P' such that $P(c_i) \neq P(c_j)$ $\forall i \neq j$ and $P(c_i) \neq Q(\ell)$ for every zero ℓ of Q' $(i = 1, \dots, k)$. Assume that there exist two functions f, $g \in \mathcal{M}_u(d(a,R^-))$ such that $P(f) = Q(g)$ and let $t = \mathrm{ram}(f)$. Then, $\widetilde{N}(r,f) \geq \frac{T(r,f)(kn-m)}{q^t n} + O(1)$. Moreover, suppose $\frac{2m}{3} < n$. Then, $k \leq 2$. Further, if $m \neq n$, then $k = 1$, c_1 is a simple zero of P' and either $n < m$ or $\mathrm{g.c.d.}(m,n) = n - m$.*

Remark. In Corollaries 60.5–60.10, when we assume that one of the functions f and g belongs to $\mathcal{M}_u(d(a,R^-))$, then by Corollary 59.2, both f and g belong to $\mathcal{M}_u(d(a,R^-))$.

Corollary 60.5. *Let P, Q be in $\mathbb{K}[x]$ with $P'Q'$ not identically 0 and such that $\deg(P) = \deg(Q)$. Assume that there exist two distinct zeros c_1, c_2 of P' such that $P(c_1) \neq P(c_2)$ and $P(c_i) \neq Q(\ell)$ for every zero ℓ of Q'. If two functions f, $g \in \mathcal{A}(d(a,R^-))$ satisfy $P(f) = Q(g)$, then f, $g \in \mathcal{A}_b(d(a,R^-))$.*

Proof. Without loss of generality, we can assume $a = 0$. Let $\deg(P) = \deg(Q)$ and assume that the conclusion is wrong, hence by Lemma 60.2, f and g are not bounded, hence by Theorem 43.10, $\lim_{r\to+\infty} T(r,f) = +\infty$. Now, by Theorem 60.4, we have $\lim_{r\to+\infty} \widetilde{N}(r,f) = +\infty$. But since $f \in \mathcal{A}(d(0,R^-))$, this is absurd. □

Corollary 60.6. *Let P, Q be in $\mathbb{K}[x]$ where $P'Q'$ is not identically zero, and let $m = \deg(P)$, $n = \deg(Q)$ with $3 \le m$ and $n \ge \min(2, \frac{m}{2})$. Assume that there exist two distinct zeros c_1, c_2 of P' such that $P(c_i) \neq Q(\ell)$ for every zero ℓ of Q' $(i = 1,2)$. Assume further that, if $m \ge 4$, then $P(c_1) \neq P(c_2)$. If two functions f, $g \in \mathcal{A}(d(a,R^-))$ satisfy $P(f) = Q(g)$, then f, $g \in \mathcal{A}_b(d(a,R^-))$.*

Proof. Without loss of generality, we can assume $a = 0$. If $m = 3$, by Lemma 60.1, we have $P(c_1) \neq P(c_2)$, hence that inequality is satisfied anyway. So, we can apply Theorem 60.4 with $k = 2$ which shows that $\widetilde{N}(r,f)$ is unbounded (because $2\deg(Q) - m > 0$), a contradiction since $f \in \mathcal{A}(d(0,R^-))$. □

Corollary 60.7. *Let P, Q be in $\mathbb{K}[x]$ with $P'Q'$ not identically zero and $\deg(P) < \deg(Q)$. Assume that there exists a zero c of P' such that $P(c) \neq Q(\ell)$ for every zero ℓ of Q'. If two functions f, $g \in \mathcal{A}(d(a,R^-))$ satisfy $P(f) = Q(g)$, then f, $g \in \mathcal{A}_b(d(a,R^-))$.*

Proof. Without loss of generality, we can assume $a = 0$. Let $m = \deg(P)$, $n = \deg(Q)$ and assume that f or g is not bounded. By Lemma 60.2, both f and g are not bounded. We can apply Theorem 59.4 with $k = 1$, $m < n$, which shows that $\widetilde{N}(r,f)$ is unbounded, a contradiction since $f \in \mathcal{A}(d(0,R^-))$. □

We now apply Theorem 59.4 to meromorphic functions in $d(a,R^-)$.

Corollary 60.8. *Let P, Q be in $\mathbb{K}[x]$ with $P'Q'$ not identically zero and $m = \deg(P)$, $n = \deg(Q)$ satisfying $m \neq n$ and $2m < 3n$. Assume that there exist two distinct zeros c_1 and c_2 of P' such that $P(c_1) \neq P(c_2)$ and $P(c_i) \neq Q(\ell)$ for every zero ℓ of Q' $(i = 1,\ 2)$. If f, $g \in \mathcal{M}(d(a,R^-))$ satisfy $P(f) = Q(g)$, then f, $g \in \mathcal{M}_b(d(a,R^-))$.*

Proof. Let k be the number of zeros c_i of P such that $P(c_i) \neq P(c_j)$ $\forall i \neq j$ and $P(c_i) \neq Q(\ell)$ for every zero ℓ of Q' $(i = 1,\ 2)$. By Lemma 60.2, if one of the functions f and g belongs to

$\mathcal{M}_u(d(a,R^-))$, then both belong to $\mathcal{M}_u(d(a,R^-))$. So, if f and g belong to $\mathcal{M}_u(d(a,R^-))$, then by Theorem 60.4 and the assumption that $2m < 3n$, we have $k = 1$, a contradiction. □

Corollary 60.9. *Let P, Q be in $\mathbb{K}[x]$ with $P'Q'$ not identically zero and $3 \leq \deg(P) \leq 4$, $\deg(Q) \geq 3$, $\deg(P) \neq \deg(Q)$. Assume that there exist two distinct zeros c_1 and c_2 of P' such that $P(c_i) \neq Q(\ell)$ for every zero ℓ of Q' ($i = 1, 2$). Moreover, if $\deg(P) = 4$, we assume that $P(c_1) \neq P(c_2)$. If $f, g \in \mathcal{M}(d(a,R^-))$ satisfy $P(f) = Q(g)$, then $f, g \in \mathcal{M}_b(d(a,R^-))$.*

Proof. Let $m = \deg(P)$, $n = \deg(Q)$. We note that the inequality $P(c_1) \neq P(c_2)$ is assured by Lemma 59.1 when $m = 3$ and is assumed when $m = 4$ and hence holds in all cases. Assume that one of the two functions f, g belongs to $\mathcal{M}_u(d(a,R^-))$, hence by Lemma 60.2, both f and g belong to $\mathcal{M}_u(d(a,R^-))$. By Theorem 60.4, we have $k = 1$, a contradiction to the hypothesis. □

Corollary 60.10. *Let P, Q be in $\mathbb{K}[x]$ with $P'Q'$ not identically 0 and $m = \deg(P)$, $n = \deg(Q)$, satisfying $m < n$ and $n - m \neq$ g.c.d.(m,n). Assume that there exists a zero c of P' such that $P(c) \neq Q(\ell)$ for every zero ℓ of Q'. If $f, g \in \mathcal{M}(d(a,R^-))$ satisfy $P(f) = Q(g)$, then $f, g \in \mathcal{M}_b(d(a,R^-))$.*

Proof. Assume that one of the two functions f, g belongs to $\mathcal{M}_u(d(a,R^-))$, hence by Lemma 60.2, both f and g belong to $\mathcal{M}_u(d(a,R^-))$. Since $m \neq n$, by Theorem 60.4, we have that either $n < m$ or g.c.d.$(m,n) = n - m$, both cases that are excluded from the hypotheses of the corollary. Hence, both f and g belong to $\mathcal{M}_b(d(a,R^-))$. □

Corollary 60.11. *Let P, Q be in $\mathbb{K}[x]$ with $P'Q'$ not identically zero and $\deg(P) = \deg(Q) \geq 4$. Assume that there exist three distinct zeros c_1, c_2, c_3 of P' such that $P(c_i) \neq P(c_j)$ $\forall i \neq j$ and $P(c_i) \neq Q(\ell)$ for every zero ℓ of Q' ($i = 1, 2, 3$). If $f, g \in \mathcal{M}(d(a,R^-))$ satisfy $f' \neq 0$ and $P(f) = Q(g)$, then $f, g \in \mathcal{M}_b(d(a,R^-))$.*

Proof. Without loss of generality, we can assume $a = 0$. Let $n = \deg(P) = \deg(Q)$ and assume that f or g belongs to $\mathcal{M}_u(d(a,R^-))$, hence by Lemma 60.2, both f and g belong to $\mathcal{M}_u(d(a,R^-))$.

Since $f' \neq 0$, we have $\mathrm{ram}(f) = 0$, hence by Theorem 60.4, $\widetilde{N}(r, f) \geq T(r, f)\left(\frac{3n-n}{n}\right) + O(1)$, therefore $\lim_{r \to +\infty} \widetilde{N}(r, f) - T(r, f) = +\infty$, which is absurd. $\square$

Theorem 60.12. *Let P, $Q \in K[x]$ with $P'Q'$ not identically zero and let $m = \deg(P)$, $n = \deg(Q)$ with $2 \leq \min(m, n)$. Assume that there exist k distinct zeros $c_1, \ldots, c_k$ of P' such that $P(c_i) \neq P(c_j)$ $\forall i \neq j$ and $P(c_i) \neq Q(\ell)$ for every zero ℓ of Q' $(i = 1, \ldots, k)$. Assume that there exist two non-constant functions f, $g \in \mathcal{M}(\mathbb{K})$ such that $P(f) = Q(g)$ and let $t = \mathrm{ram}(f)$. Then, $q \leq p$ and f satisfies $\widetilde{N}(r, f) \geq \frac{T(r,f)(kq-p)}{q^t q} + \log r + O(1)$. Moreover, if $\frac{p}{2} < q$, then $k = 1$ and c_1 is a simple zero of P'.*

Corollary 60.13. *Let P, Q be in $\mathbb{K}[x]$ with $P'Q'$ not identically zero and* $\deg(P) \leq \deg(Q)$. *Assume that there exists a zero c of P' such that $P(c) \neq Q(\ell)$ for every zero ℓ of Q'. If there exist f, $g \in \mathcal{A}(\mathbb{K})$ satisfying $P(f) = Q(g)$, then f and g are constant.*

Proof. Let $m = \deg(P)$, $n = \deg(Q)$. Assume that f and g are not constant. Then by Theorem 60.12, we have $\widetilde{N}(r, f) \geq T(r, f)\left(\frac{n-m}{q^t n}\right) + \log r + O(1) \geq \log r + O(1)$, hence $\lim_{r \to +\infty} \widetilde{N}(r, f) = +\infty$, which is absurd. $\square$

Corollary 60.14. *Let P, $Q \in \mathbb{K}[x]$ with $P'Q'$ not identically zero and let $m = \deg(P)$, $n = \deg(Q)$ with $2 \leq \min(m, n)$ and $\frac{m}{2} < n$. Assume that there exist 2 distinct zeros c_1, c_2 of P' such that $P(c_1) \neq P(c_2)$ and $P(c_i) \neq Q(\ell)$ for every zero ℓ of Q' $(i = 1, 2)$. Assume that there exist two functions f, $g \in \mathcal{M}(\mathbb{K})$ such that $P(f) = Q(g)$. Then f and g are constants.*

Proof (Theorems 60.4 and 60.12). In Theorem 60.4, without loss of generality, we can assume $a = 0$. We first assume that both f, g have a ramification index equal to 0. Let $b = \text{g.c.d.}(m, n)$ and $m = b\widehat{m}$, $n = b\widehat{n}$. We can clearly assume that $a = 0$ and that none of the meromorphic functions to be considered attains 0 or ∞ at 0. We denote by I (respectively J) an interval of the form $[u, +\infty[$ (respectively $[u, \log r[$).

Clearly, $mT(r, f) = nT(r, g) + O(1)$, hence

(1) $\quad \widehat{m}T(r, f) = \widehat{n}T(r, g) + O(1).$

Let α be a pole of f of order k. Then α is a pole of g of order l such that $k\widehat{m} = l\widehat{n}$, hence $\widehat{n}$ divides k, therefore

(2) $N(r, f) \geq \widehat{n}\widetilde{N}(r, f)$.

Since c_i is a zero of P', $P - P(c_i)$ has a factorization of the form $(x - c_i)^{s_i} R_i(x)$, with $R_i \in \mathbb{K}[x]$, $R_i(c_i) \neq 0$. Consequently,

(3) $(f - c_i)^{s_i} R_i(f) = Q(g) - P(c_i)$ $(i = 1, \dots, k)$.

We put $S = \sum_{i=1}^{t}(s_i - 1)$. Since by definition we have $s_i \geq 2$ $\forall i = 1, \dots, t$, then $S \geq t$.

Since, by hypothesis, $Q - P(c_i)$ does not vanish at any zero of Q'; this implies that it has no multiple zero and hence it has a factorization in the form $\prod_{j=1}^{q}(x - b_{i,j})$, where the $b_{i,j}$ are all distinct for each fixed i. It follows that

$$(4) \quad \widetilde{Z}(r, Q - P(c_i)) = \sum_{j=1}^{w} \widetilde{Z}(x - b_{i,j}) \ (i = 1, \dots, t).$$

On the other hand, we note that $b_{i,j} \neq b_{k,l}$ whenever $(i, j) \neq (k, l)$. Indeed, suppose that $b_{i,j} = b_{k,l}$ for some $(i, j) \neq (k, l)$. Then $i \neq k$, hence $P(c_i) \neq P(c_k)$, therefore $(Q - P(c_i)) - (Q - P(c_k))$ is a constant different from 0. But since $b_{i,j} = b_{k,l}$, the point $b_{i,j}$ is a zero of $(Q - P(c_i)) - (Q - P(c_k))$, a contradiction. Thus, all the points $b_{i,j}$ are distinct $(i = 1, \dots, t;\ j = 1, \dots, w)$.

Now, by applying Theorem 58.11 to g at the points $b_{i,j}$, for all $i = 1, \dots, t$ and $j = 1, \dots, w$, we have

$$(5) \quad (nt - 1)T(r, g) \leq \sum_{i=1}^{t} \sum_{j=1}^{w} \widetilde{Z}(r, g - b_{i,j}) + \widetilde{N}(r, g) - \log r + O(1).$$

Hence, by (4), we have

(6) $(tn - 1)T(r, g) \leq \sum_{i=1}^{t} \widetilde{Z}(r, Q - P(c_i))$.

But by (3) and (6), we obtain

$$\begin{aligned}(7) \quad \widetilde{Z}(r, Q(g) - P(c_i)) &\leq \widetilde{Z}(r, f - c_i) + \widetilde{Z}(r, R_i(f)) + O(1) \\ &\leq (m - s_i + 1)T(r, f) + O(1) \\ &\leq (m - 1)T(r, f) + O(1), \quad (i = 1, \dots, t).\end{aligned}$$

On the other hand, it is obvious that $\widetilde{N}(r,g) = \widetilde{N}(r,f)$. Hence, it follows from (2) that

$$\widetilde{N}(r,g) = \widetilde{N}(r,f) \leq \frac{1}{\widehat{n}} N(r,f) \leq \frac{1}{\widehat{n}} T(r,f).$$

Next, by (5) and (7), we have $\widetilde{N}(r,f) \geq (tn-1)T(r,g) - \sum_{i=1}^{t} \widetilde{Z}(r, f-c_i) + \widetilde{Z}(r, R_i(f)) + \log r + O(1) \geq (tn-1)T(r,g) - T(r,f) \sum_{i=1}^{t}(1+m-s_i) + \log r + O(1) = (tn-1)T(r,g) - T(r,f)$ $(tm - S) + \log r + O(1)$, hence

$$(8) \quad \widetilde{N}(r,f) \geq \left(\frac{(Sn-m}{n}\right) T(r,f) + \log r + O(1).$$

Since $t \leq S$, the inequality $\widetilde{N}(r,f) \geq \frac{T(r,f)(tn-m)}{n} + O(1)$ is then proven.

Now by (1), we can write $(tn-1)T(r,g) = (\frac{tn-1}{n})mT(r,f) + O(1)$, hence by (2), (5), (6), and (7), we obtain

$$(9) \qquad \left(\frac{tn-1}{n}\right) mT(r,f) \leq \left(\sum_{i=1}^{t}(m - s_i + 1) + \frac{1}{\widehat{n}}\right) T(r,f) \\ - \log r + O(1).$$

Therefore, since f, g belong to $\mathcal{M}_u(d(0,R^-))$ in Theorem 60.4 and in $\mathcal{M}(\mathbb{K})$ in Theorem 60.12, then by Theorem 43.10, (9) implies

$$(10) \quad (tn-1)m \leq ntm - n\left(\sum_{i=1}^{t} s_i\right) + nt + b.$$

We can then deduce $(nt-1)\widehat{m} \leq \widehat{n}tm - \widehat{n}(\sum_{i=1}^{t} s_i) + \widehat{n}t + 1$. But since $n\widehat{m} = \widehat{n}m$, we have $\sum_{i=1}^{t}(s_i - 1)\widehat{n} \leq \widehat{m} + 1$. It follows that

$$(11) \; S\widehat{n} \leq \widehat{m} + 1.$$

Thus, in Theorem 60.4, by (11), we see that either $n \leq m$ or $\widehat{n} = \widehat{m} + 1$, hence g.c.d.$(m,n) = n - m$ and then $S = 1 = t$.

In Theorem 60.12, both f and g belong to $\mathcal{M}(\mathbb{K})$. By (8), we have already shown $\widetilde{N}(r,f) \geq (\frac{nt-m}{n})T(r,f) + \log r + O(1)$.

Moreover, since $\log r$ tends to ∞ with r, by (9), we obtain $(\frac{tn-1}{n})m < \sum_{i=1}^{t}(m - s_i + 1) + \frac{1}{\widehat{n}}$, hence $S\widehat{n} < \widehat{m} + 1$, i.e.

(12) $\quad S\widehat{n} \leq \widehat{m}.$

Thus, in Theorem 60.12, by (12), we have $n \leq m$.

Now assume the hypotheses of Theorem 60.4 hold and suppose that $m < \frac{3n}{2}$. So we have $\widehat{m} < \frac{3\widehat{n}}{2}$ in Theorem 60.4. Therefore, by (11), we obtain

(13) $\quad \frac{2}{3}S\widehat{m} < \widehat{m} + 1,$

hence $S \leq 2$.

Suppose $S = 2$. By (13), we have $\widehat{m} = 1$ or 2. If $\widehat{m} = 2$, since $S\widehat{m} \leq \widehat{m} + 1$, and since g.c.d.$(\widehat{m}, \widehat{n}) = 1$, we are led to $\widehat{n} = 1$, a contradiction to the hypothesis $m \leq \frac{3n}{2}$. Consequently, $\widehat{m} = 1$, and then $m = n = 1$. Thus, if $m \neq n$, we have $S = 1$, hence $t = 1$, $s_1 = 2$, therefore c_1 is a simple zero of P'. This also completes the proof of Theorem 60.4.

Now, in Theorem 60.12, assuming $\frac{m}{2} < n$, by (12), we see that the inequality $S \geq 2$ is impossible, hence $S = t = 1$, therefore c_1 is a simple zero of P'.

We must now consider the general case when the ramification index of one of the two functions f, g is not 0. By Lemma 60.3, f and g have the same ramification index h and satisfy $P_h(\sqrt[q^h]{f}) = Q_h(\sqrt[q^h]{g})$. Moreover, one can easily verify that all hypotheses on the degrees satisfied by P and Q are also satisfied by P_h and Q_h. And of course, $\sqrt[q^h]{f}$, $\sqrt[q^h]{g}$ lie in $\mathcal{M}_u(d(0, R^-))$ in Theorem 60.4 and in $\mathcal{M}(\mathbb{K})\setminus\mathbb{K}$ in Theorem 60.12. Consequently, putting $f_{(h)} = \sqrt[q^h]{f}$, $g_{(h)} = \sqrt[q^h]{g}$, we have

$$\widetilde{N}(r, f_{(h)}) \geq \frac{T(r, f_{(h)})(tn - m)}{n} + O(1).$$

Actually, $\widetilde{N}(r, f_{(h)}) = \widetilde{N}(r, f)$ and $T(r, f_{(h)}) = \frac{T(r,f)}{q^h}$, thereby the inequality $\widetilde{N}(r, f_{(h)}) \geq \frac{T(r,f_{(h)})(tn-m)}{q^h n} + O(1)$ is established. Now, suppose $\frac{2m}{3} < n$. Then we can check that all statements on t, m, n, true for P_h, Q_h, hold for P, Q. Similarly, we can conclude in Theorem 60.12, when $\frac{m}{2} < n$. □

Henceforth, we suppose that the characteristic of $\mathbb{K}$ *is null and we consider functional equations of the form* $P(x)(g(x))^n = Q(x)(f(x))^m + S(x)$, *where* S *is a small function with respect to* f *and* g *and we show that the equations have no solutions* f, g *when* f, g *are transcendental or are unbounded inside a disk.*

Theorem 60.15. *Let* P, Q, $S \in \mathbb{K}(x)$ *and suppose that two functions* f, $g \in \mathcal{M}(\mathbb{K})$ *have finitely many poles and satisfy* $P(x)(g(x))^n = Q(x)(f(x))^m + S(x)$ $\forall x \in \mathbb{K}$ *with* $\min(m, n) \geq 2$. *Then,* f *and* g *belong to* $\mathbb{K}(x)$.

Proof. Let $F(x) = Q(x)(f(x))^m + S(x)$. Then F belongs to $\mathcal{M}(\mathbb{K})$ and has finitely many poles. Since $P(x)(g(x))^n = F(x)$, 0 is a perfectly branched value for F. On the other hand, $F(x) - S(x) = Q(x)(f(x))^m$, hence $Q(x)$ is a perfectly branched polynomial for F. But then by Theorem 49.9, F is not transcendental and hence belongs to $\mathbb{K}(x)$ and then so does $(f(x))^m$. But since the entire function f is algebraic over $\mathbb{K}(x)$, actually by Corollary 51.7, it belongs to $\mathbb{K}(x)$. But then, so does g^n and similarly, so does g. □

Corollary 60.16. *Let* P, Q, $S \in \mathbb{K}[x]$ *and suppose that two entire functions* f, $g \in \mathcal{A}(\mathbb{K})$ *satisfy* $P(x)(g(x))^n = Q(x)(f(x))^m + S(x)$ $\forall x \in \mathbb{K}$ *with* $\min(m, n) \geq 2$. *Then,* f *and* g *belong to* $\mathbb{K}[x]$.

In Theorem 29.23, we saw that the function $\sqrt[q]{.}$ is defined in $d(1, 1^-)$ when q is prime to p and that $\sqrt{.}$ is defined in $d(1, (\frac{1}{2})^-)$ if $p = 2$.

Theorem 60.17. *Let* $h \in \mathcal{A}(\mathbb{K})$ *have all its zeros of order multiple of* $q \geq 2$ *and let* $w \in \mathbb{K}[x]$ *be non-identically* 0. *Then the functional equation*

$$(\mathcal{E}) \quad (g(x))^q = h(x)(f(x))^q + w(x)$$

has no solution f, $g \in \mathcal{A}(\mathbb{K}) \setminus \mathbb{K}[x]$.

Proof. Let f, $g \in \mathcal{A}(\mathbb{K})$ satisfy $(\mathcal{E})$. If $h \in \mathbb{K}[x]$, by Corollary 60.16, both f and g are polynomials. Now suppose that $h \notin \mathbb{K}[x]$. By Theorem 29.7, there exists $\phi \in \mathcal{A}(\mathbb{K})$ such that $\phi^q = h$. Consequently, since w is not 0, we have $w(x) = (g(x))^q - (\phi(x)f(x))^q$, and hence, by Theorem 50.4, $(g(x))^q - (\phi(x)f(x))^q$ does not belong to $\mathcal{A}_f(\mathbb{K})$, a contradiction to the hypothesis $(g(x))^q - h(x)(f(x))^q = w(x)$. □

We can now consider analytic functions inside a disk $d(a, R^-)$.

Theorem 60.18. *Let $q \in \mathbb{N}^*$ be prime to p and let $R \in]0, +\infty[$. Let $h \in \mathcal{A}(d(a, R^-))$ be such that that all zeros of h are of order multiple of q and let $w \in \mathcal{A}_b(d(a, R^-))$. Then the functional equation*

$$(\mathcal{E}) \quad (g(x))^q = h(x)(f(x))^q + w(x)$$

has no solution in $\mathcal{A}_u(d(a, R^-))$.

Similarly, we can prove Theorem 60.19:

Theorem 60.19. *Suppose $\mathbb{K}$ has residue characteristic 2. Let $a \in \mathbb{K}$ and $R \in]0, +\infty[$ and let $h,\ w \in \mathcal{A}_b(d(a, R^-))$ be such that that all zeros of h are of even order. Then the equation $(\mathcal{E})$ $g^2 = hf^2 + w$ has no solution $f,\ g \in \mathcal{A}_u(d(a, R^-))$.*

Proof of Theorems 60.18 and 60.19. We can obviously suppose $a = 0$. Suppose that $(\mathcal{E})$ has solutions $f,\ g \in \mathcal{A}_u(d(0, R^-))$. By Theorem 29.23, we can apply the root function $\sqrt[q]{.}$ to the function h in both Theorems 60.18 and 60.19. Therefore, there exists $\phi \in \mathcal{A}_b(d(0, R^-))$ such that $\phi^q = h$. Consequently, we have $g^q - (\phi f)^q = w$ (with $q = p = 2$ in Theorem 60.19). But by Theorem 50.4, $g^q - (\phi f)^q$ is not bounded, which ends the proof. □

By Theorem 60.18, we can also derive the following corollary:

Corollary 60.20. *Suppose $\mathbb{K}$ has residue characteristic $p \neq 2$. Let $R \in]0, +\infty[$, and let $h,\ w \in \mathcal{A}_b(d(a, R^-))$ be such that $h(x) \neq 0$ $\forall x \in d(a, R^-)$. Then the equation $(\mathcal{E})$ $g^2 = hf^2 + w$ has no solution $f,\ g \in \mathcal{A}_u(d(a, R^-))$.*

Chapter 61

Yoshida's Equation in the Field $\mathbb{K}$

Throughout this chapter, the field $\mathbb{K}$ has characteristic $p \geq 0$.

We call Yoshida's equation a differential equation of the form $(\mathcal{E})$ $(y')^m = F(x,y)$ (with $F(x,y) \in \mathbb{K}(x,y)\setminus\mathbb{K}$). Several results were obtained in characteristic 0 and in characteristic $p \neq 0$, for meromorphic functions in the whole field $\mathbb{K}$ or inside a disk $d(a,R^-)$. In [24], it was shown that if $(\mathcal{E})$ admits solutions in $\mathcal{M}(\mathbb{K})\setminus\mathbb{K}(x)$, then $F \in \mathbb{K}(x)[y]$ and $\deg_y(F) \leq 2m$. Moreover, it was shown that if $F \in \mathbb{K}(y)$, then any solution of the equation lying in $\mathcal{M}(\mathbb{K})$ is a rational function with a very specific form. That was generalized in characteristic $p \neq 0$ in [27]. Theorems 61.1 and 61.2 were proven when the ground field is $\mathbb{C}_p$. They have an immediate generalization in any algebraically closed field complete for an ultrametric absolute value, such as $\mathbb{K}$.

Theorem 61.1. *Let F, $G \in \mathbb{K}(x,Y)$ be irreducible. If the differential equation $F(x,y^{(n)}) = G(x,y)$ admits a solution $f(x) \in \mathcal{M}(\mathbb{K})\setminus\mathbb{K}(x)$, then* $\deg(G) \leq (n+1)\deg(F)$.

Proof. Suppose the given equation admits a solution $f(x) \in \mathcal{M}(\mathbb{K})\setminus\mathbb{K}(x)$. Let $w = \deg(F)$, $\ell = \deg(G)$.

By Theorem 51.5, we have $T(r, G(x,f(x))) = \ell T(r,f) + o(T(r,f))$ and $T(F(x,f^{(n)}(x))) = wT(r,f^{(n)}) + o(T(r,f))$. Now by Corollary 43.15, we have $T(r,f^{(n)}) \leq (n+1)T(r,f) + O(1)$. Consequently, $\ell T(r,f) \leq w(n+1)T(r,f) + o(T(r,f))$. Thus, since $T(r,f)$ is unbounded, $\ell \leq (n+1)w$. □

Notation. Let $F(x,y) \in \mathbb{K}(x,y)$. We denote by $\deg_x(F)$ (respectively $\deg_y(F)$) the degree of F as a rational function that belongs to $\mathbb{K}(y)(x)$ (respectively to $\mathbb{K}(x)(y)$).

Theorem 61.2 comes from [24] in characteristic 0 and from [27] in characteristic $p \neq 0$.

Theorem 61.2 (Yoshida). *Let $F(x,y) \in \mathbb{K}(x,y)\setminus\{0\}$ and suppose that there exists a non-constant solution $f \in \mathcal{M}(\mathbb{K})\setminus\mathbb{K}(x)$ of the differential equation*

$$(\mathcal{E}) \quad (y')^m = F(x,y).$$

Then, F belongs to $\mathbb{K}(x)[y]$ and $\deg_y(F) \leq 2m$.

Proof. Let us write $G(x,y)$ in an irreducible form $\frac{P(x,y)}{Q(x,y)}$ with $P,\ Q \in \mathbb{K}[x,y]$, $\deg(P) = s$, $\deg(Q) = t$. Let $\alpha \in \mathbb{K}$ be such that $P(x,\alpha)$ is not identically zero and set $z = \frac{1}{y-\alpha}$. Equation $(\mathcal{E})$ is equivalent to

$$(\mathcal{E}') \quad (z')^n = R(x,z),$$

where $R(x,z)$ is an irreducible rational function that we may write in the form $\frac{\sum_{j=0}^{s} a_j(x) z^{2n-j}}{\sum_{j=0}^{t} b_j(x) z^{-j}}$, with $a_j \in \mathbb{K}[x], (0 \leq j \leq s),\ b_j \in \mathbb{K}[x], (0 \leq j \leq t)$. We note that $a_0(x) = P(\alpha) \neq 0$.

Suppose first $s - 2n \geq t$. Then,

$$G(x,z) = (-1)^n \frac{\sum_{j=0}^{s} a_j(x) z^{s-j}}{\sum_{j=0}^{t} b_j(x) z^{s-2n-j}}.$$

So, $\deg_z(R(x,z)) = s$. By Theorem 61.1, we have $s \leq 2n$, hence $t = 0$ and therefore G belongs to $\mathbb{K}[x,y]$.

Suppose now $s - 2n < t$. We may write $R(x,z)$ in the form

$$(-1)^n \frac{\sum_{j=0}^{s} a_j(x) z^{2n+t-j}}{\sum_{j=0}^{t} b_j(x) z^{t-j}}.$$

Thus, we can see that $\deg_z(R) = 2n + t$, hence by Theorem 61.1, we have $2n \geq 2n + t$, thereby $t = 0$ again. $\square$

Theorem 61.3. *Let $F(Y) \in \mathbb{K}(Y)\setminus\mathbb{K}$ and let $(\mathcal{E})$ be the equation $(y')^m = F(y)$. Then there exists non-constant solutions $f \in \mathcal{M}(\mathbb{K})$ of $(\mathcal{E})$ if and only if F is of the form $B(Y-c)^\ell$ $(0 \le \ell \le 2m)$, whereas $m-\ell$ divides m. If those hypotheses are satisfied, then the solutions are the functions of the form $f(x) = c + (ax + \phi(x^p))^{\frac{m}{m-\ell}}$, where a satisfies*

$\left(\frac{am}{m-\ell}\right)^m = B$ *when* $m > \ell$,
$\left(\frac{am}{m-\ell}\right)^m = (-1)^m B$ *when* $m < \ell$

and

ϕ *is a constant when* $p = 0$,
$\phi(u) \in \mathcal{M}(\mathbb{K})$ *when* $p \neq 0$.

Proof. Suppose first that a solution f is such that $f' = 0$. Then we have $F(f(x)) = 0$. Therefore, f must be equal to one of the zeros of F for every $x \in K$, and finally, f is a constant. Consequently, a solution f is such that f' is not identically 0. Therefore, $\text{ram}(f) = 0$.

We first verify that F is a polynomial of degree $\ell \le 2m$. Let $f \in \mathcal{M}(\mathbb{K})\setminus\mathbb{K}$ be a solution of the equation $(\mathcal{E})$. Let b be a pole of $F(X)$ and β be a zero of $f - b$. Then β is a pole of $F \circ f$, hence by equation $(\mathcal{E})$, β is a pole of f', a contradiction since f has no pole at β. Consequently, f must avoid any pole of $F(X)$. Therefore, by Corollary 36.23, $F(X)$ admits at most one pole b and hence is of the form $\frac{Q(X)}{(X-b)^s}$, with $Q(X) \in \mathbb{K}[X], Q(b) \neq 0$ and $s \in \mathbb{N}$. Suppose $s \ge 1$. Since f must avoid any pole of F, f is of the form $f = b + \frac{1}{h}$ with $h \in \mathcal{A}(\mathbb{K})\setminus\mathbb{K}$ and hence $h' \neq 0$. From $(\mathcal{E})$, we derive that h satisfies $(h')^m = (-1)^m h^{2m+s} Q(b + \frac{1}{h})$. Since $h \in \mathcal{A}(\mathbb{K})$ and is not a constant, h admits at least one zero γ and then we can find $M \in \mathbb{R}$ such that $M \ge \Psi(\gamma)$ and $\Psi(h, \mu) > -\Psi(b)$ $\forall \mu \ge M$. Then, $\Psi(b + \frac{1}{h}, \mu) = \Psi(b)$ $\forall \mu \ge M$, hence $\Psi(Q(b + \frac{1}{h})) = \Psi(Q(b))$ $\forall \mu \ge M$. Since $h' \neq 0$, we obtain $m\Psi(h', \mu) = (2m + s)\Psi(h, \mu) + \Psi(Q(b))$ $\forall \mu \ge M$. But by Theorem 37.10, we have $\Psi(h', \mu) \le \Psi(h, \mu) - \mu$ $\forall \mu \in \mathbb{R}$, hence $m\big(\Psi(h, \mu) - \mu)\big) \ge (2m + s)\Psi(h, \mu) + \Psi(Q(b))$ and therefore

$$(1) \quad (m + s)\Psi(h, \mu) + m\mu + \Psi(Q(b)) \le 0.$$

Now, since $M \ge \Psi(\gamma)$ and since h belongs to $\mathcal{A}(\mathbb{K})$, the function $m(\Psi(h, \mu) + \mu) + \Psi(Q(b))$ is strictly increasing and tends to $+\infty$

with μ, a contradiction to Relation (1). Consequently, $F(X)$ has no pole and hence is a polynomial $a_0 + a_1X + \cdots + a_\ell X^\ell$. By Theorem 43.8, we have $T(r, F \circ f) = \ell T(r, f) + O(1)$, $(r \in J)$, so, then, using the inequality $T(r, f') \leq 2T(r, f)$, we can derive from $(\mathcal{E})$ that $\ell \leq 2m$.

Now, let c be a zero of $F(X)$ of order n and let α be a zero of $f - c$ of order w such that $p \not| w$. We check that

$$(2) \quad w = \frac{m}{m-n} \quad \text{and} \quad m > n.$$

So, all zeros of $f - c$ are multiple, and, if their order is not a multiple of p, they have the same multiplicity order. Furthermore, we have $\frac{m}{m-n} \geq 2$.

If we consider a pole of f of order t such that $p \not| t$, we have

$$(3) \quad t = \frac{m}{\ell - m} \quad \text{and} \quad m < \ell.$$

So, all poles of f are multiple and the poles whose order is not multiple of p have the same multiplicity order.

Let $c_1, \ldots, c_k$ be the distinct zeros of F and for each $j = 1, \ldots, k$, and let n_j be the order of c_j as a zero of F. Let us apply Theorem 58.11 to f. Since $\text{ram}(f) = 0$, we have

$$(k-1)T(r, f) \leq \sum_{j=1}^{k} \widetilde{Z}(r, f - c_j) + \widetilde{N}(r, f) - \log r + O(1),$$

and then, by (2) and (3), we can derive

$$(k-1)T(r, f) \leq (Tr, f) \left(\sum_{j=1}^{k} \left(\frac{m - n_j}{m} \right) + \frac{\ell - m}{m} \right) - \log r + O(1),$$

hence

$$(k-1)T(r, f) \leq \frac{\left(\sum_{j=1}^{k} (m - n_j) \right) + \ell - m}{m} T(r, f) - \log r + O(1).$$

But $\sum_{j=1}^{k} n_j = \ell$, so we obtain

$$(k-1)T(r, f) \leq (k-1)T(r, f) - \log r + O(1),$$

which is impossible whenever $k \geq 2$. That shows that $k = 1$. Therefore, $F(X)$ is of the form $B(X - c)^\ell$. Now, by (2) and (3), we can see that $f - c$ can't have simultaneously a zero and a pole whose order is not multiple of p.

So, by (2) and (3), either $m > \ell$ and then f belongs to $\mathcal{A}(K)$ and admits at least a zero, or $\ell > m$ and then f has no zero but admits at least a pole and is of the form $c + \frac{1}{h}$ with $h \in \mathcal{A}(\mathbb{K})$. And since f' is not identically zero, we note that f must have at least one zero or one pole, therefore f is not a constant.

(A) Suppose $m > \ell$ and hence f belongs to $\mathcal{A}(K)$. Let γ be a zero of $f - c$. Then $m\omega_\gamma(f') = \ell\omega_\gamma(f - c)$. Let $\rho = \text{g.c.d.}(m, \ell)$ and set $m = \sigma\rho$, $\ell = \tau\rho$. Then $\sigma\omega_\gamma(f') = \tau\omega_\gamma(f - c)$. Now, since σ, τ are relatively prime, σ divides $\omega_\gamma(f - c)$ for every zero γ of $f - c$, and therefore, by Theorem 27.7, there exists $g \in \mathcal{A}(\mathbb{K})$ such that $g^\ell = f - c$. On the other hand, from the given equation $(f')^{\sigma\rho} = C_2(f - c)^{\tau\rho}$, the function $\theta(x) = \frac{(f')^\sigma}{C_2(f-c)^\tau}$ is a meromorphic function such that $\theta^\rho = 1$. Then by Lemma 36.1, θ is a constant. Consequently, $(f')^\sigma$ is of the form $C_3(f - c)^\tau$, with $(C_3)^\rho = C_2$, hence $(f')^\ell = C_3 g^{\tau\ell}$, therefore $f' = C_4 g^\tau$ with $C_4^\sigma = C_3$. But $f' = \sigma g' g^{\sigma-1}$, hence $\sigma g' g^{\sigma-1} = C_4 g^\tau$ and finally $g' g^{\sigma-\tau-1} = \frac{C_4}{\sigma}$. Now, since we have noted that $\ell < m$, we have $\sigma - \tau - 1 \geq 0$. Consequently, since $g \in \mathcal{A}(\mathbb{K})$, the only solution is that both g' and $g^{\sigma-\tau-1}$ are constant i.e. $g(x)$ is of the form $ax + v(x)$ with $\sigma = \tau + 1$, with $a \in \mathbb{K}^*$ and $v \in \mathcal{M}(\mathbb{K})$ has a null derivative i.e. if $p = 0$, then v is just a constant $b \in \mathbb{K}$, and if $p \neq 0$, then v is of the form $\phi(x^p)$, whereas $\phi(u)$ belongs to $\mathcal{M}(\mathbb{K})$.

More precisely, we note that $\sigma = \frac{m}{m-\ell}$ because $\sigma = \tau + 1$. Consequently, $f(x) = c + (ax + v)^{\frac{m}{m-\ell}}$ and therefore $f'(x) = \frac{am}{m-\ell}(ax + v)^{\frac{\ell}{m-\ell}}$. Now, since f is solution of $(\mathcal{E})$ and since $v'(x)$ is identically zero, we can derive

$$\left(\frac{am}{m-\ell}\right)^m (ax + v(x))^{\frac{m\ell}{m-\ell}} = B(ax + v(x))^{\frac{m\ell}{m-\ell}},$$

therefore $\left(\frac{am}{m-\ell}\right)^m = B$.

(B) Suppose $\ell > m$ and hence f is of the form $c + \frac{1}{h}$ with $h \in \mathcal{A}(\mathbb{K})$. Then h satisfies $h'^m = (-1)^m B h^{2m-\ell}$. Thus, we are lead to the same problem with $2m - \ell$ instead of ℓ and a polynomial $G(Y)$ of the form $(-1)^m BY^{2m-\ell}$ instead of F. Consequently, h is of the form $(ax + v(x))^{\frac{m}{\ell-m}}$ with v of the form $\phi(x^p)$, whereas $\phi(u)$ belongs to

$\mathcal{M}(\mathbb{K})$ and with $\left(\frac{am}{\ell-m}\right)^m = (-1)^m B$. So, $f(x) = c + (ax + v(x))^{\frac{m}{m-\ell}}$ with $\left(\frac{am}{\ell-m}\right)^m = (-1)^m B$.

Conversely, suppose that $m - \ell$ divides m and suppose that $F(Y)$ is of the form $B(Y - c)^\ell$. Let $f(x) = c + (ax + \phi(x^p))^{\frac{m}{m-\ell}}$, where a satisfies the following: $\left(\frac{am}{m-\ell}\right)^m = B$ when $m > \ell$, $\left(\frac{am}{m-\ell}\right)^m = (-1)^m B$ when $m < \ell$ and satisfies further the following: ϕ is a constant when $p = 0$, $\phi(u) \in \mathcal{M}(\mathbb{K})$ when $p \neq 0$. Then we can check that f is a solution of $(\mathcal{E})$. $\square$

Chapter 62

Yoshida's Equation Inside a Disk

Here we want to consider Yoshida's equation for meromorphic functions inside a disk $d(a, R^-)$ [23]. Several methods look like parts of the proof of Theorem 61.3. However, we do not obtain a result as general as in Chapter 61.

Throughout this chapter, $\mathbb{K}$ *has characteristic* 0.

Proposition 62.1. *Let* $F(Y)$, $G(Y) \in \mathcal{M}_b(d(a, R^-))(Y)$ *be irreducible as rational functions. If the differential equation* $F(y^{(n)}) = G(y)$ *admits a solution* $f \in \mathcal{M}_u(d(a, R^-))$, *then* $\deg(G) \leq (n+1)\deg(F)$.

Proof. Suppose there exists $f \in \mathcal{M}_u(d(a, R^-))$ such that $F(f^{(n)}) = G(f(x))$.

On one hand, by Corollary 43.15, we have $T(r, (f^{(n)})) \leq (n+1)\, T(r, f) + O(1)$. Hence, by Theorem 51.5 and Corollary 43.15, we have $T(r, F(f^{(n)})) \leq (n+1)\deg(F)T(r, f) + o(T(r, f))$. On the other hand, by Theorem 51.5, we have $T(r, G(f)) = \deg(G)T(r, f) + o(T(r, f))$. Now, since $T(r, f)$ is unbounded, the conclusion is clear. □

Theorem 62.2. *Let* $m \in \mathbb{N}^*$ *and let* $F \in \mathcal{M}_b(d(a, R^-))(Y)$. *If the differential equation* $(\mathcal{E})$ $(y')^m = F(y)$ *admits a solution* $f \in \mathcal{M}_u(d(a, R^-))$, *then* F *belongs to* $\mathcal{M}_b(d(a, R^-))[Y]$ *and* $\deg(F) \leq 2m$.

Proof. Let us write $f(Y) = \frac{\sum_{i=0}^{k} c_i(x)Y^i}{\sum_{i=0}^{l} b_i(x)Y^i}$ and suppose that $(\mathcal{E})$ admits a solution $f \in \mathcal{M}_u(d(a, R^-))$. Let us take $\alpha \in \mathbb{K}$ such that $\sum_{i=0}^{k} c_i(x)\alpha^i$ is not identically zero and let $g = \frac{1}{f-\alpha}$. Let $G \in \mathcal{M}_b(d(a, R^-))(Y)$ be defined by $F(Y) = G(\frac{1}{Y-\alpha})$. Then g satisfies an equation

$$(g')^m = G(g) \tag{1}$$

with $G(Y)$ of the form $(-1)^m \frac{\sum_{i=0}^{k} c_i(x)Y^{2m-i}}{\sum_{i=0}^{l} e_i(x)Y^{-i}}$ and $c_0 = \sum_{i=0}^{k} c_i(x)\alpha^i$.

Suppose first $k - 2m \geq l$. Then, $G(Y) = (-1)^m \frac{\sum_{i=0}^{k} c_i(x)Y^{k-i}}{\sum_{i=0}^{l} e_i(x)Y^{k-2m-i}}$. So, the degree of G is k. Then by Proposition 62.1, we have $2m \geq k$, hence $l = 0$ and hence F is a polynomial of degree $k \leq 2m$. Suppose now $k - 2m < l$. We have $G(Y) = (-1)^m \frac{\sum_{i=0}^{k} c_i(x)Y^{2m+l-i}}{\sum_{i=0}^{l} e_i(x)Y^{l-i}}$. So, G is of degree $2m + l$ and then, by applying Proposition 62.1 to (1), we obtain $l = 0$ and hence $k < 2m$ i.e. F is a polynomial of degree $k < 2m$. □

Notation. In proofs of Theorems 62.3 and 62.4, we have to consider a spherically complete extension $\widehat{\mathbb{K}}$ of $\mathbb{K}$ and the disk $\widehat{d}(a, R^-)$.

Theorem 62.3 looks like Theorem 61.3 when we assume that F is a polynomial with a unique zero.

Theorem 62.3. *Suppose $\mathbb{K}$ is of characteristic 0. Let $F(Y) \in \mathbb{K}[Y]$ be of the form $B(Y - c)^\ell$. The solutions $f \in \mathcal{M}(d(a, R^-))$ of $(\mathcal{E})$ are the functions of the form $c + (ax + b)^{\frac{m}{m-\ell}}$ such that*

either $m > \ell$ and then $B = \left(\frac{am}{m-\ell}\right)^m$

or $m < \ell$ and then $(-1)^m B = \left(\frac{am}{m-\ell}\right)^m$.

Proof. Suppose $(\mathcal{E})$ admits a solution $f \in \mathcal{M}_u(d(a, R^-))$. Let α be a zero of $f - c$ of order s. So, we have

(1) $s = \dfrac{m}{m-\ell}$ and $m > \ell$.

Consequently, all zeros of $f - c$ are multiple and of same order. Now, let β be a pole of f of order t. Then we have

(2) $t = \dfrac{m}{\ell - m}$ with $m < \ell$.

So, all poles are of same order. Moreover, we note that f cannot have both zeros and poles.

Consequently, either $m > \ell$ and f belongs to $\mathcal{A}_u(d(a, R^-))$ or $m < \ell$ and f is of the form $c + \frac{1}{h}$ with $h \in \mathcal{A}_u(d(a, R^-))$.

Suppose first that $m > \ell$. Then f belongs to $\mathcal{A}_u(d(a, R^-))$. By (1), every zero α of $f - c$ is of order $s = \frac{m}{m-\ell}$, hence α is a zero of f' of order $\frac{\ell}{m-\ell}$. By Theorem 36.12, there exists $g \in \mathcal{A}(d(a, R^-))$ such that $g^s = f - c$. Then, by Equation $(\mathcal{E})$, we have $(sg')^m (g)^{m(s-1)} = Bg^{s\ell}$ which, by (1), yields $(sg')^m = B$. Consequently, since $\mathbb{K}$ is of characteristic 0, by Lemma 36.1, g' is a constant and hence $g(x)$ is of the form $(ax + b)$. So, $f - c = (ax + b)^s$ and (similar to Theorem 61.3) we obtain $s = \frac{m}{m-\ell}$, $B = \left(\frac{am}{m-l}\right)^m$, and $f(x) = c + (ax + b)^{\frac{m}{m-\ell}}$.

Suppose now that $m < \ell$ and then f is of the form $c + \frac{1}{h}$ with $h \in \mathcal{A}_u(d(a, R^-))$. Then h satisfies

(3) $\quad (h')^m = (-1)^m Bh^{2m-\ell}$

and by (2) every zero β of h is of order $t = \frac{m}{\ell-m}$. Consequently, β is a zero of h' of order $t - 1 = \frac{2m-d}{\ell-m}$. Consider again a function $g \in \mathcal{A}(\widehat{d}(a, R^-))$ such that $g^t = h$. In the same way as in the previous case, we can check that g' is a constant and therefore, since $\mathbb{K}$ is of characteristic 0, f is of the form $c + \frac{1}{(ax+b)^t}$ and the conclusion follows as in Theorem 61.3. □

Proposition 62.4. *Suppose $\mathbb{K}$ is of characteristic 0. Let $F(Y) \in \mathbb{K}[Y]$ have at least two distinct zeros. Then the differential equation $(\mathcal{E})$ $(y')^m = F(y)$ does not admit solutions $f \in \mathcal{A}_u(d(a, R^-))$ or of the form $c + \frac{1}{h}$ with $h \in \mathcal{A}_u(d(a, R^-))$.*

Proof. Let $F(Y) = B\prod_{j=1}^{k}(Y - c_j)^{\ell_j}$ with $k \geq 2$, $c_i \neq c_j\ \forall i \neq j$ and let $\ell = \deg(F)$. Suppose first that a solution f of $(\mathcal{E})$ belongs to $\mathcal{A}_u(d(a, R^-))$. Since, by Corollary 18.3, $|f'|(r) \leq \frac{|f|}{r}\ \forall r \in]0, R[$, we note that $m \geq \ell$, therefore $\ell_j < m\ \forall j = 1, \ldots, k$.

For each $j = 1, \ldots, k$, let $t_j = \text{g.c.d.}(m, \ell_j)$ and set $m = t_j q_j$, $\ell_j = t_j w_j$. So, q_j and w_j are relatively prime. For each pair (i, j), $i \neq j$, since $c_i \neq c_j$, $f - c_i$ and $f - c_j$ have no common zero. So, if α is a zero of $f - c_j$, we have $m\omega_\alpha(f') = \ell_i \omega_\alpha(f - c_i)$ and therefore $q_i \omega_\alpha(f') = w_i \omega_\alpha(f - c_i)$. Thus, q_i divides $\omega_\alpha(f - c_i)$, $(1 \leq i \leq k)$.

Now, by Theorem 36.12, there exists $g_i \in \mathcal{A}_u(\widehat{d}(a, R^-))$ such that $f - c_i = (g_i)^{q_i}$.

We note that if $q_i = 1$, then m must divide ℓ_i, which is excluded because $\ell_j < m$. Consequently, $q_i \geq 2\ \forall i = 1, \dots, k$. Next, for any pair (i, j) of two distinct integers among $1, \dots, k$, we have $(g_i)^{q_i} - (g_j)^{q_j} = c_i - c_j \neq 0$. But since $\min(q_i, q_j) \geq 2$, by Theorem 47.11, both g_i, g_j are bounded a contradiction since f is unbounded.

Suppose now that $f = c + \frac{1}{h}$ with $h \in \mathcal{A}_u(d(a, R^-))$. At most one of the c_i is equal to c. So, we can assume $c_i \neq c\ \forall i = 1, \dots, k-1$. Set $a_j = c - c_j\ \forall j = 1, \dots, k$ and $\ell = \sum_{j=1}^{k-1} \ell_j$. Then f must satisfy

$$(1) \qquad (h')^m = (-1)^m B h^{2m-\ell} \prod_{i=1}^k \left(c - c_i + \tfrac{1}{h}\right)^{\ell_i}$$
$$= (-1)^m B h^{2m-\ell} \left(a_k + \frac{1}{h}\right)^{\ell_k} \prod_{i=1}^{k-1} \left(a_i + \frac{1}{h}\right)^{\ell_i}.$$

Now, since h is unbounded, when r is close enough to R, we have

$$(|h'|(r))^m = |B| (|h|(r))^{2m} \left(\left|a_k + \frac{1}{h}\right|(r)\right)^{\ell_k} \prod_{i=1}^{k-1} \left(\left|a_i + \frac{1}{h}\right|(r)\right)^{\ell_i}$$
$$= |B| (|h|(r))^{2m} \left(\left|a_k + \frac{1}{h}\right|(r)\right)^{\ell_k} \prod_{j=1}^{k-1} |a_j|^{\ell_j}.$$

Set $C = |B \prod_{i=1}^{k-1} (|a_i|)^{\ell_i}$. By Corollary 18.3, when r is close enough to R, we obtain

$$C (|h|(r))^{2m} \left(\left|a_k + \frac{1}{h}\right|(r)\right)^{\ell_k} \leq C \left(\frac{|h|(r)}{r}\right)^m,$$

therefore $Cr^m (|h|(r))^m (|a_k + \frac{1}{h}|(r))^{l_k} \leq 1$ and hence $Cr^m (|h|(r))^{m-l_k} \leq 1$ when r tends to R. But since $m < \ell_k$, that inequality is impossible when r tends to R. □

By Theorems 62.3 and 62.4, we have the following corollary:

Corollary 62.5. *Suppose $\mathbb{K}$ is of characteristic 0. Let $F(Y) \in \mathbb{K}[Y]$ and let $m \in \mathbb{N}^*$. Then the differential equation $(\mathcal{E})$ $(y')^m = F(y)$ does not admit any solution $f \in \mathcal{A}_u(d(a, R^-))$ or of the form $c + \frac{1}{h}$ with $h \in \mathcal{A}_u(d(a, R^-))$.*

Definition. In Theorem 62.6, we have to present Equation $(\mathcal{E})$ when it is in the form $(y')^{mq} = (F(y))^q$, where $q \in \mathbb{N}^*$. We say that $(\mathcal{E})$ is *a qth power of the equation* $(y')^m = (F(y))$ and an equation $(\mathcal{E})$ is said to be *irreducible* if it is not a qth power of another equation of the form $(\mathcal{E})$, for some $q > 1$.

Remark. According to Theorems 62.3 and 62.4, we have characterized all solutions of $(\mathcal{E})$ that either belong to $\mathcal{A}(d(a, R^-))$ or admit an exceptional value. In Theorem 62.6, we examine the form of Equation $(\mathcal{E})$ when a solution is not of the form as shown in Theorem 62.3.

Theorem 62.6. *Suppose $\mathbb{K}$ is of characteristic 0. Let $F(Y) \in \mathbb{K}[Y]$ and let $m \in \mathbb{N}^*$. Suppose the differential equation $(\mathcal{E})$ $(y')^m = F(y)$ admits a solution $f \in \mathcal{M}_u(d(a, R^-))\backslash\mathcal{A}(d(a, R^-))$ that has no exceptional value in $\mathbb{K}$. Then $(\mathcal{E})$ is a power of one of the following forms:*

$(\mathcal{E}_1)$ $(y')^2 = A(y - c_1)(y - c_2)(y - c_3)$,
$(\mathcal{E}_2)$ $(y')^2 = A(y - c_1)(y - c_2)(y - c_3)(y - c_4)$,
$(\mathcal{E}_3)$ $(y')^3 = A(y - c_1)^2(y - c_2)^2$,
$(\mathcal{E}_4)$ $(y')^3 = A(y - c_1)^2(y - c_2)^2(y - c_3)^2$,
$(\mathcal{E}_5)$ $(y')^6 = A(y - c_1)^3(y - c_2)^4(y - c_3)^5$,
$(\mathcal{E}_6)$ $(y')^4 = A(y - c_1)^2(y - c_2)^3(y - c_3)^3$,
$(\mathcal{E}_7)$ $(y')^6 = A(y - c_1)^4(y - c_2)^5$,
$(\mathcal{E}_8)$ $(y')^4 = A(y - c_1)^3(y - c_2)^3$,
$(\mathcal{E}_9)$ $(y')^6 = A(y - c_1)^3(y - c_2)^5$,
$(\mathcal{E}_{10})$ $(y')^4 = A(y - c_1)^2(y - c_2)^3$,
$(\mathcal{E}_{11})$ $(y')^6 = A(y - c_1)^3(y - c_2)^4$.

Proof. Let $F(Y) = A\prod_{i=1}^{k}(Y - c_i)^{\ell_i}$ with $c_i \neq a_j$ $\forall i \neq j$ and $\ell = \sum_{i=1}^{k} \ell_i$. We can note that each zero of $f - c_i$ is a zero of f' and hence is a zero of order $\frac{m}{m-\ell_i} \geq 2$. Particularly, we have

(1) $\dfrac{m}{2} \leq \ell_i \leq m - 1$

and hence $\frac{km}{2} \leq \sum_{i=1}^{k} \ell_i \leq k(m - 1)$ i.e. $\frac{km}{2} \leq \ell \leq k(m - 1)$. In the same way, for each pole β of f, we have $\omega_\beta(f) = \frac{m}{\ell - m} \geq 1$, hence $m+1 \leq \ell \leq 2m$, therefore

$$(2) \quad \max\left[m+1, \frac{km}{2}\right] \leq \ell \leq \min[2m, k(m-1)],$$

and hence

$$(3) \quad \frac{m+1}{m-1} \leq k \leq \frac{2\ell}{m} \leq 4.$$

Let $u = \dfrac{m}{\ell - m}$ and $u_i \dfrac{m}{m-\ell_i}$, $i = 1, \ldots, k$.

Then,

$$(4) \quad \frac{1}{m} \leq \frac{1}{u_i} \leq \frac{1}{2}.$$

By (2), we have

$$(5) \quad \max\left[\frac{1}{m}, \frac{k-2}{2}\right] \leq \frac{1}{u} \leq \min\left[1, \frac{k(m-1)}{m} - 1\right].$$

On the other hand, by definition of u and u_i, we also note that

$$(6) \quad \frac{1}{u} + \sum_{i=1}^{k} \frac{1}{u_i} = k - 1.$$

(A) First, suppose $m = 2$. By (1), (2), and (3), we have $k = \ell = 3$ or 4 and $\ell_i = 1\ \forall i = 1, \ldots, k$. This leads to Equations (1) and (2).

(B) Now, suppose $m \geq 3$. Since $m-1$ does not divide m, that yields $\ell_i \geq 2\ \forall i = 1, \ldots, k$. Consequently, we have

$$(7) \quad \max\left[\frac{m}{2}, 2\right] \leq \ell_i \leq m - 1.$$

Hence, (2) becomes

$$(8) \quad \max\left[m+1, \frac{km}{2}, 2k\right] \leq \ell \leq \min[2m, k(m-1)].$$

So, we have $\frac{m+1}{m-1} \leq k \leq \min\left(m, \frac{2d}{m}\right]$. And since $1 < \frac{m+1}{m-1} \leq 2\ \forall m \geq 3$, Relation (3) becomes

$$(9) \quad 2 \leq k \leq \min\left[m, \frac{2\ell}{m}\right] \leq 4.$$

Now, by (7), we obtain $\frac{1}{m} \le \frac{m-\ell_i}{m} \le \min\left[\frac{m-2}{m}, \frac{1}{2}\right]$ i.e.

$$(10) \quad \frac{1}{m} \le \frac{1}{u_i} \le \min\left[\frac{m-2}{m}, \frac{1}{2}\right],$$

and by (5) and (8), we can derive

$$(11) \quad \max\left[\frac{1}{m}, \frac{k-2}{2}, \frac{2k-m}{m}\right] \le \frac{1}{u} \le \min\left[1, \frac{k(m-1)}{m} - 1\right].$$

(B1) Suppose $m = 3$. By (7) and (8), we have $\ell_i = 1 \ \forall i = 1, \ldots, k$ and $\max\left[4, \frac{3k}{2}, 2k\right] \le d \le \min[6, 2k]$. Consequently, either $\ell = 4$, $k = 2$, or $\ell = 6$, $k = 3$. This leads to Equations (3) and (4).

(B2) $m \ge 4$. In that subcase, we have to consider three subcases.

(B2a) $k = 4$. By (6), (9), (10), and (11), we obtain $\ell = 2m \ge 8$, $\frac{1}{m} \le \frac{1}{u_i} \le \frac{1}{2}$, $u = 1$, $\sum_{i=1}^{4} \frac{1}{u_i} = 2$. Consequently, $u_i = 2$ and $\ell_i = \frac{m}{2} \ \forall i = 1, 2, 3, 4$. That leads to the equation $(y')^m = A(y-c_1)^{\frac{m}{2}}(y-c_2)^{\frac{m}{2}}(y-c_3)^{\frac{m}{2}}(y-c_4)^{\frac{m}{2}}$ with $m \ge 4$, m multiple of 2.

That is the mth power of Equation $(\mathcal{E}_2)$:

(B2b) $k = 3$. Thanks to (6), (7), (8), and (10), we obtain

$$\frac{m}{2} \le \ell_i \le m-1, \quad \frac{3m}{2} \le \ell \le 2m, \quad 2 \le u_i \le m,$$

$$1 \le u \le 2, \quad \frac{1}{u} + \sum_{i=1}^{4} \frac{1}{u_i} = 2.$$

We can assume that $2 \le u_1 \le u_2 \le u_3 \le m$ i.e. $\frac{1}{m} \le \frac{1}{u_3} \le \frac{1}{u_2} \le \frac{1}{u_1} \le \frac{1}{2}$.

(B2b)(i) Suppose first $u = 1$. The only possible values for $\frac{1}{u_1}, \frac{1}{u_2}, \frac{1}{u_3}$ are

$$\left(\frac{1}{2}, \frac{1}{3}, \frac{1}{6}\right), \ \left(\frac{1}{2}, \frac{1}{4}, \frac{1}{4}\right), \ \left(\frac{1}{3}, \frac{1}{3}, \frac{1}{3}\right).$$

Consequently, the only possible values for (ℓ_1, ℓ_2, ℓ_3) are respectively

$$\left(\frac{m}{2}, \frac{3m}{2}, \frac{5m}{6}\right), \ \left(\frac{m}{2}, \frac{3m}{4}, \frac{3m}{4}\right), \ \left(\frac{2m}{3}, \frac{2m}{3}, \frac{2m}{3}\right).$$

That leads to the following equations, respectively:

$$(y')^m = A(y-c_1)^{\frac{m}{2}}(y-c_2)^{\frac{m}{3}}(y-c_3)^{\frac{5m}{6}} \text{ with } m \geq 4,\ m \text{ multiple of } 6.$$

$$(y')^m = A(y-c_1)^{\frac{m}{2}}(y-c_2)^{\frac{3m}{4}}(y-c_3)^{\frac{3m}{4}} \text{ with } m \geq 4,\ m \text{ multiple of } 4.$$

$$(y')^m = A(y-c_1)^{\frac{2m}{3}}(y-c_2)^{\frac{2m}{3}}(y-c_3)^{\frac{2m}{3}} \text{ with } m \geq 4,\ m \text{ multiple of } 3.$$

Such equations are respectively the mth power of $(\mathcal{E}_5)$, $(\mathcal{E}_6)$, $(\mathcal{E}_4)$.

(B2b) (ii) $u = 2$. The only possible value for $\left(\frac{1}{u_1}, \frac{1}{u_2}, \frac{1}{u_3}\right)$ is $\left(\frac{1}{2}, \frac{1}{2}, \frac{1}{2}\right)$. Consequently, the only possible value for (ℓ_1, ℓ_2, ℓ_3) is $\left(\frac{m}{2}, \frac{m}{2}, \frac{m}{2}\right)$. That leads to the equation $(y')^m = A(y-c_1)^{\frac{m}{2}}(y-c_2)^{\frac{m}{2}}$ $(y-c_3)^{\frac{m}{2}}$ with $m \geq 4$, m multiple of 2.

Thus, that equation is the mth power of $(\mathcal{E}_1)$.

(B2c) $k = 2$. Thanks to (7), (8), (10), and (11), we can write

$$\frac{m}{2} \leq \ell_i \leq m-1, \quad m+1 \leq \ell \leq 2m-2,$$

$$2 \leq u_i \leq m, \quad (i = 1, 2)\ \frac{m}{m-2} \leq u \leq m.$$

Particularly, $2 \leq u \leq m$. By (6), we then obtain $\frac{1}{u_1} + \frac{1}{u_2} = \frac{u-1}{u}$ i.e.:

$$(12) \quad \frac{1}{u_1} + \frac{1}{u_2} + \frac{1}{u} = 1.$$

Suppose $u > 6$. By (12), we have $\frac{1}{u_1} + \frac{1}{u_2} > \frac{5}{6}$. But by (12), $u_1 = u_2 = 2$ is impossible. And if $(u_1, u_2) \neq (2, 2)$ we can see that $\frac{1}{u_1} + \frac{1}{u_2} \leq \frac{5}{6}$. Consequently, $u \leq 6$. Moreover, if $u = 5$, by (12), we have $\frac{1}{u_1} + \frac{1}{u_2} = \frac{4}{5}$. If $u_i \geq 3, i = 1, 2$, then $\frac{1}{u_1} + \frac{1}{u_2} \leq \frac{2}{3}$. Suppose $u_1 = 2$. If $u_2 = 3$, $\frac{1}{u_1} + \frac{1}{u_2} = \frac{5}{6}$. And if $u_2 \geq 4$, then $\frac{1}{u_1} + \frac{1}{u_2} \leq \frac{3}{4}$. Consequently, $u \neq 5$ and hence u can only take the values 2, 3, 4, and 6. We consider those 4 cases.

(B2c) (i) $u = 2$. The only possible values for $\left(\frac{1}{u_1}, \frac{1}{u_2}\right)$ are $\left(\frac{1}{3}, \frac{1}{6}\right)$ and $\left(\frac{1}{4}, \frac{1}{4}\right)$. Consequently, the only possible values for (ℓ_1, ℓ_2)

are respectively $\left(\frac{2m}{3}, \frac{5m}{6}\right)$ and $\left(\frac{3m}{4}, \frac{3m}{4}\right)$. They correspond to the equations

$$(y')^m = A(y - c_1)^{\frac{2m}{3}}(y - c_2)^{\frac{5m}{6}} \text{ with } m \geq 4 \text{ and } m \text{ multiple of 6,}$$

$$(y')^m = A(y - c_1)^{\frac{3m}{4}}(y - c_2)^{\frac{3m}{4}} \text{ with } m \geq 4 \text{ and } m \text{ multiple of 4,}$$

which are the mth power of $(\mathcal{E}_7)$ and $(\mathcal{E}_8)$.

(B2c) (ii) $u = 3$. The only possible values for $\left(\frac{1}{u_1}, \frac{1}{u_2}\right)$ are $\left(\frac{1}{2}, \frac{1}{6}\right)$ and $\left(\frac{1}{3}, \frac{1}{3}\right)$. Consequently, the only possible values for (ℓ_1, ℓ_2) are respectively $\left(\frac{m}{2}, \frac{5m}{6}\right)$ and $\left(\frac{2m}{3}, \frac{2m}{3}\right)$. They correspond to the equations

$$(y')^m = A(y - c_1)^{\frac{m}{2}}(y - c_2)^{\frac{5m}{6}} \text{ with } m \geq 4 \text{ and } m \text{ multiple of 6,}$$

$$(y')^m = A(y - c_1)^{\frac{2m}{3}}(y - c_2)^{\frac{2m}{3}} \text{ with } m \geq 4 \text{ and } m \text{ multiple of 3,}$$

which are the mth power of $(\mathcal{E}_9)$ and $(\mathcal{E}_3)$.

(B2c) (iii) $u = 4$. The only possible value for $\left(\frac{1}{u_1}, \frac{1}{u_2}\right)$ is $\left(\frac{1}{2}, \frac{1}{4}\right)$. Consequently, the only possible value for (ℓ_1, ℓ_2) is $\left(\frac{m}{2}, \frac{3m}{4}\right)$. This corresponds to the equation $(y')^m = A(y - c_1)^{\frac{m}{2}}(y - c_2)^{\frac{3m}{4}}$ with $m \geq 4$ and m multiple of 4 and this is the mth power of $(\mathcal{E}_9)$ and $(\mathcal{E}_{10})$.

(B2c) (ii) $u = 6$. The only possible value for $\left(\frac{1}{u_1}, \frac{1}{u_2}\right)$ is $\left(\frac{1}{2}, \frac{1}{3}\right)$. Consequently, the only possible value for (ℓ_1, ℓ_2) is $\left(\frac{m}{2}, \frac{2m}{3}\right)$. This corresponds to the equation $(y')^m = A(y - c_1)^{\frac{m}{2}}(y - c_2)^{\frac{2m}{3}}$ with $m \geq 4$ and m multiple of 6 and this is the mth power of $(\mathcal{E}_9)$ and $(\mathcal{E}_{11})$.

That finishes the proof of Theorem 62.6. □

Corollary 62.7. *Suppose $\mathbb{K}$ is of characteristic 0. Let $F(Y) \in \mathbb{K}(Y) \backslash \mathbb{K}$ and let $m \in \mathbb{N}^*$. If one of the following hypotheses is satisfied, there exists no solution $f \in \mathcal{M}_u(da, R^-)$ for the irreducible equation $(\mathcal{E})$ $(y')^m = F(y)$:*

(1) *$F(Y) \in \mathbb{K}[Y]$ and $\deg(F) \geq 2m + 1$,*
(2) *$F(Y) \in \mathbb{K}[Y]$ has at least five zeros, ignoring multiplicity,*
(3) *$m = 2$, $F(Y) \in \mathbb{K}[Y]$ has at least one multiple zero,*

(4) $m = 3$, $F(Y) \in \mathbb{K}[Y]$ *has at least four distinct zeros,*
(5) $m = 3$, $F(Y) \in \mathbb{K}[Y]$ *has at least one zero of order* $s \neq 2$,
(6) $m = 4$, $F(Y) \in \mathbb{K}[Y]$ *has four distinct zeros,*
(7) $m = 5$,
(8) $m = 6$, $F(Y) \in \mathbb{K}[Y]$ *has at least four distinct zeros,*
(9) $m \geq 7$.

Bibliography

[1] Amice, Y. *Les nombres p-adiques*, P.U.F. (1975).

[2] Amice, Y. Dual d'un espace $H(D)$ et transformation de Fourier, *Groupe d'étude d'analyse ultramétrique de l'IHP*, 1(5), 1–12 (1973–1974).

[3] An, T.T.H., Wang, J.T.Y. and Wong, P.M. Strong uniqueness polynomials: The complex case, *Complex Variables*, 49(1), 25–54 (2004).

[4] An, T.T.H. and Escassut, A. Meromorphic solutions of equations over non-Archimedean fields, *Ramanujan Journal*, 15(3), 415–433 (2008).

[5] An, T.T.H. and Nguyen Phuong Viet. *Non-Archimedean Second Main Theorem Sharing Small Functions.* ArXiv: 20111.08917v2.

[6] Araujo, J. Nonmaximal ideals and the Berkovich space of the algebra of bounded analytic function, *Journal of Mathematical Analysis and Applications*, 455(1), 221–245 (2017).

[7] Bartels, S. Meromorphic functions sharing a set with 17 elements, ignoring multiplicities, *Complex Variable Theory and Applications*, 39(1), 85–92 (1999).

[8] Bergweiler, W. and Eremenko, A. On the singularities of the inverse to a meromorphic function of finite order, *Revista Matemática Iberoamericana*, 11, 355–373 (1995).

[9] Berkovich, V. *Spectral Theory and Analytic Geometry over Non-Archimedean Fields*, AMS Surveys and Monographs, 33, (1990).

[10] Bezivin, J.-P., Wronskien et equations differentielles p-adiques, *Acta Arithmetica*, 158(1), 61–78 (2013).

[11] Bezivin, J.-P., Boussaf, K. and Escassut, A. Zeros of the derivative of a p-adic meromorphic function, *Bulletin des Sciences Mathématiques*, 136(8), 839–847 (2012).

[12] Bezivin, J.-P., Boussaf, K. and Escassut, A. Some old and new results on zeros of the derivative of a p-adic mermorphic function, *Contemporary Mathematics*, 596, 23–30 (2013).

[13] Boussaf, K. Motzkin factorization in algebras of analytic elements, *Annales Mathématiques Blaise Pascal*, 2(1), 73–91 (1995).

[14] Boussaf, K., Hemdaoui, M. and Maïnetti, N. Tree structure on the set of multiplicative semi-norms of Krasner algebras $H(D)$, *Revista Matematica Complutense*, 13(1), 85–109 (2000).

[15] Boussaf, K. Shilov boundary for algebras $H(D)$, *Italian Journal of Pure and Applied Mathematics*, 8, 75–82 (2000).

[16] Boussaf, K., Boutabaa, A. and Escassut, A. Growth of p-adic entire functions and applications, *Houston Journal of Mathematics*, 40(3), 715–736 (2014).

[17] Boussaf, K., Escassut, A. and Ojeda, J. Complex and p-adic branched functions and growth of entire functions, *Bulletin of the Belgian Mathematical Society Simon Stevin*, 22, 781–796 (2015).

[18] Boussaf, K., Boutabaa, A. and Escassut, A. Order, type and cotype of growth of p-adic entire functions, a survey with additional results, *p-adic Numbers, Ultrametric Analysis and Applications to Number Theory*, 8(4), 280–297 (2016).

[19] Boussaf, K., Ojeda, J. and Escassut, A. Primitives of p-adic meromorphic functions, *Contemporary Mathematics*, 551, 51–56 (2011).

[20] Boussaf, K. and Escassut, A. Growth of analytic functions in an ultrametric open disk and branched values, *Bulletin of the Belgian Mathematical Society, Simon Stevin*, 28, 1–16 (2021).

[21] Boutabaa, A. Théorie de Nevanlinna p-adique, *Manuscripta Mathematica*, 67, 251–269 (1990).

[22] Boutabaa, A., Escassut, A. and Haddad, L. On uniqueness of p-adic entire functions, *Indagationes Mathematicae*, 8, 145–155 (1997).

[23] Boutabaa, A. and Escassut, A. URS and URSIMS for p-adic meromorphic functions inside a disk, *Proceedings of the Edinburgh Mathematical Society*, 44, 485–504 (2001).

[24] Boutabaa, A. and Escassut, A. Applications of the p-adic Nevanlinna theory to functional equations, *Annales de l'Institut Fourier*, 50(3), 751–766 (2000).

[25] Boutabaa, A. and Escassut, A. Nevanlinna Theory in characteristic p and applications, *Italian Journal of Pure and Applied Mathematics* 23, 45–66 (2008).

[26] Boutabaa, A., Cherry, W. and Escassut, A. Unique Range sets in positive characteristic, *Acta Arithmetica*, 103(2), 169–189 (2002).

[27] Boutabaa, A. About the p-adic Yosida equation inside a disk, *Indagationes Mathematicae*, 20(3), 397–413 (2009).

[28] Carleson, L. Interpolation by bounded analytic functions and the corona problem, *Annals of Mathematics*, 76, 547–559 (1962).

[29] Charak, K.S. Value distribution theory of meromorphic functions, *Mathematics Newsletter*, 18(4), 1–35 (2009).

[30] Chen, H. and Fang, M. On the value distribution of $f^n f'$, *Science in China*, 38 A (7), 789–798 (1995).

[31] Cherry, W. and Yang, C.C. Uniqueness of non-Archimedean entire functions sharing sets of values counting multiplicities, *Proceedings of the American Mathematical Society*, 127, 967–971 (1998).

[32] Datta, S.K. and Arindam, J. On the weak type of meromorphic functions, *International Mathematical Forum*, 4(12), 569–579 (2009).

[33] Diarra, B. Ultraproduits ultramétriques de corps valués, Annales Scientifiques de l'Université de Clermont II, Série Mathématique, 22, 1–37 (1984).

[34] Diarra, B. and Escassut, A. Survey on bezout rings of p-adic analytic functions, *Southeast Asian Bulletin of Mathematics*, 39, 1–8 (2015).

[35] Escassut, A. Algèbres d'éléments analytiques en analyse non archimedienne, *Indagationes Mathematicae*, 36(4), 339–351 (1974).

[36] Escassut, A. Eléments analytiques et filtres percés sur un ensemble infraconnexe, *Annali di Matematica Pura ed Applicata*, Toma. CX, 335–352 (1976).

[37] Escassut, A. Algèbres de Krasner-Tate et algèbres de Banach ultramétriques, *Astérisque*, 10, 1–107 (1973).

[38] Escassut, A. T-filtres, ensembles analytiques et transformations de Fourier p-adique, *Annales de l'Institut Fourier*, 25(2), 35–80 (1975).

[39] Escassut, A. Eléments spectralement injectifs et générateurs universels dans une algèbre de Tate, *Collectanea Matematica*, XXVIII(2), 131–148 (1977).

[40] Escassut, A. The ultrametric spectral theory, *Periodica Mathematica Hungarica*, 11(1), 7–60 (1980).

[41] Escassut, A. Transcendence order over $\mathbb{Q}_p$ in $\mathbb{C}_p$, *Journal of Number Theory*, 16(3), 395–402 (1982).

[42] Escassut, A. and Sarmant, M.C. Mittag-Leffler series and Motzkin product for invertible analytic elements, *Rivista di Matematica Pura ed Applicata*, 2, 61–73 (1992).

[43] Escassut, A., Haddad, L. and Vidal, R. Urs, Ursim, and nonurs, *Journal of Number Theory*, 75, 133–144 (1999).

[44] Escassut, A. *Analytic Elements in p-adic Analysis*, World Scientific Publishing Co. Pte. Ltd. Singapore (1995).

[45] Escassut, A. *Ultrametric Banach Algebras*, World Scientific Publishing Co. Pte. Ltd. Singapore (2003).

[46] Escassut, A. and Yang, C.C. *The Functional Equation $P(f) = Q(g)$ in a p-adic Field, Journal of Number Theory,* 105(2), 344–360 (2004).
[47] Escassut, A. Meromorphic functions of uniqueness, *Bulletin des Sciences Mathématiques,* 131(3), 219–241 (2007).
[48] Escassut, A. Ultrametric, Corona problem and spherically complete fields, *Proceedings of the Edingburgh Mathematical Society,* (Series 2), 53(0)2, 353–371 (2010).
[49] Escassut, A. *Value Distribution in p-adic Analysis.* World Scientific Publishing Co. Pte. Ltd. Singapore (2015).
[50] Escassut, A. *p-adic Analytic Functions.* World Scientific Publishing Co. Pte. Ltd. Singapore (2021).
[51] Escassut, A. and Ojeda, J. Branched values and quasi-exceptional values for p-adic mermorphic functions, *Houston Journal of Mathematics,* 39(3), 781–795 (2013).
[52] Escassut, A. and Ojeda, J. The p-adic Hayman conjecture when $n = 2$, *Complex Variable and Elliptic Equations,* 59(10), 1451–1455 (2014).
[53] Escassut, A. *Banach Algebras of Ultrametric Functions.* World Scientific Publishing Co. Pte. Ltd. Singapore (2022).
[54] Escassut, A. and Ta Thi Hoai An. New applications of the p-adic nevanlinna theory, *p-Adic Numbers, Ultrametric Analysis and Applications,* 10(1), 12–31 (2018).
[55] Escassut, A. Survey and additional properties on the transcendence order over $\mathbb{Q}_p$ in $\mathbb{C}_p$, *p-Adic Numbers, Ultrametric Analysis and Applications,* 7(1) 17–23 (2015).
[56] Escassut, A. and Yang, C.C. A short note on two p-adic meromorphic functions sharing a few small ones, *Rendiconti del Circolo Matematico di Palermo* 70(2), 623–630 (2021).
[57] Escassut, A. Short communication, Number of zeros of exponential polynomials in zero residue characteristic, *p-Adic Numbers, Ultrametric Analysis and Applications,* 15(1), 90–92 (2023).
[58] Feneyrol-Perrin, Y. Transformations conformes dans les corps Hédériques, *Studia Scientiarum Mathematica Hungarica,* 24, 219–239 (1989).
[59] Frank, G. and Reinders, M. A unique range set for meromorphic functions with 11 elements, *Complex Variable Theory Application,* 37, 185–193 (1998).
[60] Fujimoto, H. On uniqueness of meromorphic functions sharing finite sets, *American Journal of Mathematics,* 122(6), 1175–1203 (2000).
[61] Garandel, G. Les semi-normes multiplicatives sur les algèbres d'éléments analytiques au sens de Krasner, *Indagationes Mathematicae,* 37(4), 327–341 (1975).

[62] Guennebaud, B. *Sur une notion de spectre pour les algèbres normées ultramétriques*, thèse Université de Poitiers, Poitiers, France (1973).

[63] Ha, H.K. On p-adic meromorphic functions, *Duke Mathematical Journal*, 50, 695–711 (1983).

[64] Ha, H.K. and Yang, C.C. On the functional equation $P(f) = Q(g)$, *Value Distribution Theory*, Marcel Dekker (2004).

[65] Hayman, W.K. *Meromorphic Functions.* Oxford University Press (1975).

[66] Hayman, W.K. Picard values of meromorphic functions and their derivatives, *Annals of Mathematics*, 70, 9–42 (1959).

[67] Hu, P.C. and Yang, C.C. *Meromorphic Functions over non-Archimedean Fields*, Kluwer Academic Publishers, (2000).

[68] Kaplansky, I. *Fields and Rings. Reprint of the Second (1972) Edition*, Chicago Lectures in Mathematics. University of Chicago Press, Chicago, IL, (1995).

[69] Krasner, M. *Prolongement analytique uniforme et multiforme dans les corps valués complets. Les tendances géométriques en algèbre et théorie des nombres*, Clermont-Ferrand, 94–141 (1964). Centre National de la Recherche Scientifique (1966), (Colloques internationaux de C.N.R.S. Paris, 143).

[70] Krasner, M. *Nombre d'extensions d'un degré donn'e d'un corps p-adique. Les tendances géométriques en algèbre et théorie des nombres*, Clermont-Ferrand, pp. 143–169 (1964). Centre National de la Recherche Scientifique (1966), (Colloques internationaux de C.N.R.S. Paris, 143).

[71] Lang, S. *Introduction to Transcendental Numbers.* Addison Wesley Reading, Mass.-London-Don Mills, Ont. (1966).

[72] Lazard, M. Les zéros des fonctions analytiques sur un corps valué complet, *IHES, Publications Mathématiques*, 14, 47–75 (1962).

[73] Malher, K. Ein Beweis der Transzendenz der P-adischen Exponentialfunktion, *Journal für die reine und angewandte Mathematik*, 169, 61–66 (1932).

[74] Motzkin, E. and Robba, Ph. *Prolongement analytique en analyse p-adique*, Séminaire de theorie des nombres, année 1968-69, Faculté des Sciences de Bordeaux.

[75] Motzkin, E. La décomposition d'un élément analytique en facteurs singuliers, *Annales de l'institut Fourier*, 27(1), 67–82 (1977).

[76] Mues, E. Uber ein Problem von Hayman, *Mathematische Zeitschrift*, 164, 239–259 (1979).

[77] Mues, E. and Reinders, M. Meromorphic functions sharing one value and a unique range sets, *Kodai Mathematical Journal*, 18, 515–522 (1995).

[78] Mues, E. and Reinders, M. On a question of C. C. Yang, *Complex Variables Theory Application*, 34(1–2), 171–179 (1997).

[79] Nevanlinna, R. *Le théorème de Picard-Borel et la théorie des fonctions méromorphes.* Gauthiers-Villars, Paris (1929).

[80] Ojeda, J. On Hayman's Conjecture over a p-adic field, *Taiwanese Journal of Mathematics*, 12(9), 2295–2313 (2008).

[81] Picard, E. *Mémoire Sur Les Fonctions Entières*, Annales de l'Ecole Normale Supérieure, (1880).

[82] Rivera-Letelier, J. Bi-analytic elements and partial isometries of hyperbolic space, *Contemporary Mathematics*, 319, 319–343 (2003).

[83] Robba, Ph. *Fonctions Analytiques sur les Corps Valués Ultramétriques Complets. Prolongement Analytique et Algèbres de Banach ultramétriques*, Astérisque, 10, 109–220 (1973).

[84] Robert, A. *A Course in P-Adic Analysis*, Graduate texts. Springer (2000).

[85] Rubel, Lee, A. *Entire and Meromorphic Functions*, Springer-Verlag, New York (1996).

[86] Schikhof, W.H. *Ultrametric Calculus. An Introduction to p-adic Analysis.* Cambridge University Press (1984).

[87] Serre, J.-P. Dépendance d'exponentielles p-adiques. *Séminaire Delange-Pisot-Poitou. Théorie des nombres*, 7(2), 1–14 (1965–1966), Exp. No. 15, IHP, 13 Rue Pierre et Marie Curie, Paris 5e (France).

[88] Shamseddine, K. A brief survey of the study of power series and analytic func- tions on the Levi-Civita fields, *Contemporary Mathematics*, 596, 269–279 (2013).

[89] Tate, J. Rigid analytic spaces, *Inventiones Mathematicae*, 12, 257–289 (1971).

[90] Valiron, G. *Lectures on the General Theory of Integral Functions*, Chelsea Publishing Company (1949).

[91] Van Der Put, M. The Non-Archimedean Corona Problem, Table Ronde d'Analyse Ultramétrique, *Bulletin de la Socété Mathématique de France, Mémoire*, 39–40, 287–317 (1974).

[92] Waldschmidt, M. *Nombres Transcendants*, Lecture Notes in Mathematics, 402, Springer Verlag (1974).

[93] Wang, J.T.Y. Uniqueness polynomials and bi-unique range sets, *Acta Arithmetica*, 104, 183–200 (2002).

[94] Yamanoi, K. The second main theorem for small functions and related problems, *Acta Mathematica*, 192, 225–294 (2004).

[95] Yang, C.C. On the value distribution of a transcendental meromorphic functions and its derivatives, *Indian Journal of Pure and Applied Mathematics*, 35, 1027–1031 (2004).

[96] Yang, C.C. and Hua, X. Unique polynomials of entire and meromorphic functions, *Matematicheskaia Fizika Analys Geometriye*, 4(3), 391–398 (1997).

[97] Yang, C.C. and Li, P. *Some Further Results on the Functional Equation $P(f) = Q(g)$, Series Advances in Complex Analysis and Applications*, Value distribution theory and related topics, pp. 219–231, Kluwer Academic Publishers, Dordrecht-Boston-London (2004).

[98] Yi, H.X. and Yang, C.C. *Uniqueness Theorems of Meromorphic Functions.* Science Press, China (1995).

Definitions

Chapter 1

centered similarity,
center of a centered similarity,
non-centered similarity,
translation,
affinely rigid set,
centered affinely rigid set,
non-centered affinely rigid analytic set,
Condition (F).

Chapter 2

valuation ring, valuation ideal,
residue class, residue class field, residue characteristic,
circle of center a, or radius r, classes of a circle,
diameter of a set, codiameter of a set,
hole of a set,
infraconnected set, infraconnected component,
affinoid set,
empty annulus of a set.

Chapter 3

filter thinner than another one,
sequence thinner than a filter,
filter secant with a set, with another filter,
increasing distances sequence, decreasing distances sequence,

monotonous distances sequence,
equal distances sequence,
increasing filter of center a and diameter r,
decreasing filter of center a and diameter r,
decreasing filter with no center of diameter r, of canonical basis (D_n),
monotonous filter,
body of a monotonous filter, beach of a monotonous filter,
spherically complete field,
pierced monotonous filter,
circular filter of center a and diameter r,
peripheral of a bounded set,
circular filter with no center, of diameter diameter r, of canonical basis (D_n),
large circular filter, punctual circular filter,
$\mathcal{F}$-affinoid.

Chapter 4
Gauss norm.

Chapter 5
quasi-monic polynomial.

Chapter 6
p-adic absolute value.

Chapter 7
Eisenstein polynomial,
uniformizer of an extension of $\mathbb{Q}_p$,
ramification index of an extension of $\mathbb{Q}_p$.

Chapter 8
principal ultrafilter,
incomplete ultrafilter,
immediate extension of an ultrametric field.

Chapter 9
transcendence order $\leq \tau$ over $\mathbb{Q}_p$,
weak transcendence order $\leq \tau$ over $\mathbb{Q}_p$, infinite order,
algebraic number in $\mathbb{C}_p$, transcendental number in $\mathbb{C}_p$,
denominator of an algebraic number.

Chapter 10
transcendence type $\leq \tau$ in $\mathbb{C}_p$,
infinite transcendence type.

Chapter 11
analytic element on D,
invertible element of a space $H(D)$,
pole of order q of an element $f \in H(D)$,
polynomial of poles of an element of $H(D)$ in $\overline{D} \backslash D$,
residue of an element $f \in H(D)$ at a pole a.

Chapter 12
bianalytic element from A onto B.

Chapter 13
punctual semi-norms of $H(D)$.

Chapter 14
radius of convergence,
entire function,
power series, Laurent series,
zero of multiplicity order q.

Chapter 15
f-hole, Mittag-Leffler term of f associated with a hole T_n,
principal term of f,
Mittag-Leffler series of f on the infraconnected set D,
specific circular filter of a hole, of $\widetilde{D}$,
residue of an analytic element on a hole.

Chapter 16
polynomial of zeros of an analytic element on a subset A,
semi-invertible analytic element,
quasi-invertible analytic elements,
quasi-minorated analytic element.

Chapter 17
analytic element vanishing along a filter $\mathcal{F}$,
analytic element properly vanishing along a filter $\mathcal{F}$.

Chapter 18
piercing of a subset,
well pierced subset.

Chapter 20
analytic element strictly vanishing along a monotonous filter $\mathcal{F}$,
analytic element collapsing along a monotonous filter $\mathcal{F}$.

Chapter 21
analytic set.

Chapter 23
strictly injective analytic element,
strictly injective analytic function,
bianalytic function.

Chapter 28
divisor on K, divisor on a disk $d(a, r^-)$,
bounded divisor in a disk,
divisor of a function,
Euclidean division by a polynomial.

Chapter 29
index of an analytic element,
pure factor associated to a hole,
Motzkin factor in a hole,
Motzkin index,
f-supersequence,
Motzkin factorization,
principal factor.

Chapter 30
order of growth of an entire function,
cotype of growth of an entire function.

Chapter 31
type of growth of an entire function.

Chapter 33
order of growth of an analytic function inside a disk,
type of growth of of an analytic function. inside a disk,
cotype of growth of of an analytic function inside a disk.

Chapter 34
contiguous filters,
quasi-invertible analytic function inside a disk.

Chapter 35
regular sequence,
regular ultrafilter,
coroner filter,
multibijective Banach algebra.

Chapter 36
meromorphic function in K, in a disk $d(a, R^-)$,
pole of order s of a meromorphic function,
meromorphic function holomorphic at a point a,
divisor of a meromorphic function in K, in a disk $d(a, R^-)$,
divisor of the poles of f,
exceptional value,
pseudo-exceptional value,
quasi-exceptional value.

Chapter 37
residue of a meromorphic function at a point a,
analytic element meromorphic in a hole,
singular part at a point b,
polynomial of the poles in a hole.

Chapter 39
exceptional value,
pseudo-exceptional value,
quasi-exceptional value,
Motzkin factor,
Motzkin index.

Chapter 40
circular filter surrounding another circular filter,
circular filter surrounding a monotonous filter,
monotonous surrounding a circular filter,
monotonous surrounding another monotonous filter,
D-bordering filter,
strictly D-bordering filter,
δ-topology,
Shilov set,
Shilov boundary.

Chapter 41
degree of a rational function,
absolute degree of a rational function,
segment of the tree $\Phi(D)$.

Chapter 42
shortest path.

Chapter 43
counting function of zeros $Z(r, f)$, counting multiplicity,
counting function of zeros $\overline{Z}(r, f)$, ignoring multiplicity,
counting function of poles $N(r, f)$, counting multiplicity,
counting function of poles $\overline{N}(r, f)$, ignoring multiplicity,
characteristic function or Nevanlinna function $T(r, f)$.

Chapter 47
function of uniqueness for a family of functions, polynomial of uniqueness,
function of strong uniqueness for a family of functions,
polynomial of strong uniqueness.

Chapter 48
singular point of a curve,
degenerate curve,
elliptic curve, hyperelliptic curve.

Chapter 49
perfectly branched value for f,
totally branched value for f.

Chapter 51
small function with respect to f,
perfectly branched small function with respect to f.

Chapter 52
functions f, g sharing a function h ignoring multiplicity.

Chapter 53
Laurent series of a function at a point,
regular point,
singular point,
point of high singularity,
Laurent series of $h \circ f$ at α.

Chapter 56

unique range set counting multiplicity, urscm for $\mathcal{F}$,
unique range set ignoring multiplicity, ursim for $\mathcal{T}$,
f, g share S, counting multiplicity,
f, g share S, ignoring multiplicity.

Chapter 58

ramification index.

Chapter 62

qth power of an equation,
irreducible equation.

Notations

Chapter 2

$\mathbb{N},\ \mathbb{Z},\ \mathbb{Q},\ \mathbb{R},\ \mathbb{C}$

$\overline{S},\ \overset{\circ}{S}$

$\overline{a}\ (a \in \mathbb{L}),\ \overline{P}\ (P \in \mathbb{L}[x]$

$|x|,\ (x \in E)$

$|t|_\infty,\ (t \in \mathbb{R})$

$\log,\ \theta,\ v,\ \Psi,\ |E|$

$U_E,\ M_E,\ \mathcal{L}$

$d(a,r),\ d(a,r^-),\ C(a,r)$

$\Delta(a,r',r''),\ \Gamma(a,r',r''),\ \delta(a,D),\ \delta(D,E)$

$\mathrm{diam}(D),\ \mathrm{codiam}(D)$

$\widehat{L},\ \widehat{d}(a,r),\ \widehat{d}(a,r^-)),\ \widehat{C}(a,r))$

$\widehat{D},\ \widetilde{D}$

$\mathcal{J}(\Lambda),\ \mathcal{E}(\Lambda),\ \chi(D)$

Chapter 3

$\mathcal{C}_D(\mathcal{F})$

$\mathcal{B}_D(\mathcal{F})$

$\mathrm{diam}(\mathcal{F})$

$\widehat{G}$

γ

$\widehat{\mathcal{F}},\ \widehat{\mathcal{G}}$

Chapter 4

$f'^r,\ f'^l$

$\|\ .\ \|$

$\Phi(D),\ \Phi^0(D)$

$v(h,\mu)$

$\Psi(h,\mu)$

$\Psi_a(h,\mu)$

$h_a(u)$

$\Psi_a(h,\mu)$

$\|\ .\ \|_0$

$|P|(r),\ P \in \mathbb{K}[x],\ |h|(r),\ h \in \mathbb{K}(x)$

$\varphi_{\mathcal{F}},\ \varphi_{a,r}$

φ_a

$\nu^+(f,\mu),\ \nu^-(f,\mu),\ \nu(f,\mu).$

$\frac{d_r f}{d\mu},\ \frac{d_l f}{d\mu}$

Chapter 5

(g,h)

$B(f,g)$

Chapter 6

$\mathrm{irr}(a,L)$

$\mathcal{N}$

$|n|_p$

Ψ_p

$\mathbb{Z}_p,\ \mathbb{Q}_p,\ \Omega_p,\ \mathbb{C}_p,\ \mathbb{F}_p$

Chapter 7

$u_s,\ r_s$

$A_s,\ B_s$

$F_s,\ G_s$

$W_s,\ W$

$\mathcal{H}$

Chapter 8

$\mathcal{U}$, $\mathcal{R}$

$\widehat{\mathbb{K}}$

$\widehat{d}(a,r)$, $\widehat{d}(a,r^-)$

$\widehat{C}(a,r)$

Chapter 9

$\mathcal{S}(\tau)$

$\mathcal{S}(\tau^+)$

$\mathrm{Card}(S)$

$|z|_\infty$, $z \in \mathbb{C}$

$den(a)$, $a \in \Omega$

$\mathcal{T}(a)$, $a \in \Omega$

$\overline{|a|} = \max_{j=1,\dots,n}(|a_j|_\infty)$, $a \in \Omega$

$s(a) = \max(\log(\overline{|a|}, \mathrm{den}(a))$

Chapter 10

$\mathcal{U}_D$

$R(D)$, $R_b(D)$, $R_0(D)$

$\| \, . \, \|_D$

$\mathrm{Mult}(A)$

$\mathrm{Mult}(A, \| \, . \, \|)$

$\mathrm{Mult}_m(A, \| \, . \, \|)$, $\mathrm{Mult}_1(A, \| \, . \, \|)$

$Ker(\phi)$

$\| \, . \, \|_{si}$

Chapter 11

$H(D)$, $H_b(D)$, $H_0(D)$

$\mathcal{U}_D$

Alg

$\mathrm{res}(f,a)$

Chapter 12

$\Xi(D', D)$

Chapter 13

Mult$(H(D),\ \mathcal{U}_D)$

${}_D\varphi_{\mathcal{F}}$, ${}_D\varphi_{a,r}$, φ_a, ${}_D\varphi_D$, ${}_D\varphi_D$, ${}_D\varphi_\infty$

$\mathcal{J}(\mathcal{F})$, $\mathcal{J}_0(\mathcal{F})$, $\mathcal{J}(a)$

$\Psi_a(h,\mu)$

Chapter 14

$\mathcal{A}(\mathbb{K})$, $\mathcal{A}(d(a,r^-))$, $\mathcal{A}_b(d(a,r^-))$, $\mathcal{A}_u(d(a,r^-))$, $\mathcal{A}(\mathbb{K}\setminus d(a,r))$, $\mathcal{A}_b(\mathbb{K}\setminus d(a,r))$, $\mathcal{A}_u(\mathbb{K}\setminus d(a,r))$ $\mathcal{A}(\Gamma(a,r',r''))$, $\mathcal{A}_b(\Gamma(a,r',r''))$

$|f|(r)$

$\nu^+(f,\mu)$, $\nu^-(f,\mu)$, $\nu(f,\mu)$

Chapter 15

$\overline{\overline{f_0}}$, $\overline{\overline{f_T}}$

$\widehat{H}(\widehat{D})$

res(f,T)

$E^{\circledast}$

Chapter 16

$\omega_\alpha(f)$

Chapter 17

Alg

$f^*, \overline{f}$

$\mathcal{J}(a)$

Condition A, Condition B

Chapter 19

$\lambda(a)$

$\Psi(x)$, $\Psi(f,\mu)$, $\Psi_a(f,\mu)$

Chapter 22

$f^{<n>}$

Chapter 24

$\mathcal{Q}_n(D)$, $\mathcal{Q}(D)$

Chapter 25

Log, exp, e^x

Chapter 26

$\mathbb{L}$

Ψ_p, Ψ_0

$O(\zeta(r))$, $o(\zeta(r))$

Chapter 27

$\mathcal{T}(E)$, $\mathcal{T}_{a,R}(E)$

$\mathcal{D}(f)$

Chapter 29

f^T

$mo(f, T)$

η_q, $\sqrt[q]{u}$

Chapter 30

$\rho(f)$, $\widetilde{\rho}(f)$

$\psi(f)$ $\widetilde{\psi}(f)$

$\psi(f, r)$ $\mathcal{A}^*(\mathbb{K})$, $\mathcal{A}^*(\mathbb{K}, t)$

$s(r, f)$

Chapter 31

$\sigma(f)$, $\widetilde{\sigma}(f)$

$\sigma(f, r)$,

Chapter 32

$\theta(f)$, $\psi(f)$

$\zeta(f, r)$

$\mathcal{A}^*(E)$

$\sigma(f)$

Chapter 37

res(f, α) (f meromorphic function)

Chapter 35

$\mathcal{I}_s$, $t(J)$, $\ell(J)$, $u(J)$, $m(s)$,
$\lambda(f_1, \dots, f_s)$, $w(f_1, \dots, f_s)$,
$t(J)$, $l(J)$
$\mathcal{D}(f, \epsilon)$
$\mathcal{G}_I$
$\mathcal{V}(\phi, f_1, \dots, f_q, \epsilon)$.
$B(\mathbb{N}, \mathbb{K})$
$\mathcal{W}$

Chapter 36

$\mathcal{M}\mathbb{K})$, $\mathcal{M}(d(a, R^-)))$
$\mathcal{M}_b(d(a, R^-)))$
$\mathcal{M}_u(d(a, R^-)))$
$\mathcal{W}(f, r, \epsilon)$
$\omega_a(f)$

Chapter 37

$\text{res}(\text{f}, \alpha)$

Chapter 39

$D = \mathbb{K} \setminus S$
$\mathcal{A}(D)$, $\mathcal{A}^c(D)$,
$\mathcal{M}(D)$, $\mathcal{M}^c(D)$,
$\mathcal{A}^*(D)$
$\mathcal{M}^*(D)$

Chapter 40

$\mathcal{F}_{a,r}$
$\mathcal{Q}(\mathcal{F})$.
$\mathcal{F} \preceq \mathcal{G}$
$\mathcal{F} \prec \mathcal{G}$
$\Sigma(D)$, $\Sigma_0(D)$
$\sup(\mathcal{F}, \mathcal{G})$

$\lambda(\mathcal{F}, \mathcal{G})$

$\delta(\mathcal{F}, \mathcal{G})$

$\widetilde{\delta}$

Chapter 41

$h_*(\mathcal{F})$

$\Phi(D, u)$

$\Pi(f)$

Chapter 42

$\widetilde{\delta}(\mathcal{F}, \mathcal{G})$

$\widetilde{d}(u, v)$

$\overline{f}$

$\mathcal{P}$

Chapter 43

$O(\zeta(r))$, $o(\zeta(r))$

$Z(r, f), \overline{Z}(r, f),$

$N(r, f), \overline{N}(r, f)$

$T(r, f)$

$\overline{\omega}_0$, $\overline{\overline{\omega}_0}$

Chapter 46

$Z_R(r, f)$, $\overline{Z}_R(r, f)$

$N_R(r, f)$, $\overline{N}_R(r, f)$

$T_R(r, f)$

Chapter 48

$g.c.d.(m, n)$

Chapter 50

$Z(r, f \mid {}''f(x)$ satisfying Property P$'')$

$\overline{Z}(r, f \mid {}''f(x)$ satisfying Property P$'')$

$N(r, f \mid {}''f(x)$ satisfying Property P$'')$

$\overline{N}(r, f \mid {}''f(x)$ satisfying Property P$'')$

Chapter 58

$\mathcal{E}(f, S)$

ram(f)

$\widetilde{Z}(r, f)$, $\widetilde{N}(r, f)$

Chapter 61

$\deg_x(P)$